Magnesium and Its Alloys

Metals and Alloys

Series Editor
George E. Totten

International in scope, this series focuses on vitally important areas relating to the science and engineering of the development, selection, and use of metals and alloys. Topics covered include microstructural characterization, heat treatment, surface engineering, designing for residual stresses, developments in the modeling and simulation related to material and process design, properties, and failures related to the use of metals and alloys as a class of materials. Encompassing state-of-the-art advancements, the books in this series target the practicing engineer, researcher, and student.

Encyclopedia of Iron, Steel, and Their Alloys, Five-Volume Set
Rafael Colás, George E. Totten

Encyclopedia of Aluminum and Its Alloys, Two-Volume Set
George E. Totten, Murat Tiryakioglu, Olaf Kessler

Agglomeration of Iron Ores
Ram Pravesh Bhagat

Magnesium and Its Alloys: Technology and Applications
Leszek A. Dobrzański, George E. Totten, Menachem Bamberger

For more information on this series, please visit:
https://www.crcpress.com/Metals-and-Alloys/book-series/CRCMETALL

Magnesium and Its Alloys

Technology and Applications

Edited by
Leszek A. Dobrzański
George E. Totten
Menachem Bamberger

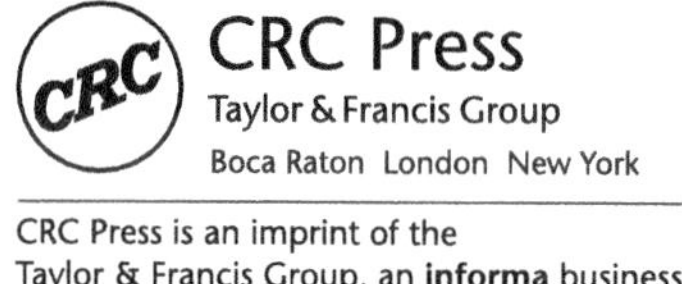

CRC Press is an imprint of the
Taylor & Francis Group, an **informa** business

CRC Press
Taylor & Francis Group
6000 Broken Sound Parkway NW, Suite 300
Boca Raton, FL 33487-2742

First issued in paperback 2021

© 2020 by Taylor & Francis Group, LLC
CRC Press is an imprint of Taylor & Francis Group, an Informa business

No claim to original U.S. Government works

ISBN 13: 978-0-367-77924-5 (pbk)
ISBN 13: 978-1-4665-9662-7 (hbk)

This book contains information obtained from authentic and highly regarded sources. Reasonable efforts have been made to publish reliable data and information, but the author and publisher cannot assume responsibility for the validity of all materials or the consequences of their use. The authors and publishers have attempted to trace the copyright holders of all material reproduced in this publication and apologize to copyright holders if permission to publish in this form has not been obtained. If any copyright material has not been acknowledged please write and let us know so we may rectify in any future reprint.

Except as permitted under U.S. Copyright Law, no part of this book may be reprinted, reproduced, transmitted, or utilized in any form by any electronic, mechanical, or other means, now known or hereafter invented, including photocopying, microfilming, and recording, or in any information storage or retrieval system, without written permission from the publishers.

For permission to photocopy or use material electronically from this work, please access www.copyright.com (http://www.copyright.com/) or contact the Copyright Clearance Center, Inc. (CCC), 222 Rosewood Drive, Danvers, MA 01923, 978-750-8400. CCC is a not-for-profit organization that provides licenses and registration for a variety of users. For organizations that have been granted a photocopy license by the CCC, a separate system of payment has been arranged.

Trademark Notice: Product or corporate names may be trademarks or registered trademarks, and are used only for identification and explanation without intent to infringe.

Library of Congress Cataloging-in-Publication Data

Names: Bamberger, Menachem, editor. | Dobrzanski, Leszek A., editor. | Totten, George E., editor.
Title: Magnesium and its alloys : technology and applications / [edited by] Menachem Bamberger, Leszek A. Dobrzanski, and George E. Totten.
Description: First edition. | Boca Raton, FL : CRC Press/Taylor & Francis Group, [2020] | Series: Metals and alloys | Includes bibliographical references and index.
Identifiers: LCCN 2019015206 | ISBN 9781466596627 (hardback : alk. paper) | ISBN 9781351045476 (ebook)
Subjects: LCSH: Magnesium--Metallurgy.
Classification: LCC TN799.M2 M24 2020 | DDC 669/.96723--dc23
LC record available at https://lccn.loc.gov/2019015206

Visit the Taylor & Francis Web site at
http://www.taylorandfrancis.com

and the CRC Press Web site at
http://www.crcpress.com

Contents

Series Preface

METALS AND ALLOYS

Although iron, aluminum, nickel, titanium, copper, and other metals and their alloys have been used for hundreds, if not thousands, of years, the understanding of these materials is certainly not complete and, in fact, often inadequate or simply not sufficiently current. There continue to be important advances in the development of metals and alloys, which continue to be critically important in every major market sector. Thus, there is an ongoing and vitally important need to provide updated coverage pertaining to the physical and mechanical metallurgy, equipment and thermal process developments, structure, properties, and modeling and simulation, and failure analysis of traditional and more recently developed metals and alloys.

In view of these ongoing challenges and their commercial and technological impacts, this series will focus on these vitally important areas relating to the science and engineering of the development, selection, and use of metals and alloys. General examples of the areas of topical coverage include: microstructural characterization, heat treatment, surface engineering, designing for residual stresses, developments in the modeling and simulation related to material, and process design, properties, and failures related to the use of metals and alloys as a class of materials. The books in this series target the practicing engineer, researcher, and student. The coverage of the books in this series will be international in scope and will encompass state-of-the-art advancements to facilitate their instructional and reference value, which will enhance their long-term utility for the reader.

George E. Totten
G.E. Totten & Associates, LLC
Seattle, WA

Editors

Leszek A. Dobrzański, Prof, DSc, PhD, MSc, Eng, Hon. Prof., M Dr. HC is currently a full professor and the director of the Science Centre ASKLEPIOS of the Medical and Dental Engineering Centre for Research, Design and Production ASKLEPIOS Ltd in Gliwice, Poland, the president of the World Academy of Materials and Manufacturing Engineering, and the editor-in-chief of the "Journal of Achievements in Materials and Manufacturing Engineering," "Archives of Materials Science and Engineering" and "Open Access Library." From 1971 to 2017, he worked at the faculty of Mechanical Engineering of the Silesian University of Technology in Gliwice, Poland as a full professor, and among other, the vice-rector of the University, three times the dean of the faculty, and multiyear the director of the Institute. His published creative scientific output includes about 2,300 scientific publications and pertains to materials science, surface engineering, nanotechnology, biomaterials and bioengineering, computer-aided engineering materials design and modelling of their structure and properties. That scientific output includes close to 50 books and monographs and 50 chapters in books and monographs, 220 papers in English in journals covered in the Web of Science Core Collection, and 52 patents. He is a creator and leader of the Scientific School promoting by him personally of a group of 62 completed PhD theses and next PhD dissertations in progress, as well as promoting ca. 1000 MSc and BSc theses.

George E. Totten, PhD, FASM, is the president and founder of G.E. Totten & Associates, LLC, Seattle, Washington, USA. He has more than 31 years of experience in the heat treating and fluids/lubricants industry. His most recent work has focused on metalworking quenchants (heat treating), hydraulic fluids technology, and lubricant formulation and testing. Widely published, he has worked extensively within SAE International, ASM International, and ASTM International committees. He is also past president of the International Federation for Heat Treatment and Surface Engineering (IFHTSE).

Menachem Bamberger, has an excellent international reputation in the field of materials engineering. He is chairman of the Casting and Solidification Committee of the Israel Metallurgical Society and a member of the Verein Deutscher Giessereifachleute and the TMS- Mg group. From 1996 to 2000 he served as the chairman of the scientific committee of the Consortium for the Development of Production and Finishing Processes of Magnesium Alloys. His group works in the fields of development of new Mg alloys based on thermodynamic simulations as well as processing of materials and their simulation. He published 65 papers in Mg casting, solidification and phase stability. In the last 20 years he has led 30 projects funded by (among others) Israel Academy of Science, German Israel research foundation (GIF) and BMBF-Germany.

Contributors

Trevor Abbott
Magontec Limited
Sydney, New South Wales, Australia

and

School of Engineering
RMIT University
Carlton, Victoria, Australia

and

Department of Materials Science and Engineering
Monash University
Melbourne, Victoria, Australia

Janusz Adamiec
Institute of Materials Science
Silesian University of Technology
Katowice, Poland

Murat Alkan
Metallurgical and Materials Engineering Department
Faculty of Engineering
Dokuz Eylül University
Izmir, Turkey

Menachem Bamberger
Department of Materials Science and Engineering
Technion University
Haifa, Israel

Mehmet Buğdayci
Metallurgical and Materials Engineering Department
Faculty of Chemical and Metallurgical Engineering
Istanbul Technical University
Istanbul, Turkey

and

Chemical and Process Engineering Department
Faculty of Engineering
Yalova University
Yalova, Turkey

Paul Burke
Digital Alloys Inc.
Burlington, Massachusetts

V.S.K. Chakravadhanula
Material Characterization Division
Materials and Metallurgy Group
Vikram Sarabhai Space Center
Trivandrum, India

Jae-Hyung Cho
Materials Processing Innovation Research
Korea Institute of Materials Science
Changwon, South Korea

Leszek A. Dobrzański
The Medical and Dental Engineering Centre for Research
Design and Production ASKLEPIOS Ltd
Gliwice, Poland

Rajendra Laxman Doiphode
Department of Mechanical Engineering
Government Polytechnic
Kolhapur, India

Sebastián Feliu Jr.
ECORR Research Group
Department of Surface Engineering, Corrosion and Durability
National Center for Metallurgical Research (CENIM-CSIC)
Madrid, Spain

Robert Jarosz
ZM WSK Rzeszów S.A.
Rzeszów, Poland

William D. Judge
Department of Materials Science & Engineering
University of Toronto
Toronto, Ontario, Canada

Suk-Bong Kang
Advanced Materials
Korea Institute of Materials Science
Changwon, South Korea

Bhagwati Prasad Kashyap
Department of Metallurgical Engineering and Materials Science
Indian Institute of Technology Bombay
Mumbai, India

Andrzej Kiełbus
Institute of Materials Science
Silesian University of Technology
Katowice, Poland

Hyoung-Wook Kim
Advanced Materials
Korea Institute of Materials Science
Changwon, South Korea

Shae K. Kim
UST and Korea Institute of Materials Science
Changwon, South Korea

Georges J. Kipouros
Department of Mechanical Engineering
Dalhousie University
Halifax, Nova Scotia, Canada

Yiannis G. Kipouros
Department of Economics
Queen's University
Kingston, Ontario, Canada

Zi-Kui Liu
Department of Materials Science and Engineering
The Pennsylvania State University
University Park, Pennsylvania

Sushant K. Manwatkar
Material Characterization Division
Materials and Metallurgy Group
Vikram Sarabhai Space Center
Trivandrum, India

S.V.S. Narayana Murty
Material Characterization Division
Materials and Metallurgy Group
Vikram Sarabhai Space Center
Trivandrum, India

P. Ramesh Narayanan
Material Characterization Division
Materials and Metallurgy Group
Vikram Sarabhai Space Center
Trivandrum, India

Filip Pastorek
Research Centre
University of Zilina
Zilina, Slovakia

George E. Totten
Department of Mechanical and
Materials Engineering
Portland State University
Portland, Oregon

Ahmet Turan
Chemical and Process Engineering
Department
Faculty of Engineering
Yalova University
Yalova, Turkey

Onuralp Yücel
Metallurgical and Materials
Engineering Department
Faculty of Chemical and Metallurgical
Engineering
Istanbul Technical University
Istanbul, Turkey

Yu Zhong
Department of Mechanical and
Materials Engineering
College of Engineering & Computing
Florida International University
Miami, Florida

1 The Importance of Magnesium and Its Alloys in Modern Technology and Methods of Shaping Their Structure and Properties

Leszek A. Dobrzański, George E. Totten, and Menachem Bamberger

CONTENTS

**"Learning naturally is a true pleasure;
how unfortunate then it is
that in most schools it is made a pain."**

Sir Humphry Davy (1778–1829)

1.1 GENERAL CHARACTERISTICS OF MAGNESIUM

How greatly amazed would have been Sir Humphry Davy (1778–1829), a Cornish chemist and inventor, the ***Fellow and President of the Royal Society, Member of the Royal Irish Academy and Fellow of the Geological Society***, to whose memory we dedicate this book, and who was the first to isolate, with electrochemical methods, in 1807–1808, a series of elements, including magnesium [1], had he known how great technical importance was gained by this chemical element after 210 years since discovery. Magnesium's atomic number is 12. It occurs in the solid state as a substance with a silver-grey colour, metallic gloss, with a close physical similarity to the other five elements in the second column of the periodic table. It belongs to alkaline earth metals, which also include Ca, Sr, Ba, Ra and Be. All the elements of the group have the same electron configuration in the outer shell of electrons and a similar crystalline structure [2–4].

Magnesium is one of the lightest metals with a density of 1.74 g/cm^3 and the atomic mass of 24.31. The melting point and the boiling point of magnesium is, respectively,

---

The sentence from 1836 is cited from the book *Memoirs of the Life of Sir Humphry Davy. Vol. 1 of 2*. London, UK, Longman, Rees, Orme, Brown, Green, & Longman.

The portrait of Sir Humphry Davy was made by Thomas Phillips (died 1845). User: Dcoetzee from the National Portrait Gallery, London, UK.

650°C and 1107°C. A linear coefficient of thermal expansion of magnesium at room temperature is equal to 26 10^{-6} 1/K [2–4]. The strength and plastic properties of pure magnesium are relatively low and dependent upon its purity. Tensile strength Rm in the cast state is 80–120 MPa, the yield point Re = 20 MPa, elongation A = 4%–6%, and hardness 30 HBW [2–6]. It should be noted that a relatively high specific strength of magnesium, which is 4.6–6.8 km, is very favourable, and in the case of structural alloys of this element, reaches up to 19 km in comparison with the non-alloy steel, which reaches 6–7 km, and after quenching and tempering 11.5 km, and over 19 km only after patenting in the form of very fine wires having a diameter of not more than 5 mm. The properties of pure magnesium are summarised in Table 1.1.

Magnesium crystallises in a compact hexagonal lattice A3 [7–10], with the parameters a = 0.321 nm and c = 0.521 nm. Below 225°C, slide in the magnesium structure is only possible in the base plane {0001} <1120>, together with twinning in the plane {1012} <1011>. Pure magnesium and conventional cast magnesium alloys have a tendency to brittleness, characterised by an intercrystalline fracture and a local transcrystalline fracture in the twin zones or in the base plane {0001} with large grains. A new base plane is activated above 225°C {1011}, causing good magnesium deformability. The temperature range, in which magnesium can be most easily subjected to plastic working, is 350°C–450°C [4].

Magnesium rather easily oxidises in air, but, as in the case of aluminium, the corrosion process is inhibited by passivation. In contrast to aluminium (PBR = 1.28), magnesium, however, has an unfavourable Pilling–Bedworth ratio PBR = 0.80 [11,12], as a result of which the passivation coating is less effective. Magnesium is also passivated in

TABLE 1.1
Comparison of Selected Properties of Pure Magnesium

Properties	Mg
Atomic number	12
Atomic mass	24.3056
Crystalline structure	A3
Density in 20°C, g/cm^3	1.738
Melting point, °C	650
Boiling point, °C	1107
Thermal expansion coefficient, 10^{-6} 1/K	25.2
Specific heat capacity, kJ/(kg·K)	1.025
Heat conductivity, W/(m·K)	418
Resistivity, nΩ·m	44.5
Tensile strength, *Rm*, MPa	180 ÷ 220[a]
Yield point *Rp*, 0.2, MPa	115 ÷ 140[a]
Elongation, %	2 ÷ 10[a]
Elastic modulus, GPa	40
Hardness, HBW	45 ÷ 47[a]

Source: Dobrzański, L.A., *OAlib. Annal VII*, 2, 1–982, 2017.
[a] Rolling

concentrated (98%) sulphuric acid and in the presence of iodine vapours. A passivation layer of the poorly soluble magnesium fluoride protects it to the temperature of 600°C, and also against the activity of hydrofluoric acid [13]. Magnesium reacts slowly with hot water with temperature of >70°C, forming magnesium hydroxide. It is completely resistant to the activity of alkali, and vigorously reacts with acids, by creating corresponding salts and releasing hydrogen [10]. Magnesium is a flammable substance, the flash point is approx. 760°C. Magnesium dust is pyrophoric, its flash point is approx. 470°C. Magnesium in the air burns with a blinding white flame, whose temperature is 3000°C–3100°C. The main product is magnesium oxide, accompanied by magnesium nitride. Incineration is also maintained in an atmosphere of water vapour and carbon dioxide. Magnesium dissolves when heated in methanol and ethanol by producing relevant magnesium alcoholates. Such reactions are initiated by iodine and inhibited by water with the fraction of over 1%. They are used to produce alcoholates and to obtain the so-called absolute ethanol, i.e., the product with a very small fraction of water.

Magnesium is the eleventh most abundantly occurring element in the human body by mass [14], and is essential for all cells and for about 300 enzymes [15–19]. Magnesium ions play an important role in maintaining the osmotic pressure of blood and other tissues and in maintaining the proper structure of ribosomes. It is a component of bone, reduces the degree of hydration of cellular colloids, participates in the transmission of signals in the nervous system. The demand for magnesium in adults is 300–400 mg per day [15] and although in the natural environment it is richly present in foods consumed by man, there is less and less of it due to chemical fertilisation of soil with compounds containing potassium, and due to the use of excessive amounts of food preservatives. Other causes of magnesium deficiency are alcohol abuse, the use of hormonal contraceptives, stress, excessive consumption of fats and renal failure. One of the most common symptoms of magnesium deficiency is muscle spasms. Aches, poor digestion, worries and difficulty in sleeping are also common. It can also be a foundational cause of the restless leg syndrome [20–24]. Magnesium ions interact with polyphosphate compounds, such as deoxyribonucleic acid (DNA), adenosine triphosphate (ATP) and ribonucleic acid (RNA). The activity of magnesium ions is required for hundreds of enzymes [19]. Magnesium compounds are used in medicine as conventional laxatives, antacids and for stabilising the abnormal stimulation of nerves or vasoconstriction.

Magnesium is the ninth element most abundant in the universe and the eighth most abundant in the earth crust [25], but the fourth most commonly occurring element on Earth, after iron, oxygen and silicon [26], constituting 13% of the planet mass and constitutes about 2% of the crust of the Earth. It is the third element most commonly dissolved in sea water, after soda and chlorine [25], which averages about 0.13% magnesium by weight. When it is harvested from this resource, it becomes the third most-plentiful element on our planet [25,27].

1.2 ACQUISITION OF MAGNESIUM FROM NATURAL SOURCES AND SCALE OF MAGNESIUM PRODUCTION

Magnesium occurs naturally only in combination with other elements, where invariably it has the oxidation state^{+2}. It is present in large deposits of magnesite, dolomite and other minerals and in mineral waters, where the magnesium ion is soluble. Although magnesium is present in over 60 minerals, only dolomite [$MgCO_3$*$CaCO_3$], magnesite [$MgCO_3$], brucite [Mg $(OH)_2$], carnallite [$MgCl_2$ * KCl * $6H_2O$], talc [$Mg_3Si_4O_{10}(OH)_2$] and olivine [$(Mg, Fe)_2SiO_4$] and also sea water [Mg^{2}+(aq)] have commercial significance. Resources from which magnesium compounds can be recovered range from large to virtually unlimited and are globally widespread. Identified world magnesite and brucite resources total 12 billion tons and several million tons, respectively. Resources of dolomite, forsterite, magnesium-bearing evaporite minerals and magnesia-bearing brines are estimated to constitute a resource of billions of tons. Magnesium hydroxide can be recovered from seawater. Australia holds the highest levels of magnesium reserves in the world today, accounting for 30.4% of the total available raw mineral. China has the second-most reserves, with 28.3% of the available raw mineral. They are followed by North Korea and Russia, with 19.6% and 17.4% of the available raw mineral, respectively [29]. Dolomite is used for the production of magnesium in the United States, Canada, France, China, Serbia, India and Brazil. Magnesite is used for magnesium production in Australia and Canada. Asbestos waste may also be the source of magnesium. As serpentine could be used as a source of magnesia, Canada was exploring a method to produce magnesia from serpentine in tailings of an asbestos mine. Brine is used for magnesium production in Russia and Kazakhstan. Solar energy can be used for concentrating the water containing magnesium, as was done in the Great Salt Lake in Utah in the United States. Magnesium can be produced from evaporite deposits, as in the case of the Dead Sea in Israel. Carnallite salts are used in Perm, Russia. For many years, the leading and largest manufacturer of magnesium were the United States. However, these days China leads the world in terms of magnesium production. They are currently responsible for about 80% of the current supply of this needed metal [26,28,29].

At least 60% of the world production of primary magnesium is manufactured in thermal processes. In China, dominating in magnesium production, the silicothermic Pidgeon process is used almost exclusively, consisting of magnesium oxide reduction at high temperature by means of silicone, often in the form of ferrosilicon alloy [30]. However, starting with 2017, some plants producing magnesium using the Pidgeon process in China were expected to shut down, owing to energy cost increases and to comply with environmental regulations [31,32] obligated in China [29]. The process can also be carried out with carbon at a temperature of about 2300°C. It should be noted though that the silicothermic reduction of dolomite to produce magnesium was first developed by Dr Amati in 1938 at the University of Padua. Industrial production with this method was first started in the world by the Societe Anonima Italiana per il Magnesio e Leghi in Bolzano, yet in 1938. Externally heated retorts had been used in the process, identical as discovered later at the beginning of the 1940s by Lloyd Montgomery Pidgeon (1903–1999) of the Canadian National Research Council (NRC) also in a doctoral thesis. The first factory on the North American continent was built only in 1941 in Haley, Ontario, Canada. Despite this, traditionally the name of this technology was well established as the Pidgeon process.

Many large plants producing magnesium in the world currently use electrolytic processes requiring a charge in the form of magnesium chloride and apply electric current to the dissociation of the metal to magnesium and chlorine gas, which can also be traded [33,34]. In the United States, natural brine water and sea water account for about two-thirds of the total magnesium industry compound production values. Magnesium is mainly produced in the Dow process, by electrolysis of fused magnesium chloride from brine and sea water. The Dow Chemical Company, as one of the predecessors of DowDuPont Inc., was set up by Herbert Henry Dow (1866–1930), who used electric current in 1891 to separate bromides from brine, with the method known as the Dow process. For magnesium, electrolysis processes comprise two stages: preparing a raw material containing magnesium chloride and dissociation of this compound to the magnesium and the chlorine gas in electrolytic cells. This process used calcined dolomite and 75% FeSi ground, blended and briquetted into pellets that were charged into closed end, externally heated retorts. The salt solution containing ions of Mg^{2+} is first treated with lime (calcium oxide) to precipitate magnesium hydroxide. Magnesium hydroxide is then converted to a partial hydrate of magnesium chloride by subjecting hydroxide to the activity of hydrochloric acid, and heating the product. The salt is then subjected to electrolysis in a molten state. Mg^{2+} ion in a cathode is reduced by two electrons to magnesium. In industrial processes, charges consist of various molten salts containing anhydrous or partially dehydrated magnesium chloride or anhydrous carnallite. In order to avoid impurities present in carnallite ores, dehydrated artificial carnallite is produced by controlled crystallisation from heated solutions containing magnesium and potassium. Partially dehydrated magnesium chloride can be obtained in the Dow process, in which seawater is mixed in a flocculator with lightly burned reactive dolomite. The insoluble magnesium hydroxide is precipitated on the bottom of the settling tank, from where it is pumped as a suspension, filtered, and converted to magnesium chloride, by reaction with hydrochloric acid, and then it is subject to the multi-stage drying to 25% of water. The final dehydration takes place during melting. Anhydrous magnesium chloride is produced by two main methods, by dehydration of magnesium chloride brines or chlorination of magnesium oxide.

One of the newer processes of magnesium production is the solid oxide membrane technology including electrolytic reduction of MgO to magnesium involving the yttria-stabilized zirconia (YSZ) as the electrolyte and the anode in the form of molten metal. O^{2-} anode is being oxidised. The graphite layer borders the anode of molten metal, and carbon and oxygen react to form carbon oxide. When silver is used as a liquid metal anode, a carbon reducer or hydrogen are not required, and oxygen gas is only released on the anode [35–37]. This method is more environmentally friendly than others, as it emits much less carbon dioxide and provides a 40% reduction in costs in relation to the electrolytic reduction method [36,37].

The global primary production of magnesium is steadily growing in the last decade by at least 50% compared to the level of 2010 (Figure 1.1) [38–41]. At the same time, manufacturing capacities of primary magnesium have been increased many times in the world, mainly caused by higher production capacity in China (Figure 1.2). The sole US producer of primary magnesium temporarily shut down some capacity at the end of 2016 citing the shutdown of a titanium sponge plant that had been a major customer, and this capacity was not expected to restart in the foreseeable future. In China, a new 100,000-ton-per-year plant that would

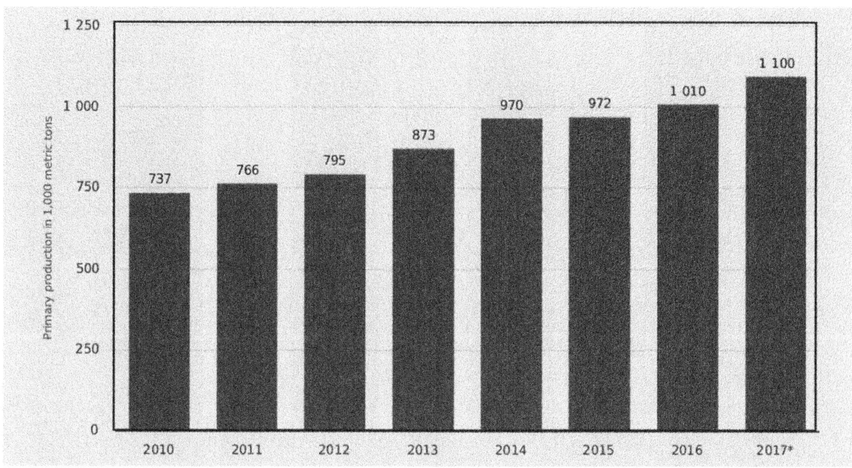

FIGURE 1.1 Primary magnesium production worldwide from 2010 to 2017 (in kmetric tons) (From Statista—The Statistics Portal, Primary magnesium production worldwide 2010–2017, https://www.statista.com/statistics/569515/primary-magnesium-production-worldwide/; Accessed February 23, 2019).

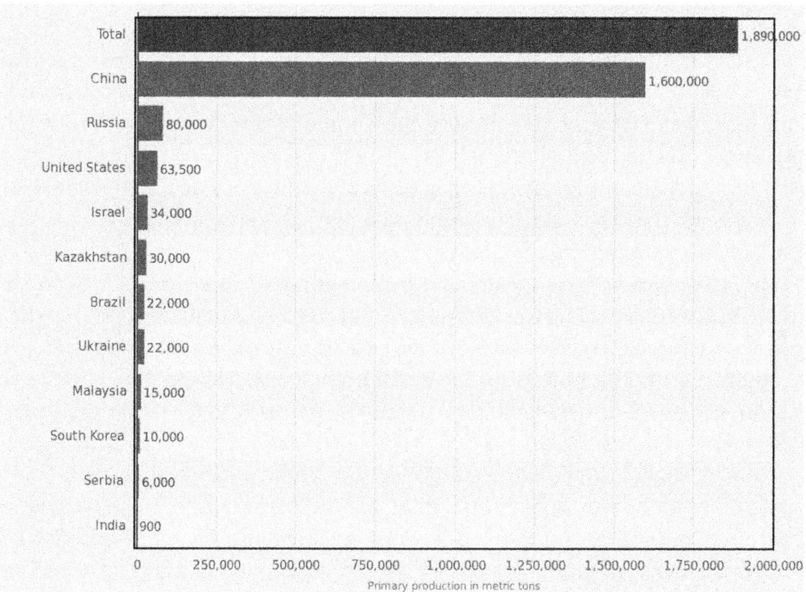

FIGURE 1.2 Primary magnesium production capacity worldwide with example of 2014 (in metric tons) (Statista—The Statistics Portal, Primary magnesium production worldwide in 2016, https://www.statista.com/statistics/569535/primary-magnesium-production-capacity-by-country/; Accessed February 23, 2019).

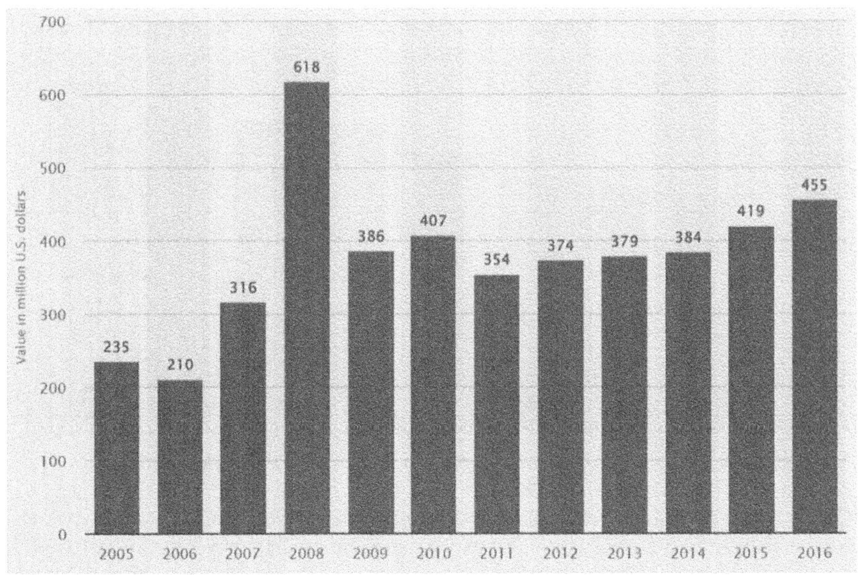

FIGURE 1.3 Value of magnesium recycled from scrap between 2005 and 2016 years (in million USD) (From Statista—The Statistics Portal, U.S. value of magnesium recycled from scrap between 2005 and 2016 (in million U.S. dollars), https://www.statista.com/statistics/209425/recycled-value-of-magnesium-in-the-us/; Accessed February 23, 2019).

produce magnesium from lake brines was completed in 2017 and was expected to ramp up to full capacity in early 2018. In the recent years, the turnover of magnesium scrap has stabilised, mainly in the United States, where it totals approx. 380 million USD per year (Figure 1.3).

However, global consumption of magnesium products has declined slightly since 2015. This is an indication that the steel production rates around the world are beginning to stabilise once again.

Products for non-ferrous metals and glass within this market have seen sharp declines. Magnesium, as a pure element, is not used as a constructional material, as it is characterised by low strength and plastic properties. The advantages of magnesium, i.e., high combustion heat, high chemical activity and low density, contribute to its use in pyrotechnics, chemical industry, nuclear power industry and metallurgy (Table 1.2).

In 2017 in United States the leading use for primary magnesium metal, which accounted for 34% of reported consumption, was in aluminium-base alloys that were used for packaging, transportation and other applications. Castings accounted for 30% of primary magnesium metal consumption, desulphurisation of iron and steel, 22%; wrought products, 6%; and other uses, 8%. Consumption of magnesium as a reducing agent for metals production decreased dramatically compared with that in 2016 because of the shutdown of a titanium sponge producer in Utah at the end of 2016. About 52% of the secondary magnesium was consumed for structural uses and about 48% was used in aluminium alloys [38].

TABLE 1.2
Estimated Share of Various Applications in Annual Consumption of Magnesium

Application	Mass Fraction of Consumed Magnesium,%
Constructional	
Die castings	approx. 11.4
Gravity castings	0.8
Wrought products	2.9
Non-constructional	
Alloy additive in Al alloys	53.5
Desulphurisation	11.5
Cast iron modification	6.3
Reduction of metals	4.1
Chemicals	3.2
Electrochemicals	3.2
Others	3.2

Source: Dobrzański, L.A. et al., *OAlib.*, 5, 1–319, 2012.

1.3 GENERAL CHARACTERISTICS OF MAGNESIUM ALLOYS AND THEIR APPLICATION AREAS

Magnesium is used most frequently in aluminium alloys as an alloy additive, which is added to improve strength and corrosion resistance. It is also used as an additive in zinc alloys to improve mechanical properties and dimensional stability [4,41,42]. Magnesium gained its largest application area by creating alloys in combination with other elements. Magnesium alloys used as construction materials contain various alloy additives as shown in Table 1.3. Such elements have different effects on the structure and properties of magnesium alloys, and many of them are useful in various applications, but some of them are used for very specific cases. Five basic groups of alloys, which are currently produced for commercial purposes, based mainly on main alloy elements, are distinguished according to their chemical composition and they fall into the following subgroups [4,41–47]:

- Mg-Mn,
- Mg-Al-Mn,
- Mg-Al-Zn-Mn,
- Mg-Zr,
- Mg-Zn-Zr,
- Mg-RE-Zr,
- Mg-Ag- RE-Zr,
- Mg-Y-RE-Zr.

Until recently, thorium was also a component in the given groups of alloys [4]:

- Mg-Th-Zr,
- Mg-Th-Zn-Zr,
- Mg-Ag-Th-RE-Zr.

TABLE 1.3
The Effects of Various Alloying Elements in Magnesium (Compiled According to Data Available from the Website of the International Magnesium Association)

Alloying Element	Effects of Addition	Additional Information
Al	• Increases hardness, strength and castability while only increasing density minimally • Increased amount of aluminium decreases the ductility of the alloy	• Average alloy contains about 2–9 weight percent of aluminium and can be heat treated with >6 weight percent
Be	• Significantly reduces surface melt oxidation during processing • Grain coarsening can occur • Can be a carcinogenic material and is being rejected by some companies for use	• Included only in very small quantities
Ca	• Improves thermal and mechanical properties as well as assists in grain refinement and creep resistance • Reduces oxidation during processing when added to cast alloys • Allows for better rollability of sheet metal • Reduces surface tension	• Additions exceeding 0.3 weight percent, increases the risk of cracking during welding
Ce	• Improves corrosion resistance • Increases plastic deformation capability, magnesium elongation, and work hardening rates • Reduces yield strength	
Cu	• Assists in increasing both room and high temperature strength • Negatively impacts ductility and corrosion resistance	
Mn	• Increases saltwater corrosion resistance within some aluminium containing alloys • Reduces the adverse effects of iron, usually present in 2–4 weight percent	
Ni	• Increases both yield and ultimate strength at room temperature • Negatively impacts ductility and corrosion resistance	
Ne	• Improves material strength	
REM	• Increase in high temperature creep and corrosion resistance and strength • Allows lower casting porosity and weld cracking in processing	
Si	• Can increase molten alloys' fluidity • Improves elevated temperature properties, especially creep resistance	• Only used in pressure die casting
Sr	• It enhances creep performance	• Used in conjunction with other elements

(Continued)

TABLE 1.3 (*Continued*)
The Effects of Various Alloying Elements in Magnesium (Compiled According to Data Available from the Website of the International Magnesium Association)

Alloying Element	Effects of Addition	Additional Information
Sn	• It improves ductility, and reduces tendency to crack during processing	• Used with aluminium
Y	• Enhances high temperature strength and creep performance	• Combined with other rare earth metals
Zn	• Increases the alloys' fluidity in casting • It can improve corrosion resistance • Additions of 2 weight percent or greater tend to be prone to hot cracking	• Second most commonly used alloying metal with magnesium • Added to magnesium alloys with nickel and iron impurities
Zr	• Refines grain size in sand and gravity castings	• Not combined with aluminium

Source: International Magnesium Association, Magnesium alloys overview, https://www.intlmag.org/page/design_mag_all_ima; Access February 23, 2019.

Currently such alloys are withdrawn from use due to adverse environmental impact of the element.

The chemical composition of modern magnesium alloys has been constantly improved, to enhance various properties and find new fields of applications [4,26,48–51]. Figure 1.4 shows schematically the future directions of magnesium alloys development for industrial applications.

Some magnesium alloys can be produced using non-conventional technologies, for example by the rapid-solidification process (RSP) [52–54]. This provides up to a 20-fold increase in resistance to corrosion of the material in comparison with the conventional alloys, and improvement in creep resistance. Cast strips made of these materials are then mechanically ground into powders, which—when put in a metal container—are pressed to form rods. The structure of MgAl5Zn5Nd5 alloy, fabricated by RSP, has a grain size of $0.3 \div 5$ μm, providing tensile strength Rm of approx. 500 MPa, which is almost double that of conventional alloys, corresponding to the specific strength of over 28 km. Some of magnesium alloys produced by RSP exhibit superplastic properties [8,55–57].

Enhanced mechanical properties of magnesium alloys can also be obtained by making metallic glass from them [2,3,58–60]. The most promising properties are seen for triple alloys of Mg–M–Ln systems, where M is Ni or Cu, whereas Ln are lanthanides, for example Y. Tensile strength of some metallic glasses made of Mg–M–Ln alloys may reach $Rm = 610 \div 850$ MPa, and elasticity modulus $E = 40 \div 60$ GPa. Properties of metallic Mg65Cu25Y10 glasses are similar to those obtained by RSP. Nanocrystallisation—during annealing for 20 s at 100°C—of the cast strips with the amorphous structure may produce

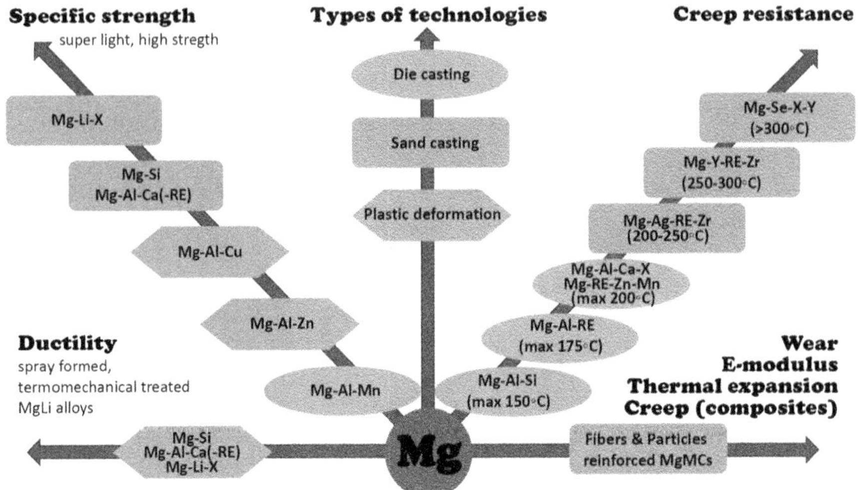

FIGURE 1.4 Scheme of future directions of magnesium alloys development for industrial applications. (Based on Dobrzański, L.A. et al., *OAlib.*, 5, 1–319, 2012; Kumar, D.S. et al., *Am. J. Mater. Sci. Technol.*, 4, 12–30, 2015; Mordike, B.L. and Ebert, T. *Mater. Sci. Eng.*, A302, 37–45, 2001; Blawert, C. et al., *Trans. Indian Inst. Met.*, 57, 397–408, 2001; Asadi, P. et al., *Welding of Magnesium Alloys, New Features on Magnesium Alloys*, W.A. Monteiro (Ed.), InTech, 2012.)

(e.g., in Mg85Zn12Ce3 alloy) nanocrystalline grains with the size of 3 ÷ 20 nm and tensile strength of Rm = 930 MPa, which corresponds to the specific strength of ca. 53.5 km, when it is thought that beryllium has the highest specific strength of 38 km for metals and alloys. Works are in progress over further development of alloys of this group and on their practical applications.

Magnesium and its alloys were produced almost exclusively for military purposes until the first half of the twentieth century. They could not compete with the much cheaper construction materials in the civil industry, due to the complicated manufacturing process and high production costs, despite better mechanical properties. Now, with the advancements achieved in the technology [2–4,10,52–81] formation, heat treatment, technological improvements and corrosion prevention, magnesium alloys are increasingly used in many areas of life [2]. Advanced technologies of magnesium alloys fabrication include casting in solid-ductile state, re-casting, thixocasting, thixopressing and the mentioned rapid solidification process (RSP) and Magnesium Matrix Composites (MMCs). The cast and plastically worked components made of magnesium alloys have a number of common applications. Due to low density with very favourable mechanical properties, components can be manufactured from them with the method of casting, plastic working, machining or welding. High-strength casting alloys with properties identical or very similar to the properties of alloys for plastic working are of particular interest in the modern machine building industry. Accurate and complex shapes can be achieved by producing constructional components with the casting technology, with a high degree of material homogeneity and with high strength, at the same time with good alloy ductility, and with lower work inputs to produce them and relatively low machining costs.

Magnesium alloys are constructional materials widely used in various industries due to low density and high specific strength (relative to the density) [2–4]. In addition, they exhibit good corrosion resistance, no aggressiveness with respect to the mould material and small melting heat. A high vibration damping ability and low inertia allow to use magnesium alloys for rapidly moving parts in places where there are rapid changes in speed. Growth in the consumption of magnesium alloys was also a result of progress in the production of new high-strength alloys, among others, with additions of Zr, Ce and Cd, very light alloys from Li used include mainly in aircraft constructions and space vehicles. [2,82].

Magnesium alloys with low density and relatively high specific strength are used worldwide in many different applications. One of the broad areas of application of magnesium, and mainly its alloys, is the automotive and land transport industry [26, 83–86]. The use of magnesium in automobile parts has continued to increase as automobile manufacturers sought to decrease vehicle weight in order to comply with fuel-efficiency standards. Magnesium castings have substituted for aluminium, iron and steel in some automobiles. The substitution of aluminium for steel in automobile sheet was expected to increase consumption of magnesium in aluminium alloy sheet. Although some magnesium sheet applications have been developed for automobiles, these were generally limited to expensive sports cars and luxury vehicles, automobiles where the higher price of magnesium is not a deterrent to its use. Table 1.4 shows a wide range of automotive parts which are made or can be made of magnesium and its alloys.

There is quite a long history of magnesium usage in the aerospace industry for multiple applications, both civil and military ones [87,88]. The idea is to reduce the weight of aircrafts and aviation constructions, space and missile vehicles, in order to reduce fume gas emissions and increase fuel efficiency, which has also a major impact on reducing operating costs. The competitiveness of magnesium in these applications is the greater, the more limited is the trend to reduce the weight of these devices by using light aluminium alloys, and the relatively higher are the costs of

TABLE 1.4
Magnesium Alloys Applications in Car Components

Group of Car Parts	Car Components
Engine and parts transmission	Engine block, gear box, crank case, oil pump housing, cylinder head cover, transfer case, covers, cams, bed plate, engine cradle, clutch and engine parts
Interior parts	Steering wheel covers, seat components, instrument panel, brake and clutch pedal, air bag retainer, door inner
Chassis components	Wheels, suspension arms, engine cradle, rear support, tailgate, bumper, brake system, fuel storage system
Body components	Cast components, radiator support, sheet components, extruded components, exterior and interior components, seats, instruments and controls

Source: Kumar, D.S. et al., *Am. J. Mater. Sci. Technol.*, 4, 12–30, 2015.

using laminates containing metallic fibres or composites containing a carbon fibre. Polymeric materials with low density, when subjected to a load under extremely high temperature, feature especially poor impact and damage properties. Magnesium can be used in thrust reversals, and also in fans of jet engines and housings of aircraft engines and helicopters. Recent changes in the standards allow currently to use magnesium alloys that meet, among others, flammability criteria, in seats of passenger aircraft, and research is pending to use magnesium alloys more widely in a cabin. Space ships and rockets also contain magnesium and its alloys. Magnesium is able to withstand short-wave electromagnetic radiation, exposure to ozone and influence of high-energy particles and matter. Dimensional stability is also a key factor, as magnesium is used in imaging optical devices carried by satellites.

Magnesium and its alloys, mainly, are very attractive materials for biomedical applications, especially in implants [89–109]. Since the natural bone density is 1.8–2.1 g/cm^3, and magnesium alloys have a density of 1.74–2.0 g/cm^3, and magnesium is comparable to a natural bone in terms of resistance to brittle cracking, modulus of elasticity and compressive yield strength, for these reasons, magnesium was first introduced in the medical industry as an orthopaedic biomaterial in the first half of the twentieth century. Magnesium occurs naturally as an ion in the human body, half of which is stored in the bone tissue, and has good biocompatibility and is non-toxic. Implants made of magnesium can be biodegradable in body fluids, which can be used to eliminate the re-operation necessary to remove implants. With a variety of coating deposition technologies, magnesium alloys gain more and more uses for non-biodegradable implants. Many research centres in the world conduct extensive research in this area. Metallic magnesium and its alloys can be used for various types of devices for orthopaedic surgeries, mainly plates, screws and fasteners. Current studies have demonstrated that magnesium implants show nominal changes in the blood composition in a six-month examination after implantation, but without causing damage to the excretory organs such as liver or kidney. Because of the required mechanical properties and biocompatibility, magnesium and its alloys are also used as temporary bio-implants. The limitation may, however, be a significant corrosion rate after implantation in an environment of tissues and body fluids, as implants can potentially corrode even before the completion of the healing process. There are many technological possibilities of preventing this problem, involving the selection of the chemical composition of the alloy, surface treatment, among others, by coating with corrosion resistant layers. However, the corrosion behaviour of magnesium implants with surface treatment and without such treatment has been tested in vitro studies so far, and tests in vivo conditions need to be continued, which is the immediate cause why such implants have not been commonly implemented in clinical practice. Recently, research and implementation works have been intensified over new generations of bioactive, biocompatible and biodegradable metallic materials for orthopaedic applications, and generally biomedical applications, among which pure magnesium, magnesium alloys and composites based on magnesium alloys have a leading role. Such materials are characterised by substantial biocompatibility and the required mechanical properties, however, due to an excessive corrosion rate of pure magnesium and its alloys under physiological conditions and loss of properties before bone healing, they cannot be used—even as temporary—for too long.

The developing composites with the MgMCs magnesium matrix are a solution to the problem [102–104, 110–117]. Magnesium matrix composites, as biocompatible materials, show the required mechanical properties, including tensile strength, ductility, elasticity modulus and corrosion resistance in physiological conditions. In MgMC-type composites, the matrix consists of magnesium alloys, calcium polyphosphate (CPP) and the molecules of β-tricalcium phosphate (β-TCP). Table 1.5 shows examples of manufacturing technologies of MgMC composites systematised by increasing costs. Apart from those given in the table, Physical Vapour Deposition (PVD) can also be distinguished as vapour processing for the manufacture of composite MgMC materials.

The market of portable and miniaturised electronic devices is constantly expanding. Magnesium alloys, often replacing the polymeric materials due to the greater strength properties, and the consequent greater durability, more favourable thermal conductivity and the ability to protect the electromagnetic and radio interferences, have been used for housings and for other minor constructional components including cameras, mobile phones, laptops and portable media devices [118].

TABLE 1.5
Review of Fabrication Processes of MgMC Composites with Magnesium Alloy Matrix

Group	Route	Application	Comments	Cost
Liquid processing	Stir casting	Basic process for MgMCs used in automotive and aerospace industry	Applicable for mass production, low volume fractions up to 30%	Low
	Melt deposition technique	Used to produce structural shapes such as rods, beams	Uniform distribution, high strengths	Low/ medium
	Squeeze casting	Widely used in automotive industry like connecting rods	Lower porosity, expensive, moulds needed, large capacity presses needed	Medium
Solid state processing	Powder metallurgy	Used to produce small objects bolts, pistons, valves	High volume fractions of particulate are possible (better properties), powders are expensive, not for near-net-shape parts	Medium
	Diffused bonding	Used to make sheets, blades, vane shafts	Lower temperatures than hot pressing but not capable of complex parts, slow, expensive fibre damage can occur	High

Source: Kumar, D.S. et al., *Am. J. Mater. Sci. Te chnol.*, 4, 12–30, 2015.

Sports equipment uses the advantageous specific strength of magnesium, its impact resistance and mouldability mainly for the manufacture of golf clubs, tennis rackets and arch handles. The same properties, and moreover the ability to reduce vibrations, point to the use of magnesium for bicycle frames and chassis of in-line rolls. It is advantageous to manufacture the parts of poles using magnesium alloys due to high torsional strength [119].

There are also many other applications of magnesium alloys, among others, for portable power tools and for renovation equipment, for printing and textile machines, and as a replacement for some of the engineering polymer materials [120].

1.4 OVERVIEW OF ISSUES RELATING TO MAGNESIUM AND ITS ALLOYS DISCUSSED IN THIS BOOK

Magnesium is currently the subject of interest of numerous publishers and authors. Numerous new publications are released all the time, mainly containing detailed technical solutions, and possibly an explanation of the physiochemical basics of the occurring phenomena crucial to the structure and, consequently, to the properties and potential applications of these materials. Many of such publications present a new approach to the technological processes, which—in addition to the chemical and phase composition—are dictating the applicational prospects of either magnesium alloys and pure metal, as well as magnesium-matrix MgMCs composites. A topic of biomedical and biomimetic engineering materials is interesting, both biodegradable ones developed for a long time, and non-biodegradable ones, where focus is now put more often. It is important to distinguish between engineering materials, intentionally and artificially produced by man, and natural materials, which can be obtained from nature, such as wood or rocks, which do not require further processing in addition to providing the shape and geometric form. Both, engineering materials and natural materials, are considered to be technical materials that are used in modern technology [2,3]. Reports in the literature regarding pure magnesium, magnesium alloys and magnesium matrix MgMCs composite materials, as a specific group of metal matrix composites (MMCs), are disseminated comprehensively in various journals and conference papers, but rarely in specially published guides, handbooks or books. Hence, the intention of CRC Press/Taylor & Francis Publisher was to develop and publish a comprehensive handbook on the subject, which was finally decided to be entitled "Magnesium and Its Alloys: Technology and Applications." After establishing the thematic scope, requests were received from different authors to participate in this publishing project, although, as usual, not everyone managed to be involved fully and on time, so ultimately the content was accepted containing 12 main sections prepared by authors from different countries, including the present Preface. Such chapters include both, extensive monographs, usually based on long-term own research experiments of the Authors, but also elaborations similar to research papers, outlining a single specific technological or research problem. For obvious reasons, the thematic scope of such chapters covers selected aspects only. In the editors' opinion, the so prepared elaborations, representing individual chapters, present a sufficiently broad spectrum of new information on pure magnesium, magnesium alloys and magnesium matrix MgMCs composites, to draw the attention of even a demanding Reader of scientific literature. Undoubtedly, such information will be useful to researchers and technologists, as well as PhD students and regular students of engineering, mainly materials engineering, mechanical engineering, manufacturing engineering and metallurgy.

This Preface is followed by the second chapter titled "Overview on Magnesium and Its Alloys, Casting Technologies, Microstructure, and Properties." It was created by Trevor Abbott from Magontec Limited in Glen Waverley, Victoria, Australia. The author points out that, unlike the steel and aluminium alloys, the majority of magnesium alloys is produced as casting, with a very small share of wrought products. In addition, gravity casting is not very common, with the result that the vast majority of magnesium alloys is produced by pressure die casting. The author therefore focussed on high pressure die casting (HPDC) processes with respect to magnesium alloys. An equilibrium system concerning Mg-Al alloys was discussed, as the one used practically most often, as well as dependencies between the diagrams of phase equilibrium, microstructure, castability and properties.

The third chapter is entitled "Magnesium production from calcined dolomite via the pidgeon process." The team of authors: Mehmet Buğdayci, Ahmet Turan, Murat Alkan and Onuralp Yücel, represents the Istanbul Technical University in Istanbul, Turkey, Yalova University in Yalova, Turkey and Dokuz Eylül University in Izmir, Turkey. A vacuum silicothermic production process of magnesium, called the Pidgeon process, is discussed in this chapter, as well as its historical development, basic principles, theoretical basis, industrial applications and neutralisation of wastes from this process. These issues have been fully explained by theoretical examples and industrial applications.

A group of authors: Rajendra Laxman Doiphode from the Government Polytechnic in Kolhapur, India, S.V.S. Narayana Murty from Vikram Sarabhai Space Center in Trivandrum, India and Bhagwati Prasad Kashyap from the Indian Institute of Technology Bombay in Mumbai, India, has prepared the fourth chapter entitled "Hot Deformation Characteristics of Magnesium Alloys." The purpose of this study is to revise the characteristics of hot deformations of magnesium alloys, including superplasticity. An example of such magnesium alloy is a commercial grade for plastic processing, i.e., AZ31, containing 3% of Al and 1% of Zn, very popular in industry. It was found that changes in mechanical properties are related to the course of dynamic recrystallisation, although the effect is not uniform with respect to the grain size depending on the flow stress at high temperature. Deviations from the expected trends are associated with plastic deformation by twinning and the texture and their impact on flow stress and ductility.

Leszek A. Dobrzański from The Medical and Dental Engineering Centre for Research, Design and Production ASKLEPIOS Ltd in Gliwice, Poland, a co-editor of this book, is an author of the next, fifth chapter, which is an extensive monograph concerning "Effect of Heat and Surface Treatment on the Structure and Properties of the Mg-Al-Zn-Mn Casting Alloys." The elaboration concerns the results of own research of the Mg-Al-Zn-Mn casting magnesium alloys. The effect of the cooling rate and mass concentration of Al on the kinetics of alloys crystallisation was described. Another aspect outlined is represented by the effect of heat treatment conditions and different cooling mediums on the structure of the investigated alloys. The results of own investigations using very advanced materials science research methodology were presented. The investigations concerned surface layers deposition by physical (PVD) and chemical (CVD) vapour deposition techniques on the substrates of Mg-Al-Zn-Mn alloys. The results of investigations concerning the surface treatment of Mg-Al-Zn-Mn alloys were also presented. The paper also analyses the effect of the surface treatment process carried out through rapid heating, and then rapid crystallisation and the embedding of hard ceramic particles into the laser remelted surface zone using carbides of titanium, tungsten, vanadium, niobium, tantalum and aluminium oxide with the use of high-performance diode lasers (HPDL) on the structure and properties of the investigated materials. Interdisciplinary investigations with technological foresight methods concerning the surface engineering of magnesium alloys were also added. The elaboration also contains a computer analysis using neural networks, enabling to determine and/or predict the phenomena taking place in the examined Mg-Al-Zn-Mn alloys during the heat and surface treatment.

The sixth chapter concerns "Microstructure and Mechanical Properties of the Alloys of Mg-Zn-Zr System." The chapter was prepared by Jae-Hyung Choa, Hyoung-Wook Kima and Suk-Bong Kanga from the Korea Institute of Materials Science in Seongsan-gu, South Korea and the University of Science and Technology in Daejeon, South Korea. A variety of research has been conducted to enhance the mechanical properties of such alloys, both by careful alloy design and by thermomechanical processing, and thus to develop wrought Mg alloys with high strength and formability. The addition of Zn to Mg alloys contributes to their strength by solid solution hardening and precipitate hardening. Additional doping of small amounts of Zr in Mg alloys also improves mechanical properties by grain refinement. The effect of Zn on rollability and mechanical properties of Mg alloys was investigated. Rollability and mechanical properties were improved with Zn content. The effect of ageing heat treatment on microstructure, texture and mechanical properties was also examined. After the ageing, many precipitates were found to be formed and affected mechanical behaviours.

"Effect of Heat Treatment Parameters on the Microstructure of Mg-9Al Magnesium Alloy" is the title of another, seventh chapter prepared by Andrzej Kiełbus, Janusz Adamiec and Robert Jarosz from the Silesian University of Technology, Institute of Materials Science in Katowice, Poland and ZM WSK Rzeszów S.A. in Rzeszow, Poland, respectively. This is a research paper. The results were presented of experimental studies concerning changes in the structure of magnesium alloy with 9% of Al after heat treatment. The examined alloy, in the cast state, is characterised by a constant structure of the α-Mg solution with discontinuous and continuous precipitates of $Mg_{17}Al_{12}$ phase at grain boundaries. Moreover, the occurrence of Laves' phase Mg_2Si and precipitates of Mn_5Al_8 phase, has been provided. The relationship between the as-cast structure, heat treatment conditions and quantity of intermetallic phases was specified. After supersolution treatment a reduction of the contents of $Mg_{17}Al_{12}$ phase precipitations was observed. The application of ageing treatment caused the precipitation of discontinuous $Mg_{17}Al_{12}$ phase.

Filip Pastorek from the University of Zilina, Slovakia, prepared a research paper on "Biodegradable magnesium alloys with aluminium, lithium and rare earth additions" published as the eighth chapter. The chapter describes the interaction between various magnesium alloys and environments of the human body or simulated body fluids, in vitro and in vivo corrosion behaviour tests of Mg alloys with Al, Li and RE addition. These materials could be applied as temporary and biodegradable implants in the human body. Attention was drawn to gas cavities formation as a result of the interactions between magnesium alloys and body fluids.

The ninth chapter entitled "Air-Formed Oxide Films on Magnesium Alloys, Composition, Structure and Protective Properties" was prepared by Sebastián Feliu Jr. from the National Center for Metallurgical Research in Madrid, Spain. The author notes that despite the growing application of magnesium alloys in corrosive environments, their corrosion resistance was not properly explained. The author focusses on the nature and properties of a thin oxide layer with a thickness of approx. 3 nm, which is formed on the outer surface of the alloy by contact with an oxidising environment. A review is made of the current knowledge on the air-formed thin oxide layers on the surface of magnesium alloys and the possible relationship between the composition of the surface and degradation of the material. There is a critical discussion on the protection ability of some of these thin layers in different corrosive environments. The protection properties of Mg alloys in the corrosive environment depend on differences in the thickness, homogeneity and chemical composition of such thin layers associated with a concentration of an alloy element in the magnesium alloy or physiochemical properties of the substrate surface of this alloy.

Shae K. Kim from the Korea Institute of Industrial Technology and the University of Science and Technology in Incheon, South Korea, is also showing interest in a thin layer of oxides on the surface of magnesium alloys, but presents a completely different approach in the prepared chapter ten entitled "Importance of Cleanliness for Magnesium Alloys." The author draws attention to three essential aspects, i.e., SHE (Safety, Health, Environment), which, when followed during the production of magnesium alloys, can bring substantial environmental benefits. The author links complications in this respect with non-chronic porous oxides on the surface of Mg alloys. These surface oxides are fused together with the entire charge during the technological processes, casting for example, thus contaminating these alloys, with such contamination growing with each next remelting of the alloy. Industrial cooperation is therefore important, to create technological solutions ensuring compliance with the conditions of SHE. The author suggests a solution associated with the introduction of ECO-Mg alloys by adding CaO powders to all the new and previously produced types of Mg alloys. The improvement of castability of ECO-Mg alloys is of strategic importance, especially in case of pressure-cast thin parts with a complex shape with ensuring high efficiency. On the other hand, ECO-Mg alloys for plastic processing should be subject to plastic working at the temperature higher than until now, with a higher speed of deformation and higher efficiency, and the Author also devotes attention to this aspect.

Yu Zhong from Florida International University in Miami, Florida, and Zi-Kui Liu from the Pennsylvania State University in Pennsylvania, prepared the eleventh chapter titled "Alloy Development—Application of Computational Thermodynamics." The fundamentals of computational thermodynamics including the software CALPHAD (CALculation of PHAse Diagram) and the first-principles calculations are reviewed in this chapter. Furthermore, the applications of computational thermodynamics to Mg-based alloys are discussed in detail including calculations of phase equilibria diagrams and thermochemical properties, simulations of solidification and hydrogen embrittlement, predictions of bulk metallic glass, and the linkage with mechanical properties. In addition, the first-principles predictions on new phases are illustrated. Finally, the future trend of computational thermodynamics is discussed, including the additional capabilities of first-principles calculations, thermodynamic database automation, and kinetic simulations.

The form of the last, twelfth chapter differs, by the rule, from all others. A team of authors including S.V.S. Narayana Murty, Sushant K Manwatkar, Venkata Sai Kiran Chakravadhanula and P. Ramesh Narayanan from the Vikram Sarabhai Space Centre in Trivandrum, India created a chapter titled "Atlas of Microstructures." The state-of-the-art magnesium alloys necessitate a thorough understanding of the structure-property correlation imagined by the tetrahedron bonding structure, processing, properties and applications. The chapter presents the examples of the structure obtained through optical, scanning and transmission electron microscopes. The detailed methodology of these investigations of magnesium alloys has been described along with a list of etching reagents to delineate the microstructures for optical microscopic examination. An atlas of structures of several, commercial and experimental magnesium alloys in various conditions has been included.

1.5 GENERAL REMARKS

The book is addressed to the honourable readers. The authors are in the hope that the efforts made to prepare the book will be regarded as useful in everyday acquisition of knowledge by different persons. In closing this message, the Editors want to recall a few thoughts of eminent scholars, which may be a signpost in the acquisition of knowledge, also by reading this book.

Sir Prof. Harry Kroto, a Nobel Prize winner in Chemistry in 1996, pointed out that the undertaken tasks should be carried out with full determination, using the words "When choosing something worth doing, never give up and try not to disappoint anyone." We remain deeply convinced that, as Editors, we just so acted.

Wisława Szymborska, a Nobel Prize winner in Literature in 1996, drew attention to the need for constant deepening of knowledge, saying in her Nobel speech that "Any knowledge that does not lead to new questions becomes in a short time dead." This is the purpose of this book and the book is to be a proof that the right things have been done so far, and at the same time it defines a certain intellectual boundary, which requires the adoption of new efforts to set out a new scope of cognition in this field.

Finally, we quote Prof. Hiroshi Amano, a Nobel Prize winner in Physics in 2014, who formulated a very important expectation in his Nobel speech, saying "I would like to see the younger generation attempting to tackle subjects which will greatly contribute to improving the quality of human lives. By doing so, the younger generation can develop a much better world for themselves." It is evident that we have prepared this book about magnesium and its alloys, as the older generation of researchers, primarily in order to make it easier to recognise in what place on the path of cognition we are on. It is up to the younger generation to decide what will be soon done in this regard.

Above all, however, is the thought formulated by Prof. Albert Einstein, a Nobel Prize winner in Physics in 1921, saying "Imagination is more important as knowledge. For knowledge is limited to all we now know and understand, while imagination embraces the entire world, and all there ever will be to know and understand." However, one should have the knowledge and recognition of the current state of affairs, because it is the starting point for the imagination, which by itself will not function to create the progress. The purpose of the book is to convey the current knowledge on magnesium alloys to young researchers, what can engage their imagination, inspiring new discoveries and achievements.

Hence, there is nothing left but to propagate the thoughts of the greatest representatives of our civilisation. As the editors we wish you an interesting reading and new associations and inspirations from studying the content of the book, which—in this way and now—we deliver to your hands.

REFERENCES

1. H. Davy, Electrochemical researches, on the decomposition of the earths; with observations in the metals obtained from the alkaline earths, and on the amalgam procured from ammonia, *Philosophical Transactions of the Royal Society*, 98 (1808) pp. 333–370, doi:10.1098/rstl.1808.0023.
2. L.A. Dobrzański, Metals and alloys, *Open Access Library, Annal VII*, 2 (2017) pp. 1–982 (in Polish).
3. L.A. Dobrzański, *Materiały inżynierskie i projektowanie materiałowe. Podstawy nauki o materiałach i metaloznawstwo*, Wydawnictwa Naukowo-Techniczne, Wydanie II zmienione i uzupełnione, Warszawa, Poland (2006) pp. 1–1600.
4. L.A. Dobrzański, T. Tański, A.D. Dobrzańska-Danikiewicz, M. Król, S. Malara & J. Domagała-Dubiel, Mg-Al-Zn alloys structure and properties, *Open Access Library*, 5, 11 (2012) pp. 1–319 (in Polish).
5. H. Friedrich & S. Schumann, Research for a "new age of magnesium" in the automotive industry, *Journal of Materials Processing Technology* 117 (2001) 276–281.
6. D. Ghosh, R. Carnahan, R. Decker, C. Van Schilt, P. Frederick & N. Bradley, *Magnesium Alloys and Their Applications*, Deutsche Gesellschaft für Metallkunde (1992).
7. R.V. Mises, Mechanik der plastichen formänderung von kristallen, *Zeitschrift für Angewandte Mathematik und Mechanik*, 8 (1928) pp. 161–185. doi:10.1002/zamm.19280080302.
8. G.I. Finch & A.G. Quarrell, Structure of magnesium, zinc and aluminium films, *Proceedings of the Royal Society; Mathematical, Physical And Engineering Sciences*, 141, 844 (1933) pp. 398–414. doi:10.1098/rspa.1933.0126.
9. Z. Trojanová, Z. Drozd & P. Lukáč, Superplastic behaviour of selected magnesium alloys. In T. Tanski (Ed.), *Magnesium Alloys*, Intech (2018), doi:10.5772/intechopen.79752.
10. X.J. Wang, D.K. Xub,, R.Z. Wuc, X.B. Chend, Q.M. Penge, L. Jinf, Y.C. Xing et al., What is going on in magnesium alloys? *Journal of Materials Science & Technology*, 34 (2018) pp. 245–247. doi:10.1016/j.jmst.2017.07.019.

11. N. Pilling & R.J. Bedworth, The oxidation of metals at high temperatures, *Journal of the Institute of Metals*, 29 (1923) p. 529.
12. L.M. Sheppard, Using "corrosion" to make ceramics. Form a metal precursor into a complex, shape, oxidize it, and get a ceramic with the same shape and dimensions, *Chemical Innovation*, 31 (2001) pp. 23–30.
13. E.L. Dreizin, Ch.H. Berman & E.P. Vicenzi, Condensed-phase modifications in magnesium particle combustion in air, *Scripta Materialia*, 122 (2000), pp. 30–42. doi:10.1016/S0010-2180(00)00101-2.
14. A.M.P. Romani, Magnesium in health and disease. In: A. Sigel, H. Sigel & R.K.O. Sigel, *Interrelations Between Essential Metal Ions and Human Diseases. Metal Ions in Life Sciences*, Springer, Dordrecht, the Netherlands, 13 (2013) pp. 49–79. doi:10.1007/978-94-007-7500-8_3.
15. Standing Committee on the Scientific Evaluation of Dietary Reference Intakes; Food and Nutrition Board; Institute of Medicine (IOM). *Dietary Reference Intakes for Calcium, Phosphorus, Magnesium, Vitamin D, and Fluoride*, National Academies Press, Washington, DC (1997).
16. U. Gröber & U. Magnesium. In U. Gröber (Ed.), *Micronutrients: Metabolic Tuning-Prevention-Therapy*, 1st ed., MedPharm Scientific Publishers, Stuttgart, Germany (2009) pp. 159–166.
17. R.K. Rude & R.K. Magnesium. In P.M. Coates, J.M. Betz (Eds.), *Encyclopedia of Dietary Supplements*, 2nd ed., Informa Healthcare, New York (2010) pp. 527–537.
18. R.K. Rude & R.K. Magnesium. In A.C. Ross, B. Caballero, R.J. Cousins, K.L. Tucker & T.R. Ziegler (Eds.), *Modern Nutrition in Health and Disease*, 11th ed., Lippincott Williams & Wilkins, Baltimore, MD (2012) pp. 159–175.
19. U. Gröber, J. Schmidt & K. Kisters, Magnesium in prevention and therapy, *Nutrients*, 7, 9 (2015) pp. 8199–8226. doi:10.3390/nu7095388.
20. P.C. Pham, P.M. Pham, S.V. Pham, J.M. Miller & P.T. Pham, Hypomagnesemia in patients with type 2 diabetes, *Clinical Journal of the American Society of Nephrology*, 2 (2007) pp. 366–373.
21. W. Vierling, D.H. Liebscher, O. Micke, B. von Ehrlich & K. Kisters, Magnesium deficiency and therapy in cardiac arrhythmias: Recommendations of the german society for magnesium research, *Deutsche Medizinische Wochenschrift*, 138 (2013) pp. 1165–1171.
22. B. Chernow, S. Bamberger, M. Stoiko, M. Vadnais, S. Mills, V. Hoellerich & A.L. Warshaw, Hypomagnesemia in patients in postoperative intensive care,. *Chest*, 95 (1989) pp. 391–397.
23. R. Whang & K.W. Ryde, Frequency of hypomagnesemia and hypermagnesemia. Requested vs. routine, *JAMA*, 263 (1990) pp. 3063–3064.
24. L. Spätling, H.G. Classen, W.R. Külpmann, F. Manz, P.M. Rob, H.F. Schimatschek, W. Vierling, J. Vormann, A. Weigert & K. Wink, Diagnosing magnesium deficiency. Current recommendations of the society for magnesium research, *Fortschritte der Medizin Originalien*, 118 (2000) pp. 49–53.
25. United States Geological Survey. Magnesium. Statistics and Information, https://minerals.usgs.gov/minerals/pubs/commodity/magnesium/; Accessed February 23, 2019.
26. D.S. Kumar, C.T. Sasanka, K. Ravindra & K.N.S. Suman, Magnesium and its alloys in automotive applications: A review, *American Journal of Materials Science and Technology*, 4, 1 (2015) pp. 12–30. doi:10.7726/ajmst.2015.1002.
27. G.J. Simandl, H. Schultes, J. Simandl & S. Paradis, Magnesium: Raw materials, metal extraction and economics: Global picture, *Proceedings of the Ninth Biennial SGA Meeting*, Dublin, ireland (2007), pp. 827–830.
28. United States Geological Survey, 2016 *Minerals Yearbook, Magnesium*, [Advance Release], August 2018, https://minerals.usgs.gov/minerals/pubs/commodity/magnesium/myb1-2016-mgmet.pdf; Accessed February 24, 2019.

29. United States Geological Survey. *Magnesium Compounds*, https://minerals.usgs.gov/minerals/pubs/commodity/magnesium/mcs-2018-mgcom.pdf; Accessed February 23, 2019.
30. M. Halmann, A. Frei & A. Steinfeld, Magnesium production by the pidgeon process involving dolomite calcination and Mg) silicothermic reduction: Thermodynamic and environmental analyses, *Industrial & Engineering Chemistry Research*, 47 (2008) pp. 2146–2154. doi:10.1021/ie071234v.
31. S. Ramakrishnan & P. Koltun, Global warming impact of the magnesium produced in China using the pidgeon process, *Resources, Conservation and Recycling*, 42 (2004) pp. 49–64.
32. H. Li & S. Xie,. Research on development of dolomitesferrosilicon thermal reduction process of magnesium production, *Journal of Rare Earths*, 23 (2005) (Suppl. Dec), pp. 606–610.
33. G.J. Kipouros & D.R. Sadoway, The chemistry and electrochemistry of magnesium production in G. Mamantov, C.B. Mamantov and J. Braunstein (Eds.), *Advances in Molten Salt Chemistry*, Vol. 6, Elsevier, Amsterdam, the Netherlands (1987) pp. 127–209.
34. P. Duhaime, P. Mercille & M. Pineau, Electrolytic process technologies for the production of primary magnesium, *Journal Mineral Processing and Extractive Metallurgy Transactions of the Institutions of Mining and Metallurgy: Section C*, 111, 2 (2002) pp. 53–55. doi:10.1179/mpm.2002.111.2.53.
35. U.B. Pal & A. Powell, The use of solid-oxide-membrane technology for electrometallurgy, *The Journal of the Minerals, Metals & Materials Society*, 59, 5 (2007) pp. 44–49. doi:10.1007/s11837-007-0064-x.
36. X. Guan & U.B. Pal, Design of optimum solid oxide membrane electrolysis cells for metals production, *Progress in Natural Science: Materials International*, 25 (2015) pp. 591–594. doi:10.1016/j.pnsc.2015.11.004.
37. X. Guan, U.B. Pal, Y. Jiang & S. Su, Clean metals production by solid oxide membrane electrolysis process, *Journal of Sustainable Metallurgy*, 2 (2016) pp. 152–166. doi:10.1007/s40831-016–0044-x.
38. Statista—The Statistics Portal, Primary magnesium production worldwide 2010–2017, https://www.statista.com/statistics/569515/primary-magnesium-production-worldwide/; Accessed February 23, 2019.
39. Statista—The Statistics Portal, Primary magnesium production worldwide in 2016 https://www.statista.com/statistics/569535/primary-magnesium-production-capacity-by-country/; Accessed February 23, 2019.
40. Statista—The Statistics Portal, U.S. value of magnesium recycled from scrap between 2005 and 2016 (in million U.S. dollars), https://www.statista.com/statistics/209425/recycled-value-of-magnesium-in-the-us/; Accessed February 23, 2019.
41. S. Housh, B. Mikucki & A. Stevenson, Selection and application of magnesium and magnesium alloys and Properties of magnesium alloys in ASM International Handbook Committee, ASM Handbook, Vol. 2, Properties and selection: Nonferrous alloys and special-purpose materials, Fifth printing, ASM International (1998) p. 455–516.
42. M. Avedesian & H. Baker (Eds.), *ASM Specialty Handbook: Magnesium and Magnesium Alloys*, ASM International, Materials Park, OH (1999).
43. H. Baker, *Physical Properties of Magnesium and Magnesium Alloys*, The Dow Chemical Company, Midland, MI (1997).
44. B.L. Mordike & F. Bush, Development of high temperature creep resistant alloys. In K.U. Kainem (Ed.), *Magnesium Alloys and Their Applications*, Wiley-VCH, Weinheim, Germany (2000) pp. 35–39.
45. Z. Koren, H. Rosenson, E.M. Gutman, Ya. Unigovski & A. Eliezer, Development of semisolid casting for AZ91 and AM50 magnesium alloys, *Journal of Light Metals*, 2 (2002) pp. 81–87.

46. Y.Z. Lu, Q.D. Wang, W.J. Ding, X.Q. Zeng & Y.P. Zhu, Fracture behavior of AZ91 magnesium alloy, *Materials Letters*, 44 (2000) pp. 265–268.
47. S. Lun Sin & D. Dubé, Influence of process parameters on fluidity of investment-cast AZ91D magnesium alloy, *Materials Science and Engineering A*, 386 (2004) pp. 34–42.
48. International Magnesium Association, Magnesium alloys overview, https://www.intlmag.org/page/design_mag_all_ima; Accessed February 23, 2019.
49. B.L. Mordike & T. Ebert, Magnesium: Properties—applications—potential, *Materials Science and Engineering*, A302 (2001) pp. 37–45. doi:10.1016/S0921-5093(00)01351-4.
50. C. Blawert, N. Hort & K.U. Kainer, Automotive applications of magnesium and its alloys, *Transactions of the Indian Institute of Metals*, 57, 4 (2001) pp. 397–408.
51. P. Asadi, K. Kazemi-Choobi & A. Elhami. In W.A. Monteiro (Ed.), *Welding of Magnesium Alloys, New Features on Magnesium Alloys*, InTech (2012). doi:10.5772/47849.
52. M. Sugamata, S. Hanawa & J. Kaneko, Structures and mechanical properties of rapidly solidified Mg-Y based alloys, *Materials Science and Engineering: A*, 226–228 (1997) pp. 861–866. doi:10.1016/S0921-5093(97)80089-5.
53. J. Cai, G.C. Ma, Z. Liu, H.F. Zhang & Z.Q. Hu, Influence of rapid solidification on the microstructure of AZ91HP alloy, *Journal of Alloys and Compounds*, 422, 1–2 (2006) pp. 92–96. doi:10.1016/j.jallcom.2005.11.054.
54. X.F. Guo & D. Shechtman, Reciprocating extrusion of rapidly solidified Mg–6Zn–1Y–0.6Ce–0.6Zr alloy, *Journal of Materials Processing Technology*, 187–188 (2007) pp. 640–644. doi:10.1016/j.jmatprotec.2006.11.056.
55. A. Uoya, T. Shibata, K. Higashi & A. Inoue, Superplastic deformation characteristics and constitution equation in rapidly solidified Mg–Al–Ga alloy, *Journal of Materials Research*, 11, 11 (1996) pp. 2731–2737. doi:10.1557/JMR.1996.0346.
56. W.J. Kim, W.Chung, S. Chung & D. Kum, Superplasticity in thin magnesium alloy sheets and deformation mechanism maps for magnesium alloys at elevated temperatures, *Acta Materialia*, 49, 16 (2001) pp. 3337–3345. doi.org/10.1016/S1359-6454(01)00008-8.
57. K. Kubota, M. Mabuchi & K. Higashi, Review processing and mechanical properties of fine-grained magnesium alloys, *Journal of Materials Science*, 34, 10 (1999) pp. 2255–2262.
58. T.B. Matias, G.H. Asato, B.T. Ramasco, W.J. Botta, C.S. Kiminami & C. Bolfarini, Processing and characterization of amorphous magnesium based alloy for application in biomedical implants, *Journal of Materials Research and Technology*, 3, 3 (2014) pp. 203–209. doi.org/10.1016/j.jmrt.2014.03.007.
59. J. S.-C. Jang C.-T. Tseng L.-J. Chang J. C.-C. Huang Y.-C. Yeh & J.-L. Jou, Thermoplastic forming properties and microreplication ability of a magnesium-based bulk metallic glass, *Advanced Engineering Materials*, 10, 11 (2008) pp. 1048–1052. doi.org/10.1002/adem.200800104.
60. H. Men, Z.Q. Hu & J. Xu, Bulk metallic glass formation in the Mg–Cu–Zn–Y system, *Scripta Materialia*, 46, 10 (2002) pp. 699–703. doi.org/10.1016/S1359-6462(02)00055-6.
61. M.S, Yong & A.J. Clegg, Process optimisation for a squeeze cast magnesium alloy, *Journal of Materials Processing Technology*, 145, 1 (2004) pp. 134–141. doi:10.1016/j.jmatprotec.2003.07.006.
62. H. Hu, Squeeze casting of magnesium alloys and their composites, *Journal of Materials Science*, 33, 6 (1998) pp. 1579–1589.
63. D.G. Eskin, J. Zuidema Jr., V.I. Savran & L. Katgerman, Structure formation and macrosegregation under different process conditions during DC casting, *Materials Science and Engineering: A*, 384, 1–2 (2004) pp. 232–244. doi:10.1016/j.msea.2004.05.066.

64. Z. Shao, Q. Le, Z. Zhang &, J. Cui, A new method of semi-continuous casting of AZ80 Mg alloy billets by a combination of electromagnetic and ultrasonic fields, *Materials & Design*, 32, 8–9 (2011) pp. 4216–4224. doi:10.1016/j.matdes.2011.04.035.
65. F. Pan, X. Chen, T. Yan, T. Liu, J. Mao, W. Luo, Q. Wang, J. Peng, A. Tang & B. Jiang, A novel approach to melt purification of magnesium alloys, *Journal of Magnesium and Alloys*, 4 (2016) pp. 8–14. doi:10.1016/j.jma.2016.02.003.
66. D.C. Hofmann, K.S. Wecchio, Submerged friction stir processing (SFSP): An improved method for creating ultra-fine-grained bulk material, *Materials Science and Engineering A*, 402, 1–2 (2005) pp. 234–241. doi:10.1016/j.msea.2005.04.032.
67. P. Cavaliere & P.P. De Marco, Superplastic behavior of friction stir processed AZ91 magnesium alloy produced by high pressure die cast, *Journal of Materials Processing Technology*, 184 (2007) pp. 77–83. doi:10.1016/j.jmatprotec.2006.11.005.
68. D. Zhang, S. Wang, C. Qiu & W. Zhang, Superplastic tensile behavior of a fine grained AZ91 magnesium alloy prepared by friction stir processing, *Materials Science and Engineering A*, 556 (2012) pp. 100–106. doi:10.1016/j.msea.2012.06.063.
69. F. Chai, D. Zhang, Y. Li & W. Zhang, High strain rate superplasticity of a fine-grained AZ91 magnesium alloy prepared by submerged friction stir processing, *Materials Science and Engineering A*, 568 (2013) pp. 40–48. doi:10.1016/j.msea.2013.01.026.
70. A. Mohan, W. Yuan & R.S. Mishra, High strain rate superplasticity in friction stir processed ultrafine-grained Mg-Al-Zn alloys, *Materials Science and Engineering A*, 562 (2013) pp. 69–76. doi:10.1016/j.msea.2012.11.026.
71. A. Raja, P. Biswas & V. Pancholi, Effect of layered microstructure on the superplasticity of friction stir processed AZ91 magnesium alloy, *Materials Science and Engineering A*, 725 (2018) pp. 492–502. doi:10.1016/j.msea.2018.04.028.
72. J.A. del Valle, F. Carreño & O.A. Ruano, Influence of texture and grain size on work hardening and ductility in magnesium-based alloys processed by ECAP and rolling, *Acta Materialia*, 54, 16 (2006) pp. 4247–4259. doi:10.1016/j.actamat.2006.05.018.
73. R.B. Figueired & T.G. Langdon, Principles of grain refinement and superplastic flow in magnesium alloys processed by ECAP, *Materials Science and Engineering: A*, 501, 1–2 (2009) pp. 105–114, doi:10.1016/j.msea.2008.09.058.
74. K. Bryła, M. Krystian, J. Horky, B. Mingler, K. Mroczka, P. Kurtyka & L. Lityńska-Dobrzyńska, Improvement of strength and ductility of an EZ magnesium alloy by applying two different ECAP concepts to processable initial states, *Materials Science and Engineering: A*, 737 (2018) pp. 318–327; doi.org/10.1016/j.msea.2018.09.070.
75. Z. Zeng, N. Stanford, Ch. Huw, J. Davies, J.-F. Nie & N. Birbilis, Magnesium extrusion alloys: A review of developments and prospects, *Journal International Materials Reviews*, 64,1 (2019) pp. 27–62. doi:10.1080/09506608.2017.1421439.
76. S. You, Y. Huang, K.U. Kainer & N. Hort, Recent research and developments on wrought magnesium alloys, *Journal of Magnesium and Alloys*, 5 (2017) pp. 239–253. doi:10.1016/j.jma.2017.09.001.
77. J. Jayakumar, B.K. Raghunath & T.H. Rao, Recent development and challenges in synthesis of magnesium matrix nano composites: A review, *International Journal of Latest Research in Science and Technology*, 1, 2 (2012) pp. 164–171.
78. H. Singh, N.J. Sarabjit & A.K Tyagi, An overview of metal matrix composite: Processing and sic based mechanical properties, *Journal of Engineering Research and Studies*, II, IV (2011), pp. 72–78.
79. P.-Ch. Lin, S.-J. Huang & P.-S. Hong, Formation of magnesium metal matrix composites Al2o3p/Az91d and their mechanical properties after heat treatment. *Acta Metallurgica Slovaca*, 16, 4 (2010) pp. 237–245.
80. M. Gupta, M.O. Lai & D. Saravanaranganathan, synthesis, microstructure and properties characterization of disintegrated melt deposited Mg/SiC composites, *Journal of Materials Science*, 35, 9 (2010) pp. 2155–2165.

81. M. Gupta & N.M.L. Sharon, *Magnesium, Magnesium Alloys and Magnesium Composites*, John Wiley & Sons, Hoboken, NJ (2011).
82. Y.A. Filatov, V.I. Yelagin & V.V. Zakharow, New Al-Mg-Sc alloys, *Materials Science and Engineering A*, 280 (2000) pp. 97–101.
83. M.K. Kulekci, Magnesium and its alloys applications in automotive industry, *The International Journal of Advanced Manufacturing Technology*, 39, 9–10 (2008) pp. 851–865.
84. E. Aghion, B. Bronfin & D. Eliezer, The role of the magnesium industry in protecting the environment, *Journal of Materials Processing Technology*, 117, 3 (2001) pp. 381–385; doi:10.1016/S0924-0136(01)00779-8.
85. A. H. Musfirah & J.A. Ghani, Magnesium and aluminum alloys in automotive industry, *Journal of Applied Sciences Research*, 8, 10 (2012) pp. 4865–4875.
86. Klaumünzer D. et al., Magnesium process and alloy development for applications in the automotive industry. In V. Joshi, J. Jordon, D. Orlov, N. Neelameggham. (Eds.), *Magnesium Technology The Minerals, Metals & Materials Series*, Springer, Cham, Switzerland (2019). doi:10.1007/978-3-030-05789-3_3.
87. A. Wendt, K. Weiss, A. Ben-Dov, M. Bamberger & B. Bronfin, Magnesium castings in aeronautics applications: Special requirements. In S.N. Mathaudhu, A.A. Luo, N.R. Neelameggham, E.A. Nyberg & W.H. Sillekens (Eds.), *Essential Readings in Magnesium Technology*, Springer, Cham, Switzerland (2016). doi:10.1007/978-3-319-48099-2_9.
88. M. Toozandehjani, N. Kamarudin, Z. Dashtizadeh, E.Y. Lim, A. Gomes & Ch. Gomes, Conventional and advanced composites in aerospace industry: Technologies revisited, *American Journal of Aerospace Engineering*, 5, 1 (2018) pp. 9–15. doi:10.11648/j.ajae.20180501.12.
89. Y. Wang, D. Tie, R. Guan, N. Wang, Y. Shang, T. Cui & J. Li. Microstructures, mechanical properties, and degradation behaviors of heat-treated Mg-Sr alloys as potential biodegradable implant materials, *Journal of the Mechanical Behavior of Biomedical Materials*, 77 (2018) pp. 47–57.
90. D. Lin, F. Hung, T. Lui & M. Yeh, Heat treatment mechanism and biodegradable characteristics of ZAX1330Mg alloy, *Materials Science & Engineering C-Materials for Biological Applications*, 51 (2015) pp. 300–308.
91. Y. Dai, Y. Lu, D. Li, K. Yu, D. Jiang, Y. Yan, L. Chen & T. Xiao, Effects of polycaprolactone coating on the biodegradable behavior and cytotoxicity of Mg-6%Zn-10%Ca3(PO4)2 composite in simulated body fluid, *Materials Letters*, 198 (2017) pp. 118–120.
92. Y. Chen, Z. Xu, C. Smith & J. Sankar, Recent advances on the development of magnesium alloys for biodegradable implants, *Acta Biomaterialia*, 10 (2014) p. 4561.
93. A. McGoron & D. Persaud-Sharma, Biodegradable magnesium alloys: A review of material development and applications, *Journal of Biomaterials and Tissue Engineering*, 12 (2011) pp. 25–39.
94. D. Tie, R. Guan, H. Liu, A. Cipriano, Y. Liu, Q. Wang, Y. Huang & N. Hort, An in vivo study on the metabolism and osteogenic activity of bioabsorbable Mg-1Sr alloy, *Acta Biomaterialia*, 29 (2016) pp. 455–467.
95. S. Zhang, X. Zhang, C. Zhao, J. Li, Y. Song, C. Xie, H. Tao, Y. Zhang, Y. He, Y. Jiang & Y. Bian, Research on an Mg-Zn Alloy as a degradable biomaterial, *Acta Biomaterialia*, 6 (2010) pp. 626–640.
96. Z. Li, X. Gu, S. Lou & Y. Zheng, The development of binary Mg-Ca alloys for use as biodegradable materials within bone, *Biomaterials*, 29 (2008) pp. 1329–1344.
97. K. Yu, L. Chen, J. Zhao, S. Li, Y. Dai, Q. Huang & Z. Yu, In vitro corrosion behavior and in vivo biodegradation of biomedical β-Ca3(PO4)2/Mg–Zn composites, *Acta Biomaterialia*, 8 (2012) pp. 2845–2855.

98. M.B. Kannan & R.K. Raman, In vitro degradation and mechanical integrity of calcium-containing magnesium alloys in modified-simulated body fluid, *Biomaterials*, 29 (2008) pp. 2306.
99. G. Song, Control of biodegradation of biocompatible magnesium alloys, *Corrosion Science*, 49 (2007) pp. 1696–1701.
100. S. Farè, Q. Ge, M. Vedani, G. Vimercati, D. Gastaldi, F. Migliavacca, L. Petrini & S. Trasatti, Evaluation of material properties and design requirements for biodegradable magnesium stents, *Matéria*, 15, 2 (2010) pp. 96–103.
101. X. Li, C. Chua & P. K. Chu, Effects of external stress on biodegradable orthopedic materials: A review, *Bioactive Materials* (2016). doi:10.1016/j.bioactmat.2016.09.00.
102. M. Haghshenas, Mechanical characteristics of biodegradable magnesium matrix composites: A review. *Journal of Magnesium and Alloys*, 5 (2017) pp. 189–120, doi:10.1016/j.jma.2017.05.001.
103. V. K. Bommala, M. Krishna & Ch. T. Rao, Magnesium matrix composites for biomedical applications: A review, *Journal of Magnesium and Alloys*, (2019) in press. doi:10.1016/j.jma.2018.11.001.
104. P. Ch. Banerjee, S. Al-Saadi, L. Choudhary, S.E. Harandi & R. Singh, Magnesium implants: Prospects and challenges, *Materials*, 12, 136 (2019) pp. 1–21. doi:10.3390/ma12010136.
105. R. Radha & D. Sreekanth. Insight of magnesium alloys and composites for orthopedic implant applications: A review, *Journal of Magnesium and Alloys*, 5 (2017) pp. 286–312. doi:10.1016/j.jma.2017.08.003.
106. F. Witte, J. Fischer, J. Nellesen, H.A. Crostack, V. Kaese, A. Pisch, F. Beckmann & H. Windhagen. In vitro and in vivo corrosion measurements of magnesium alloys, *Biomaterials*, 27 (2006) pp. 1013–1018.
107. M. Erinc, W.H. Sillekens, R. Mannens & R.J. Werkhoven, Applicability of existing magnesium alloys as biomedical implant materials, magnesium technology. In San Francisco, E.A. Nyberg, S.R. Agnew, N.R. Neelameggham, and M.Q. Pekguleryuz (Eds.), *Minerals, Metals and Materials Society*, Warrendale, PA (2009) pp. 209–214.
108. M. Bamberger & G. Dehm. Trends in the development of new Mg alloys, *Annual Review of Materials Research*, 38 (2008) pp. 505–533.
109. N. Sezer, Z. Evis, S. M. Kayhanb, A. Tahmasebifar & M. Koç. Review of magnesium-based biomaterials and their applications, *Journal of Magnesium and Alloys*, 6 (2018) pp. 23–43.
110. X. Du, W. Du, Z. Wang, K. Liu & S. Li, Ultra-high strengthening efficiency of graphene nanoplatelets reinforced magnesium matrix composites, *Materials Science and Engineering: A*, 711 (2018) pp. 633–642. doi:10.1016/j.msea.2017.11.040.
111. R. Purohit, Y. Dewang, R.S. Rana, D. Koli & S. Dwivedi, Fabrication of magnesium matrix composites using powder metallurgy process and testing of properties, *Materials Today: Proceedings*, 5, 2, 1 (2018) pp. 6009–6017. doi:10.1016/j.matpr.2017.12.204.
112. M.E. Turan, Y. Sun, F. Aydin, H. Zengin, Y. Turen & H. Ahlatci, Effects of carbonaceous reinforcements on microstructure and corrosion properties of magnesium matrix composites, *Materials Chemistry and Physics*, 218 (2018) pp. 182–188. doi:10.1016/j.matchemphys.2018.07.050.
113. H. Zhi, Y. Xing & Y. Liu, Review of recent studies in magnesium matrix composites, *Journal of Materials Science*, 39, 20 (2004) pp. 6153–6171.
114. M.E. Turan, Y. Sun & Y. Akgul, Mechanical, tribological and corrosion properties of fullerene reinforced magnesium matrix composites fabricated by semi powder metallurgy, *Journal of Alloys and Compounds*, 740 (2018) pp. 1149–1158. doi:10.1016/j.jallcom.2018.01.103.

115. S. Wu, S. Wang, D. Wen, G. Wang & Y. Wang, Microstructure and mechanical properties of magnesium matrix composites interpenetrated by different reinforcement, *Applied Science*, 8 (2018) 2012. doi:10.3390/app8112012.
116. D. Liu, Y. Zuo, W. Meng, M. Chen & Z. Fan, Fabrication of biodegradable nano-sized β-TCP/Mg composite by a novel melt shearing technology, *Materials Science and Engineering: C*, 32, 5 (2012) pp. 1253–1258. doi:10.1016/j.msec.2012.03.017.
117. D. Ahmadkhaniha, Fedel, M. Heydarzadeh Sohi, A. Zarei Hanzaki & F. Deflorian, Corrosion behavior of magnesium and magnesium–hydroxyapatite composite fabricated by friction stir processing in Dulbecco's phosphate buffered saline, *Corrosion Science*, 104 (2016) pp. 319–329. doi:10.1016/j.corsci.2016.01.002.
118. International Magnesium Association, Electronic application, https://www.intlmag.org/page/app_electronic_ima; Accessed February 23, 2019.
119. International Magnesium Association, Sports applications, https://www.intlmag.org/page/app_sports_ima; Accessed February 23, 2019.
120. International Magnesium Association, Other applications, https://www.intlmag.org/page/app_other_ima; Accessed February 23, 2019.

2 Casting Technologies, Microstructure and Properties

Trevor Abbott

CONTENTS

2.1 INTRODUCTION

Magnesium alloys constitute about one-third of total magnesium metal consumption worldwide. The other two-thirds include alloying additions to aluminium, use in the production of steel, and as a reducing agent for titanium and zirconium production (Figure 2.1). The vast majority of magnesium alloy consumption is in high-pressure die-cast applications. Small volumes of alloys are used in gravity cast applications while wrought applications are even less. This product mix is in marked contrast to aluminium and steel where the majority of alloy consumption is in wrought and semi finished products.

In this chapter, the high-pressure die casting (HPDC) processes are described, as well as the common magnesium alloys. The magnesium—aluminium alloy

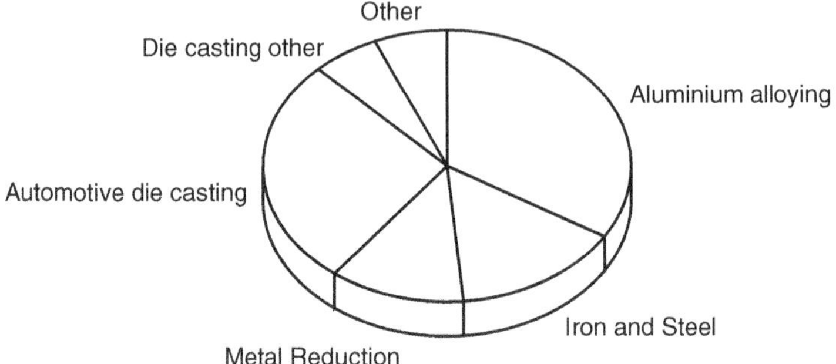

FIGURE 2.1 Breakdown of primary magnesium metal consumption. (Adapted from Abbott, T.B., *Corrosion*, 71, 120–127, 2015.)

system is discussed as it forms the basis of all common die cast magnesium alloys. Relationships between phase equilibrium diagrams, microstructure, castability and properties are explained.

2.2 CASTING TECHNOLOGIES

2.2.1 Comparison of Wrought and Cast Alloy Properties

The dominance of high-pressure diecasting for magnesium alloy applications [1] can be readily understood by comparing the properties of magnesium and aluminium alloys. Magnesium and aluminium are both recognised as light alloys. Aluminium is well established and conversion from an existing aluminium component to a magnesium component requires a property or price advantage.

Two of the most important properties to consider when comparing materials are specific strength (yield strength or tensile strength divided by density) and tensile elongation or ductility. Figure 2.2 compares the properties of wrought and gravity cast aluminium and magnesium alloys [2,3] and shows that, even with the density differences included, wrought magnesium alloys are generally inferior to aluminium. The situation is actually worse for magnesium than Figure 2.2 suggests because processing productivity factors, such as extrusion speed, are generally lower for magnesium, and magnesium also suffers from other property limitations such as tension—compression anisotropy and inferior corrosion resistance.

Figure 2.2 illustrates that for gravity castings (sand castings and permanent mould castings), magnesium alloys are also at a disadvantage. Magnesium alloys are typically more costly than aluminium alloys [1] and so wrought and gravity cast magnesium applications are limited to niche applications where other characteristics such as dent resistance (body panels), electrochemical properties (sacrificial anodes), sound dampening, machinability, castability and other properties are important.

Figure 2.3 shows, that for high-pressure diecastings (HPDC), magnesium alloys do have significant property advantages over aluminium [2,3]. The benefits of magnesium are in fact greater than the strength and ductility comparisons

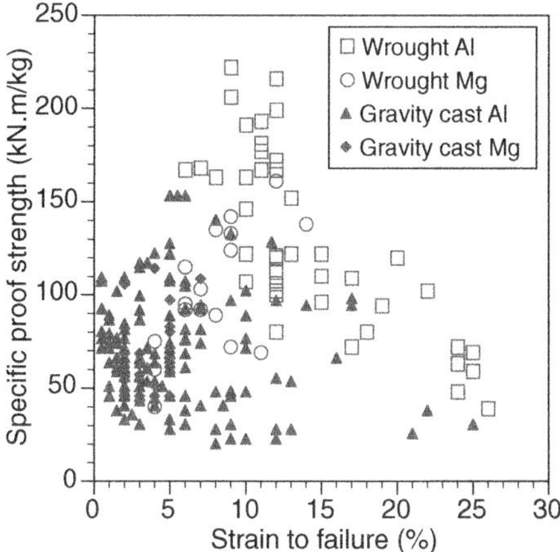

FIGURE 2.2 Comparison of specific proof strength and ductility for wrought and gravity cast magnesium and aluminium alloys. (From Polmear, I.J., *Light Alloys: From Traditional Alloys to Nanocrystals*, 4th Ed, Butterworth Heinemann, Norwood MA, 2005; Kearney, A. and Rooy, E.L., *Aluminum Foundry Products, Properties and Selection: Nonferrous Alloys and Special-Purpose Materials*, Vol 2, ASM Handbook, ASM International, 1990, 123–151.)

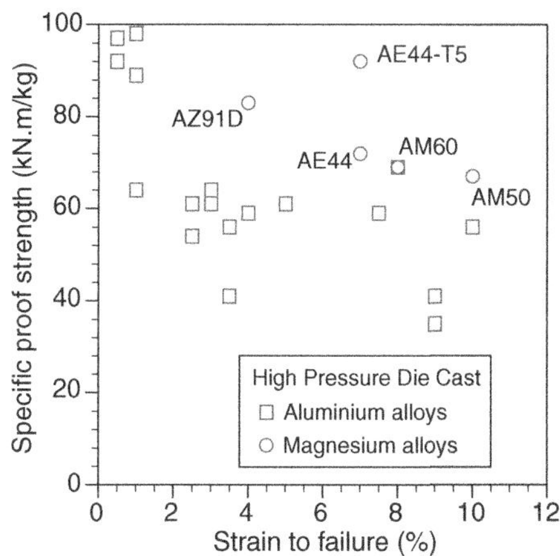

FIGURE 2.3 Comparison of specific proof strength and ductility for high pressure die cast magnesium and aluminium alloys. (From Polmear, I.J., *Light Alloys: From Traditional Alloys to Nanocrystals*, 4th Ed, Butterworth Heinemann, Norwood MA, 2005; Kearney, A. and Rooy, E.L., *Aluminum Foundry Products, Properties and Selection: Nonferrous Alloys and Special-Purpose Materials*, Vol 2, ASM Handbook, ASM International, 1990, 123–151.) T5 ageing treatment for AE44 was 30 hours at 200°C.

suggest. Molten aluminium is much more aggressive towards steel, which reduces die lifetimes compared to magnesium alloys and prevents aluminium alloys from being cast with hot chamber die casting machines. Magnesium alloys can be cast into thinner sections and more complex parts and the lower volumetric specific heat increases productivity.

These differences, in comparative advantages of magnesium and aluminium alloys in wrought, gravity cast and HPDC processing routes, are clearly reflected in the relative volumes of commercial consumption [1].

2.2.2 Magnesium Casting Alloy Designations

Magnesium alloys are commonly named according to a code defined by the ASTM standard, ASTM-B275, which indicates the two highest alloying elements and their concentrations. For example the alloy AZ91D contains approximately 9% aluminium (A), 1% zinc (Z) and the letter D indicates the fourth alloy specification within the series.

This system suffers from an inadequate number of letters in the alphabet, especially as rare earth elements, which were previously all covered by the letter E, increasingly need to be specified individually. Also adding to confusion is the use of proprietary designations that provide little or no indication of the composition. An alternative naming system is to use chemical element symbols as used in the ISO standard (ISO 16220).

In Table 2.1 the common magnesium high-pressure casting alloys are listed according to both their ASTM and/or proprietary naming together with their equivalents according to the ISO system. A typical composition is given for each alloy, which, in most cases, is the centre of the specified composition ranges.

2.2.3 High-Pressure Die Casting Processes

High-pressure die casting involves the rapid injection of molten metal into a die under high pressure. There are two types of die casting machines. Cold chamber die casting

TABLE 2.1
Magnesium High Pressure Die Casting Alloys

Common/ASTM and/ or Proprietary Name	ISO Equivalent (ISO 16220)	Typical Composition
AE42	Mg Al4RE2[a]	Mg-4%Al-2.5%RE-0.2%Mn[a]
AE44-4	Mg Al4RE4[a]	Mg-4%Al-4%RE-0.3%Mn[a]
AE44-2	Mg Al4Ce3La1	Mg-4%Al-2.6%Ce-1.4%La-0.3%Mn
AJ62	Mg Al6Sr2	Mg-6%Al-2%Sr-0.3%Mn
AM20	Mg Al2Mn	Mg-2.1%Al-0.4%Mn
AM50A	Mg Al5Mn	Mg-4.9%Al-0.4%Mn
AM60B	Mg Al6Mn	Mg-6%Al-0.4%Mn
AS21	Mg Al2Si1	Mg-2.2%Al-1%Si-0.2%Mn
AS31	Mg Al3Si1	Mg-3.2%Al-1%Si-0.2%Mn
AS41B	Mg Al4Si1	Mg-4.2%Al-1%Si-0.2%Mn
AZ91D	Mg Al9Zn1	Mg-9%Al-0.7%Zn-0.15%Mn
MRI 230D	Mg Al7Ca2	Mg-6.8%Al-2.2%Ca-0.7%Sn0.3%Sr-0.3%Mn
MRI 153M	Mg Al8Ca1	Mg-8%Al-1%Ca-0.3%Sr-0.25%Mn

[a] RE designates a mixture of rare earths with ~50% cerium, ~25% lanthanum and the remainder predominantly neodymium and praseodymium.

separates the casting machine and the furnace and uses a dosing system to transfer the required amount of molten metal for each shot. Figure 2.4 shows a schematic diagram of the main features of a cold chamber die casting machine. All HPDC magnesium alloys are suitable for cold chamber casting. Cold chamber die casting is especially used for larger components and for components requiring higher integrity such as for the automotive industry. Figure 2.5 shows and example of a cold chamber die cast magnesium component, an engine torque support unit cast in AE44 alloy.

The second process is hot chamber die casting and is shown in Figure 2.6. Hot chamber machines directly couple the furnace and the casting machine and have higher productivity than cold chamber machines. They are generally limited to magnesium alloys with low liquidus temperatures such as AZ91D to maintain reasonable lifetimes for the steel components, such as the gooseneck, which are immersed in the molten metal. Some higher melting point alloys, such as AM60B, are occasionally cast by the hot chamber process. Hot chamber casting is generally used for small components where higher

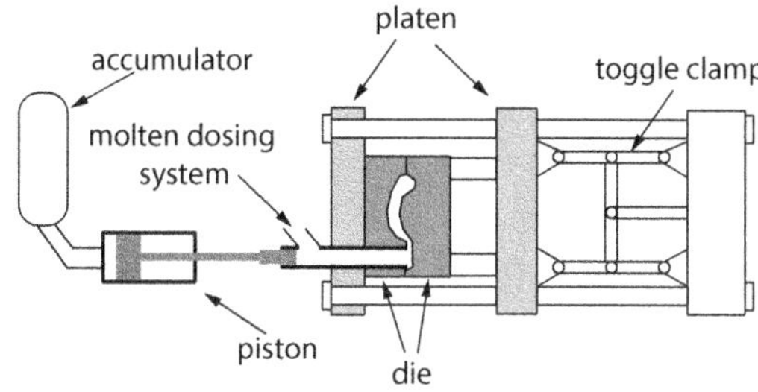

FIGURE 2.4 Schematic illustration of cold chamber high-pressure die casting.

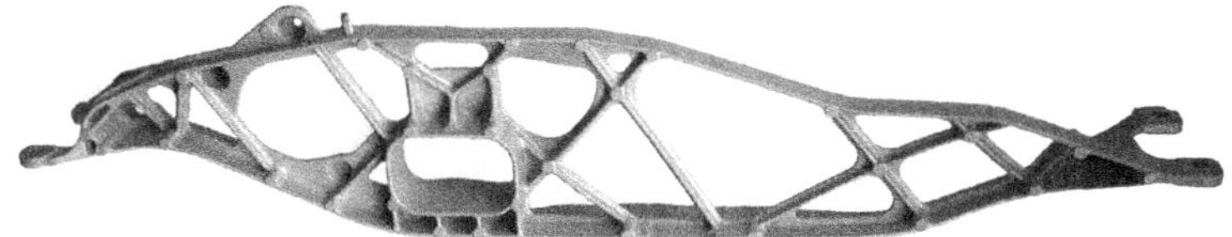

FIGURE 2.5 Example of a cold chamber die cast component. An engine torque support unit cast in AE44 magnesium alloy.

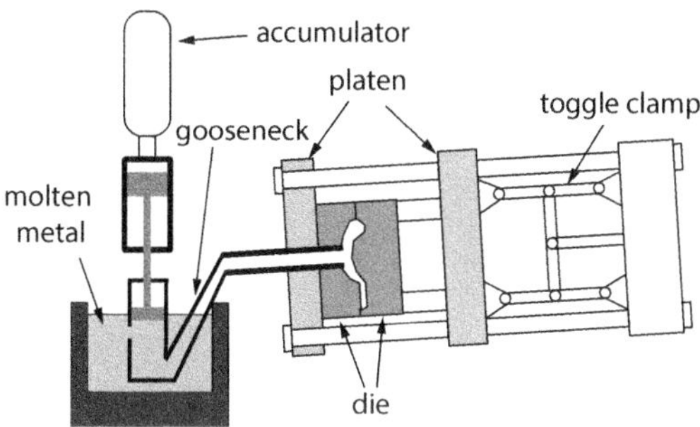

FIGURE 2.6 Schematic illustration of hot chamber high-pressure die casting.

FIGURE 2.7 Example of a hot chamber die cast component. A mobile phone component cast in AZ91D magnesium alloy.

productivities are needed and structural requirements are less important. They are widely used for casting housings for electronic devices such as shown in Figure 2.7.

2.3 PHASE EQUILIBRIA, MICROSTRUCTURE AND CASTABILITY

2.3.1 Magnesium—Aluminium Alloys

The magnesium—aluminium system, shown in Figure 2.8, forms the basis of magnesium alloys used for high-pressure die casting. The three most common magnesium alloys, AZ91D, AM60B and AM50A, are essentially magnesium—aluminium

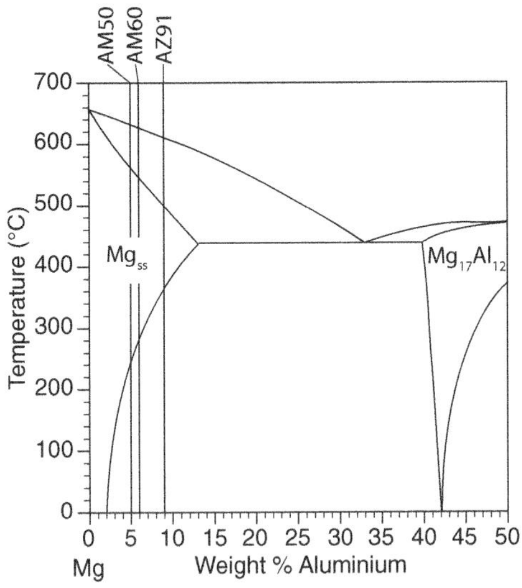

FIGURE 2.8 The magnesium rich end of the magnesium—aluminium phase diagram showing the compositions of the common alloys.

alloys with minor additions of manganese, and in the cases of AZ91D, zinc. Manganese is a necessary addition in the production of aluminium containing alloys. Additions in the range of 0.1%–0.5% manganese are added to reduce iron content and to protect from the harmful effects of residual iron on corrosion performance. Approximately 1% zinc is present in AZ91D to provide solid solution strengthening.

Figure 2.9 shows the microstructures of AM60B and AZ91D alloys. The dominant phase is primary magnesium, which appears darkest. Surrounding the primary magnesium is a network of magnesium saturated in aluminium that was the last to solidify (medium gray). Within this network is $Mg_{17}Al_{12}$ intermetallic particles,

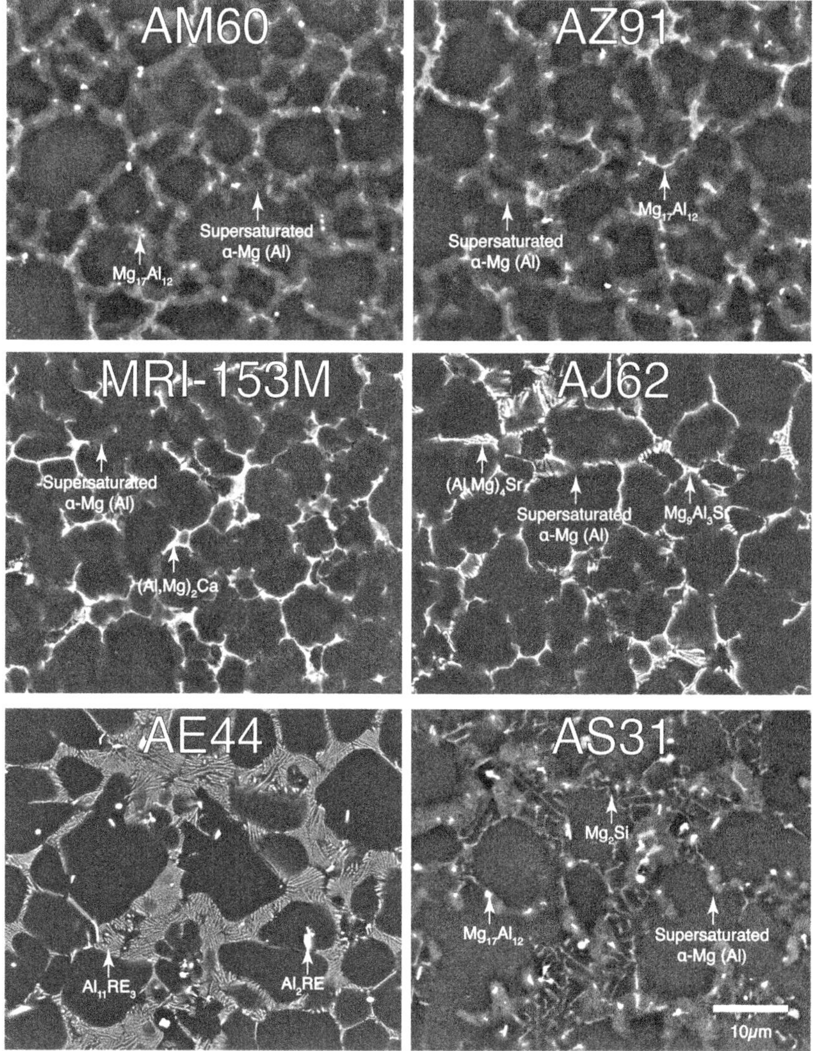

FIGURE 2.9 Microstructures of selected high-pressure die cast magnesium alloys.

which appear white. The main difference between AM60B and AZ91D is that in AM60B the intermetallic fraction is much smaller and exists mostly as isolated particles. In AM50A (not shown) the faction of $Mg_{17}Al_{12}$ is reduced further. In the case of AZ91D the network of $Mg_{17}Al_{12}$ extends over a larger proportion of the primary Mg cell boundaries. This leads to both an increase in yield strength and a decrease in ductility. If the aluminium content is raised significantly above 9% then the $Mg_{17}Al_{12}$ network becomes continuous leading to brittleness.

There is a basic difference between the Mg-Al system, and the Al-Si system that forms the basis of aluminium casting alloys. In the Al-Si system the aluminium phase makes up the bulk of the eutectic and, as a consequence, eutectic, and even hyper-eutectic Al-Si alloys have sufficient ductility to be commercially useful.

In the Mg-Al system the eutectic is dominated by $Mg_{17}Al_{12}$ phase, which renders near eutectic alloys too brittle to be commercially useful. All of the commercial alloys are located at the magnesium end of the diagram where to solidus slope is steep. This has important ramifications for castability, as is discussed in the next section.

2.3.2 Castability and Hot Tearing

Hot tearing and hot cracking phenomena have been studied extensively for aluminium alloy castings [4]. It is recognised that many alloy systems exhibit a maximum in cracking susceptibility with alloying additions. Figure 2.10 shows an idealised example of how cracking susceptibility varies with alloying additions. The curve is described as a lambda curve due to its similarity to the shape of the Greek letter

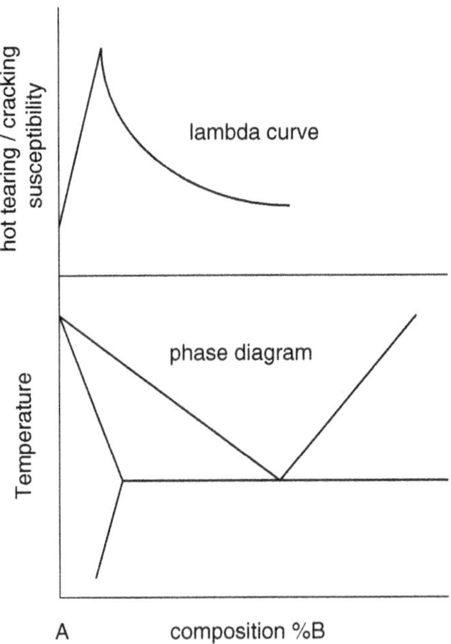

FIGURE 2.10 Schematic illustration of the variation in hot tearing susceptibility with composition.

Λ [5]. Hot tearing modelling has recently been extended to high-pressure die cast magnesium—rare earth alloys [6].

The common magnesium alloys shown in Figure 2.8 would appear to be located at the most crack prone compositions; however, cracking in the Mg-Al system is relatively low. Cracking susceptibility occurs in alloys that have wide freezing ranges, especially when the last stages of solidification occur over a wide range. The most crack-sensitive stage occurs when the solidified phase exhibits sufficient continuity to transfer stress but liquid films remain that are easy paths for cracks. The reason for the lower than expected cracking in Mg-Al alloys, appears to be related to their extremely low creep resistance [7]. The solid phase is extremely weak at high temperatures and so transfers stress poorly. Evidence for this is seen when additions are made to Mg-Al alloys that improve creep performance and also dramatically increase cracking susceptibility.

2.3.2.1 Effect of Zinc

It has long been known that the addition of zinc to Mg-Al alloys increases cracking [8]. Figure 2.11 shows the relative severity of cracking in HPDC tests bars. The contour values are related to the percentage of castings that broke into separate pieces on ejection from the die [9]. The cracking severity reaches a maximum at around 6% zinc and then declines at higher zinc levels. At higher levels the increased fraction of intermetallic phases makes the alloys brittle.

Zinc has a two-fold effect on crack sensitivity. Zinc provides some solid solution strengthening and improves creep resistance. Zinc also extends the final freezing interval because the binary Mg-Zn eutectic temperature is at 340°C or roughly 100°C lower than the Mg-Al eutectic.

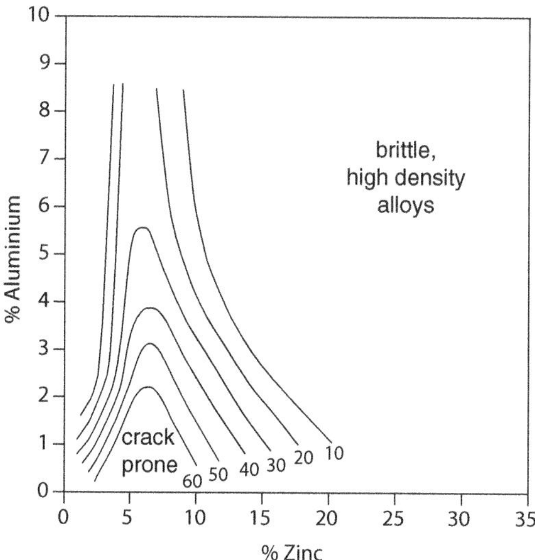

FIGURE 2.11 Sensitivity to hot tearing of die cast alloys in the Mg-Al-Zn system. The contours with high values indicate high severity of hot tearing.

2.3.2.2 Effects of Calcium and Strontium

The alloy MRI-153M (Mg Al8Ca1) is an Mg-Al alloy modified by the addition of small amounts of calcium and strontium. It has been in use commercially for a number of years due to its enhanced creep resistance relative to AM60B and AZ91D, but is prone to hot cracking [10]. A common practice for reducing cracking is to apply a pattern to the die such as shown in Figure 2.12. This has the effect of distributing stress in the part during solidification to prevent stress localisation leading to cracking.

The microstructure of MRI-153M is shown in Figure 2.9. It shows a similar proportion of second phases to AZ91D but as $(Mg, Al)_2Ca$ rather than $Mg_{17}Al_{12}$[7]. The high cracking tendency appears to be linked to the freezing range of Mg-Al alloys combined with the high temperature strength enhancement arising from calcium and strontium additions.

AJ62 is an Mg-Al based alloy with 2% Strontium. Its microstructure, shown in Figure 2.9, appears similar to MRI-153M but with a slightly higher proportion of second phase intermetallics. Some studies [11] have shown AJ62 to have poor castability (see also Section 4.2), although it has been used successfully for serial production of engine blocks.

2.3.2.3 Effects of Rare Earths and Silicon

The tendency for alloying additions to Mg-Al alloys to initially increase and then decrease hot tearing has been shown clearly in the case of rare earth additions. Figure 2.13 shows the results of an investigation (see acknowledgements) where progressive additions of rare earths (most Ce and La) where made to an Mg-4%Al based alloy. High-pressure die casting was made using a ribbed box die designed to encourage defects such as hot tears. A small addition (<0.5%) of rare earths caused a dramatic increase in defects. The level of defects remained high until the rare

FIGURE 2.12 High-pressure die cast magnesium oil pan. The cross-hatch pattern is commonly employed to minimise hot tearing.

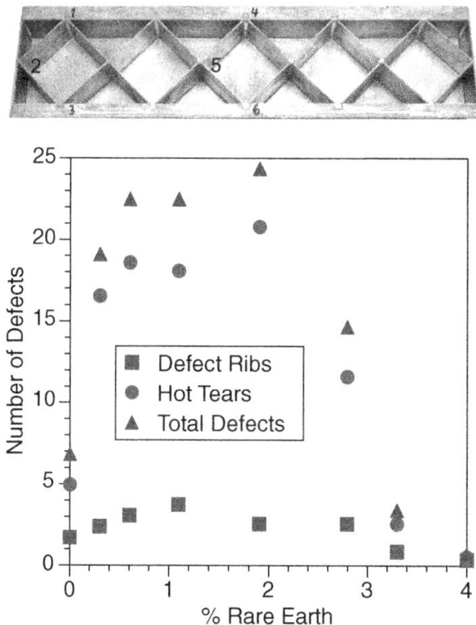

FIGURE 2.13 Variation of hot tearing and other casting defects with rare earth content added to an Mg-4%Al base alloy. A typical casting is shown above with inspected areas marked by numbers.

earth content exceeded 3%. At the level of 4% rare earths the frequency of defects had fallen to close to zero and the casting performance was better than the original Mg-4%Al base alloy.

The observed behaviour with rare earth additions is consistent with the lambda curve phenomena for hot tearing originally developed for aluminium alloys and illustrated in Figure 2.10. The microstructure of AE44 (Mg-4%Al-4%RE), shown in Figure 2.9, has a high-area fraction of eutectic, much higher than other alloys shown (with the possible exception of AS31), and so a reduced hot tearing susceptibility is to be expected.

The alloy AS31 contains approximately 1% silicon and is used in the production of transmission housings. The Mg-Si system forms a eutectic consisting of Mg and Mg_2Si at just over 1% silicon therefore AS31 lies close to the eutectic composition. The microstructure of AS31 seen in Figure 2.9 shows the Mg-Mg_2Si eutectic taking up a large area fraction. AS31 has been reported to have good castability [11], at least in terms of its resistance to cracking, and this is consistent with the observed microstructure.

2.3.2.4 Overview of Effects of Alloying Additions to Mg-Al Alloys

A consistent pattern emerges with the addition of many elements to Mg-Al alloys. When additions are made at low levels (relative to the eutectic compositions) the tendency for casting defects, such as hot tears, increases rapidly. At higher levels of additions the frequency of defects decreases but the higher fraction of intermetallic phases often leads to unacceptably low ductility. Alloy systems in which the eutectic contains a majority of magnesium phase, such as Mg-$Al_{11}RE_3$ and Mg-Mg_2Si offer the best opportunities for alloys with resistance to hot tearing combined with good ductility.

2.4 MECHANICAL PROPERTIES

2.4.1 Tensile Properties

The commonly determined tensile properties are elastic modulus, 0.2% proof stress, tensile strength and elongation to failure. For aluminium alloys the determination of these properties is straight forward, but in the case of magnesium alloys serious issues arise with the consistent determination of elastic modulus and 0.2% proof strength. Magnesium alloys exhibit pseudo elastic behaviour (non linear reversible strain). As a consequence the apparent elastic modulus changes with stress and values obtained by 0.2% offset method are not equivalent to the stress required to impart 0.2% residual plastic strain.

Figure 2.14 shows a loading—unloading tensile test for HPDC AM60. If the initial modulus is used to measure a 0.2% offset stress then a value of 125MPa is obtained. At higher stress and strain values, the apparent modulus on unloading decreases to below 30 GPa. By interpolating the appropriate modulus value, the stress level for 0.2% residual plastic strain is determined to be 145 MPa.

The difference between 0.2% offset and 0.2% residual plastic is rarely acknowledged and leads to ambiguity and inconsistency. Another source of variation occurs when the apparent modulus is drawn manually from the tensile curve, as there is often no clearly defined straight section. The standard for

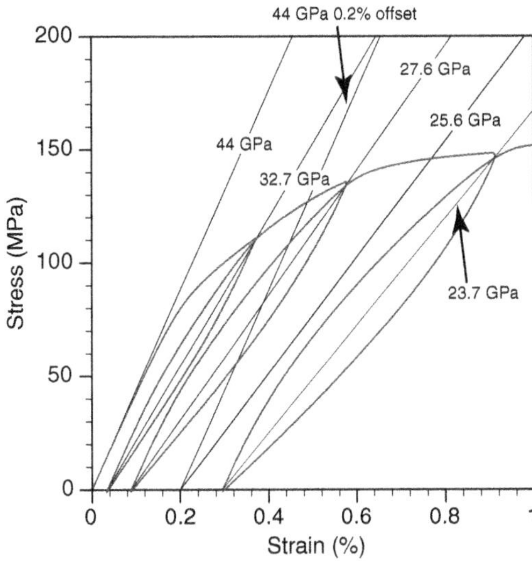

FIGURE 2.14 Cyclic tensile test curve for AM60 alloy demonstrating the variation in elastic modulus and the associated problems with consistent determination of proof stress.

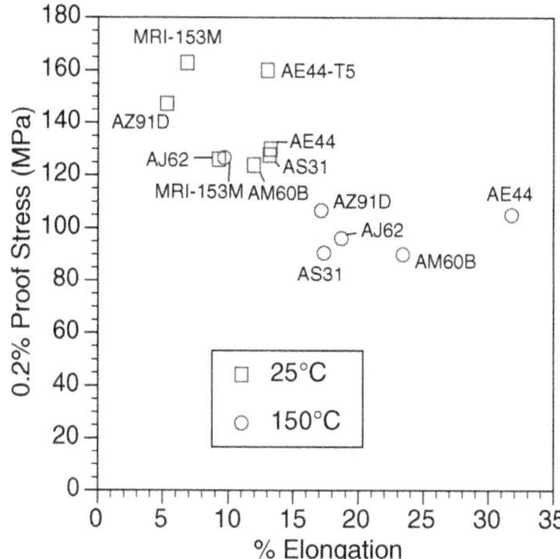

FIGURE 2.15 Proof stress and ductility for various high-pressure die cast magnesium alloys at room temperature and 150°C [7]. T5 ageing treatment for AE44 was 30 hours at 200°C.

tensile testing, ISO 6892-1:2009, allows for multiple methods despite the fact that they produce inconsistent results. In this chapter the 0.2% offset values are used wherever possible, including in Figure 2.3, these values are lower than 0.2% residual plastic.

Figure 2.15 shows 0.2% offset stresses for various alloys plotted against total elongation for room temperature and 150°C. The results are from cast tensile bars, which give generally higher elongation values than specimens cut from plates (Figure 2.3). Generally there is an inverse relationship between strength and ductility, although, in the case of AE44, a T5 age hardening is possible (30 hours at 200°C) which produces both high strength and ductility. AS31 and AE44 alloys both have relatively high elongation values, at least at room temperature. This is despite their high eutectic fractions. This is presumed to be due to the low proportion of intermetallic phase within their respective eutectics.

2.4.2 Mechanical Properties of Castings

The tensile properties described in the previous section were from test bars. They provide information about the properties of relatively defect-free materials and so overlook issues related to castability in complex castings. Four-point bending tests with box castings (as shown in Figure 2.13) have been conducted to assess the effects of castability on mechanical properties (see acknowledgements).

The alloys AZ91D, AM60B, AM40 (Mg-4%Al), AE44, AE35 (Mg-3%Al-5%RE) and AJ62 were tested. Figure 2.16 shows the testing geometry, and the average displacement and load attained by each alloy on yielding (average of five casting per alloy). Figure 2.17 shows macrographs from location 2 (according to labels shown in Figure 2.13) for each alloy.

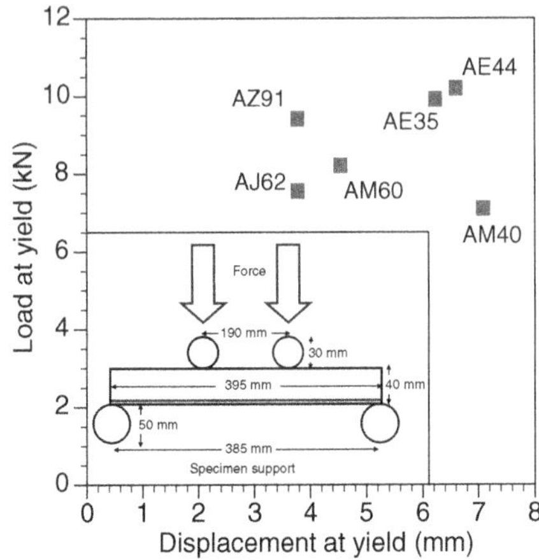

FIGURE 2.16 Loading and displacement at yield of box die samples (see Figure 2.13) cast in various alloys.

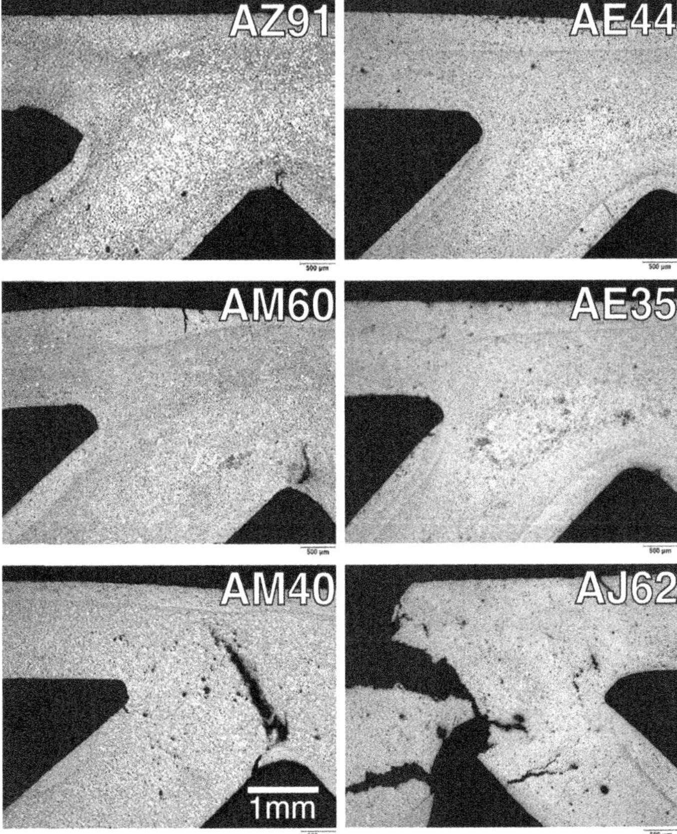

FIGURE 2.17 Macrographs of location 2 shown in Figure 2.13 for various alloys.

The alloys AZ91D, AM60B and AM40 follow a trend from high load and low displacement at yield (AZ91D) to high displacement and low load (AM40). AJ62 performed poorly in both load and displacement and the AE alloys (AE35 and AE44) performed well in both measures. Figure 2.17 shows that castability factors play an important role with AJ62 showing severe hot tearing while the AE alloys were virtually free of cracks. The AM and AZ alloys showed intermediate levels of cracking with castability (resistance to tearing) deteriorating at lower aluminium levels.

2.4.3 Creep Properties

All of the alloys described in this chapter, beyond the AM and AZ alloys, where originally developed to overcome the poor high temperature creep resistance of existing alloys. The poor creep strength of AZ91D and AM60B are evident in Figure 2.18 where the creep strengths of various alloys at 150°C are shown. Creep strength is defined as the stress required to produce 0.1% creep strain after 100 hours.

The poor creep strength in AZ91 has been attributed to the poor thermal stability of the $Mg_{17}Al_{12}$ phase and the high diffusion rate of Al in Mg. The microstructures shown in Figure 2.9 illustrate that the addition of elements Ca, Sr and RE reduce or eliminate $Mg_{17}Al_{12}$ and regions of α-Mg supersatured with Al. This is due to these elements high affinity for Al. In the case of AS31 the Si combines with Mg to form Mg_2Si. The main benefit of Si appears to be in improving castability and strength over that of an Mg-3%Al alloy.

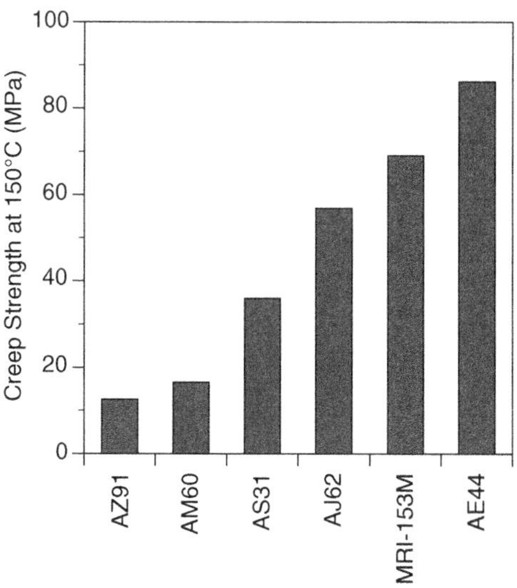

FIGURE 2.18 Creep strength at 150°C (stress required to induce 0.1% strain after 100 hours) for various alloys. (From Zhu, S. et al., *Metall. Mater. Trans. A.*, 46A, 3543–3554, 2015.)

2.5 CONCLUSIONS

The overwhelming majority of all magnesium alloy consumption is for high-pressure die cast components made with the alloys AM50A, AM60B and AZ91D. The poor creep performance of these alloys lead to the development of more complex alloys where elements such as Ca, Sr, rare earths and Si were added.

In many cases the addition of these elements lead to problems with castibility, especially hot tearing. Most of these alloy systems exhibit a maximum in hot tearing susceptibility at a particular level of alloy addition. Alloys with higher levels of additions show reduced hot tearing susceptibility but in many cases other factors then impinge such as low ductility and die soldering.

Alloys with high fractions of eutectic are generally less prone to hot tearing, but commercially useful near-eutectic magnesium alloys are only possible if the eutectic contains magnesium as the major component and intermetallic phases as the minor component. Eutectics with intermetallic phases as the major components are too brittle. The systems Mg-Al-RE and Mg-Al-Si both contain magnesium dominated eutectics and alloys from these systems, AE44 and AS31 are currently used in commercial production.

ACKNOWLEDGEMENTS

The castability studies using the box die described in Sections 3.2.3 and 4.2 were originally conducted in 2006 by Per Bakke at Hydro Magnesium (now Magontec). Suming Zhu supplied the original micrographs, used in Figure 2.9 and Hua Qian Ang supplied the data used to create Figure 2. 14, both from RMIT University.

REFERENCES

1. Abbott, T.B., Magnesium: Industrial and research developments over the last 15 years. *Corrosion* 2015, 71(2), 120–127.
2. Polmear, I.J., *Light Alloys: From Traditional Alloys to Nanocrystals.* 4th Ed. Butterworth Heinemann, Norwood MA, 2005.
3. Kearney, A., Rooy, E.L., *Aluminum Foundry Products, Properties and Selection: Nonferrous Alloys and Special-Purpose Materials,* Vol 2, ASM Handbook, ASM International, 1990, 123–151.
4. Eskin, D.G., Katgerman, L., Mechanical properties in the semi-solid state and hot tearing of aluminium alloys. *Progress in Materials Science* 2004, 49, 629–711.
5. Upadhya, G., Cheng, S., Chandra, U., A mathematical model for prediction of hot tears in castings. *Light Metals* 1995, 1101–1106.
6. Easton, M.A., Gibson, M.A., Zhu, S., Abbott, T.B., An a priori hot-tearing indicator applied to die-cast magnesium-rare earth alloys. *Metallurgical and Materials Transactions A* 2014, 45A, 3586–3595.
7. Zhu, S., Easton, M.A., Abbott, T.B., Nie, J.F., Dargusch, M.S., Hort, N., Gibson, M.A., Evaluation of magnesium die-casting alloys for elevated temperature applications: Microstructure, tensile properties, and creep resistance. *Metallurgical and Materials Transactions A* 2015, 46A, 3543–3554.
8. Foerster, G., Improved magnesium die casting alloys. In *Eighth SDCE International Die Casting Exposition and Congress* Detroit, MI 1–6, 1975.

9. Easton, M.A., Abbott, T.B., Nie, J.F., Savage, G., An assessment of high pressure die cast Mg-Zn-Al alloys. *Magnesium Technology*, Pekguleryuz, M.O., Neelameggham, N.R., Beals, R.S., Nyberg, E.A. (Eds.), TMS (The Minerals, Metals and Materials Society), 2008.
10. Easton, M.A., Gibson, M.A., Gershenzon, M., Savage, G., Tyagi, V., Abbott, T.B., Hort, H., Castability of some magnesium alloys in a novel castability die. *Materials Science Forum* 2011, 690, 61–64.
11. Zhu, S., Gibson, M., Easton, M., Zhen, Z., Abbott, T., Creep resistant magnesium alloys and their properties. *Metal Casting Technologies* 2012, 58(2), 20–24.

3 Magnesium Production from Calcined Dolomite via the Pidgeon Process

Mehmet Buğdayci, Ahmet Turan, Murat Alkan, and Onuralp Yücel

CONTENTS

3.1 INTRODUCTION

3.1.1 Lloyd Montgomery Pidgeon and the History of Pidgeon Process

The Pidgeon Process is a widely used vacuum silicothermic reduction process to produce primary magnesium (Mg) metal and, the process was developed by Lloyd Montgomery Pidgeon in the beginning of 1940s in Canada.[1,2]

L. M. Pidgeon was born in 1903 in Ontario, Canada (Figure 3.1). He graduated in chemistry from the University of Manitoba in 1925 and, he received his M.Sc. and Ph. D. degrees in the field of chemistry from the McGill University in 1927 and 1929. After his Ph. D., he attended to Oxford University to work on "anti-knock compounds" under the supervision of Sir Alfred Egerton (1929–1931). Since his undergraduate and postgraduate degrees were in chemistry, the period at Oxford was the first time he dealt with the processes at high temperatures. Dr. Pidgeon got back to Canada and joined to National Research Council in Ottawa (1931–1941) where he began to work in metallurgy. During that period, Pidgeon developed his path-breaking primary magnesium production process because of high magnesium demand in WWII. In 1941, he was appointed the director of research in Dominion Magnesium Ltd where the first founded magnesium production plant by using the Pidgeon Process. He worked in the department of metallurgical engineering at the

FIGURE 3.1 Lloyd Montgomery Pidgeon. (From Canadian Mining Hall of Fame Online, http://mininghalloffame.ca/inductees/p-r/lloyd_m._pidgeon, Accessed August 2017.)

University of Toronto as a professor between 1943 and 1969 as well as served as the head of department. He led the department to be a worldwide known and strong department. Particularly he contributed to the department to expand in physical metallurgy and materials science. He was made an officer of the Order of Canada because of his distinguishing works. Pidgeon died in 1999.[2–4]

Several patents on magnesium production were published by Dr. Pidgeon. But the milestone patent was released in 1943. It was explained the basic principles of magnesium production from magnesium oxide containing raw materials by using ferrosilicon as a reductant under vacuum. Molten salt electrolysis was the dominant primary magnesium production method until Pidgeon's invention. This process is called the Pidgeon Process, in honour of L. M. Pidgeon.[2,5]

3.1.2 Basic Principles of Pidgeon Process

Calcined dolomite is the most remarkable raw material to use as magnesium source in the Pidgeon process and, it theoretically contains 58.2 mass% CaO and 41.8 mass% MgO, respectively.[6] The fundamental principle of the Pidgeon Process is the reduction of magnesium from calcined dolomite by using silicon (or any silicon containing material like ferrosilicon) under vacuum. A mixture, containing calcined dolomite, ferrosilicon and a slight amount of fluorspar (CaF_2, about 2.5wt.%) as catalyst, is charged into a high alloy steel retort to heat up to 1200°C–1250°C and to keep under vacuum (about 1 mbar) for about 6 hours.[7,8] The general flowchart of the Pidgeon Process is shown in Figure 3.2.

The general reaction which occurs during the Pidgeon Process is given by Eq. 3.1. However, in industry, ferrosilicon is favorably used instead of metallic silicon due to its lower price and greater availability.[1] Furthermore, when FeSi consists of higher than 65% Si by mass, its activity is close to the activity of pure silicon (0.97°C at 1,200°C).[5,9] Reduction reaction of Mg from a dolomite ore by using ferrosilicon (with a composition of 75 mass% Si) as a reductant is shown by Eq. 3.2.

$$2MgO{\cdot}CaO + Si \rightarrow 2Mg + Ca_2SiO_4 \quad (3.1)$$

$$2MgO{\cdot}CaO + \text{ferrosilicon } (Si + 0.167Fe) \rightarrow 2Mg + Ca_2SiO_4 + 0.167Fe \quad (3.2)$$

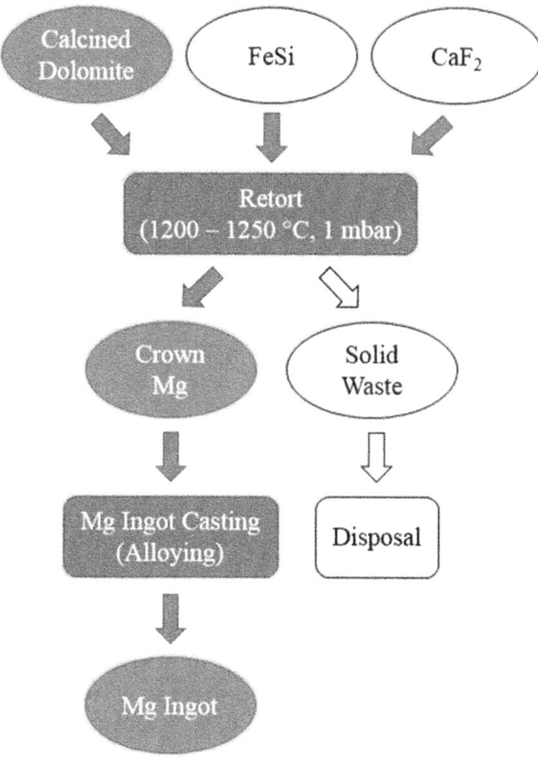

FIGURE 3.2 General flow chart of the Pidgeon Process.

Magnesium is in the gaseous phase at the reduction temperature. It is therefore collected and condensed in the cooling part of reaction retort whereas the rest of charge remains in as-charged zone.[10] A schematic sketch of cross-section of a magnesium reduction retort can be seen in Figure 3.3. A retort is basically in the form of one-closed-end tube. Other side is closed by using an air-tight cover (using gasket sealant) that is made of high alloy steel, same as the retort. Retorts are placed in a retort furnace where reduction zone stays in the furnace while condensation zone (cooling zone) stays out. A heat shield is utilized in the retort for the separation of reduction zone and condensation zone. The heat shield is formed as a disc having a hole in its center. The main reasons to use of the heat shield are heat barrier, sodium-potassium condensation and blocking the scattering of magnesium back. Without a heat shield, earth alkaline metals could be collected in condensation zone with magnesium, and it might cause to the deflagration and explosion of condensed metal when cover is open to remove products with the release of air into the retort. In this case, reduced crown magnesium, containing earth alkaline metals, has a microstructure of tiny and non-uniform crystals as well. Optimum condenser temperature is reported to be at 485°C with an acceptable vacuum value for the process.[11–15] Selected steps of primary magnesium production in semi-pilot scale are given in Figure 3.4.

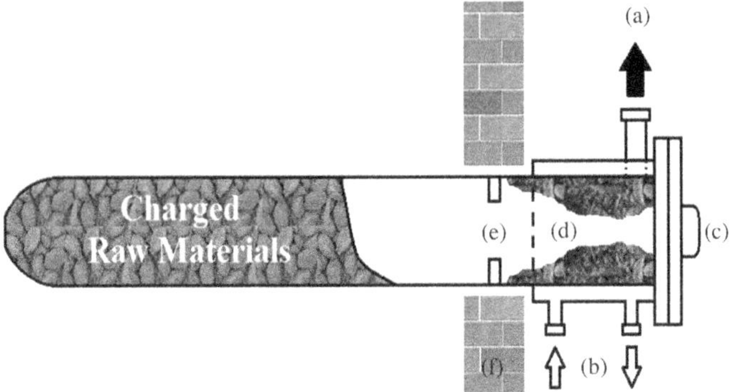

FIGURE 3.3 Schematic sketch of reduction retort cross-section; (a) vacuum outlet, (b) chilled water inlet and outlet, (c) cover, (d) crown magnesium, (e) heat shield, (f) retort furnace wall.

FIGURE 3.4 Mg production process through the Pidgeon Process in semi-pilot scale (a) placing of a retort in a retort furnace, (b) and (c) as-reduced crown Mg.

3.1.3 Thermodynamical Background

Metallothermic reduction reactions are mostly highly exothermic that allow reactions to continue in a self-sustaining mode without any additional heat.[1] However, the Pidgeon Process is based on an endothermic reaction ($\Delta H°_{298K}$ is about 183.0 kJ/mol Si, Eq. 3.2). Heat supply is a critical part in industrial retorts. Thermodynamically, reaction-beginning temperatures decrease with the application of vacuum for both MgO and calcined dolomite [Table 3.1, Figure 3.5a]. Reaction beginning temperature is 1,098°C for the conditions where ferrosilicon (with 75 mass% Si) is used as reductant under 1 mbar vacuum.

Reaction of calcined dolomite with ferrosilicon causes the formation of crown Mg in the cooling stage of the retort and a slag mixture mainly consists of Ca_2SiO_4 (higher than 80 mass%) and metallic iron [Figure 3.5b]. The slag may also

TABLE 3.1
Minimum Reduction Temperatures (°C) of Mg in MgO and Calcined Dolomite by Using Si and FeSi Reductants under Vacuum and Atmospheric Conditions

	MgO		Calcined Dolomite	
Reductant	**1 bar**	**1 mbar**	**1 bar**	**1 mbar**
Si	2,143	1,325	2,489	1,318
FeSi	1,789	1,155	1,870	1,098

Source: Bugdayci, M. et al., *High Temperature Materials and Processes*, 2017.

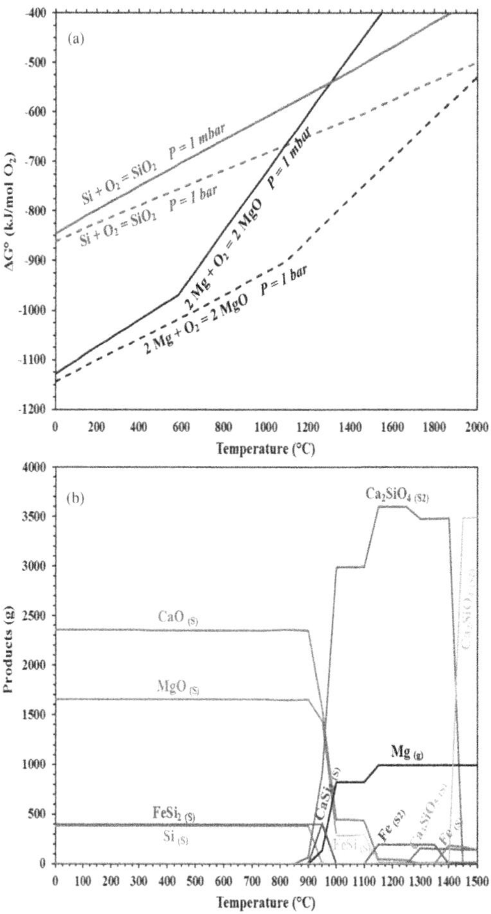

FIGURE 3.5 (a) The change of mg reduction temperature by using Si under different atmosphere pressures and, (b) the change of reaction products for the reaction of calcined dolomite and 100% stoichiometric FeSi (75wt.%) with the increase in reaction temperature.

contain a slight amount of MgO and Fe_xO_y in practice due to insufficient reduction conditions in terms of vacuum value, temperature and calcined dolomite content.

The use of CaC_2 as a cheap reductant is also reported in the literature. Although CaC_2 slightly decreases the recovery ratio of magnesium, its price makes it a promising alternative reductant than ferrosilicon. Moreover, the use of CaC_2 with ferrosilicon is possible to reduce process costs. Furthermore, it has been reported that aluminum is a more effective reductant than ferrosilicon, but its price does not allow using it for mass production.[1]

3.1.4 Industrial Application

Volume and dimensions of retorts and number of retorts in a retort furnace change in accordance with fuel type in industry. Detailed information can be seen in Table 3.2. Retorts are produced via centrifugal casting to minimize defects and porosities that might cause to decrease vacuum level. Alloy type of retorts is chosen to show high oxidation resistance at high temperatures such as high alloy steel having a Cr content between 24%–30% and a Ni content up to 30%.[16]

An industrial plant working through Pidgeon Process consists of dolomite calcination, reduction and magnesium foundry units. Reduction units contain retort furnaces where the retorts are inserted in (Figure 3.6). Number of retorts in a retort furnace can change from 11 to 64 and, heavy oil, coal gas, coal or natural gas can be used as fuel to heat up a retort furnace. Some adjunct systems and workshops (such as retort casting, mechanical and electrical workshops) are also take place in a Pidgeon Process plant. Primary Mg, which is produced in the form of crown (Mg > 99.5% by wt.), is casted into ingots with or without alloying additions in a casthouse [Figure 3.7a and b].[16,17]

A relatively newer production method was described and studied by Bettanini, Zanier and Enrici.[18] They established the vertical thermal reduction technology. The first application of this technology was realized in a magnesium plant near the town of Bolzano (Italy), so, this process is called as the Bolzano Process. The differences of the Bolzano Process than the Pidgeon Process are: the heating process is realized by electric heating conductors, and retorts are placed vertically into large blocks.[19]

TABLE 3.2
Selected Industrial Retort Properties in Accordance with Fuel Type

	Fuel Type	
Retort Properties	**Heavy Oil, Coal Gas**	**Coal**
Quantity in a retort furnace	16–64	11–22
Dimensions, ϕ*length, mm	325–370 * 2600–2800	325–339 * 2600–2700
Retort quantity connected to the same vacuum system	4–7	6–11

Source: Esan Magnezyum, http://www.esanmagnezyum.com/(Accessed August 2017).

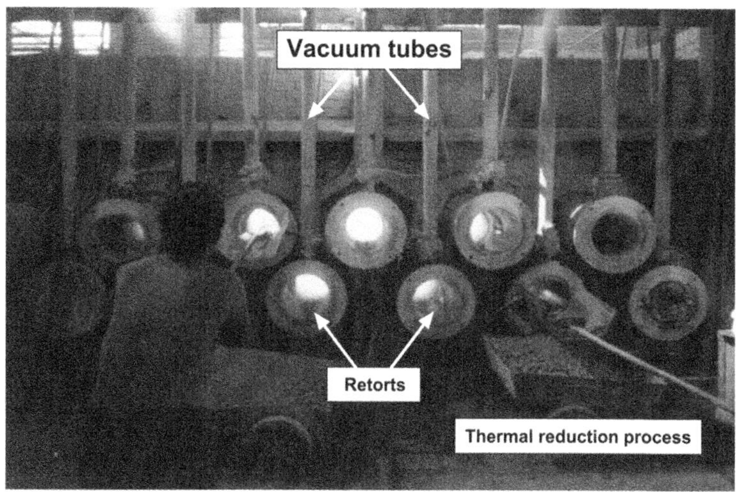

FIGURE 3.6 Industrial-scale Mg production through the Pidgeon Process; retorts which are connected to vacuum system and retort furnace, (From Aghion, E. and Golub, G. Production technologies of magnesium, In *Magnesium Technology: Metallurgy, Design, Applications*, Friedrich, H.E., Mordike, B.L. (Eds.), Springer, Heidelberg, Germany, 29–62, 2006.)

FIGURE 3.7 (a) Mg casting unit, (b) casting of Mg ingots, (c) slag removal from industrial retorts and (d) cooled slag pellets.

TABLE 3.3
Technology Comparison of the Pidgeon Process and Electrolytic Reduction

	Pidgeon Process	Electrolytic Reduction
Capital intensity ($/ton Mg annual capacity)	3250 (±650)	16667 (±2500)
Production costs ($/tonne Mg)	2447	3108
Energy intensity (GJ/tonne Mg)	300 (±60)	270 (±27)
Environmental impact (kg $CO_{2\ eq.}$/kg Mg)	43.30	27.93

Source: Prentice, L.H. and Nawshad, H., Magsonic™ carbothermal technology compared with the electrolytic and Pidgeon processes, In *Magnesium Technology 2012*, Springer International Publishing, 37–41, 2012; D'Errico et al., *JOM.*, 61, 14–18, 2009; Yücel, O. *Evaluation of Reduction Wastes which Arises in Magnesium Production (in Turkish: Magnezyum Uretiminde Olusan Reduksiyon Atiklarinin Degerlendirilmesi*), Istanbul, Turkey, 2016; Das, S. *JOM*, 60, 63–69, 2008.

The Brasmag company in Brazil and POSCO in South Korea are also developed their own vertical thermal reduction technologies.[20,21]

Comparison of the Pidgeon Process and Electrolytic Reduction is given in Table 3.3.

3.1.5 Disposal of Process Wastes

An inert waste mainly consists of CaO, SiO_2 and a slight amount of MgO, Fe_2O_3 and Al_2O_3 arises during the Pidgeon Process. The sum of CaO and SiO_2 is minimum 80% by weight in waste and, these compounds are in the form of Ca_2SiO_4 structure. That compound is very stable and hard to break by using thermal and leaching operations. Approximately 2 metric tons of waste is produced for per 1 metric ton of magnesium. Thus, it is a necessity to apply a disposal method for waste although it has an inert characteristic. It is reported that it is possible to add the waste to cement up to 6% by weight. That solution can help the protection of natural resources as well as CO_2 emissions in cement industry.[25] Furthermore, another suggestion is the use of waste for CO_2 capture for the CO_2 emissions of the Pidgeon Process. That waste disposal process comprises waste activation by leaching, carbon capture, and the use of carbonated waste for mining field reclamation steps.[26]

3.2 CONCLUSION

Between primary magnesium mass production methods, the outstanding technology is the Pidgeon Process. This process is based on the reduction of magnesium in calcined dolomite ores by using silicon based materials (mainly ferrosilicon) as reductant. Charge mixture comprises a slight amount of CaF_2 as catalyst. Reduction is performed for about 6 hours and at the temperatures between 1200°C and 1250°C. As a thermodynamical requirement, process is conducted in retorts to allow working under vacuum atmosphere. Vacuum value is wanted to be as low as possible. It is considered to be 1 mbar for theoretical studies. In industrial applications, insufficient vacuum value which is due to weak air-sealing is the primary reason to work with low magnesium recovery ratios. Magnesium is in gaseous phase at its reduction temperature, it thereby collected in the cooling zone of retorts. It is called crown magnesium due to its shape and have a magnesium content of higher than 99.5% by weight. Crown magnesium is casted in the form of ingots with or without alloying and, ingots are introduced into the market as a semi-product.

REFERENCES

1. Bugdayci, M., Turan, A., Alkan, M., Yücel, O. Effect of reductant type on the metallothermic magnesium production process. *High Temperature Materials and Processes* 2017. doi:10.1515/htmp-2016-0197.
2. Brown, R.E., Lloyd, M. Pidgeon: Magnesium Pioneer. In *Essential Readings in Magnesium Technology*, Mathaudhu, S.N., Luo, A.A., Neelameggham, N.R., Nyberg, E.A., Sillekens, W.H. (Eds.), Springer: Cham, Switzerland, 2016, 89–91.
3. The Canadian Encyclopedia Online, http://www.thecanadianencyclopedia.ca/en/article/lloyd-montgomery-pidgeon/(Accessed August 2017).
4. Canadian Mining Hall of Fame Online, http://mininghalloffame.ca/inductees/p-r/lloyd_m._pidgeon (Accessed August 2017).
5. Pidgeon, L.M. Method and apparatus for producing magnesium. US Patent 2,330,143, September 21, 1943.
6. Shahraki, B.K., Mehrabi, B., Dabiri, R. Thermal behavior of Zefreh dolomite mine (Central Iran). *Journal of Mining and Metallurgy B: Metallurgy* 2009, 45 (1), 35–44.
7. Wulandari, W., Brooks, G.A., Rhamdhani, M.A., Monaghan, B.J. Kinetic analysis of silicothermic process under flowing argon atmosphere. *Canadian Metallurgical Quarterly* 2014, 53 (1), 17–25.
8. Wang, C., Zhang, C., Zhang, S.J., Guo, L.J. The effect of CaF_2 on the magnesium production with silicothermal process. *International Journal of Mineral Processing* 2015, 142, 147–153.
9. Pidgeon, L.M., Alexander, W.A. Thermal production of magnesium-Pilot plant studies on the retort ferrosilicon process. *Transactions AIME* 1944, 159, 315–352.
10. Wynnyckyj, J.R., Rao, D.B., Mueller, G.S. Reaction kinetics in the silicothermic magnesium process. *Canadian Metallurgical Quarterly* 1977, 16 (1), 73–81.
11. Çetin, A. M.Sc. Thesis. *Istanbul Technical University, Graduate School of Science Engineering and Technology*, ITU, İstanbul, Turkey, 2005.
12. Yücel, O., Yiğit, S., Derin, B. Production of magnesium metal from Turkish calcined dolomite using vacuum silicothermic reduction method. *Materials Science Forum: Trans Tech Publications* 2005, 488, 39–42.
13. Demiray, Y. Refined magnesium metal ingot production of domestic resource. M.Sc. Thesis. Istanbul Technical University, Graduate School of Science Engineering and Technology. ITU; İstanbul, Turkey, 2008.
14. Demiray, Y., Yucel, O. Production and refining of magnesium metal from Turkey originating dolomite. *High Temperature Materials and Processes* 2012, 31 (3), 251–257.
15. Turan, A., Alkan, M., Tavsanoglu, T., Sahin, F.C., Goller, G., Yücel, O. (2013). *Abstract Books XII. International Symposium on Self-Propagating High Temperature Synthesis*, South Padre Island, TX, Oct. 21–24, 127–128, 2013.
16. Esan, E., *China Office. Mg Metal Production*, Esan, Eskisehir, Turkey, 2013.
17. Esan Magnezyum, http://www.esanmagnezyum.com/(Accessed August 2017).

18. Bettanini, C., Zanier, S., Enrici, M. Method of extracting magnesium from magnesium oxides. *US Patent* No. 4,238,223, December 9, 1980.
19. Aghion, E., Golub, G. Production technologies of magnesium. In *Magnesium Technology: Metallurgy, Design, Applications*, Friedrich, H.E., Mordike, B.L. (Eds.), Springer, Heidelberg, Germany, 2006, 29–62.
20. Cherubini, F., Raugei, M., Ulgiati, S. LCA of magnesium production: Technological overview and worldwide estimation of environmental burdens. *Resources, Conservation and Recycling* 2008, 52 (8), 1093–1100.
21. Dae, K.P., Han, GG.S. Vertical type thermal reduction apparatus for magnesium production. *Korean Patent* 20120074972A, December 28, 2010.
22. Das, S. Primary magnesium production costs for automotive applications. *JOM Journal of the Minerals, Metals and Materials Society* 2008, 60 (11), 63–69.
23. Prentice, L.H., Nawshad, H. Magsonic™ carbothermal technology compared with the electrolytic and Pidgeon processes. In *Magnesium Technology 2012*, Mathaudhu, S., Sillekens, W., Neelameggham, N., Hort, N. (Eds.), Springer International Publishing, 2012, Warrendale, PA, 37–41.
24. D'Errico, F., Perricone, G., Oppio, R. A new integrated lean manufacturing model for magnesium products. *JOM Journal of the Minerals, Metals and Materials Society* 2009, 61 (4), 14–18.
25. Yücel, O. *Evaluation of Reduction Wastes which Arises in Magnesium Production (in Turkish: Magnezyum Uretiminde Olusan Reduksiyon Atiklarinin Degerlendirilmesi*). Istanbul, Turkey, 2016.
26. Turan, A. *The Use of Pidgeon Process Residue on the CO_2 Capture which is Generated in the Calcination Step of Dolomite Ores and, the Evaluation of the Products (In Turkish: Pidgeon Prosesi Atıklarının Dolomit Kalsinasyonunda CO_2 Yakalama İçin Kullanımı ve Ortaya Çıkan Ürünlerin Değerlendirilmesinin Araştırılması).* Tubitak, Turkey, 2017.

4 Hot Deformation Characteristics of Magnesium Alloys

Rajendra Laxman Doiphode, S.V.S. Narayana Murty, and Bhagwati Prasad Kashyap

CONTENTS

4.1 INTRODUCTION

In the aerospace industry, due to its basic requirement of light-weight metals and alloys, Mg has attracted large attention. The density of Mg (1.74 gm/cm^3) is only two-third that of aluminum (2.70 gm/cm^3) and one quarter that of steel (7.85 gm/cm^3). These alloys are even popular in other sectors like automobile, biomedical, architecture and electronic industries (1). In order to produce components by various manufacturing processes, the material should possess good ductility. Various microstructural aspects of material play a very important role in influencing flow properties (2). Mg has excellent machining characteristics; it is estimated that Mg alloys may be safely machined at about ten times the rate possible for steel and twice that possible for aluminum. But, from production point of view, there is a need for good forming properties in a material. If the formability of Mg-alloys is compared to that of other materials like steel and aluminum, it has limitations. It has been attributed to its Hexagonal Close Packed (HCP) type of crystal structure. It has only three slip systems available for deformation i.e. basal, prismatic and pyramidal and, at room temperature (RT); only two of them are active. The third slip system becomes active only at elevated temperature. So, it is necessary to carry out high temperature deformation of Mg to activate other slip systems and mechanisms for deformation. However, high temperature deformation may affect some of the mechanical and metallurgical properties unfavorably (3).

Study of high temperature flow properties is important for optimizing the manufacturing processes (4). Superplasticity is the ability of a polycrystalline material to exhibit, in a generally isotropic manner, very high elongations, and prior to failure. If the material is superplastic, then by using the required air/gas flow pressure, the products can be manufactured easily. This will improve the productivity. Under some limited experimental conditions, the material can be pulled out to exhibit a very large neck free elongation prior to failure. These high tensile elongations are examples of the occurrence of superplasticity.

4.1.1 Prerequisites for the Development of Superplasticity

- Fine and equiaxed grain size with less than 10 μm for metals
- It requires a relatively high testing temperature typically at or above ~$0.5T_m$, where T_m is the absolute melting temperature
- Presence of second phase particles
- Strain rate sensitivity index m more than 0.3
- High angle grain boundaries

There are various methods of processing the metal for grain refinement, and the mode of deformation and related micro mechanisms can be influenced by the selected method (5). It is observed that along with grain size, other microstructural aspects also play a significant role, and it could be possible that all these aspects of microstructures have synergistic effects on properties.

4.2 DEFORMATION BEHAVIOR IN HCP METALS

4.2.1 Dislocations and Plastic Deformation

It is well known that dislocations can move under adequately applied external stresses. Cumulative movement of dislocations leads to the gross plastic deformation. At microscopic level, dislocation motion involves breaking and reformation of inter-atomic bonds (6). One-dimensional crystal defects and dislocations play an important role in plastic deformation of crystalline solids. Their importance in plastic deformation is relevant to their characteristic nature of motion in specific directions (slip-directions) on specific planes (slip-planes), where the edge dislocation moves by glide and climb whilst the screw dislocation can be moved by glide and cross-slip. The onset of plastic deformation initiates the motion of existing dislocations in real crystal whilst, in the crystal having negligible dislocations, it can be attributed to both the generation and subsequent motion of dislocations. During the motion, the dislocations tend to interact among themselves. Dislocation interaction is very complex as the number of dislocations move on a number of slip planes in various directions. When they are in the same plane and have the same sign, they repel each other but they annihilate each other if they have opposite signs (leaving behind as a crystal of extremely low dislocation density).

4.2.2 Hindrances to Dislocation Motion

In general, when dislocations are close and their strain fields add to a larger value, they repel, because being closer increases the potential energy (it takes energy to strain a region of the material). When unlike dislocations are on closely spaced neighboring slip planes, complete annihilation cannot occur. In this situation, they combine to form a row of vacancies or an interstitial atom. An important consequence of the interaction of dislocations that are not on parallel planes is that they intersect each other or inhibit each other's motion. Intersection of two dislocations results in a sharp break in the dislocation line. These breaks can be of two kinds: (a) a jog breaks out from the dislocation line moving it out of slip plane; (b) a kink forms in dislocation line that remains in slip plane. Other hindrances to dislocation motion include interstitial and substitutional atoms, foreign particles, grain boundaries (GBs), external grain surface, and change in structure due to phase change. Important practical consequences of hindrance of dislocation motion are that dislocations are still movable but at higher stresses (or forces) and, in most instances, that leads to generation of more dislocations. Dislocations can initiate from existing dislocations, and from defects, GBs and surface irregularities. Thus, the number of dislocations increases dramatically during plastic deformation. As further motion of dislocations requires increase of stress, the material can be said to be strengthened i.e. materials can be strengthened by controlling the motion of dislocation.

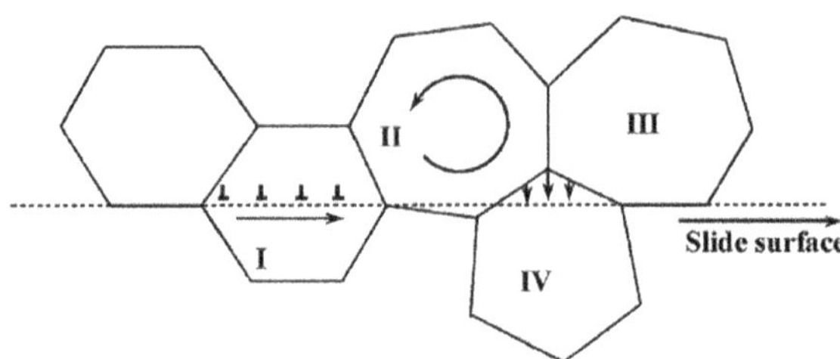

FIGURE 4.1 Possible mechanisms of the formation of the sliding surface for GBS in microcrystalline materials: I – dislocation glide; II – grain rotation; III – grain boundary sliding; IV – grain boundary migration. (From Sergueeva, A. et al., *J. Mater. Sci.*, 42, 1433–1438, 2006.)

4.2.3 Dislocation and Grain Boundary Interaction

Slip in polycrystalline material involves generation, movement and (re-)arrangement of dislocations. Because of dislocation motion on different planes in various directions, they may interact as well. This interaction can cause dislocations to be immobile or mobile at higher stresses. Figure 4.1 shows the possible mechanisms of the formation of the sliding surface for grain boundary sliding (GBS) in microcrystalline materials.

During deformation, mechanical integrity and coherency are maintained along the GBs; that is, the GBs are constrained, to some degree, in the shape it may assume by its neighboring grains. Once the yielding has occurred, continued plastic deformation is possible only if enough slip systems are simultaneously operative so as to accommodate grain shape changes while maintaining grain boundary integrity.

There are two prominent mechanisms of plastic deformation, namely *slip* and *twinning*. HCP metals deform plastically via the simultaneous activation of these modes (8).

4.3 DEFORMATION BY SLIP SYSTEM

Slip is the prominent mechanism of plastic deformation in metals. It involves sliding of blocks of crystal over one another along definite crystallographic planes, called slip planes. Slip occurs when shear stress applied exceeds a critical value. During slip each atom usually moves same integral number of atomic distances along the slip plane producing a step, but the orientation of the crystal remains the same. Steps observable under microscope as straight lines are called slip lines.

Slip occurs most readily in specific directions (slip directions) on certain crystallographic planes. This is due to limitations imposed by the fact that single crystal remains homogeneous after deformation. Generally, slip plane is the plane of greatest atomic density, and the slip direction is the close packed direction within the slip plane. It turns out that the planes of the highest atomic density are the most widely spaced planes, while the close packed directions have the smallest translation distance. Feasible combination of a slip plane together with a slip direction is considered as a slip system. Mg can be deformed easily within their basal plane at room temperature (RT) but, according to von Mises theory minimum five independent slip systems are required to be activated for deformation. An independent slip system is defined as one producing a crystal shape change that cannot be reproduced by any combination of other slip system. So, it is essential for Mg and its alloys to be deformed at elevated temperature to activate the minimum required slip systems.

4.3.1 Slip Modes

In general, the four most observed slip modes that can be activated are:

1. Basal <a> (0 0 0 1) slip – 3 possible (2 independent), activated at RT.
2. Prismatic <a> $\{1\ 0\ \bar{1}\ 0\}$ slip – 3 possible (2 independent), activated at RT.
3. First-order pyramidal <c + a> $\{1\ 0\ \bar{1}\ 1\}$ and
4. Second-order pyramidal <c + a> $\{2\ \bar{1}\ \bar{1}\ 1\}$ slip

From pyramidal slip mode 6 slip modes are possible (4 independent). But this mode can be activated at high temperature, normally more than 250°C. At RT only four slip modes are active, fifth independent slip system is still needed. Therefore, to activate pyramidal slip the high temperature deformation is carried out in HCP system. Figure 4.2a shows the slip planes in the HCP crystal.

4.3.2 Single Crystal Slip and Schmidt's Law

In a single crystal, plastic deformation is accomplished by the process called slip, and sometimes by twinning. The extent of slip depends on many factors including external load and the corresponding value of shear stress produced by it, the geometry of crystal structure, and the orientation of active slip planes with the direction of shearing stresses generated. Schmidt first recognized that single crystals at different orientations but of same material require different stresses to produce slip (9). If a stress is applied at an

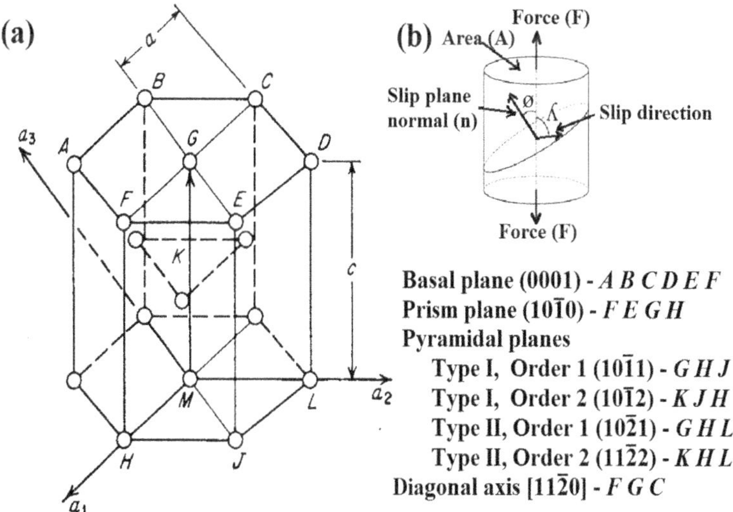

FIGURE 4.2 (a) Slip planes in HCP crystal; (b) geometry of slip during tensile testing of a single crystal.

arbitrary angle to a single crystal, as shown in Figure 4.2b, each available slip system will experience Resolved Shear Stress (RSS) acting in the associated slip plane in the slip direction. If the normal to the slip plane makes an angle ø with the tensile axis then the area of the plane is $A/\cos\phi$. If the slip direction is at an angle of λ to the tensile axis then the resolved component of the applied force F, parallel to the slip direction, is $F * \cos\lambda$. Since the slip systems are usually oriented differently with respect to the tensile axis, for a given applied stress, force F will give different values of RSS (τ_{RSS}), and one of these will experience the highest RSS. This is the one which, if sufficiently high, can initiate plastic deformation. This stress is called as Critical Resolved Shear Stress (CRSS) and is designated as τ_{CRSS}, and is related to the tensile yield stress (σ_y) through:

$$\tau_{CRSS} = \sigma y * \cos\phi \cos\lambda \tag{4.1}$$

where $\cos\phi\cos\lambda$ = Schmidt factor (SF). CRSS for basal planes is approximately one hundred times less than for pyramidal and Prismatic planes (10). So, Mg can be deformed easily only within their basal plane at RT. However, when the temperature increases, the CRSS for non-basal slip systems will decrease rapidly, thus non-basal slip systems will be also activated at elevated temperature (11). Thus, Taylor's criteria will be satisfied for plastic deformation (12). CRSS decreases with increase in temperature and by addition of solute atoms for all the slip systems. The Al and Zn reduce the CRSS of the Mg crystal (10).

Almost all engineering alloys are polycrystalline. Gross plastic deformation of these materials corresponds to the comparable distortion of the individual grains by means of slip. Although some grains may be oriented favorably for slip, yielding cannot occur unless slip occurs in the unfavorably oriented neighboring grains also. That is, to initiate plastic deformation of polycrystalline metals higher stresses are required than for equivalent single crystals, where the stress needed depends on orientation of the crystal. Much of this increase is attributed to geometrical reasons (13).

4.4 DEFORMATION BY TWINNING

The second important mechanism of plastic deformation is twinning. Each atom in the twinned region moves by a homogeneous shear a distance proportional to its distance from the twin plane. The lattice strains involved in twinning are small, usually in order of fraction of inter-atomic distance, thus resulting in very small gross plastic deformation. The roles of twinning in HCP metals have been considered to be limited as:

1. A complementary deformation mechanism to provide additional independent slip systems;
2. Makes the change in plane orientation so that further slip can occur;
3. A source of work hardening; and
4. A source of dynamic recovery or the accommodation of stress concentrations.

Twinning generally occurs when slip is restricted, because the stress necessary for twinning is usually higher than that for slip (14). Thus, some HCP metals with limited number of slip systems may preferably twin. There may be formation of early or delayed (after plastic deformation by slip) twinning. Delayed twinning usually has a rather small effect on the actual stress verses strain curve, whereas immediate twinning causes very rapid formation of twinned regions, giving large load drops. This twinning is very sensitive to temperature of deformation and to strain rate. The relative contribution of twinning to the overall strain increases as the temperature is lowered or the strain rate increased. Hence there will be contribution from slip and twinning to the overall deformation. Twin structures may form during nucleation and growth processes such as crystal growth, phase transformation or recrystallization of the solid and deformation during processing.

There is presence of multiple variants of twins inside a grain, and the twins exhibit the tendency to either initiate or terminate at the GBs (15). When twinning is caused by an external stress, it is clearly necessary that the applied forces do work during the formation of the twin. That means the stress across the twinning plane and resolved in the twinning direction should be positive. The plastic deformation is accommodated by twinning; when twins get exhausted the accommodation is achieved by <c + a> slip (16). The difference between twinning and slip deformation is that twinning is polarized, i.e. reversal of the stress direction will not produce a twin. This is the reason that for a single crystal of given orientation with respect to a uniaxially applied stress, some variants of a particular twin mode may operate only in tension, whereas others may operate in compression. Therefore, twinning mode of deformation is very sensitive to orientation.

4.4.1 Deformation/Mechanical Twins

A twin formed during mechanical deformation is called deformation/mechanical twinning. Due to the lack of independent slip systems for Mg during deformation, mechanical twinning becomes an important deformation mechanism. The dominant twinning mode observed in Mg is $\{10\bar{1}2\}$ $<1011>$. The formation and growth of the mechanical twinning is also related to the grain size and the texture of polycrystalline materials (17). The average stress of mechanical twinning for a fine grain size material is higher than the stress for a coarse grain material, because of the greater interaction value between the crystals in a fine grain size (18). Even though mechanical twinning can be activated during RT deformation of Mg, this will result in large internal stress, thus leading to reduced ductility (19). At elevated temperature, mechanical twinning becomes less important due to the activation of non-basal slip systems. Twinning in the initial stage of plastic deformation may reorient the basal planes, which become more favorable for slip (20).

The activation of each type of twinning mode is governed by the combined effect of the Schmidt factor (SF) and grain size, and this controls the texture evolution characteristics developed during deformation process (21). Twin nucleation is not thermally activated but rather occurs at places of high stress concentration. Therefore, there is slow variation of twinning stress with temperature. Once nucleated, there is evidence that over an appreciable temperature range twins can grow more readily than slip can propagate. Therefore, twinning is a much more regulated process than slip (22).

4.4.2 Twin–Twin Interactions

The effect of twin–twin interactions on the nucleation and propagation of $<10\bar{1}2>$ twinning in AM30 Mg-alloy was studied (22). It was found that twin nucleation and twin propagation rates are strongly dependent on the number of activated twins in a given grain. This behavior was identified by comparing the twin growth evolution in two grains with roughly the same high Schmidt factor for twinning.

4.4.3 Twinning and Grain Size

There is a strong dependence of mechanical twinning on grain size (23) and ductility (24) of material. Twin formation was enhanced by grain coarsening (25). The work-hardening rate in plastic flow stage is strongly affected by twinning. In easy glide stage and linear hardening stage, twin boundaries, similar to GBs, increase the work-hardening rate. In metals that yield as a consequence of mechanical twinning, the yield stress is a function of the grain size (d) in much the same way as it is for dislocation glide. At a critical grain size, there appears a transition from yielding dominated by glide to yielding controlled by twinning as shown in Figure 4.3.

Ability of a metal to deform plastically depends on ease of dislocation motion under applied external stresses. As mentioned in earlier section, strengthening of a metal consists of hindering dislocation motion. Dislocation motion can be hindered in many ways, thus are strengthening mechanisms in metals (26).

4.4.4 Grain Boundary Strengthening

This strengthening mechanism is based on the fact that crystallographic orientation changes abruptly in passing from one grain to the next across the grain boundary. Thus, it is difficult for a dislocation moving on a slip plane in one crystal to pass over to a similar slip plane in another grain, especially if the orientation is highly misaligned (27). In addition, the crystals are separated by a thin non-crystalline region, which is the characteristic structure of a large angle grain boundary. Atomic disorder

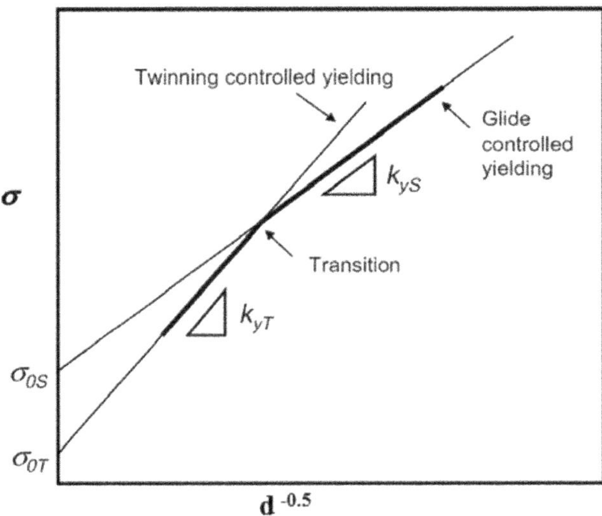

FIGURE 4.3 Transition from yielding dominated by glide to control by twinning. (From Barnett, M.R., *Scripta Mater.*, 59, 696–698, 2008.)

at the boundary causes discontinuity in slip planes. Hence dislocations are stopped by a grain boundary and they pile up against it. The smaller the grain size, the more frequent is the pile up of dislocations.

4.4.5 Low/High-Angle Grain Boundary

A grain boundary can hinder the dislocation motion in two ways: (1) by forcing the dislocation to change its direction of motion, and (2) by discontinuity of slip plane because of disorder. Effectiveness of grain boundary depends on its characteristic misalignment, represented by an angle (28). The ordinary high-angle grain boundary (misorientation angle > 15°) represents a region of random misfit between the grains on each side of the boundary. This structure contains grain-boundary dislocations which are immobile. However, they group together within the boundary to form a step or grain boundary ledge. These ledges can act as effective sources of dislocations as the stress at the end of slip plane may trigger new dislocations in adjacent grains. Low angle GBs (misorientation angle < 15°) are considered to be composed of a regular array of dislocations and are not effective in blocking the dislocations.

4.4.6 Accommodation Mechanism

Zelin and Mukherjee (27) summarized the mechanisms for superplastic deformation. During superplastic deformation, grain boundary sliding (GBS) dominates deformation mechanism. It is clear that during superplastic deformation an "accommodation mechanism" has to be active which is able to postpone for some time the formation of cavities (29). To allow sliding to continue, material must be transported, usually by diffusion creep from regions where the grains are trying to overlap (regions under compression) and be deposited in regions where the voids are trying to form (boundaries under tensile stress). Figure 4.4 shows consequences of grain boundary sliding.

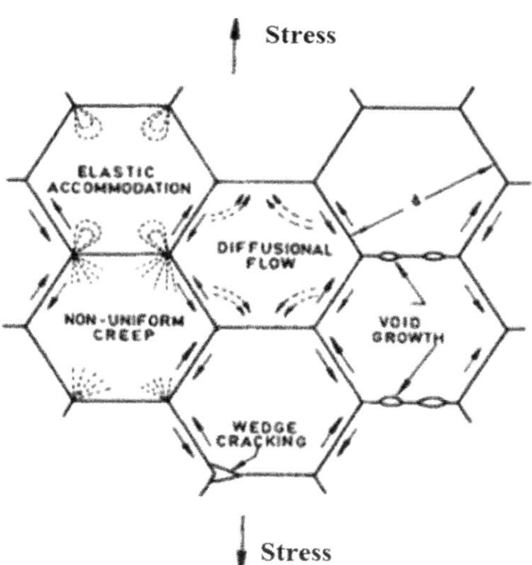

FIGURE 4.4 Consequences of GBS accommodated by slip mechanism.

4.5 GRAIN REFINEMENT OF Mg ALLOYS

Grain refinement can be achieved by various methods of mechanical processing like casting, thixo casting, mechanical/thermo mechanical processing using (forging, rolling, extrusion, hydrostatic extrusion, high pressure torsion (HPT) and severe mechanical processing like Equal Channel Angular Pressing (ECAP), Equal Channel Angular Rolling (ECAR), Friction stir processing (FSP) (30).

The processing temperature has more important role on the grain refinement (31). The grain refinement does not considerably progress after certain passes of severe processing. As the grain size becomes smaller, the yield and tensile strengths at higher temperatures decrease, while micro-hardness and fracture elongation increase. AZ31 can be successfully rolled to strains as high as 80% reduction at 300°C (32). There is effect of processing on microstructural aspects like shear bands, tensile twins and double twins which contribute to texture weakening during differential speed warm rolling of Mg alloy AZ31 (33).

4.5.1 GRAIN REFINEMENT THROUGH CALIBER ROLLING

Caliber rolling is a rolling process, initially used for grain refinement of steel in Japan (34). The process is suitable for mass production of long bars and rods of various cross sections. The process is used for other materials (35) including Mg alloys (36). Caliber rolling to obtain bar products is different from conventional plate/sheet rolling in different aspects. They are, (a) large strain can be introduced in to the work piece since it is cross section reduction against thickness reduction in plate/sheet rolling; (b) hydrostatic compressive stresses are present throughout the work piece when the material is passing through the die in caliber rolling, which reduces the tendency for crack initiation against open tensile forces in plate/sheet rolling in the transverse direction in plate/sheet rolling as alligatoring; (c) ability to rotate the work piece during caliber rolling, which is not possible in conventional rolling, which helps in reduction from all sides leading to microstructural homogeneity; (d) ability of the work piece to retain the temperature as it is a bar product compared to sheet/plate product, reducing the number of re-heats required for achieving the desired reduction.

The development of high strength bulk ultrafine-grained Mg-alloy AZ31 by multi-pass warm rolling was carried out (37). The initial microstructure composed of a very coarse grains with a large number of deformation twins. However, after multi-pass warm rolling, uniform ultrafine-grained microstructure consisting of very few twins was obtained. The yield strength, ultimate tensile strength as well as elongation markedly improved due to ultrafine grain refinement (38). Figure 4.5 shows the microstructures before and after caliber rolling along with the change in misorientation angle.

The strengthening of Mg–Al–Zn alloy was obtained by repetitive oblique shear strain with caliber rolling (39). Caliber rolling refined the grain structure effectively at a commercial processing speed with the formation of fine subgrains on a sub-micrometer scale. This resulted in high yield strength of more than 400 MPa. A simultaneous operation of oblique shear strain weakened the basal texture

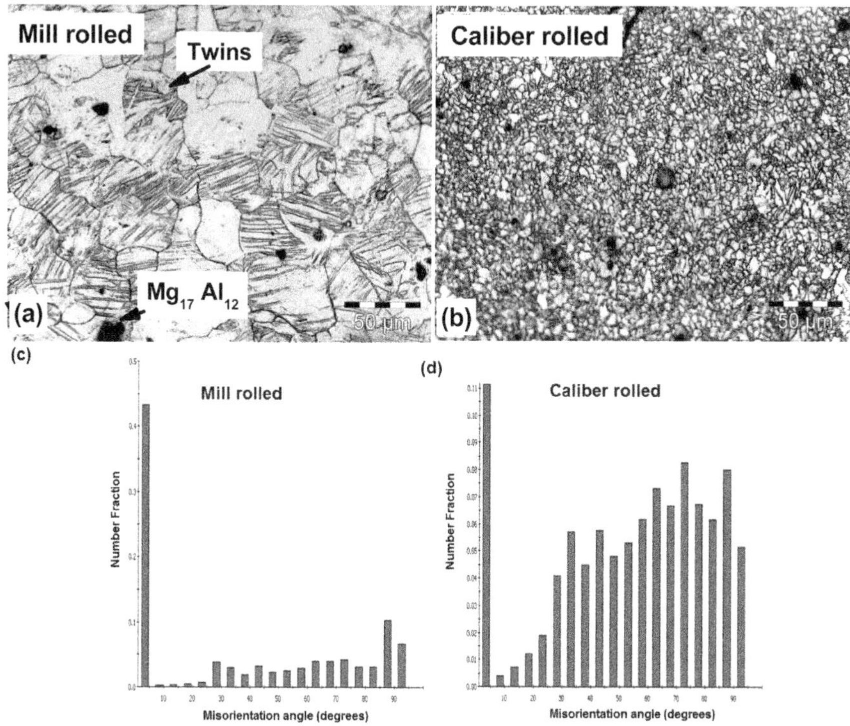

FIGURE 4.5 Optical micrographs and misorientation angle of AZ31 Mg-alloy: (a,c) mill rolled plate, (b, d) Caliber rolled at 300°C. (From Doiphode, R.L. et al., *J. Magn. Alloys*, 1, 169–175, 2013.)

compared to that of the initial as-extruded alloy. This resulted in tensile ductility comparable to that of the commercially extruded alloy.

There occurred changes in strain distribution and microstructural evolution in multi-pass warm caliber rolling (40). The microstructure has inhomogeneous distribution in the cross section of warm caliber rolled bar, which affected the various aspects of microstructure.

4.6 EFFECT OF ANNEALING

Annealing was carried out after mechanical processing to restore the ductility. Annealing affects the microstructure and subsequently changes the properties of metals (41). Microstructure becomes homogenous having equiaxed grains. Figure 4.6 shows the Optical micrographs for Mg alloy AZ31 after annealing for 5 minutes at temperature (a) 300°C (b) 450°C; and for 6000 minutes at temperature (c) 300°C (d) 450°C.

Although grain growth occurs owing to the higher annealing temperature, it is important to note that the homogeneous and equiaxed grain structure provides good ductility. In order to enhance the ductility, it is important that fine and equiaxed

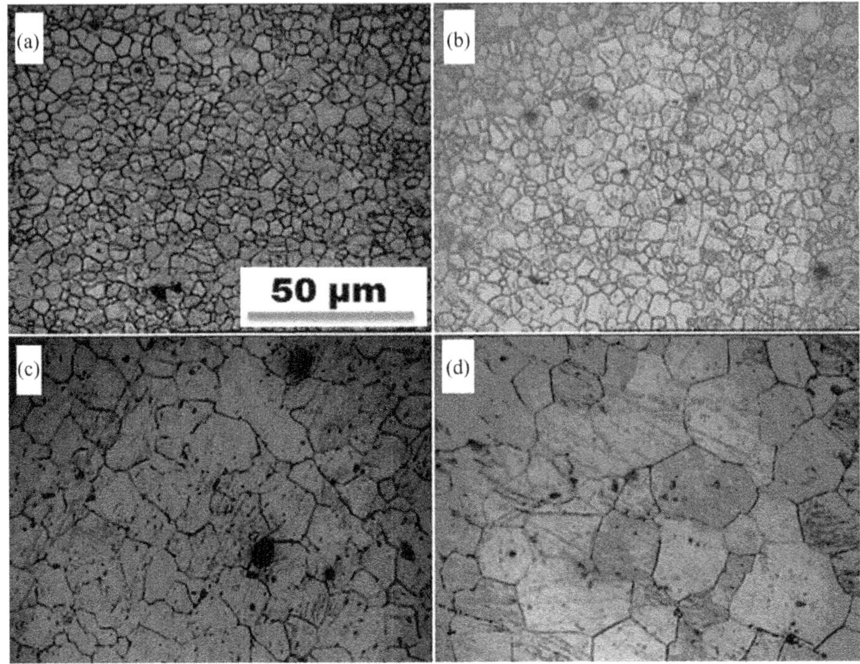

FIGURE 4.6 Optical micrographs after annealing for 5 minutes at temperatures (a) 300°C (b) 450°C; and for 6000 minutes at temperatures (c) 300°C (d) 450°C. (From Doiphode, R.L. et al., *J. Magn. Alloys*, 3, 322–329, 2015.)

grains should be distributed homogeneously. There exist very few studies on annealing of Mg-alloy AZ31 as summarized in Table 4.1 from the literature. It exhibits a wide variation in the activation energy for grain growth Q_g (29–200 kJ mol^{-1}) as compared to that expected on the basis of grain boundary Q_{gb} (92 kJ mol^{-1}) or lattice Q_l (135 kJ mol^{-1}) diffusion. Table 4.2 shows grain sizes after annealing for 5–6000 minutes.

TABLE 4.1
Studies on Annealing of Mg Alloy AZ31 from Literature

	Annealing		Grain Size (µm)				
Processing Type	Temp. (°C)	Time (min)	Initial d_o	Final d	Activation Energy Q_g (kJ mol^{-1})	Growth Law Index n_g	References
Rolling	375	1440	3	7–39	–	–	(43)
Compression	350–500	50	11	29	200	1.27	(44)
ECAP	250–500	60	1.9	4–32	94.2	0.5	(45)
Caliber rolling	300–450	5–6000	3	6–20	12.3–76.5	0.11	(42)

ECAP = Equal channel angular pressing.

TABLE 4.2
Grain Size d after Annealing of Mg Alloy AZ31 Undergone Caliber Rolling at 300°C (Initial Grain Size $d_0 = 3 \pm 0.1$ μm)

	Grain Size d (μm) at Various Annealing Temperatures T			
Annealing Time t (minutes)	300°C	350°C	400°C	450°C
5	6.2 ± 1.5	6.3 ± 1.5	7.8 ± 1.5	8.9 ± 1.5
10	7.5 ± 1.5	8.3 ± 1.5	8.3 ± 1.5	9.8 ± 2.0
20	9.4 ± 2.0	10.0 ± 2.0	10.5 ± 2.0	10.7 ± 2.0
30	10.0 ± 2.0	10.3 ± 2.0	10.7 ± 2.0	10.8 ± 2.0
60	12.3 ± 2.2	12.6 ± 2.5	12.6 ± 2.5	12.8 ± 2.5
240	13.3 ± 2.2	13.6 ± 2.5	13.9 ± 2.5	14.0 ± 2.5
600	14.0 ± 3.0	14.3 ± 3.0	16.2 ± 3.0	16.4 ± 3.0
1440	14.8 ± 3.8	15.5 ± 3.5	16.8 ± 3.5	16.9 ± 3.5
2880	16.4 ± 4.0	16.5 ± 4.0	17.8 ± 4.0	18.1 ± 4.0
6000	16.8 ± 4.2	17.4 ± 4.2	18.0 ± 4.5	20.4 ± 4.5

It is now well established that grain refinement process with the plastic deformation distort the microstructure. And, it causes the thermodynamically unstable microstructure containing dislocations, sub-grains and grains where energy is stored and tends to revert to a stable state (46). The tendency to revert to the stable state is accelerated by an increase in temperature (44).

The processes contributing to the production of the stable state called as static annealing consists of three phenomena; recovery, recrystallization and grain growth. The recovery is thermally activated re-arrangement and/or annihilation of dislocations and other defects. The driving force is the reduction in the total energy associated with these defects. Since the actual number of dislocations removed is quite small, the change in mechanical properties is limited. The recrystallization is nucleation and growth of new, relatively strain-free grains within a previously deformed material. The driving force for recrystallization is the removal of dislocations and their associated energies. Recrystallization is complete when all the new grains have impinged on one another and the old grains have been consumed. At this point, the mechanical properties of the material are restored to those prior to deformation. The grain growth is increase in the average grain size of a polycrystalline material that occurs at elevated temperatures. The energy stored during the plastic deformation of polycrystals has been utilized as the driving force for boundary migration in recrystallization. It is usually performed at temperature range of 0.4–0.5 T_m, where T_m is absolute melting temperature for metals with solute content (alloys). The free energy of GBs itself provides the driving force for grain growth in polycrystalline materials. The elimination of grain boundary area as the average grain size increases results in a reduction of the total free energy.

There is effect of time and temperature of annealing on hardness penetration. The decrease in hardness was observed as annealing temperature increases.

The reduction in hardness from surface to core is faster initially up to 20 min and becomes slower after more than 30 min of annealing (47).

4.7 DYNAMIC RECRYSTALLIZATION (DRX)

The understanding of the fundamentals of dynamic recrystallization (DRX) is very important. The phenomenon occurs during deformation, where energy is stored in the material mainly in the form of dislocations. This energy is released in three main processes, those of recovery, re-crystallization, and grain coarsening (48–49). The latter process involves the growth of a few grains which become much larger than the average. The presence of second phase particles may influence the re-crystallization process by creating nucleus sites through intense local deformation. Figure 4.7 shows the optical micrograph (ED plane) of the extruded ZWEK1000 alloy, which reveals one example of numerous fine-grained areas associated with second-phase particles. There is influence of texture on the re-crystallization mechanisms (50). The DRX mechanisms and kinetics depend on the operative deformation mechanisms and thus vary for different loading modes (tension, compression) as well as for different relative orientations between the loading axis and the c-axes of the grains (51). In particular, DRX is enhanced by the operation of <c + a> slip (52), since cross slip and climb take place more readily than for other slip systems. Thus, the formation of high angle boundaries is easier. DRX is also found to be promoted by twinning. DRX is plasticity enhancing structural development. DRX in its operation does not produce any strain. It is restoration or softening process, which facilitates straining with respect to ease and extent. DRX provides grain refinement and enhances ductility by GB sliding, thus relaxing the stress concentrations (53).

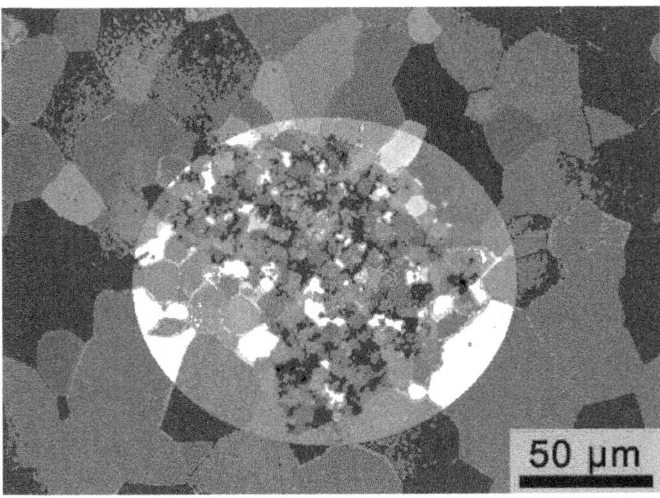

FIGURE 4.7 Optical micrograph (ED plane) of the extruded ZWEK1000 alloy showing one example of numerous fine-grained areas associated with second-phase particles. (From Hirsch, J. et al., *Acta Mater.*, 61, 818–843, 2013.)

The changes in texture that occur during the recrystallization process can be dramatic in the sense that the previous texture of the deformed state is, in some cases, replaced by an entirely different texture (55). A strong deformation texture is replaced with a different but strong recrystallization texture (56). This is consistent with the understanding of recrystallization as a process of nucleation and growth; new grains with (possibly) new orientations grow into the deformed structure, thereby eliminating the stored energy developed from plastic deformation (57). As in a solid-state phase transformation, growth of recrystallized grains can lead to drastic changes in microstructure and texture. Grain boundary mobility is affected due to the orientation dependence in metals. There are very large directly measured mobility differences of 100–1000 times, between low-angle (2°C–5°C) and high-angle (>15°) GBs (58). The fine grains with high-angle boundaries are capable of grain boundary sliding during superplastic deformation. During DRX, subgrains are first developed in the vicinity of the serrated GBs and, as deformation progresses, sub grain structure will form over the whole volume of the grain through the conversion of dislocation cell walls into sub GBs (59). There is possibility of recrystallization without grain growth at elevated temperatures and moderate strain rates. Also, necklace formation takes place as a DRX mechanism, which could be inside the twins. The texture of recrystallized twins may be much weaker than the texture of the parent grains. The mechanism of DRX in twins was found to be of continuous nature, involving the formation of low angle boundaries and their conversion to HAGBs forming new fine grains. Figure 4.8 shows the optical and SEM micrographs of AZ31 Mg alloy deformed at 200°C, showing dynamic recrystallization inside the deformation twins (60).

The literature shows that a portion of the grains from DRX satisfy the $\{10\bar{1}2\}$ $\langle 10\bar{1}1\rangle$ twin orientation relationship and indicate that twinning plays an important

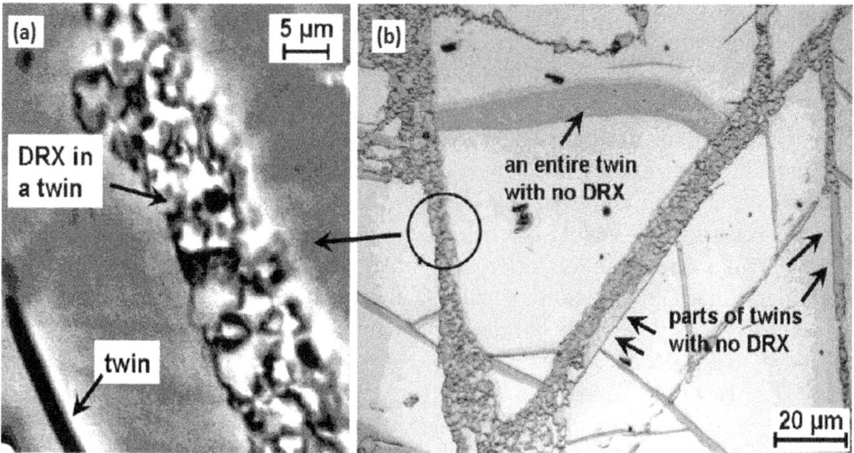

FIGURE 4.8 Optical (a) and SEM micrograph (b) of AZ31 Mg alloy deformed at 200°C/10^{-2} s^{-1} up to $\varepsilon = -1.2$ showing dynamic recrystallization inside the deformation twins. The micrographs are perpendicular to CD. (From Al-Samman, T. and Gottstein, G., *Mater. Sci. Eng. A*, 490, 411–420, 2008.)

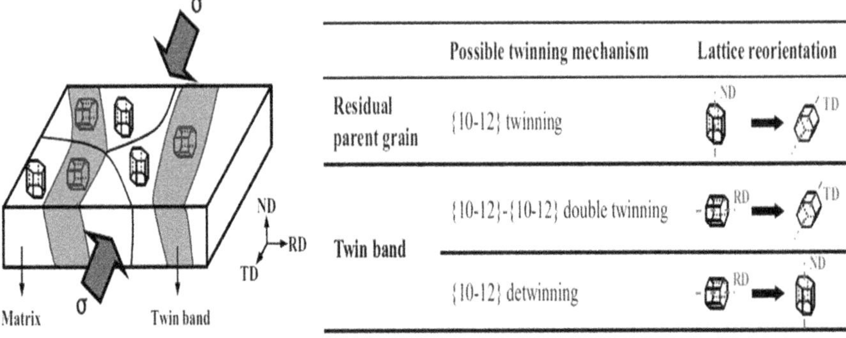

FIGURE 4.9 A schematic diagram illustrating the crystallographic lattice orientations of residual parent grains and twin bands in the Mg alloy AZ31 compressed to 6% along the RD, and the possible twinning mechanisms under subsequent compression along the TD. (From Park, S.H. et al., *Mater. Sci. Eng. A*, 532, 401–406, 2012.)

role in DRX (61). Deformation mechanisms, twinning, initial textures and DRX are very much related. Texture has a great effect on the twinning behavior; it affects DRX in the AZ31 Mg alloy. The effect of texture on DRX is caused by its effect on the character of dislocation slip. Figure 4.9 shows a schematic diagram illustrating the crystallographic lattice orientations of residual parent grains and twin bands in the Mg alloy AZ31 compressed to 6% along the Rolling Direction (RD), Normal Direction (ND), and the possible twinning mechanisms under subsequent compression along the Transverse Direction (TD).

4.8 ORIENTATION DEPENDENCE OF FLOW PROPERTIES

Complete isotropic characteristics are difficult to achieve since a certain level of mechanical fibering tends to occur in most deformed materials, inducing orientation dependence (63). However, in most materials, there is a pattern in the orientations caused firstly during crystallization from a melt or amorphous solid state and secondly by further thermo-mechanical processing involving deformation, recrystallization and phase transformation. The orientation does have effect on deformation behavior through slip and twinning in HCP metals. Magnesium has the ability to develop strong textures during mechanical processing (18). Development of texture affects the number of twin formation in the material due to misorientation.

Although mechanical twinning induces an abrupt change in orientation of the twinned material, large plastic strains are obtained usually by slip. During the slip process, the crystal lattice also rotates so that the active slip direction (in uniaxial tension) or the active slip-plane normal (in uniaxial compression) moves toward alignment with the direction of the applied stress. As a consequence, the orientation of the crystal changes during deformation. In a polycrystalline specimen, even though the grains are initially oriented at random, after sufficient deformation most of the grains are realigned into a preferred orientation, hence the development of a texture. The nature of the deformation texture depends essentially on the crystal structure of the metal, and on the nature and extent of plastic flow. Many other

TABLE 4.3
Theories of Deformation Texture Development as Reviewed from Literature

Name	Description	References
Taylor theory	All grains undergo the same shape change, Taylor factor $(M_f) = \tau_c/\sigma$	(65)
Max. work Principle	Grains with a low value of M are favorably oriented for deformation	(66)
Self consistent model	Use a mean field approach in which the deformation of each crystallite is considered individually within the surrounding of a homogeneous matrix	(67)
Crystal plasticity finite element model	Use finite element methods, include constitutive crystal plasticity equations, but are very computer intensive	(68)
Grain neighbor interaction model	Considers the deformation of pairs of grains	(69)
Relaxed constraints model	Allows the operation of less than 5 independent slip systems, it assumes that slip is homogeneous within a grain, all grains of the same orientation	(58)
Sach's theory	Each grain deforms independently of its neighbors and on the slip system that has greatest RSS	(58)

factors, such as composition, initial texture, thermal and mechanical history, the temperature, rate, and physical constraints during deformation, may all affect the resulting texture. Many attempts were made to predict the evolution of texture during deformation. Here some of the approaches are outlined, based on theories of polycrystalline plasticity, Table 4.3.

For even distribution of stress over the structure, three simple internal architectures are possible (64) as shown in Figure 4.10. Black regions correspond to grains

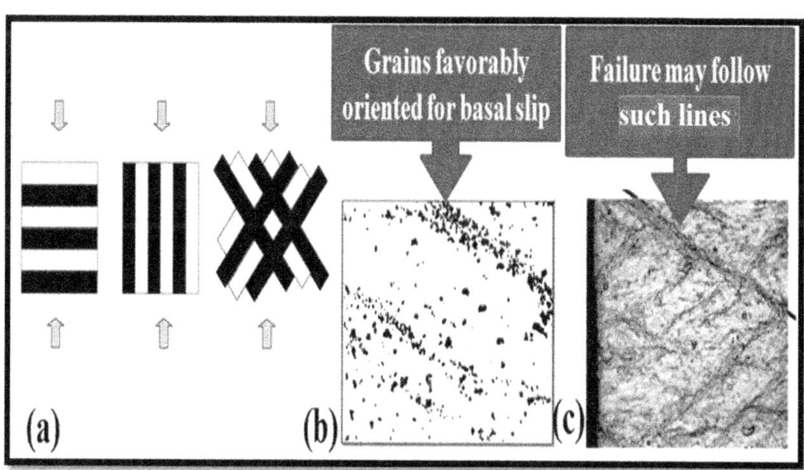

FIGURE 4.10 (a) Different idealized internal architectures (b) EBSD map showing grains favorably oriented for basal slip (c) band marked corresponding to a grouping of favorably oriented soft grains. (From Barnett, M.R., *Mater. Sci. Forum*, 618–619, 227–232, 2009.)

favorably aligned for basal slip and the material can be said to have a strong micro-texture. This speeds up the failure because of high local strain in the softer phase. Figure 4.10 also shows the electron backscatter diffraction (EBSD) map of grains favorably oriented for basal slip. It leads to strain heterogeneities. In tensile tests, failure may follow these lines of strain concentration.

4.9 ANISOTROPY AND ASYMMETRY IN Mg ALLOYS

Anisotropy is the different behavior of material in different directions under the same loading condition. Crystallographic materials are no more isotropic (same properties in all directions). Asymmetry is the different behavior of material under the different loading conditions. It is important to study the anisotropy and asymmetry present in the material. There are two main causes of these properties:

- Mechanical fibering – it is the elongation and alignment of microstructural features such as inclusions, GBs, twins and second phase particles.
- Preferred orientations of grains or crystallographic texture.

4.9.1 Effect of Twinning

The tensile-compressive mechanical anisotropy was closely related to generation of twin. The (0002) texture strongly affected the mechanical anisotropy because twin generation gets enhanced when the (0002) plane is strongly distributed parallel to the compressive testing directions. Anisotropy is also more prominent upon lower temperature or higher strain rate, which is due to twinning (70). The specific crystallographic orientation of the twin texture in rolling direction, caused by the $\{10\bar{1}2\}$ twinning during the pre-compression, gives rise to a change of active plastic deformation mechanisms. These are the basal slip, prismatic slip, $\{10\bar{1}2\}$ twinning, and $\{10\bar{1}2\}$ detwinning in twinned regions with the loading direction (i.e. in-plane anisotropic activities of slip and twinning) by affecting their SF. This led to anisotropy in in-plane deformation characteristics of the pre-compressed material (71). The amount of the pre-compression also plays a critical role in the deformation, as it is directly related to the twin volume fraction which governs the practical contribution of active deformation modes in twinned and un-twinned regions to the whole deformation. The $\{10\bar{1}2\}$ twins are formed to minimize strain incompatibility caused by the greater activity of basal slip and the less activity of prismatic slip in the twinning grain than in the surrounding grains. The $\{10\bar{1}2\}$ twinning tendency was found to increase with increasing the SF of the basal slip (72). When the loading is perpendicular to the c-axis of the HCP lattice, the specimen yields at a low stress because twinning is favored at relatively low strain, the deformed shape is strongly anisotropic owing to the slipping at high stress along $\langle 11\bar{2}3 \rangle$, and fracture occurs along $\langle 11\bar{2}3 \rangle$. On the other hand, when the loading is parallel to c-axis, there is no twinning process and the plastic deformation is dominated by slip which requires higher stress; the yield stress is higher, and the deformation is isotropic.

4.9.2 Effect of Texture

The textures cause anisotropy of mechanical properties along transverse and longitudinal directions, and the crystal grain shape may be another reason for anisotropy of mechanical properties. A strong fiber texture in the specimen with the basal plane parallel to the loading axis is a favorable orientation for $\{10\bar{1}2\}$ twinning in compression but not in tension. The very large difference in the stress-strain curves between tension and compression is attributable to the utilization of the $\{10\bar{1}2\}$ twinning mode to assist in the compressive deformation. The dependence of flow asymmetry due to $\{10\bar{1}2\}$ twinning was discussed in the literature (73).

There exists a distinctive anisotropy in their mechanical behavior which is independent of the process parameters. There is effect of microstructure (grain size, twins and texture) on the asymmetry and anisotropy of AZ31 Mg-Alloy (74). Many studies were carried out specifically on asymmetry and anisotropy. A pronounced anisotropy in the mechanical properties of the extruded sheets was observed. This anisotropy is caused by the developed texture, which allows the activation of specified deformation modes in defined directions (75). The equal channel angular rolled specimens exhibited a stronger planar anisotropy (76). The stronger anisotropy was probably due to the modified texture, which transformed from the strong basal texture for the as-received specimen to the non-basal texture for the equal channel angular rolled specimens. After annealing the textures became weaker, and the differences of crystal diameters between transverse and longitudinal directions became smaller, then the anisotropy of mechanical properties gets reduced.

4.9.3 Reduction in Asymmetry and Anisotropy

The mechanical anisotropy gets reduced by grain refinement. This is because twin generation was suppressed due to activation of non-basal slip by grain refinement (77). There occurs drop in anisotropy from a change in the relative slip activity of the prismatic and pyramidal slip systems. The mechanical anisotropy decreases with increasing temperature. This was attributed to reduction in twin generation at elevated temperature. The randomization of texture may reduce asymmetry and anisotropy (14). Texture randomization can be achieved by various mechanical processing methods. Even it is possible within the standard rolling process without the use of special techniques (78).

4.10 DEFORMATION IN-HOMOGENEITIES

The deformation obtained in the wrought Mg-alloys is not homogeneous (23); it can display deformation in-homogeneities, as influenced by variation in the ease of basal slip amongst grains, micro-textures, shear banding, twinning and GBS. Figure 4.11 shows the micrographs where SBs are formed on the surfaces of compressed UFG AZ31 specimens.

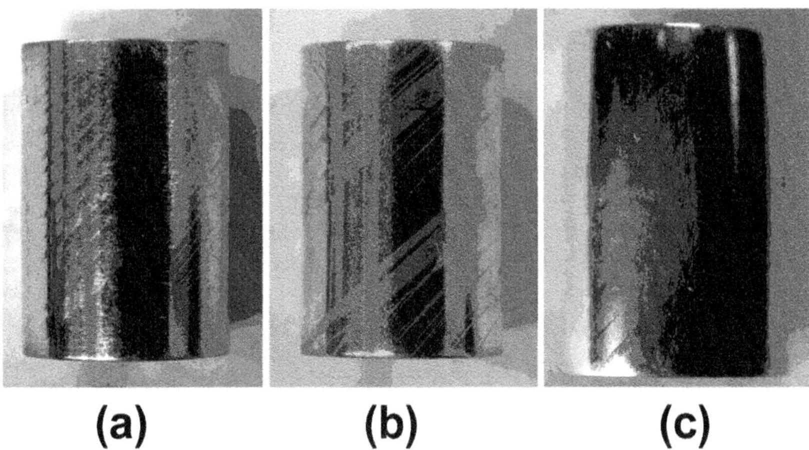

FIGURE 4.11 Micrographs showing the formation of SBs on the surfaces of compressed UFG Mg alloy AZ31 specimens: (a) 0.46-μm grain size, (b) 2.87-μm grain size, and (c) 3.22-μm grain size. Strain = 3 pct. (From Lee, W.T. et al., *Mater. Trans.*, 42, 2909–2916, 2011.)

Basal slip is the most readily activated slip mode in wrought Mg-alloys. It is homogeneous at the grain scale. In grains, if it is active, forms closely spaced and evenly dispersed slip lines. This is effectively softer region or easier deformation mode than the surrounding material. This leads to grain to grain heterogeneity. Processed materials have a considerable residual stress present. It may sometimes amount to ~50 MPa, and it may have impact on the Bauschinger effect (stress-strain characteristics change as a result of the microscopic stress distribution). This is favorable condition for twin formation.

Another important source of heterogeneity is deformation twinning. It leads to redistribution of local stresses. If there is considerable Bauschinger effect, there exists the possibility of double twinning in Mg-alloys as shown in the image quality map from a smaller area of the 45 μm grain size material after a compressive strain of 0.09 with indexable twin boundaries highlighted as shown in Figure 4.12. It may lead to formation of local regions with basal planes favorably aligned for slip. This can result in high local strain formation and may lead to the failure of material.

As studied earlier, Mg undergoes non-basal slip also. It occurs at prismatic and pyramidal plane. These slip modes occur near GBs and they create deformation heterogeneity where deformation gradients form near GBs. A high level of deformation heterogeneity on the cross-sectional planes of Mg-alloy AZ31 processed by HPT was observed (81). This heterogeneity was revealed through the nature of the flow patterns, through the distributions of grain sizes, and by comprehensive micro hardness measurements.

There are many reasons for the occurrence of asymmetry and anisotropy in the flow behavior which is another cause of in-homogeneity in flow behavior.

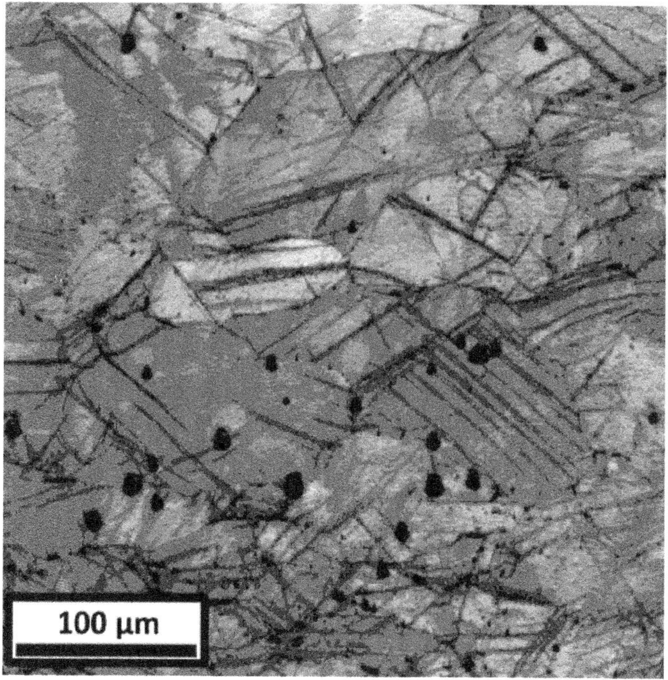

FIGURE 4.12 Image quality map from a smaller area of the 45 μm grain size of Mg alloy AZ31 after a compressive strain of 0.09. Indexable twin boundaries highlighted: blue for extension; green for contraction; red for double twins. (From Levinson, A. et al., *Acta Mater.*, 2013, 61, 5966–5978.)

4.11 HIGH TEMPERATURE FLOW PROPERTIES AND SUPERPLASTICITY

There is an increase in ductility, when the material is processed at high temperature and low strain rate as shown in the Figure 4.13; on the other hand, when analyzing the mechanical parameters related to formability, processing at high temperature (2) and strain rate assures better stretchability and drawability (82) even for compressive deformation (83). It is possible to improve the superplasticity through grain refinement (84) along with other microstructural aspects.

The constitutive relation for flow behavior at elevated temperature (86) is given by:

$$\dot{\varepsilon} = A\left[\frac{D_0 G b}{kT}\right]\left[\frac{b}{d}\right]^p\left[\frac{\sigma}{G}\right]^n \exp\left[\frac{-Q_d}{RT}\right] \tag{4.2}$$

This involves the effects of temperature T, strain rate and grain size d on flow stress σ in terms of strain rate sensitivity index $m = 1/n$, activation energy for deformation Q_d and exponent of inverse grain size p, respectively. Here, b is Burgers vector, D_o is the pre-exponential frequency factor for diffusion coefficient $\left(D = D_0 \exp(-\frac{Q_d}{RT})\right)$, G

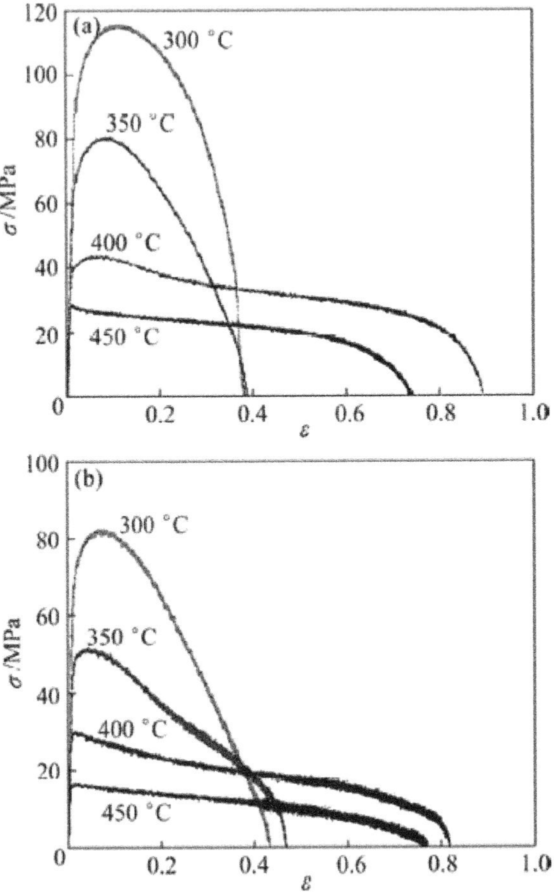

FIGURE 4.13 True stress versus true strain curves of AZ80 Mg alloy at various temperatures in elongation-to-failure tests: (a) strain rate = 10^{-2} s^{-1}; (b) strain rate = 10^{-3} s^{-1}. (From Qiao, J. et al., *Trans. Nonferrous Met. Soc. China*, 23, 2857–2862, 2013.)

is the shear modulus, k is Boltzmann's constant, A is a material and mechanism dependent constant, and R is gas constant. These parameters not only explain the mechanisms operating during deformation but also suggest the ranges of the variables employed in above equation that are suitable for material processing. Deformation mechanism map and processing map are developed based on such factors, respectively. A careful examination of the nature of stress-strain curves reveals the variation in flow stress as a function of strain ε, which causes a deviation from the steady state behavior.

Summary of various parameters for high temperature tensile deformation of Mg alloy from literature is presented in Table 4.4 along with the grain size d used. An exceptional elongation of 3050% was achieved after extrusion plus ECAP at RT under tensile testing at 473 K and strain rate of 10^{-4} s^{-1} as shown in Figure 4.14.

TABLE 4.4
Summary of High Temperature Deformation of Mg Alloy from Literature

Material and Processing	Temp. (°C)	σ (MPa)	$\dot{\varepsilon}$ (s^{-1})	e_f (%)	d (μm)	References
AZ31 Commercial alloy	500	8	10^{-3}	320	25–30	(89)
AZ31 Commercial alloy	250	125	10^{-3}	40	16.7	(59)
AZ31 Commercial alloy	350	48	10^{-2}	110	10	(59)
AZ31 Commercial alloy	400	10	10^{-4}	475	–	(90)
AZ31 Commercial alloy	400	12	5×10^{-5}	175	–	(82)
AZ31 Commercial alloy	400	–	10^{-4}	270	15	(91)
AZ31 Commercial alloy	300	45	2×10^{-2}	80	12	(92)
AZ31 Commercial alloy	300	85	5×10^{-4}	45	30	(27)
AZ31 Annealing	400	–	0.7×10^{-3}	362	15	(93)
AZ31 Annealing	450	–	10^{-5}	596	–	(94)
AZ31 ECAP	375	10	10^{-4}	325	17	(95)
AZ31 ECAP	350	32	10^{-2}	150	3	(31)
AZ31 ECAP	350	12	10^{-4}	960	2.2	(96)
Mg 0.6Zr Extrusion + ECAP	300	–	3.3×10^{-4}	400	1.4	(87)
AZ61 Hot extrusion	350	–	10^{-4}	920	12	(97)
Mg-Y-Zn Hot extrusion	400	–	10^{-3}	260	–	(98)
WE 54 Hot extrusion	520	–	0.9	130	20	(99)
AZ80 Extrusion + Rolling	400	–	10^{-2}	161	16	(85)
AZ31 Deep drawing	220	120	–	95	–	(100)
AZ31 Caliber rolling	450	18	10^{-3}	182	13	(101)
AZ31 Hot rolling	400	–	5×10^{-4}	140	29	(102)
AZ61 Hot rolling	450	–	10^{-3}	321	15	(103)
AZ31 ABRC 3 passes	200	–	5×10^{-5}	312	3–4	(104)
AZ 91 HPT at RT	300	–	10^{-4}	1308	30	(105)
AZ31 FSP	350	–	5×10^{-4}	1050	11.4	(106)
AZ91 Submerged FSP	350	–	2×10^{-2}	990	1.2	(107)

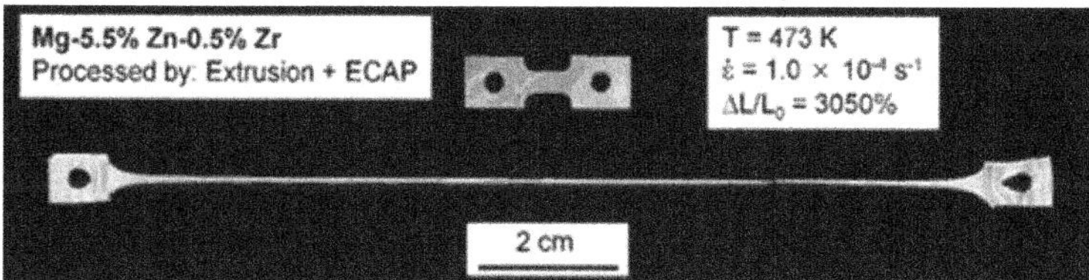

FIGURE 4.14 An exceptional elongation of 3050% is achieved after extrusion plus ECAP at RT and tensile testing at 473 K and strain rate of 10^{-4} s^{-1}. (From Langdon, T.G., *J. Mater. Sci.*, 44, 5998–6010, 2009.)

The magnitudes of parameters m and Q_d of constitutive relationship (Eq. 4.2), as obtained from the literature on AZ31 Mg-alloy for different processing methods and temperature are summarized in Table 4.5. The values of m are noted to vary from 0.33 to 0.55 at intermediate strain rates, which correspond to superplastic behavior. Figure 4.15 shows the values of m for Mg alloy AZ31 caliber rolled at 450°C at various strain rates.

TABLE 4.5
Strain Rate Sensitivity Factor *m* and Activation Energy Q_d for Mg Alloy from Literature

Material/Processing/Treatment	*m*	Q_d (kJ mol^{-1})	Deformation Mechanism	References
Tensile deformation				
AZ31 Equal Channel Angular Extrusion (ECAE)	0.55	–	–	(108)
AZ31 ECAP and annealed	0.5	84	GBS, accommodated by grain boundary diffusion	(109)
AZ31 B –O sheet – for intermediate $\dot{\varepsilon}$	0.47	~92	GBS, accommodated by grain boundary diffusion	(93)
— for high $\dot{\varepsilon}$	0.17			
AZ31 B –O sheet For strain of > 150%	0.33	~135	Lattice diffusion	(59)
AZ31 Hot rolling	0.4	–	Dislocation creep	(102)
AZ61 Hot rolling	0.33	143–160	Glide controlled Dislocation creep	(103)
AZ31 B –O sheet rolled and annealed at 345°C for 2 Hrs	–	80	GBS, accommodated by grain boundary diffusion	(110)
AZ31 B –O sheet rolled (coarse grained)	0.38	145	Dislocation creep accommodated by lattice diffusion	(89)
AZ31 B sheet caliber rolled	0.33	88	GBS, accommodated by grain boundary diffusion	(101)
Mg 0.6Zr Extrusion and ECAP	0.3	–	GBS	(87)
AZ61 Hot extrusion	0.45	–	GBS and dislocation creep	(97)
WE 54 Hot extrusion	0.23	237	–	(99)
AZ80 Extrusion followed by rolling	–	149.6	GBS and dislocation climb	(85)
AZ 91 HPT at RT	0.52	80.34	–	(105)
Compressive deformation				
AZ31 B –O	0.33	148	Lattice diffusion	(111)
As cast + Annealed at 400°C	0.1	196	Dislocation creep	(112)
AZ31 plate	0.22	143	Self diffusion	(113)
AZ31 rolled plate	0.15	164	Higher than self diffusion	(114)
AZ31 cast ingot, homogenized	–	136	Lattice diffusion	(115)
ZK60	0.2	140.3	Lattice self diffusion	(116)
Mg-Gd-Y-Zn	0.2	135	Climb controlled lattice self diffusion	(117)

For a fixed temperature, decreasing the strain rate produces a more observable ductility-enhancement compared to the case when the strain rate is held fixed and the temperature is raised. It can be inferred that $\dot{\varepsilon}$ of about 10^{-3} s^{-1} is the threshold for superplasticity (200% elongation) for all the considered high temperature tests (>325°C). It is observed that a 200% elongation is not feasible at 225°C, even at the

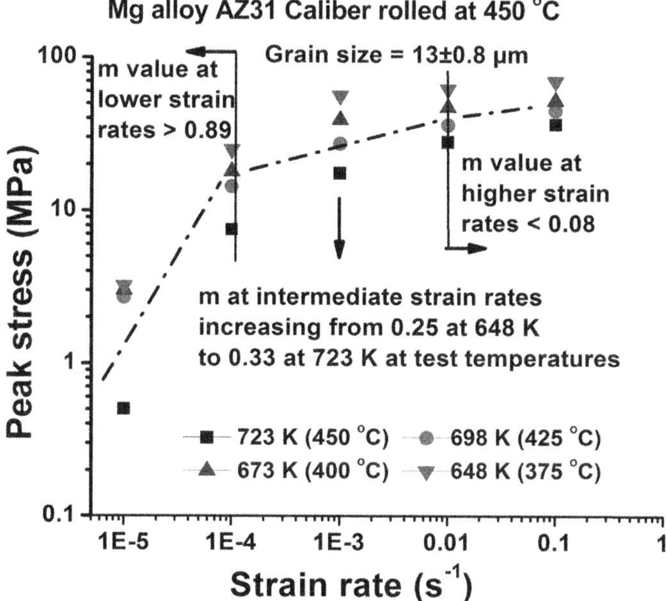

FIGURE 4.15 Log stress vs. log strain rate plot from DSR tests conducted at various temperatures for Mg alloy AZ31 caliber rolled at 450°C at various strain rates. (From Doiphode, R.L. et al., *Metall. Trans.* A, 46, 3028–3042, 2015.)

smallest strain rates considered. For each strain rate, there is a limiting temperature beyond which no further ductility-enhancement can be achieved.

The ductility is improved at elevated temperature with low strain rate due to softening originated from DRX. The mechanical properties are sensitive to strain rate at constant elevated temperature. Compared with strain rate, the temperature plays a more important role in improving ductility.

4.12 SYNERGIZED EFFECT OF VARIOUS MICROSTRUCTURAL ASPECTS IN Mg ALLOYS

Most of the available literature has studied the effects of various microstructural aspects separately. However, the various aspects of material like microstructure, texture and twinning play a combined role. It is well known that high angle grain boundaries affect the deformation behavior. However, there is combined effect of misorientation angle and twin density, which is true because $\{10\bar{1}2\}$ twins are helping in the tensile deformation (17) and higher the twin density higher is the effect. The synergistic effect is observed between PF intensity, number fraction of misorientation angle >15° and twin density. As shown schematically in Figure 4.16, for the same orientation, the two types of twins formed are giving increase in strength in tension and compression. But, if there is different orientation and the same kind of twins, there will be different behavior. Further study is necessary to quantify the relative effectiveness of these aspects in deformation.

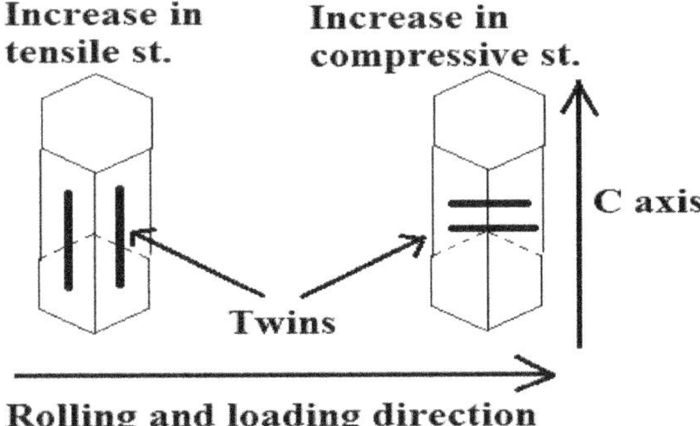

FIGURE 4.16 Schematic showing the effect of two types of twins formed at constant orientation in HCP system.

4.13 PROCESSING MAPS FOR Mg ALLOYS

The concept of processing maps is based on the dynamic materials model proposed by Prasad (118). Here the work piece is considered to be a dissipater of power and the efficiency of power dissipation (η) occurring through microstructural changes during hot deformation is calculated. The variation of η with deformation temperature and strain rate constitutes the power dissipation map. It also shows the instability of flow based on the extremism principles of irreversible thermodynamics as applied to continuum mechanics of large plastic flow. These maps can be plotted for various strains. These maps are useful for selecting the processing parameters for maximum efficiency. Figure 4.17 shows the processing map of Mg–3Al alloy developed at the strain 0.5. The contour values

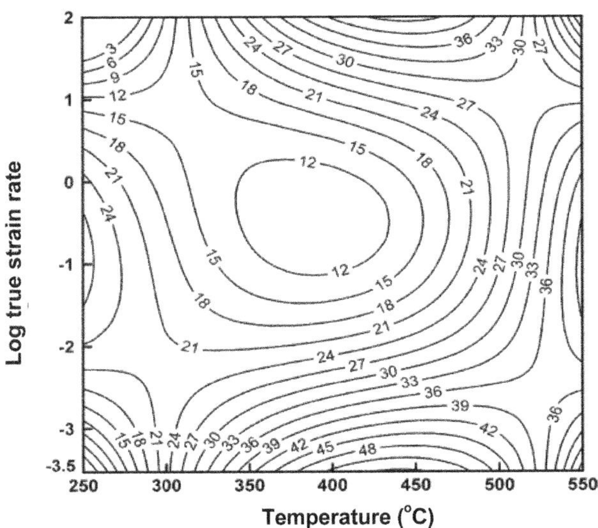

FIGURE 4.17 Processing map of Mg–3Al alloy developed at the strain 0.5. The contour values represent iso-efficiency of power dissipation (η) in percent. (From Srinivasan, N. et al., *Mater. Sci. Eng. A*, 476, 146–156, 2008.)

TABLE 4.6
Processing Conditions for Various Magnesium Alloys

Material and Processing	Temperature (°C)	Strain Rate (s^{-1})	Peak Efficiency (%)	References
AZ31 as Cast	350–400	0.003	56	(119)
Tx32-0.4al-0.6 Si Gravity casted	300–450	0.0003–0.001	40	(120)
ZM31 + 3.2y as Extruded	340–500	0.0001–0.03	42.5	(121)
ZACM8100 Gravity casted	312–350	0.001–0.22	37.5	(122)
AZ91 Powder compacts	275–325	0.001–0.01	43	(123)
WE43 as Extruded	380–430	0.003–0.0025	37	(124)
AZ80 as Extruded	350–400	0.1	30	(125)
ZK60 as Extruded	340–380	0.001–0.003	48	(126)
AZ41 as Extruded	390–450	0.005–0.015	36	(127)

represent iso-efficiency of power dissipation (η) in percent. Table 4.6 shows the parameters of processing with the efficiency of power dissipation for Mg alloys from literature.

4.14 CONCLUSIONS

It can be summarized that the Mg-alloys have been studied extensively. There are many severe plastic deformation processes which include rolling, extrusion, ECAP, HPT and FSP. The combinations of these processes are also used to enhance the mechanical properties and superplasticity. Conventional rolling process has limitations with respect to process parameters, microstructural aspects and resultant mechanical properties; the modified rolling processes like differential speed rolling, equal channel angular rolling and caliber rolling (CR) are developed. There occurs grain refinement after the processing of material but it also affects the microstructure and texture, which causes change in the properties. Also, there occurs formation of twins during deformation, which is the reason for many changes in the properties. For superplasticity, ultra fine grains, uniform distribution of second phase particles, high angle GBs and homogeneous structure are the important factors. It is further noted that twinning and texture play a very important role in the deformation of HCP metals. The changes in the flow properties were found to be the result of DRX but not so systematic effect of grain size could be revealed even in elevated temperature flow behavior. The deviation from the systematic effect of grain size was attributed to the vital role played by twinning and texture in influencing the flow stress and ductility. The synergistic effect of twins, texture and grain size could be a necessary step towards rationalizing the deviation from the known effects of grain size on flow properties in the material like magnesium alloys.

REFERENCES

1. Kainer, K.U. *Magnesium: Magnesium Alloys and Their Applications, Proceedings of the 6th International Conference*, Orlando, FL, 2003.
2. Kashyap, B.P., Arieli, A., Mukherjee, A.K. Review: Microstructural aspects of superplasticity. *J. Mater. Sci.* 1985, *20*, 2661–2686.

3. Dieter, G.D. *Mechanical Metallurgy*. McGraw-Hill Book Company, London, UK, 1988.
4. Langdon, T.G. Seventy-five years of superplasticity: Historic developments and new opportunities. *J. Mater. Sci.* 2009, *44*, 5998–6010.
5. Hertzberg, R. *Deformation and Fracture Mechanics of Engineering Materials*. John Wiley & Sons, New York, 1995.
6. Hull, D., Bacon, D. *Introduction to Dislocations* (4th ed.). Butterworth Heinemann, Oxford, 2001.
7. Sergueeva, A., Mara, N., Mukherjee A.K. Grain boundary sliding in nanomaterials at elevated temperatures. *J. Mater. Sci.* 2006, *42*, 1433–1438.
8. Capolungo, L. Dislocation junction formation and strength in magnesium. *Acta Mater.* 2011, *59*, 2909–2917.
9. Courtney, T.H. *Mechanical Behavior of Materials* (2nd ed.). Wavelande press, Long Grove, IL, 2006.
10. Friedrich, H., Mordike, B. *Magnesium Technology, Metallurgy, Design Data, Applications*. Springer, Berlin, Germany, 2006.
11. Hutchinson, W.B., Barnett, M.R. Effective values of critical resolved shear stress for slip in polycrystalline magnesium and other hcp metals. *Scripta Mater.* 2010, *63*, 737–740.
12. Hosford, W. *Mechanical Behaviour of Materials*. Cambridge University Press, Cambridge, 2005.
13. Meyers, M., Chawla, K. *Mechanical Behavior of Materials* (2nd ed.). Cambridge University Press, Cambridge, UK, 2009.
14. Christian, J., Mahajan, S. Deformation twinning. *Prog. Mater. Sci.* 1995, *39*, 1–157.
15. Balogh, L., Niezgoda, S., Kanjarla, A., Brown, D., Clausen, B., Liu, W., Tomé, C. Spatially resolved in situ strain measurements from an interior twinned grain in bulk polycrystalline AZ31 alloy. *Acta Mater.* 2013, *61*, 3612–3620.
16. Brown, D., Agnew, S., Bourke, M., Holden, T., Vogel, S., Tomé, C. Internal strain and texture evolution during deformation twinning in magnesium. *Mater. Sci. Eng.* A 2005, *399*, 1–12.
17. Hong, S.G., Park, S.H., Lee, C.S. Role of {10–12} twinning characteristics in the deformation behavior of a polycrystalline magnesium alloy. *Acta Mater.* 2010, *58*, 5873–5885.
18. Ecob, N., Ralph, B. The effect of grain size on deformation twinning in a textured zinc alloy. *J. Mater. Sci.* 1983, *18*, 2419–2429.
19. Guo, L., Chen, Z., Gao, L. Effects of grain size, texture and twinning on mechanical properties and work-hardening behavior of AZ31 magnesium alloys. *Mater. Sci. Eng. A* 2011, *528*, 8537–8545.
20. Munroe, N., Tan, X. Orientation dependence of slip and twinning. *Scripta Mater.* 1997, *36*, 1383–1386.
21. Park, J.S., Kim, J.M., Song, Y.H., Lim, D.W., Yang, W.S., Shin, H.T., Koo, Y.S. Mechanical behaviors and texture development of the extruded magnesium AZ31 alloys. *Mater. Sci. Forum* 2011, *695*, 275–278.
22. Haitham, E., Kapil, J., Oppedal, A., Hector, L., Agnew, S., Cherkaoui, M., Vogel, S. The effect of twin–twin interactions on the nucleation and propagation of $\{10\bar{1}2\}$ twinning in magnesium. *Acta Mater.* 2013, *61*, 3549–3563.
23. Barnett, M.R. A rationale for the strong dependence of mechanical twinning on grain size. *Scripta Mater.* 2008, *59*, 696–698.
24. Barnett, M.R. Twinning and the ductility of magnesium alloys Part I: "Tension" twins. *Mater. Sci. Eng.* A 2007, *464*, 1–7.
25. Chino, Y., Kimura, K., Mabuchi, M. Deformation characteristics at room temperature under biaxial tensile stress in textured AZ31 Mg alloy sheets. *Acta Mater.* 2009, *57*, 1476–1485.

26. Cahn, R., Haasen, P. *Physical Metallurgy Vol. 1* (4th ed.). Elsevier Science B.V., Cambridge, UK, 1996.
27. Zelin, M., Mukherjee, A. Cooperative phenomena at grain boundaries during superplastic flow. *Acta Mater.* 1995, *43*, 2359–2372.
28. Sergueeva, A., Mara, N., Mukherjee, A. Grain boundary sliding in nanomaterials at elevated temperatures. *J. Mater. Sci.* 2006, *42*, 1433–1438.
29. Ashby, M., Shercliff, H., Cebon, D. *Engineering, Science, Processing and Design* (1st ed.). Elsevier Linacre House, Jordan Hill, Oxford, UK, 2007.
30. Langdon, T.G. Twenty-five years of ultrafine-grained materials: Achieving exceptional properties through grain refinement. *Acta Mater.* 2013, *61*, 7035–7059.
31. Kang, S., Lee, Y., Lee, J. Effect of grain refinement of magnesium alloy AZ31 by severe plastic deformation on material characteristics. *J. Mater. Process. Technol.* 2008, *201*, 436–440.
32. Stanford, N. Fine grained AZ31 by conventional thermo-mechanical processing. *Mater. Sci. Forum* 2009, *618–619*, 239–244.
33. Cho, J., Kim, H., Kang, S., Han, T. Bending behavior, and evolution of texture and microstructure during differential speed warm rolling of AZ31B magnesium alloys. *Acta Mater.* 2011, *59*, 5638–5651.
34. Torizuka, S., Muramatsu, E., Murty, S.V.S.N., Nagai, K. Microstructure evolution and strength-reduction in area balance of ultrafine-grained steels processed by warm caliber rolling. *Scripta Mater.*, 2006, 55 (8), 751–754.
35. Lee, T., Park, C.H., Lee, S.Y., Son, I.H., Lee, D.L., Lee, C.S. Mechanisms of tensile improvement in caliber-rolled high-carbon steel. *Met. Mater. Int.* 2012, *18*, 391–396.
36. Tanno, Y., Mukai, T., Asakawa, M., Kobayashi, M. Study on warm caliber rolling of magnesium alloy. *Mater. Sci. Forum* 2003, *419–422*, 359–364.
37. Murty, S.V.S.N., Mittal, M., Sinha, P. Development of high strength bulk ultrafine grained Mg-alloy AZ31 by multi pass war rolling, *Proceedings of the International Conference Aerospace Science and Technology*, Bangalore, India, 2008.
38. Doiphode, R.L., Narayana Murty, S.V.S., Prabhu, N., Kashyap, B.P. Effects of caliber rolling on microstructure and room temperature tensile properties of Mg-3Al- 1Zn alloy. *J. Magn. Alloys* 2013, *1*, 169–175.
39. Mukai, T., Somekawa, H., Inoue, T., Singh, A. Strengthening Mg–Al–Zn alloy by repetitive oblique shear strain with caliber roll. *Scripta Mater.* 2010, *62*, 113–116.
40. Inoue, T., Yin, F., Kimura, Y. Strain distribution and microstructural evolution in multi-pass warm caliber rolling. *Mater. Sci. Eng. A* 2007, *466*, 114–122.
41. Hornbogen, E., Koster, U. *Recrystallization of Metallic Materials.* (F. Haessner et al., Editors.) 1978, 159–194.
42. Doiphode, R.L., Narayana Murty, S.V.S., Prabhu, N., Kashyap, B.P. Grain growth in calibre rolled Mg–3Al–1Zn alloy and its effect on hardness. *J. Magn. Alloys* 2015, *3*, 322–329.
43. Tsai, M., Chang, C. Grain size effect on deformation twinning in Mg–Al–Zn alloy. *Mater. Sci. Technol.* 2013, *29*, 759–763.
44. Beer, A.G., Barnett, M.R. The post-deformation recrystallization behaviour of magnesium alloy Mg–3Al–1Zn. *Scripta Mater.* 2009, *61*, 1097–1100.
45. Su, C., Lu, L., Lai, M. Mechanical behavior and texture of annealed AZ31 Mg alloy deformed by ECAP. *Mater. Sci. Technol.* 2007, *23* (3), 290–296.
46. Gottstein, G., Shvindlerman, L. *Grain Boundary Migration in Metals.* CRC Press, Boca Raton, FL, 1999.
47. Doiphode, R.L., Narayana Murty, S.V.S., Prabhu, N., Kashyap, B.P. Study of hardness penetration depth of caliber rolled Mg-3Al-1Zn alloy. *Mater. Sci. Forum* 2015, *830*, 177–180.

48. Doherty, R.D., Hughes, D.A., Humphreys, F.J., Jonas, J.J., Jensen, D.J., Kassner, M.E., Rollett, A.D. Current issues in recrystallization: A review. *Mater. Sci. Eng. A* 1997, *238*, 219–274.
49. Humphreys, F.J. A new analysis of recovery recrystallization and grain growth, *Mater. Sci. Technol.* 1999, *15*, 37–44.
50. Del Valle, J., Ruano, O. Influence of texture on dynamic recrystallization and deformation mechanisms in rolled or ECAPed AZ31 magnesium alloy. *Mater. Sci. Eng. A* 2008, *487*, 473–480.
51. Dudamell, N.V., Ulacia, I., Gálvez, F., Yi, S., Bohlen, J., Letzig, D., Pérez-Prado, M. T. Influence of texture on the recrystallization mechanisms in an AZ31 Mg sheet alloy at dynamic rates. *Mater. Sci. Eng. A* 2011, *532*, 528–535.
52. Kang, F., Li, Z., Wang, J.T., Cheng, P., Wu, H.Y. The activation of <c + a> non-basal slip in Magnesium alloys. *J. Mater. Sci.* 2012, *47*, 7854–7859.
53. Mcqueen, H.J., Imbert, C.A. Dynamic recrystallization. *J. Alloys Compd.* 2004, *378*, 35–43.
54. Hirsch, J., Al-Samman, T. Superior light metals by texture engineering: Optimized aluminum and magnesium alloys for automotive applications. *Acta Mater.* 2013, *61*, 818–843.
55. Beck, P.A., Hu, H. *Recrystallization, Grain Growth and Textures.* (H. Margolin, Editor) ASM International, 1966.
56. Doiphode, R.L., Narayana Murty, S.V.S., Prabhu, N., Kashyap, B.P. Microstructure and texture evolution during annealing of caliber rolled Mg-3Al-1Zn alloy. *Trans. Indian Inst. Met.* 2015, *68*, 317–321.
57. Gordon, P., Vandermeer, R. *Grain Boundary Migration, Recrystallization, Grain Growth and Texture.* ASM, Materials Park, OH, 1966.
58. Humphreys, F.J., Hatherly, M. *Recrystallization and Related Annealing Phenomena* (2nd ed.). Elsevier, Amsterdam, the Netherlands, 2004.
59. Tan, J., Tan, M. Dynamic continuous recrystallization characteristics in two stage deformation of Mg-3Al-1Zn alloy sheet. *Mater. Sci. Eng. A* 2003, *339*, 124–132.
60. Al-Samman, T., Gottstein, G. Dynamic recrystallization during high temperature deformation of magnesium. *Mater. Sci. Eng. A* 2008, *490*, 411–420.
61. Ma, Q., Li, B., Marin, E., Horstemeyer, S. Twinning-induced dynamic recrystallization in a Mg alloy extruded at 450°C. *Scripta Mater.* 2011, *65*, 823–826.
62. Park, S.H., Hong, S.G., Lee, J.H., Lee, C.S. Multiple twinning modes in rolled Mg-3Al-1Zn alloy and their selection mechanism. *Mater. Sci. Eng. A* 2012, *532*, 401–406.
63. Wang, Y.N., Huang, J.C. Texture analysis in hexagonal materials. *Mater. Chem. Phy.* 2003, *81*, 11–26.
64. Barnett, M.R. The challenge of inhomogeneous deformation in Mg and its alloys. *Mater. Sci. Forum* 2009, 618–619, 227–232.
65. Taylor, G.I. Plastic strain in metals. *J. Inst. Met.* 1938, *62*, 307–324.
66. Bishop, J.F.W., Hill, R. A theoritical derivation of the plastic properties of a polycrystalline face centered metal. *Philos. Mag.* 1951, *42*, 1298–1307.
67. Molinari, A., Canova, G., Ahzi, S. A self consistent approach of the large deformation polycrstal viscoplasticity. *Acta Metall.* 1987, *35*, 2983–2994.
68. Kalidindi, S.R., Bronkhorst, C.A., Anand, L. Crystallographic texture evolution in bulk deformation processing of FCC metals. *J. Mech. Phys. Solids* 1992, *40*, 537–569.
69. Houtte, V., Delannay, P., Samajdar, I. Quantitative prediction of cold rolling textures in low carbon steels by means of the LAMEL model. *Texture Microst.* 1999, *31*, 109–149.
70. Dai, Q., Zhang, D., Chen, X. On the anisotropic deformation of AZ31 Mg alloy under compression. *Mater. Des.* 2011, *32*, 5004–5009.
71. Park, S.H., Hong, S., Lee, S.C. In-plane anisotropic deformation behavior of rolled Mg–3Al–1Zn alloy by initial {10–12} twins. *Mater. Sci. Eng. A* 2013, *570*, 149–163.

72. Koike, J., Sato, Y., Ando, D. Origin of the anomalous {10–12} twinning during tensile deformation of Mg alloy sheet. *Mater. Trans.* 2008, *49A*, 2792–2800.
73. Haitham E.K., Kapil, J., Oppedal, A.L., Hector Jr., L.G., Agnew, S.R., Cherkaoui, M., Vogel, S.C. The effect of twin–twin interactions on the nucleation and propagation of $\{10\bar{1}2\}$ twinning in magnesium. *Acta Mater.* 2013, *61*, 3549–3563.
74. Doiphode, R.L., Narayana Murty, S.V.S., Prabhu, N., Kashyap, B.P. Flow-asymmetry and-anisotropy in caliber rolled Mg-3Al-1Zn alloy. *3rd Asian Symposium Material Processing.* IIT Madras, India, 2012.
75. Gall, S., Müller, S., Reimers, W. Microstructure and mechanical properties of magnesium AZ31 sheets produced by extrusion. *Int. J. Mater. Form.* 2011, *6*, 187–197.
76. Cheng, Y.Q., Chen, Z.H. Anisotropy of deformation behavior of AZ31 magnesium alloy sheets. *Adv. Mater. Res.* 2011, *314–316*, 1121–1125.
77. Chino, Y., Kimura, K., Hakamada, M., Mabuchi, M. Mechanical anisotropy due to twinning in an extruded AZ31 Mg alloy. *Mater. Sci. Eng. A* 2008, *485*, 311–317.
78. Kohzu, M., Kii, K., Nagata, Y., Nishio, H., Higashi, K., Inoue, H. Texture randomization of AZ31 magnesium alloy sheets for improving the cold formability by a combination of rolling and high-temperature annealing. *Mater. Trans.* 2010, *51*, 749–755.
79. Lee, W.T., Ding, S.X., Sun, D.K., Hsiao, C.I., Chang, C.P., Chang, L., Kao, P.W. Deformation structure of unidirectionally compressed ultrafine-grained Mg-3Al-1Zn Alloy. *Mater. Trans.* 2011, *42*, 2909–2916.
80. Levinson, A., Mishra, R.D.K., Doherty, R., Kalidindi, S. Influence of deformation twinning on static annealing of AZ31 Mg alloy. *Acta Mater.* 2013, *61*, 5966–5978.
81. Figueiredo, R., Aguilar, M., Cetlin, P., Langdon, T. Deformation heterogeneity on the cross-sectional planes of a magnesium alloy processed by high-pressure torsion. *Metall.Trans. A*, 2011, *42*, 3013–3021.
82. Pellegrini, D., Ghiotti, A., Bruschi, S. Effect of warm forming conditions on AZ31B flow behaviour and microstructural characteristics. *Int. J. Mater. Form.* 2011, *4*, 155–161.
83. Barnett, M.R., Keshavarz, Z., Beer, A.G., Atwell, D. Influence of grain size on the compressive deformation of wrought Mg–3Al–1Zn. *Acta Mater.* 2004, *52*, 5093–5103.
84. Mishra, R., Valiev, R., Mukherjee, A. The observation of tensile superplasticity in nanocrystalline materials. *Nanostruct. Mater.* 1997, *9*, 473–476.
85. Qiao, J., Bian, F., He, M., Wang, Y. High temperature tensile deformation behavior of AZ80 magnesium alloy. *Trans. Nonferrous Met. Soc. China* 2013, *23*, 2857–2862.
86. Mukherjee, A.K., Bird, J.E., Dorn, J.E. The role of climb in creep processes: (B.L. Eyre, Editor). United Kingdom Atomic Energy Authority, Harwell, UK, *II*, 422–495, 1968.
87. Langdon, T.G. Seventy-five years of superplasticity: Historic developments and new opportunities. *J. Mater. Sci.* 2009, *44*, 5998–6010.
88. Doiphode, R.L., Narayana Murty S.V.S., Prabhu, N., Kashyap, B.P. High temperature tensile flow behaviour of caliber rolled Mg-3Al-1Zn alloy. *Metall. Trans. A* 2015, *46*, 3028–3042.
89. Wu, X., Liu, Y. Superplasticity of coarse-grained magnesium alloy. *Scripta Mater.* 2002, *46*, 269–274.
90. Panicker, R., Chokshi, A., Mishra, R., Verma, R., Krajewski, P. Microstructural evolution and grain boundary sliding in a superplastic magnesium AZ31 alloy. *Acta Mater.* 2009, *57*, 3683–3693.
91. Sakamoto, H., Ohbuchi, Y., Kuramae, H., Shi, J. Deep drawing formability analysis of AZ31 Mg-alloy. *Adv. Mater. Res.* 2011, *337*, 701–704.
92. Frost, H., Ashby, M. *Deformation Mechanism Maps: The Plasticity and Creep of Metals and Ceramics.* Pergamon Press, Oxford, New York, 1982.

93. Yin, D., Zhang, K., Wang, G., Han, W. Warm deformation behavior of hot-rolled AZ31 Mg alloy. *Mater. Sci. Eng. A* 2005, 392, 320–325.
94. Watanabe, H., Fukusumi, M. Mechanical properties and texture of a superplastically deformed AZ31 magnesium alloy. *Metall. Trans. A* 2008, *477*, 153–161.
95. Del Valle, J., Penalba, F., Ruano, O. Optimization of the microstructure for improving superplastic forming in magnesium alloys. *Metall. Trans. A* 2007, *467*, 165–171.
96. Abu-Farha, F., Khraisheh, M. Mechanical characteristics of superplastic deformation of AZ31 magnesium alloy. *J. Mater. Eng. Perform.* 2007, *16*, 192–199.
97. Wang Y., Huang J. Transition of dominant diffusion process during superplastic deformation in AZ61 magnesium alloys. *Metall. Trans. A* 2003, *35*, 555–562.
98. Oñorbe, E., Garcés, G., Dobes, F., Pérez, P., Adeva, P. High-temperature mechanical behavior of extruded Mg-Y-Zn alloy containing LPSO phases. *Metall. Mater. Trans. A*. 2013, *44*, 2869–2883.
99. Carsi, M., Jesús Bartolome, M., Rieiro Félix Peñalba, I., Ruano, O. A., *Mater. Des.* 2014, *28*, 30–35.
100. Zhong, X., Huang, G., He, F., Liu, Q. Mechanical properties and microstructure of AZ31 magnesium alloy under uni-axial tensile loading. *Mater. Sci. Forum* 2011, *686*, 219–224.
101. Doiphode, R.L., Kulkarni R., Narayana Murty, S.V.S., Prabhu, N., Kashyap, B.P. Effect of severe caliber rolling on superplastic properties of Mg-3Al-1Zn (AZ31) alloy. *Mater. Sci. Forum* 2013, *735*, 327–331.
102. Deng, J., Lin, Y.C., Li, S., Chen, J., Ding, Y. Hot tensile deformation and fracture behaviors of AZ31 magnesium alloy. *Mater. Des.* 2013, *49*, 209–219
103. Olguín-González, M.L., Hernández-Silva, D., García-Bernal, M.A., Sauce-Rangel, V.M. Hot deformation behavior of hot-rolled AZ31 and AZ61 magnesium alloys. *Mater. Sci.Eng. A* 2014, *597*, 82–88.
104. Yang, Q., Ghosh, A.K. Formability of ultrafine-grain Mg alloy AZ31B at warm temperatures. *Metall. Trans. A* 2008, *39*, 2781–2796.
105. Al-Zubaydi, A.S.J., Zhilyaev, A.P., Wang, S.C., Reed, P.A.S. Superplastic behaviour of AZ91 magnesium alloy processed by high-pressure torsion. *Mater. Sci. Eng. A* 2015, *637*, 1–11.
106. Zhang, D., Xiong, F., Zhang, W., Qiu, C. Superplasticity of AZ31 magnesium alloy prepared by friction stir processing. *Trans. Nonferrous Met. Soc. China* 2011, *21*, 1911–1916.
107. Chai, F., Zhang, D., Li, Y., Zhang, W. High strain rate superplasticity of a fine-grained AZ91 magnesium alloy prepared by submerged friction stir processing. *Mater. Sci. Eng. A* 2013, *568*, 40–48.
108. Furushima, T., Shimizu, T., Manabe, K. Grain refinement by combined ECAE/extrusion and dieless drawing processes for AZ31 magnesium alloy tubes. *Mater.Sci. Forum* 2010, *654–656*, 735–738.
109. Figueiredo, R., Langdon, T.G. Developing superplasticity in a magnesium AZ31 alloy by ECAP. *J. Mater. Sci.* 2008, *43*, 7366–7371.
110. Koike, J., Ohyama, R., Kobayashi, T., Suzuki, M., Maruyama, K. Grain-boundary sliding in AZ31 magnesium alloys at room temperature to 523 K. *Mater. Trans.* 2003, *44*, 445–451.
111. Lee, B., Shin, K., Lee, C. High temperature deformation behavior of AZ31 Mg alloy. *Mater. Sci. Forum* 2005, *475–479*, 2927–2930.
112. Sun, S., Zhang, M., He, W., Zhou, J., Sun, G. Hot deformation behavior during the hot compression of AZ31 alloy. *Adv. Mater. Res.* 2010, *139–141*, 545–548.
113. Prasad, Y., Rao, K. Effect of crystallographic texture on the kinetics of hot deformation of rolled Mg–3Al–1Zn alloy plate. *Mater. Sci. Eng. A* 2006, *432*, 170–177.

114. Fatemi-Varzaneh, S., Zarei-Hanzaki, A., Haghshenas, M. A study on the effect of thermo-mechanical parameters on the deformation behavior of Mg–3Al–1Zn. *Mater. Sci. Eng. A* 2008, *497*, 438–444.
115. Kwon, S., Song, K., Shin, K., Kwun, S. Low cycle fatigue properties and an energy-based approach for as-extruded AZ31 magnesium alloy. *Met. Mater. Int.* 2011, *17*, 207–213.
116. Mirzadeh, H. Constitutive behaviors of magnesium and MgeZneZr alloy during hot deformation. *Mater. Chem. Phy.* 2015, *152*, 123–126.
117. Mirzadeh, H. Quantification of the strengthening effect of rare earth elements during hot deformation of Mg-Gd-Y-Zr magnesium alloy. *J. Mater. Technol.* 2015, *5*, 1–4.
118. Prasad, Y.V.R.K., Gegel, H.L., Doraivelu, S.M., Malas, J.C., Morgan, J.T., Lark, K.A., Barker, D.R. Modeling of dynamic material behavior in hot deformation: Forging of Ti-6242. *Metall. Trans. A* 1984, *15*, 1883–1892.
119. Srinivasan, N., Prasad, Y.V.R.K, Rama Rao, P. Hot deformation behaviour of Mg–3Al alloy – A study using processing map. *Mater. Sci. Eng. A* 2008, *476*, 146–156.
120. Dharmendra, C., Rao, K.P., Zhao, F., Prasad, Y.V.R.K., Hort, N., Kainer, K.U. Effect of silicon content on hot working, processing maps, and microstructural evolution of cast TX32–0.4Al magnesium alloy. *Mater. Sci. Eng.* A 2014, *606*, 11–23.
121. Tahreen, N., Zhang, D.F., Pan, F.S, Jiang, X.Q., Li, D.Y., Chen, D.L. Hot deformation and processing map of an as-extruded Mg–Zn–Mn–Y alloy containing I andW phases. *Mater.Des.* 2015, *87*, 245–255.
122. Zhu, S., Luo, T., Zhang, T., Yang, Y. Hot deformation behavior and processing maps of as-cast Mg–8Zn–1Al–0.5Cu–0.5 Mn alloy. *Trans. Nonferrous Met. Soc. China* 2015, *25*, 3232–3239.
123. JabbariTaleghani, M.A., Torralba, J.M. Hot deformation behavior and workability characteristics of AZ91 magnesium alloy powder compacts – A study using processing map. *Mater. Sci. Eng. A* 2013, *580*, 142–149.
124. Avadhani, G.S., Tapase, S., Satyam S. Hot deformation processing and texture in magnesium alloy WE43. *16th IFAC Symposium on Automation in Mining, Mineral and Metal Process.* San Diego, CA, 2013.
125. Li, H., Wei, X., Ouyang, J., Jiang, J., Li, Y. Hot deformation behavior of extruded AZ80 magnesium alloy. *Trans. Nonferrous Met. Soc. China* 2013, *23*, 3180–3185.
126. Li, J., Liu, J., Cui, Z. Characterization of hot deformation behavior of extruded ZK60 magnesium alloy using 3D processing maps. *Mater. Des.* 2014, *56*, 889–897.
127. Cai, Z., Chen, F., Ma, F., Guo, J. Dynamic recrystallization behavior and hot workability of AZ41M magnesium alloy during hot deformation. *J. Alloy. Compd.* 2016, *670*, 55–63.

5 Effect of Heat and Surface Treatment on the Structure and Properties of the Mg-Al-Zn-Mn Casting Alloys

Leszek A. Dobrzański

Science Centre ASKLEPIOS of the Medical and Dental Engineering Centre for Research, Design and Production ASKLEPIOS in Gliwice, Poland

cooperation in research:

Anna D. Dobrzańska-Danikiewicz, Justyna Domagała-Dubiel, Mariusz Król, Szymon Malara, Tomasz Tański, Jacek Trzaska, Lubomir Čižek, and Jerry H. Sokolowski

"When choosing something worth doing, never give up and try not to disappoint anyone."

Sir Prof. Harry Kroto (1939–2016), Nobel Prize Winner in Chemistry in 1996

This chapter was created in a very special period, when after invocation to human kindness delivered from the stage during the great charity event, in front of millions of people gathered in Gdańsk and in many Polish Cities and before television screens, **the late**

Paweł Adamowicz

Mayor of the City of Gdańsk in the years of 1998 to 2019,

and previously the Chairman of the Gdańsk City Council

and the Vice-Rector of the University of Gdańsk was brutally murdered in the 54th year of life

and at night, when in the operating room, there was a hopeful struggle for his life and over the next nights and days, when millions of people of good will in Poland and in many places in the world plunged into contemplation, silence and mourning, and joined in the protest against hate speech and such acts.

To his Memory and ideals, which he adhered to throughout his life, which I personally identify completely with, i.e. goodness, openness, tolerance, solidarity, freedom and democracy and humanism and positive activities for others

I dedicate my work

Prof. Leszek A. Dobrzański

CONTENTS

The Author thanks Springer Nature for permission to use Tables 5.1 and 5.2 and Figure 5.3 and International OCSCO Word Press in Gliwice, Poland for permission to use the figures and photos used by Author in his earlier publications.

5.1 GENERAL CHARACTERISTICS OF THE INVESTIGATED Mg-Al-Zn-Mn ALLOYS AND OF THE METHODOLOGY OF THE INVESTIGATIONS CARRIED OUT

5.1.1 General Characteristics of Magnesium Alloys

The basic expectation in relation to engineering materials nowadays is a need for their on-demand delivery [1–3]. This means that it is not the chemical composition of the material that is the basic criterion of suitability for designers and manufacturers of machines and other equipment, not even the structure, but the achievable functional properties [4–9]. On the other hand, light alloys, which include, among others, magnesium alloys and their production and processing technologies, are of particular importance in the current search for materials with low density, alternative to steels and other ferroalloys [10–14]. Magnesium alloys, except for titanium and aluminium alloys, are currently becoming a good quality and modern material for manufacturing various components of machines and equipment [1–3]. Lightweight alloys (with low density) are applied mainly in the transport resources construction industry [15], which is closely linked to energy efficiency, especially liquid fuels in a number of the initiated research programmes, both in Europe and in other countries, including the US and Japan. In this sense, by saving energy, it will be easier to carry the means of transport and the goods and people transported with such means, and this usually serves the purpose of satisfying the needs, and if associated with tourism – also pleasure. Investigations into the structure and properties of light alloys, including magnesium, serve to improve the conditions and quality of life in the modern world. Magnesium is constantly gaining an increasing importance in modern technology [14,16–79], as evidenced by, among others, its rising global production [80]. Primary magnesium production capacity worldwide in 2016 reached altogether 2 million of metric tonnes (Figure 5.1). The analysts

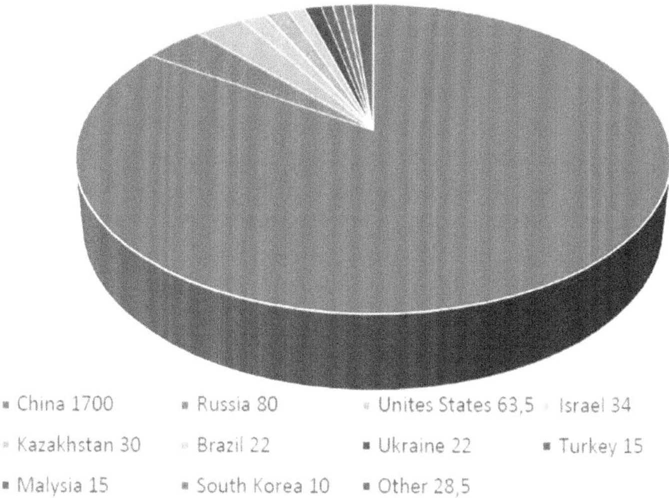

FIGURE 5.1 Primary magnesium production capacity worldwide in 2016 by country (1,000 metric tons).

forecast the global magnesium market to grow at a Compound Annual Growth Rate CAGR of 5.92% during the period 2018–2022 [80].

The demand for small-density magnesium alloy castings is to a large extent related not only to the growth of the automotive industry, but also, among others, sports industry. The properties of the core and of the surface layer of the part produced can be customised most advantageously by selecting the part's material and its structure and properties formation processes appropriately along with the surface layer type and technology ensuring the required functional properties [81–85]. As a consequence, a material is designed and delivered meeting all the requirements set by the customer.

The strength and plastic properties of pure magnesium are relatively low and for this reason it is not used as a structural material. Magnesium is most commonly used, however, as an alloy additive to produce aluminium alloy elements, because it constitutes approx. 53.5% of the produced magnesium [1], and to produce zinc alloy elements. Die castings represent ca. 11.3%, gravity castings ca. 0.8% and rolled products from constructional alloys of magnesium with other metallic elements represent approx. 2.9% [1]. Five basic groups of alloys, which are currently produced for commercial purposes, based mainly on such alloy elements as Al, Zn, Mn, Zr and RE, are distinguished according to their chemical composition. They fall into the following subgroups [86,87]: Mg-Mn, Mg-Al-Mn, Mg-Al-Zn-Mn, Mg-Zr, Mg-Zn-Zr, Mg-RE-Zr, Mg-Ag- RE-Zr and Mg-Y-RE-Zr. Other groups of alloys such as Mg-Th-Zr, Mg-Th-Zn-Zr and Ag-Mg-Th-Zr-RE were also fabricated previously, but they were withdrawn due to the detrimental effects of thorium. This chapter pertains to a selected group of such alloys only, i.e. Mg-Al-Zn-Mn.

The three basic alloy elements in magnesium alloys are Al, Zn and Mn [1–3,87,88]. The largest addition in Mg alloys is Al. Aluminium increases the hardness, the tensile strength Rm and the elongation A in Mg alloys, whilst the highest tensile strength is seen at a concentration of approx. 5% of Al, and the highest elongation at a concentration of approx. 6% of Al. Hardness is increased as a result of precipitation of the intermetallic phase $\gamma - Mg_{17}Al_{12}$ and maintained only to a temperature of 120°C. A similar increase in the strength can be obtained in the presence of zinc and manganese [89,90]. For this reason, the physical properties of, among others, these Mg alloys are enhanced through dispersion hardening. With the increasing concentration of Al, so is growing the yield point and castability (at a concentration of approx. 10% of Al, magnesium alloys have a very good castability). Aluminium also reduces the cast shrinkage but causes brittleness to heat [88]. In casting alloys, the concentration of Al is within 3%–11%, for the maximum concentration of approx. 9% in alloys for machining [1–3,91]. Magnesium, together with aluminium, creates a phase system with eutectics with a concentration of 32.3% of Al at a temperature of 437°C (Figure 5.2a.). Eutectics consists of the solid solution α and the intermetallic phase γ with a concentration of 40.2% of Al. A solubility limit of aluminium in magnesium is 12.7% at the eutectic temperature, falling significantly with the decreasing temperature down to 1.5% at room temperature. Mg-Al alloys are hence susceptible to ageing [1–3,90]. Such elements as Zn and Mn, as well as Cu play an important role in aluminium-magnesium alloys [1–3,86–89].

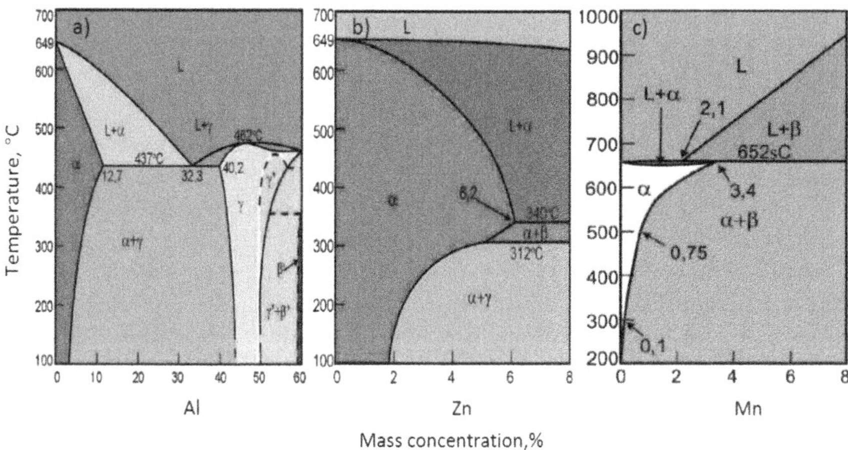

FIGURE 5.2 Phase equilibrium chart: (a) Mg-Al; (b) Mg-Zn; (c) Mg-Mn. (From Dobrzański, L.A., *OAlib.*, Annal VII, 2017, 1–982, 2017; Dobrzański, L.A., *Materiały inżynierskie i projektowanie materiałowe. Podstawy nauki o materiałach i metaloznawstwo*, Wydawnictwa Naukowo-Techniczne, Wydanie II zmienione i uzupełnione, Warszawa, Poland, pp. 1–1600, 2006; Dobrzański, L.A., *Metalowe materiały inżynierskie*, Wydawnictwa Naukowo-Techniczne, Warszawa, Poland, pp. 1–888, 2004.)

Zinc is the second component, after Al, having the greatest influence on properties of magnesium alloys. Similar like Al, it increases the strength and yield point of magnesium alloys, but only in the presence of Al and Mn. Alloys with a concentration of approx. 5% of Zn reach the greatest strength and plasticity. Magnesium alloys with a concentration of approx. 3% of zinc have very good castability, while a higher concentration of Zn causes hot brittleness and the microporosity effect. Elongation is reduced in the alloys with a concentration of above 2% of Zn. Zinc is dissolved in magnesium and forms also part of the hardening phase of the Mg-Al alloy, and it causes the presence of granular eutectics in this system. Zinc also counteracts adverse effects on the corrosion resistance of impurities comprising Fe and Ni, which may be present in Mg alloys [1–3,12,85–88,91–93]. Magnesium creates, together with zinc, a phase system with eutectics with a temperature of 340°C, which consists of the solid solution α with maximum solubility of zinc in magnesium of 6.2% and of the secondary solid solution β on the intermetallic phase matrix Mg_7Zn_3 (Figure 5.2b). At a temperature of 312°C, the phase is subject to eutectoid decomposition to the mixture of the solution α and the secondary solid solution β on the matrix of the intermetallic phase Mg_7Zn_3. The solubility of zinc in magnesium decreases with the temperature falling from 6.2% at a temperature of 340°C and to approx. 2% at 100°C.

Manganese is present in almost all magnesium alloys and, as a basic component, is added at a concentration of about up to 2.5%, although commercial alloys containing Mn rarely contain more than 1.5% of this element, and in the presence of Al, the solubility of Mn in the solid state is reduced to 0.3%. Manganese does not have much influence on the tensile strength (growth occurs at a concentration of over 1.5%), but slightly improves the yield strength. Its primary function is to improve the corrosion resistance of Mg-Al and Mg-Al-Zn alloys in salt water, as it reduces the adverse effect of iron, which causes corrosion. In Mg-Al alloys, manganese is included in various intermetallic phases, whose hardness is rising together with the fraction of Al. It also allows to weld Mg alloys [1–3,86–89,94]. Mg-Mn alloys, unlike Mg-Al alloys, are characterised by a small range of the solid solution α (Figure 5.2c). The solubility of Mn in Mg is 2.1% at the peritectic temperature of 652°C, and as such temperature drops, reaching 0.75% at 500°C, 0.1% at 300°C and at room temperature – the concentration of close to zero [92].

In case of three-component alloys, there are many more structural components, which are given in Table 5.1, according to the list in the work [95], prepared on the basis of numerous publications from the years of 1913–2003, and in Table 5.2, where the complexity of phase transformations in such alloys is presented. The selected cross-sections of the calculated diagrams of equilibrium of this system are shown in Figure 5.3.

5.1.2 General Characteristics of Investigation into Mg-Al-Zn-Mn Alloys Covered by This Elaboration

The aim of this elaboration is to present the results of long-term own research under 6 large projects led by me [96–101], summarising the experience concerning casting magnesium alloys [84,87,102–224]. The research was conducted in particular as part of the FORSURF project [101] in collaboration with Prof. A. D. Dobrzańska-Danikiewicz, within the framework of the habilitation dissertation then prepared by her [102], and under other projects [96–99], in cooperation with the following current PhDs: J. Domagała-Dubiel, M. Król, Sz. Malara, and the current Associated Professor T. Tański, as part of their doctoral dissertations [103–106] and a habilitation dissertation [107] created under my supervision and scientific custody, and with the current Associated Professor, J. Trzaska, whose doctoral dissertation was also created under my supervision and concerned another topic [108], however, the experience assembled there was used for implementation of the present research. Associated Professor L. Čižek collaborated with me as part of the international CEEPUS project [100], and Professor J. Sokolowski cooperated under long-term bilateral liaison.

An analysis of crystallisation kinetics of the examined alloys was covered by the scope of the research performed. The influence of mass concentration of Al and of the cooling rate on the kinetics of phase transformations during the cooling of the investigated alloys was studied. The temperatures of individual transformations were also identified. Heat treatment conditions were selected with cooling in different mediums and structural examinations of heat-treated alloys were performed. Coatings were deposited onto the surface of the investigated alloys by Physical (PVD) and Chemical (CVD) Vapour Deposition techniques. The structure of the deposited surface coatings was investigated and the dependence between functional properties of the investigated alloys and the coating type and deposition technology was examined. A composite structure, i.e. MMCs (Metal Matrix Composites), with the gradient of phase composition and functional properties, was produced on the surface as a result of surface laser treatment by laser imbed of hard ceramic

TABLE 5.1
Crystallographic Data of Solid Phases in Al-Mg-Zn System

Phase Temperature Range, °C	Pearson Symbol/Space Group/Prototype	Lattice Parameters, pm
(Al); <660.45	*cF*4 *Fm3m* Cu	*a* = 404.96
(Mg); <650	*hP*2 *P6₃/mmc* Mg	*a* = 320.94 *c* = 521.07
(Zn); <419.58	*hP*2 *P6₃/mmc* Mg	*a* = 266.50 *c* = 494.70
β, Mg_2Al_3; ≤452	*cF*1168 *Fd3m* Mg_2Al_3	*a* = 2816–2824
γ, $Mg_{17}Al_{12}$; <458	*cI*58 *I43m* αMn	*a* = 1054,38
ε, $Mg_{23}Al_{30}$; 410–250	*hR*159 *R3* $Mn_{44}Si_9$	*a* = 1282.54 *c* = 2174,78
δ, $Mg_{51}Zn_{20}$; 342–325	*oI*142 *Immm* $Mg_{51}Zn_{20}$	*a* = 1408.3 *b* = 1448.6 *c* = 1402.5
MgZn; <347	–	–
ζ, Mg_2Zn_3; <416	*mC*110 *B2/m*	–
η, $MgZn_2$; <590	*hP*12 *P6₃/mmc* $MgZn_2$	*a* = 522.1 *c* = 856.7
Θ, Mg_2Zn_{11}; <381	*cP*39 *Pm3* Mg_2Zn_{11}	*a* = 855.2
*τ_1, $Mg_{32}(Zn, Al)_{48}$	*cI*160 *Im3* $Mg_{32}(Zn, Al)_{48}$	*a* = 1413–1471
*φ, $Mg_{21}(Zn, Al)_{17}$	*oP*152 *Pbcm* $Mg_{21}(Zn, Al)_{17}$	*a* = 897.9 *b* = 1698.8 *c* = 1934
*τ_2, $Mg_{43}Zn_{42}Al_{15}$ or $Mg_{46}Zn_{37}Al_{17}$	*cP*640(?) *Pa3*	a = 2291 or a = 2310
*q, $Mg_{44}Zn_{41}Al_{15}$	Quasicrystalline, icosahedral	

Source: Petrov, D. et al., Materials Science International Team MSIT: Aluminium – Magnesium – Zinc, *Landolt-Börnstein, Numerical Data and Functional Relationships in Science and Technology*, G. Effenberg and S. Ilyenko, Eds., New Series/Editor in Chief: W. Martienssen, Group IV: Physical Chemistry, Vol. 11, Ternary Alloy Systems, Phase Diagrams, Crystallographic and Thermodynamic Data, critically evaluated by MSIT, Subvolume A, Light Metal Systems, Part 3, Selected Systems from Al-Fe-V to Al-Ni-Zr, vol 11A3. Springer, Berlin, Germany. pp. 191–209.

TABLE 5.2
Phase Transformations in the Three-Component System of Mg-Al-Zn Phase Equilibrium

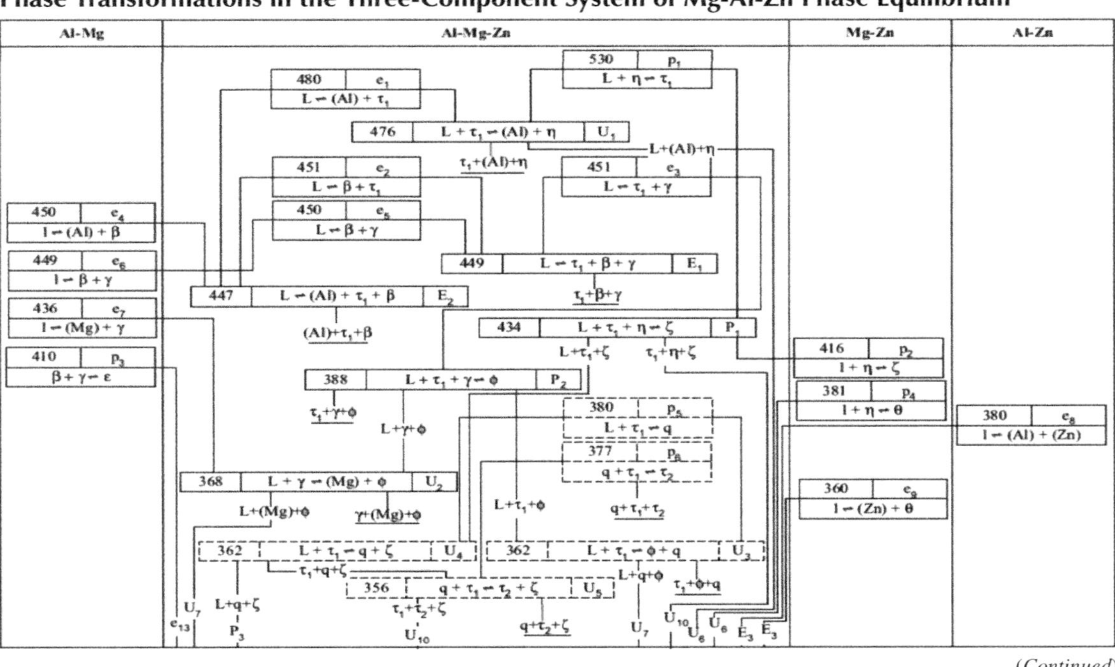

(Continued)

TABLE 5.2 (*Continued*)
Phase Transformations in the Three-Component System of Mg-Al-Zn Phase Equilibrium

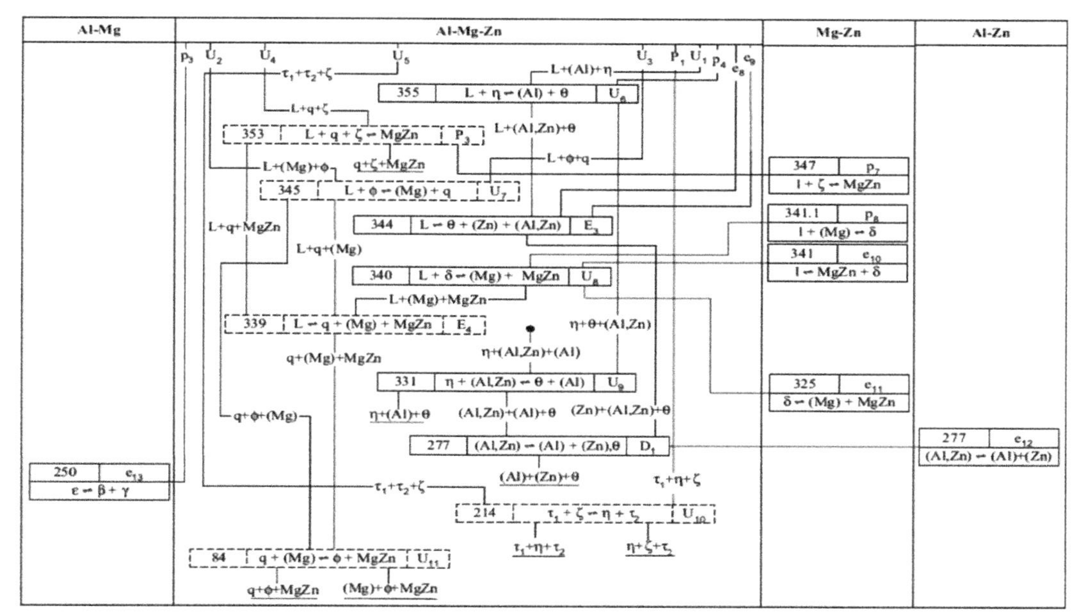

Source: Petrov, D. et al., Materials Science International Team MSIT: Aluminium – Magnesium – Zinc, *Landolt-Börnstein, Numerical Data and Functional Relationships in Science and Technology*, G. Effenberg and S. Ilyenko, Eds., New Series/Editor in Chief: W. Martienssen, Group IV: Physical Chemistry, Vol. 11, Ternary Alloy Systems, Phase Diagrams, Crystallographic and Thermodynamic Data, critically evaluated by MSIT, Subvolume A, Light Metal Systems, Part 3, Selected Systems from Al-Fe-V to Al-Ni-Zr, vol 11A3. Springer, Berlin, Germany. pp. 191–209.

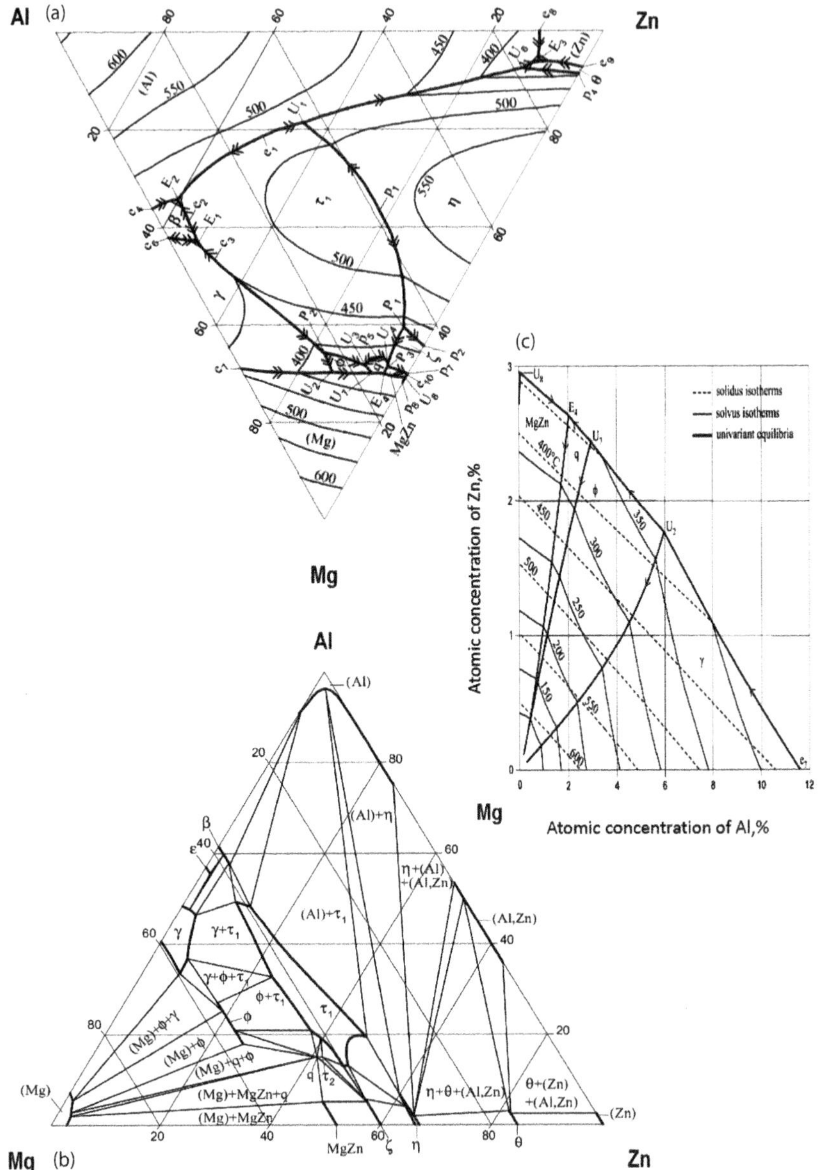

FIGURE 5.3 Examples of calculated cross-sections of the Mg-Al-Zn equilibrium system: (a) isothermal section at 335°C; (b) liquidus surface; (c) solidus and solvus isotherms of the phase (Mg); atomic concentration of elements is given on the axes. (From Petrov, D. et al., updated by H. L. Lukas–Materials Science International Team MSIT: Aluminium–Magnesium–Zinc, *Landolt-Börnstein, Numerical Data and Functional Relationships in Science and Technology*, G. Effenberg and S. Ilyenko, Eds., New Series/Editor in Chief: W. Martienssen, Group IV: Physical Chemistry, Vol. 11, Ternary Alloy Systems, Phase Diagrams, Crystallographic and Thermodynamic Data, critically evaluated by MSIT, Subvolume A, Light Metal Systems, Part 3, Selected Systems from Al-Fe-V to Al-Ni-Zr, vol 11A3. Springer, Berlin, Germany. pp. 191–209.)

particles into remelted surface zone. Before surface treatment investigations, heuristic studies were carried out, aimed at forecasting the development of materials surface engineering and at evaluating the strategic position for the relevant analysed surface technologies. Structural investigations into the surface and materials with the macro- and microscopy methods were carried out using light, transmission and scanning electron microscopy and X-ray phase analysis.

The investigations of physical properties of magnesium alloys after heat and surface treatment were also undertaken.

The investigations were carried out on experimental casting meltings of magnesium MCMgAl12Zn1, MCMgAl9Zn1, MCMgAl6Zn1 and MCMgAl3Zn1 alloys with the chemical composition given in Table 5.3.

The alloys were cast in an induction melting furnace, with the use of the protective bath Flux 12, equipped with two ceramic filters with the applied melting point appropriate for the produced material of 750°C ± 10°C. Refining with inert gas with the industrial name Emgesalem Flux 12 was performed in order to maintain metallurgical purity of the melted metal. A protective layer, Alkon M62, was applied to improve the surface quality of the alloy. The material was cast in dies with a bentonite binder, because of its favourable sorption properties, and was formed in the shape of plates with dimensions of 250 × 150 × 25 mm.

In order to enable the simulation of casting processes of the investigated Mg-Al-Zn-Mn alloys and the prediction of particular temperature values, neural network models have been developed, allowing to calculate individual values of the phase α-Mg nucleation temperature, the maximum temperature of the determined α-Mg phase growth and the solidus temperature of the casting magnesium alloys based on the mass concentration of Al and the cooling rate. Neural networks were also used to develop models of dependency between the hardness of the Al concentration, temperature and supersaturation time, cooling medium, and, accordingly, the temperature and time of ageing. The simulations, based on such developed models, allow to select the optimum heat treatment conditions of the investigated Mg-Al-Zn-Mn casting alloys.

A unidirectional MLP (Multi-layer Perceptron) network with one hidden layer and an appropriate number of neurons in this layer (Table 5.4) with, respectively, 2 or 4 neurons in the input layer, was considered as the optimum one from among the several considered options for the calculation of the assumed values, based on

TABLE 5.3
Chemical Composition of the Examined Alloys

Mass Concentration of Alloy Elements in the Examined Alloys, %						
Al	Zn	Mn	Si	Fe	Mg	Remainder
12.1	0.62	0.17	0.047	0.013	86.96	0.09
9.09	0.77	0.21	0.037	0.011	89.79	0.092
5.92	0.49	0.15	0.037	0.007	93.33	0.066
2.96	0.23	0.09	0.029	0.006	96.65	0.035

TABLE 5.4
The Architecture of the Created MLP Networks and Indicators for Assessing the Quality of Neural Networks Used to Determine the Properties of the Modelled Values

		Training Set			Validation Set			Test Set		
Size	MLP Network Architecture	Mean Absolute Error	Deviation Quotient	Pearson's Correlation	Mean Absolute Error	Deviation Quotient	Pearson's Correlation	Mean Absolute Error	Deviation Quotient	Pearson's Correlation
T_N	2-8-1	3.56	0.21	0.97	3.06	0.25	0.97	3.01	0.14	0.99
T_G	2-8-1	2.15	0.16	0.98	1.2	0.07	0.99	1.89	0.12	0.99
Tsol	2-8-1	4.54	0.15	0.98	4.6	0.15	0.98	2.83	0.1	0.99
Q_c	2-7-1	67.3	0.63	0.77	67.56	0.72	0.69	65.6	0.51	0.87
Z	2-13-1	21.36	0.56	0.83	11.09	0.39	0.92	17.83	0.61	0.8
HRF	2-6-1	1.37	0.09	0.99	1.77	0.11	0.99	1.58	0.1	0.99
R_c	2-8-1	7.88	0.39	0.92	5.66	0.32	0.94	6.92	0.4	0.91
$R_{c0,2}$	2-4-1	8.72	0.39	0.91	6.74	0.34	0.94	13.15	0.37	0.96
HRF_1	4-2-1	5.35	0.43	0.90	6.49	0.44	0.90	5.90	0.46	0.89
R_{a1}	4-5-1	1.99	0.42	0.91	1.68	0.41	0.91	1.53	0.42	0.91

the assembled results of own research described in this article. Data relating to the modelling of hardness and roughness after laser treatment is shown with index l.

The networks were trained with the backward error propagation and the conjugate gradient methods. Each of the respective sets of data, concerning the mentioned modelled dependencies of the individual alloys, was randomly divided into three subsets: learning, validation and test subset. For example, in case of a network calculating hardness after supersaturation, the number of cases was, respectively: 68, 20, 20, while for the network calculating hardness after ageing, respectively: 231, 100, 101.

The data from the training set was used for modifying the weights of networks, the data from the validation set was used for network evaluation in the training process, and the remaining part of values (test set) was used to determine, independently, network efficiency after completing fully the network creation procedure. The results used in the network training and test process were subjected to normalisation. The relative scaling of the deviation from the minimum value according to the mini-max function was applied.

The mini-max function transforms the variables domain to the interval (0,1). The network type, the number of neurons in the hidden layer (layers), the training method and parameters were determined by observing the influence of such values on the adopted factors of network quality evaluation. The following values were used as the basic indicators of model quality evaluation:

- Average network forecast error
- Quotient of standard deviations
- Pearson's correlation coefficient

An important quality indicator of the models created with a neural network is the quotient of standard deviations for errors and for data (Table 5.3). The correctness of the model adopted by a network can only be considered if the results presented by the forecast network involve a smaller error than the simple estimation of an unknown output value. The simplest way to estimate an output value is to take on an average value from the output data for a training set and present it as a forecast for the data not presented during the training process. An average error in such case equals the standard deviation of the output value in a training set, while the quotient of standard deviations takes on a value one. The smaller a network prediction error, the smaller values assumed by a quotient of standard deviations, reaching zero for an 'ideal' forecast. For the created networks, the applied error function and the logistic activation function was assumed as the sum of squares using the training method based on the conjugate gradient algorithm, which allowed for presentation of examples from a training set for example for 101 training epochs for a network calculating hardness after supersaturation and for 195 epochs for a network calculating hardness after ageing. The triaxial charts, shown later in this chapter, were created based on the developed neural network models.

5.2 THERMAL-DERIVATIVE ANALYSIS OF Mg-Al-Zn-Mn ALLOYS

5.2.1 Methodology of Thermal-Derivative Studies

The development of the magnesium and magnesium alloys casting technology is currently associated with the safety of the process, the lowering of costs, with ensuring product repeatability and with achieving the highest possible rate of metal injection into the mould. Magnesium alloys are very often used in the form of castings produced by the solidification of liquid metal in sand or metal moulds, which in turn are a semi-product in the manufacture of machined products. Crystallisation kinetics can be characterised with the following values:

- Metal temperature
- Heat loss rate
- Cooling rate
- Generation rate of hidden crystallisation heat
- Density of grains which is equivalent to the density of the nuclei formed
- The share of the solid fraction of crystallised metal
- Concentration of components in the residual liquid
- Characteristic distances and values describing the shape and size of the structural components

All these values are the variables of crystallisation time and geometric coordinates of the casting. Crystallisation kinetics can be characterised fully by combining crystallisation equations with heat exchange equations, and the crystallisation heat being released is a factor linking these equations and it depends on the proportion of the crystallised structural components. Details concerning the basis of the analysed research methods are given in the paper [87].

From among several types of thermal analyses, applied in the research and industrial practice, including simplified thermal analysis and differential thermal analysis (DTA) [225–236], the thermal derivative analysis (ATD) of solidifying metals and alloys, employed in this work, has gained particular importance. A particular advantage of the ATD method is not only the ability to evaluate the alloy in terms of the chemical composition, but also the ability to evaluate, in the same measurement process, numerous details of the primary or secondary crystallisation process kinetics. By obtaining so much information about the material, such as, e.g. the degree of liquid supercooling, temporary cooling rate, nucleation temperature, it is possible, within a short time (2–5 minutes) to make an immediate decision about process quality improvement, thus allowing to improve the quality of casting and metallurgical production, which dictates the reliability class of machines and equipment. The development of this method relates primarily to determination of crystallisation processes' kinetics and the related possibility of determining the concentration of certain elements, properties of the alloy and structure as well as heat loss and heat exchange processes taking into account internal sources. The thermal-derivative (ATD) analysis method consists of recording at the same time a cooling curve $T = f(t)$ and the first derivative of temperature after the time T′ = dT/dt. The curve dT/dt is determined in a mathematical way. All operations are carried out simultaneously with temperature measurement. The experience necessary to carry out such investigations was gathered as a result of long-term collaboration with Prof. J. Sokolowski from the University of Windsor, Canada, with the example of other Mg and Al alloys, and this in a widely understood range of evaluation of the quality of castings and construction of the Universal Metallurgical Simulator and Analyzer device – UMSA [237–255].

The chart of thermal analysis and derivative analysis shows three curves: cooling, crystallisation and base curve. The cooling curve reflects the kinetics of alloy crystallisation. For continuous cooling, all intervals or breaks inform about internal heat sources. The characteristic points on the curve T = f(t) inform about the temperature of phase transformations of the examined alloys. The crystallisation curve gives a special picture of temperature changes during solidification of alloys. These changes are caused by internal heat sources, and the magnitude of changes is proportional to the thermal effects. Specific heat (thermal capacity) has also significant impact on the progress of the curve. Characteristic points can be determined in the diagram: LM, L, SM and S, which, in conjunction with the cooling curve T = f(t), determine, successively, characteristic temperatures of crystallisation stages: TLM, TL, TSM, TS. The temperatures TL and TS stand for, respectively, the liquidus and solidus temperature during solidification in conditions of stable equilibrium, and the temperatures TLM and TSM are the temperatures of the same transformations taking place during supercooling (in conditions of unstable equilibrium) [146,162,172]. In addition to the characteristic points described above, there are also maxima and minima on the cooling derivative chart. The point K should be distinguished, which identifies the end of metal crystallisation in all points of the probe; this point also identifies the characteristic temperature TK – the temperature of end of eutectics crystallisation.

The curves T(t) and T′(t) describe exactly metal crystallisation. It can be generally stated that the ‘stops’ existing on the cooling are the result of the acting heat sources. This of course results from the fact that the examined probe (metal) releases heat continuously to the surrounding. If temperature is maintained on the constant level, this means that heat is released inside the material. The temperature rise observed occasionally on the temperature ‘stop’ signifies very intensive release of heat. The liquid supercooling temperature is considered to be the lowest temperature. Thus, all these events indicate that the solidification heat is released, and the crystallisation curve describes the kinetics of these processes.

When the three values are known (dT/dt)c, the curve Tc′(t) is determined, according to which the temporary thermal effect of the crystallising alloy can be calculated, and after integrating in the entire time interval – the total heat effect of the crystallisation stage can be calculated.

The melting heat, Qt, determines the amount of heat required to melt a unit of mass at the melting temperature. For the alloys crystallising in the range of the liquidus and solidus temperature, this is a sum of melting heat and of thermal capacity of the mixture of liquid phase and of crystals in the range of the liquidus and solidus temperature.

A thermal-derivative analysis of the investigated Mg-Al-Zn-Mn alloys was made using the UMSA device [234–236]. The experiments consisted of melting, annealing and cooling a specimen with a suitably predetermined cooling rate. The specimens with the shape and dimensions shown in Figure 5.3 were heated inductively to a temperature of 700 ± 1°C using an induction coil supplied by an induction generator with the maximum power of 5 kW. Remelting was performed in a sealed chamber into which a protective gas, argon, was supplied with the flow rate of 2.4 l/min. to prevent self-ignition of the magnesium alloy. To achieve an adequate cooling rate of the tested material, the specimens were cooled using argon fed through nozzles arranged in the coil. The intensity rate of the flowing argon cooling the specimen was controlled using a control and measuring system measuring the temperature of the flowing gas at the inlet and outlet of the valve. The gas flow rate was chosen experimentally and was applied throughout the entire experiment.

The experiments were performed at the following rates of cooling:

- 0.6°C/s – the specimens were cooled without forced cooling
- 1.2°C/s – argon flew through the cooling system with the rate of 30 l/min
- 2.4°C/s – argon flew through the cooling system with the rate of 125/min.

The relationship between the temperature and the time of structural changes, the liquidus and solidus temperature, the nucleation temperature of eutectics and the phase containing Mn and Al and Mg_2Si phase, for the given alloy, was determined by performing three full loops of specimen melting and solidification. The first remelting provided a very good contact of the thermocouple with the metal of the specimen, the second and third remelting is the main experiment guaranteeing the repeatability of tests results.

5.2.2 Results of Thermal-Derivative Studies

Characteristic points describing the thermal effects occurring during alloy crystallisation, defining the temperature and time values, read from the thermal-derivative analysis curves, were determined based on the ATD analysis. When analysing the progress of alloys crystallisation, on the basis of cooling curves, those selected with the marked points describing the heat phenomena are presented in Figure 5.4, and the interpretation of the points obtained is given in Table 5.5.

The temperature and time of formation and precipitation of the individual phases during crystallisation was respectively determined by projecting the designated points along the cooling curve, and then on the axis of coordinates and abscissa. In addition, the latent heat of crystallisation of the individual phases was calculated based on the selected values characterising the differential curve. It was assumed that thermal capacity – $c_p(t)$, alloy crystallisation is the function of time, thus it is dependent on the fraction of the solid phase, thermal capacity in the liquid state c_{pLiq} and thermal capacity in the solid state c_{pSol}. The thermal capacity of the alloy in time t, thermal capacity in the liquid state c_{pLiq} and thermal capacity in the solid state c_{pSol} were calculated based on chemical compositions of the examined alloys using Thermo-Calc Software [236], and the total

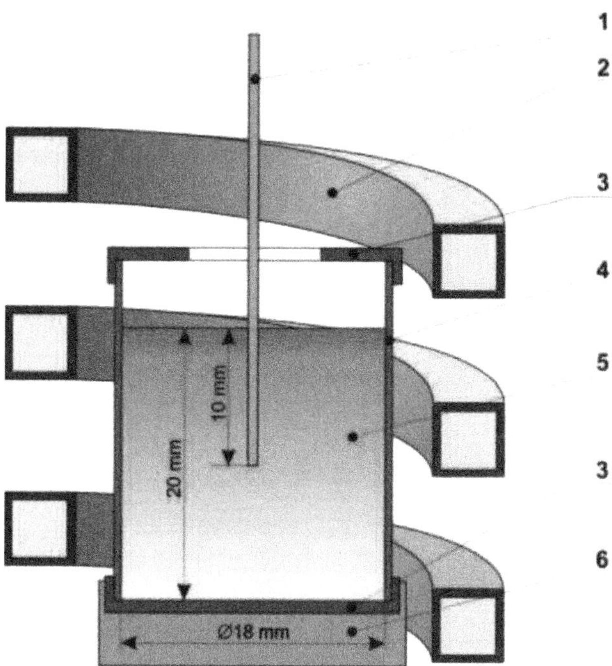

FIGURE 5.4 Diagram of a heating and cooling device UMSA and arrangement of the thermocouple and dimensions of specimens for thermoanalysis, 1 – thermocouple, 2 – induction coil with cooling nozzles, 3 – safety caps, 4 – protective film, 5 – specimen, 6 – tester insulation.

latent heat of the examined alloys' crystallisation was calculated based on the relationships given in the works [87,256].

Table 5.6 presents the examples of crystallisation temperature of particular phases and the solid phase fraction for the selected MCMgAl6Zn1 alloy depending on the cooling rate of 0.6°C/s; 1.2°C/s; 2.4°C/s. The effect of cooling rate and magnesium concentration on the nucleation temperature of the phase $\alpha - T_N$ is shown in Figure 5.5. Primary importance for the nucleation temperature can be assigned both to Al concentration (reduces the nucleation temperature of the phase α according to the liquidus line) as well as cooling rate (causes the increase of nucleation temperature of the phase α). For example, for MCMgAl3Zn1 alloy, if the cooling rate is increased from 0.6°C/s to 1.2°C/s, the nucleation temperature of the phase α will rise from 633.16°C to 635.39°C, but a further increase in the cooling rate to 2.4°C/s will increase the temperature from 640.32°C. The degree of liquid supercooling during the crystallisation of the solid solution α depends primarily on the applied cooling rate. A change in the cooling rate has no effect on the change of the maximum temperature T_G of phase α crystallisation. The value of temperature T_G is reduced, however, as the concentration of Al rises. For instance, for alloys cooled with the rate of 0.6°C, growth in the concentration from 3% to 6% of Al decreases the maximum temperature T_G of the crystallisation of the phase α from 630.85°C to 611.92°C, and further growth in concentration to 9% Al decreases the

TABLE 5.5

Determination of Temperature and the Time Determined Based on the Chart of Thermal Analysis and Derivative Analysis and Abridged Methodology of Determination of Characteristic Points from the Thermal-Derivative Curve

Point Determination	Temperature	Discussion	Method of Determination on the Thermal-Derivative Curve
I	T_N	start of nucleation of the phase α (liquidus temperature)	the intersection point of the tangent led to the straight section of the crystallisation curve with the base curve
II	T_{Dmin}	growth of crystals of the phase α upon reaching the critical value	the value of the first derivative equal to zero
III	T_{DKP}	coherence of dendrites of the phase α	reaching the maximum value by the first derivative
IV	T_G	determined growth of the phase α	first derivative again reaches the value zero
V	$T_{(Mg+Si+Al+Mn)}$	crystallisation of the phase α, phase Mg_2Si and the phase containing Al and Mn	
VI	$T_{(Mg+Si+Al+Mn)f}$	end of crystallisation of the phase Mg_2Si and the phase containing Al and Mn	
VII	$T_{E(Mg+Al)N}$	nucleation of eutectics $\alpha + \gamma$ ($Mg + Mg_{17}Al_{12}$)	
VIII	$T_{E(Mg+Al)min}$	growth of nuclei of the phase γ upon reaching the critical value and growth of eutectics $\alpha + \gamma$ ($Mg + Mg_{17}Al_{12}$)	the first derivative reaches the value zero
IX	$T_{E(Mg+Al)G}$	determined growth of eutectics $\alpha + \gamma$ ($Mg + Mg_{17}Al_{12}$)	the value of the first derivative of the cooling curve again reaches the value zero
X	T_{sol}	end of alloy crystallisation (solidus temperature)	the crystallisation curve coincides with the base curve

temperature value to 592.91°C. The lowest value of the maximum temperature T_G of crystallisation of the phase α of 576.03°C was found for the alloy containing 12% of Al. If the mass concentration of Al and the cooling rate are increased, this also reduces the value of the temperature T_{sol}. For example, for MCMgAl6Zn1 alloy, if the cooling rate is increased from 0.6°C/s to 1.2°C/s, the solidus temperature drops from 419.47°C to 415.44°C, and a further increase in the cooling rate to 2.4°C/s will decrease the temperature T_{sol} to 401.66°C. A variable concentration of Al has on the other hand the highest effect on increasing the value of the heat Q_c generated during crystallisation of the investigated Mg-Al-Zn-Mn alloys. A higher concentration of Al

TABLE 5.6
Thermal Characteristics of MCMgAl6Zn1 Alloy Cooled with the Rate of 0.6;1.2; 2.4 °C/s

Temperature, T or Percentage of Solid Fraction, f_s	Cooling Rate, °C/s					
	0.6		1.2		2.4	
	Average Value	Standard Deviation	Average Value	Standard Deviation	Average Value	Standard Deviation
T_N, °C	615.88	3.4	615.74	2.95	619.77	6.35
T_{Dmin}, °C	611.51	1.93	not recorded		not recorded	
f_s, %	1.96	0.49				
T_{DKP}, °C	611.75	1.94	not recorded		not recorded	
f_s, %	3.16	0.6				
T_G, °C	611.92	1.99	610.33	2.46	608.14	3.21
f_s, %	5.44	0.87	5.92	1.51	8.9	0.49
T(Mg + Si + Al + Mn), °C	533.65	4.81	532.77	5.28	536.37	15.74
f_s, %	85.28	1.97	83.45	1.92	81.03	6.52
T(Mg + Si + Al + Mn)f, °C	520.18	5.54	509.72	5.21	511.99	21.35
f_s, %	88.08	1.89	88.13	2.03	86.30	5.55
TE(Mg+Al)N, °C	429.45	2.33	431.69	3.04	432.99	3.9
f_s, %	95.02	0.95	94.2	1.31	93.85	0.82
TE(Mg+Al)min, °C	426.59	2.23	not recorded		not recorded	
f_s, %	96.03	1.05				
TE(Mg+Al)G, °C	427.17	2.49	not recorded		not recorded	
f_s, %	97.96	0.46				
T_{sol}, °C	419.47	3.26	415.44	3.71	401.66	8.57

in the examined alloys causes an increase in the value of the heat Q_c. If the cooling rate is increased, the generated heat of crystallisation is slightly increased, except for the MCMgAl3Zn1 alloy (Figure 5.6). The examples of information concerning the crystallisation heat of the MCMgAl6Zn1 alloy are given in Table 5.7.

Figure 5.7 shows three-dimensional charts of the effect of cooling rate and Al concentration in alloys on the nucleation temperature of the phase (T_N) phase α, crystallisation (T_G) phase α and solidus (T_{sol}) and on the value of crystallisation heat (Q_c), determined using the models of neural networks.

5.2.3 Structure and Mechanical Properties of Mg-Al-Zn-Mn Alloys after Thermal-Derivative Studies

The investigated MCMgAl3Zn1, MCMgAl6Zn1, MCMgAl9Zn1 and MCMgAl12Zn1 alloys, in the cast state, are characterised by a structure of the solid solution α with precipitates of the phase γ-$Mg_{17}Al_{12}$ along the boundaries of grains and by areas of the plate mixture $\alpha + \gamma$. The zones of so-called abnormal eutectics (pseudo eutectics)

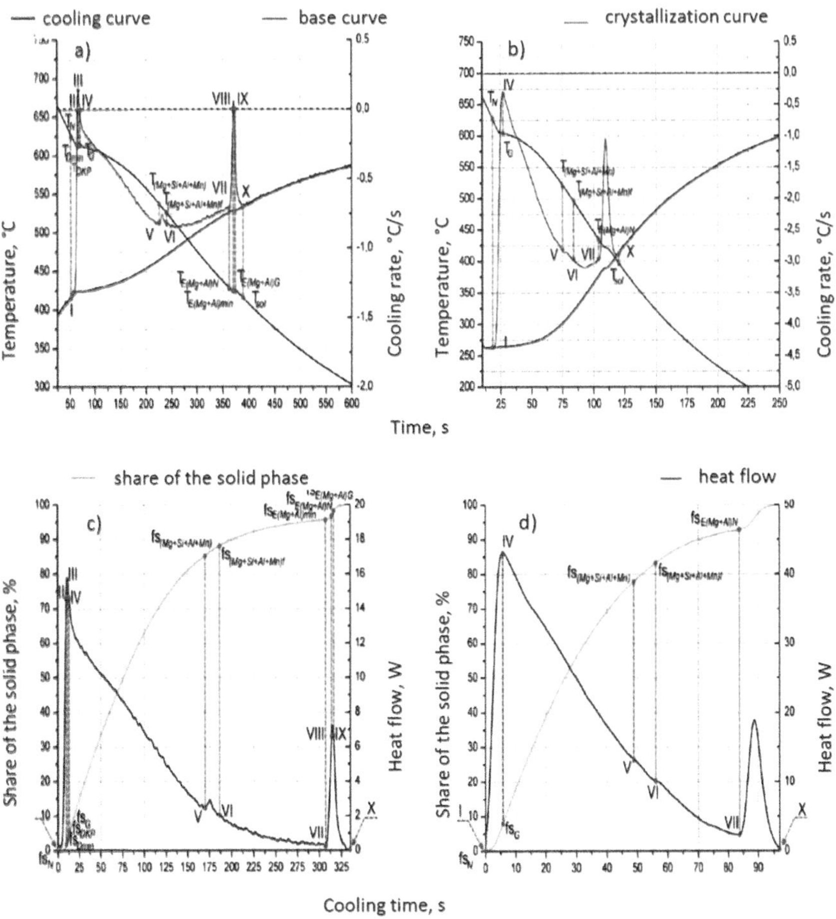

FIGURE 5.5 Results of investigations of the MCMgAl6Zn1 alloy cooled with the rate of (a, c) 0.6 °C/s (b, d) 2.4 °C/s; (a, b) cooling curve, crystallisation curve and base curve; (c, d) changes during the stream of heat and fraction of the solid phase.

also exist, of precipitation of the phases containing Al and Mn, and of the Mg_2Si phase, whose fraction depends on the mass concentration of Al, and increases along with its growth. The shape of individual precipitates changes along with the varying cooling rate. Figure 5.8 shows for example the results of metallographic investigations, with a light microscope, of MCMgAl6Zn1 and MCMgAl9Zn1 alloys, cooled with a varied rate.

It was found by examining thin foils in a transmission electron microscope (Figure 5.9) that the structure of MCMgAl3Zn1, MCMgAl6Zn1, MCMgAl9Zn1 and MCMgAl12Zn1 casting magnesium alloys, cooled with the cooling rates used in the experiment, is represented by the solid solution α-Mg with precipitates of the phase γ-$Mg_{17}Al_{12}$. The phase precipitates mainly form part of eutectics and occur as large particles located mainly at grain boundaries.

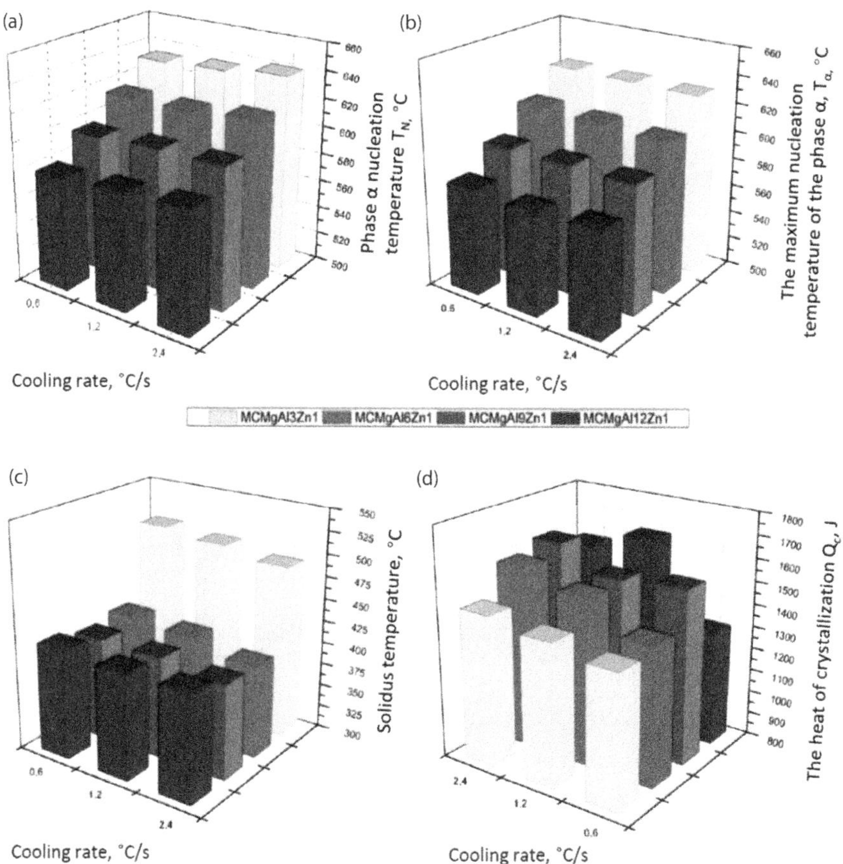

FIGURE 5.6 The effect of cooling rate of Mg-Al-Zn-Mn alloys on temperature of (a) crystallisation (T_N) of the phase α, (b) crystallisation (T_G) of the phase α, (c) solidus (T_{sol}), (d) on the value of crystallisation heat (Q_c).

An X-ray qualitative phase analysis confirms that the structure of all the examined alloys contains the phase γ-$Mg_{17}Al_{12}$ and phase α-Mg being the matrix of the alloys, which is shown for example in Figure 5.9 for MCMgAl6Zn1 alloy. An X-ray qualitative phase analysis did not confirm the presence of the Mg_2Si-type Mg phases and phases containing Al and Mn, which may prove that the fraction of such precipitates is below the minimum detection limit appropriate for the X-ray method. The investigations results of the phase composition and the chemical composition of the studied Mg-Al-Zn-Mn casting alloys confirm that the rate of cooling does not affect the phase composition of these alloys (Figure 5.10).

The structure observations undertaken using a scanning electron microscope and an X-ray quantitative and qualitative microanalysis examination, made using an EDS scattered X-ray radiation spectroscope, confirm the presence of α-Mg + γ-$Mg_{17}Al_{12}$ eutectics in the examined alloys with precipitates of the Mg_2 Si phase with angular shapes as well as the precipitates of phases containing Al

TABLE 5.7
Latent Heat of Crystallisation Released by the Crystallising Phases and Its Portion in the Overall Crystallisation Heat of MCMgAl6Zn1 Alloy Cooled with the Rate of 0.6; 1.2; 2.4°C/s

Reaction	Cooling Rate, °C/s					
	0.6		1.2		2.4	
	Latent Heat of Unit Crystallisation Specimen Mass, J/g	Fraction of Phases, %	Latent Heat of Unit Crystallisation Specimen Mass, J/g	Fraction of Phases, %	Latent Heat of Unit Crystallisation Specimen Mass, J/g	Fraction of Phases, %
L→ α (Mg)	139.96	86.96	140.28	86.96	154.61	83.79
L→ α (Mg) + Mg_2Si + (Al + Mn)	15.98	9.09	14.67	9.09	17.92	9.71
L→ α (Mg) + γ (Mg + $Mg_{17}Al_{12}$)	6.23	3.93	6.35	3.93	11.98	6.49
In total	162.16	100	161.31	100	184.52	100

(a)

Phase α nucleation temperature T_N, °C

Cooling rate, °C/s

Mass concentration Mg,%

> 640
< 640
< 630
< 620
< 610
< 600
< 590
< 580

(b)

The maximum nucleation temperature of the phase α, $T_α$, °C

Cooling rate, °C/s

Mass concentration Mg,%

> 630
< 630
< 620
< 610
< 600
< 590
< 580
< 570

(c)

Solidus temperature, °C

Cooling rate, °C/s

Mass concentration Mg,%

> 540
< 540
< 520
< 500
< 480
< 460
< 440
< 420
< 400

(d)

The heat of crystallization Q_c, J

Cooling rate, °C/s

Mass concentration Al,%

> 1600
< 1600
< 1550
< 1500
< 1450
< 1400
< 1350
< 1300

Al and Mn with

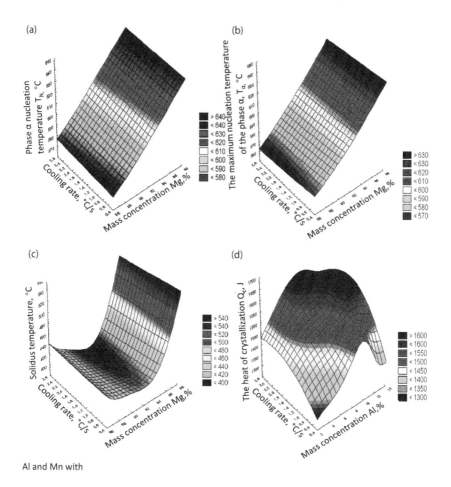

FIGURE 5.7 The effect of cooling rate and concentration of Al in Mg-Al-Zn-Mn alloys on temperature, determined using models of neural networks, of (a) nucleation (T_N) of the phase α, (b) crystallisation (T_G) of the phase α, (c) solidus (T_{sol}), (d) on the value of crystallisation heat (Q_c).

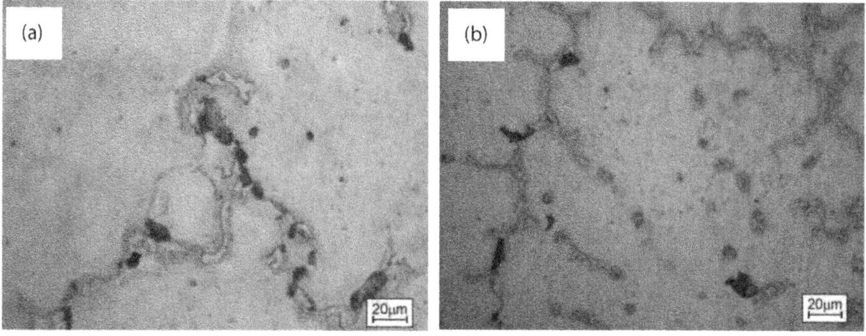

FIGURE 5.8 Structure of magnesium alloys (a) MCMgAl6Zn1 cooled with the rate of 0.6°C/s; (b) MCMgAl9Zn1 cooled with the rate of 2.4°C/s.

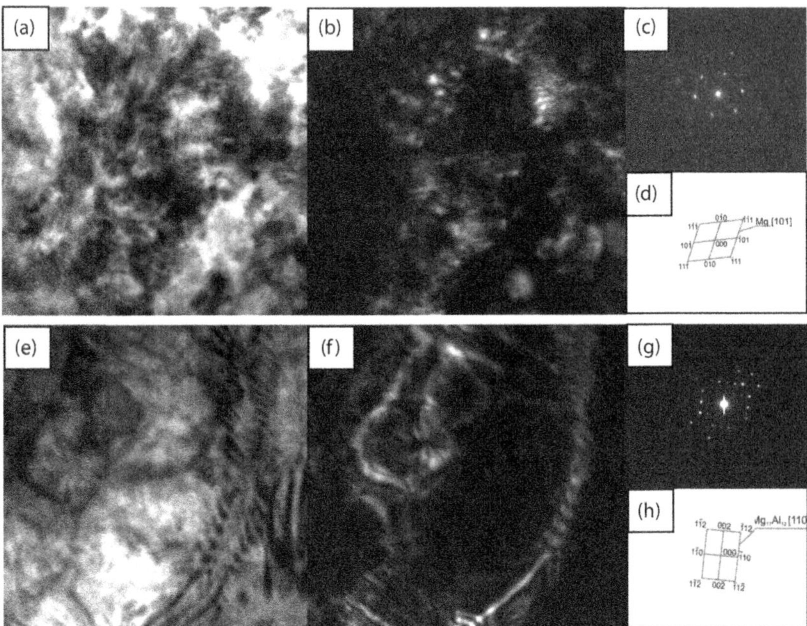

FIGURE 5.9 Structure of thin foils made of MCMgAl9Znl magnesium casting alloy cooled at a rate of 0.6°C/s: (a) the image of the matrix, the α-Mg solutions of the alloy in bright field, (b) image in the dark field from the area as on drawing (a); (c) diffraction pattern from the area as in Figure (a) and (b), (d) diffraction pattern solution from Figure (c); (e) image of precipitates of the phase γ-$Mg_{17}Al_{12}$ in the bright field, (f) image in the dark field from the area as in Figure (a); (f) diffraction pattern from the area as in figure (e) and (f); (h) diffraction pattern solution from Figure (g).

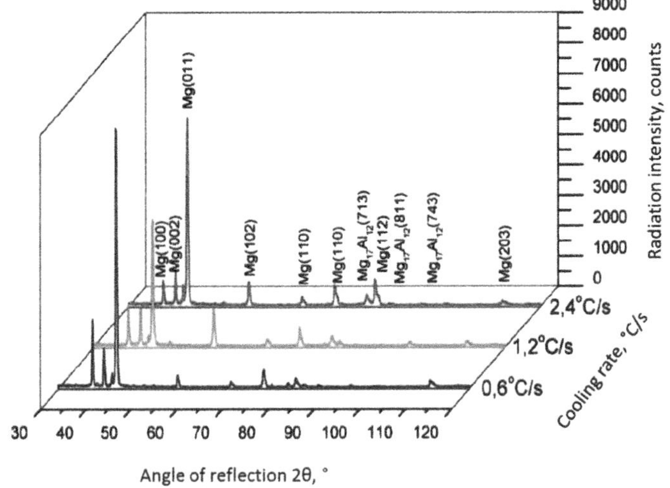

FIGURE 5.10 X-ray diffraction pattern of MCMgAl6Zn1 casting alloy cooled with the rate of 0.6;1.2; 2.4°C/s.

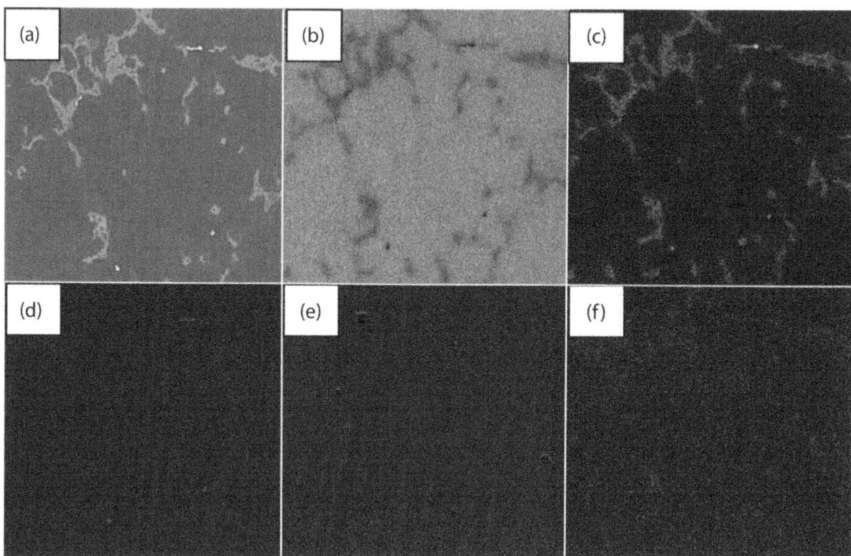

FIGURE 5.11 Structure of MCMgAl6Zn1 casting alloy cooled with the rate of 1.2°C/s; (a) an image obtained with back scattered electrons; (b–f) maps of surface distribution of elements: (b) Mg; (c) Al; (d) Mn; (e) Si; (f) Zn.

and Mn with irregular shapes, occurring mostly in the needle form (Figure 5.11). A chemical analysis of the distribution of surface elements and a quantitative microanalysis performed on lateral fractures of casting magnesium alloys by means of the EDS system (Figure 5.12) confirms in the examined micro-areas a heightened concentration of Mg, Si, as well as Al and Mn, which indicates the presence of precipitates containing Mg and Si in the structure of alloys, as well as phases with a high concentration of Al and Mn.

As the cooling rate is rising, so is decreasing the grain size for each of the analysed Mg-Al-Zn-Mn alloys. The MCMgAl6Zn1 alloy has the largest grain (190.3 μm) (Figure 5.12). The grain size is decreased two times due to a change in the cooling rate from 0.6°C/s to 2.4°C/s. The similar relationships were identified for the other examined alloys.

The increased mass concentration of Al causes a slight decrease in grain size. The increased mass concentration of Al influences however the linear increase of hardness, the same as higher cooling rate in case of MCMgAl6Zn1, MCMgAl9Zn1 and MCMgAl12Zn1 alloys. The hardness increased to 26 HRF for the cooling rate of 1.2°C/s was found for the MCMgAl3Zn1 alloy. If the cooling rate is increased to 2.4°C/s, hardness is decreased to 19.6 HRF. The highest hardness of 74.2 HRF was achieved for the MCMgAl12Zn1 alloy

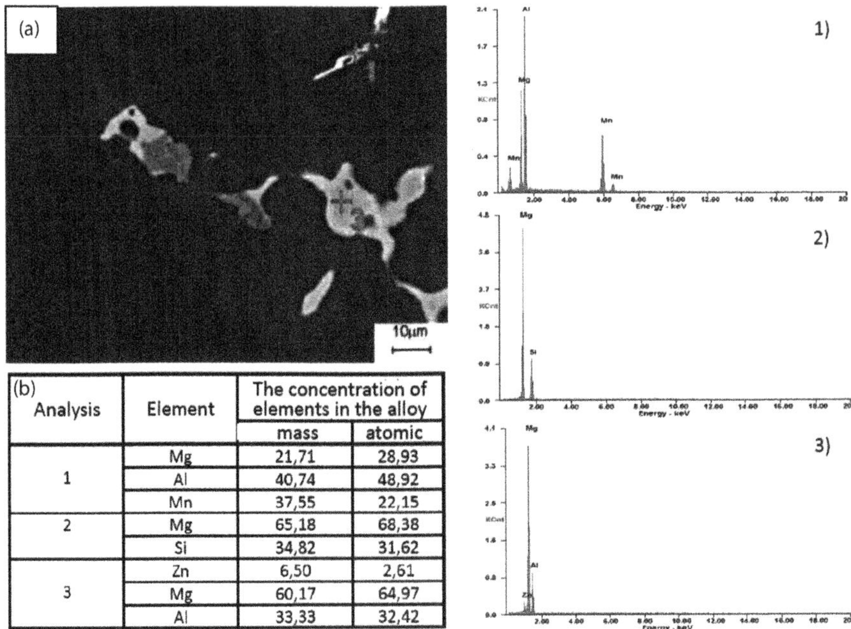

(b)

Analysis	Element	The concentration of elements in the alloy	
		mass	atomic
1	Mg	21,71	28,93
	Al	40,74	48,92
	Mn	37,55	22,15
2	Mg	65,18	68,38
	Si	34,82	31,62
3	Zn	6,50	2,61
	Mg	60,17	64,97
	Al	33,33	32,42

FIGURE 5.12 Structure (a), the results of the quantitative analysis of chemical composition (b) and intensity charts as a function of X-ray energy (c) of the MCMgAl6Zn1 casting alloy cooled at a rate of 0.6°C/s; (1) analysis (1, 2) analysis (2, 3) analysis 3, from the points indicated in Figure (a).

cooled at a rate of 2.4°C/s. The MCMgAl6Zn1 alloy has the highest compressive strength value of 296.7 MPa, and the MCMgAl3Zn1 alloy has the lowest strength of 245.9 MPa (the alloys cooled with the rate of 0.6°C/s). A change in the cooling rate of the examined alloys leads to higher compressive strength. The MCMgAl3Zn1 and MCMgAl9Zn1 alloys have the highest increase in the value Rc for the rising cooling rate.

If the cooling rate is increased maximally, this leads to higher compressive strength for MCMgAl3Zn1, MCMgAl9Zn1 alloys to the value of, respectively, 275.8 and 316 MPa, and for MCMgAl6Zn1, MCMgAl12Zn1 alloys by the average of 10–15 MPa.

A higher concentration of Al in the examined alloys has the largest impact on increased yield strength for compression. The MCMgAl3Zn1 alloy has the smallest yield strength value of 73.9 MPa, and the MCMgAl9Zn1 alloy has the highest strength value of 133.8 MPa. A change in the cooling rate from 0.6°C/s to 2.4°C/s for MCMgAl12Zn1 alloy increases yield strength from 125.9 to 164 MPa.

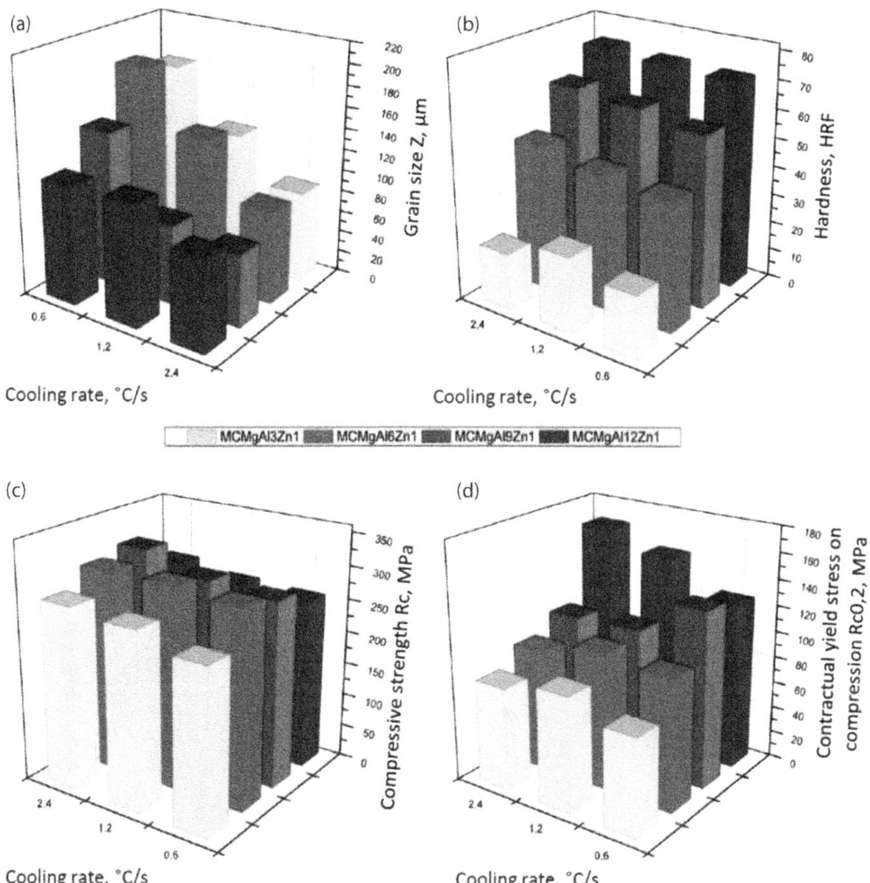

FIGURE 5.13 The effect of cooling rate of Mg-Al-Zn-Mn alloys on (a) the grain size of the phase α, (b) hardness, (c) compressive strength R_c, (d) yield strength on compression $R_{c0.2}$.

The opposite situation occurs in the case of the MCMgAl9Zn1 alloy. As the cooling rate is rising, so is decreasing the yield strength from 133.8 to 104.9 MPa. A change in the cooling rate does not change the yield strength of MCMgAl3Zn1 and MCMgAl6Zn1 alloys (Figure 5.13).

Figure 5.14 shows three-dimensional charts of the effect of cooling rate and Al concentration in alloys on the grain size of the phase α, hardness, compressive strength R_c and yield strength $R_{c0.2}$ determined with the use of models of neural networks.

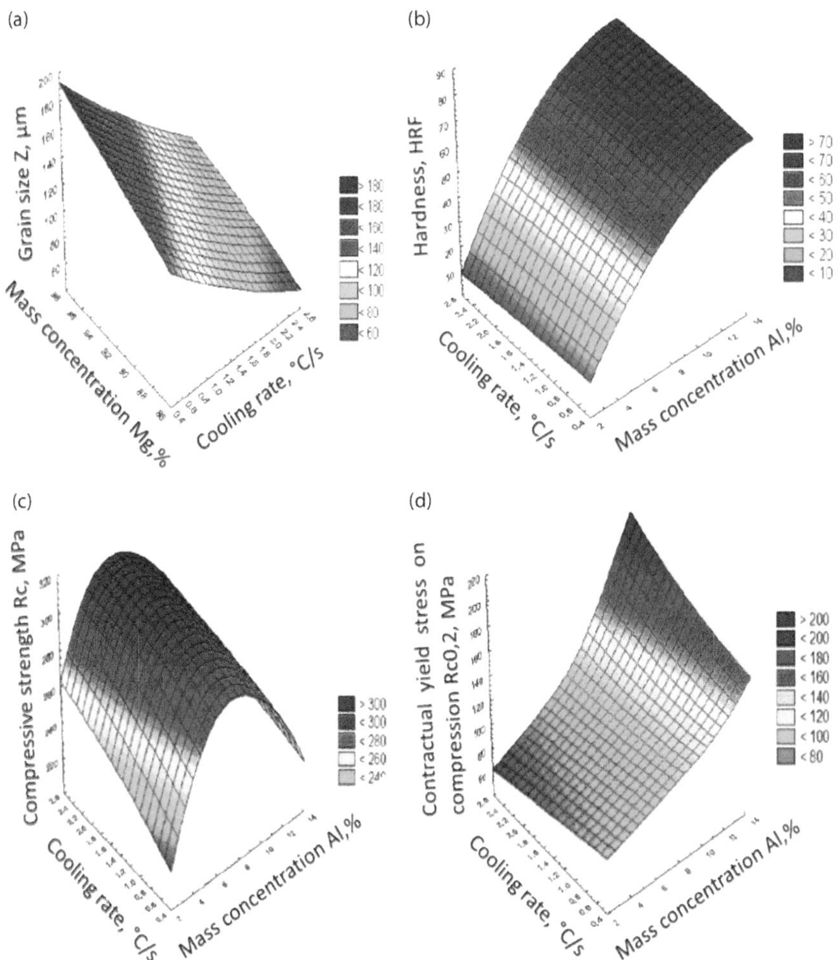

FIGURE 5.14 The effect of cooling rate and Al concentration in Mg-Al-Zn-Mn alloys on the grain size of the phase α, hardness, compressive strength R_c and yield strength on compression $R_{c0.2}$ determined with the use of models of neural networks.

5.3 STRUCTURE AND PROPERTIES OF HEAT-TREATED Mg-Al-Zn-Mn ALLOYS

5.3.1 Heat Treatment Conditions of Mg-Al-Zn-Mn Alloys

The investigated alloys underwent heat treatment, the conditions of which, i.e. annealing time and temperature during supersaturation and ageing and cooling rate after supersaturation, were optimised based on the results of hardness tests and metallographic observations. Supersaturation was performed with cooling in water or only after prior supersaturation in water. Ageing was carried out in a temperature

TABLE 5.8
Heat Treatment Conditions for Examined Alloys

Determination of Heat Treatment State	Heat Treatment Conditions		
	Temperature, °C	Annealing Time, h	Cooling Method
0 – cast, no treatment	–	–	–
1 – supersaturated	430	10	water
2 – supersaturated	430	10	air
3 – after heat treatment with cooling with furnace	430	10	furnace
4 – after ageing	190	15	air

range of 150°C to 210°C every 20°C and during 5, 10 and 15 hours. The conditions of heat treatment of the examined alloys listed in Table 5.8 were selected for further analysis based on preliminary trials.

5.3.2 The Results of Structural Examinations of the Investigated Heat-Treated Alloys

The results of metallographic examinations made with a light and scanning microscope (Figure 5.15) and also based on analyses of the distribution of surface elements and X-ray quantitative micro-analysis and X-ray phase analysis indicate that MCMgAl12Zn1, MCMgAl9Zn1, MCMgAl6Zn1, MCMgAl3Zn1 casting magnesium alloys in the cast state are characterised by a structure of the solid solution α-Mg being the matrix of the alloy and the intermetallic phase γ-$Mg_{17}Al_{12}$ in a plate-like form, located mainly at grain boundaries. Furthermore, the presence of needle eutectics was found near the intermetallic phase precipitates ($\alpha + \gamma$) (Figure 5.15).

Phases exist in the structure of the investigated casting magnesium alloys, apart from precipitates of the intermetallic phase γ-$Mg_{17}Al_{12}$, and such phases also are coloured grey and have angular contours with smooth edges in the form of hexagonal particles (Figure 5.15b).

It was pointed out based on the results of examinations of chemical composition with an EDS X-ray scattered radiation energy spectrometer, as well as the literature data, that this is the phase Mg_2Si, which, when precipitating, increases the hardness of castings.

After heat treatment with heating at the temperature of 430°C and with cooling with the furnace, in MCMgAl12Zn1, MCMgAl9Zn1, MCMgAl6Zn1 alloys there is a structure of the solid solution α-Mg with numerous areas of precipitation of the secondary phase γ-$Mg_{17}Al_{12}$ (areas with the morphology similar to eutectics) and phase precipitates and on the boundaries of grains and the light-grey phase situated mostly within the limits of grains of the phase γ (Figures 5.16 through 5.18).

The MCMgAl3Zn1 alloy, after heat treatment with cooling with a furnace, has a structure with few precipitates of the phase γ distributed unevenly in the matrix.

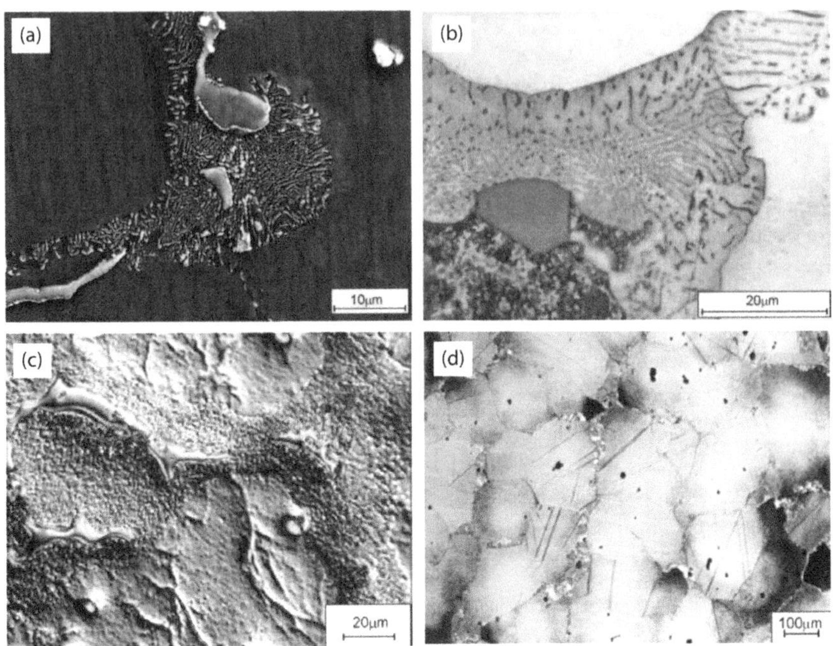

FIGURE 5.15 Structure of casting magnesium alloys in the cast state: (a) MCMgAl6Zn1 (SEM); (b) MCMgAl9Zn1 (LM); (c) observation with the phase contrast technique (LM); (d) observation in polarised light; (LM).

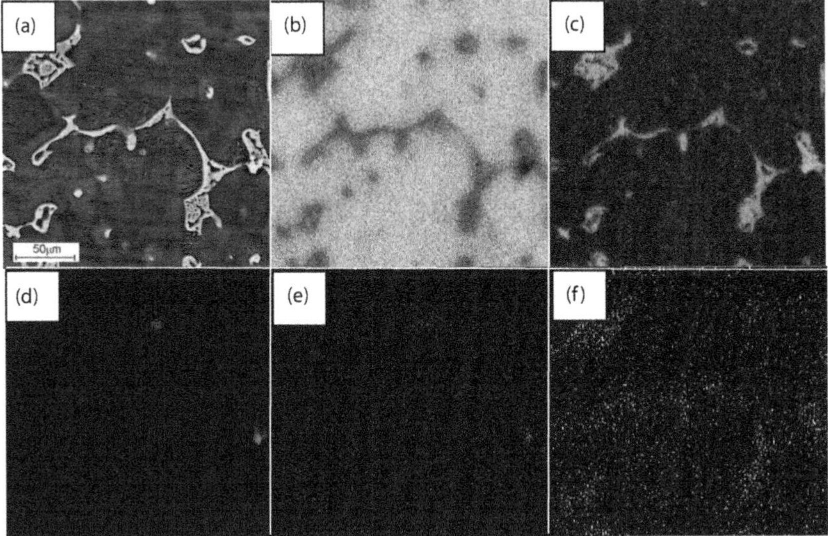

FIGURE 5.16 Structure of MCMgAl12Zn1 casting alloy in the cast state: (a) an image obtained with back scattered electrons; (b–f) maps of surface distribution of the following elements: (b) Mg; (c) Al; (d) Mn; (e) Fe; (f) Zn.

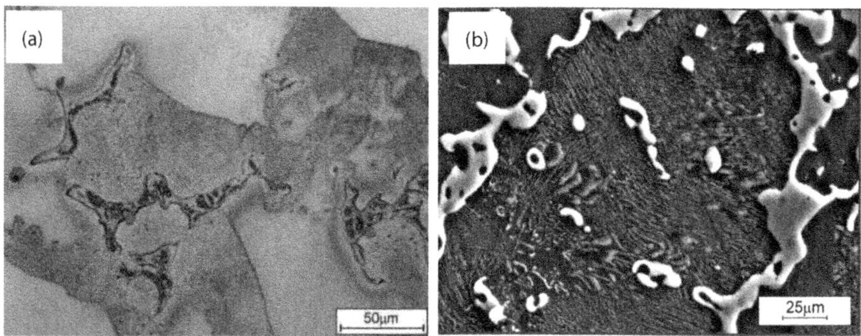

FIGURE 5.17 Structure of MCMgAl9Zn1 casting alloy after heat treatment with cooling with an furnace; (a) LM; (b) SEM.

The predominant fraction of Mg and Al, and also a small concentration of Zn (Figures 5.17 and 5.18), was found in the alloy matrix as well as in the area of eutectics and in the area of large precipitates formed on the boundaries of phases identified as $Mg_{17}Al_{12}$.

A chemical analysis of the distribution of surface elements and a quantitative microanalysis performed on lateral fractures of casting magnesium alloys by means of the EDS system also reveal that in some areas there is clearly a heightened concentration of Mg, Si, as well as Al., Mn and Fe, which indicates the presence of precipitates containing Mg and Si with angular contours in the structure of alloys, as well as of phases with a high concentration of Al and Mn, having an irregular shape, occurring frequently in the spheroidal or needle-like form.

After supersaturation with cooling in water, as well as in air (Figure 5.19), in MCMgAl9Zn1, MCMgAl6Zn1, MCMgAl3Zn1 alloys, apart from a matrix of the solid solution α-Mg, the phase γ-$Mg_{17}Al_{12}$ exists in the structure, with a trace fraction, and single precipitates of Mg-Si phases exist, as well as Mn-Al-Fe phases occurring frequently in the spheroidal or needle-like form (Figure 5.20). Eutectic precipitates do not exist in the structure. Numerous areas exist in the MCMgAl12Zn1 alloy with the phase γ-$Mg_{17}Al_{12}$ unsolved in a solid solution and areas of eutectics (Figure 5.19b).

It was found, by examining thin foils in a transmission electron microscope, that the matrix of MCMgAl12Zn1, MCMgAl9Zn1, MCMgAl6Zn1, MCMgAl3Zn1 casting magnesium alloys in a state after supersaturation consists of the supersaturated solid solution α-Mg (Figure 5.20). Low density dislocations are presented in the supersaturated solid solution (Figure 5.21).

It was pointed out with X-ray qualitative and quantitative phase analysis methods that – after supersaturation and ageing – the intermetallic phase γ-$Mg_{17}Al_{12}$ and the solid solution α-Mg – being a matrix of the examined Mg-Al-Zn-Mn alloys – exist in the investigated alloys (Figure 5.22). The volumetric fractions of precipitates of the phase γ-$Mg_{17}Al_{12}$ in the structure of MCMgAl12Zn1, MCMgAl9Zn1, MCMgAl6Zn1 alloys depend on the concentration of Al introduced as the main alloy additive, assuming the maximum value of 11.9% for the MCMgAl12Zn1 alloy in the state after supersaturation and ageing and the

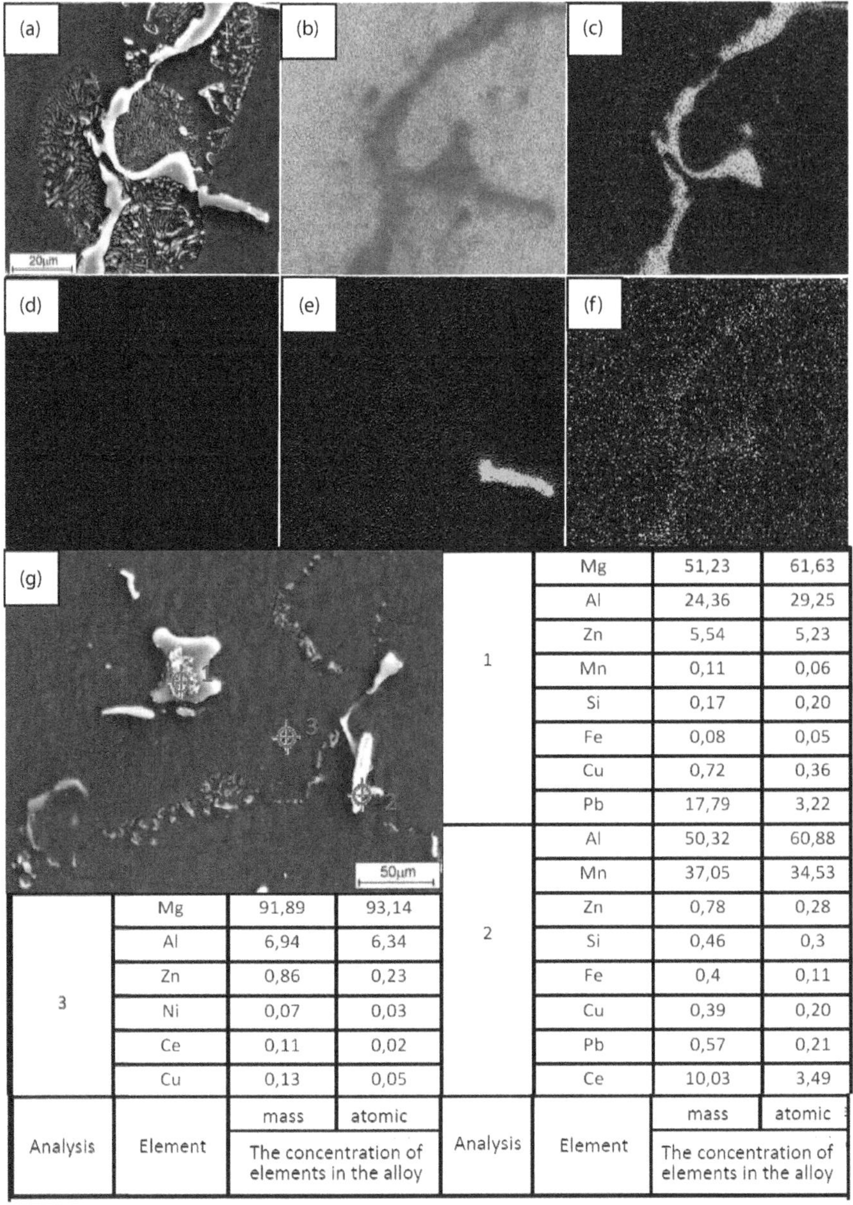

Analysis	Element	The concentration of elements in the alloy: mass	The concentration of elements in the alloy: atomic
1	Mg	51,23	61,63
	Al	24,36	29,25
	Zn	5,54	5,23
	Mn	0,11	0,06
	Si	0,17	0,20
	Fe	0,08	0,05
	Cu	0,72	0,36
	Pb	17,79	3,22
2	Al	50,32	60,88
	Mn	37,05	34,53
	Zn	0,78	0,28
	Si	0,46	0,3
	Fe	0,4	0,11
	Cu	0,39	0,20
	Pb	0,57	0,21
	Ce	10,03	3,49
3	Mg	91,89	93,14
	Al	6,94	6,34
	Zn	0,86	0,23
	Ni	0,07	0,03
	Ce	0,11	0,02
	Cu	0,13	0,05

FIGURE 5.18 Structure of MCMgAl6Zn1 casting alloy after heat treatment with cooling with the furnace: (a) an image obtained with back scattered electrons; (b–f) maps of surface distribution of the elements: (b) Mg; (c) Al; (d) Mn; (e) Si; (f) Zn; (g) the structure with the marked places of EDS point analysis and the results of the quantitative analysis of chemical composition.

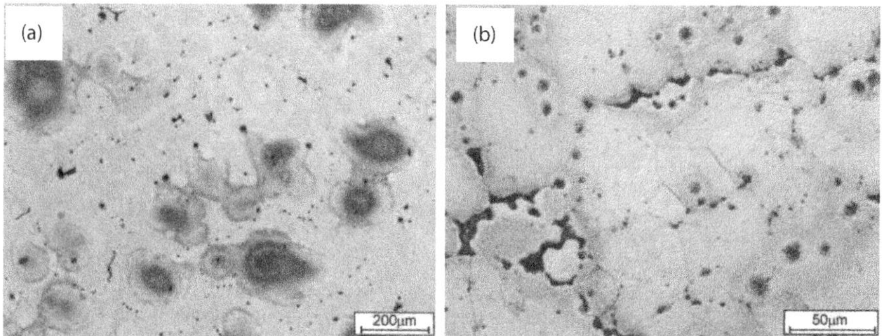

FIGURE 5.19 Structure of casting magnesium alloys in the supersaturated state: (a) MCMgAl3Zn1 after cooling in air; (b) MCMgAl12Zn1 after cooling in water.

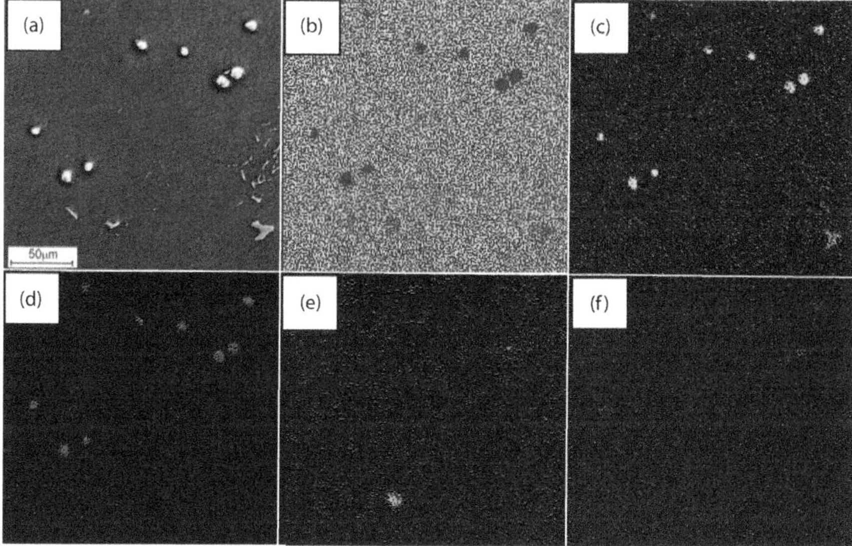

FIGURE 5.20 Structure of MCMgAl9Zn1 casting alloy in the supersaturated state after cooling in air: (a) an image obtained with back scattered electrons; (b–f) maps of surface distribution of the elements: (b) Mg; (c) Al; (d) Mn; (e) Si; (f) Fe.

minimum value – 1.6% for MCMgAl9Zn1 alloys in the state after supersaturation and ageing and MCMgAl6Zn1 in the cast state (Table 5.9).

The examinations of thin foils of the investigated alloys after ageing show that the structure of casting magnesium alloys in this state consists of the solid solution α-Mg with uniformly arranged precipitates of the secondary phase γ-$Mg_{17}Al_{12}$ (Figure 5.23). In the solid solution, being a matrix of Mg-Al-Zn-Mn casting alloys,

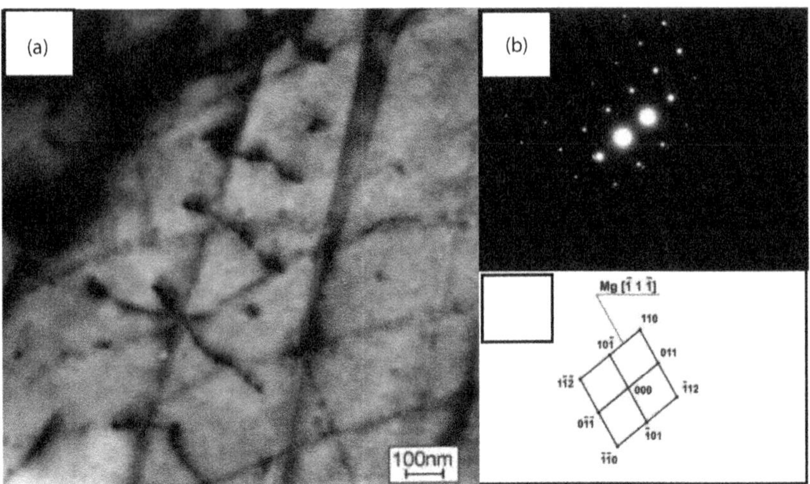

FIGURE 5.21 Structure of thin foils made of MCMgAl9Zn1 magnesium casting alloy in the supersaturated state: (a) image in bright field, (b) diffraction pattern from the area as in Figure (a,c) diffraction pattern solution from Figure (c).

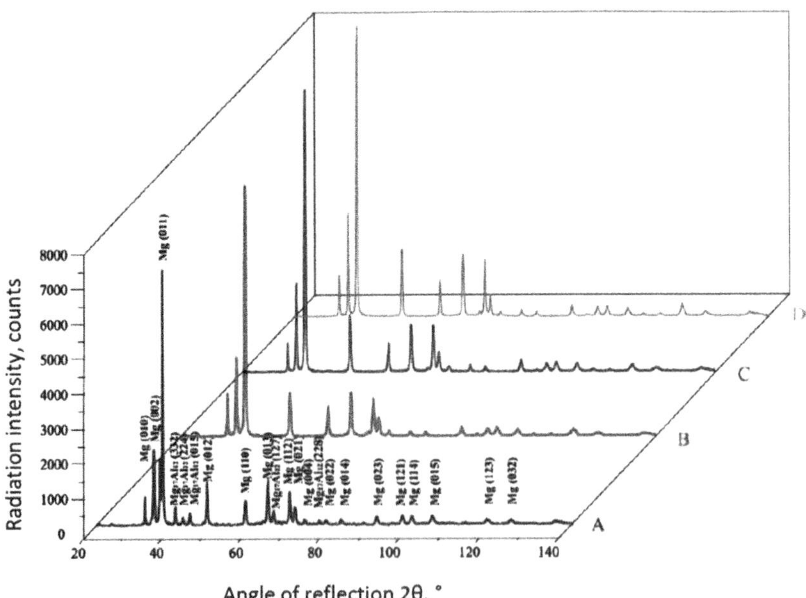

FIGURE 5.22 X-ray diffraction pattern of the following casting magnesium alloys: A – MCMgAl12Zn1, B – MCMgAl9Zn1, C – MCMgAl6Zn1, D – MCMgAl3Zn1 in the state after supersaturation and ageing.

TABLE 5.9
Results of X-ray Quantitative Analysis of Mg-Al-Zn-Mn Casting Magnesium Alloys in the Cast State and in Different State of Heat Treatment

Stop	State of Heat Treatment	Mass Fraction, [%]	
		α-Mg	γ-$Mg_{17}Al_{12}$
MCMgAl12Zn1	Cast state	93.99	6.01
	Supersaturated state after cooling in water	98.1	1.9
	Supersaturated state after cooling in air	96.6	3.4
	State after heat treatment with cooling with furnace	92.6	7.4
	State after supersaturation and ageing	88.1	11.9
MCMgAl9Zn1	Cast state	93.6	6.4
	State after heat treatment with cooling with furnace	93.1	6.9
	State after supersaturation and ageing	98.4	1.6
MCMgAl6Zn1	Cast state	98.4	1.6

dislocations occur after supersaturation and ageing, which create clusters and tangled lattices connected with the precipitation of the intermetallic phase γ-$Mg_{17}Al_{12}$ particles. The precipitates of the phase γ-$Mg_{17}Al_{12}$ usually have the shape of rods or plates, and the dominant direction of their growth are directions from the family <110> α-Mg. The dispersion precipitates contained in the solid solution in the aged magnesium alloys have in most of the cases a privileged crystallographic orientation with the matrix. Some of them exhibit the following relations:

$$(1\bar{1}01)\ \alpha\text{-Mg} \parallel (10\bar{1})\ Mg_{17}Al_{12}$$
$$[11\bar{2}0]\ \alpha\text{-Mg} \parallel [111]\ Mg_{17}Al_{12}$$

according to those given by S. Guldberg and N. Ryum [257] for those existing, among others, in the eutectic structure of Mg alloys containing 33% of Al. Certain precipitates in the investigated supersaturated and aged magnesium alloys show an orientation in which planes from the family {110} γ-$Mg_{17}Al_{12}$ are deflected by approx. 10° from the planes from the family {1 1 01} of the solid solution α-Mg, while others are even more deflected from the given relationship. For example, a morphology and crystallographic orientation of precipitates of particles of the intermetallic phase γ-$Mg_{17}Al_{12}$ in relation to the matrix of the solid solution α-Mg are presented for the MCMgAl9Zn1 alloy (Figure 5.23). However, no cases were found confirming the crystallographic relationships presented by Y. Liu, J. Zhou, D. Zha and S. Tang in the paper [258] with reference to Mg-Al-Zn-Sn alloys. It was found by examining the areas of thin foils of the MCMgAl9Zn1 alloy with the X-ray quantitative microanalysis method, with an EDS X-ray scattered radiation energy spectrometer, that most of all Mg (over 84% atomically), as well as Al (15% atomically) is contained in the phase causing the hardening of magnesium alloys after ageing. A high concentration of Mg and Al is also maintained in the matrix of alloys, by assuming, respectively, the values Mg ~90% atomically and Al ~10% atomically.

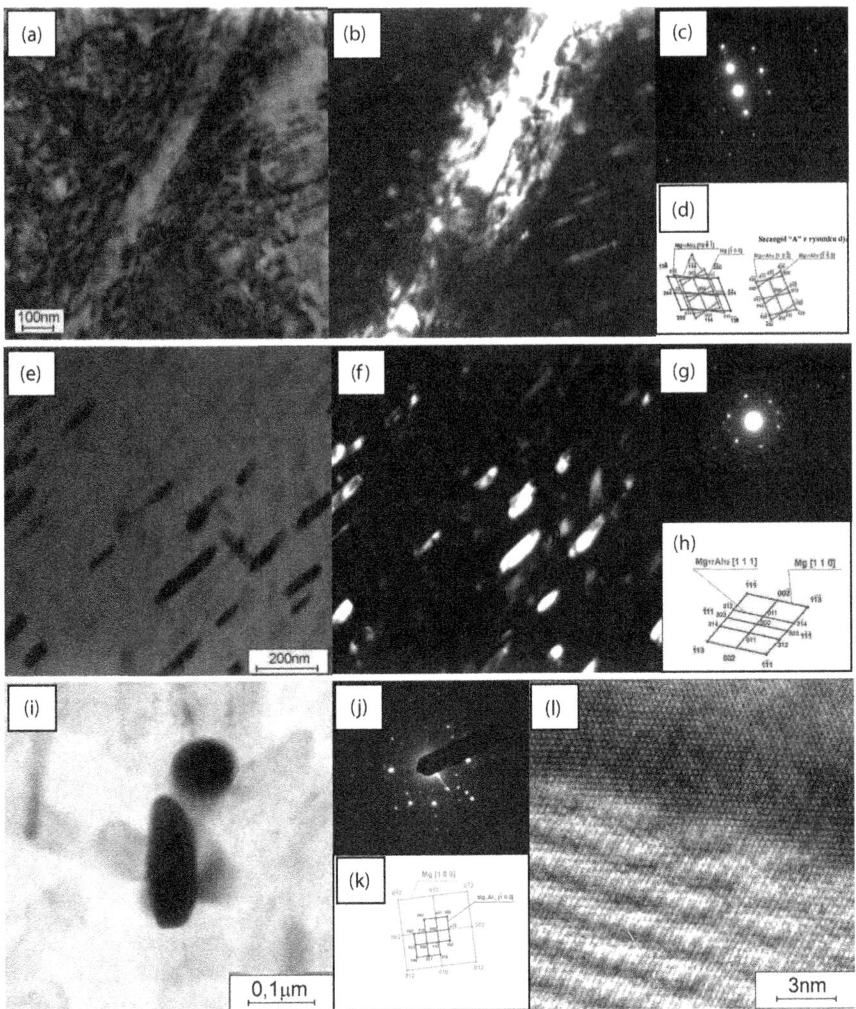

FIGURE 5.23 The structure of thin foils made of the MCMgAl9Zn1 magnesium casting alloy after supersaturation and ageing at 190°C for 15 hours and cooling in air: (a) image in the bright field, (b) image in the dark field from the reflex {402} of the phase γ-$Mg_{17}Al_{12}$; (c) diffraction pattern from the area as in Figure (a) and (b), (d) diffraction pattern solution from Figure (c); (e) image of precipitates of the phase $\gamma\gamma$-$Mg_{17}Al_{12}$ in the bright field, (f) image in the dark field from the reflex {314} of the phase γ-$Mg_{17}Al_{12}$ of the area as in Figure (a); (b) diffraction pattern from the area as in Figure (e) and (f); (h) diffraction pattern solution from Figure (g); (i) image of precipitates of the phase γ-$Mg_{17}Al_{12}$ in the bright field; (j) diffraction pattern from the area as in Figure (i); (k) diffraction pattern solution from Figure (i); (l) high-resolution image of the area of the matrix limit α-Mg and precipitates of the phase γ-$Mg_{17}Al_{12}$.

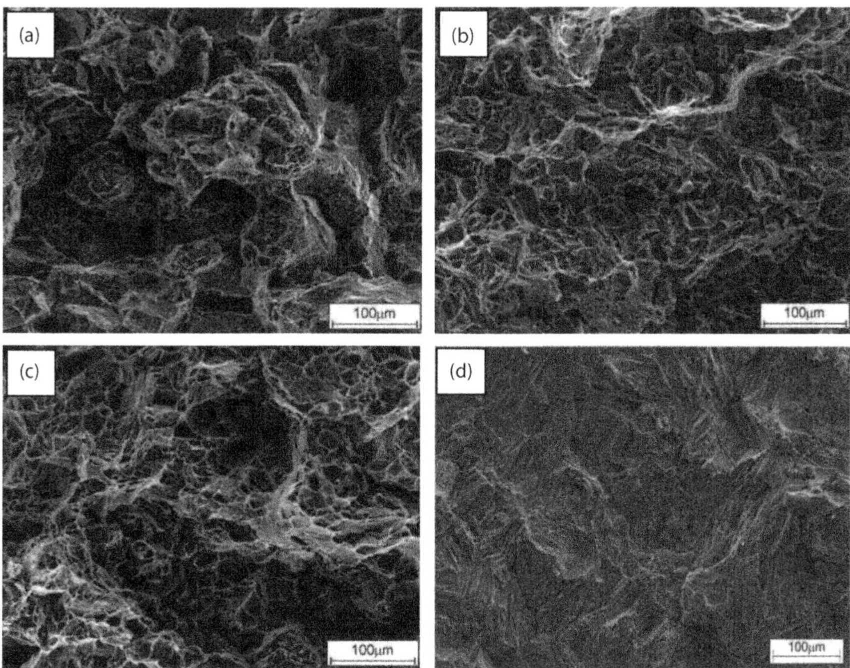

FIGURE 5.24 The fracture structure of Mg-Al-Zn-Mn casting alloys: (a) ductile MCMgAl3Zn1 alloy in the cast state; (b) mixed MCMgAl6Zn1 alloy in the state after heat treatment with cooling with furnace; (c) ductility of MCMgAl9Zn1 alloy in the supersaturated state after cooling in air; (d) brittle MCMgAl12Zn1 alloy in the state after supersaturation and ageing.

The structure of fractures after a static tensile test (Figure 5.24) was examined to characterise more completely the effect of heat treatment and a concentration of Al on the properties of Mg-Al-Zn-Mn casting alloys.

5.3.3 Mechanical and Operational Properties of the Investigated Alloys

MCMgAl12Zn1, MCMgAl9Zn1 and MCMgAl6Zn1 alloys in the cast state are characterised by a mixed fracture, whereas a ductile fracture exists for MCMgAl3Zn1 alloys (Figure 5.24). After heat treatment consisting of supersaturation with cooling in water and air, the plasticity of alloys increases, as may be signified in majority of cases by an increase in the narrowing and elongation value (Figure 5.25) and by a ductile character of fractures. On the other hand, the MCMgAl12Zn1 alloy, after heat treatment with cooling in air and ageing, and the MCMgAl9Zn1 alloy, after heat treatment with cooling in an furnace, where hardness increased in relation to the cast state and the narrowing and elongation value has fallen slightly, exhibit a brittle

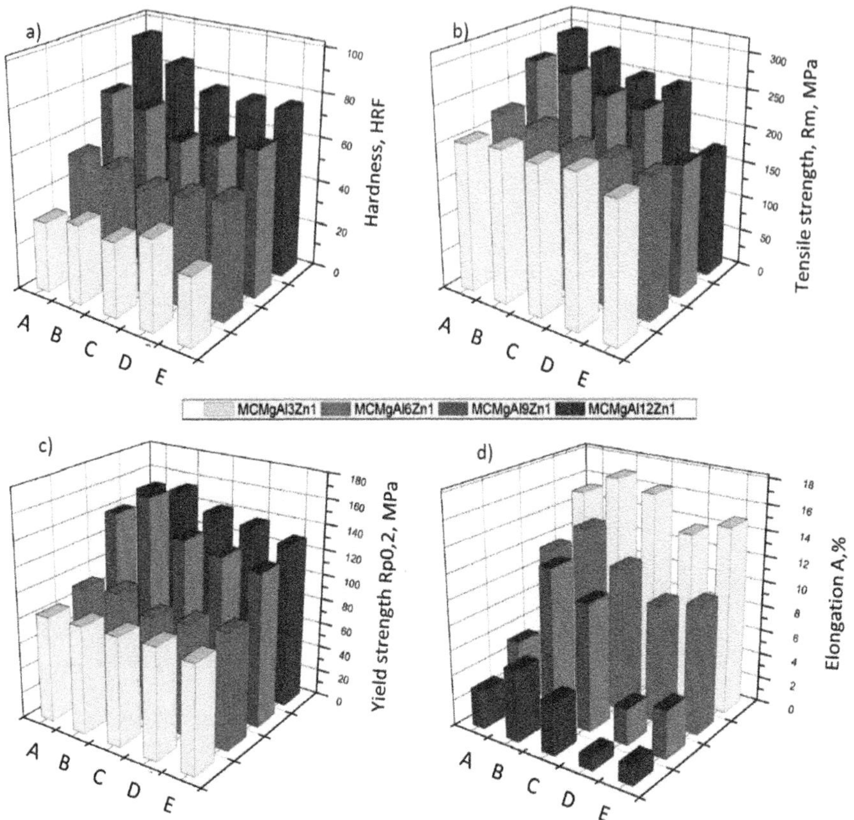

FIGURE 5.25 Results of measurement of (a) hardness; (b) tensile strength Rm HRF; (c) yield point $R_{p0,2}$; (d) elongation A of the investigated Mg-Al-Zn-Mn casting alloys after different heat treatment operations; A – after supersaturation and ageing; B – after heat treatment with cooling with an furnace; C – after supersaturation with cooling in the air; D – after supersaturation with cooling in water; E – cast state.

fracture (Figure 5.24). A mixed fracture occurs in castings made of MCMgAl6Zn1 and MCMgAl3Zn1 alloys (Figure 5.24).

Three-dimensional graphs of supersaturation conditions (Figure 5.26) and aging conditions (Figure 5.27) for the hardness of the investigated Mg-Al-Zn-Mn casting alloys were created using calculation models prepared using neural networks. The results obtained clearly indicate that the most favourable type of heat treatment according to the optimum working conditions and energy consumed and time needed to carry out supersaturation and ageing, and to achieve the most advantageous mechanical properties, is the supersaturation of the investigated Mg-Al-Zn-Mn casting alloys at 430°C for 10 hrs and ageing at 190°C for 15 hrs, regardless the concentration of Al. The conditions for heat treatment shown in Table 5.8 were chosen for this reason.

The effect of Al concentration and the type of heat treatment on the hardness and other mechanical properties of the Mg-Al-Zn-Mn alloys being cast are

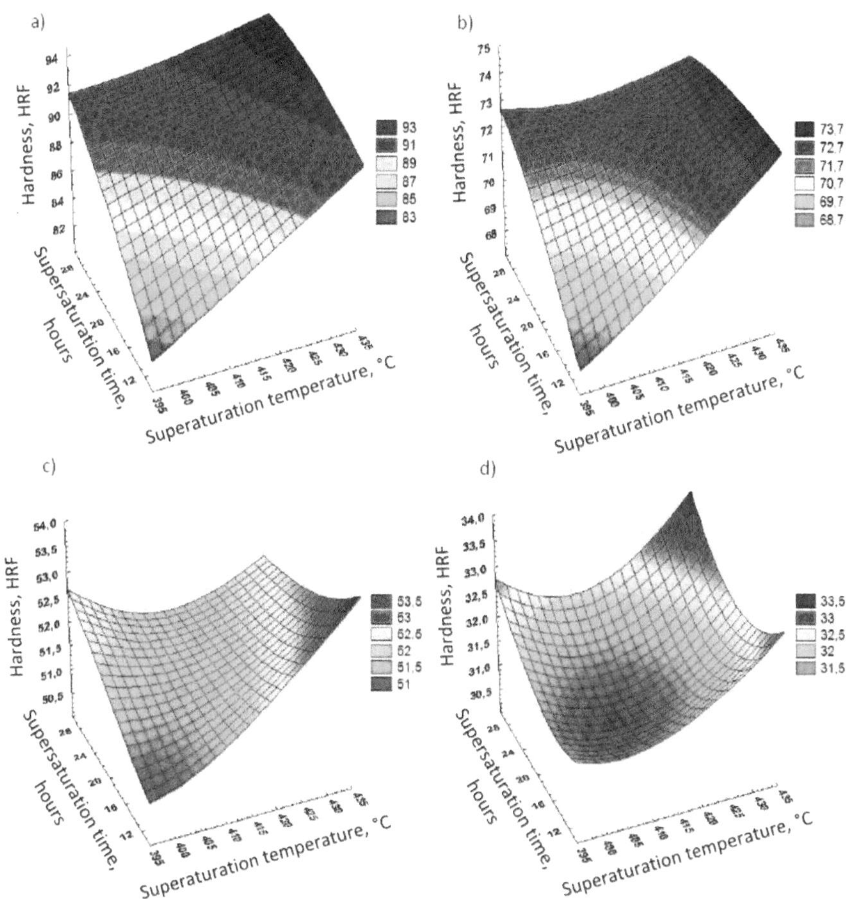

FIGURE 5.26 The effect of temperature and supersaturation time on the hardness of Mg-Al-Zn-Mn al-Zn-Mn casting alloys determined by using models of neural networks with the assumed ageing temperature of 190°C and ageing time of 15 hours: (a) MCMgAl12Zn1; (b) MCMgAl9Zn1; (c) MCMgAl6Zn1; (d) MCMgAl3Zn1.

shown in Figures 5.25 through 5.27. The hardness increases as the Al concentration rises from 3% to 12%. The MCMgAl12Zn1 alloy exhibits the highest hardness of 75.4 HRF in the cast state. The hardness is two times higher than for the MCMgAl3Zn1 alloy of 30.6 HRF. Hardness is increasing each time when the investigated alloys are subject to heat treatment consisting of successive supersaturation and ageing. The MCMgAl12Zn1, MCMgAl9Zn1 and MCMgAl6Zn1 alloys reach the hardness of, respectively, 85.1; 71.2 and 51.84 HRF after heat treatment with cooling, and the highest hardness after supersaturation and ageing of, respectively: 94.6; 75.1 and 53.2 HRF. The hardness after supersaturation is however slightly smaller than after ageing. The MCMgAl3Zn1 alloys exhibit the highest hardness of 40.7 HRF after supersaturation with cooling in water. The hardness in other cases is similar.

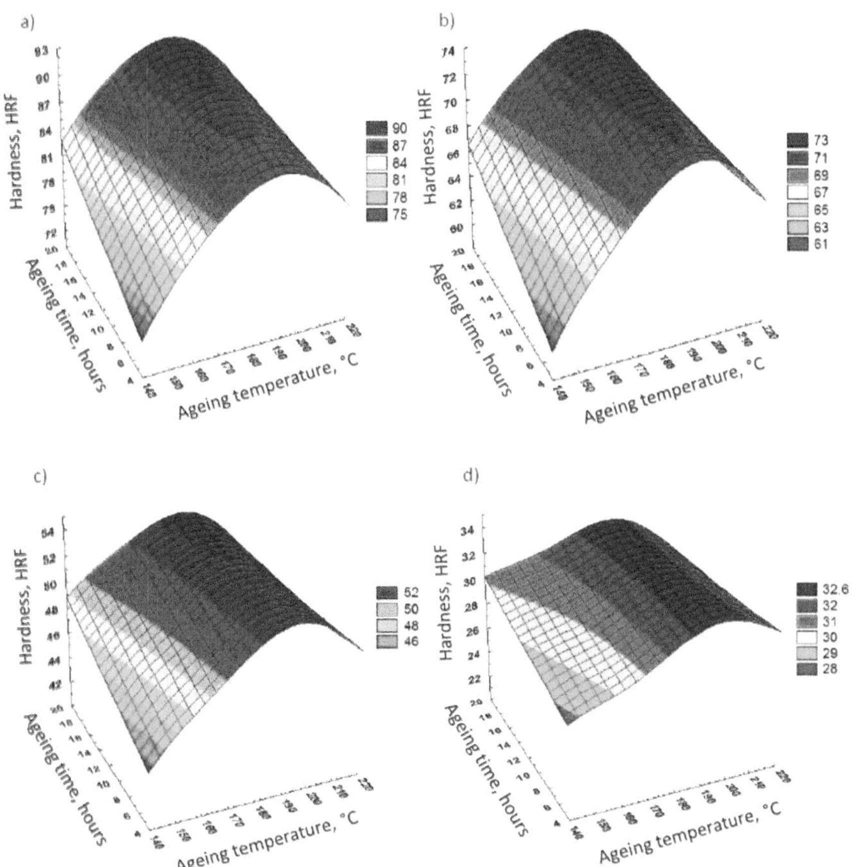

FIGURE 5.27 The effect of temperature and ageing time on the hardness of Mg-Al-Zn-Mn casting alloys determined by using models of neural networks with the assumed supersaturation temperature of 430°C and supersaturation time of 10 hours: (a) MCMgAl12Zn1; (b) MCMgAl9Zn1; (c) MCMgAl6Zn1; (d) MCMgAl3Zn1.

MCMgAl6Zn1 and MCMgAl3Zn1 alloys enjoy the highest tensile strength in the cast state of, respectively, 192.1 and 191.3 MPa (Figures 5.25 through 5.27), with the highest elongation in the cast state of, respectively, 11.6% and 15.2%. If the concentration of Al is increased from 6% to 12%, this affects reduction in elongation in the cast state to 170.9 MPa. Heat treatment with cooling with a furnace and supersaturation, followed by ageing, elevates the tensile strength. The maximum tensile strength of 294.8 MPa was obtained after ageing the MCMgAl12Zn1 alloy. A substantial increase in the tensile strength of MCMgAl9Zn1 specimens by 50% after ageing was also found. The smallest increase in tensile strength after heat treatment of, respectively, 30.3 and 12.4 MPa, is exhibited by MCMgAl6Zn1 and MCMgAl3Zn1 alloys. Variations in the tensile strength of the alloys subjected to supersaturation with cooling in water and in the air account for a maximum of 6 MPa.

The MCMgAl12Zn1 alloys show the maximum yield point of 129.4 MPa in the cast state (Figures 5.25 through 5.27). After applying heat treatment, the highest yield point values of respectively 153.7; 158.9 and 96.3 MPa in the heat-treated state with cooling with an furnace, are reached for MCMgAl12Zn1, MCMgAl9Zn1 and MCMgAl6Zn1 alloys. The yield point after supersaturation and ageing is only slightly higher and is, respectively, 149.6, 143.8 and 92.03 MPa.

If the Al concentration rises to 12%, the elongation of the investigated Mg-Al-Zn-Mn alloys in the cast state is decreased to approx. 3%, five times lower than elongation of MCMgAl3Zn1 alloys (Figures 5.25 through 5.27). Supersaturation with cooling in water and air increases the elongation value even two times for MCMgAl12Zn1 and MCMgAl9Zn1 alloys. Alloys after heat treatment with cooling in a furnace and after supersaturation and ageing have a slightly decreased elongation in relation to the cast state.

The influence of chemical composition and heat treatment technology on the applications and functional properties of the investigated Mg-Al-Zn-Mn alloys was established with abrasion resistance tests and corrosion tests.

Abrasion tests with the metal-to-metal configuration were carried out to compare the abrasion resistance under conditions simulating the operating conditions of Mg-Al-Zn-Mn casting alloys. The results of the executed abrasion test reveal that the lowest mean weight loss in the cast state and after the heat treatment, with the load rising from 6 to 12 N, occurs for the MCMgAl12Zn1 alloy (Figure 5.28). This alloy is characterised by the best tribological properties among the studied

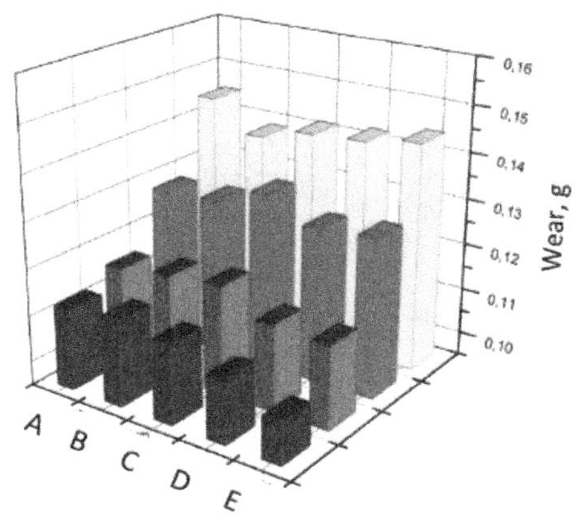

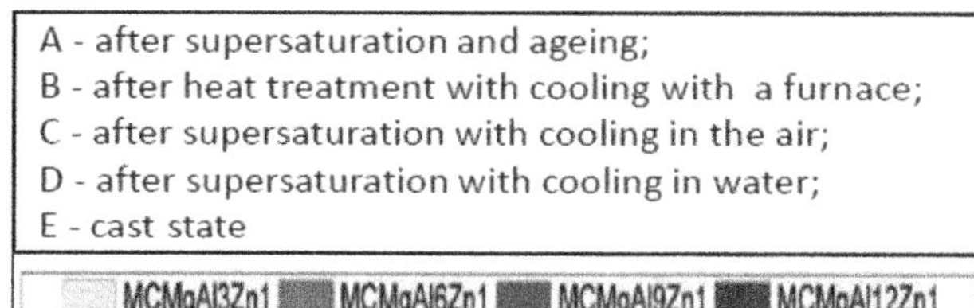

FIGURE 5.28 The results of wear of magnesium alloys wear for the load of 12 N.

Mg-Al-Zn-Mn alloys. The MCMgAl3Zn1 alloy exhibits the highest mass loss for the analysed cases in the cast state and after heat treatment. The abrasion resistance of the investigated alloys, represented by a mean mass loss, corresponds to their hardness. The heat treatment of the investigated alloys increases their wear resistance.

Corrosion tests using the electrochemical potentiodynamic method in a 3% NaCl water solution were performed in order to determine the effect of Al concentration as well as heat treatment on the corrosion resistance of the investigated Mg-Al-Zn-Mn alloys. The corrosive wear of the tested materials' surface depending on the mass fraction of Al, as well as the cast state and supersaturated state, annealed state and in the hardened-by-precipitation state, was determined as a result. Polarisation curves (Figure 5.29) and anode loops for the Mg-Al-Zn-Mn casting alloys in the cast state and after heat treatment were obtained based on

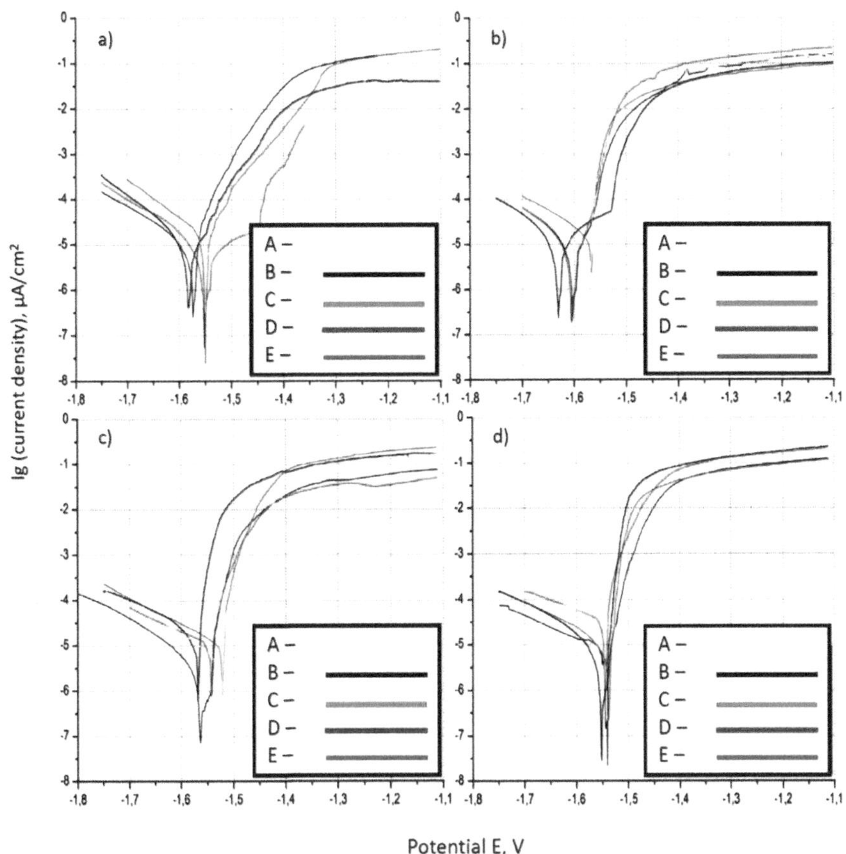

FIGURE 5.29 Charts of anode polarisation of Mg-Al-Zn-Mn casting alloys: (a) MCMgAl3Zn1; (b) MCMgAl6Zn1; (c) MCMgAl9Zn1; (d) MCMgAl12Zn1; A – after supersaturation and ageing; B – after heat treatment with cooling with furnace; C – after supersaturation with cooling in the air; D – after supersaturation with cooling in water; E – cast state.

the executed potentiodynamic tests. These curves indicate that the tested alloys undergo pitting corrosion, and the examined alloys are particularly susceptible to such corrosion. Quantitative data was established describing the phenomenon of electrochemical corrosion of the investigated alloys on the basis of the obtained anode polarisation curves, using the Tafel extrapolation method near a corrosion potential: corrosion potential value E_{kor} (mV), polarisation resistance R_p (kΩ/cm^2), corrosion current density i_{kor} (μA/cm^2), corrosion rate V_p (mm/year) and mass loss V_c (g/m^2) (Figure 5.29).

The anode polarisation curves as well as the corrosion current density value signify the rate of quench annealing of the investigated alloys. The MCMgAl3Zn1 alloys, whose corrosion potential is 1578.4 mV, corrosion resistance is 1.29 kΩ/cm^2, and corrosion current density in the passive range is 3.43 μA/cm^2, show the best corrosive resistance in the cast state. In case of the MCMgAl3Zn1 alloy, the corrosion current density, except for the condition after supersaturation and ageing, is smaller than for alloys with the higher concentration of Al. The MCMgAl6Zn1 alloy has a slightly lowered value describing corrosion in relation to the MCMgAl3Zn1 alloy. A significant deterioration of corrosion resistance, when the polarisation resistance decreases, with increased corrosion current density at the same time, occurs for MCMgAl12Zn1 and MCMgAl9Zn1 alloys (Figure 5.29).

The smallest corrosion current density, hence the smallest anodic quench annealing for the investigated Mg-Al-Zn-Mn alloys, and the related best corrosion resistance after heat treatment, is exhibited by the MCMgAl3Zn1 alloy, whereas the worst one by the MCMgAl12Zn1 alloy (Figure 5.30). This is equivalent to the acceleration of surface corrosion in a corrosive solution, which was shown based on the calculated values of the corrosion rate V_p and mass loss V_c of the investigated alloys.

It is seen by analysing the results obtained for alloys with a 12%, 9% and 6% concentration of Al that corrosion resistance of alloys after precipitation hardening takes place in relation to alloys in the cast state, as well as those subject to supersaturation only. The investigated alloys exhibit the worst corrosion resistance in all the analysed cases after heat treatment with cooling with a furnace.

The pitting potential value, E_n, at which pits start to appear on the surface of the investigated alloys, and of the repassivation potential, E_{cp}, below which no active pits exist on the surface any more, was determined also from the course of potentiodynamic curves. It was found by comparing the width of corrosion loops of the investigated alloys (E_n and E_{cp} parameters within the range where no new pits are formed, however corrosive processes may still occur in the existing ones), as well as the tilt angle and height of such loops, that the worst parameters (potting and repassivation potential) have the alloys subject to heat treatment with cooling with the furnace, whereas the best results are obtained for alloys after precipitation hardening. The charts of corrosive loops also reveal that a corrosive factor in MCMgAl3Zn1 casting magnesium alloys requires longer time to puncture the passive layer and to penetrate the material, whereas above the puncture potential value, where pits are started to be formed, the alloy is subject to faster quench annealing in the NaCl solution.

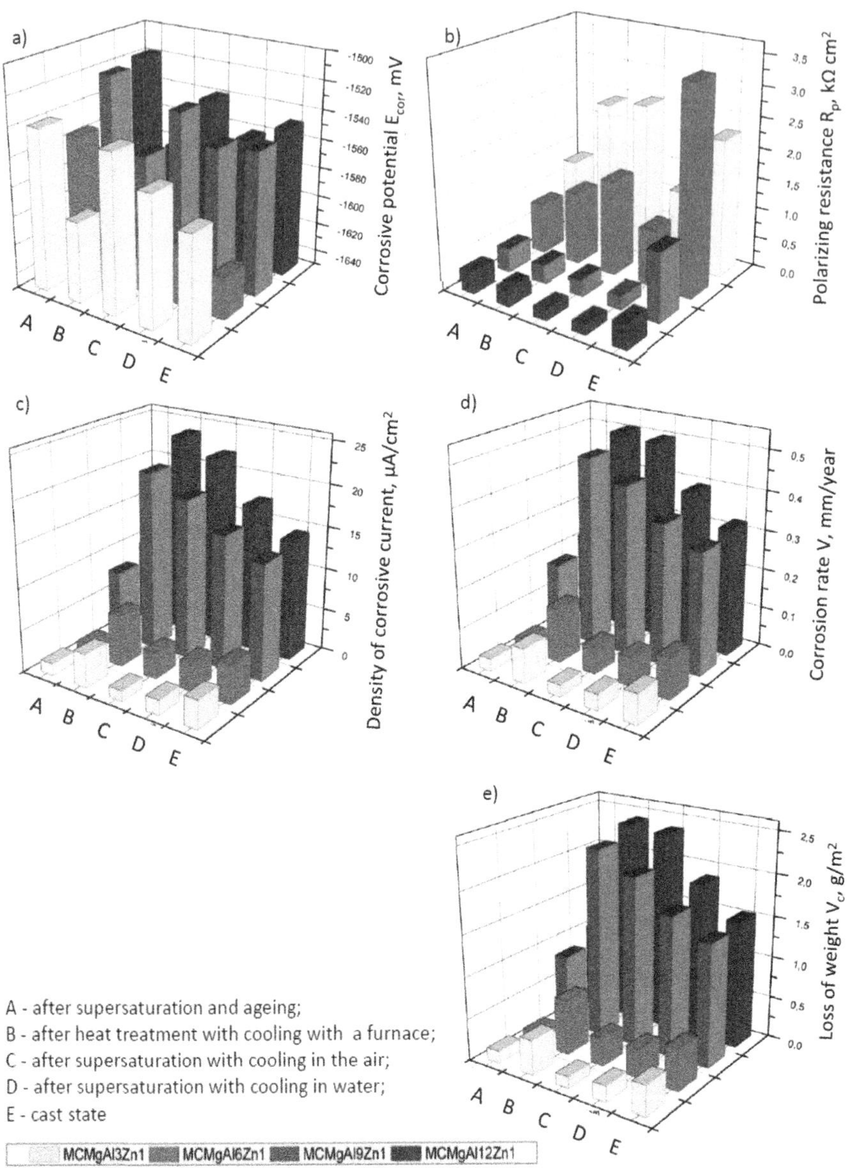

FIGURE 5.30 The results of measurements of values characterising the course of corrosion using the electrochemical potentiodynamic method of the investigated Mg-Al-Zn-Mn alloys; (a) corrosion potential E_{kor} (mV); (b) polarisation resistance R_p (kΩ/cm^2); (c) corrosion current density i_{kor} (μA/cm^2); (d) corrosion rate V_p (mm/year); (e) mass loss V_c (g/m^2).

5.4 APPLICATION SUITABILITY AND FORECAST DEVELOPMENT OF SURFACE TREATMENT OF Mg-Al-Zn-Mn ALLOYS

5.4.1 Methodological Scope of Foresight Research with Respect to Mg-Al-Zn-Mn Alloys

The design, technological and operational experience of numerous products show that it is not reasonably substantiated in engineering calculations and in real expectations to ensure the expected properties, as required in the beginning, equally for the whole section of the product. Different requirements are many times imposed on the core of the material, most often the product or its element, and different on the surface or the surface layer. The properties of the core and surface layer of the part produced can be customised most advantageously by selecting the part's material and its structure and properties formation processes appropriately along with the surface layer type and technology ensuring the required functional properties. Extensive investigations were carried out to confirm such regularities also in relation to Mg-Al-Zn-Mn casting alloys. Scientific technology foresight methods were applied in order to justify the desirability of employing the surface treatment of Mg-Al-Zn-Mn alloys and to forecast its development. Technology foresight consists of looking regularly, over a long-term prospective, into the future of science and technology, economy and societies, in conjunction with an ability to select strategic technologies aimed at bringing considerable economic and social benefits. The activities performed under the author's FORSURF project [101] are concentrated around the externalisation of knowledge based on the transformation of implicit knowledge, available only to experts and specialists in a given field, into explicit knowledge available to the broadly conceived public, which – in a long-term perspective – should lead to the strengthening of a knowledge- and innovation-based economy. Priority technologies, referred to as critical technologies, were compared from among 500 detailed technologies of structure and surface properties formation of engineering materials, with the best development outlooks and/or of key significance for industry within the analysed time horizon of 20 years. A set of 140 critical materials surface engineering technologies was generated based on the research performed.

Nearly 500 independent foreign and domestic experts representing scientific, business and public administration circles have taken part at various stages of works in the FORSURF [for] technology foresight implemented as part of own works, who have completed approx. 800 multi-question surveys. A collection of 140 critical technologies was thoroughly analysed according to three iterations of the e-Delphix method performed according to the author's idea of e-foresight [102,259–262]. This paper presents the outcomes of foresight research [87,101], based on source data, pertaining to the position of physical (PVD) and chemical (CVD) vapour deposition and the technology of laser formation of the surface layer of Mg-Al-Zn-Mn casting alloys, against the overall materials surface engineering. The outcomes of materials science and heuristic research are presented, relating to eleven groups of detailed technologies of surface treatment of MCMgAl12Zn1, MCMgAl9Zn1, MCMgAl6Zn1, MCMgAl3Zn1 casting magnesium alloys, by using – as a classification criterion – the physical processes resulting in changes to the surface layer properties, the type of the deposited coatings and/or the type of powder deposited onto the substrate. Such technologies relate in particular to the deposition of:

- The following coatings applied in PVD processes:
  A – Ti/Ti(C, N)/(Ti, Al)N
  B – Ti/Ti(C, N)/CrN
  C – Cr/CrN/CrN
  D – Cr/CrN/TiN
  E – Ti/(Ti, Si)N/(Ti, Si)N
- The following coatings applied in CVD processes:
  F – Ti/DLC/DLC
- Laser surface treatment by laser imbed into remelted surface zone of hard ceramic particles of:
  G – Titanium carbides
  H – Tungsten carbides
  I – Vanadium carbides
  J – Silicone carbides
  K – Aluminium carbides

Particular attention in the deliberations conducted was devoted to: Cathodic Arc Deposition (CAD), being a physical vapour deposition method, to Plasma Assisted Chemical Vapour Deposition (PACVD) and to alloying/laser imbed of carbide or oxide hard ceramic particles into the substrate remelted surface zone, and such technologies were used during the performed materials science experiments.

A universal scale of relative states, being a single-pole positive scale without zero, where 1 is a minimum rate and 10 an extraordinarily high rate, was used in the research heuristic undertaken (Figure 5.31).

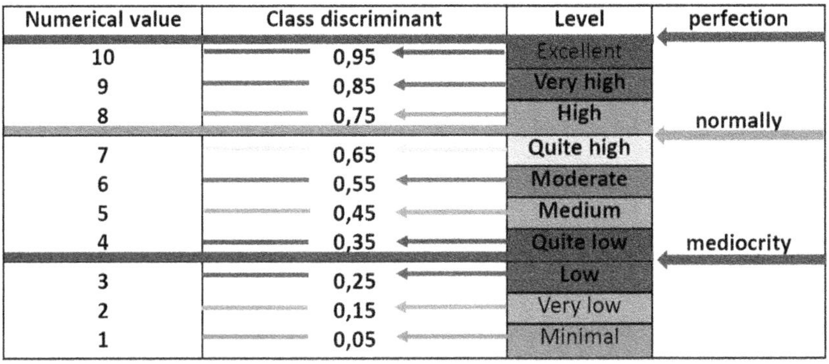

Numerical value	Class discriminant	Level	perfection
10	0,95	Excellent	
9	0,85	Very high	
8	0,75	High	normally
7	0,65	Quite high	
6	0,55	Moderate	
5	0,45	Medium	
4	0,35	Quite low	mediocrity
3	0,25	Low	
2	0,15	Very low	
1	0,05	Minimal	

FIGURE 5.31 Universal scale of relative states. (From Dobrzańska-Danikiewicz, A., *Arch. Mater. Sci. Eng.*, 44/1, 43–50, 2010.)

The positions of technologies are ranked with a set of author's contextual matrices, containing in particular: dendrological matrices of technology value, metrological matrices of environment influence and matrices of strategies for technologies. The matrices represent a tool of a graphical comparative analysis of the individual technologies or their groups.

A dendrological matrix of technology value was applied in order to determine the objectivised values of the individual, separate technologies or their groups, and a meteorological matrix of environment influence was used to determine the intensity of positive and negative influence of the micro- and macroenvironment on the specific technologies.

5.4.2 Results of Foresight Analyses for Mg-Al-Zn-Mn Alloys

The dendrological matrix of technology value (Figure 5.32) presents graphically the evaluation results of the individual technologies for their potential representing a real, objective value of the given technology and for its attractiveness reflecting the subjective perception of the given technology by its potential users.

The analysis made has shown that in the considered case, all the technologies (A)–(K) were classified to the most promising quarter called wide-stretching oak, encompassing the technologies with a high potential and attractiveness, as shown graphically in Figure 5.33. The best results were achieved by: a technology of laser

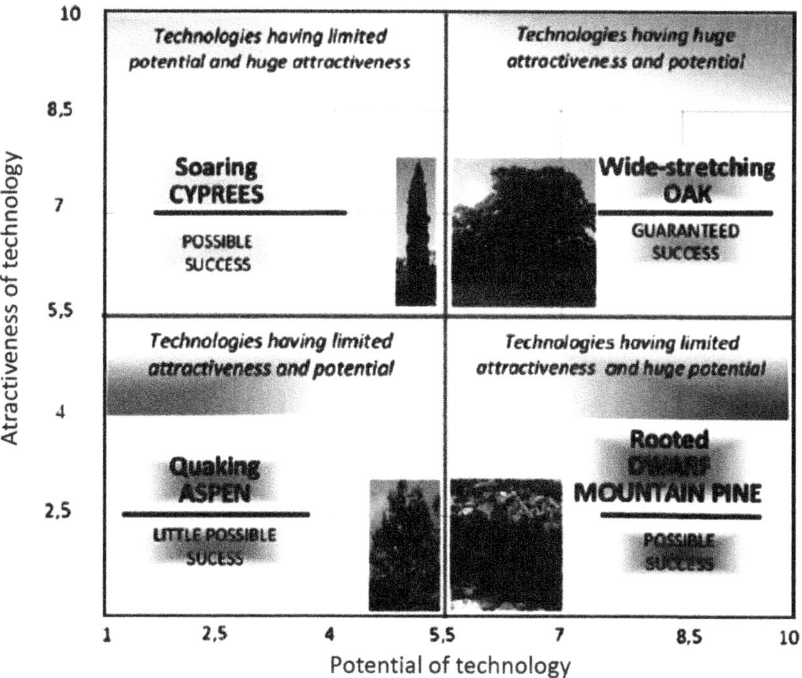

FIGURE 5.32 Dendrological matrix of technology value; presentation of approach.

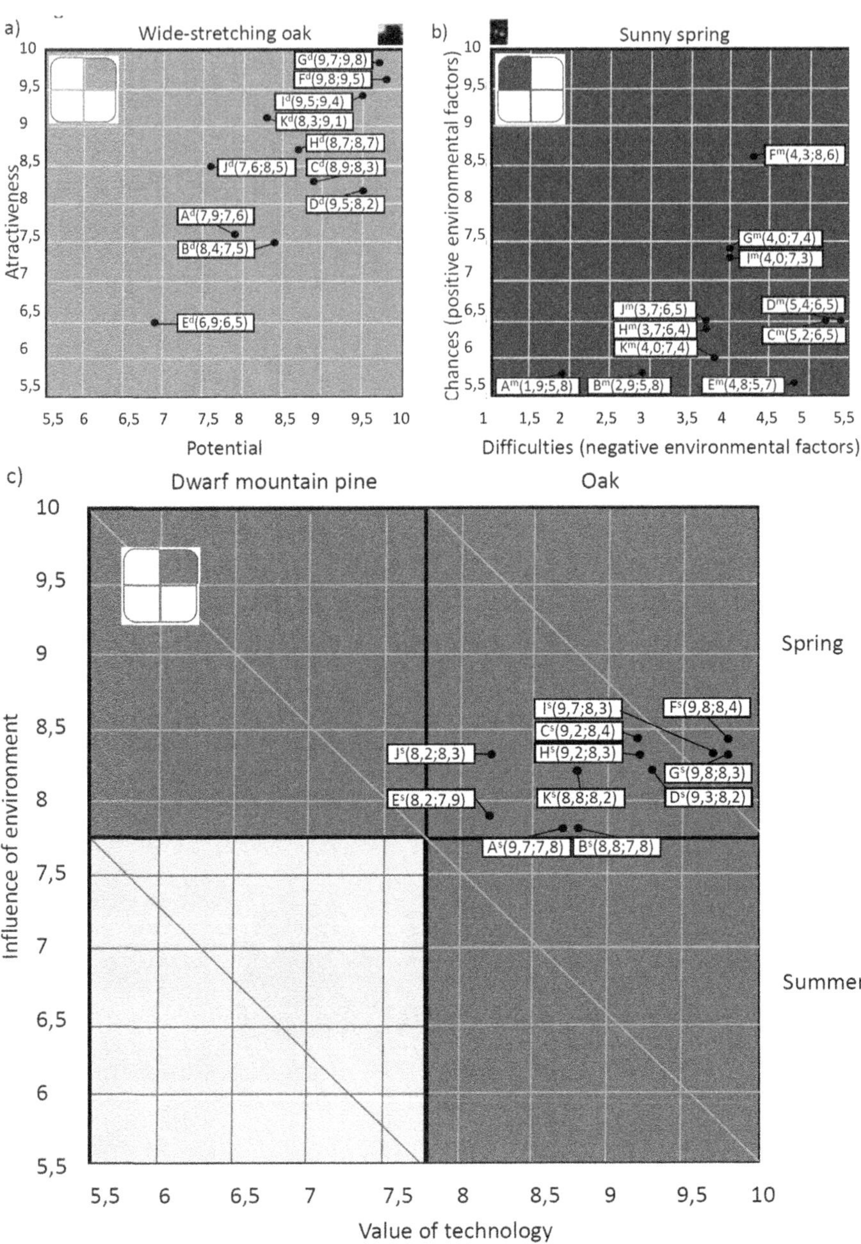

FIGURE 5.33 The corresponding quarters of contextual matrices to determine the objectivised values with positioning of particular selected technologies used for surface treatment of Mg-Al-Zn-Mn alloys: (a) top-right quarter of the dendrological matrix of technology value; (b) bottom-left quarter of metrological matrix of environment influence; (c) top-right quarter of the matrix of strategies for technologies.

surface treatment using titanium carbide G^{d}[1] (9,7; 9,8) and vanadium carbide I^{d} (9,5; 9,4) and Plasma Assisted Chemical Vapour Deposition (PACVD) of Diamond-Like Coatings Ti/DLC/DLC F^{d} (9,8; 9,6). Taking into consideration the eleven analysed technologies of surface treatment of Mg-Al-Zn-Mn casting alloys, the lowest value E^{d} (6,9; 6,5), being the resultant of the technology potential and attractiveness, is seen for physical vapour deposition of Ti/(Ti, Si)N/(Ti, Si)N coatings.

The meteorological matrix of environment influence (Figure 5.34) presents the results of how the external positive factors impacting the specific groups of technologies are evaluated according to classification into difficulties having negative influence and opportunities having positive influence on the technologies analysed. The examination of experts' opinions regarding positive and negative factors influencing the given technologies took place with the use of a survey consisting of several dozens of questions concerning the micro- and macroenvironment, according to social, technological, economic, environmental, political and legal environment, in strictly defined proportions.

The results of the research conducted, concerning Mg-Al-Zn-Mn casting alloys subject to surface treatment, presented in Figure 5.33, show that for all the investigated groups of technologies (A)–(K), the environment is very supportive, bringing multiple opportunities and few difficulties. All the analysed technologies were

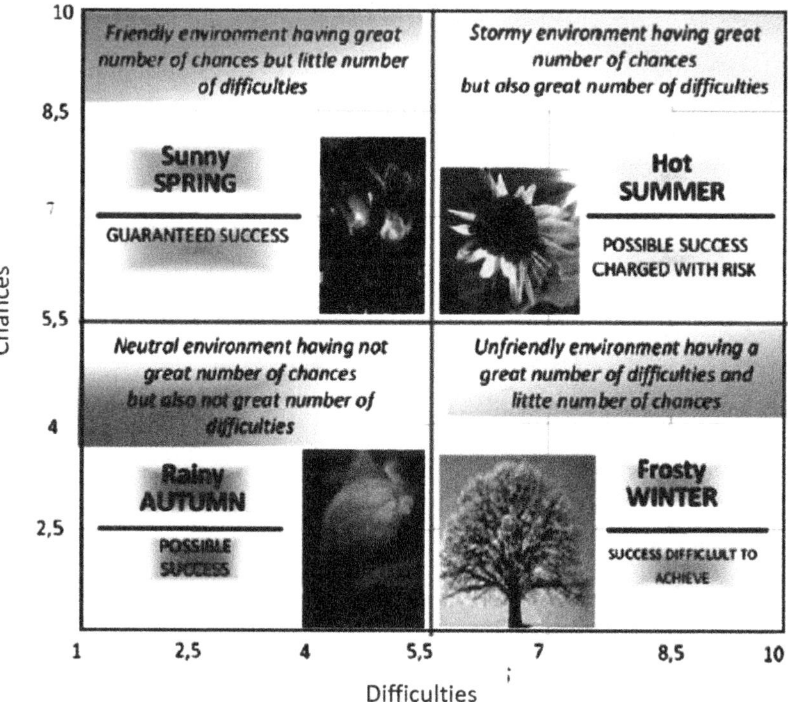

FIGURE 5.34 Meteorological matrix of technology value; presentation of the approach.

[1] *d* in the index means that a position of a given technology related to a dendrological matrix.

therefore placed in the quarter corresponding to sunny spring, which is tantamount to good development perspectives over the nearest 20 years. The Assisted Chemical Vapour Deposition technology for hard, abrasion resistant (sometimes called 'self-lubricating') and biocompatible diamond-like Ti/DLC/DLC F^{m}[2] coatings (4,3; 8,6) offer greatest opportunities, but also relatively many difficulties, which is associated with comprehensive current and future applicational capabilities in the tool, automotive, aviation, microelectronic, medical and biomedical industry, but is accompanied however by strong competition from laser surface treatment, pulsed laser deposition, ion implantation, thermal spraying and hybrid technologies.

To be able to transfer the relevant numerical values from the four-field dendrological and meteorological matrices to the sixteen-field matrix of strategies for technologies (Figure 5.35), mathematical dependencies were formulated allowing to rescale and objectivise the research results, and a computer programme was developed on their basis enabling the fast calculation of the values

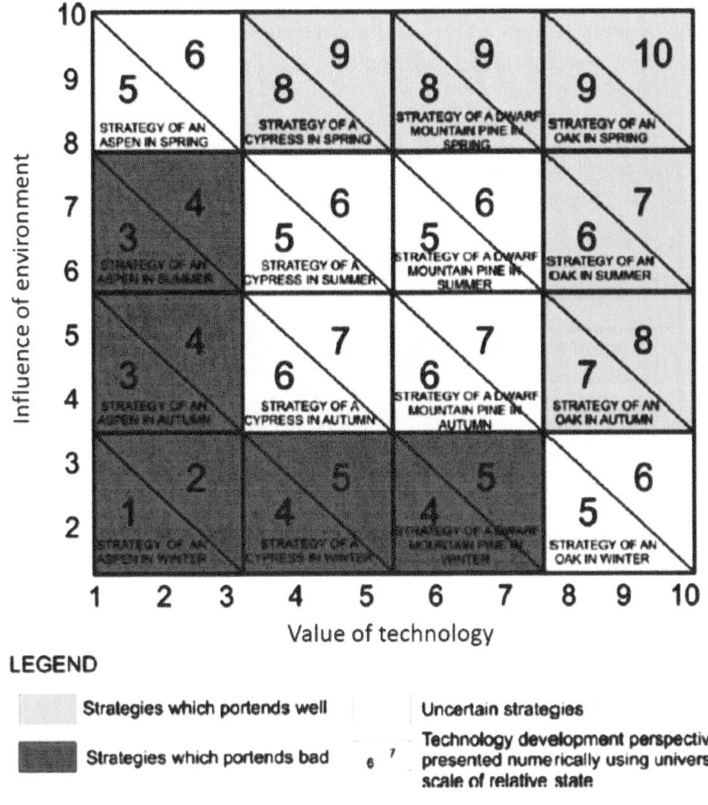

FIGURE 5.35 General form of the matrix of strategies for technologies.

[2] *d* in the index means that a position of a given technology relates to a meteorological matrix.

sought and to generate graphically the matrix of strategies for technologies. The following concepts were introduced: the relative value of the technology V_n and the relative value of environment influence E_n.

In case of surface treatment of Mg-Al-Zn-Mn casting alloys, it is the most advantageous quarter of the matrix of strategies for technologies. The matrix shown in Figure 5.33c presents graphically the place of individual surface treatment technologies according to their value and the environment influence force, indicating the relevant action strategy. The best possible prospects for development rated as outstanding (10 points) are seen for the Assisted Chemical Vapour Deposition technology for Ti/DLC/DLC F^s[3] coatings (9,8; 8,4) and for laser imbed into remelted surface zone of titanium carbide powders G^s (9,8; 8,3) and vanadium carbide powders I^s (9,7; 8,3) into the surface of Mg-Al-Zn-Mn casting alloys, which is connected with the possibility of ensuring the use of the best structure and mechanical and functional properties of the investigated alloys for such surface treatment technologies. Taking into consideration the examined technologies of surface treatment of Mg-Al-Zn-Mn casting alloys, the most advantageous environment conditions correspond to the deposition of diamond-like coatings in the process of Assisted Chemical Vapour Deposition F^s, as well as Physical Vapour Deposition of Cr/CrN/CrN coatings by Cathode Arc Evaporation (CAD) C^s (9,2; 8,4), which is mainly influenced by a wide range of possible uses of the products produced with these technologies, possessing unique mechanical, tribological and anticorrosive properties.

All the analysed technologies are therefore placed respectively in the triangles 9 and 10 of the most advantageous quarter of the matrix of strategies for technologies (Figure 5.33c), corresponding to the oak in spring strategy, connected with achieving the success and consisting of developing, strengthening and implementing attractive technologies with a large potential in the industrial practise to achieve a spectacular success.

The results of the foresight research conducted served as a reason for selecting the specific technologies used in these investigations for the surface treatment of Mg-Al-Zn-Mn casting alloys and for performing a full set of investigations of the structure and properties of the so fabricated layers and surface coatings, presented in the successive parts of this paper.

5.5 NANOSTRUCTURED SURFACE LAYERS APPLIED BY VAPOUR DEPOSITION ONTO THE INVESTIGATES ALLOYS

5.5.1 General Characteristic of CVD (Chemical Vapour Deposition) and PVD (Physical Vapour Deposition) Techniques

From among a myriad of techniques enhancing the life of engineering materials, this work uses two coating deposition techniques in relation to the investigated Mg-Al-Zn-Mn alloys, in particular due to their low melting temperature up to 180°C, namely: Cathodic Arc Evaporation (PVD CAE) and Plasma Assisted CVD (PACVD) (Figure 5.36).

[3] *s* in the index means that a position of a given technology relates to a matrix of strategies for technologies.

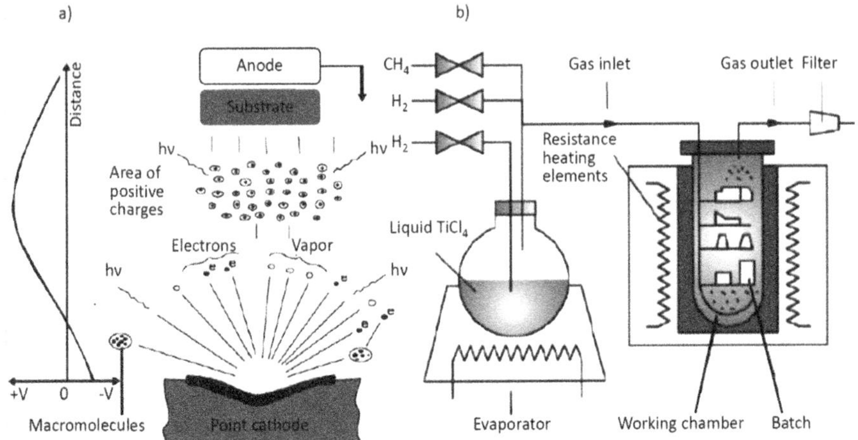

FIGURE 5.36 Diagram of coatings deposition by the following methods: (a) Arc evaporation and distribution of potential by the cathode in the CAE method; (b) APCVD process; 1 – work chamber, 2 – charge, 3 – resistive heating elements (according to T. Burakowski and T. Wierzchonia).

Table 5.10 presents the fabricated surface coatings and the conditions and the set conditions of the both coatings deposition methods on the substrate of the examined Mg-Al-Zn-Mn casting alloys.

The substrates, before deposition of coatings, were cleaned chemically by way of washing and rinsing in ultrasound washes and cascade washers and dried in a stream of hot air. They were also cleaned with ions for 20 min. using Ar ions with the substrate polarisation voltage of 800/200 V.

Coatings in the PVD CAE process were deposited using a DREVA ARC400 device by German company Vakuumtechnik with the CAE-PVD method. The device is equipped with three independent sources of metal vapours. Targets with the diameter of 65 mm cooled with water were used for deposition of coatings by PVD, containing pure metals of (Cr, Ti) and TiAl and Ti-Si alloys. The coatings were deposited in the atmosphere of Ar inert gas and N_2 reactive gases to obtain nitrides and a mixture of N_2 and C_2H_2 to obtain a layer of carbonitrides. A gradient change in the concentration of chemical composition on the cross section of the coatings was obtained by changing the dosage ratio of reactive gases or by changing the intensity of target evaporation current on arc sources. During the PVD process of coatings deposition, the substrates made of casting magnesium alloys were displaced relative to the sources of vapours, by performing rotary movements in order to obtain a uniform thickness of coatings and thus to counteract the generation of the so-called shadow effect on the deposited surfaces.

The plasma-assisted chemical vapour deposition process (PACVD), in which a relatively low surface treatment temperature can be achieved, was used for producing DLC carbon coatings at the set pressure and in the atmosphere of C_2H_2 acetylene and silicon, whose variable concentration in the central coating constituted the applied gradient.

TABLE 5.10
Types, Conditions and Techniques of Coatings Deposition onto the Substrate of Mg-Al-Zn-Mn Alloys

	Type and Technique of Coatings Deposition					
	PVD					PACVD
Process Conditions	**Ti/Ti(C, N)-gradient/CrN**	**Ti/Ti(C, N)-gradient/ (Ti, Al)N**	**Cr/CrN-gradient/CrN**	**Cr/CrN-gradient/TiN**	**Ti/(Ti, Si) N-gradient/(Ti, Si)N**	**Ti/DLC/ DLC**
Base pressure, Pa	5×10^{-3}	5×10^{-3}	5×10^{-3}	5×10^{-3}	5×10^{-3}	1×10^{-3}
Working pressure, Pa	0.9/1.1 to 1.9/2.2	0.9/1.1 to 1.9/2.8	1.0/1.4 to 2.3/2.2	1.0/1.4 to 2.3/2.2	0.89/1.5 to 2.9/2.9	2
Flow of argon, cm³/min	80[a]	80[a]	80[a]	80[a]	80[a]	80[a]
	10[b]	10[b]	80[b]	80[b]	20[b]	–
	10[c]	10[c]	20[c]	20[c]	20[c]	–
Flow of nitride, cm³/min	225→0[b]	0→225[b]	0→250[b]	0→250[b]	0→300[b]	–
	250[c]	350[c]	250[c]	250[c]	–	–
Flow of acetylene, cm³/min	0→170[b]	140→0[b]	–	–	–	230
Voltage on the substrate, V	70[a]	70[a]	60[a]	60[a]	70[a]	500
	70[b]	70[b]	60[b]	60[b]	100[b]	
	60[c]	70[c]	60[c]	100[c]	100[c]	
Current intensity in cathode, A	60	60	60	60	60	–
Process temperature, °C	<150	<150	<150	<150	<150	<180

[a] During deposition of metallic layer.
[b] During deposition of graded layer.
[c] During deposition of ceramic layer.

5.5.2 The Structure of Coatings Deposited by PVD CAE and PACVD Methods onto the Substrate of Mg-Al-Zn-Mn Alloys

A qualitative phase composition analysis carried out with the X-ray diffraction method by the Bragg-Brentano technique has confirmed the correctness of the produced TiN, (Ti, Al)N, (Ti, Si)N, Ti(C, N), CrN coatings (Figure 5.37). Due to the overlapping reflexes of the substrate material and of the coating, a relatively small thickness of individual layers of up to 3.5 om, as well as the identical values of Miller's indicators (hkl) for Ti(C, N) and (Ti, Al)N coatings, it was difficult to identify the individual phases. Reflexes coming from the phases existing in the substrate were also identified, i.e. a solid solution of α-Mg with precipitates of the secondary phase γ-$Mg_{17}Al_{12}$c-Mg. The insufficient volumetric fraction of other phases existing in the substrate material does not allow for their unequivocal identification on the X-ray diffraction patterns made. The presence of reflexes from the substrate was determined for all diffraction patterns, which is due to the thickness of coatings, smaller than the penetration depth of X-ray beams inside the material. The grazing incidence X-ray diffraction (GIXD) method for the primary X-ray beam was employed in the further investigations to obtain more accurate information from the layer of surface coatings. Reflexes from the thin surface layers were only registered for different incidence angles of the primary beam. For example, portions of respective diffraction patterns are provided as d-e) in Figure 5.37.

The tests of chemical composition made with Glow Discharge Optical Emission Spectrometry (GDOES) confirm that chemical elements are present in the coatings produced forming part of the analysed layers (Figure 5.38). The character of changes in the joint zone, i.e. a higher concentration of elements forming part of the substrate, with an accompanying abrupt reduction in the concentration of elements forming the coatings, may signify that a transition layer exists between the substrate material and the coating, improving the adhesion of the deposited coatings to the substrate, despite the fact that the results cannot be interpreted unequivocally as such, due to the inhomogeneous evaporation of the material from the specimens' surface. Moreover, the existence of a zone with linearly changing contents of elements forming part of the investigated coatings was confirmed with an optical spectrometer, which indicates their gradient character (Figure 5.38).

The surface morphology of the coatings produced with the PVD CAE technique is characterised by inhomogeneity because numerous droplet-shaped particles are present in the structure, the presence of which, confirmed by images from an electron scanning microscope (Figure 5.39), results from the essence of the cathodic arc evaporation process. The highest surface inhomogeneity as compared to the surface of the other examined coatings, is exhibited by Ti/Ti(C, N)/(Ti, Al)N and Ti/Ti(C, N)/CrN coatings, in which numerous solidified droplets were found. The individual droplets are varied according to their size and shape, depending on the conditions of the process and type of the applied sources of metal vapours. Depressions are also formed, created by solidified droplets being ejected. In case of the DLC coating obtained in the PACVD process on the surface, small droplets also exist with the spheroidal shape (Figure 5.40). The results of examinations

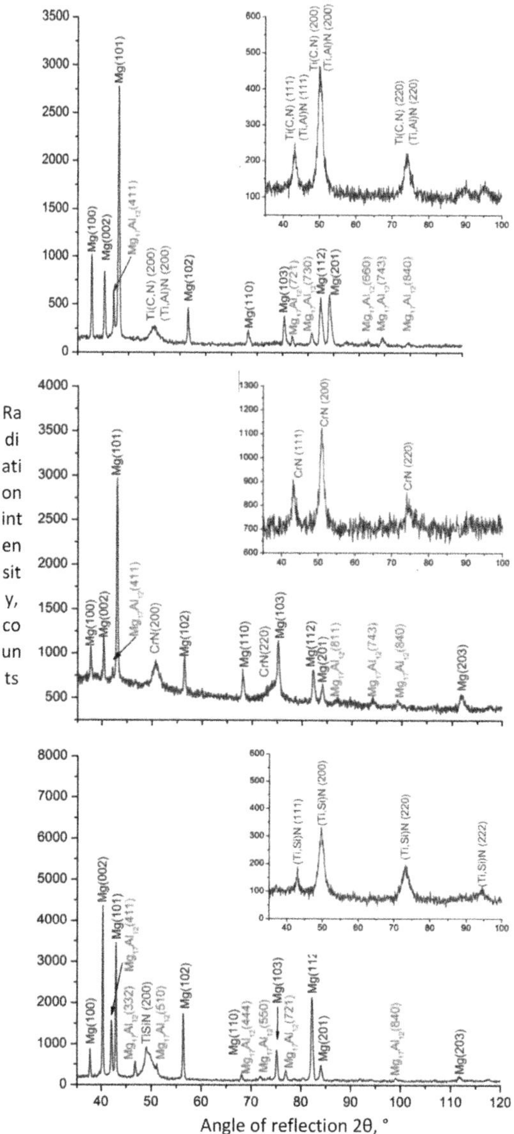

FIGURE 5.37 X-ray diffraction patterns of Ti/Ti(C, N)/(Ti, Al)N (a, d), Cr/CrN/CrN (b, e) and Ti/(Ti, Si)N/(Ti, Si)N (c, f) coatings deposited onto MCMgAl9Zn1 (a, d), MCMgAl6Zn1 (b, e) and MCMgAl12Zn1 (c, f) casting magnesium alloys obtained with the Bragg-Brentano technique (a–c) and made with the grazing incidence X-ray diffraction method c = 4° (d–f).

carried out with the use of an EDS X-ray scattered radiation energy spectrometer indicate that carbon is contained in the droplets existing on the surface in ~95% atomically (Figure 5.40). A lattice of microcracks, waved surfaces or surfaces with a circular shape, distinct for other classical high-temperature CVD processes, does not exist on the surface of the DLC coating.

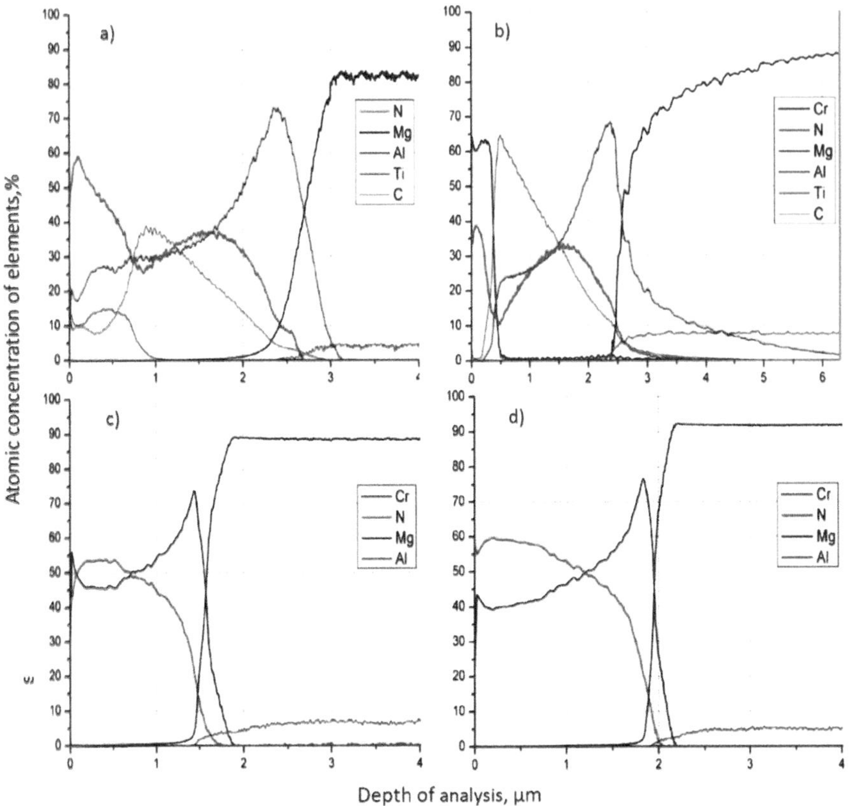

FIGURE 5.38 Changes in a concentration of components of the coatings deposited onto the substrate of the examined Mg-Al-Zn-Mn casting alloys: (a) Ti/Ti(C, N)/(Ti, Al)N on the substrate of MCMgAl9Zn1 alloy; (b) Ti/Ti(C, N)/CrN on the substrate of MCMgAl9Zn1 alloy; (c) Cr/CrN/CrN on the substrate of MCMgAl9Zn1 alloy; (d) Cr/CrN/CrN on the substrate of MCMgAl6Zn1 alloy.

The fractographic tests of fractures in a scanning electron microscope indicate that the coatings deposited onto the substrate of the examined Mg-Al-Zn-Mn alloys have a compact structure, without clear stratifications and defects, and are deposited uniformly and tightly adhere to the substrate (Figure 5.41). Ti/Ti(C, N)/(Ti, Al)N and Ti/Ti(C, N)/CrN coatings have a layered structure, with a clearly marked transition zone between a graded coating and an antiwear coating, achieved as a result of using separate sources of metal vapours for each of the layers (Figure 5.42). In the case of Cr/CrN/CrN i Ti/(Ti, Si)N/(Ti, Si)N coatings, no visible differences were identified on the cross section (Figure 5.42).

At the fractures to the examined coatings, their thickness was measured with a scanning electron microscope (Figure 5.42). Such measurements, in the different places of fractures, confirmed the uniformity of the layers applied. In case of Ti/Ti(C, N)/(Ti, Al)N and Ti/Ti(C, N)/CrN coatings with a clear layered structure, it

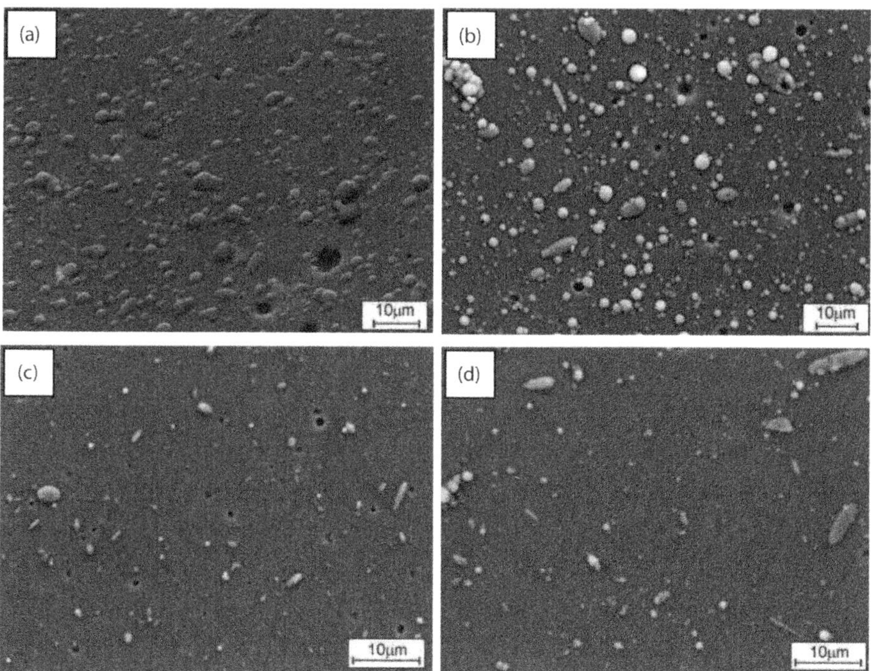

FIGURE 5.39 Surface topography of PVD coatings deposited onto the substrate of the examined Mg-Al-Zn-Mn casting alloys: (a) Ti/Ti(C, N)/(Ti, Al)N on the substrate of MCMgAl6Zn1 alloy; (b) Ti/Ti(C, N)/CrN on the substrate of MCMgAl6Zn1 alloy; (c) Cr/CrN/CrN on the substrate of MCMgAl9Zn1 alloy; (d) Ti/(Ti, Si)N/(Ti, Si)N on the substrate of MCMgAl12Zn1 alloy.

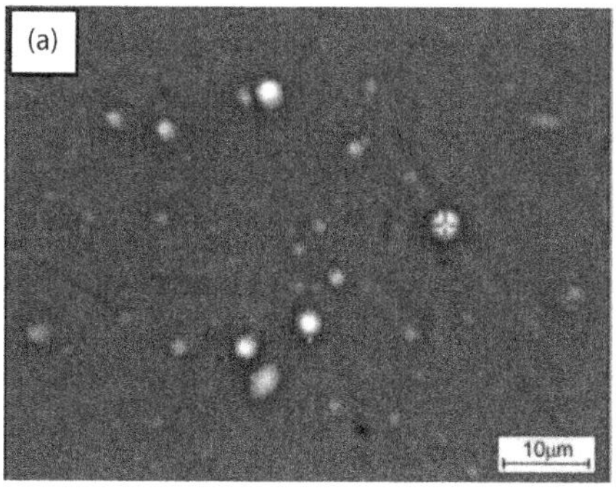

(a)

Chemical element	Concentration of the elements in selected point, %	
	mass	atomic
C	84,11	95,21
Mg	4,38	2,11
Ti	10,31	2,68

FIGURE 5.40 Surface topography of the Ti/DLC/DLC coating deposited onto the substrate made of the MCMgAl12Zn1 casting alloy (a), the results of the quantitative analysis of chemical composition of the selected solidified droplet on the coating surface in the place indicated in Figure (a).

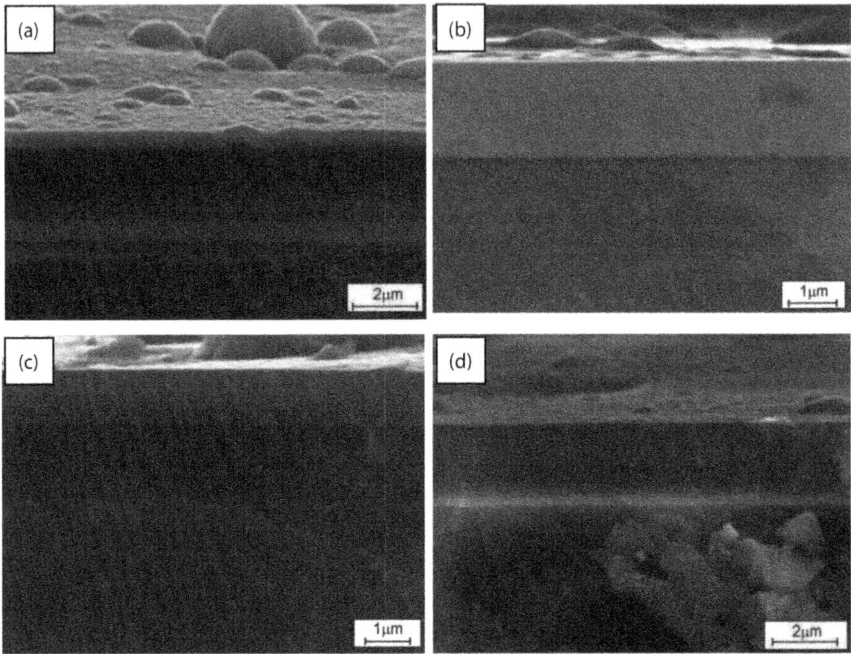

FIGURE 5.41 The structure of fractures of the coatings deposited onto the substrate of the examined Mg-Al-Zn-Mn casting alloys: (a) Ti/Ti(C, N)/CrN on the substrate of MCMgAl6Zn1 alloy; (b) Cr/CrN/CrN on the substrate of MCMgAl6Zn1 alloy; (c) Ti/(Ti, Si)N/(Ti, Si)N on the substrate of MCMgAl9Zn1 alloy; (d) Ti/DLC/DLC on the substrate of MCMgAl9Zn1 alloy.

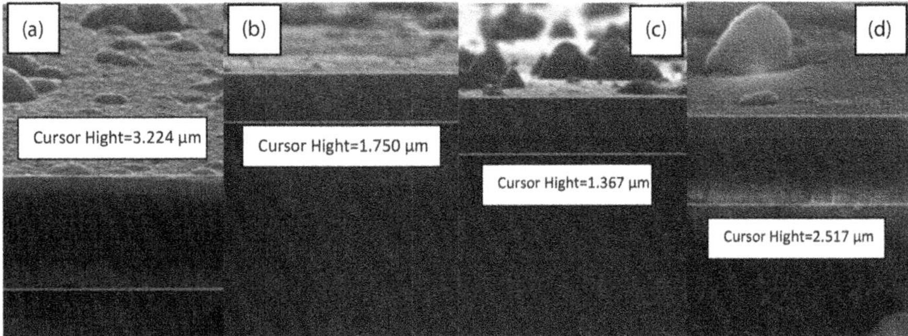

FIGURE 5.42 Measurements of the thickness of coatings deposited onto the substrate of the examined Mg-Al-Zn-Mn casting alloys: (a) Ti/Ti(C, N)/CrN on the substrate of MCMgAl6Zn1 alloy; (b) Cr/CrN/CrN on the substrate of MCMgAl3Zn1 alloy; (c) Ti/(Ti, Si)N/(Ti, Si)N on the substrate of MCMgAl9Zn1 alloy; (d) Ti/DLC/DLC on the substrate of MCMgAl6Zn1 alloy.

was possible to measure the thickness of individual layers. The graded layer thickness is not larger than 2.5 μm, whereas the top, antiwear coating has the thickness of up to approx. 0.6 μm (Figure 5.42). The thickness of other layers, i.e. Cr/CrN/CrN, Cr/CrN/TiN, Ti/(Ti, Si)N/(Ti, Si)N and Ti/DLC/DLC, does not exceed 2.6 μm and is, respectively, ~1.8; ~1.8; ~1.4; ~2.5 μm (Figure 5.42). Multilayer Ti/DLC/DLC carbon coatings fabricated with the PACVD method, given the similarity of the phase composition of the graded and antiwear coating, do not exhibit a clear transition zone between the particular coatings. In the case of Ti/DLC/DLC and Cr/CrN/CrN coatings deposited onto the investigated Mg-Al-Zn-Mn alloys, the graded DLC and CrN layers possess a structure classified as the T zone according to Thornton's model (Figure 5.42). In the area where a thin adhesion coating exists, the purpose of which is to improve layer adhesion to the substrate, it is possible to identify a characteristic, light, continuous titanium layer, as confirmed with an EDS analysis (Figure 5.43).

No clear differences were identified for the structure of the coatings deposited onto the investigated Mg-Al-Zn-Mn alloys depending on the applied substrates of the examined Mg-Al-Zn-Mn alloys, differing mainly in the concentration of 3% to 12% Al. An EDS analysis confirms in all the examined cases the presence of the main elements, i.e. Mg, Al, Zn, Ti, Cr, C, N, Si, in the composition of the examined Mg-Al-Zn-Mn casting alloys, as well as in the deposited coatings (Figure 5.43). Information about the mass and atomic concentration of elements in the microareas of the matrix and deposited coatings, tested locally, should only be regarded as estimated, as they involve systematic errors, in relation to the measurements of a concentration of elements, so-called Light elements with energy <1 keV (C, N) due to strong absorption. A measuring error for the mass concentration in the range of 5% to 20% is approximately 4%, and above 20%, the error is 2%. Moreover, the value of particular structural components is often smaller than the diameter of the analysing beam, the result obtained is averaged for this component and for the partly excited matrix. A qualitative analysis of the surface distribution of elements, carried

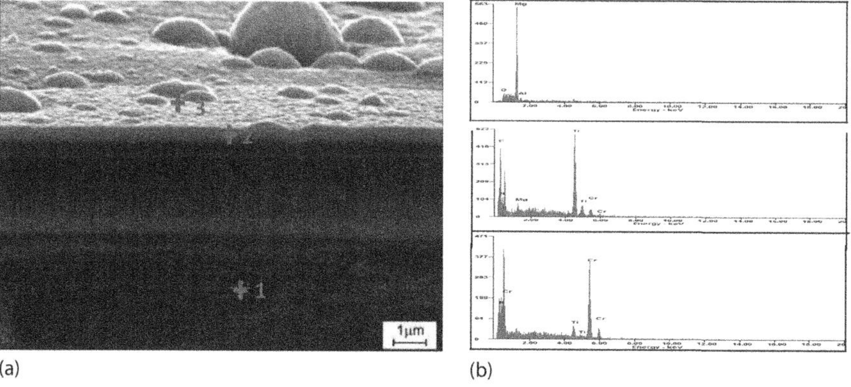

FIGURE 5.43 The structure of fracture of the Ti/Ti(C, N)/CrN coating deposited onto the substrate made of the MCMgAl6Zn1 casting alloy: (a) and intensity charts as a function of X-ray energy (b); (1) analysis (1, 2) analysis (2, 3) analysis 3, from the points indicated in Figure (a).

out on the cross section of the examined alloys with the deposited coatings, clearly confirms, however, a higher concentration of elements on the following boundaries: graded/multi-component boundary, of the coatings being constituted.

The results of diffraction tests using a high-resolution transmission electron microscope confirm that TiN and CrN phases (Figure 5.44) and graphite phases (Figure 5.45) – as expected – are present, respectively, in the surface layers of Ti/Ti(C, N)/(Ti, Al)N, Ti/Ti(C, N)/CrN, Cr/CrN/CrN, Ti/(Ti, Si)N/(Ti, Si)N, Cr/CrN/TiN and Ti/DLC/DLC coatings deposited onto the substrate surface of the examined Mg-Al-Zn-Mn alloys.

For the phase TiN, its regular lattice structure of the spatial group Fm3m(225) with identicality periods of a = b = c = 0,424173 nm was confirmed. The isomorphism of the TiN phase, and the possibility of substitution of atomic positions occupied by Ti by Al and/or Si, respectively in (Ti, Al)N and (Ti, Si)N, prevents the diffractive differentiation of such phases (Figure 5.46). The CrN phase also, in the Ti/Ti(C, N)/CrN and Cr/CrN/CrN coating, has a regular structure of the spatial group Fm3m(225) with the identicality periods of a = b = c = 0.414 nm. The presence of crystalling graphite in a hexagonal cell of the spatial group P63mc(186) with the lattice parameters of a = b = 0.2 nm, c = 0.679 nm, identified by diffraction, was

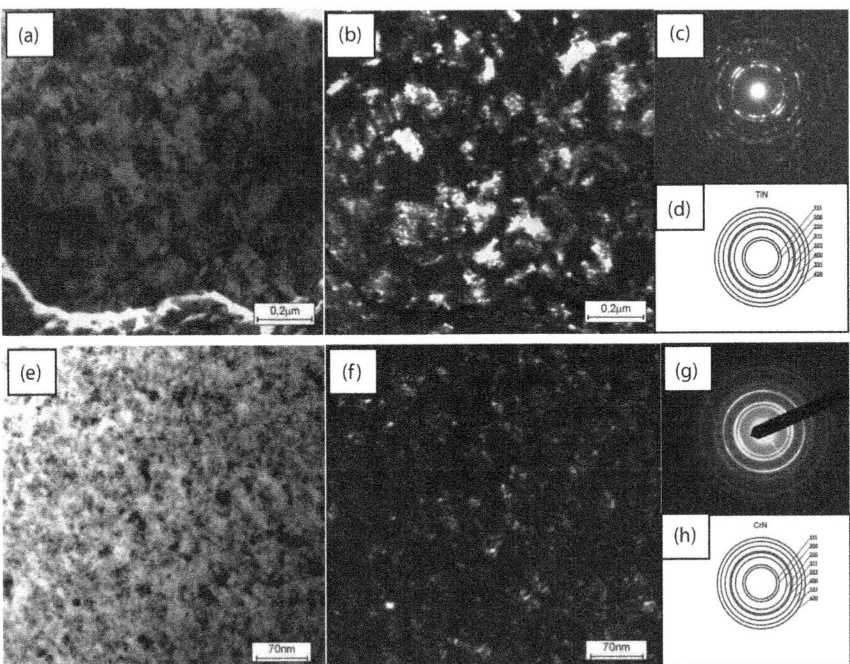

FIGURE 5.44 Structure of thin foils from the layers of, respectively, TiN (a–d) and CrN (e–h) in Cr/CrN/TiN (a–d) and Cr/CrN/CrN (e–h) coatings on the substrates of MCMgAl9Zn1 (a–d) and MCMgAl3Zn1 (e–h) casting alloys: (a) and (e) images in the bright field, (b) and (f) images in the dark field; (c) and (g) diffraction patterns from areas as, respectively, in Figure (a) and (e); (d) and (h) diffraction pattern solutions from figures, respectively, (c) and (g).

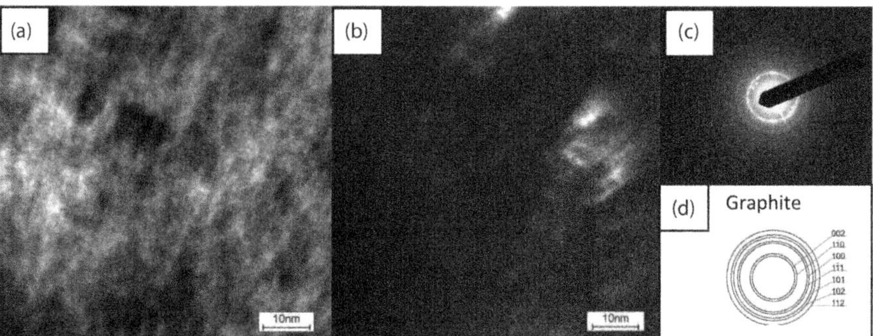

FIGURE 5.45 A structure of thin foils from a DLC layer in the Ti/DLC/DLC coating on the MCMgAl6Zn1 casting alloy substrate: (a) image in the bright field, (b) image in the dark field, (c) diffraction pattern from the area as in Figure (a) and (b).

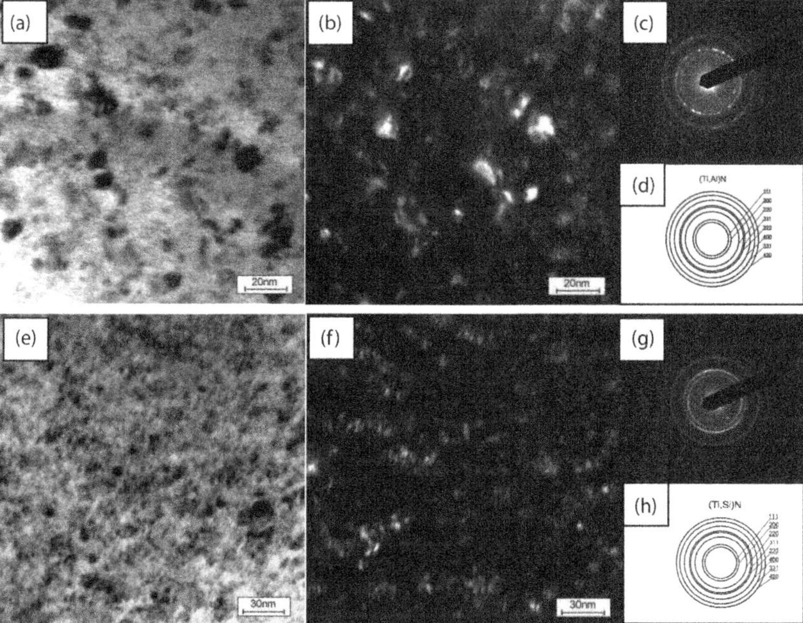

FIGURE 5.46 Structure of thin foils from the layers of, respectively, Ti(Al, N) (a–d) and Ti, Si)N (e–h) in Ti/Ti(C, N)/(Ti, Al)N (a–d) and Ti/(Ti, Si)N/(Ti, Si)N (e–h) coatings on the substrates of MCMgAl6Zn1 (a–d) and MCMgAl9Zn1 (e–h) casting alloys: (a,e) images in the bright field, (b,f) images in the dark field; (c,g) diffraction patterns from areas as, respectively, in Figures (a) and (e); (d,h) diffraction pattern solutions from figures, respectively, (c,g).

confirmed in the Ti/DLC/DLC coating fabricated in the PACVD process. In all the cases, the analysed coatings show a nanocrystalline structure (Figures 5.44 through 5.46). The size and dispersion of crystallites of the deposited coatings was determined using the dark field technique. The coatings are characterised by a compact structure with a high homogeneity of crystallites, as well as small distribution in

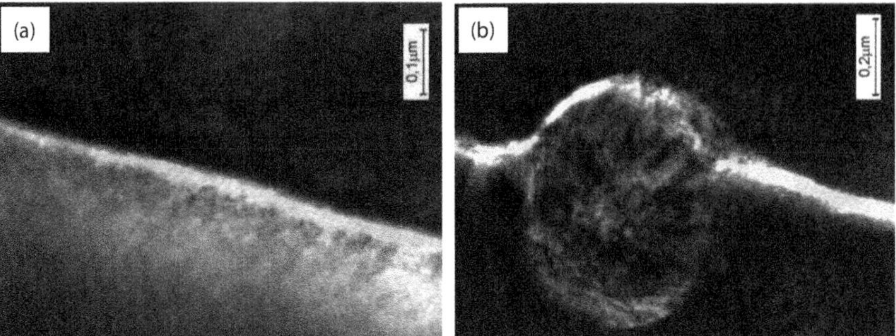

FIGURE 5.47 Structure of a thin foil from a Ti/(Ti, Si)N/(Ti, Si)N coating deposited onto the substrate made of MCMgAl9Zn1 casting alloy, bright field, TEM: (a) boundary of surface layer; (b) structure of droplet solidified in the PVD CAE process.

terms of their size in the range of 10–20 nm, for example, in the Ti(C, N)/CrN coating with the diameter of crystallites not larger than ø10 nm. The only exception is the TiN phase in the Cr/CrN/TiN coating, where the size of crystallites is approx. 200 nm. Droplets of solidified metal exist inside the coatings deposited by PVD CAE, for example such presented in the structure of thin foils in a Ti/(Ti, Si)N/(Ti, Si)N coating (Figure 5.47).

5.5.3 Mechanical and Operational Properties of Coatings Deposited by PVD CAE and PACVD Methods onto the Substrate of Mg-Al-Zn-Mn Alloys

The deposition of Ti/Ti(C, N)/CrN, Ti/Ti(C, N)/(Ti, Al)N, Cr/CrN/CrN, Cr/CrN/TiN and Ti/(Ti, Si)N/(Ti, Si)N layers with the PVD CAE method and of Ti/DLC/DLC layers with the PACVD method onto the substrate of Mg-Al-Zn-Mn alloys causes a significant increase in microhardness of the examined coated alloys versus uncoated alloys (Figure 5.48).

The measured microhardness of MCMgAl3Zn1, MCMgA16Znl, MCMgAl9Zn1 and MCMgA112Znl casting alloys not coated with the examined coatings was, respectively, 82, 98, 133 and 153 HV. Two groups of coatings can be differentiated based on the microhardness tests conducted. The first group has the microhardness of up to 2000 HV, including CrCrN/CrN, Cr/CrN/TiN and Ti/(Ti, Si)N/(Ti, Si)N, Ti/DLC/DLC coatings and the other group of Ti/Ti(C, N)/CrN and Ti/Ti(C, N)/(Ti, Al)N coatings with the measured hardness of above 2000 HV (Figure 5.48).

The Ra roughness tests of the surface of casting magnesium alloys with the deposited coatings indicate the lack of significant impact of the substrate type on the value of the surface roughness parameter, because differences in roughness parameter values depending on the type of alloy amount to, respectively, 0.05 μm (Figure 5.48). Coatings with the graded CrN layer show the smallest values of surface roughness in the range of 0.12 to 0.18 μm, when the overall range of surface roughness of the examined coatings is between 0.12 and 0.32 μm (Figure 5.48).

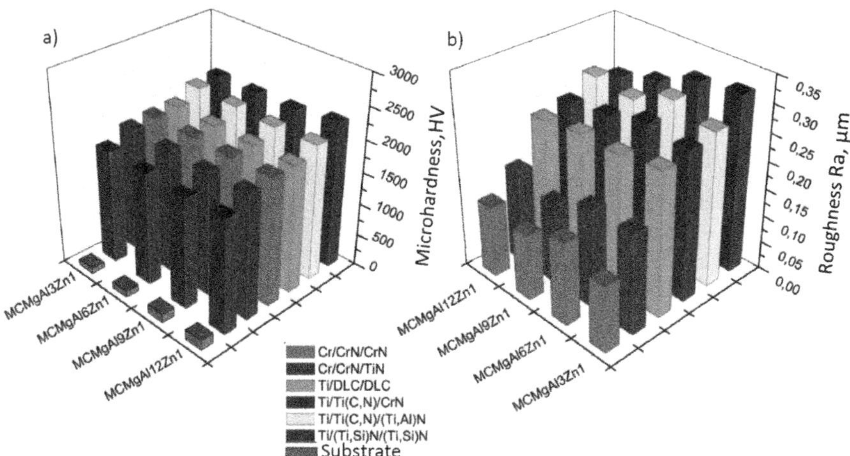

FIGURE 5.48 Results of measurements of (a) microhardness; (b) roughness of Ti/Ti(C, N)/CrN, Ti/Ti(C, N)/(Ti, Al)N, Cr/CrN/CrN, Cr/CrN/TiN and Ti/(Ti, Si)N/(Ti, Si)N coatings deposited by PVD CAE and Ti/DLC/DLC coatings deposited by PACVD onto the substrate of Mg-Al-Zn-Mn alloys.

Wear resistance tests with the ball-on-disk method of the investigated hybrid coatings on the substrate of Mg-Al-Zn-Mn alloys were carried out in dry friction conditions in a horizontal configuration of the disc rotation axis using a tungsten carbide ball as a counter-specimen. Resistance to abrasion wear and the friction coefficient of all the tested coatings on all the substrates were examined, according to the displacement rate, surface pressure, atmospheric conditions and other factors (Figures 5.49 and 5.50). The diagrams of dependency of the friction factor and/or counter-specimen displacement were recorded in the vertical axis depending on the number of disc rotations or the travelled friction path through the counter-specimen until wearing the examined coating. Both parts reveal the characteristics of the friction coefficient depending on the number of revolutions or the wear path (Figure 5.49). In the first unsteady state of friction, the friction coefficient rises suddenly as the friction path is increasing (Figure 5.49b). The second part of the diagram is close to a steady state. The cause of sudden changes in the friction coefficient are chippings on the surface of the examined coatings and counter-specimen.

The values of the friction path, until wearing appears of the tested coatings, range between 6 and 630 m (Figure 5.49). Carbon DLC coatings have the best abrasion resistance. With the load of 5 N, the average friction coefficient of DLC coatings for the slide speed of 0.05 m/s is within the range of 0.08–0.15 µm, smaller by the order of magnitude than in the case of other coatings (Figure 5.50). The cause are the lubricating properties of the graphite present in the DLC type coatings, which is favoured by an increase in temperature resulting from the essence of friction. The wear path values until wearing the DLC coatings exceed, by even 70 times, the values of this friction path for other coatings, e.g. for the Cr/CrN/CrN coating (Figure 5.49).

The smallest force at which the coating is damaged, referred to as the critical load L_C, measuring the adhesion of coatings to a substrate, was determined with the

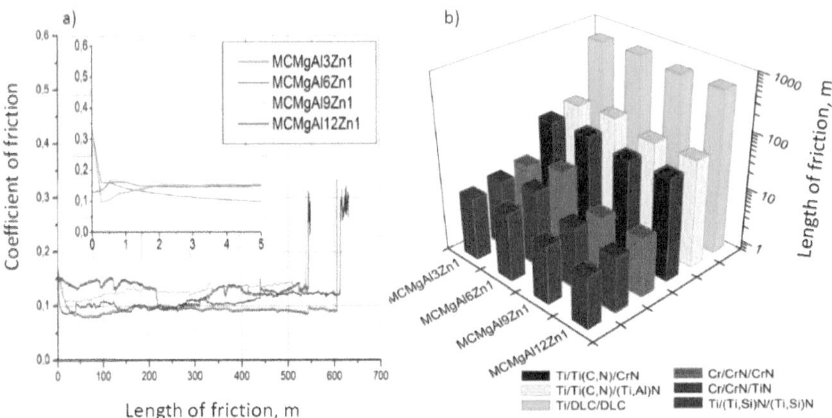

FIGURE 5.49 The results of tribological examinations with the ball-on-disk method for the investigated coatings deposited onto a substrate from all the investigated Mg-Al-Zn-Mn casting alloys: (a) characteristics of the friction coefficient depending on the friction path of Ti/DLC/DLC coatings deposited onto all the examined Mg-Al-Zn-Mn alloys; (b) comparison of the friction path until wearing appears of all the tested coatings deposited onto all the investigated Mg-Al-Zn-Mn casting alloys.

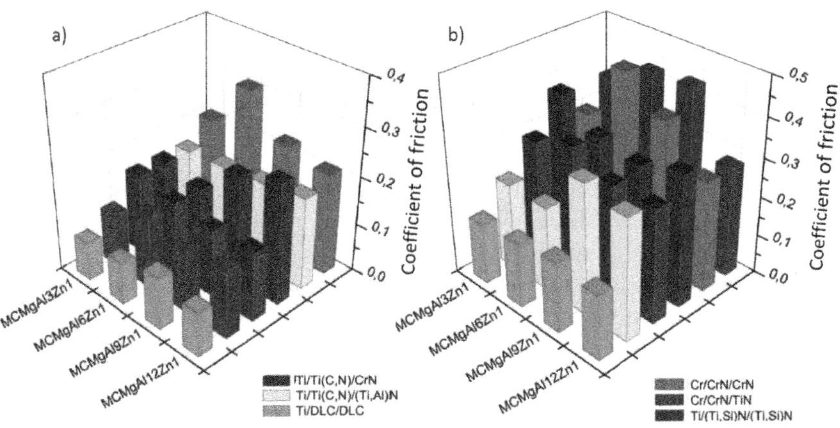

FIGURE 5.50 The results of friction coefficient measurements in an abrasion resistance test with the ball-on-disk method of the tested coatings deposited onto Mg-Al-Zn-Mn casting alloys: (a) minimum; (b) maximum.

scratch-test method (Figures 5.51 and 5.52). The coefficients L_{C1} and L_{C2} were determined based on variations in the acoustic emission values registered during a measurement, created at the indenter – tested specimen interface and by measuring the friction force of a diamond indenter (Figure 5.51) and through metallographic observations in a light microscope connected to a measuring instrument (Figure 5.52). The critical load L_{C1} is recorded on the chart of dependency of the friction force and

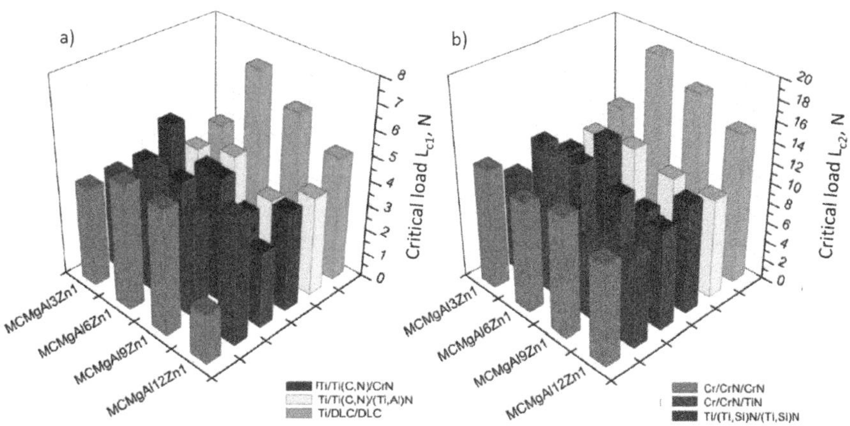

FIGURE 5.51 The results of critical load measurements of the examined coatings deposited onto Mg-Al-Zn-Mn casting alloys: (a) L_{C1}; (b) L_{C2}.

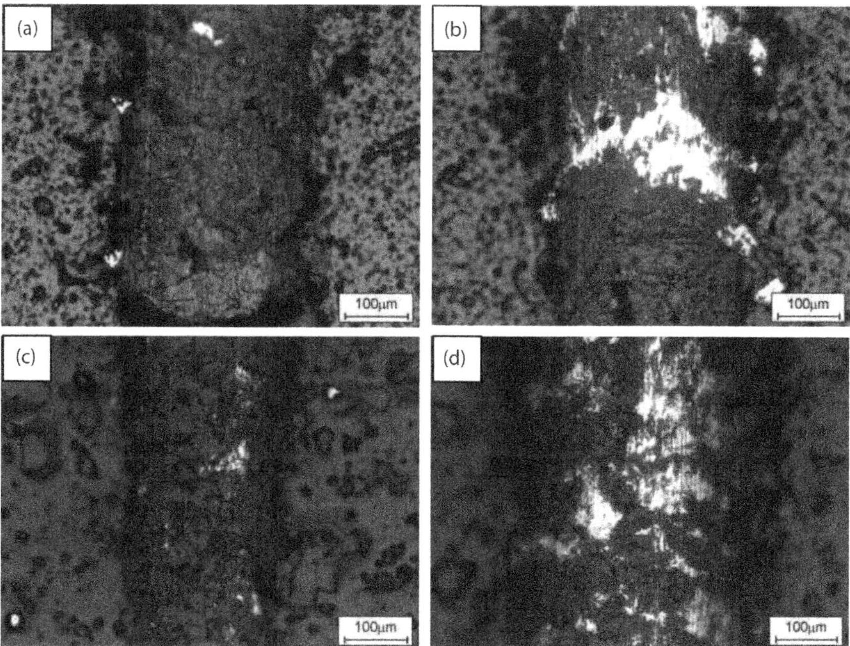

FIGURE 5.52 The examples of the scratch mark with a diamond indenter of the surface of selected coatings on the substrate of Mg-Al-Zn-Mn casting alloys with the scratch test method for the critical load of: (a,c) LC1, (b,d) LC2: (a,b) Ti/Ti(C, N)/(Ti, Al)N coatings on the substrate of MCMgAl6Zn1 alloy; (c,d) Ti/DLC/DLC coatings on the substrate of MCMgAl6Zn1 alloy.

acoustic emission from the load, as the first small stroke of the acoustic emission signal. In contrast, the critical load L_{C2} refers to the point at which the delamination of the coating occurs, and where cracks, chippings and deliminations appear outside and inside the scratch path along with exposure of the substrate material and where a sound signal is amplified. The highest values of the critical load L_{C1} and L_{C2} are, respectively, 7 and 19 N, therefore, the best coating adhesion to the substrate was achieved for a Ti/DLC/DLC coating produced on an MCMgAl9Zn1 substrate. The other critical load values measured, signifying coating adhesion to the substrate, do not exceed 14 N (Figure 5.51).

The pitting corrosion resistance of the investigated coatings for the tested Mg-Al-Zn-Mn casting alloys was evaluated by recording anode polarisation curves with the potentiodynamic method in 1 M water NaCl solution (Figure 5.53). The characteristic values describing pitting corrosion resistance were determined based on the registered anode polarisation curves,

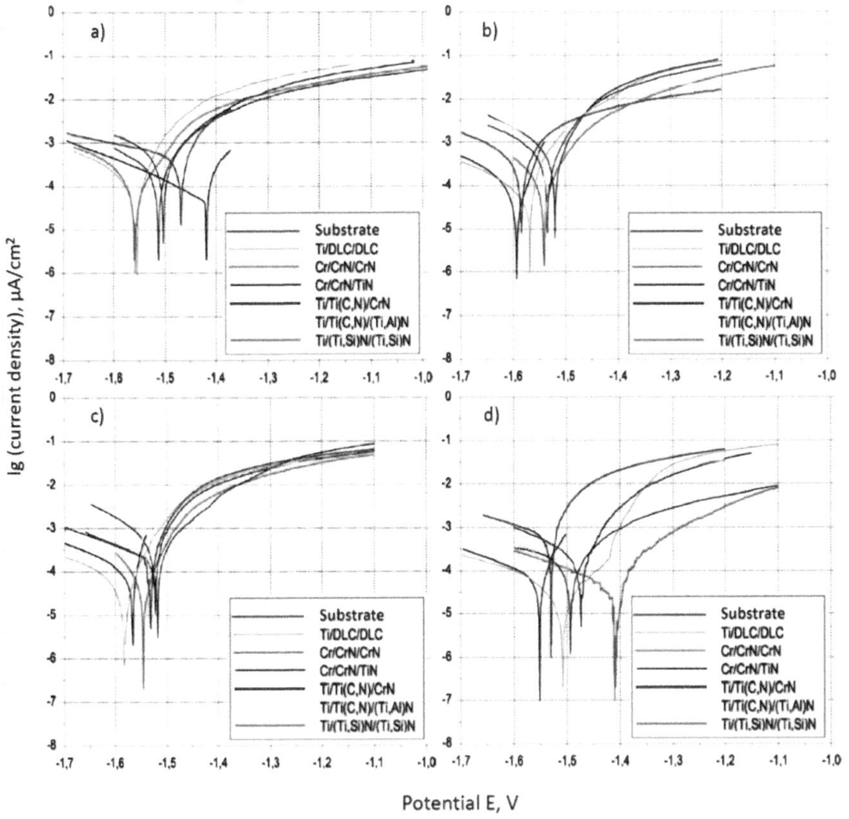

FIGURE 5.53 The charts of anode polarisation of the tested coatings deposited onto substrates of Mg-Al-Zn-Mn casting alloys: (a) MCMgAl3Zn1; (b) MCMgAl6Zn1; (c) MCMgAl9Zn1; (d) MCMgAl12Zn1.

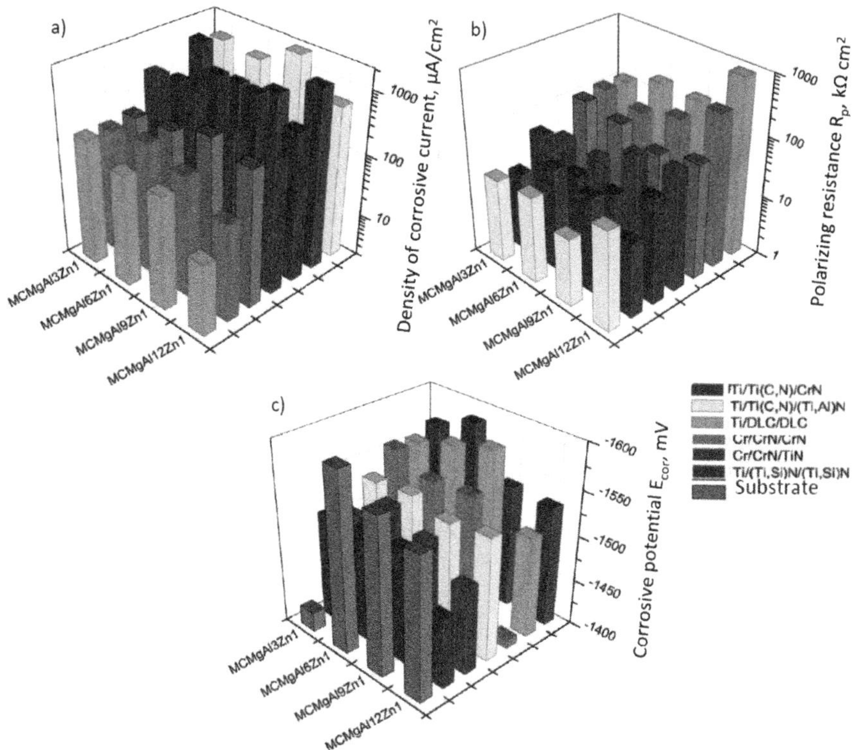

FIGURE 5.54 The results of measurements of the values characterising the course of corrosion using the electrochemical potentiodynamic method of the investigated coatings deposited onto substrates of Mg-Al-Zn-Mn casting alloys: (a) corrosion current density i_{kor} ($\mu A/cm^2$); (b) polarisation resistance R_p ($k\Omega/cm^2$); (c) corrosion potential E_{kor} (mV).

i.e.: corrosion potential E_{kor} (mV), polarisation resistance R_p (Ωcm^2), corrosion current density i_{kor} ($\mu A/cm^2$) (Figure 5.54).

It was found that there is no effect of repassivation in the passive range and the coatings applied are subject to pitting corrosion in the examined environment.

Ti/DLC/DLC and Cr/CrN/CrN coatings have the best electrochemical properties, for which the lowest corrosion current density values were measured (Figure 5.54).

Open pits with different shapes – from semisphere-shaped to roll-shaped ones – were identified in the structure of the examined coatings and in a substrate of Mg-Al-Zn-Mn casting alloys after the corrosion test on the basis of metallographic tests, depending on the applied substrate, coating, corrosion environment and polarisation conditions, which are least visible for the investigated Cr/CrN/CrN and Ti/DLC/DLC coatings (Figure 5.55). Corrosion products created when the tested material dissolves exist in the bottom of the pits.

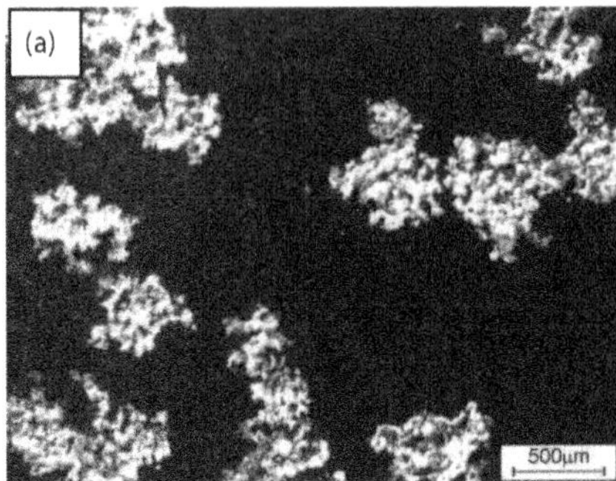

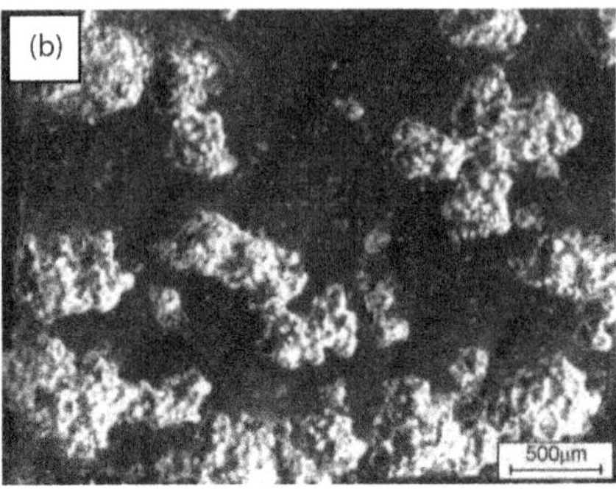

FIGURE 5.55 The surface structure of examples of the selected coatings on the substrate of Mg-Al-Zn-Mn casting alloys after a corrosion test: (a) Cr/CrN/CrN coatings on the substrate of MCMgAl9Zn1 alloy; (b) Ti/DLC/DLC coatings on the substrate of MCMgAl6Zn1 alloy.

5.6 LASER SURFACE TREATMENT OF INVESTIGATED Mg-Al-Zn-Mn ALLOYS

5.6.1 General Characteristics of Laser Surface Treatment of the Investigated Mg-Al-Zn-Mn Alloys

In this work, laser surface treatment was performed with a high-power diode laser (HPDL) in relation to the examined Mg-Al-Zn-Mn alloys, and encompassed the laser imbed into remelted surface zone of each of investigated alloys - the carbides of, respectively, TiC, WC, VC, NbC, SiC, TaC and Al_2O_3 oxide, listed in Table 5.11.

The laser treatment of Mg-Al-Zn-Mn casting alloys, consisting in the laser imbed of hard ceramic particles into remelted surface zone, was performed by the technique of feeding the powder continuously to the area of the pool remelted with a laser, by dosing the powder using a fluidising feeder, supplied with inert gas transporting the

TABLE 5.11
Characteristics of Particles of the Phases Applied for Laser Imbed of Hard Ceramic Particles into Remelted Surface Zone of the Investigated Mg-Al-Zn-Mn Alloys

		Types of Particles of the Phases Applied for Laser Imbed of Hard Ceramic Particles into Remelted Surface Zone of the Investigated Mg-Al-Zn-Mn Alloys into the Surface						
Property		**WC**	**TiC**	**VC**	**NbC**	**SiC**	**TaC**	**Al_2O_3**
Density, kg/m^3		15.69	4.25	5.36	7.6	3.44	15.03	3.97
Hardness, HV		3400	1550	2850	2100	1600	1725	2300
Melting point, °C		2870	3140	2830	3500	1900	3880	2047
Grain size, μm	min.	0.7–0.9	<1.0		<10	<10	<10	1–5
	max.	>5	>6.4	>1.8	<45	<75	<45	80

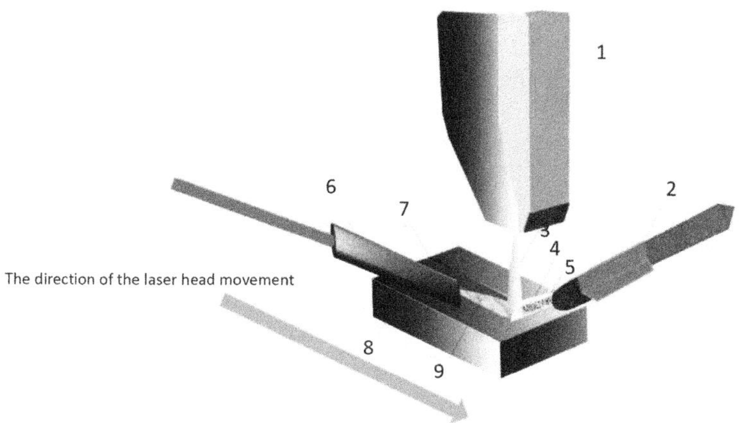

FIGURE 5.56 Diagram of laser imbed of hard ceramic particles into remelted surface zone of Mg alloys: 1 – laser head, 2 – nozzle transporting shield gas and powder, 3 – laser beam, 4 – gas, 5 – powder, 6 – nozzle with gas, 7 – remelted area, 8 – material enhanced on the surface, 9 – blow of shielding gas.

powder of the melt injected particles with the rate of 5 l/min and with a nozzle supplying shield gas (Figure 5.56).

The laser imbed of hard ceramic particles into remelted surface zone process was carried out using a continuous-mode operation laser, because the reinforcing material is fed to the remelted zone during laser heating only. The reinforcing material is only slightly dissolved in the matrix of the quasi-composite layer or is completely insoluble. The surface obtained by laser imbed of hard ceramic particles into remelted surface zone is relatively smooth, and by obtaining dimensions similar to final dimensions, finishing after such laser treatment is usually limited to a minimum. The investigations presented in this chapter were performed using an HPDL laser, Rofin DL 020, with the specifications given in Table 5.12. HPDL lasers consist of many single diode emitters with GaAs, where so-called diode bars with the section of 0.6 × 0.115 mm and length of 11 mm are provided [244–246]. Diode bars are assembled into packets, which are then mounted into a copper body intensively cooled with water. Standard diode packets with the dimensions of 182 × 130 × 272 mm provide the laser power of up to 2.0 kW. GaAs doped with Al, In or P is the semiconducting material in high-performance diodes.

TABLE 5.12
Technical Data of HPDL Rofin DL 020 Laser

Length of laser radiation waves, nm	808–940
Smooth adjustable range of power, W	100–2000
Focal length of laser beam, mm	82
Power density range in laser beam focal plane, kW/cm²	0.8–36.5
Focal point dimensions of laser beam, mm	1.8 × 6.8

After initial tests, laser power was selected in the range of 1.2–2.0 kW and the laser imbed of hard ceramic particles into remelted surface zone rate of 0.25; 0.50; 0.75; 1.0 m/min. The optimum geometric features of a single laser path were achieved with the laser imbed of hard ceramic particles into remelted surface zone rate of 0.75 m/min. with the exception of Al_2O_3 and NbC powders, for which the optimum laser imbed of hard ceramic particles into remelted surface zone rate is, respectively, 0.50 m/min and 0.25 m/min. Laser imbed of hard ceramic particles into remelted surface zone was performed in an argon shield to protect the substrate against oxidisation. The alloy subjected to laser imbed of hard ceramic particles into remelted surface zone was in a shield of a shielding gas blown from two nozzles, one directed axially to the material processed by the laser, and the other directed perpendicular to the pool area (Figure 5.56). The flow rate of the shield gas (Argon 5.0) was 12 L/min. Nozzle distance to the processed material does not exceed 20 mm. One or two paths after laser imbed of hard ceramic particles into remelted surface zone were made on one surface of rectangular specimens of the tested materials with varied laser power and varied rate of laser imbed of hard ceramic particles into remelted surface zone rate. The surface structure of the investigated Mg-Al-Zn-Mn alloys subjected to laser treatment by laser imbed of hard reinforcing particles into remelted surface zone.

A quasi-composite layer with a matrix with initially laser-remelted and then again crystallised alloy, being a substrate, and with appropriately melt injected carbides or aluminium oxide, is achieved on the surface as a result of the applied laser treatment. As a consequence of an X-ray qualitative phase analysis of surface layers of Mg-Al-Zn-Mn casting magnesium alloys subjected to laser imbed of hard ceramic particles into remelted surface zone, the presence of the α-Mg phase and the precipitates of the intermetallic phase were found, as well as, respectively, WC, TiC, NbC, VC, SiC carbides and Al_2O_3 oxide (Figure 5.57).

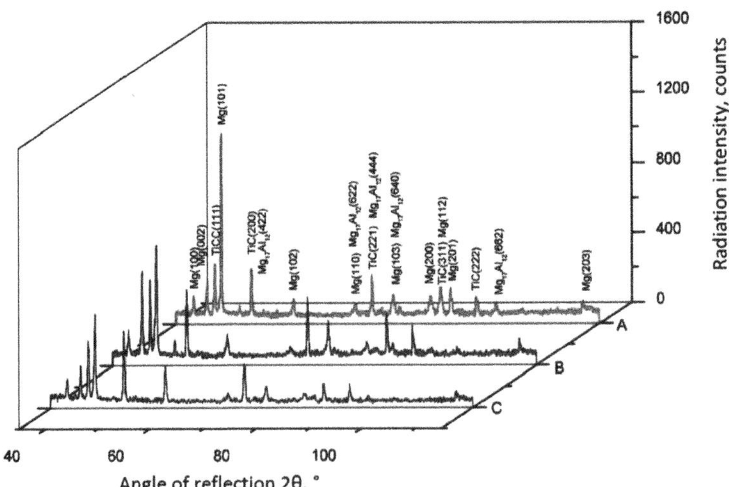

FIGURE 5.57 Example of the X-ray diffraction pattern of MCMgAl12Zn1 casting alloy after laser imbed of into remelted surface zone of TiC powder at a rate of 0.75 m/min for the laser power of, respectively: A – 1.2 kW, B – 1.6 kW, C – 2.0 kW.

The process conditions, and in particular the laser beam power and the type of powder particles used and of the substrate, influence the shape of the face and the surface topography (Figure 5.58). Depending on these conditions, the surface is non-uniformly heated by a laser, which has a direct effect on the formation of the molten material in the remelting pool. Moreover, some of the substrate material is sublimed under a high temperature, as recesses in the middle part of the remelting. After the laser imbed into remelted surface zone of TiC and WC powders, the face surface has a high regularity, without visible cracks and burrs on the sides of the bead. In the case of using the VC powder, the remelting surface is characterised by a flat shape, but with visible discontinuities in the surface layer.

The laser imbed into remelted surface zone of SiC particles means that there are clear protrusions of the remelted zone above the surface of the substrate. In the case of laser imbed into remelted surface zone of NbC carbide powders, the face has a high irregularity and visible flashes of the material on the sides of the bead. The laser imbed into remelted surface zone of Al_2O_3 powder particles causes small recesses in the central part of the bead face for the applied laser power of 1.6 and 2.0 kW.

The shape of the remelted zone (RZ) and of the heat-affected zone (HAZ) on the cross-section of the remelting bead of the investigated alloys (Figure 5.59) depends on laser treatment conditions and the type of the substrate.

The thickness of RZ and HAZ zones established based on the computer analysis of images with a light microscope are the function of four variables, namely: laser beam power, laser imbed of hard ceramic particles into remelted surface zone

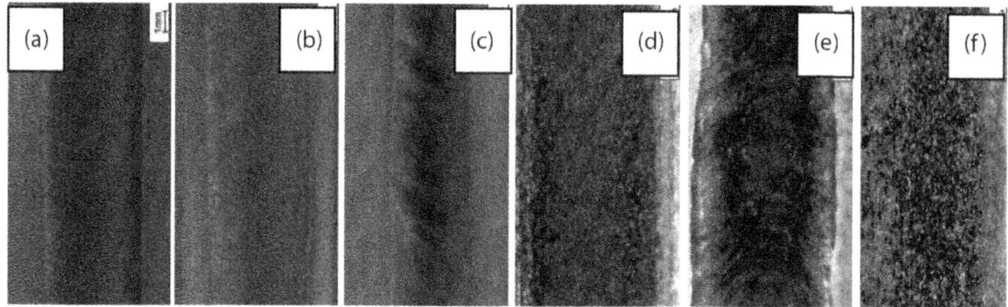

FIGURE 5.58 Examples of views of the face of remelting the surface layer of Mg-Al-Zn-Mn casting alloys after the laser imbed into remelted surface zone of the investigated powders: (a) TiC into the substrate of MCMgAl3Zn1 alloy with the laser power of 1.2 kW and the laser imbed of hard ceramic particles into remelted surface zone rate of 0.75 m/min; (b) TiC into the substrate of MCMgAl3Zn1 alloy with the laser power of 1.6 kW and the laser imbed of hard ceramic particles into remelted surface zone rate of 0.75 m/min; (c) VC into the substrate of MCMgAl3Zn1 alloy with the laser power of 1.6 kW and laser imbed of hard ceramic particles into remelted surface zone rate of 0.75 m/min; (d) SiC into the substrate of MCMgAl12Zn1 alloy with the laser power of 2.0 kW and laser imbed of hard ceramic particles into remelted surface zone rate of 0.75 m/min; (e) NbC into the substrate of MCMgAl12Zn1 alloy with the laser power of 2.0 kW and laser imbed of hard ceramic particles into remelted surface zone rate of 0.25 m/min; (f) Al_2O_3 into the substrate of MCMgAl9Zn1 alloy with the laser power of 2.0 kW and laser imbed of hard ceramic particles into remelted surface zone rate of 0.50 m/min.

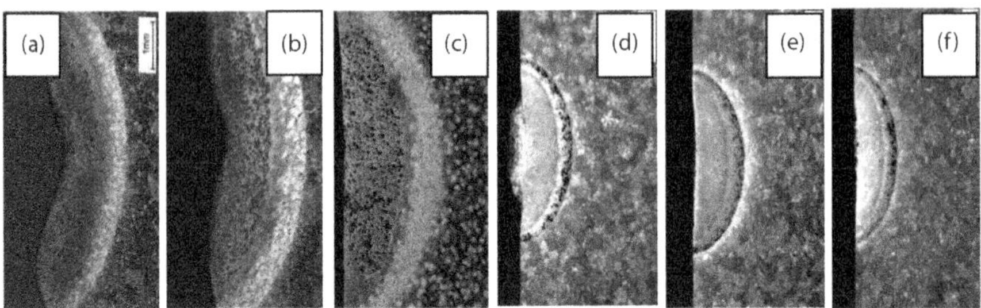

FIGURE 5.59 Examples of surface layers of Mg-Al-Zn-Mn alloys after the laser imbed into remelted surface zone of the investigated powders: (a) TiC into the substrate of MCMgAl9Zn1 alloy with the laser power of 1.2 kW and the laser imbed of hard ceramic particles into remelted surface zone rate of 0.75 m/min; (b) TiC into the substrate of MCMgAl12Zn1 alloy with the laser power of 1.6 kW and the laser imbed of hard ceramic particles into remelted surface zone rate of 0.75 m/min; (c) WC into the substrate of MCMgAl9Zn1 alloy with the laser power of 2.0 kW and the laser imbed of hard ceramic particles into remelted surface zone rate of 0.75 m/min; (d) SiC into the substrate of MCMgAl12Zn1 alloy with the laser power of 2.0 kW and the laser imbed of hard ceramic particles into remelted surface zone rate of 0.75 m/min; (e) Al_2O_3 into the substrate of MCMgAl12Zn1 alloy with the laser power of 2.0 kW and the laser imbed of hard ceramic particles into remelted surface zone rate of 0.50 m/min; (f) Al_2O_3 into the substrate of MCMgAl12Zn1 alloy with the laser power of 1.6 kW and the laser imbed of hard ceramic particles into remelted surface zone rate of 0.50 m/min.

rate, type of the melt injected powder, as well as the substrate type (Figure 5.60). Remelted zones (RZ) and heat-affected zones (HAZ) with a different thickness and shape, depending on the laser power and type of ceramic powder, are present in each surface layer after laser surface treatment of MCMgAl12Zn1 and MCMgAl9Zn1 alloys. In case where TiC, WC and VC powders are melt injected into the substrate of the MCMgAl6Zn1 alloy, a minimum heat-affected zone exists, which increases along with higher power of the laser. In case where powders are melt injected into the surface of MCMgAl3Zn1 alloy, a remelted zone is only created with a marked boundary between the remelted zone and the substrate material. If the laser power is changed at the constant remelting speed, this will increase the thickness of both zones in the surface layer. The laser power used also influences the shape and convexity of the remelted zone projecting above the substrate surface. The highest thickness of the surface layer is seen after laser imbed into remelted surface zone of SiC and NbC powders, with the laser power of 2.0 kW into the substrate of MCMgAl12Zn1 alloy and is, respectively, 3590 μm and 3950 μm. For the other powders applied, the highest achieved thickness of the surface layer for MCMgAl12Zn1 and MCMgAl9Zn1 alloys is within the range of 2340–2470 μm. MCMgAl3Zn1 alloy has the smallest thickness of the surface layer. The remelted zone in these alloys, after the laser imbed into remelted surface zone of WC, TiC and VC carbides with the laser power 1.2 kW, is within the range of 450–720 μm. Similarly, the width of the remelted face with the laser imbeded particles of carbide and Al_2O_3 oxide, depends on the applied laser power, laser imbed of hard ceramic particles into remelted surface zone

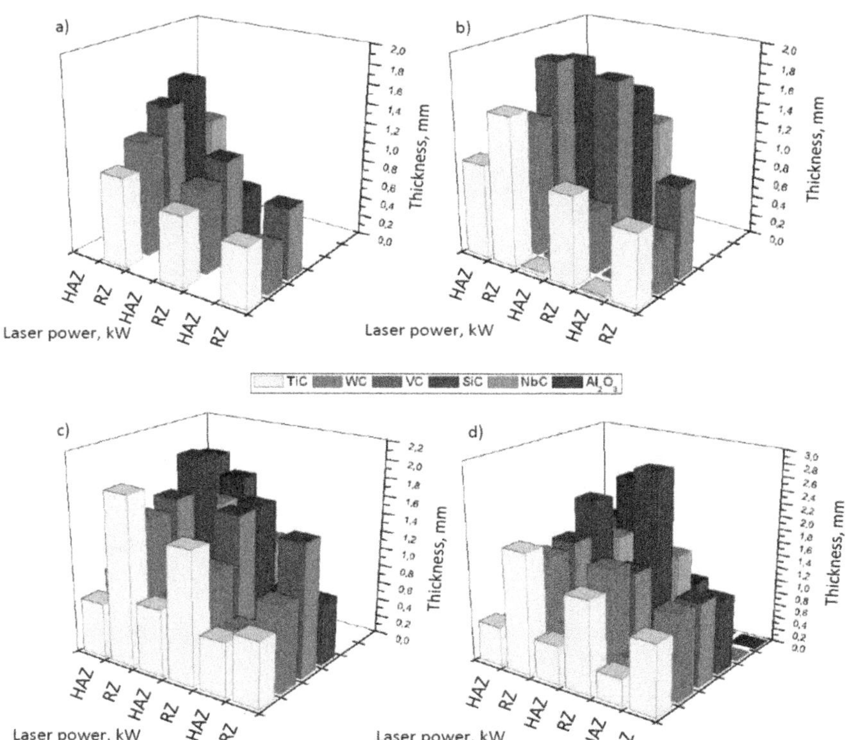

FIGURE 5.60 The effect of laser power on the thickness of remelted zone RZ and heat affected zone HAZ in the surface layer of Mg-Al-Zn-Mn casting alloys after the laser imbed into remelted surface zone of the investigated powders: (a) MCMgAl3Zn1; (b) MCMgAl6Zn1; (c) MCMgAl9Zn1; (d) MCMgAl12Zn1.

rate, as well as the powder and substrate type and assumes the values of 3540–8320 μm (Figure 5.61). The biggest width of remelting is recorded for treatment with the laser power of 2.0 kW.

The structure of the solidifying material after laser imbed of hard ceramic particles into remelted surface zone, examined with the use of a metallographic light microscope and an electron scanning microscope, is characterised by areas with a varied morphology connected with crystallisation of Mg-Al-Zn-Mn alloys (Figures 5.62 through 5.64). Multiple changes in the growth direction of crystals are seen in such areas. Small dendrites, whose main axes are oriented towards the heat evacuation direction, exist in the area on the boundary between the solid and liquid phase. The crystals are much smaller in this zone, compared to the central remelted zone, because solidification is initiated on the unsolved phases contained in the matrix and partly remelted grains of the native material. The next stages of crystals' growth are closely related to the behaviour of the privileged orientation. The growth direction of the crystals corresponds to the direction of the highest temperature gradient. As a consequence of laser imbed of hard ceramic particles into remelted surface zone, a structure is created which is free of defects, with the dispersed grain of the native material containing mainly dispersion particles of the

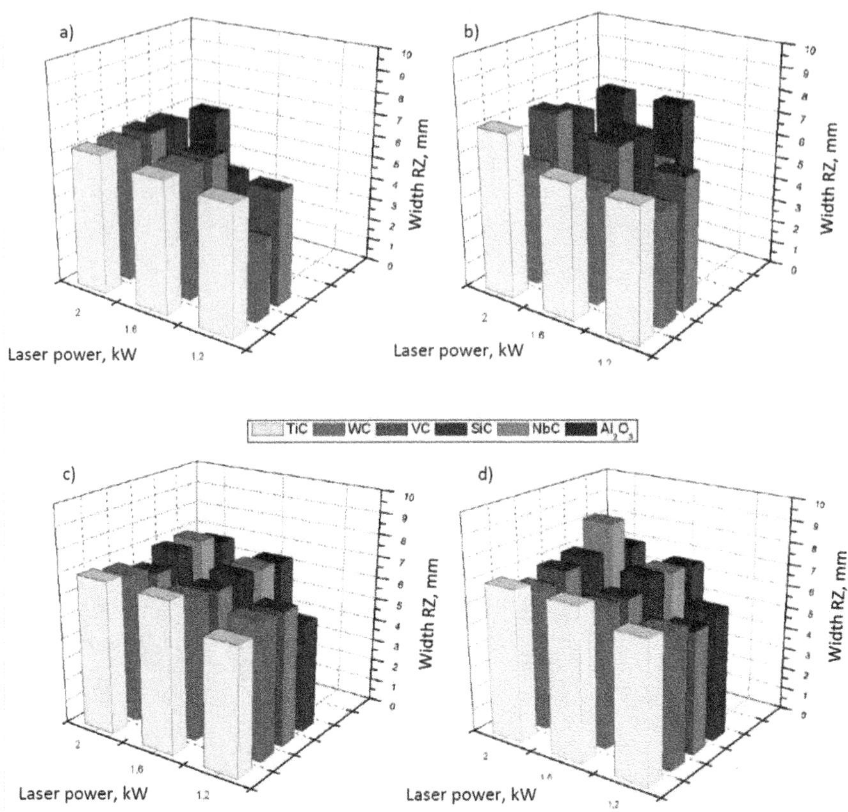

FIGURE 5.61 The effect of laser power on the width of the remelted zone of Mg-Al-Zn-Mn casting alloys after the laser of the investigated powders: (a) MCMgAl3Zn1; (b) MCMgAl6Zn1; (c) MCMgAl9Zn1; (d) MCMgAl12Zn1.

melt injected carbide, respectively, TiC, WC, VC, SiC and NbC or Al_2O_3 oxide. Laser remelting and/or laser imbed of hard ceramic particles into remelted surface zone influence the refinement of the structure in the surface layer within the entire applied range of laser power of 1.2 to 2.0 kW. The particles of TiC, WC and Al_2O_3 powders are distributed evenly throughout the remelted zone. If SiC particles are melt injected with the laser power of 1.2 kW, carbides are arranged mainly at the surface layer. For the power rate of 2.0 and 1.6 kW in MCMgAl12Zn1 and MCMgAl9Zn1 alloy, the particles are scattered throughout the remelted zone owing to the rapid stirring of the molten metal. The exception are the Mg-Al-Zn-Mn casting alloys with the laser imbeded particles of VC, whose fraction in the remelted zone is insignificant. NbC particles are not identified in the structure of the remelted zone or just a small fraction of them occurs in the produced quasi-composite layer. Moreover, small cracks in the surface layer exist in the investigated alloys treated on the surface with the use of NbC powder.

The results of the linear, point and surface X-ray qualitative and quantitative phase analysis with the use of an EDS X-ray scattered radiation energy spectrometer (Figure 5.65) on the cross section of surface layers of Mg-Al-Zn-Mn casting

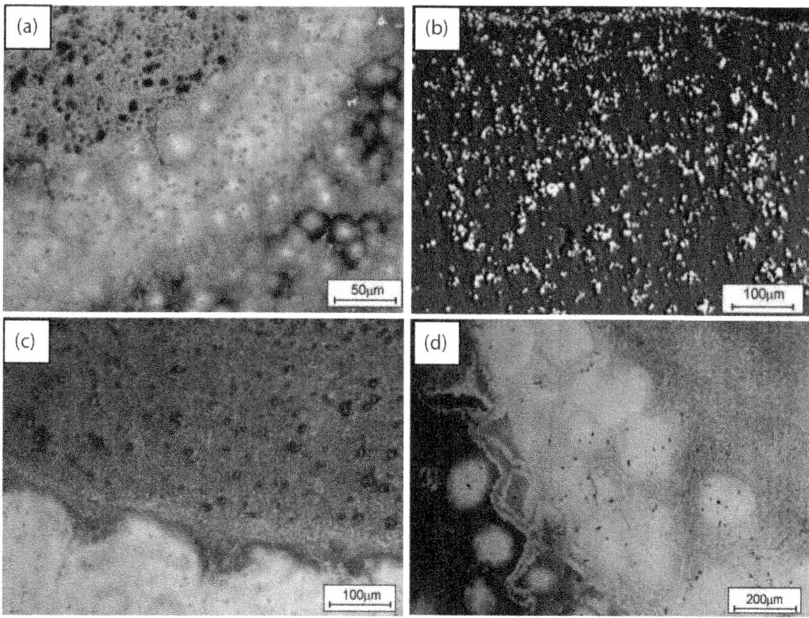

FIGURE 5.62 The examples of remelted zones of Mg-Al-Zn-Mn alloys after the laser imbed into remelted surface zone of the investigated powders: (a) WC into the substrate of MCMgAl12Zn1 alloy with the laser power of 2.0 kW and the laser imbed of hard ceramic particles into remelted surface zone rate of 0.75 m/min (LM); (b) WC into the substrate of MCMgAl6Zn1 alloy with the laser power of 1.6 kW and the laser imbed of hard ceramic particles into remelted surface zone rate of 0.75 m/min (SEM); (c) TiC into the substrate of MCMgAl6Zn1 alloy with the laser power of 1.6 kW and the laser imbed of hard ceramic particles into remelted surface zone rate of 0.75 m/min (LM); (d) Al_2O_3 into the substrate of MCMgAl9Zn1 alloy with the laser power of 1.6 kW and the laser imbed of hard ceramic particles into remelted surface zone rate of 0.50 m/min (LM).

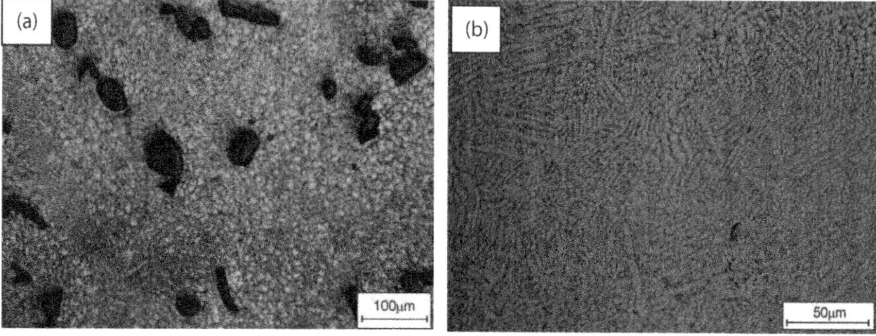

FIGURE 5.63 The examples of the central remelted zone of Mg-Al-Zn-Mn alloys after the laser imbed into remelted surface zone of the investigated powders: (a) SiC into the substrate of MCMgAl6Zn1 alloy with the laser power of 2.0 kW and the laser imbed of hard ceramic particles into remelted surface zone rate of 0.75 m/min (LM); (b) NbC into the substrate of MCMgAl6Zn1 alloy with the laser power of 2.0 kW and the laser imbed of hard ceramic particles into remelted surface zone rate of 0.25 m/min (LM).

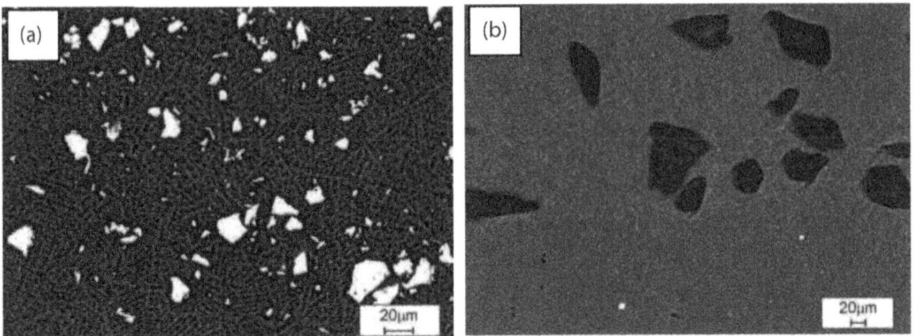

FIGURE 5.64 The examples of the central remelted zone of Mg-Al-Zn-Mn alloys after the laser imbed into remelted surface zone of the investigated powders: (a) TiC into the substrate of MCMgAl3Zn1 alloy with the laser power of 1.2 kW and the laser imbed of hard ceramic particles into remelted surface zone rate of 0.75 m/min (SEM); (b) Al_2O_3 into the substrate of MCMgAl6Zn1 alloy with the laser power of 2.0 kW and the laser imbed of hard ceramic particles into remelted surface zone rate of 0.50 m/min (SEM).

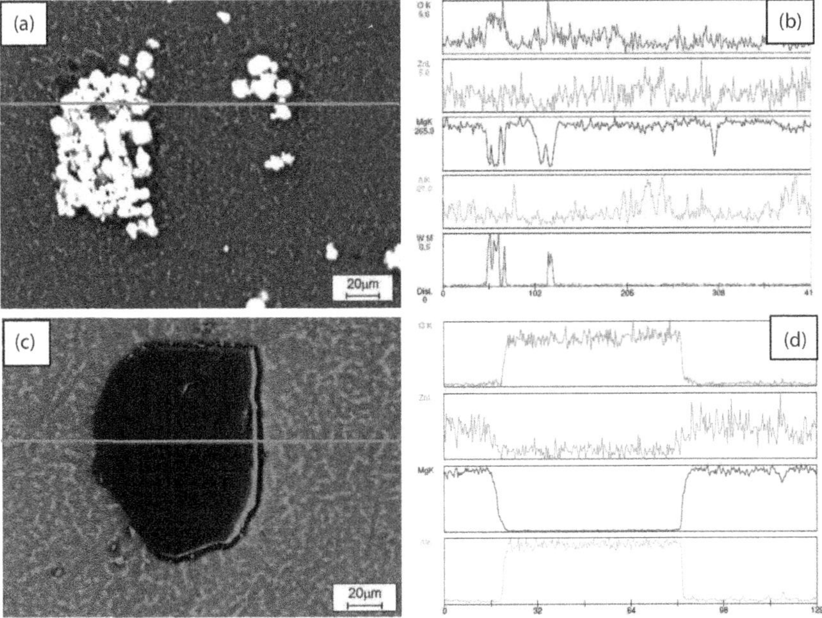

FIGURE 5.65 The example of linear distribution of elements in the central remelted zone of Mg-Al-Zn-Mn alloy after laser imbed into remelted surface zone of the investigated powders: (a) WC in MCMgAl9Zn1 alloy with the laser power of 2.0 kW and the laser imbed of hard ceramic particles into remelted surface zone rate of 0.75 m/min; (b) linear analysis of changes in chemical composition from Figure (a) acc. to the sequence from the top: O, Zn, Mg, Al, W; (c) Al_2O_3 into the substrate of MCMgAl9Zn1 alloy with the laser power of 2.0 kW and the laser imbed of hard ceramic particles into remelted surface zone rate of 0.50 m/min; (d) linear analysis of changes in chemical composition from Figure (c) acc. to the sequence from the top: O, Zn, Mg, Al: the lines along which analyses were made are marked.

magnesium alloys treated with laser with the use of TiC, WC, VC, SiC, Al_2O_3 powders confirm the existence of both, the main alloy elements contained in the substrate and of the phases existing there, i.e. Mg, Al, Zn, Mn, Si, as well as of elements introduced into the alloys in the laser imbed of hard ceramic particles into remelted surface zone process, i.e. Ti, W, V, Si, Si and O, thereby confirming the insolubility of the melt injected particles in the remelted substrate. Due to structure refinement and the size of its components smaller than the diameter of the beam of electrons, the chemical composition of particles may be averaged with the chemical composition of the matrix.

It was found based on the examinations of thin foils in an electron microscope that the structure of casting magnesium alloys with the laser imbeded particles of carbide and Al_2O_3 oxide are represented by very fine grains of the solid solution α-Mg with a hexagonal lattice from the spatial group P63/mmc, with a high density of dislocation, with precipitates of the inter-metallic phase γ-$Mg_{17}Al_{12}$ (regular network, spatial group 143 m), analogous as discussed in the state of the heat treated Mg-Al-Zn-Mn alloys in part 3.1 of this paper. The presence of phases of hard particles used for laser imbed into remelted surface zone was confirmed by diffraction, and the examples are given in Figure 5.66.

5.6.2 The Mechanical and Operational Properties of the Investigated Mg-Al-Zn-Mn Alloys after the Laser Surface Treatment

The conditions of laser imbed of hard ceramic particles into remelted surface zone, i.e. laser power and laser imbed of hard ceramic particles into remelted surface zone rate, as well as the type of the substrate and type of the melt injected powders influence the hardness of the Mg-Al-Zn-Mn casting alloys with the laser imbeded particles of carbides and Al_2O_3 oxide. The measured hardness of the obtained layers accounts for 32.4 HRF in case of MCMgAl3Zn1 alloy with the laser imbeded particles of Al_2O_3, up to 105.06 HRF for the MCMgAl3Zn1 alloy with the laser imbeded particles of WC (Figure 5.67). The largest increase in hardness was observed in case of MCMgAl3Zn1 and MCMgAl6Zn1 casting alloys with the laser imbeded ceramic particles. For MCMgAl9Zn1 and MCMgAl12Zn1 alloys, the hardness is comparable to, respectively, the hardness of each of the alloys not subjected to laser treatment, and in some laser imbed of hard ceramic particles into remelted surface zone conditions, even deteriorates insignificantly in relation to the alloys which were not laser-treated.

Triaxial charts of the effect of laser power, Al concentration, as well as the type of imbeded powder on the hardness of such alloys were created based on the prepared models of neural networks for forecasting the properties of the tested Mg-Al-Zn-Mn alloys treated on the surface (Figure 5.68). The results obtained clearly indicate that the highest hardness is shown by MCMgAl12Zn1 alloy, into the surface of which TiC and WC powders were laser melt injected with the laser power of 2.0 kW and the laser imbed of hard ceramic particles into remelted surface zone rate of 0.75 m/min. The results of forecasts correspond to the previously presented results of experimental studies, which indicates the adequacy of the developed computer model.

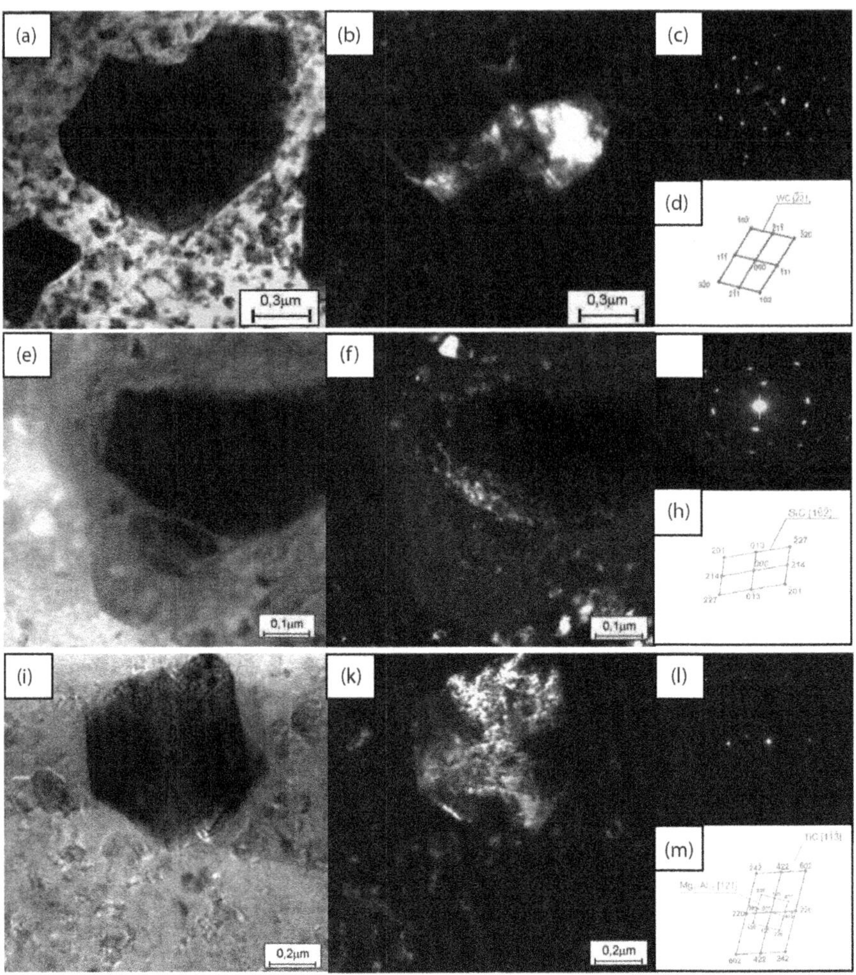

FIGURE 5.66 The structure of thin foils of Mg-Al-Zn-Mn casting alloys after the laser imbed into remelted surface zone of carbide particles: (a–d) WC into the substrate of MCMgAl6Zn1 alloy with the laser power of 2.0 kW, (a) image in the bright field, (b) image in the dark field from the reflex <320> WC; c) diffraction pattern from the area as in the Figure (a); (d) diffraction pattern solution from the Figure (c); (e–h) SiC into the substrate of MCMgAl9Zn1 alloy with the laser power of 1.6 kW; (e) image in the bright field, (b) image in the dark field from the reflex <013> SiC; (c) diffraction pattern from the area as in the Figure (e); (d) diffraction pattern solution from the Figure (g); (i–l) TiC into the substrate of MCMgAl12Zn1 alloy with the laser power of 1.6 kW, (i) image in the bright field, (b) image in the dark field from the reflex <220> TiC, (k) diffraction pattern from the area as in the Figure (i); (l) diffraction pattern solution from the Figure (k).

The microhardness of the surface layer of MCMgAl3Zn1 and MCMgAl6Zn1 casting alloys with laser melt injected particles of carbides and Al_2O_3 oxide is, on average, two times higher than the microhardness of such alloys without laser treatment performed (Figure 5.69). The native material has the hardness of >70 to 140 HV0.1, whereas the hardness of the remelted zone ranges between 80

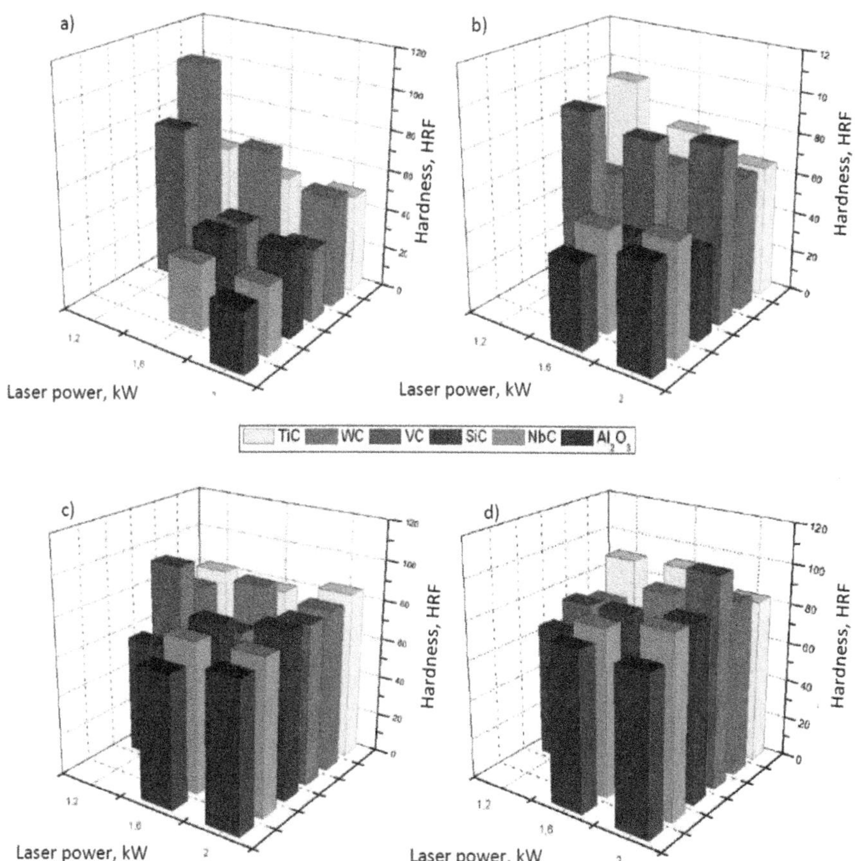

FIGURE 5.67 The effect of laser power on the hardness of the surface layer after laser imbed of the investigated powders into remelted surface zone of Mg-Al-Zn-Mn casting alloys: (a) MCMgAl3Zn1; (b) MCMgAl6Zn1; (c) MCMgAl9Zn1; (d) MCMgAl12Zn1.

and 700 HV0.1. The microhardness values show such large disparities on the cross section of the remelted zone due to fluctuation of chemical composition in the laser treated zone.

The results of measurements of roughness of Mg-Al-Zn-Mn casting alloys after the laser imbed into remelted surface zone of TiC, WC, VC, SiC, Al_2O_3 powders with the laser power in the range of 1.2–2.0 kW reveal its growth relative to roughness of the untreated surface. The roughness is in the range of Ra = 6.4–42.5 μm (Figure 5.70). For each substrate, regardless the concentration of Al in the alloy, the highest roughness exists after laser imbed of hard ceramic particles into remelted surface zone with the laser power of 2.0 kW with the rate of 0.5 m/min. For the constant rate of laser imbed of hard ceramic particles into remelted surface zone and unchangeable intensity of powder feeding, surface roughness decreases along with higher laser power. The smallest roughness of, respectively, 4.0 and 5.6 μm,

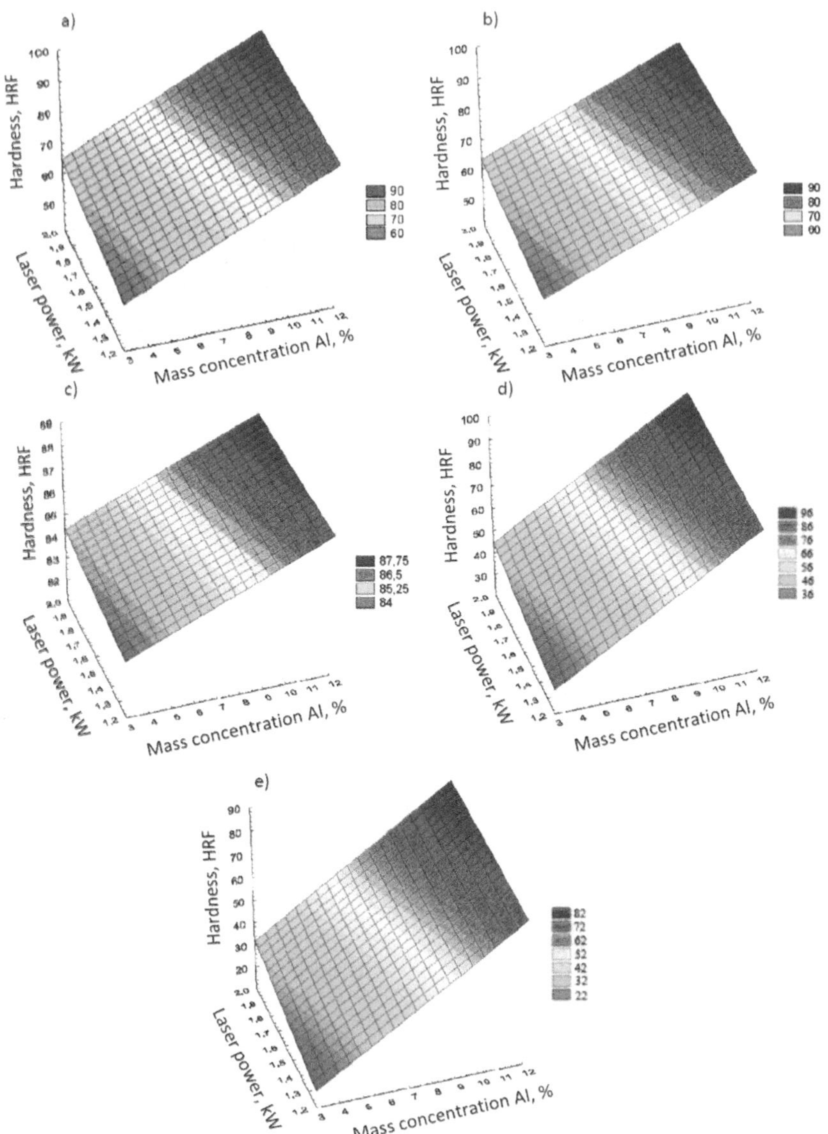

FIGURE 5.68 The effect of Al concentration and laser power on the hardness of Mg-Al-Zn-Mn casting alloys after laser imbed into remelted surface zone of powders of carbides and Al_2O_3 oxide with the laser imbed of hard ceramic particles into remelted surface zone rate of 0.75 m/min, determined by using models of neural networks: (a) TiC; (b) WC; (c) VC; (d) SiC; (e) Al_2O_3.

is exhibited by MCMgAl9Zn1 and MCMgAl12Zn1 alloys after laser imbed into remelted surface zone of VC power with the laser power of 2.0 kW. The maximum measured surface roughness of Ra = 42.5 μm exists in case of the surface layer of MCMgAl9Zn1 alloy after laser imbed into remelted surface zone of SiC power with the laser power of 1.2 kW.

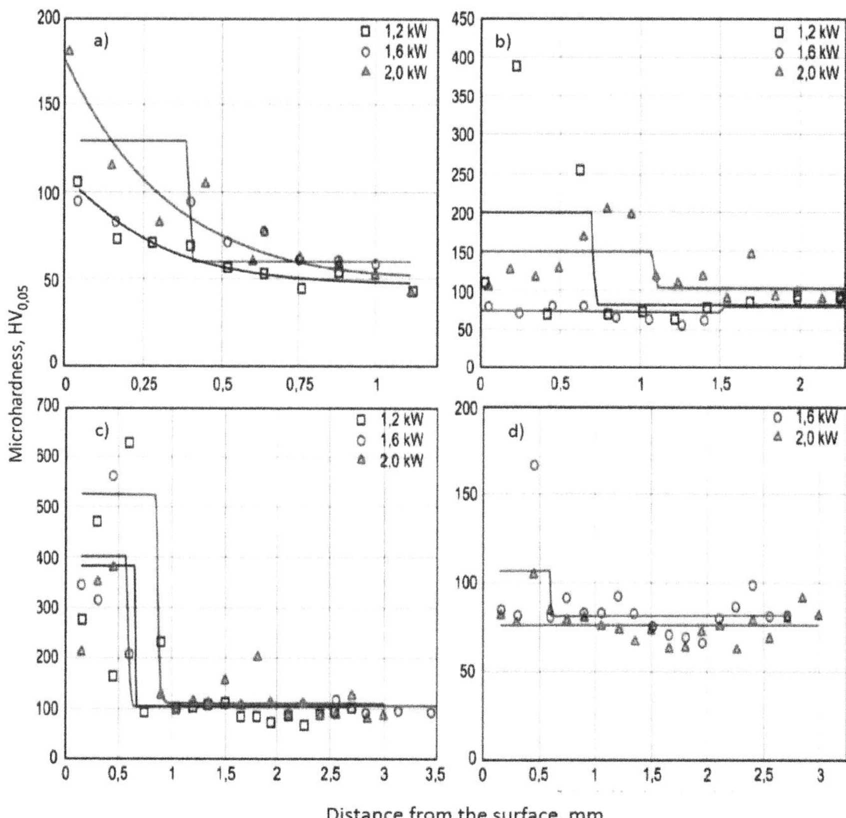

FIGURE 5.69 Curve of changes in microhardness of the surface layer of Mg-Al-Zn-Mn casting alloys after laser imbed into remelted surface zone of powders of carbides and Al_2O_3 oxide: (a) MCMgAl6Zn1 after the laser imbed into remelted surface zone of WC with the laser imbed of hard ceramic particles into remelted surface zone rate of 0.75 m/min, (b) MCMgAl6Zn1 after the laser imbed into remelted surface zone of TiC with the laser imbed of hard ceramic particles into remelted surface zone rate of 0.75 m/min, (c) MCMgAl12Zn1 after the laser imbed into remelted surface zone of SiC with the laser imbed of hard ceramic particles into remelted surface zone rate of 0.75 m/min, (d) MCMgAl9Zn1 after the laser imbed into remelted surface zone of Al_2O_3 with the laser imbed of hard ceramic particles into remelted surface zone rate of 0.50 m/min.

Similar to the network to calculate the hardness of the surface layer after laser imbed of hard ceramic particles into remelted surface zone, neural networks were also developed to predict surface roughness of Mg-Al-Zn-Mn alloys after laser imbed into remelted surface zone of TiC, WC, VC, SiC and Al_2O_3 powders. Experimental data was employed to build a network model, taking into account the type of powder used, the concentration of Al in the alloy, laser power and laser imbed of hard ceramic particles into remelted surface zone rate – as input variables – and the roughness Ra as the output variable. On the basis of the created neural network model, charts were developed of the effect of laser power,

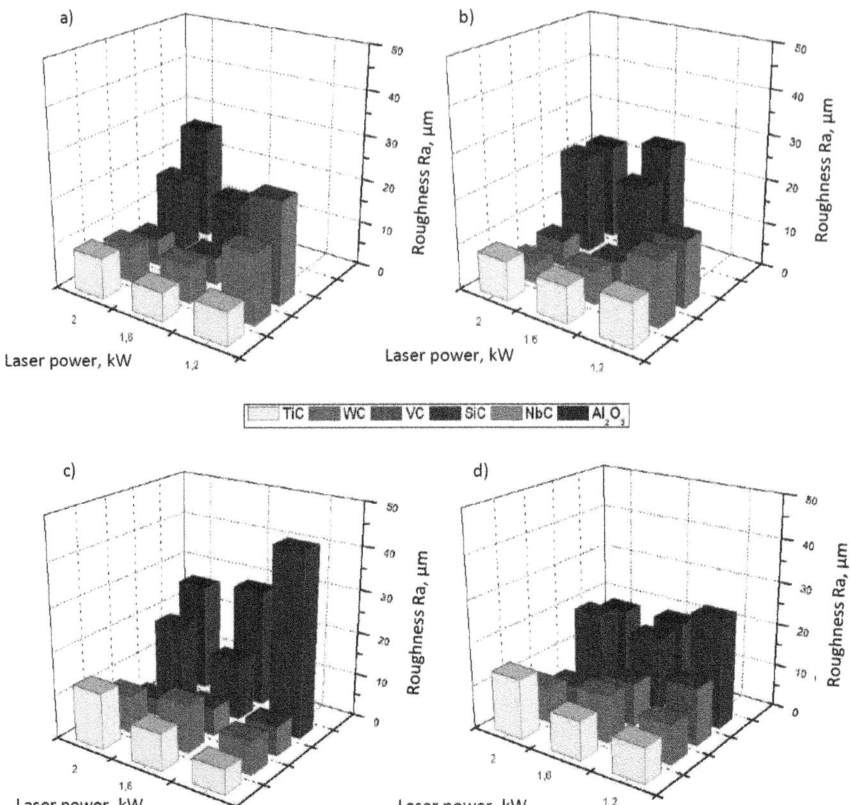

FIGURE 5.70 The effect of laser power on the roughness of the surface layer after laser imbed of the investigated powders into remelted surface zone of Mg-Al-Zn-Mn casting alloys: (a) MCMgAl3Zn1; (b) MCMgAl6Zn1; (c) MCMgAl9Zn1; (d) MCMgAl12Zn1.

concentration of Al in the alloy, as well as the type of the melt injected powder on the roughness of casting magnesium alloys after laser imbed into remelted surface zone of the studied ceramic particles (Figure 5.71). The outcomes obtained clearly point out that the highest roughness of the produced composite surface layer is shown by MCMgAl12Zn1 casting alloys, into the surface of which VC, WC powders were melt injected with the laser power of 2.0 kW and laser imbed of hard ceramic particles into remelted surface zone rate of 0.75 m/min.

The abrasive wear resistance of Mg-Al-Zn-Mn casting alloys after the laser imbed into remelted surface zone of TiC, WC, VC, SiC, Al_2O_3 powders is measured by the rate Δm of loss mass as a result of abrasive wear of the alloy after laser imbed of hard ceramic particles into remelted surface zone to Δm of the loss mass as a result of abrasive wear of the heat treated alloy. The highest

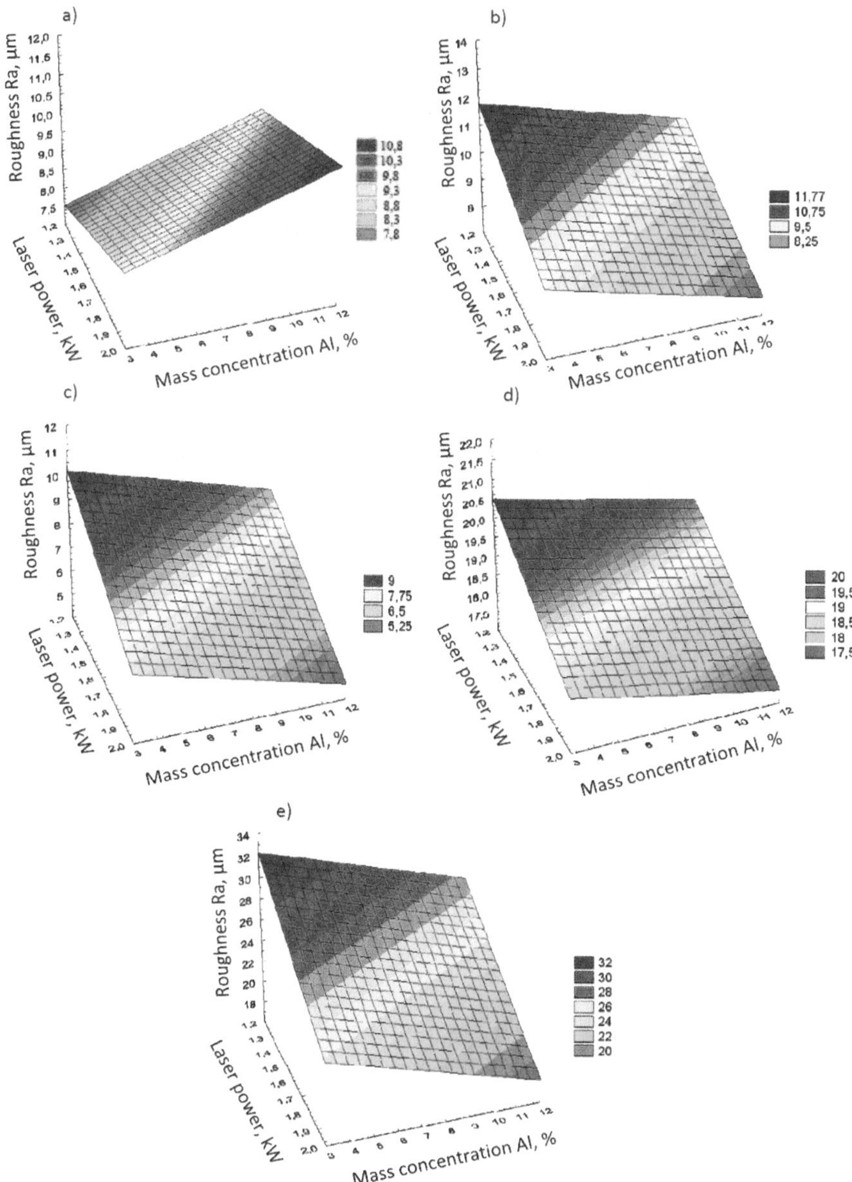

FIGURE 5.71 The effect of Al concentration and laser power on the roughness of Mg-Al-Zn-Mn casting alloys after laser imbed into remelted surface zone of powders of carbides and Al_2O_3 oxide with the laser imbed of hard ceramic particles into remelted surface zone rate of 0.75 m/min, determined by using models of neural networks: (a) TiC; (b) WC; (c) VC; (d) SiC; (e) Al_2O_3.

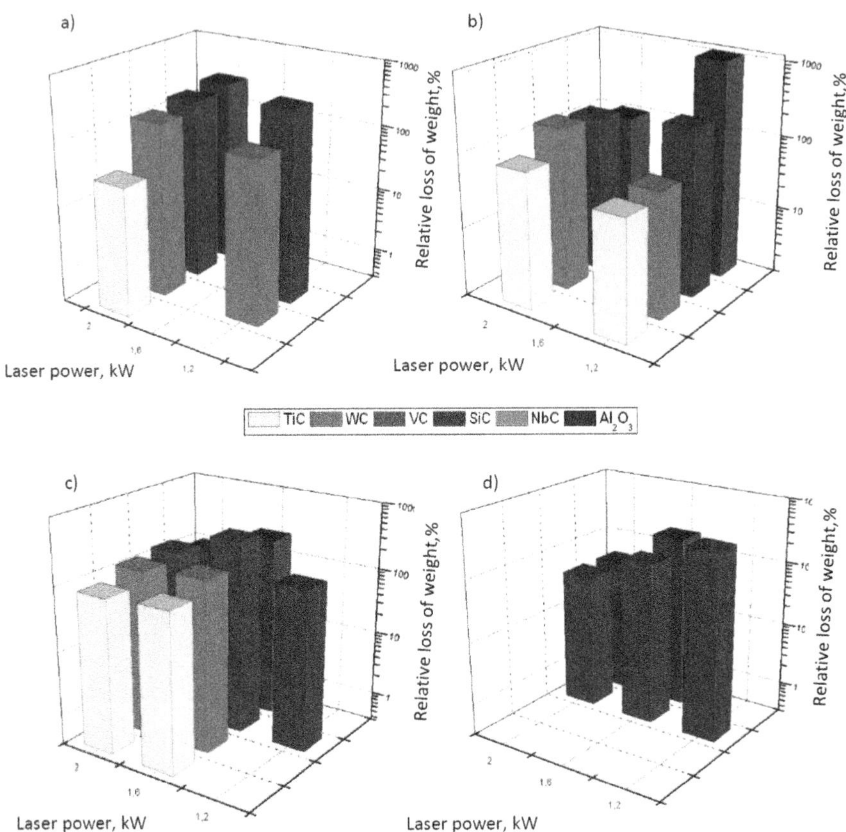

FIGURE 5.72 The effect of laser power on the relative loss in the mass as a result of abrasion wear of the surface layer after laser imbed of the investigated powders into remelted surface zone of Mg-Al-Zn-Mn casting alloys: (a) MCMgAl3Zn1; (b) MCMgAl6Zn1; (c) MCMgAl9Zn1; (d) MCMgAl12Zn1.

abrasive wear resistance is recorded for MCMgAl3Zn1 and MCMgAl6Zn1 casting alloys with laser melt injected particles of TiC carbides (Figure 5.72). It was found that there is nearly a proportional effect of wear of the surface layer of Mg-Al-Zn-Mn casting alloys with laser melt injected particles of TiC carbides, depending on the type of the substrate and the thickness of the laser treated surface layer.

Corrosion tests using the electrochemical potentiodynamic method in a 3% NaCl water solution were performed in order to determine the effect of laser imbed into remelted surface zone of TiC, WC, VC, SiC, Al_2O_3 powders into the substrate of Mg-Al-Zn-Mn casting alloys. The corrosive wear of the examined alloys' surface depending on the mass concentration of Al, the powder applied as well as the laser power was determined by examining Mg-Al-Zn-Mn alloys after laser imbed of hard ceramic particles into remelted surface zone. The determined values of resistance polarisation Rp (Figure 5.73) and

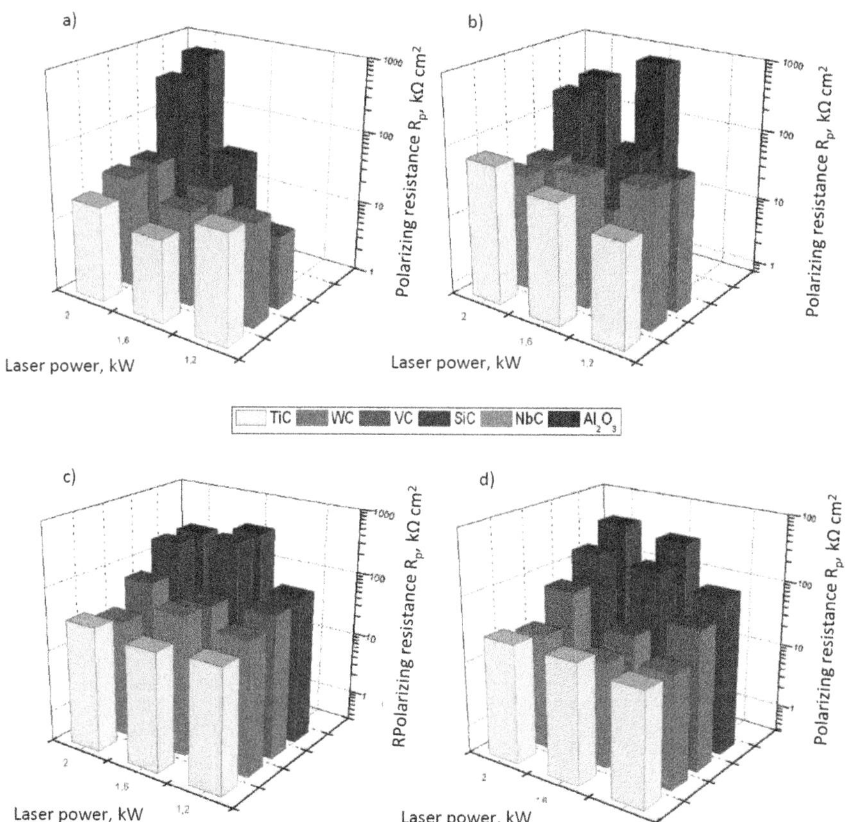

FIGURE 5.73 The effect of laser power on changes in the average value of polarisation resistance of the surface layer after laser imbed of the investigated powders into remelted surface zone of Mg-Al-Zn-Mn casting alloys: (a) MCMgAl3Zn1; (b) MCMgAl6Zn1; (c) MCMgAl9Zn1; (d) MCMgAl12Zn1.

corrosive power density i_{cor} (Figure 5.74) indicate the disadvantageous effect of the melt injected particles on the corrosive resistance of Mg-Al-Zn-Mn casting alloys. The value of corrosive current measures corrosive resistance as a proportion between the corrosion rate and corrosive power density. The higher values of corrosive currents and smaller values of corrosive potential obtained for Mg-Al-Zn-Mn casting alloys with laser melt injected TiC, WC, SiC, VC and Al_2O_3 particles in relation to the values determined for the substrate indicate the weaker corrosive resistance of the alloys laser treated by means of laser imbed of hard ceramic particles into remelted surface zone. The corrosive current density i_{cor} is between 0.009 and 1.913 mA/cm², respectively, for the surface layer of MCMgAl6Zn1 alloy after the laser imbed into remelted surface zone of Al_2O_3 powder with the power of 1.6 kW and MCMgAl3Zn1 alloy after laser imbed into remelted surface zone of TiC powder with the laser power of 1.6 kW (Figure 5.74). The polarisation resistance value R_p is within the range of

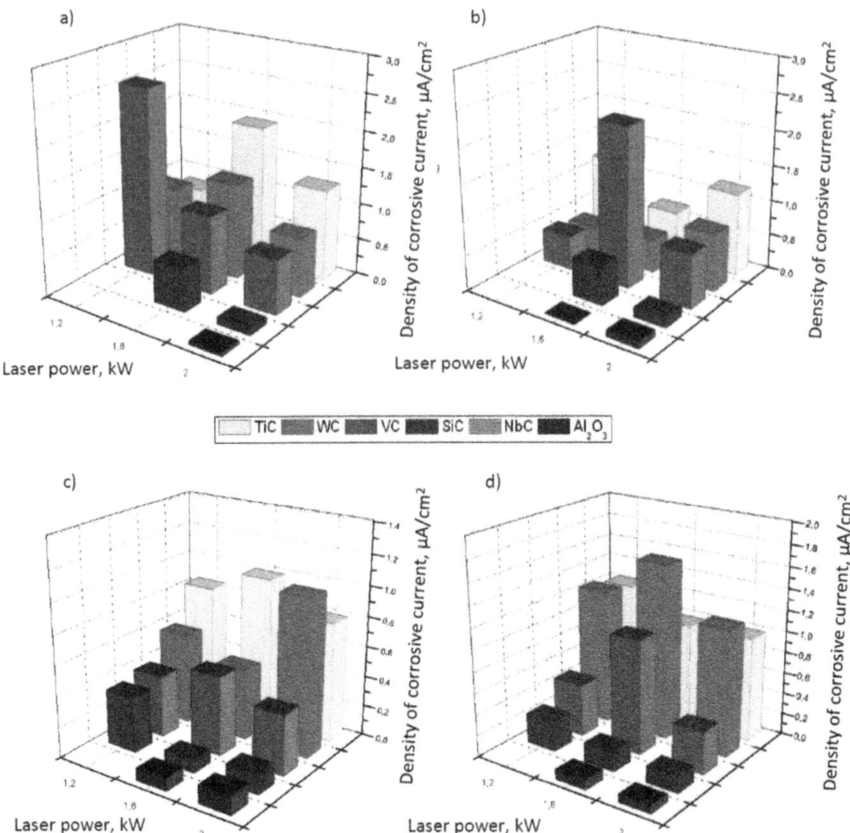

FIGURE 5.74 The effect of laser power on changes in the average value of corrosive power density of the surface layer after laser imbed of the investigated powders into remelted surface zone of Mg-Al-Zn-Mn casting alloys: (a) MCMgAl3Zn1; (b) MCMgAl6Zn1; (c) MCMgAl9Zn1; (d) MCMgAl12Zn1.

13.59–650 Ωcm^2, respectively for the surface layer of the MCMgAl6Zn1 alloy after laser imbed into remelted surface zone of Al_2O_3 powder with the power of 1.6 kW and MCMgAl3Zn1 alloy after laser imbed into remelted surface zone of TiC powder with the laser power of 1.6 kW (Figure 5.73). Polarisation resistance R_p falls in all the examined alloys after laser treatment in relation to the alloys not treated by laser (Figure 5.73). A slight shift in anodic polarisation curves by approx. 0.1 V, determined for all the investigated surface layers in the positive direction, in relation to the curves determined for alloys without the executed laser treatment, shows a slight increase in corrosive resistance of laser treated Mg-Al-Zn-Mn alloys as compared to the same alloys, but not treated by laser. The existence of pitting corrosion is however indicated by an increase in the course of polarisation curves in the anode area, coming from surface layers of laser-treated alloys.

5.7 FINAL REMARKS AND SCOPE OF FURTHER WORKS

The conditions imposed by modern technology and economic requirements indicate very favourable development prospects of Mg, and especially its alloys. The introduction to this paper points to a number of areas of such applications, although note-worthy is also a very broad and still open area of the possible uses of such group of engineering materials for biomedical purposes, both as biodegradable materials [263–276], as well as non-biodegradable materials [276–280]. Undoubtedly, low density, at least 60% lower than that of titanium, considered to be a lightweight alloy in such applications, opens up broad prospects, especially in implantology, in view of the excellent biocompatibility of Mg and its alloys. Attention is drawn to the diversity of these alloys, and the Figure 5.75 shows the effect of individual elements on the properties and biocompatibility of these materials [276].

The use of additive technologies, e.g. selective laser sintering, to fabricate components using such materials, is very attractive, despite the need to observe special safety requirements. It is very tempting to combine these two tendencies, to produce prosthetic and implantological components using Mg alloys with the method of selective laser sintering, and such idea was presented in relation to Ti alloys in the own works [281–283] and was broadly presented at numerous prestigious global scientific conferences [284–300]. The author is pursuing works to utilise such achievements with respect to Mg alloys in the framework of the current own projects [301,302]. Figure 5.76 shows the scope of scientific and research objectives and works expected to be pursued for the undertaken projects [301] in connection with the use of magnesium alloys for the manufacture of dental restorations by means of additive technologies.

This paper is a synthesis of the author's research into the group of Mg-Al-Zn-Mn alloys, carried out for more than last ten few years. The research was conducted under the Author's guidance as part of own projects, including doctoral and

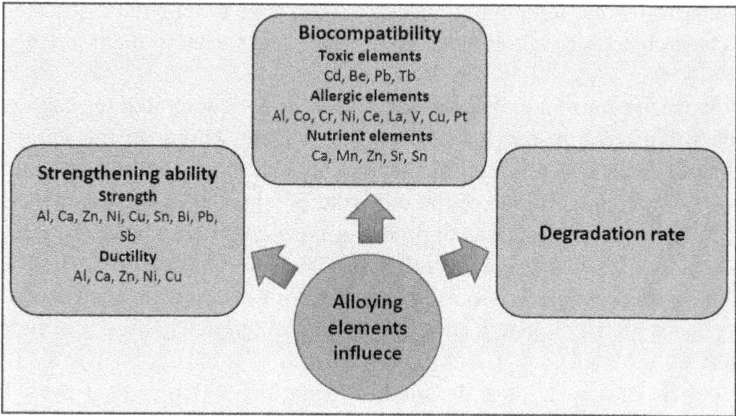

FIGURE 5.75 The effect of individual elements on the properties and biocompatibility of Mg alloys. (Prepared by Author using information from Radha, R. and Sreekanth, D., *J. Magnes. Alloys*, 5, 286–312, 2017.)

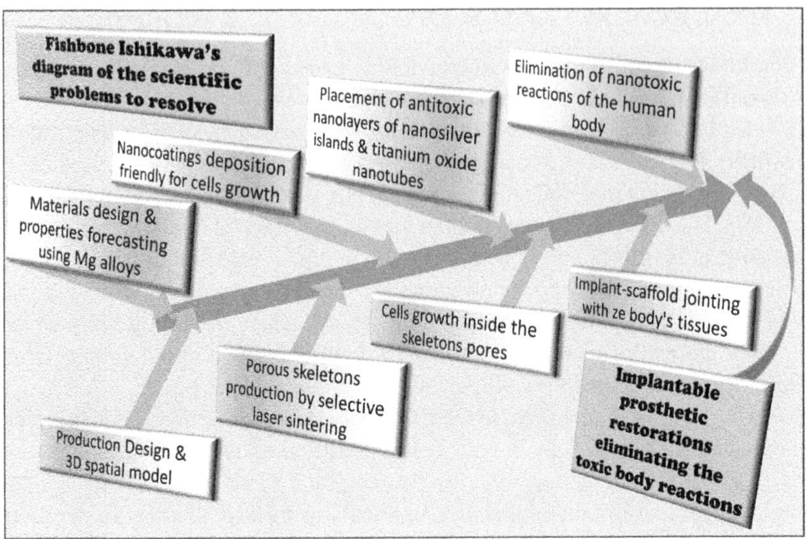

FIGURE 5.76 The scope of scientific and research objectives and works expected to be pursued for the undertaken projects in connection with the use of magnesium alloys for the manufacture of dental restorations by means of additive technologies. (From Dobrzański, L.A., Project proposal DENTANANO-PLFRO–The explanation of the interaction in nanoscale of the human body in connection zone with the surface of the metallic porous or solid implanted devices applied in regenerative dentistry coated by the nanostructural layers in order to avoid postoperative surgical complications and for the ensure proliferation of living cells and progress of the osseointegration, *Medical and Dental Engineering Centre for Research, Design and Production ASKLEPIOS in Gliwice*, Poland, EURONANOMED2019–2022.)

habilitation dissertations created under the Author's direction and supervision. It is a fully authorial work, although, of course, could have never been elaborated without the research activity of the listed Associates, and I want to thank them warmly for such efforts.

This paper presents an extensive, multi-faceted and integrated research project. Comprehensive basic research has been carried out, based on the paradigm of materials engineering and materials science, which can be expressed by the 6xE rule [1–4]. A graphical image of the 6xE rule is represented by an octahedron in Figure 5.76. The utility functions of products expected by the client will be ensured if the expected engineering material is used to fabricate them, treated using a technology with the expected quality, in order to obtain an expected shape of the final product with the expected structure and expected mechanical and functional properties.

This paper contains a lot of detailed information, which can certainly help in the proper selection of technologies for specific products within their engineering design. Engineering design, where the design of a manufacturing system and of products can be differentiated, is not a standalone activity, because it influences all the other phases of introducing a given product to the market, which are

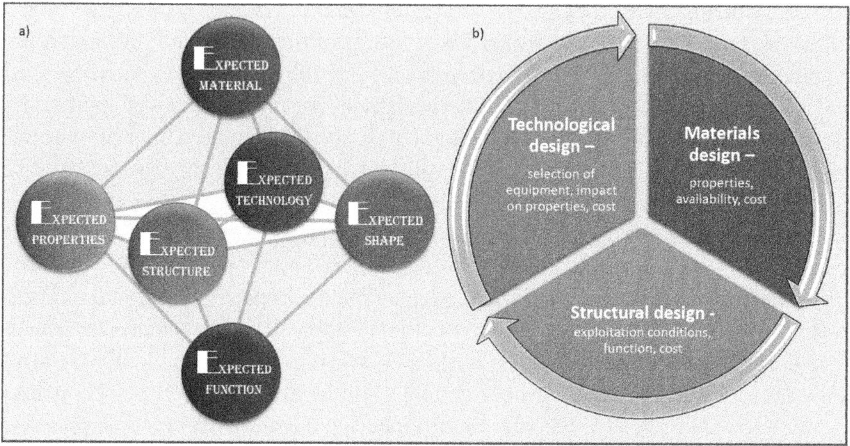

FIGURE 5.77 (a) Paradigm of materials engineering and materials science – an octahedron of the rule of 6 expectations, 6xE; (b) Diagram of inter-relation between the components of the product's engineering design, i.e. constructional design, materials design and technological design. (Prepared by Author using the idea given by Dieter, G.E., *ASM Handbook: Materials Selection and Design*, Vol. 20, ASM International, Materials Park, OH, 1997.)

also preconditioning it [1–4]. Product design combines three equally vital and inseparable components (Figure 5.77):

- Structural design whose purpose is to develop a shape and geometrical features of a product satisfying human needs
- Materials design to guarantee the required durability of a product or product components made of engineering materials with the required physiochemical and technological properties
- Technological process design enabling to achieve the required geometrical features and properties for particular components of the product and also their correct interaction after assembly, considering the volume of production, level of automation and computer aids, while ensuring also the lowest possible product costs

The research programme comprised, therefore, the selection of the chemical composition of selected Mg-Al-Zn-Mn alloys and investigations into the effect of volumetric heat treatment and into surface treatment by deposition of thin coatings by PVD and CVD and by the laser imbed of powders of multiple high-melting phases with very high hardness into remelted surface zone, on the structure and the resulting mechanical properties and the related selection of functional properties including corrosion resistance and abrasion wear resistance.

The collected information meets the requirements of the modern trend of materials design, consisting in the production of engineering materials on demand [303–305]. As part of the 5th and 6th Framework Programmes of the European Communities and as part of the then announced implementation reports for the projects 'The future of manufacturing in Europe – FutMan' and 'Manufacturing visions the futures project – ManVis', as a generalisation of the results of the European Foresight, it was agreed that the manufacturing of materials – having the properties ordered by product users – is expected. This fundamentally changes the methodology of engineering design, especially materials and technological design. The manufacture of materials on demand is setting a new role for Mg alloys. Materials ensuring the required set of physical and chemical properties have to be delivered on demand of manufacturers. Previously, manufacturers of products were adapting to their needs the material closest to the expectations with the offered structure and properties, made available on the market by suppliers of materials. Meanwhile, the reports show that the type of materials used is less important than their functionality. This means that the market of material manufacturers no longer dominates, and much more competitive requirements are set for them. New engineering materials and their production and processing processes are linked to customer requirements and features of practical products.

It is worth noting that an important part of the investigations performed was very extensive and careful research with the use of knowledge engineering methods, especially technology foresight. The results of such investigations, involving approx. 500 eminent experts from many countries, allowed to nominate the most attractive and pro-development technologies of surface formation and relate them to the investigated group of Mg-Al-Zn-Mn alloys. Both, the deposition of thin coatings with PVD/CVD methods, as well as laser imbed of powders of hard ceramic particles into remelted surface zone are among the most promising technologies with the highest projected growth rate and growing application areas. The completed research used a very large collection of cutting-edge research methods, with high-resolution transmission electron microscopy inclusive. The results of these multi-faceted studies are extremely attractive and indicate potential applications, and also explain phase transformations and structural mechanisms, the control of which in technological processes enables to put into life the idea of the material on demand.

It should be noted that the use of additive technology at the same time with advanced design of technologies and products manufactured from Mg alloys, used in relation to prosthetic restorations and implants, ensures consistency with the avant-garde development trend of the modern technology, referred to as Industry 4.0 [306–314].

Without going into the substance of whether it is substantiated to claim that after the era of steam 1.0, electricity 2.0, and computers 3.0, the next stage of integration of people and machines, described below, can be actually considered to be level 4.0 (Figure 5.78), or only 3.5, and whether a shift between the successive, so perceived levels is equally important in terms of progress of technology, an analysis is provided below to what extent the process of manufacturing and designing prosthetic restorations and implants in the project [314] fulfils the requirements of the definition of Industry 4.0. As noted before, the fact of mastering the technological process – corresponding to the requirements and assumptions of this concept – will then be implemented with respect to Mg alloys [302].

Industry 4.0 is connected with integration of the Internet of Things and Internet of Services in the manufacturing process [307,308]. The Internet of Things is both, an internal as well as cross-organisational network, where services are offered and used

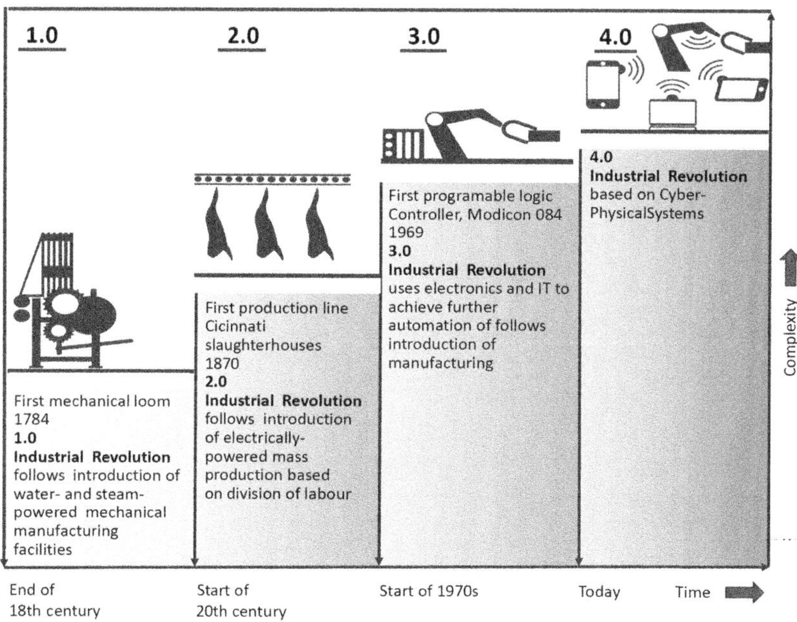

FIGURE 5.78 Four stages of the Industrial Revolution. (Prepared by Author using the idea from Kagermann, H. et al., Recommendations for implementing the strategic initiative Industrie 4.0: Final report of the Industrie 4.0 Working Group, Frankfurt a. Main, 2013, pp. 1–84, http://forschungsunion.de/pdf/industrie_4_0_final_report.pdf, access 30 January 2019.)

by participants of the chain of values [309]. Cyber-Physical Systems include things and objects (in particular such as radio-frequency identification, sensors, actuators, cell phones), which are aiding the people and machines contextually in the execution of such tasks in real time via the Internet of Things, based on information coming from the physical and virtual world, and ensure the ability to communicate and interact with them, through unique addressing schemes (Figure 5.79) [310,311].

Cyber-Physical Systems interact with each other mutually, but also cooperate with the adjacent intelligent components to achieve the set objectives [306]. The independent components of Industry 4.0 do not include machine to machine communication as the basis of the Internet of Things, nor intelligent products being a sub-component of Cyber-Physical Systems. Each Cyber-Physical System is monitoring physical processes, creates a virtual copy of the physical world and makes decentralised decisions as part of the modular structure of Smart Factories (Figure 5.80), being a key component of Industry 4.0 [307].

An example of Smart Factory is a system of fabricating prosthetic restorations and implants. This technology is now developed for Co-Cr and Ti-Al-V alloys, commonly used for this purpose [314], with a close perspective of applications for Mg alloys [301]. If the Internet acquires information about anatomical features of patients from different clinics, in the form of electronically digitalised results of diagnosis using, in particular, X-ray radiation, and after individualised design of constructional and material features of prosthetic restorations using CAD/CAM techniques, it is

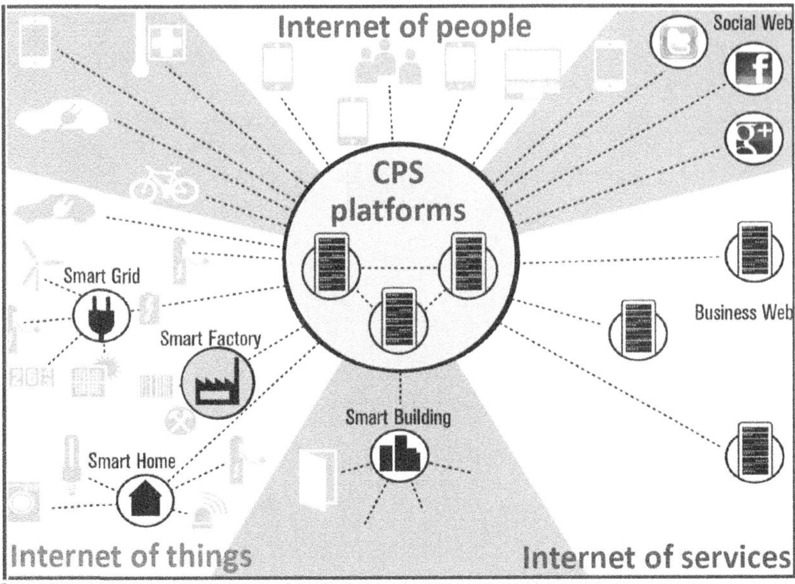

FIGURE 5.79 The Internet of Things and Services – Networking people, objects and systems. (From Kagermann, H. et al., Recommendations for implementing the strategic initiative Industrie 4.0: Final report of the Industrie 4.0 Working Group, Frankfurt a. Main, 2013, pp. 1–84, http://forschungsunion.de/pdf/industrie_4_0_final_report.pdf, access 30 January 2019.)

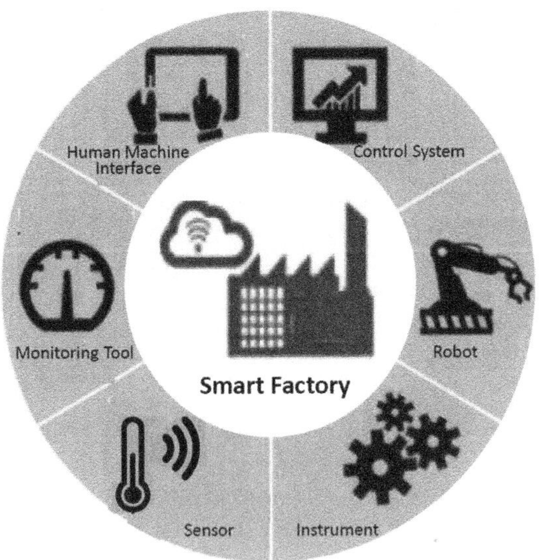

FIGURE 5.80 Scheme of the main elements included in the Smart Factory.

feasible to manufacture on such basis such components with the method of selective laser sintering integrated with finish treatment by milling, with the possible use of atomic layers deposition, ensuring the required structural and material features and appropriate properties and biocompatibility.

This paper contains a lot of detailed information, which can certainly help in the proper selection of technologies for specific products manufactured using the investigated Mg-Al-Zn-Mn alloys within engineering design of such products. I remain deeply convinced that it was just such a task about which once spoke Sir Prof. Harry Kroto (1939–2016), a Nobel Prize winner in Chemistry in 1996 (whom I was honoured to meet personally), which had to be done trying not to disappoint anyone, and I used this sentence as a motto of this paper. This work is also a tribute to the murdered mayor of Gdańsk Paweł Adamowicz and his idea.

A general conclusion of this paper is a conviction demonstrated herein about very attractive properties of the investigated Mg-Al-Zn-Mn alloys for many practical applications, and the materials science basis, widely explained herein, of the analysed heat and surface treatment technologies of these alloys.

REFERENCES

1. L.A. Dobrzański: Metals and alloys, *Open Access Library*, Annal VII (2) 2017, pp. 1–982 (in Polish).
2. L.A. Dobrzański: *Materiały inżynierskie i projektowanie materiałowe. Podstawy nauki o materiałach i metaloznawstwo*, Wydawnictwa Naukowo-Techniczne, Wydanie II zmienione i uzupełnione, Warszawa, Poland, 2006, pp. 1–1600.
3. L.A. Dobrzański: *Metalowe materiały inżynierskie*, Wydawnictwa Naukowo-Techniczne, Warszawa, Poland, 2004, pp. 1–888.
4. L.A. Dobrzański: *Podstawy metodologii projektowania materiałowego*, Wydawnictwo Politechniki Śląskiej, Gliwice, Poland, 2009, pp. 1–324.
5. M.F. Ashby: *Materials Selection in Mechanical Design*, 4th ed., Elsevier/Butterworth-Heinemann, Amsterdam, the Netherlands, 2011.
6. M.F. Ashby, D.R.H. Jones: *Engineering Materials. 1. An Introduction to Properties, Applications and Design*, 4th ed., Elsevier/Butterworth-Heinemann, Amsterdam, the Netherlands, 2012.
7. M.F. Ashby, D.R.H. Jones: *Engineering Materials. 2. An Introduction to Microstructures, Processing and Design*, 4th ed., Elsevier/Butterworth-Heinemann, Amsterdam, the Netherlands, 2013.
8. G.E. Dieter (Ed.): *ASM Handbook – Materials Selection and Design*, Vol. 20, ASM International, Materials Park, OH, 1997.
9. M. Rühle, M. Dosch, E.J. Mittemeijer, M.H. Van de Voorde: *European White Book on Fundamental Research in Materials Science*, Max-Planck-Institute für Metall- forschung, Stuttgart, Germany, 2001.
10. J.R. Davis, P. Allen (Eds.): *ASM Handbook – Properties and Selection: Nonferrous Alloys and Special-Purpose Materials*, Vol. 2, ASM International, Materials Park, OH, 1990.
11. K.U. Kainer (Ed.): *Magnesium Alloys and Technology*, John Wiley & Sons–VCH Verlag, Weinheim, Germany, 2003.
12. I. Polmear, D. StJohn, J. Nie, M. Qian: Metallurgy of the Light Metals, 5th ed., Butterworth-Heinemann, Oxford, 2017.
13. M. Avedesian, H. Baker (Eds.): *ASM Specialty Handbook: Magnesium and Magnesium Alloys*, ASM International, The Materials Information Society, OH, 1999.

14. F. Czerwinski (Ed.): *Magnesium Alloys: Design, Processing and Properties*, In- Tech, Rijeca, Croatia, 2011.
15. D. Sameer Kumar, C. Tara Sasanka, K. Ravindra, K.N.S. Suman: Magnesium and its alloys in automotive applications: A review, *American Journal of Materials Science and Technology*, 2015, 4(1), 12–30.
16. D. Jiang, Y. Dai, Y. Zhang, Y. Yan, J. Ma, D. Li, K. Yu: Effects of heat treatment on microstructure, mechanical properties, corrosion resistance and cytotoxicity of ZM21 magnesium alloy as biomaterials, *Journal of Materials Engineering and Performance*, 2019, 28(1), 33–43.
17. X.J. Wang, D.K. Xub, R.Z. Wuc, X.B. Chend, Q.M. Penge, L. Jinf, Y.C. Xing et al.: What is going on in magnesium alloys? *Journal of Materials Science & Technology*, 2018, 34(2), 245–247.
18. X. Chen, L. Liu, J. Liu, F. Pan: Microstructure, electromagnetic shielding effectiveness and mechanical properties of Mg-Zn-Y-Zr alloys, *Materials and Design*, 2015, 65, 360–369.
19. F. Pan, X. Chen, T. Yan, T. Liu, J. Mao, W. Luo, Q. Wang, J. Peng, A. Tang, B. Jiang: A novel approach to melt purification of magnesium alloys, *Journal of Magnesium and Alloys*, 2016, 4(1), 8–14.
20. L.Y. Chen, J.Q. Xu, H. Choi, M. Pozuelo, X. Ma, S. Bhowmick, J.M. Yang, S. Mathaudhu, X.C. Li: Processing and properties of magnesium containing a dense uniform dispersion of nanoparticles, *Nature*, 2015, 528(7583), 539–543.
21. H.L. Shi, X.J. Wang, C.L. Zhang, C.D. Li, C. Ding, K. Wu, X.S. Hu: A novel melt processing for Mg matrix composites reinforced by multiwalled carbon nanotubes, *Journal of Materials Science and Technology*, 2016, 32(12), 1303–1308.
22. Wu, G., Chan, K.-C., Zhu, L., Sun, J. Lu: Dual-phase nanostructuring as a route to high-strength magnesium alloys, *Nature*, 2017, 545(7652), 80–83.
23. Z. Shao, Q. Le, Z. Zhang, Z., J. Cui: A new method of semi-continuous casting of AZ80 Mg alloy billets by a combination of electromagnetic and ultrasonic fields, *Materials and Design*, 2011, 32(8–9), 4216–4224.
24. M. Esmaily, J.E. Svensson, S. Fajardo, N. Birbilis, G.S. Frankel, S. Virtanen, R. Arrabal, S. Thomas, L.G. Johansson: Fundamentals and advances in magnesium alloy corrosion, *Progress in Materials Science*, 2017, 89, 92–193.
25. A. Atrens, G.L. Song, F. Cao, Z. Shi, P.K. Bowen: Advances in Mg corrosion and research suggestions, *Journal of Magnesium and Alloys*, 2013, 1, 177–200.
26. Z. Shi, F. Cao, G.L. Song, A. Atrens: Low apparent valence of Mg during corrosion, *Corrosion Science*, 2014, 88, 434–443.
27. B.J. Wang, S.D. Wang, D.K. Xu, E.H. Han: Recent progress in fatigue behavior of Mg alloys in air and aqueous media: A review, *Journal of Materials Science and Technology*, 2017, 33, 1075–1086.
28. Y. Liu, H. Ren, W.C. Hu, D.J. Li, X.Q. Zeng, K.G. Wang, J. Lu: First-principles calculations of strengthening compounds in magnesium alloy: A general review, *Journal of Materials Science and Technology*, 2016, 32(12), 1222–1231.
29. T. Cain, L.G. Bland, N. Birbilis, J.R. Scully: A compilation of corrosion potentials for magnesium alloys, *Corrosion*, 2014, 70, 1043–1051.
30. G. Williams, H. ap Llwyd Dafydd, R. Grace: The localised corrosion of Mg alloy AZ31 in chloride containing electrolyte studied by a scanning vibrating electrode technique, *Electrochimica Acta*, 2013, 109, 489–501.
31. R. Wu, Y. Yan, G. Wang, L.E. Murr, W. Han, Z. Zhang, M. Zhang: Recent progress in magnesium-lithium alloys, *International Materials Reviews*, 2015, 60(2), 65–100.
32. W. Xu, N. Birbilis, G. Sha, Y. Wang, J.E. Daniels, Y. Xiao, M. Ferry: A high-specific-strength and corrosion-resistant magnesium alloy, *Nature Materials*, 2015, 14(12), 1229–1235.

33. S. Al-Saadi, P.C. Banerjee, M. Anisur, R.S. Raman: Hexagonal boron nitride impregnated silane composite coating for corrosion resistance of magnesium alloys for temporary bioimplant applications. *Metals*, 2017, 7, 518.
34. Q. Jiang, L. Yang, B. Hou: The effect of deep cryogenic treatment on the corrosion behavior of Mg-7Y-1.5Nd magnesium alloy. *Metals*, 2017, 7, 427.
35. V. Vijayaraghavan, A. Garg, L. Gao, R. Vijayaraghavan: Finite element based physical chemical modeling of corrosion in magnesium alloys. *Metals*, 2017, 7, 83.
36. H. Li, S. Rong, P. Sun, Q. Wang: Microstructure, residual stress, corrosion and wear resistance of vacuum annealed TiCN/TiN/Ti films deposited on AZ31, *Metals*, 2017, 7, 5.
37. A. Ma, F. Lu, Q. Zhou, J. Jiang, D. Song, J. Chen, Y. Zheng: Formation and corrosion resistance of micro-arc oxidation coating on equal-channel angular pressed AZ91D Mg alloy, *Metals*, 2016, 6, 308.
38. S.H. Salleh, S. Thomas, J.A. Yuwono, K. Venkatesan, N. Birbilis: Enhanced hydrogen evolution on Mg (OH)2 covered Mg surfaces, *Electrochimica Acta*, 2015, 161, 144–152.
39. S. Fajardo, C.F. Glover, G. Williams, G.S. Frankel: The evolution of anodic hydrogen on high purity magnesium in acidic buffer solution, *Corrosion*, 2017, 73, 482–493.
40. S.Z. Zhu, T.J. Luo, T.A. Zhang, Y.T. Liu, Y.S. Yang: Effects of extrusion and heat treatments on microstructure and mechanical properties of Mg–8Zn–1Al–0.5Cu–0.5Mn Alloy, *The Transactions of Nonferrous Metals Society*, 2017, 27, 73–81.
41. H. Zhang, J. Fan, L. Zhang, G. Wu, W. Liu, W. Cui, S. Feng: Effect of heat treatment on microstructure, mechanical properties and fracture behaviors of sand-cast Mg-4Y-3Nd-1Gd-0.2Zn-0.5Zr Alloy, *Materials Science and Engineering: A Structural Materials*, 2016, 677, 411–420.
42. D. Lysne, S. Thomas, M.F. Hurley, N. Birbilis: On the Fe enrichment during anodic polarization of Mg and its impact on hydrogen evolution, *Journal of the Electrochemical Society*, 2015, 162, C396–C402.
43. T. Cain, S.B. Madden, N. Birbilis, J.R. Scully: Evidence of the enrichment of transition metal elements on corroding magnesium surfaces using rutherford backscattering spectrometry, *Journal of the Electrochemical Society*, 2015, 162, C228–C237.
44. D.G. Eskin, J. Zuidema, V.I. Savran, L. Katgerman: Structure formation and macrosegregation under different process conditions during DC casting, *Materials Science and Engineering A*, 2004, 384(1–2), 232–244.
45. Z. Trojanová, K. Halmešová, J. Džugan, P. Palček, P. Minárik, P. Lukáč: Influence of strain rate on deformation behaviour of an AX52 alloy processed by equal channel angular pressing (ECAP), *Letters on Materials*, 2018, 8(4), 517–523.
46. M. Curioni: The behaviour of magnesium during free corrosion and potentiodynamic polarization investigated by real-time hydrogen measurement and optical imaging, *Electrochimica Acta*, 2014, 120, 284–292.
47. T.W. Cain, I. Gonzalez-Afanador, N. Birbilis, J.R. Scully: The role of surface films and dissolution products on the negative difference effect for magnesium: Comparison of Cl^- versus Cl^- free solutions, *Journal of the Electrochemical Society*, 2017, 164, C300–C311.
48. S. Fajardo, C.F. Glover, G. Williams, G.S. Frankel: The source of anodic hydrogen evolution on ultra high purity magnesium, *Electrochimica Acta*, 2016, 212, 510–521.
49. Y. Yan, H. Cao, Y. Kang, K. Yu, T. Xiao, J. Luo, Y. Deng, H. Fang, H. Xiong, Y. Dai: Effects of Zn concentration and heat treatment on the microstructure, mechanical properties and corrosion behavior of as-extruded Mg-Zn alloys produced by powder metallurgy, *Journal of Alloys and Compounds*, 2017, 693, 1277–1289.
50. Z. Shi, A. Atrens: An innovative specimen configuration for the study of Mg corrosion, *Corrosion Science*, 2011, 53, 226–246.

51. Y. Zhou, Y. Li, D. Luo, Y. Ding, P. Hodgson: Microstructures, mechanical and corrosion properties and biocompatibility of as extruded Mg–Mn–Zn–Nd alloys for biomedical applications, *Materials Science & Engineering C-Materials for Biological Applications*, 2015, 49, 93–100.
52. X. Wang, P. Zhang, L.H. Dong, X.L. Ma, J.T. Li, Y.F. Zheng: Microstructure and characteristics of interpenetrating β-TCP/Mg–Zn–Mn composite fabricated by suction casting, *Materials and Design*, 2014, 54, 995–1001.
53. Y. Song, E.-H. Han, D. Shan, C.D. Yim, B.S. You: The effect of Zn concentration on the corrosion behavior of Mg–xZn alloys, *Corrosion Science*, 2012, 65, 322–330.
54. N. Birbilis, T. Cain, J.S. Laird, X. Xia, J.R. Scully, A.E. Hughes: Nuclear microprobe analysis for determination of element enrichment following magnesium dissolution, *ECS Electrochemistry Letters*, 2015, 4, C34–C37.
55. D. Hoche, C. Blawert, S.V. Lamaka, N. Scharnagl, C. Mendis, M.L. Zheludkevich: The effect of iron re-deposition on the corrosion of impurity-containing magnesium, *Physical Chemistry Chemical Physics*, 2016, 18, 1279–1291.
56. S.V. Lamaka, D. Höche, R.P. Petrauskas, C. Blawert, M.L. Zheludkevich: A new concept for corrosion inhibition of magnesium: Suppression of iron re-deposition, *Electrochemistry Communications*, 2016, 62, 5–8.
57. X. Li, J.-H. Jiang, Y.-H. Zhao, A.-B. Ma, D.-J. Wen, Y.-T. Zhu: Effect of equal-channel angular pressing and aging on corrosion behavior of ZK60Mg alloy, *The Transactions of Nonferrous Metals Society*, 2015, 25, 3909–3920.
58. M. Jamesh, S. Kumar, T.S.N.S. Narayanan: Corrosion behavior of commercially pure Mg and ZM21Mg alloy in ringer's solution–Long term evaluation by EIS, *Corrosion Science*, 2011, 53, 645–654.
59. D. Zhang, X. Hao, D. Fang, Y. Chai: Effects of heat treatment on microstructure and mechanical properties of as-extruded Mg-9Sn-1.5Y-0.4Zr magnesium alloy, *Rare Metal Materials and Engineering*, 2016, 45, 2208–2213.
60. D. Mercier, J. Światowska, S. Zanna, A. Seyeux, P. Marcus: Role of segregated Iron at grain boundaries on Mg corrosion, *Journal of the Electrochemical Society*, 2018, 165, C42–C49.
61. M. Curioni, L. Salamone, F. Scenini, M. Santamaria, M. Di Natale: A mathematical description accounting for the superfluous hydrogen evolution and the inductive behaviour observed during electrochemical measurements on magnesium, *Electrochimica Acta*, 2018, 274, 343–352.
62. X. Gu, W. Zhou, Y. Zheng, L. Dong, Y. Xi, D. Chai: Microstructure, mechanical property, bio-corrosion and cytotoxicity evaluations of Mg/HA composites, *Materials Science & Engineering C-Materials for Biological Applications*, 2010, 30, 827–832.
63. A. Atrens, M. Liu, N.I.Z. Abidin, G.L. Song: 3–Corrosion of magnesium (Mg) alloys and metallurgical influence, *Corrosion of Magnesium Alloys*, G.L. Song, Ed., Woodhead, Cambridge, UK, 2011, pp. 117–165.
64. S. Bagherifard, D.J. Hickey, S. Fintová, F. Pastorek, I. Fernandez-Pariente, M. Bandini, T.J. Webster, M. Guagliano: Effects of nanofeatures induced by severe shot peening (SSP) on mechanical, corrosion and cytocompatibility properties of magnesium alloy AZ31, *Acta Biomaterialia*, 2017. doi:10.1016/j.actbio.2017.11.032.
65. N. Birbilis, A.D. King, S. Thomas, G.S. Frankel, J.R. Scully: Evidence for enhanced catalytic activity of magnesium arising from anodic dissolution, *Electrochimica Acta*, 2014, 132, 277–283.
66. K. Gusieva, C.H.J. Davies, J.R. Scully, N. Birbilis: Corrosion of magnesium alloys: The role of alloying, *International Materials Reviews*, 2015, 60, 169–194.
67. A.D. Südholz, N.T. Kirkland, R.G. Buchheit, N. Birbilis: Electrochemical properties of intermetallic phases and common impurity elements in magnesium alloys, *Electrochemical and Solid State Letters*, 2011, 14, C5–C7.

68. X.-N. Gu, S.-S. Li, X.-M. Li, Y.-B Fan: Magnesium based degradable biomaterials: A review, *Frontiers of Materials Science*, 2014, 8(3), 200–218.
69. A. Atrens, G.L. Song, M. Liu, Z. Shi, F. Cao, M.S. Dargusch: Review of recent developments in the field of magnesium corrosion, *Advanced Engineering Materials*, 2015, 17, 400–453.
70. G.L. Song: 1–Corrosion electrochemistry of magnesium (Mg) and its alloys, *Corrosion of Magnesium Alloys*, G.L. Song, Ed., Woodhead, Cambridge, UK, 2011, pp. 3–65.
71. E. Michailidou, H.N. McMurray, G. Williams: Quantifying the role of transition metal electrodeposition in the cathodic activation of corroding magnesium, *Journal of the Electrochemical Society*, 2018, 165, C195–C205.
72. S. Fajardo, O. Gharbi, N. Birbilis, G.S. Frankel: Investigating the effect of ferrous ions on the anomalous hydrogen evolution on magnesium in acidic ferrous chloride solution, *Journal of the Electrochemical Society*, 2018, 165, C916–C925.
73. Y. Song, E.H. Han, D. Shan, D.Y. Chang, B.S. You: The role of second phases in the corrosion behavior of Mg–5Zn Alloy, *Corrosion Science*, 2012, 60, 238–245.
74. X.-B. Liu, D.-Y. Shan, Y.-W. Song, E.-H. Han: Effects of heat treatment on corrosion behaviors of Mg-3Zn magnesium alloy, *The Transactions of Nonferrous Metals Society*, 2010, 20, 1345–1350.
75. X. Liu, D. Shan, Y. Song, R. Chen, E. Han: Influences of the quantity of Mg 2 Sn phase on the corrosion behavior of Mg–7Sn magnesium alloy, *Electrochimica Acta*, 2011, 56, 2582–2590.
76. G.S. Frankel, S. Fajardo, B.M. Lynch: Introductory lecture on corrosion chemistry: A focus on anodic hydrogen evolution on Al and Mg, *Faraday Discussions*, 2015, 180, 11–33.
77. S. Fajardo, G.S. Frankel: A kinetic model explaining the enhanced rates of hydrogen evolution on anodically polarized magnesium in aqueous environments, *Electrochemistry Communications*, 2017, 84, 36–39.
78. L.G. Bland, A.D. King, N. Birbilis, J.R. Scully: Assessing the corrosion of commercially pure magnesium and commercial AZ31B by electrochemical impedance, mass-loss, hydrogen collection, and inductively coupled plasma optical emission spectrometry solution analysis, *Corrosion*, 2015, 71, 128–145.
79. S. Fajardo, J. Bosch, G.S. Frankel: Anomalous hydrogen evolution on AZ31, AZ61 and AZ91 magnesium alloys in unbuffered sodium chloride solution, *Corrosion Science*, 2019, 146, 163–171.
80. Global Magnesium Market 2018–2022 ID 45 41163 Report May 2018 Dublin, pp. 1–111, Tech Navio; https://www.statista.com/.
81. S. Zhang (Ed.): *Thin Films and Coatings: Toughening and Toughness Characterization*, CRC Press, Boca Raton, FL, 2015.
82. L.A. Dobrzański, A.D. Dobrzańska-Danikiewicz: Materials surface engineering; Compendium of knowledge and academic textbook, *Open Access Library*, Annal VIII (1) 2018, pp. 1–1138 (in Polish)
83. L.A. Dobrzański, A.D. Dobrzańska-Danikiewicz: Engineering materials surface treatment, *Open Access Library*, 5, 2011, pp. 1–480 (in Polish).
84. A.D. Dobrzańska-Danikiewicz (Ed.): Materials surface engineering development trends, *Open Access Library*, 6, 2011, pp. 1–594.
85. A.D. Dobrzańska-Danikiewicz: Księga technologii krytycznych kształtowania struktury i wáasności powierzchni materiałów inżynierskich, *Open Access Library*, 8(26), 2013, pp. 1–823.
86. A. Fajkiel, P. Dudek, G. Sęk-Sas: Odlewnictwo XXI w. Kierunki rozwoju metalurgii i odlewnictwa stopów metali lekkich, Wydawnictwo Instytutu Odlewnictwa, Kraków, 2002.
87. L.A. Dobrzański, T. Tański, A.D. Dobrzańska-Danikiewicz, M. Król, S. Malara, J. Domagała-Dubiel: Struktura i własności stopów Mg-Al-Zn, International OCSCO World Press, Gliwice, *Open Access Library*, Vol. 5(11), 2012, pp. 1–319.

88. Z. Górny: Odlewnictwo metali i stopów, Tom III. Specjalne metody odlewania, Instytut Odlewnictwa, Kraków, 1997.
89. K.U. Kainem: *Magnesium Alloys and Technology*, Willey-VH, Weinheim, Germany, 2003.
90. K.U. Kainem (Ed.): *Magnesium Alloys and Their Applications*, Willey-VH, Weinheim, Germany, 2000.
91. H. Baker: *Physical Properties of Magnesium and Magnesium Alloys*, The Dow Chemical Company, Midland, MI, 1997.
92. S. Housh, B. Mikucki, A. Stevenson: Properties and selection, Nonferrous alloys and special purpose materials, ASM Handbook, ASM International, Materials Park, OH, 1991, 455–479.
93. A.K. Dahle, Y. Lee, M. Nave, P. Schaffer, D. StJohn: Development of the as-cast microstructure in magnesium-aluminium alloys, *Journal of Light Metals* 2001, 1, 61–72.
94. B.L. Mordike, Creep-resistant magnesium alloys, *Materials Science and Engineering A*, 2002, 324, 103–112.
95. D. Petrov, A. Watson, J. Gröbner, P. Rogl, J.-C. Tedenac, M. Bulanova, V. Turkevich, updated by H. L. Lukas – Materials Science International Team MSIT: Aluminium – Magnesium – Zinc, *Landolt-Börnstein, Numerical Data and Functional Relationships in Science and Technology*, G. Effenberg and S. Ilyenko, Eds., New Series/Editor in Chief: W. Martienssen, Group IV: Physical Chemistry, Vol. 11, Ternary Alloy Systems, Phase Diagrams, Crystallographic and Thermodynamic Data, critically evaluated by MSIT, Subvolume A, Light Metal Systems, Part 3, Selected Systems from Al-Fe-V to Al-Ni-Zr, vol 11A3. Springer, Berlin, Germany, 2005, pp. 191–209. doi:10.1007/10915998_1.
96. L.A. Dobrzański (contractor–project manager): Research project N N507 468837, Forming the functional properties of elements made of light metal alloys by depositing hybrid PVD coatings consisting of a graded transient layer and a multicomponent outer layer, 2009–2012, Silesian University of Technology, Poland - unpublished source materials.
97. L.A. Dobrzański (contractor–project manager): Development project R15 007 02, Improving the functional properties of elements of light casting magnesium and aluminium alloys heat treated by optimizing their chemical composition and by laser remelting and/or alloying the surface with carbides and/or ceramic particles, 2007–2009, Silesian University of Technology, Poland - unpublished source materials.
98. L.A. Dobrzański (contractor–project manager): INFONANO project UDA-POKL.04.01.01-00-003/09-00, Opening and development of engineering and PhD studies in the field of nanotechnology and materials science, 2009–2014, Silesian University of Technology, Poland - unpublished source materials.
99. L.A. Dobrzański (contractor–project manager): Research project KBN Nr 3 T08A 022 27, Opracowanie metodyki zautomatyzowanej oceny jakości i wad strukturalnych w stopach Al i Mg z wykorzystaniem narzędzi sztucznej inteligencji, 2004–2006, Silesian University of Technology, Poland - unpublished source materials.
100. L.A. Dobrzański (contractor–project manager): Serial Project in the framework Central European Exchange for Universities Studies (CEEPUS), 1998–2010, Silesian University of Technology, Poland - unpublished source materials.
101. L.A. Dobrzański (contractor–project manager): FORSURF Research project UDA-POIG.01.01.01-00.23/08-00, Foresight of surface properties formation leading technologies of engineering materials and biomaterials, FORSURF), 2009–2012, Silesian University of Technology, Poland - unpublished source materials.
102. A.D. Dobrzańska-Danikiewicz: Computer integrated development prediction methodology in materials surface engineering, *Open Access Library*, 1(7), 2012, pp. 1–289 (in Polish).

103. T. Tański (advisor L.A. Dobrzański): Kształtowanie struktury i własności odlewniczych stopów Mg-Al-Zn obrabianych cieplnie, PhD thesis, Library of Silesian University of Technology, Gliwice, 2006, unpublished manuscript.
104. M. Król (advisor L.A. Dobrzański): Wpływ stężenia Al oraz szybkości chłodzenia na strukturę i własności odlewniczych stopów Mg-Al-Zn, PhD thesis, Library of Silesian University of Technology, Gliwice, 2010, unpublished manuscript.
105. J. Domagała-Dubiel (advisor L.A. Dobrzański): Struktura i własności laserowo przetapianych i wtapianych warstw na stopach odlewniczych Mg-Al-Zn, PhD thesis, Library of Silesian University of Technology, Gliwice, 2010, unpublished manuscript.
106. Sz. Malara (advisor L.A. Dobrzański): Struktura i własności powierzchni odlewniczych stopów Mg-Al-Zn z wtapianymi laserowo cząstkami ceramicznymi, PhD thesis, Library of Silesian University of Technology, Gliwice, 2010, unpublished manuscript.
107. T. Tański: Forming the structure and surface properties of Mg-Al-Zn alloys, *Open Access Library*, 2 (8), 20121–20158 (in Polish).
108. J. Trzaska (advisor L.A. Dobrzański): Metodyka komputerowego modelowania kinetyki przemian austenitu przechłodzonego stali konstrukcyjnych; PhD thesis, Library of Silesian University of Technology, Gliwice, 2002, unpublished manuscript.
109. L.A. Dobrzański, T. Tański, A.D. Dobrzańska-Danikiewicz, E. Jonda, M. Bonek, A. Drygała: Structures, properties and development trends of laser surface treated hot-work steels, light metal alloys and polycrystalline silicon, in: J. Lawrence, D. Waugh, Eds., *Laser Surface Engineering. Processes and Applications*, Woodhead Publishing Series in Electronic and Optical Materials, Elsevier, Amsterdam, the Netherlands, 2015, pp. 3–32.
110. L.A. Dobrzański, T. Tański, S. Malara, M. Król, J. Domagała-Dubiel: Contemporary Forming methods of the structure and properties of cast magnesium alloys, in: F. Czerwinski, Ed., *Magnesium Alloys: Design, Processing and Properties*, InTech, Rijeka, Croatia, 2011, pp. 321–350.
111. T. Tański, E. Jonda, K. Labisz, L.A. Dobrzański: *Toughness of Laser-Treated* Surface Layers Obtained by Alloying and Feeding of Ceramic Powders, in: S. Zhang, Ed., *Thin Films and Coatings. Toughening and Toughness Characterization*, CRC Press, Boca Raton, FL, 2015, pp. 225–314.
112. L.A. Dobrzański, D. Pakuła, M. Staszuk: Chemical Vapor Deposition in Manufacturing, in: A.Y.C. Nee, Ed., *Handbook of Manufacturing Engineering and Technology*, Springer-Verlag, London, UK, 2015, pp. 2755–2803.
113. L.A. Dobrzański, K. Gołombek, K. Lukaszkowicz: Physical Vapor Deposition in Manufacturing, in: A.Y.C. Nee, Ed., *Handbook of Manufacturing Engineering and Technology*, Springer-Verlag, London, UK, 2015, pp. 2719–2754.
114. L.A. Dobrzański, A.D. Dobrzańska-Danikiewicz, T. Tański, E. Jonda, A. Drygała, M. Bonek: Laser Surface Treatment in Manufacturing, in: A.Y.C. Nee, Ed., *Handbook of Manufacturing Engineering and Technology*, Springer-Verlag, London, UK, 2015, pp. 2677–2717.
115. L.A. Dobrzański, A.D. Dobrzańska-Danikiewicz: Foresight of the Surface Technology in Manufacturing, in: A.Y.C. Nee, Ed., *Handbook of Manufacturing Engineering and Technology*, Springer-Verlag, London, UK, 2015, pp. 2587–2637.
116. L.A. Dobrzański, J. Trzaska, A.D. Dobrzańska-Danikiewicz: Use of neural networks and artificial intelligence tools for modeling, characterization, and forecasting in material engineering, in: S. Hashmi, Ed., *Comprehensive Materials Processing*, G.F. Batalha, Ed., Vol. 2: *Materials Modeling and Characterization*, Elsevier, Amsterdam, the Netherlands, 2014, pp. 161–198.
117. L.A. Dobrzański: The outstanding achievements in the scientific activity of the Institute of Engineering Materials and Biomaterials of the Silesian University of Technology in Gliwice, Poland, in: Y.I. Shalapko and L.A. Dobrzański, Ed., *Scientific Basis of*

Modern Technology: Experience and Prospects. Monograph, Department of Principles of Engineering Mechanics of Khmelnitsky National University, Khmelnitsky, Ukraine, 2011, pp. 545–600.

118. L.A. Dobrzański, T. Tański, L. Čížek, Z. Brytan: Structure and properties of magnesium cast alloys, *Journal of Materials Processing Technology*, 2007, 192–193, 567–574.
119. L. Čížek, M. Greger, L. Pawlica, L.A. Dobrzański, T. Tański: Study of selected properties of magnesium alloy AZ91 after heat treatment and forming, *Journal of Materials Processing Technology*, 2004, 157–158, 466–471.
120. A.D. Dobrzańska-Danikiewicz, L.A. Dobrzański, A. Sękala: Results of technology foresight in the surface engineering area, *Applied Mechanics and Materials*, 2014, 657, 916–920.
121. L.A. Dobrzański, T. Tański, S. Malara, J. Domagała, A. Klimpel: Laser surface treatment of Mg-Al-Zn alloys, *Strojarstvo*, 2011, 53(1), 5–10.
122. L.A. Dobrzański, T. Tański, S. Malara, M. Król: Structure and properties investigation of a magnesium alloy processed by heat treatment and laser surface treatment, *Materials Science Forum*, 2011, 674, 11–18.
123. L.A. Dobrzański, T. Tański, J. Trzaska: Optimization of heat treatment conditions of magnesium cast alloys, *Materials Science Forum*, 2010, 638–642, 1488–1493.
124. L.A. Dobrzański, T. Tański: Influence of aluminium content on behaviour of magnesium cast alloys in bentonite sand mould, *Solid State Phenomena*, 2009, 147–149, 764–769.
125. T. Tański, L.A. Dobrzański, L. Čížek: Influence of heat treatment on structure and properties of the cast magnesium alloys, *Advanced Materials Research*, 2007, 15–17, 491–496.
126. L.A. Dobrzański: Applications of newly developed nanostructural and microporous materials in biomedical, tand mechanical engineering, *Archives of Materials Science and Engineering*, 2015, 76(2), 53–114.
127. L.A. Dobrzański: Overview and general ideas of the development of constructions, materials, technologies and clinical applications of scaffolds engineering for regenerative medicine, *Archives of Materials Science and Engineering*, 2014, 69(2), pp. 53–80.
128. T. Tański, L.A. Dobrzański, S. Rusz, W. Matysiak, M. Kraus: Characteristic features of fine-grained coatings deposited on magnesium alloys, *Archives of Materials Science and Engineering*, 2014, 66(1), 13–20.
129. S. Rusz, L. Čížek, L.A. Dobrzański, S. Tylšar, J. Kedroň: ECAP methods application on selected non-ferrous metals and alloys, *Archives of Materials Science and Engineering*, 2010, 43(2), 69–76.
130. L.A. Dobrzański, S. Malara, T. Tański, J. Konieczny: Effect of high power diode laser surface alloying on structure of MCMgAl12Zn1 alloy, Archives *of Materials Science and Engineering*, 2010, 43(1), 54–61.
131. L.A. Dobrzański, T. Tański, L. Čížek, J. Madejski: The influence of the heat treatment on the microstructure and properties of Mg-Al-Zn based alloys, *Archives of Materials Science and Engineering*, 2009, 36(1), 48–54.
132. L.A. Dobrzański, J. Domagała, T. Tański. A. Klimpel, D. Janicki: Laser surface treatment of cast magnesium alloys, *Archives of Materials Science and Engineering*, 2009, 35(2), 101–106.
133. L.A. Dobrzański, S. Malara, J. Domagała, T. Tański, K. Gołombek: Influence of the laser modification of surface on properties and structure of magnesium alloys, *Archives of Materials Science and Engineering*, 2009, 35(2), 95–100.
134. L.A. Dobrzański, J. Domagała, S. Malara, T. Tański, W. Kwaśny: Structure changes and mechanical properties of laser alloyed magnesium cast alloys, *Archives of Materials Science and Engineering*, 2009, 35(2), 77–82.

135. L.A. Dobrzański, S. Malara, T. Tański, A. Klimpel, D. Janicki: Laser surface treatment of magnesium alloys with silicon carbide powder, *Archives of Materials Science and Engineering*, 2009, 35(1), 54–60.
136. L.A. Dobrzański, M. Król, T. Tański, R. Maniara: Thermal analysis of the MCMgAl9Zn1 magnesium alloy, *Archives of Materials Science and Engineering*, 2008, 34(2), 113–116.
137. L.A. Dobrzański, J. Domagała, T. Tański, A. Klimpel, D. Janicki: Characteristic of Mg-Al-Zn alloys after laser treatment, *Archives of Materials Science and Engineering*, 2008, 34(2), 69–74.
138. L.A. Dobrzański, J. Domagała, T. Tański, A. Klimpel, D. Janicki: Laser surface treatment of magnesium alloy with WC powder, *Archives of Materials Science and Engineering*, 2008, 30(2), 113–116.
139. M. Greger, R. Kocich, L. Čížek, L.A. Dobrzański, M. Widomská: Influence of ECAP technology on the metal structures and properties, *Archives of Materials Science and Engineering*, 2007, 28(12), 709–716.
140. M. Greger, R. Kocich, L. Čížek, L.A. Dobrzański, I. Juřička: Possibilities of mechanical properties and microstructure improvement of magnesium alloys, *Archives of Materials Science and Engineering*, 2007, 28(2), 83–90.
141. L.A. Dobrzański, T. Tański, J. Trzaska: Modeling of the optimum heat treatment conditions of Mg-Al-Zn magnesium cast alloys, *International Journal of Computational Materials Science and Surface Engineering*, 2007, 1(5), 540–554.
142. T. Tański, M. Król, L.A. Dobrzański, S. Malara, J. Domagała-Dubiel: Precipitation evolution and surface modification of magnesium alloys, *Journal of Achievements in Materials and Manufacturing Engineering*, 2013, 61(2), 97–149.
143. L.A. Dobrzański, M. Król: Structure and properties investigation of MCMgAl12Zn1 magnesium alloy, *Archives of Foundry Engineering*, 2013, 13(1), 9–14.
144. L.A. Dobrzański: Report on the main areas of the materials science and surface engineering own research, *Journal of Achievements in Materials and Manufacturing Engineering*, 2011, 49(2), 514–549.
145. L.A. Dobrzański, M. Król: Thermal and structure analysis of the MA MgAl6Zn3 magnesium alloy, *Journal of Achievements in Materials and Manufacturing Engineering*, 2011, 46(2), 189–195.
146. T. Tański, K. Labisz, L.A. Dobrzański: Effect of Al additions and heat treatment on corrosion properties of Mg-Al based alloys, *Journal of Achievements in Materials and Manufacturing Engineering*, 2011, 44(1), 64–72.
147. L.A. Dobrzański, M. Hetmańczyk, E. Łągiewka: Current state and development perspectives of Materials Science and Engineering in Poland, *Journal of Achievements in Materials and Manufacturing Engineering*, 2010, 43(2), 782–789.
148. L.A. Dobrzański, M. Król, T. Tański: Effect of cooling rate and aluminum contents on the Mg-Al-Zn alloys' structure and mechanical properties, *Journal of Achievements in Materials and Manufacturing Engineering*, 2010, 43(2), 613–633; *Open Access Library*, 1, 2011, 9–54.
149. T. Tański, L.A. Dobrzański, K. Labisz: Investigations of microstructure and dislocations of cast magnesium alloys, *Journal of Achievements in Materials and Manufacturing Engineering*, 2010, 42(1–2), 94–102.
150. L.A. Dobrzański, M. Król, T. Tański: Influence of cooling rate on crystallization, structure and mechanical properties of MCMgAl6Zn1 alloy, *Archives of Foundry Engineering*, 2010, 10(3), 105–110.
151. L.A. Dobrzański, M. Król, T. Tański: Application a neural networks in crystallization process of Mg-Al-Zn alloys, *Archives of Computational Materials Science and Surface Engineering*, 2010, 2(3), 149–156.

152. L.A. Dobrzański, M. Król, T. Tański: Thermal analysis, structure and mechanical properties of the MCMgAl3Zn1 cast alloy, *Journal of Achievements in Materials and Manufacturing Engineering*, 2010, 40(2), 167–174.
153. L.A. Dobrzański, M. Król: Thermal and mechanical characteristics of cast Mg-Al-Zn alloy, *Archives of Foundry Engineering*, 2010, 10(1), 27–30.
154. T. Tański, L.A. Dobrzański, R. Maniara: Microstructures of Mg-Al-Zn and Al-Si-Cu cast alloys, *Journal of Achievements in Materials and Manufacturing Engineering*, 2010, 38(1), 64–71
155. L.A. Dobrzański, M. Król: Application of the neural network for Mg-Al-Zn mechanical properties modelling, *Journal of Achievements in Materials and Manufacturing Engineering*, 2009, 37(2), 549–555: *Archives of Computational Materials Science and Surface Engineering*, 2010, 2(4), 181–188.
156. L.A. Dobrzański, T. Tański, J. Domagała, S. Malara, M. Król: Effect of high power diode laser surface melting and cooling rate on microstructure and properties of magnesium alloy, *Journal of Achievements in Materials and Manufacturing Engineering*, 2009, 37(2), 238–257: *Archives of Computational Materials Science and Surface Engineering*, 2010, 2(1), 24–43.
157. L.A. Dobrzański, S. Malara, T. Tański: Laser surface treatment of magnesium alloys with aluminium oxide powder, *Journal of Achievements in Materials and Manufacturing Engineering*, 2009, 37(1), 70–77.
158. L.A. Dobrzański, M. Król, T. Tański, R. Maniara: Effect of cooling rate on the solidification behaviour of MCMgAl6Zn1 alloy, *Journal of Achievements in Materials and Manufacturing Engineering*, 2009, 37(1), 65–69.
159. L.A. Dobrzański, J. Domagała-Dubiel, K. Labisz, E. Hajduczek, A. Klimpel: Effect of laser treatment on microstructure and properties of cast magnesium alloys, *Journal of Achievements in Materials and Manufacturing Engineering*, 2009, 37(1), 57–64.
160. L. Čížek, A. Hanus, O. Blahož, T. Tański, L.A. Dobrzański, M. Prażmowski, L. Pawlica: Structure and mechanical properties of Mg-Si alloys at elevated temperatures, *Journal of Achievements in Materials and Manufacturing Engineering*, 2009, 35(1), 37–46.
161. L.A. Dobrzański, M. Król, T. Tański, R. Maniara: Effect of cooling rate on the solidification behavior of magnesium alloys, *Archives of Computational Materials Science and Surface Engineering*, 2009, 1(1), 21–24.
162. L.A. Dobrzański, T. Tański, L. Čížek, J. Madejski: Selection of heat treatment condition of the Mg-Al-Zn alloys, *Journal of Achievements in Materials and Manufacturing Engineering*, 2009, 32(2), 203–210.
163. L. Čížek, M. Greger, I. Juřička, R. Kocich, L. Pawlica, L.A. Dobrzański, T. Tański: Structure and properties of alloys of the Mg-Al-Zn system, *Journal of Achievements in Materials and Manufacturing Engineering*, 2009, 32(2), 179–187.
164. L.A. Dobrzański, T. Tański, J. Domagała, M. Bonek, A. Klimpel: Microstructure analysis of the modified casting magnesium alloys after heat and laser treatment, *Journal of Achievements in Materials and Manufacturing Engineering*, 2009, 32(1), 7–12.
165. L.A. Dobrzański, T. Tański, J. Domagała, M. Król, S. Malara, A. Klimpel: Structure and properties of the Mg alloys in as-cast state and after heat and laser treatment, *Journal of Achievements in Materials and Manufacturing Engineering*, 2008, 31(2), 123–147.
166. L.A. Dobrzański, T. Tański, L. Čížek, J. Domagała: Mechanical properties and wear resistance of magnesium casting alloys, *Journal of Achievements in Materials and Manufacturing Engineering*, 2008, 31(1), 83–90.
167. W. Kasprzak, J.H. Sokolowski, M. Sahoo, L.A. Dobrzański: Thermal characteristics of the AM50 magnesium alloy, *Journal of Achievements in Materials and Manufacturing Engineering*, 2008, 29(2), 179–182.

168. L.A. Dobrzański, T. Tański, L. Čížek: Investigation of the MCMgAl12Zn1 magnesium alloys structure after heat treatment, *Journal of Achievements in Materials and Manufacturing Engineering*, 2008, 29(1), 23–30.
169. L.A. Dobrzański, J. Domagała, T. Tański, A. Klimpel, D. Janicki: Laser surface treatment of magnesium alloy with WC and TiC powders using HPDL, *Journal of Achievements in Materials and Manufacturing Engineering*, 2008, 28(2), 179–186.
170. W. Kasprzak, J.H. Sokolowski, M. Sahoo, L.A. Dobrzański: Thermal and structural characteristics of the AM50 magnesium alloy, *Journal of Achievements in Materials and Manufacturing Engineering*, 2008, 28(2), 131–138.
171. L.A. Dobrzański, T. Tański, J. Trzaska, L. Čížek: Modelling of hardness prediction of magnesium alloys using artificial neural networks applications, *Journal of Achievements in Materials and Manufacturing Engineering*, 2008, 26(2), 187–190.
172. L. Čížek, M. Greger, L.A. Dobrzański, R. Kocich, T. Tański, M. Praźmowski: Fracture analysis of selected magnesium alloys after different testing methods, *Journal of Achievements in Materials and Manufacturing Engineering*, 2007, 24(2), 131–134.
173. L.A. Dobrzański, T. Tański, J. Domagała, L. Čížek: Mechanical properties of magnesium casting alloys, *Journal of Achievements in Materials and Manufacturing Engineering*, 2007, 24(2), 99–102.
174. L.A. Dobrzański, T. Tański, J. Domagała: Microstructure analysis of MCMgAl16Zn1 M CMgAl8Zn1 MCMgAl4Zn1 cast magnesium alloys, *Archives of Foundry Engineering*, 2007, 7(4), 33–38.
175. L.A. Dobrzański, T. Tański, L. Čížek: Heat treatment impact on the structure of die-cast magnesium alloys, *Journal of Achievements in Materials and Manufacturing Engineering*, 2007, 20(1–2), 431–434.
176. L.A. Dobrzański, T. Tański, L. Čížek: Microstructure of MCMgAl12Zn1 magnesium alloy, *Archives of Foundry Engineering*, 2007, 7(1), 179–182.
177. L.A. Dobrzański, T. Tański, L. Čížek: Influence of Al addition on structure of magnesium casting alloys, *Journal of Achievements in Materials and Manufacturing Engineering*, 2006, 19(2), 49–55.
178. L. Čížek, M. Greger, L.A. Dobrzański, I. Juřička, R. Kocich, L. Pawlica, T. Tański: Mechanical properties of magnesium alloy AZ91 at elevated temperatures, *Journal of Achievements in Materials and Manufacturing Engineering*, 2006, 18(1–2), 203–206.
179. M. Greger, R. Kocich, L. Čížek, L.A. Dobrzański, M. Widomská, B. Kuřetová, A. Silbernagel: The structure and properties of chosen metals after ECAP, *Journal of Achievements in Materials and Manufacturing Engineering*, 2006, 18(1–2), 103–106.
180. L.A. Dobrzański, T. Tański, L. Čížek: Influence of Al addition on structure of magnesium casting alloys, *Journal of Achievements in Materials and Manufacturing Engineering*, 2006, 17(1–2), 221–224.
181. L.A. Dobrzański, T. Tański, K. Labisz, W. Pakieła: Structure and properties of AlMg5Si2Mn after laser feeding, Herald of Khmelnytskyi National University. *Technical Sciences* (Вiсник Хмельницького нацiонального унiверситету. Технiчнi науки), 2014, 3(213), 125–128.
182. L.A. Dobrzański, T. Tański, S. Malara: Effect of the heat and surface laser treatment on the corrosion degradation of the Mg-Al alloys, *Materials Engineering–Materiálové inžinierstvo*, 2011, 18(3), 85–92.
183. L.A. Dobrzański, A. Śliwa, T. Tański: Numerical simulation model for the determination of hardness for casting the magnesium alloys MCMgAl6Zn1, *Archives of Materials Science*, 2008, 29(3), 118–124.
184. L.A. Dobrzański, M. Król, T. Tański, R. Maniara: Effect of cooling rate on the solidification behavior of MC MgAl6Zn1 alloy, *Archives of Materials Science*, 2008, 29(3), 110–117.

185. L.A. Dobrzański, T. Tański, L. Čížek: Characteristics of MCMgAl9Zn1 and MCMgAl6Zn1 magnesium alloys structure, *Inżynieria Materiałowa*, 2007, 28(3–4) (157–158), 381–386.
186. L.A. Dobrzański, T. Tański: Zastosowanie sieci neuronowych do optymalizacji warunków obróbki cieplnej stopów Mg-Al, *Czasopismo Techniczne, Mechanika*, 2011, 108(7/4-M), 81–86.
187. L.A. Dobrzański, M. Król: Predykcja własności odlewniczych stopów Mg na podstawie wyników analizy termiczno-derywacyjnej, *Czasopismo Techniczne, Mechanika*, 2011, 108 (7/4-M), 73–79.
188. L.A. Dobrzański, J. Domagała, T. Tański, A. Klimpel, D. Janicki: Modyfikowanie warstwy wierzchniej odlewniczych stopów magnezu przy użyciu lasera diodowego dużej mocy (HPDL), *Inżynieria Materiałowa*, 2008, 29(6), 580–584.
189. L.A. Dobrzański, T. Tański, L. Čížek: Mikrostruktura i własności mechaniczne odlewniczych stopów magnezu Mg-Al, *Archiwum Odlewnictwa*, 2006, 6(22), 614–619.
190. L.A. Dobrzański, T. Tański, L. Čížek, J. Domagała: Mikrostruktura odlewniczych stopów magnezu Mg-Al, *Archiwum Odlewnictwa*, 2006, 6(21), 133–140.
191. L.A. Dobrzański, T. Tański, L. Čížek, L. Pawlica, J. Hubáčková: Analiza mikrostruktur modyfikowanych, odlewniczych stopów magnezu, *Zeszyty Naukowe* Nr 298, *Mechanika/Politechnika Opolska*, 2004, 78, 53–56.
192. L.A. Dobrzański, M. Piec, M. Bilewicz: Materiały kompozytowe o osnowie stopu magnezu EN-MCMgA19Zn1 wzmacniane cząstkami Al_2O_3, *Inżynieria Materiałowa*, 2003, 24(6), 605–608.
193. L.A. Dobrzański, A. Dobrzańska-Danikiewicz, T. Tański, W. Sitek: Surface engineering development perspectives identified on the basis of technology foresight results, *2nd Mediterranean Conference & New Challenges on Heat Treatment and Surface Engineering, Proceedings*, Dubrovnik, Croatia, 2013, pp. 61–68.
194. M. Król, L.A. Dobrzański: Thermal analysis of magnesium alloy, *Proceedings of the International Forum of Young Researchers*, (Problemy Nedropol'zovaniâ, Meždunarodnyj Forum-Konkurs Molodyh Učenyh, Sbornik Naučnyh Trudov, Sankt-Peterburg), St. Petersburg, Russia, 2012, Part II, pp. 12–13.
195. L.A. Dobrzański, M. Król, M. Bonek, M. Chalupová: Definition of magnesium alloy structure by quantity evaluation, Degradácia Konštrukčných Materiálov 2011, XII. vedecká konferencia so zahraničnou účasťou, Zborník prednasok, Terchová: Biely Potok, Žilina, Slovakia, 2011, pp. 38–43.
196. L.A. Dobrzański, T. Tański, S. Malara: Corrosion resistance of the Mg-Al alloys after the heat and surface laser treatment, Degradácia Konštrukčných Materiálov 2011, XII. vedecká konferencia so zahraničnou účasťou, Terchová: Biely Potok, Žilina, Slovakia, 2011, pp. 31–37.
197. T. Tański, L.A. Dobrzański, K. Labisz: Investigation of the microstructure characteristic of chosen magnesium alloys, *Proceedings of 13th International Materials Symposium, IMSP'2010*, Denizli, Turkey, 2010, pp. 761–768.
198. T. Tański, L.A. Dobrzański, K. Labisz: Investigation of the microstructure characteristic of chosen magnesium alloys, *Modern Achievements of Science and Education. IV International Conference*, Budva, Montenegro, 2010, pp. 761–768.
199. L.A. Dobrzański, T. Tański, J. Domagała, S. Malara, A. Klimpel: Laser surface treatment of Mg-Al-Zn alloys, *Proceedings of the International Conference of IFHTSE 2009 "New Challenges in Heat Treatment and Surface Engineering"*, Dubrovnik: Cavtat, Croatia, 2009, pp. 179–184.
200. L.A. Dobrzański, J. Domagała, T. Tański, A. Klimpel, D. Janicki: Laser surface treatment of magnesium alloy with side injection of titanium carbide, *Proceedings of the 2th International Conference on Modern Achievements of Science and Education*, Netanya, Israel, 2008, pp. 67–71.

201. L.A. Dobrzański, T. Tański: Influence of the chemical composition and precipitation processes on the structure and properties of the magnesium cast alloys, *24th International Manufacturing Conference*, IMC 24, Manufacturing – Focus on the Future, Waterford, Ireland, 2007, Vol. 1, pp. 499–505.
202. L.A. Dobrzański, T. Tański, L. Čížek, Z. Brytan: Structure and properties of magnesium cast alloys, *7th Asia Pacific Conference on Materials Processing*, APCMP 2006, Singapore, 2006, pp. 159–167.
203. T. Tański, L.A. Dobrzański, L. Čížek: Influence of heat treatment on structure and properties of the cast magnesium alloys, *International Conference on Processing & Manufacturing of Advanced Materials, Processing, Fabrication, Properties, Applications, THERMEC'2006*, Vancouver, Canada, 2006, CD-ROM.
204. L.A. Dobrzański, M. Król: Structure and properties investigation of MCMgAl12Zn1 magnesium alloy, Krzepnięcie i krystalizacja metali–2012. 53 Międzynarodowa Konferencja Naukowa, Cedzyna: Kielce, 2012, CD-ROM, pp. 1–6.
205. M. Greger, R. Kocich, L. Čížek, LA. Dobrzański, I. Juřička: Mechanical properties and microstructure of Mg-Al alloys after forming, *Proceedings of the 11th International Scientific Conference on the Contemporary Achievements in Mechanics, Manufacturing and Materials Science, CAM3S'2005*, Gliwice: Zakopane, 2005, pp. 370–375.
206. L.A. Dobrzański, T. Tański, L. Čížek: Influence of heat treatment on structure and property of the casting magnesium alloys, *Proceedings of the 11th International Scientific Conference on the Contemporary Achievements in Mechanics, Manufacturing and Materials Science, CAM3S'2005*, Gliwice: Zakopane, 2005, pp. 283–288.
207. L.A. Dobrzański, T. Tański, L. Čížek: Influence of modification with chemical elements on structure of magnesium casting alloys, *Proceedings of the 13th International Scientific Conference on Achievements in Mechanical and Materials Engineering, AMME'2005*, Gliwice: Wisła, 2005, pp. 199–202.
208. L. Čížek, M. Greger, R. Kocich, S. Rusz, I. Juřička, L.A. Dobrzański, T. Tański: Structure characteristics after rolling of magnesium alloys, *Proceedings of the 13th International Scientific Conference on Achievements in Mechanical and Materials Engineering, AMME'2005*, Gliwice: Wisła, 2005, pp. 87–90.
209. L. Čížek, M. Greger, L. Pawlica, L.A. Dobrzański, T. Tański: Study of selected properties of magnesium alloy AZ91 after heat treatment and forming, *Proceedings of the 12th International Scientific Conference on Achievements in Mechanical and Materials Engineering, AMME'2003*, Gliwice: Zakopane, 2003, pp. 165–169.
210. L.A. Dobrzański, M. Król: Wpływ szybkości chłodzenia na proces krystalizacji stopu MB MgAl6Zn3, VIII Ukraïns'ko-Pol's'ka Konferencìâ Molodih Naukovcìv, Mehanìka ta Ìnformatika, Tezi naukovih prac', (VIII Ukraińsko-Polska Konferencja Młodych Naukowców, Mechanika i Informatyka, Materiały Ukraińsko-Polskiej Konferencji Naukowej), Chmielnicki, Ukraina, 2011, pp. 160–161.
211. L.A. Dobrzański, T. Tański, L. Čížek, J. Domagała: Wpływ zawartości aluminium na własności odlewniczych stopów magnezu, *Polska metalurgia w latach 2002–2006*, Komitet Metalurgii PAN, Kraków, 2006, 693–698.
212. L.A. Dobrzański, T. Tański, W. Sitek, L. Čížek: Modelowanie własności mechanicznych stopu magnezu EN: MCMgAl9Zn1, *Proceedings of the 12th International Scientific Conference on Achievements in Mechanical and Materials Engineering AMME'2003*, Gliwice: Zakopane, 2003, 289–292.
213. T. Tański, L.A. Dobrzański: Laser surface treatment of cast magnesium alloys, 16. Internationales Dresdner Leichtbausymposium. Erfolgsfaktor Systemleichtbau im globalen Wettbewerb: Internationalisierung der Vernetzung von Leichtbau-Kompetenzclustern, Dresden, Germany, 2012, poster p. 1.

214. L.A. Dobrzański, T. Tański, M. Bonek, J. Domagała, A. Klimpel: The analysis of microstructure modified the casting magnesium alloys after heat and laser treatment, Innovative engineering practices for smart & susteinable manufacturing, *9th Global Congress on Manufacturing & Management, GCMM 2008*, Abstracts, Surfers Paradise, Australia, 2008, p. 106.
215. T. Tański, L.A. Dobrzański: Study of the structure and mechanical properties of magnesium casting alloys, *The 16th International Federation for Heat Treatment and Surface Engineering (IFHTSE) Congress, Proceedings*, Brisbane, Australia, 2007, p. 98.
216. T. Tański, L.A. Dobrzański, K. Labisz: Corrosion behaviour of die-casting magnesium alloys, *Mechatronics Systems and Materials,6th International Conference*, Abstracts, Opole, Poland, 2010, pp. 191–192.
217. L.A. Dobrzański, M. Król, K. Gołombek: Ocena struktury stopów Mg na podstawie analizy termiczno-derywacyjnej, *Proceedings of the Eighteenth International Scientific Conference on Contemporary Achievements in Mechanics, Manufacturing and Materials Science, CAM3S'2012*, Gliwice: Ustroń, 2012, p. 36.
218. L.A. Dobrzański, M. Król: Analiza ATD oraz komputerowa analiza obrazu w ocenie struktury stopów Mg, *Proceedings of the Seventeenth International Scientific Conference on Contemporary Achievements in Mechanics, Manufacturing and Materials Science, CAM3S'2011*, Gliwice: Wrocław, 2011, p. 35.
219. T. Tański, L.A. Dobrzański: Utwardzanie wydzieleniowe stopów Mg-Al-Zn, *Proceedings of the Fourteenth International Scientific Conference on Contemporary Achievements in Mechanics, Manufacturing and Materials Science, CAM3S'2008*, Gliwice: Ryn, 2008, 56.
220. M. Król, L.A. Dobrzański: Kinetyka krystalizacji odlewniczych stopów Mg-Al-Zn, *Proceedings of the Fourteenth International Scientific Conference on Contemporary Achievements in Mechanics, Manufacturing and Materials Science, CAM3S'2008*, Gliwice: Ryn, 2008, 42.
221. L.A. Dobrzański, T. Tański, L. Čížek: Mikrostruktura i własności mechaniczne odlewniczych stopów magnezu Mg-Al, Archiwum Odlewnictwa 6/22 (2006) 614–619.
222. L.A. Dobrzański, T. Tański, J. Domagała, L. Čížek: Mechanical properties of magnesium casting alloys, *Journal of Achievements in Materials and Manufacturing Engineering*, 2007, 24/2, 99–102.
223. L.A. Dobrzański, T. Tański, L. Čížek: Influence of aluminium on properties of magnesium cast alloys, *Proceedings of the 15th International Metallurgical & Material Conference*, Hradec nad Moravicí, Czech Republic, 2006 (CD ROM).
224. L.A. Dobrzański, T. Tański, L. Čížek: Influence of Al addition on microstructure of die casting magnesium alloys, *Journal of Achievements in Materials and Manufacturing Engineering*, 2006, 19/1, 49–55.
225. W.T. Kierkus, J.H. Sokolowski: Recent Advances in CCA: A new method of determining baseline equation, *AFS Transactions*, 1999, 66, 161–167.
226. M.B. Djurdjevis, W.T. Kierkus, G.E. Byczynski, T.J. Stockwell, J.H. Sokolowski: Modeling of fraction solid for 319 aluminum alloy, *AFS Transactions*, 1999, 14, 173–179.
227. L.A. Dobrzański, W. Kasprzak, M. Kasprzak, J.H. Sokolowski: A novel approach to the design and optimization of aluminum cast component heat treatment processes using advanced UMSA physical simulations, *Journal of Achievements in Materials and Manufacturing Engineering*, 2007, 24/2, 139–142.
228. W. Kasprzak, J.H. Sokolowski, W. Sahoo, L.A. Dobrzański: Thermal and structural characteristics of the AZ50 magnesium alloy, *Journal of Achievements in Materials and Manufacturing Engineering*, 2008, 29/2, 179–182.
229. R. Maniara, L.A. Dobrzański, J.H. Sokolowski, W. Kasprzak, W.T. Kierkus: Influence of cooling rate on the size of the precipitates and thermal characteristic of Al-Si cast alloys, *Advanced Materials Research*, 2007, 15–17, 59–64.

230. L.A. Dobrzański, W. Kasprzak, J. Sokolowski, R. Maniara, M. Krupiński: Applications of the derivation analysis for assessment of the ACAlSi7Cu alloy crystallization process cooled with different cooling rate, *Proceedings of the 13th Scientific International Conference "Achievements in Mechanical and Materials Engineering" AMME'2005*, Gliwice, Poland, 2005, 147–150.
231. L.A. Dobrzański, R. Maniara, J.H. Sokolowski: The effect of cast Al-Si-Cu alloy solidification rate on alloy thermal characteristics, *Journal of Achievements in Materials and Manufacturing Engineering*, 2006, 17, 217–220.
232. L.A. Dobrzański, R. Maniara, J.H. Sokolowski: The effect of cooling rate on microstructure and mechanical properties of AC AlSi9Cu alloy, *Archives of Materials Science and Engineering*, 2007, 28/2, 105–112.
233. L.A. Dobrzański, R. Maniara, J. Sokolowski, W. Kasprzak, Effect of cooling rate on the solidification behavior of AC AlSi7Cu2 alloy, *Journal of Materials Processing Technology*, 2007, 191, 317–320.
234. J. Sokolowski et al.: Method and Apparatus for Universal Metallurgical Simulation and Analysis–United States Patent, Patent No.: US 7,354,491 B2, Date of patent: April 8,2008, Canadian patent, Patent No.: CA 2 470 127, Date of patent: February17, 2009.
235. UMSA, Universal Metallurgical Simulator and Analyzer, University of Windsor, Canada, 2012, web4.uwindsor.ca/umsa.
236. Centre for Research in Computational Thermochemistry, École Polytechnique de Montréal, Canada, 2012, www.crct.polymtl.ca.
237. L.A. Dobrzański, R. Maniara, J.H. Sokolowski, M. Krupiński: Modelling of mechanical properties of Al-Si-Cu cast alloys using the neural Network, *Journal of Achievements in Materials and Manufacturing*, 2007, 20, 347–350.
238. M. Krupiński, L.A. Dobrzański, J.H. Sokołowski: Microstructure analysis of the automotive Al-Si-Cu castings, *Archives of Foundry Engineering*, 2008, 8/1, 71–74.
239. L.A. Dobrzański, R. Maniara, J. Sokolowski, W. Kasprzak, M. Krupiński, Z. Brytan: Applications of the artificial intelligence methods for modeling of the ACAlSi7Cu alloy crystallization process, *Journal of Materials Processing Technology*, 2007, 192–193, 582–587.
240. L.A. Dobrzański, M. Krupiński, R. Maniara, J.H. Sokolowski: Computer aided method for quality control of automotive Al-Si-Cu cast components, *Journal of Achievements in Materials and Manufacturing Engineering*, 2007, 24/2, 151–154.
241. L.A. Dobrzański, M. Krupiński, J.H. Sokołowski, P. Zarychta: The use of neural networks for the classification of casting defect, *International Journal of Computational Materials Science and Surface Engineering*, 2007, 1/1, 18–27.
242. L.A. Dobrzański, M. Krupiński, P. Zarychta, R. Maniara: Analysis of influence of chemical composition of Al-Si-Cu casting alloy on formation of casting defects, *Journal of Achievements in Materials and Manufacturing Engineering*, 2007, 21/2, 53–56.
243. L.A. Dobrzański, M. Krupiński, J.H. Sokolowski: Methodology of automatic quality control of aluminium castings, *Journal of Achievements in Materials and Manufacturing Engineering*, 2007, 20, 69–78.
244. M. Krupiński, L.A. Dobrzański, J.H. Sokolowski, W. Kasprzak, G. Byczynski: Methodology for automatic control of automotive Al-Si cast components, *Materials Science Forum*, 2007, 539–543, 339–344.
245. L.A. Dobrzański, M. Krupiński, J.H. Sokolowski, P. Zarychta, A. Włodarczyk-Fligier: Methodology of analysis of casting defects, *Journal of Achievements in Materials and Manufacturing Engineering*, 2006, 18, 267–270.
246. L.A. Dobrzański, M. Krupiński, J.H. Sokolowski: Methodology of automatic quality control of Al-Si casting alloys, *Proceedings of the 9th International Conference on Advances in Materials and Processing Technologies, AMPT 2006*, Las Vegas, NV, 2006, CD-ROM.

247. L.A. Dobrzański, M. Krupiński, J.H. Sokolowski: Methodology of quality assessment of castings from Al-Si, *3rd International Conference on Thermal Process Modelling and Simulation*, Budapest, Hungary, 2006, CD-ROM.
248. L.A. Dobrzański, M. Krupiński, J.H. Sokolowski: Zastosowanie metod sztucznej inteligencji do klasyfikacji wad w odlewach ze stopów Al-Si-Cu, *Archiwum Odlewnictwa*, 2006, 6/22, 598–605.
249. L.A. Dobrzański, M. Krupiński, J. Sokolowski: Metodyka oceny wad odlewniczych w stopach aluminium, Polska metalurgia w latach 2002–2006, Komitet Metalurgii PAN, Kraków, Poland, 2006, 699–704.
250. L.A. Dobrzański, M. Krupiński, J.H. Sokolowski, P. Zarychta, W. Kasprzak: The use of artificial intelligence methods for the identification of casting defects, *Proceedings of the 11th International Scientific Conference on the Contemporary Achievements in Mechanics, Manufacturing and Materials Science, CAM3S'2005*, Gliwice: Zakopane, Poland, 2005, 229–233.
251. L.A. Dobrzański, M. Krupiński, P. Zarychta, J. Sokolowski, W. Kasprzak: Computer assisted classification of flaws identified with the radiographical methods in castings from the aluminium alloys, *Proceedings of the 13th International Scientific Conference on Achievements in Mechanical and Materials Engineering, AMME'2005*, Gliwice: Wisła, Poland, 2005, 155–160.
252. L.A. Dobrzański, M. Krupiński, J.H. Sokolowski: Computer aided classification of flaws occurred during casting of aluminum, *Journal of Materials Processing Technology*, 2005, 167/2-3, 456–462.
253. L.A. Dobrzański, M. Krupiński, P. Zarychta, J. Sokolowski: Metodyka zautomatyzowanej oceny jakości w odlewach ze stopów aluminium, VI Międzynarodowa Konferencja Naukowa Wydziału Inżynierii Procesowej, Materiałowej i Fizyki Stosowanej WIPMiFS'55, Częstochowa, Poland, 2005, 544–547.
254. L.A. Dobrzański, M. Krupiński, R. Maniara, J. Sokolowski, W. Kasprzak: Komputerowe wspomaganie klasyfikacji wad identyfikowanych metodami radiograficznymi w odlewach ze stopów aluminium, *Proceedings of the 3rd Scientific Conference on Materials, Mechanical and Manufacturing Engineering, M3E'2005*, Gliwice: Wisła, Poland, 2005, 101–108.
255. L.A. Dobrzański, M. Krupiński, R. Maniara, J. Sokolowski, W. Kasprzak: Komputerowe wspomaganie klasyfikacji wad identyfikowanych metodami radiograficznymi w odlewach ze stopów aluminium, *Rudy i metale nieżelazne*, 2005, 50/5, 229–236.
256. L. Bäckerud, G. Chai: *Solidification Characteristics of Aluminum Alloys*, Vol. 3: Foundry Alloys, AFS SkanAluminium, Stockholm, Sweden, 1992.
257. S. Guldberg, N. Ryum: Microstructure and crystallographic orientation relationship in directionally solidified Mg–$Mg_{17}Al_{12}$-eutectic, *Materials Science and Engineering: A*, 2000, 289(1–2), 143–150.
258. Y. Liu, J. Zhou, D. Zhao, S. Tang: Investigation of the crystallographic orientations of the β- $Mg_{17}Al_{12}$ precipitates in an Mg-Al-Zn-Sn alloy, *Materials Characterization*, 2016, 118, 481–485.
259. A. Dobrzańska-Danikiewicz: E-foresight of materials surface engineering, *Archives of Materials Science Engineering*, 2010, 44/1, 43–50.
260. A.D. Dobrzańska-Danikiewicz: Foresight methods for technology validation, roadmapping and development in the surface engineering area, *Archives of Materials Science Engineering*, 2010, 44/2, 69–86.
261. A.D. Dobrzańska-Danikiewicz, T. Tański, S. Malara, J. Domagała-Dubiel: Technology foresight results concerning laser surface treatment of casting magnesium alloys, in: W.A. Monteiro Ed, *New Features on Magnesium Alloys*, InTech, Rijeka, 2012, pp. 1–30.

262. D. Dobrzańska-Danikiewicz, T. Tański, S. Malara, J. Domagała-Dubiel: Assessment of strategic development perspectives of laser treatment of casting magnesium alloys, *Archives of Materials Science Engineering*, 2010, 45/1, 5–39.
263. Y. Wang, D. Tie, R. Guan, N. Wang, Y. Shang, T. Cui, and J. Li: Microstructures, mechanical properties, and degradation behaviors of heat-treated Mg-Sr alloys as potential biodegradable implant materials, *The Journal of the Mechanical Behavior of Biomedical*, 2018, 77, 47–57.
264. D. Lin, F. Hung, T. Lui, and M. Yeh: Heat treatment mechanism and biodegradable characteristics of ZAX1330Mg alloy, *Materials Science & Engineering C-Materials for Biological Applications*, 2015, 51, 300–308.
265. Y. Dai, Y. Lu, D. Li, K. Yu, D. Jiang, Y. Yan, L. Chen, T. Xiao: Effects of polycaprolactone coating on the biodegradable behavior and cytotoxicity of Mg-6%Zn-10%Ca3(PO4)2 composite in simulated body fluid, *Materials Letters*, 2017, 198, 118–120.
266. Y. Chen, Z. Xu, C. Smith, J. Sankar: Recent advances on the development of magnesium alloys for biodegradable implants, *Acta Biomaterialia*, 2014, 10, 4561.
267. A. McGoron, D. Persaud-Sharma: Biodegradable magnesium alloys: A review of material development and applications, *Journal of Biomaterials and Tissue Engineering*, 2011, 12, 25–39.
268. D. Tie, R. Guan, H. Liu, A. Cipriano, Y. Liu, Q. Wang, Y. Huang, N. Hort: An in vivo study on the metabolism and osteogenic activity of bioabsorbable Mg-1Sr alloy, *Acta Biomaterialia*, 2016, 29, 455–467.
269. S. Zhang, X. Zhang, C. Zhao, J. Li, Y. Song, C. Xie, H. Tao, Y. Zhang, Y. He, Y. Jiang, Y. Bian: Research on an Mg-Zn alloy as a degradable biomaterial, *Acta Biomaterialia*, 2010, 6, 626–640.
270. Z. Li, X. Gu, S. Lou, Y. Zheng: The development of binary Mg-Ca alloys for use as biodegradable materials within bone, *Biomaterials*, 2008, 29, 1329–1344.
271. K. Yu, L. Chen, J. Zhao, S. Li, Y. Dai, Q. Huang, Z. Yu: In vitro corrosion behavior and in vivo biodegradation of biomedical β-Ca3(PO4)2/Mg–Zn composites, *Acta Biomaterialia*, 2012, 8, 2845–2855.
272. M.B. Kannan, R.K. Raman: In vitro degradation and mechanical integrity of calcium-containing magnesium alloys in modified-simulated body fluid, *Biomaterials*, 2008, 29, 2306.
273. G. Song: Control of biodegradation of biocompatable magnesium alloys, *Corrosion Science*, 2007, 49, 1696–1701.
274. S. Farè, Q. Ge, M. Vedani, G. Vimercati, D. Gastaldi, F. Migliavacca, L. Petrini, S. Trasatti: Evaluation of material properties and design requirements for biodegradable magnesium stents, *Matéria*, 2010, 15(2), 96–103.
275. X. Li, C. Chua, P.K. Chu, Effects of external stress on biodegradable orthopedic materials: A review, *Bioactive Materials*, 2016, doi:10.1016/j.bioactmat.2016.09.00.
276. R. Radha, D. Sreekanth: Insight of magnesium alloys and composites for orthopedic implant applications – A review, *Journal of Magnesium and Alloys*, 2017, 5, 286–312. doi:10.1016/j.jma.2017.08.003.
277. F. Witte, J. Fischer, J. Nellesen, H.A. Crostack, V. Kaese, A. Pisch, F. Beckmann, H. Windhagen: In vitro and in vivo corrosion measurements of magnesium alloys, *Biomaterials*, 2006, 27, 1013–1018.
278. M. Erinc, W.H. Sillekens, R. Mannens, R.J. Werkhoven: Applicability of existing magnesium alloys as biomedical implant materials, *Magnesium Technology*. E.A. Nyberg, S.R. Agnew, N.R. Neelameggham, and M.Q. Pekguleryuz, Eds, Minerals, Metals and Materials Society, San Francisco, CA, 2009, p 209–214.
279. M. Bamberger, G. Dehm: Trends in the Development of New Mg Alloys, *The Annual Review of Materials Research*, 2008, 38, 505–533.

280. N. Sezer, Z. Evis, S.M. Kayhanb, A. Tahmasebifar, M. Koç: Review of magnesium-based biomaterials and their applications, *Journal of Magnesium and Alloys*, 2018, 6, 23–43.
281. L.A. Dobrzański, The concept of biologically active microporous engineering materials and composite biological-engineering materials for regenerative medicine and dentistry, *Archives of Materials Science and Engineering*, 2016, 80/2, 64–85.
282. L.A. Dobrzański, A.D. Dobrzańska-Danikiewicz, Z.P. Czuba, L.B. Dobrzański, A. Achtelik-Franczak, P. Malara, M. Szindler, L. Kroll: Metallic skeletons as reinforcement of new composite materials applied in orthopaedics and dentistry, *Archives of Materials Science and Engineering*, 2018, 92(2), 53–85.
283. L.A. Dobrzański, A.D. Dobrzańska-Danikiewicz, Z.P. Czuba, L.B. Dobrzański, A. AchtelikFranczak, P. Malara, M. Szindler, L. Kroll, The new generation of the biological-engineering materials for applications in medical and dental implant-scaffolds, *Archives of Materials Science and Engineering*, 2018, 91(2), 56–85.
284. L.A. Dobrzanski: Research on the new generation of biological-engineering materials for medical and dental application (keynote lecture), *2nd Global Summit on Chemistry & Chemical Engineering GSCCE'2019*, 18–19 February 2019, Frankfurt, Germany.
285. L.A. Dobrzanski: Role of nanoscale surface treatment of the skeleton implant-scaffolds fabricated by selective laser sintering in dentistry (invited lecture), *14th Global Congress on Manufacturing & Management GCMM'2018*, 5–7 December 2018, Brisbane, Australia.
286. L.A. Dobrzanski: Surface nanocoatings inside pores of metallic skeletons improving nesting and proliferation human living cells (invited opening lecture), *International Conference Materials Research and Development*, 29–30 October 2018, Prague, Czechia.
287. L.A. Dobrzanski: New obtained nanostructural coatings on sintered tool materials and inside pores of a new generation microporous materials for medical application (invited plenary lecture), *The 9th International Conference on Technological Advances of Thin Films and Surface Coatings THINFILMS2018*, 17–20 July 2018, Shenzhen, China.
288. L.A. Dobrzanski: The selected applications (including medicine) of composite and nanocomposite materials (invited plenary lecture), *Multidisciplinary International Conference Advances in Metallurgical Processes and Materials AdMet'2018*, 10–13 June 2018, Lviv, Ukraine.
289. L.A. Dobrzanski: Effect of the biocompatible thin films on the properties of the skeleton materials fabricated by selective laser sintering for medical and dental application (plenary invited lecture), *International Conference on Materials Science and Graphene Technology 2018*, 9–11 April 2018, Dubai, UAE.
290. L.A. Dobrzanski: Okolicznościowy wykład specjalny z okazji 70. *Rocznicy Urodzin LAD, Politechnika Śląska*, 27 September 2017, Gliwice, Poland.
291. L.A. Dobrzanski: The skeleton microporous materials with coatings inside the pores for medical and dental application (keynote lecture), *11th International Conference on Advanced Materials & Processing*, 7–8 September 2017, Edinburgh, UK.
292. L.A. Dobrzanski: The application of the concept of the implant-scaffolds and hybrid multilayer biological-engineering composite materials (invited lecture), The workshop about application of 3D printing for manufacturing of dental constructions, 22 June 2017, Varna, Bulgaria.
293. L.A. Dobrzanski: Innovative technologies of a new obtained nanostructural and microporous materials (invited lecture), *III Scientific Congress INNOVATIONS'2017*, 19–22 June 2017, Varna, Bulgaria.

294. L.A. Dobrzanski: Structure and innovative technologies of a new obtained nanostructural materials (invited lecture), *The second International Conference Frontiers in Materials Processing Applications, Research and Technology, FiMPART'2017*, 4–7 June 2017, Bordeaux, France.
295. L.A. Dobrzanski: The application of the concept of the implant-scaffolds and hybrid multilayer biological-engineering composite materials (opening lecture), *16th International Materials Symposium, IMSP'2016*, 12–14 October 2016, Denizli, Turkey.
296. L.A. Dobrzanski: Original concept of the biological-engineering composites (special lecture), *Seminar in Universidade Estadual Paulista "Júlio de Mesquita Filho", Dental Faculty*, 31 August 2016, São José dos Campos, Brazil.
297. L.A. Dobrzanski: Metallic implants-scaffolds for dental and orthopedic application (keynote lecture), *9° COLAOB – Congresso Latino-Americano de Orgãos Artificiais e Biomateriais*, 24–27 August 2016, Foz do Iguaçu, PR, Brazil.
298. L.A. Dobrzanski: Application of the additive manufacturing by selective laser sintering for constituting implant-scaffolds and hybrid multilayer biological and engineering composite materials (keynote lecture), *International Conference on Processing & Manufacturing of Advanced Materials THERMEC'2016, Processing, Fabrication, Properties, Applications*, 29.05–3 June 2016, Graz, Austria.
299. L.A. Dobrzanski, P. Malara: Porous titanium scaffolds for cell grooving in dental application produced by additive manufacturing methods (invited lecture), *BIT's 6th Annual World Gene Convention-2015, "More Advanced, More Healthy and More Safety", WGC-2015*, 13–15 November 2015, Qingdao, China.
300. L.A. Dobrzanski: Additive manufacturing of metallic scaffolds for orthopedic application (invited lecture), *XXIV International Materials Research Congress, IMRC 2015*, 16–20 August 2015, Cancun, Mexico.
301. L. A. Dobrzański (coordinator): Project proposal DENTANANO-PLFRO – The explanation of the interaction in nanoscale of the human body in connection zone with the surface of the metallic porous or solid implanted devices applied in regenerative dentistry coated by the nanostructural layers in order to avoid postoperative surgical complications and for the ensure proliferation of living cells and progress of the osseointegration, *Medical and Dental Engineering Centre for Research, Design and Production ASKLEPIOS in Gliwice*, Poland, EURONANOMED2019–2022.
302. L. A. Dobrzański (coordinator): Project proposal IMEMLICE Interaction of microporous engineering materials and live cells in a newly developed category of composite medical and dental engineering-biological materials, *Medical and Dental Engineering Centre for Research, Design and Production ASKLEPIOS in Gliwice*, Poland National Science Centre 2018/31/B/ST8/00004.
303. L.A. Dobrzanski, Editorial, *Journal of Achievements in Materials and Manufacturing Engineering*, 34/1, 2009, p. 4; http://jamme.acmsse.h2.pl/papers_vol34_1/editorial.pdf; access 3 February 2019.
304. http://ec.europa.eu/research/industrial_technologies/pdf/pro-futman-doc3a.pdf; the archival document currently unavailable.
305. http://manufacturing-visions.org/download/Final_Report_final.pdf; the archival document currently unavailable.
306. M. Hermann, T. Pentek, B. Otto, Working Paper No. 01/2015, Design Principles for Industrie 4.0 Scenarios: A Literature Review, Technische Universität Dortmund Fakultät Maschinenbau Audi Stiftungslehrstuhl Supply Net Order Management, Dortmund 2015, p. 1–15; http://www.iim.mb.tu-dortmund.de/cms/de/forschung/Arbeitsberichte/Design-Principles-for-Industrie-4_0-Scenarios.pdf; access 30 January.2019.

307. H. Kagermann, W. Wahlster, J. Helbig, Eds.: Recommendations for implementing the strategic initiative Industrie 4.0: Final report of the Industrie 4.0 Working Group, Frankfurt a. Main, 2013, pp. 1–84, http://forschungsunion.de/pdf/industrie_4_0_final_report.pdf, access 30 January 2019.
308. H. Kagermann: Chancen von Industrie 4.0 nutzen. In: Bauernhansl, T., M. ten Hompel and B. Vogel-Heuser, Eds, 2014: *Industrie 4.0 in Produktion, Automatisierung und Logistik*, Anwendung, Technologien und Migration, 603–614.
309. G. Misra, V. Kumar, A. Agarwal, K. Agarwal: Internet of things (IoT) – A technological analysis and survey on vision, concepts, challenges, innovation directions, technologies, and applications (An upcoming or future generation computer communication system technology), *American Journal of Electrical and Electronic Engineering*, 2016, 4(1), 23–32. doi:10.12691/ajeee-4-1-4.
310. D. Giusto, A. Iera, G. Morabito, L. Atzori Eds, "The Internet of Things", Springer, Berlin, Germany, 2010.
311. D. Lucke, C. Constantinescu, E. Westkämper, 2008: Smart Factory – A Step towards the Next Generation of Manufacturing. In: M. Mitsuishi, K. Ueda F. Kimura, Eds, *Manufacturing Systems and Technologies for the New Frontier, the 41st CIRP conference on manufacturing systems*, Tokyo, Japan, 115–118.
312. J. Gracel, M. Stoch, A. Biegańska: Inżynierowie Przemysłu 4.0 (Nie)gotowi do zmian?, ASTOR WHITEPAPER, Kraków, 2017, pp. 1–84.
313. T. Iwański, J. Gracel: Przemysl 4.0; Rewolucja już tu jest. Co o niej wiesz? ASTOR WHITEPAPER, Kraków, 2016, pp. 1–44.
314. L.A. Dobrzański, L.B. Dobrzański (coordinators): the IMSKA-MAT Project number POIR.01.01.00-0397/16-00 Innovative dental and maxillo-facial implant-scaffold manufactured using the innovative technology and additive computer-aided materials design ADD-MAT, Medical and Dental Engineering Centre for Research, Design and Production ASKLEPIOS in Gliwice, Poland granted by Polish National Centre of Research and Development, Warsaw, Poland.

NOTICE

The Author takes part in the Project POIR.01.01-00-0485/16-00 on 'IMSKA-MAT Innovative dental and maxillofacial implants manufactured using the innovative additive technology supported by computer-aided materials design ADD-MAT' realized by the Medical and Dental Engineering Centre for Research, Design and Production ASKLEPIOS in Gliwice, Poland and co-financed by the National Centre for Research and Development in Warsaw, Poland.

6 Microstructure and Mechanical Properties of Mg-Zn Based Alloys

A Review of Mg-Zn-Zr Systems

Jae-Hyung Cho, Hyoung-Wook Kim, and Suk-Bong Kang

CONTENTS

6.1 INTRODUCTION

Magnesium alloys have drawn attention from many industrial sectors, particularly transportation, electronics, and sports, because of their lightness and great strength. A great deal of research has been carried out to fabricate better Mg products for industrial applications. Replacement of cast-aluminum components with magnesium in automobile and electronic parts has been used successfully to reduce mass and costs. Wrought Mg alloys fabricated by thermo-mechanical processing (extrusion, drawing or rolling); however, are hindered for wider applications because of their low formability. Low crystal symmetry and limited slip systems of Mg alloys inherently result in low formability. Therefore, fabrication of Mg sheets with high strength and high formability has been a crucial goal. In order to enhance the mechanical properties and formability of Mg sheets, a variety of alloy designs and thermo-mechanical processes have been tried.

Many magnesium alloys such as Mg-Al, Mg-Zn, Mg-Sn, and Mg-Ca based systems achieve their improved mechanical properties by age hardening [1]. The age hardening involves solution treatment at a relatively high temperature within the α

Mg region, and subsequent aging at a relatively low temperature to achieve a decomposition of the supersaturated solid solution into fine precipitates in the Mg matrix.

Mg-Zn based system is one of the well-known precipitate hardening alloys [2–5]. The eutectic temperature is about 613–615 K (340°C–342°C), and the maximum solid solubility of Zn in Mg is about 6.2 wt% (or 2.4 at.%) at the eutectic temperature [6–8]. Mg-Zn binary phase diagram and intermetallics are presented in Figure 6.1. The equilibrium solid solubility of Zn in Mg varies with temperatures, and the supersaturated solid solution of Zn in Mg can produce a remarkable age-hardening effect. Aging hardening can be carried out by solution heating (T4) around at 600 K, and subsequent precipitation aging (T6). It is generally accepted that for Mg-Zn alloys containing 4–9 wt% Zn, aged isothermally at 393–533 K (120°C–260°C) the strengthening precipitate phases develop. Age-hardening responses of two Mg-Zn systems are presented in Figure 6.2. Depending on the alloy composition and aging temperatures, the decomposition of the supersaturated solid-solution matrix phase involves the formation of G.P. zones, β_1 ($MgZn_2$), β_2 ($MgZn_2$), and β (Mg_2Zn_3). Hardness varies with aging time.

During isothermal aging, an evolution of precipitates occurs. Precipitate β_1 was formed as a rod or lath shape that was aligned with the long axis parallel to the c-axis of the α-matrix. Disc-shaped β_2 was located on the 0002 plane of the α-matrix. Several bulk-like particles formed at the casting stage were also found. It is known that variations in the Zn contents, from 3% to 6% in weight, affected the age hardening responses [9]. As Zn content increased, more precipitates were formed. After over-aging, the micro-hardness gradually decreased. The rod and disc shaped precipitates, β_1 and β_2, can act as obstacles that prevent dislocation movements. In particular, it is known that β_1 precipitates vertically pinned on basal planes can block the basal slip deformation effectively.

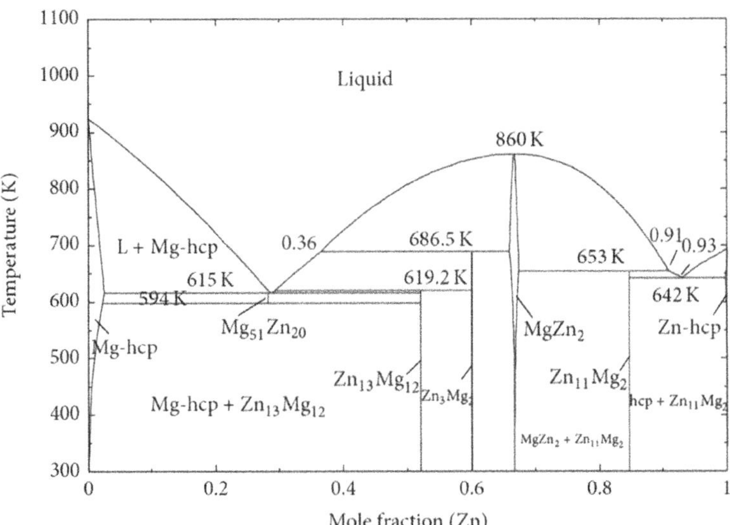

FIGURE 6.1 Mg-Zn phase diagram. (From Mezbahul-Islam, M. et al., *J. Mater.*, 704283, 1–33, 2014.)

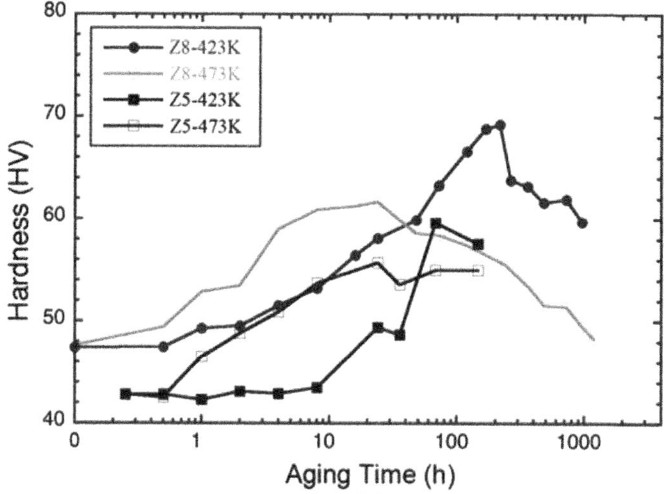

FIGURE 6.2 Age-hardening response of Mg-8 wt% Zn and Mg-5 wt% Zn alloys at 423 K and 473 K. (From Nie, J., *Metall. Mater. Trans. A*, 43, 3891–3939, 2012.)

Mg alloys with hexagonal crystal structure possess limited slip systems. In addition, rolled Mg alloys usually have a strong basal intensity because of easy activation of the basal slip. This is related to higher critical resolved shear stresses (CRSS) of non-basal slip systems than to those of basal slip systems at room temperature. Limited slip systems also result in twinning activation to accommodate plastic deformation. Various research has been conducted to better understand the plastic deformation and texture evolution of hexagonal materials [10–12].

Zr is frequently added to Mg-Zn system in order to refine the grain structure. The ZK (Mg-Zn-Zr) system possessed improved its mechanical properties due to grain refinement and aging hardening. ZK60 (overall chemical composition of Mg – 5.5wt% Zn – 0.5wt% Zr), in particular, exhibits high strength and elongation [13,14]. The mechanisms of the plastic deformation of magnesium alloys of ZK60 were investigated under compression or tension at various temperatures and strain rates [15–17]. Deep drawing was carried out on two different ZK60 magnesium sheets fabricated by ingot and twin-roll casting [18]. Microstructure and texture effect on draw-ability were examined under various temperature and drawing speed. Microstructure and mechanical properties of ZK60 sheets fabricated by twin-roll casting were also investigated in references [19–24].

The microstructure of ZK60 fabricated by twin-roll casting, and followed by warm rolling, was investigated using optical microscopy (OM) and transmission electron microscopy (TEM) after annealing treatment [22]. Tensile tests at room temperature were performed to show the influence of the annealing treatment on mechanical properties. Static recrystallization (SRX) was observed during annealing treatment of the ZK60 alloy sheets at and above 573 K (300°C). Shear bands, dislocations, and twins played important roles during static recrystallization and served as nucleation sites. Microstructure, texture and mechanical properties of the ZK60 sheets processed by differential speed rolling (DSR) were compared with those by conventional speed rolling [21,24]. The sheets processed by DSR showed significantly greater elongation than those processed by the conventional rolling sheets. Increased elongation was associated with deviation and weakening of the basal fiber texture. Below, we further summarize microstructure, hardening, and mechanical responses of Mg and Zn alloys with added Zr (ZK alloys) during aging, compression, conventional rolling, and differential speed rolling.

6.2 Zn VARIATION OF Mg-Zn-Zr (ZK) SYSTEM

Addition of some amount of Zn to Mg results in solid solution hardening of Mg alloys, and aging heat treatment of Mg-Zn alloys invokes various precipitates of MgxZny. In this part of the study, variation of rollability and mechanical properties of Mg-Zn alloys, with respect to Zn, was investigated. In order to refine the grain size, a small amount of Zr (about 0.5 wt%) was also added, and thus various Mg-Zn-Zr (ZK) systems with Zn content were obtained. Each sample contained approximately 1, 2, 3.8, or 5.8 wt% Zn, and these alloys were named ZK10, ZK20, ZK40, and ZK60, respectively.

Ingot-cast billets with a thickness of 22 mm were hot-rolled to 4 mm at a temperature of 573 K (300°C). The reduction in area for each rolling pass was about 5-15%. Figure 6.3 shows optical micrographs of the cross-sections of various ZK sheets.

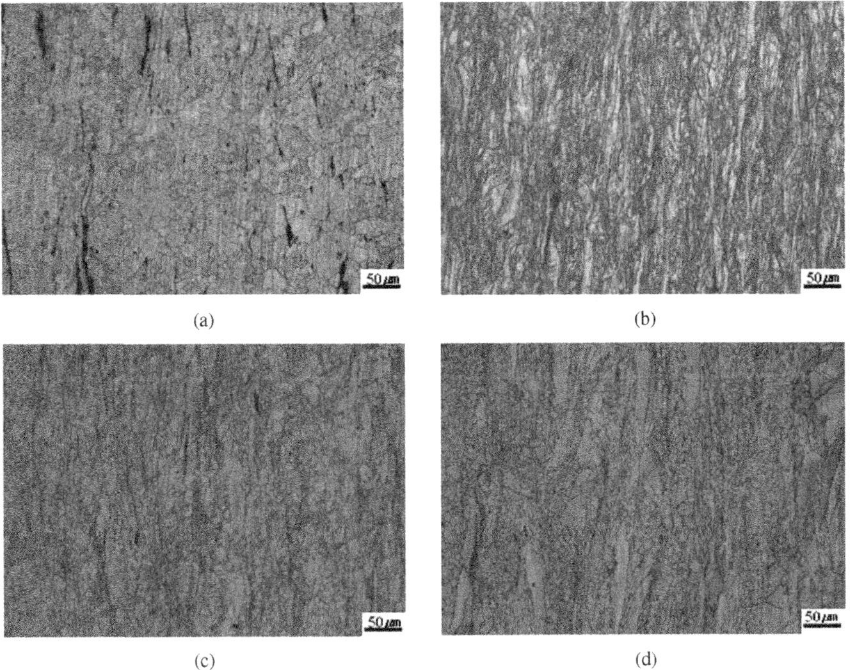

FIGURE 6.3 Optical micrographs across the thickness (4 mm) ZK sheets. (a) ZK10, (b) ZK20, (c) ZK40, and (d) ZK60.

ZK10 contained many large and equi-axial grains, in addition to cracks, as shown in Figure 6.3a. There were few cracks in ZK20, ZK40, and ZK60 (Figure 6.3b–d). The most shear bands were found in ZK20. Small grains formed by dynamic recrystallization during hot-rolling were most frequently found in ZK60. It seemed that the hardening effect caused by addition of Zn, promoted dynamic recrystallization. Small grains formed by dynamic recrystallization improved rollability and reduced interior cracks.

Further rolling of the 4-mm sheets was carried out at a temperature of 523 K (250°C) to achieve a thickness of 1 mm. Reduction in area per rolling pass was about 50%. ZK10 and ZK20 (relatively small additions of Zn) had some cracks inside during warm rolling, as shown in Figure 6.4a and b. Shear bands, and irregular dynamically recrystallized grains between shear bands, were found in the ZK20. This inhomogeneous microstructure degraded rollability of the ZK20 during warm rolling. ZK40 and ZK60 (high Zn contents) possessed a sound microstructure without cracks, as shown in Figure 6.4c and d. As Zn content increased, shear bands decreased and the number of dynamically-recrystallized grains increased. Actually, high Zn content resulted in more dynamic recrystallization.

The ZK40 and ZK60 alloys possessed better rollability. For this reason, different reductions in area (10% and 30%) were also carried out with the ZK40 and

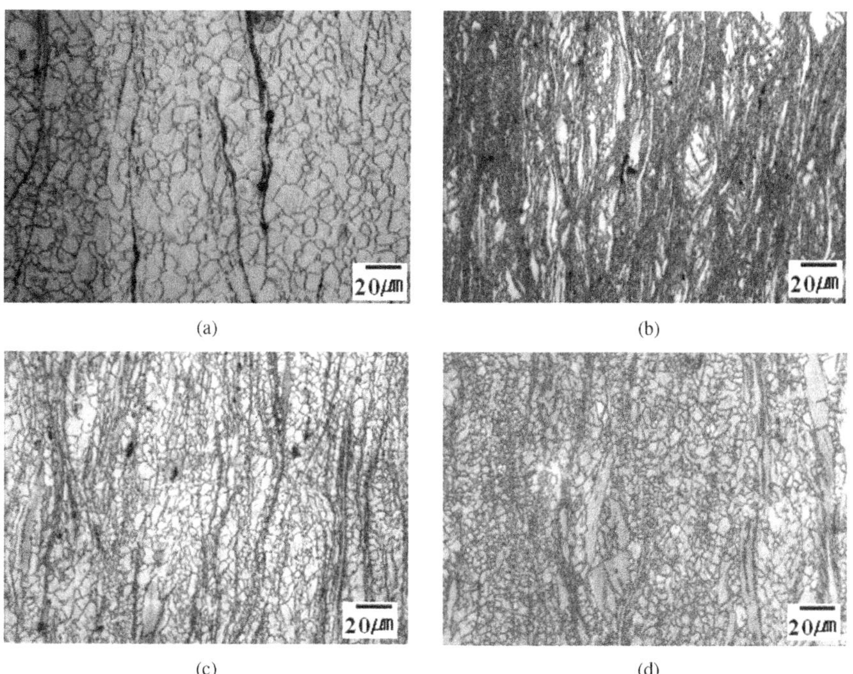

FIGURE 6.4 Optical micrographs across the thickness (1 mm) ZK sheets. (a) ZK10, (b) ZK20, (c) ZK40, and (d) ZK60.

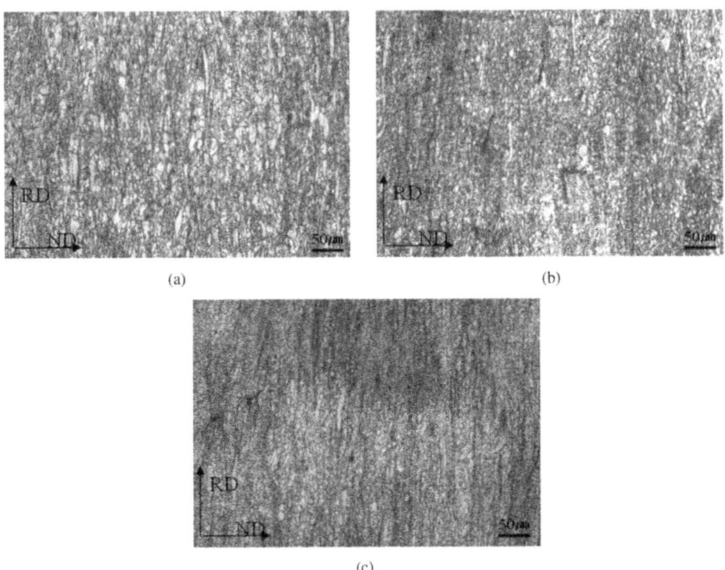

FIGURE 6.5 Optical micrographs of as-rolled ZK40 across its thickness (1 mm). Different reduction in area of (a) 10%, (b) 30%, and (c) 50% was applied.

ZK60 sheets (initially 4 mm thick). In Figure 6.5, optical micrographs of as-rolled ZK40 alloys 1-mm thick subjected to reduction in area of 10%, 30%, or 50% were compared. The temperature of the warm rolling process was 523 K (250°C). With greater reduction in area, the volume fraction of dynamic recrystallization increased and overall grain size was further refined.

Uniaxial tension tests of the as-rolled ZK40 1-mm thick were carried out for three different directions, the rolling direction (RD), 45° to the RD, and the transverse direction (TD) (Figure 6.6). Tensile strength along the RD, 45°, and TD was similar, regardless of the sample direction, and little effect from reduction in area on tensile strength was observed. Yield strength along each direction showed some difference, and the effect of reduction in area on yield strength was noticeable. Yield strength along the RD and 45° was not so different. Yield strength of the TD, however, was greater than those along the RD and 45°. The highest reduction in area (50%) resulted in less variation in yield strength in each direction. The most decreased shear bands, and most increased volume fraction of dynamic recrystallization, were observed for reduction in area of 50% (Figure 6.5). The microstructural feature seemed to decrease yield strength anisotropy.

The warm-rolled specimens mentioned above were heat-treated at a temperature of 523 K (250°C) for 10 min. Optical micrographs of annealed ZK40 alloys were observed with different reduction in area, as shown in Figure 6.7. Annealing heat-treatment removed most shear bands and the number of equi-axial grains increased. The overall grain size of specimens processed with a reduction in area of 50% was smaller than that of specimens processed with reduction in areas of 10% or 30%.

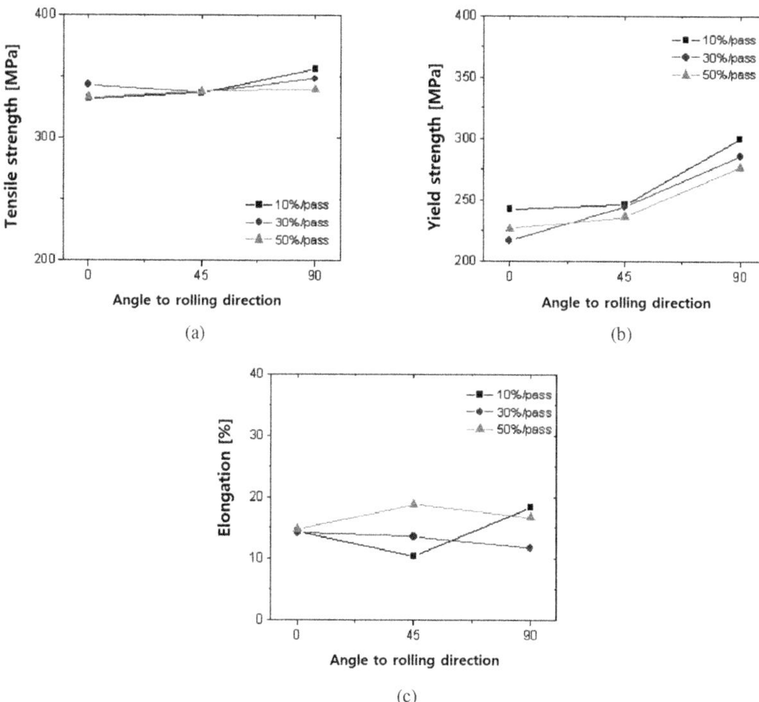

FIGURE 6.6 Tensile properties of as-rolled ZK40 with different reduction in area (1 mm sheets). (a) Ultimate tensile strength (UTS), (b) Yield strength (YS), and (c) Elongation (El).

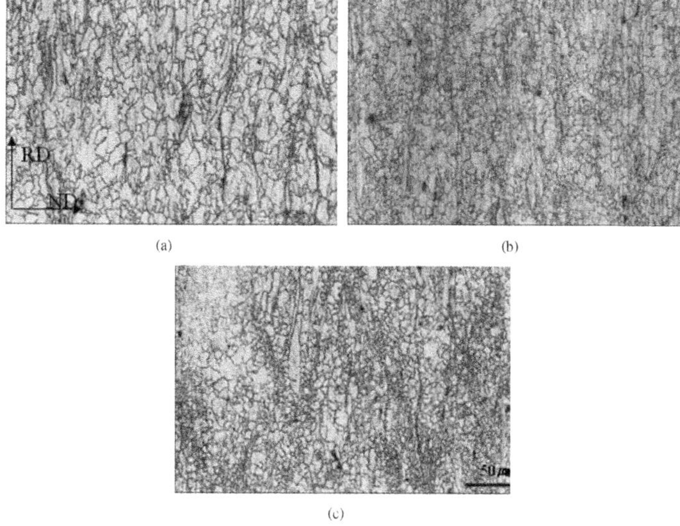

FIGURE 6.7 Optical micrographs of annealed ZK40 across its thickness (1 mm sheets) with annealing conditions of 523 K (250°C) for 10 min and different reduction in area of (a) 10%, (b) 30%, and (c) 50%.

Tensile properties of annealed ZK40 with different reduction in area are given in Figure 6.8. Note that each value of UTS, YS, and El decreased due to relaxation of deformed microstructure during annealing, compared with the rolled ones in Figure 6.6.

Warm rolling at a temperature of 523 K (250°C) was also carried out with ZK60 sheets with an initial thickness of 4 mm. Different reduction in areas of 10%, 30%, and 50% were applied to obtain sheets with a final thickness of 1 mm. Figure 6.9 shows optical micrographs of as-rolled sheets of ZK60. Some grains were elongated in the rolling direction. Dynamically recrystallized grains less than 5 μm also were observed. The number of dynamically recrystallized grains increased with increase in reduction in area.

Uniaxial tension tests of as-rolled ZK60 sheets 1-mm thick were carried out for three different directions, RD, 45° to RD, and TD (Figure 6.10). Tensile strength and yield strength in the RD, 45°, and TD were similar, regardless of the sample direction and degree of reduction in area.

However, elongation revealed some variation according to each direction and reduction in area. Planar anisotropy in elongation was associated with microstructural inhomogeneity.

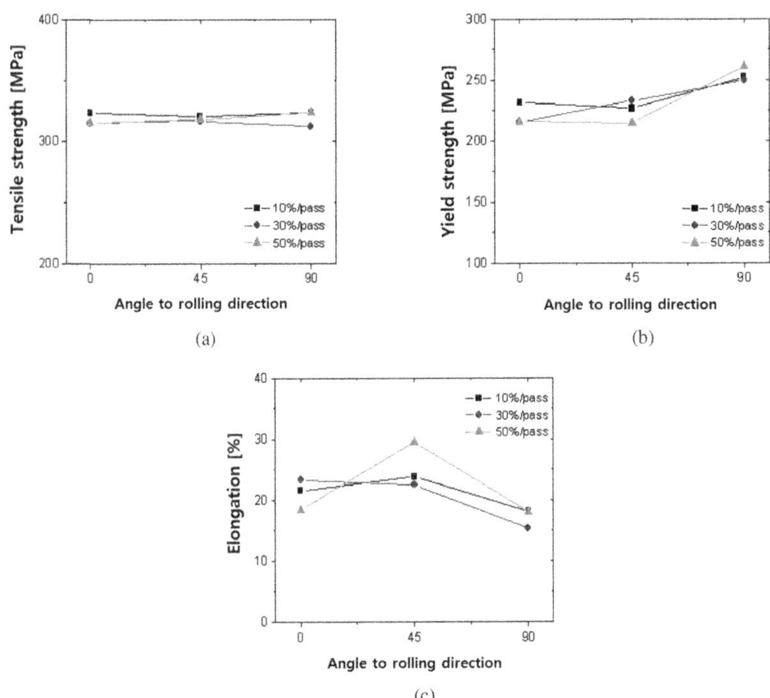

FIGURE 6.8 Tensile properties of annealed ZK40 with different reduction in area after annealing at 523 K (250°C) for 10 min (1 mm sheets). (a) Ultimate tensile strength (UTS), (b) Yield strength (YS), and (c) Elongation (El).

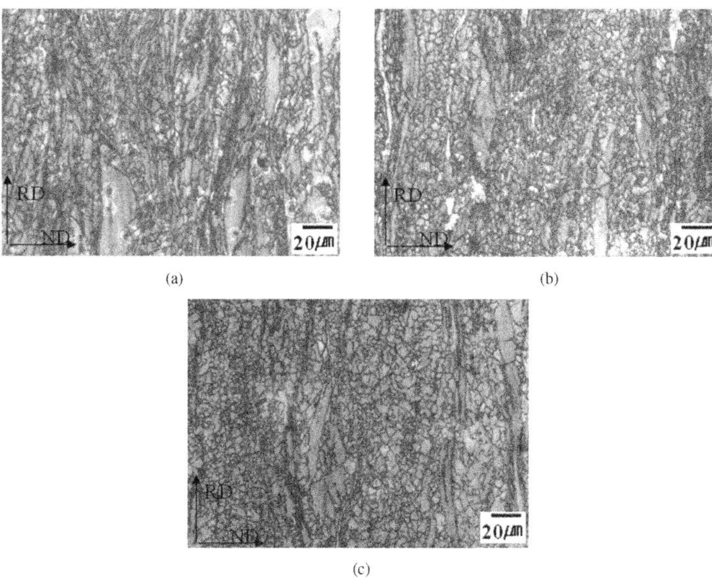

FIGURE 6.9 Optical micrographs of as-rolled ZK60 across its thickness (1 mm sheets) with different reduction in areas of (a) 10%, (b) 30%, and (c) 50%.

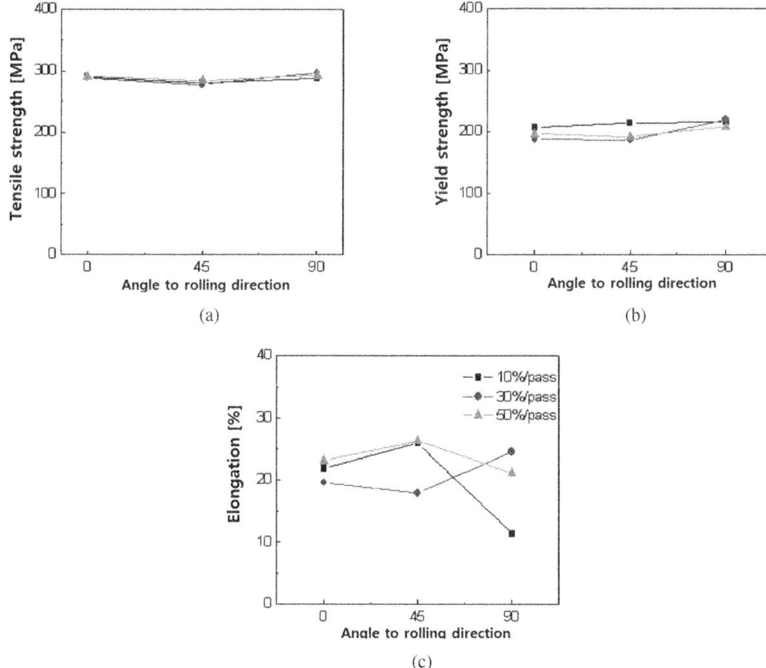

FIGURE 6.10 Tensile properties of the as-rolled ZK60 with different reduction in area (1 mm sheets). (a) Ultimate tensile strength (UTS), (b) Yield strength (YS), and (c) Elongation (El).

Dynamically recrystallized grains with homogeneous distribution revealed increased elongation (Figure 6.10c). However, mixtures of elongated (deformed) grains and refined (DRX) grains, such as in TD samples, resulted in little elongation. With increase in reduction in area, overall elongation increased. Increased volume fraction of recrystallized grains seemed to cause more elongation.

Optical micrographs of the as-annealed ZK60 alloys were observed with different degrees of reduction in area, as shown in Figure 6.11. The as-rolled sheets 1-mm thick were heat-treated at a temperature of 523 K (250°C) for 10 min. Compared with the as-rolled sheets in Figure 6.9, the number of equi-axial grains increased after annealing. As annealing temperature increased, equi-axial grains became larger.

The tensile properties of ZK60 with different degrees of reduction in area are given in Figure 6.12. The tensile strength of annealed ZK60 was greater than that of rolled sheets. Planar anisotropy of tensile strength was not common, and each direction possessed similar tensile strength. The yield strength of annealed ZK60 revealed some deviation in each direction, and that along the TD was particularly large. The overall yield strength of annealed sheets was also greater than that of rolled sheets. This trend is different from the case of ZK40, as indicated in Figures 6.6 and 6.8. In addition, possible relaxation of deformed microstructure during annealing was not observed in ZK60.

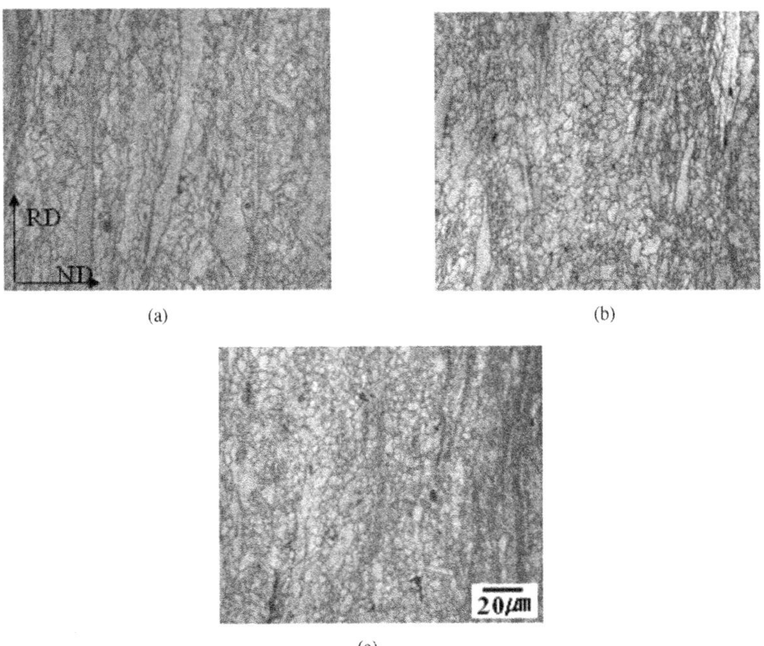

FIGURE 6.11 Optical micrographs of as-annealed ZK60 across its thickness (1 mm sheets) with annealing conditions of 523 K (250°C) for 10 min and different reduction in areas of (a) 10%, (b) 30%, and (c) 50%.

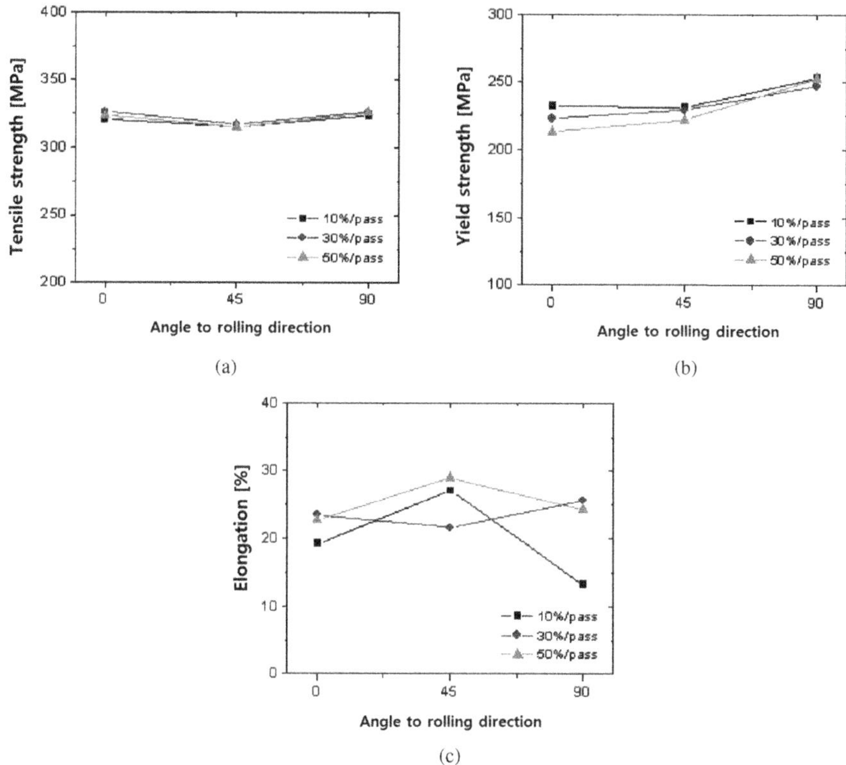

FIGURE 6.12 Tensile properties of ZK60 with different reduction area after annealing at 523 K (250°C) for 10 min (1 mm sheets). (a) Ultimate tensile strength (UTS), (b) Yield strength (YS), and (c) Elongation (El).

As mentioned in the introduction, Mg-Zn binary system possesses hardening behavior by MgxZny precipitation. Precipitation caused by annealing at 523 K (250°C) for 10 min could increase the strength of ZK alloys, although precipitate hardening effect can be maximized by solid solution followed by aging. The aging behavior of ZK60 alloys with high Zn content will be discussed further.

Overall, strength and elongation of ZK60 were greater than those of ZK40. Addition of Zn seems to strengthen Mg alloys by the solid solution mechanism. In addition, a great reduction in area was more effective for reducing planar anisotropy and increasing elongation. Greater reduction in area was associated with fewer shear bands and more dynamic recrystallization. Grain size was further refined with large reduction in area. It seems that both Zn content (alloying) and reduction in area (thermo-mechanical processing) mainly affected microstructural features including grain size, shear band, and dynamic recrystallization. Working temperature, or annealing heat treatment also can affect microstructural relaxation, grain growth, and even precipitation behavior. Crystallographic texturing also usually affects mechanical properties. The mechanical properties, including strength and elongation, reflected all those microstructural features.

6.3 AGING HEAT TREATMENT OF ZK60

Here aging behavior of ZK60 with high strength and elongation was further discussed. The approximate chemical composition of the ZK60 magnesium alloy used was of Mg-5.47Zn- 0.58Zr (wt%). The alloys were originally fabricated by ingot casting (IC), and then they were solution heat-treated at 673 K for 15 hr (T4). This was followed by artificial aging (T6) at 448 K for 12 hr. More detailed conditions for the solid solution and aging hardening heat treatments of ZK60 alloys were provided in [13,14].

ZK60 alloys show variation in hardness with aging time. After peak hardness was attained, hardness gradually decreased. The hardness value dropped after 100 hr because of overaging, as shown in Figure 6.13. Rod-like β_1 precipitates were most found in specimens with aging for 6–100 hr. The increased β_1 precipitates contributed to the increase in hardness as Mg-Zn alloys.

The density and shape of the precipitates changed with aging. The drop in hardness after 100 hr was related to the decrease of β_1 precipitates. Increase and decrease in hardness was related to variation in precipitates with aging.

When observing the microstructure of rolled ZK60 alloy sheets, various precipitates were present. Rectangular particles (T-precipitate) usually formed during casting. Both rod-like β_1 and disc-shaped β_2 precipitates were also found in the matrix. They mainly consisted of Mg and Zn. The precipitates had various concentrations of Mg and Zn. Zr was rarely found in the β_1, but was occasionally found in the β_2. β_1 has a unique growing direction of [0001]. This directionality of the precipitates was frequently destroyed by complex lattice misorientation caused by deformation during thermo-mechanical processing. Even after solution heat treatment at 673 K (400°C) for 1.5 hr, T and β_2 precipitates were found undissolved in some regions, while most β_1 precipitates were dissolved.

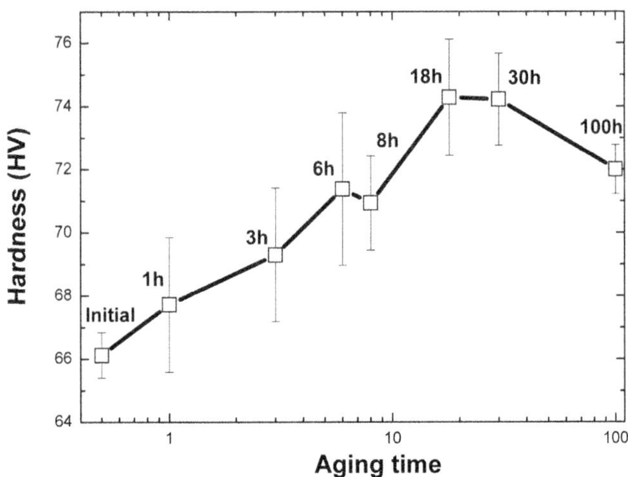

FIGURE 6.13 Hardness variation in ZK60 alloys with aging at 448 K (175°C). (From Cho, J.H. et al., *Mater. Sci. Forum*, 558–559, 159–164, 2007.)

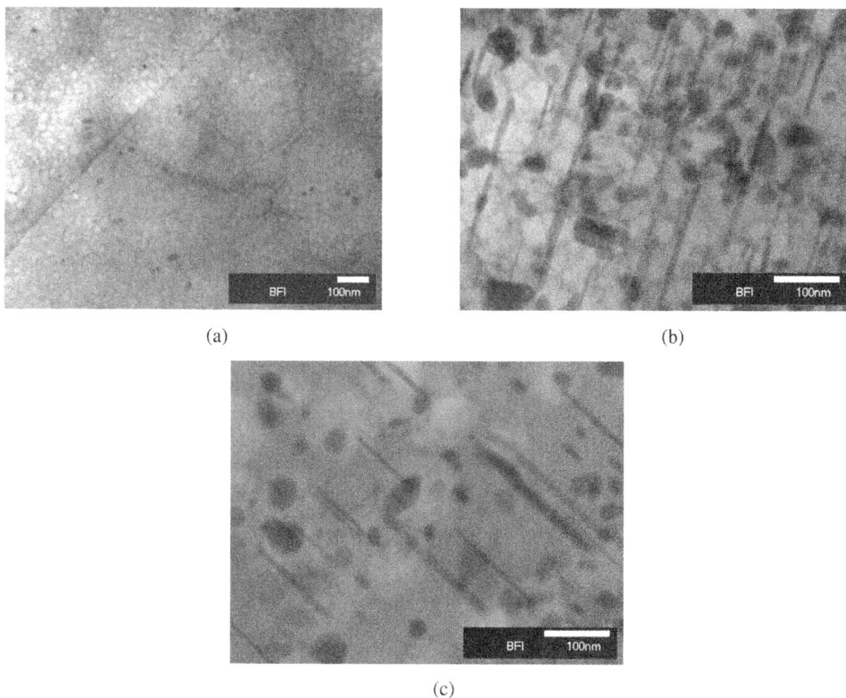

FIGURE 6.14 Precipitation evolution of ZK60 alloys with aging time: (a) Initial, (b) 8 hr, and (c) 100 hr.

TEM micrographs taken of aged ZK60 specimens are shown in Figure 6.14. Rod-like β_1 particles appeared after aging of about 6 hr. Morphologies and spatial distributions of precipitates of ZK60 revealed similarity to Mg-Zn alloys during aging [3]. The main effect of the addition of Zr to Mg is grain refinement, which is caused by the peritectic reaction at around 926 K (653°C) for the Mg-Zr binary alloy [25]. The amount of Zr is very small, and precipitates containing Zr are less often found than are precipitates of Zn.

Bar-shaped precipitates without Zr, which were shorter and thicker than rod-shaped β_1 were analyzed with SAED patterns, and they were found to be MgZn2. Most β_1 precipitates contain Zn only. While, Zr is rarely found in the β_1 precipitates. Zr is usually found in the disc-shaped or irregularly-shaped precipitates. Figure 6.15 shows Mg-Zn-Zr precipitates, which seem to be formed during cast solidification. Each precipitate (pt1, pt2, and pt3) contains all three elements (Mg, Zn, and Zr), and their average compositions were 53 wt% Mg, 32 wt% Zn, and 13 wt% Zr, respectively. Precipitate clusters containing Zr were also found in the matrix, which seem to be the result of shattering of T-precipitates containing Zr, by rolling process. Mg is the richest element in the ZK60 alloys and is mostly found in the EDS mapping.

In conclusion, solid solution and subsequent artificial aging of ZK60 alloys critically affected precipitation and mechanical properties. Variation in shape and

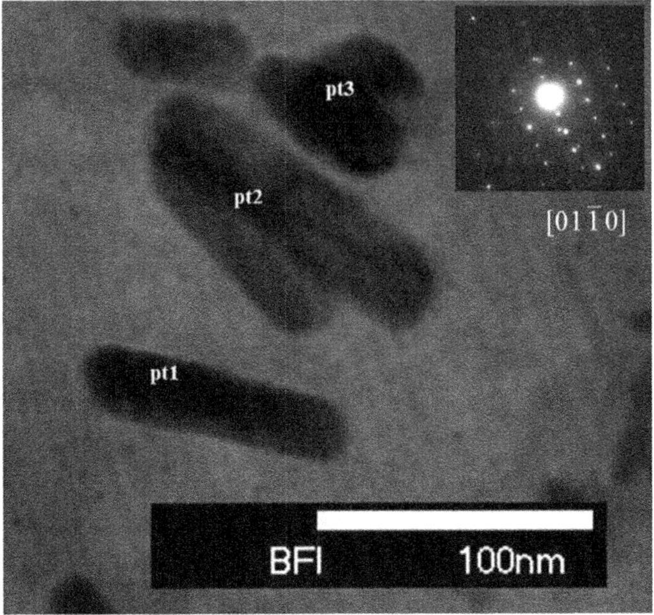

FIGURE 6.15 EDS of Mg-Zn precipitation of ZK60.

frequency of the precipitates with aging was closely related to hardness. Near the peak hardness, homogeneously-distributed rod-shaped β_1 precipitates were predominant. The rod-shaped β_1 precipitates were oriented with a growth direction of [0001] of Mg. When over-aged, the rod-shaped β_1 and disc-shaped β_2 precipitates coarsened, and their density decreased. Detailed discussion of the crystal structure of β_1 and β_2 can be found elsewhere [1,4,5,26].

6.4 UNIAXIAL COMPRESSION AND TENSION OF ZK60

Aged ZK60 alloys had improved mechanical properties because of precipitates. A comparison study of ZK60 alloys with and without aging heat-treatment helped understand the strengthening mechanism of ZK60 alloys caused by precipitates. Two ZK60 specimens were prepared: one by solid solution heat treatment (T4), and the other by solid-solution heat treatment followed by artificial aging (T6). Variation in texture and microstructure of ZK60 during deformation was also investigated. The aged samples were heat-treated to have hardness peak by aging condition, as mentioned in the previous section. High resolution scanning electron microscopy (HR-SEM) with electron backscatter diffraction (EBSD) system was effectively used to examine deformation and recrystallization of magnesium alloys [27–29].

The slip system was first activated during plastic deformation. Twinning also quickly started at the beginning of the deformation. Figure 6.16 displays inverse

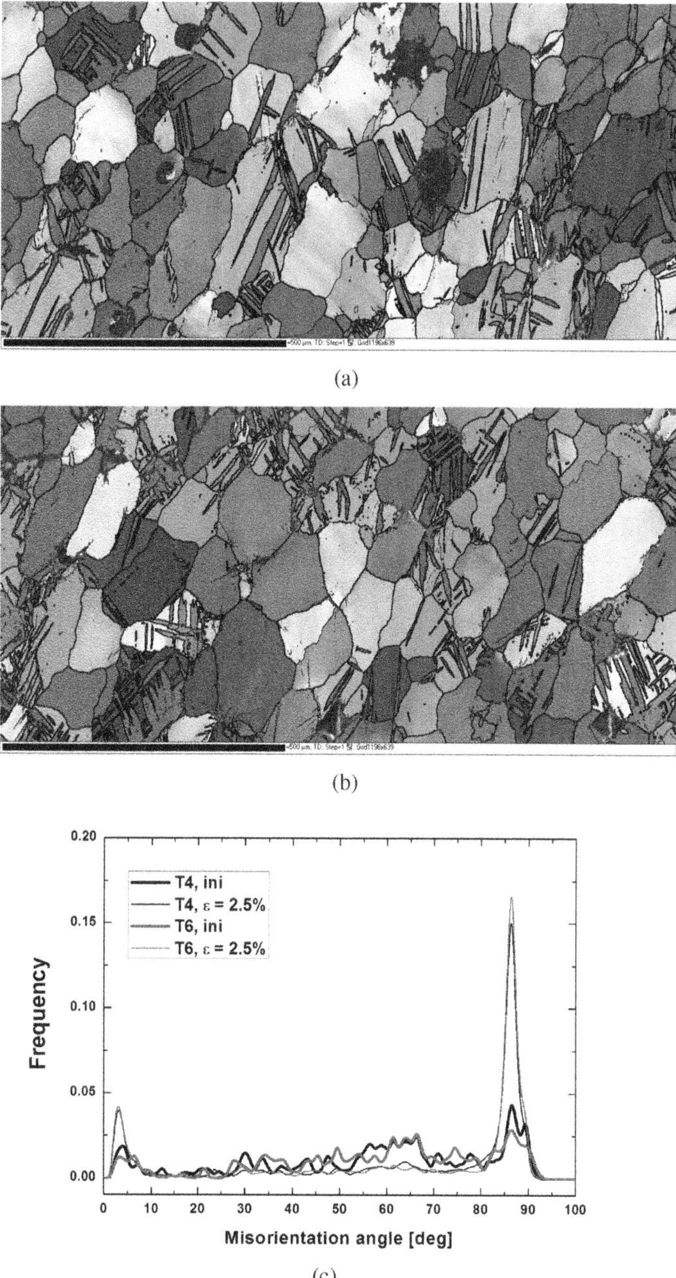

FIGURE 6.16 IPF maps during uniaxial compression ($\varepsilon = 2.5\%$) of solid solution and peak-aged ZK60 and the misorientation angle distribution. Thick lines represent a grain identification (GID) angle of 15°, and thin lines, a GID of 2°: (a) Solid solution (T4), (b) Peak-aged (T6) samples, and (c) Misorientation angle distribution. (From Cho, J.-H. et al., *Microsc. Microanal.*, 19, 8–12, 2013.)

pole figure (IPF) maps at a strain of 2.5% for both the solid solution (T4) and the aged (T6) ZK60 alloys. Some grains revealed the initial stage of twinning, and both sharp- and narrow-shaped twin regions were observed in the parent matrix. Other grains contained wide twinning regions, which implies twinning propagation into the parent matrix in progress. Most of the twinned regions have one twin variant, and a few grains had two twin variants. The other grains just exhibited slip deformation without twinning. Most of the twins were of the tensile twinning type (i.e., 86° $11\bar{2}0$), and twin variants in the same matrix had some relationship as discussed in [28]. The highest frequency of a misorientation angle of approximately 86° is associated with activation of tensile twinning for both the solid solution and the aged ZK60 alloys. The difference in the misorientation angle distribution, between the solid solution and the aged samples during compression, was not evident.

Many precipitates formed by the aging process contributed to an increase in the hardness of ZK60 alloys because of an increase in the interaction between the precipitates and dislocations. In addition, the aging process could change the chemical composition in the matrix, such as that of Zn and Zr, and the variations in the proportions of the elements could affect the SFE (stacking fault energy) of ZK60 alloys. With a decrease in the SFE, easy dissociation of a dislocation into partials, and a higher strain hardening rate during slip deformation, were usually observed [30,31]. The initiation of a partial dislocation was enhanced by low SFE [30]. A few studies reported SFE variation in Mg alloys containing alloy elements. The SFE of the Mg-Y system significantly decreased with Zn doping [32] according to a first-principles study. With the observation of a large number of stacking faults and wide splitting of the partial dislocations, the SFE of the Mg-Y system was found to be lower than that of pure Mg [33]. The kinetics of deformation twinning was also related to variation in the SFE, and thus some difference in the deformed microstructure, and the texture between the T4 and T6 samples, was expected.

When considering aging effect on the flow curves of ZK60 alloys, it was found that there existed some difference in the yield strength, strain hardening rates, and tensile strength between solution heat-treated (T4) and artificially aged (T6) samples. The flow curves of solution heat- treated (T4) ZK60 alloys have lower yield stresses and a higher strain hardening rate than those of artificially aged (T6) samples (Figure 6.17).

In conclusion, the texture and microstructure evolution of wrought magnesium alloys of ZK60 (Mg-Zn-Zr) were examined during uniaxial compression using EBSD. ZK60 is a type of precipitate hardening material, and its strength and hardness change with aging time. The grain orientations were a dominant factor in the activation of twinning when the concerned grain size was sufficiently large for twinning. In most cases, twinning was found to occur with the appearance of tensile twins. Grains with non-basal orientations underwent twinning, followed by slip deformation. The grain orientations were a dominant factor in the activation of twinning when the concerned grain size was sufficiently large for twinning.

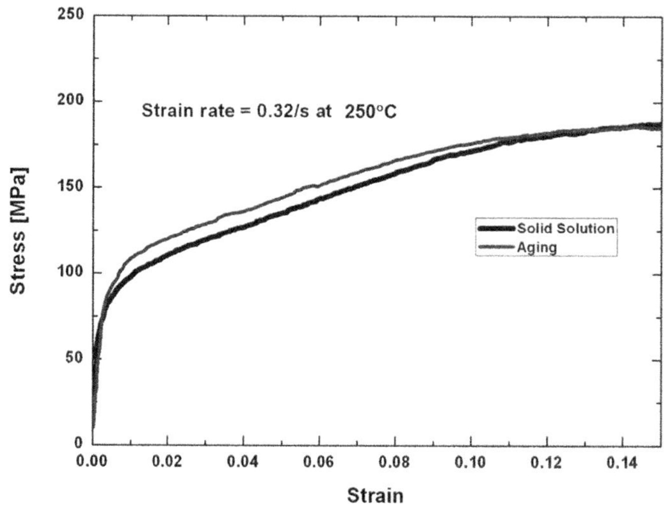

FIGURE 6.17 Flow curves for solid-solution heat-treated and aged samples at a temperature of 250°C and a strain rate of 0.32/s. (From Cho, J.-H. et al., *Metall. Mater. Trans. A*, 41, 2575–2583, 2010.)

6.5 SYMMETRIC AND ASYMMETRIC ROLLING OF ZK60

The manufacturing cost of the wrought aluminum or magnesium sheets by conventional direct-chill (DC) casting is usually high, because numerous thermo-mechanical steps of DC casting, such as scalping, homogenizing, hot rolling, cold rolling, and heat treatment should be applied. One way to reduce high manufacturing cost is to use twin-roll casting (TRC) [34–36]. The process possesses the significant merit of one-step processing from the liquid melt to the solid wrought strips by combining the casting and hot rolling processes. In addition, fast cooling during the TRC frequently improves mechanical properties due to refined grains and particles. Figure 6.18 shows the optical microstructure of various magnesium alloys including ZK60 (Mg- 5.5Zn-0.5Zr), AZ41 (Mg-4Al-1Zn), and AM31 (Mg-3Al-1Mn). The large and equi-axial grains were usually found in the DC samples, while the microstructure of the TRC samples revealed refined grains with a dendritic shape. Rapid solidification and subsequent hot deformation also critically affected the texture and microstructure of the TRC strips [19]. The fast solidification rate resulted in a microstructural gradient between the surface and center regions of the strips. The grain size in the surface region was relatively finer than that in the center due to a thermal gradient.

Variation of mechanical properties of ZK60 fabricated by the TRC process with temperature is shown in Figure 6.19. It is evident that overall strength decreases and elongation increases, with increasing temperature.

In this section, the effect of roll-speed ratio and deformation temperature on the texture evolution of solution heat-treated (T4) and aged (T6) samples of ZK60 sheets

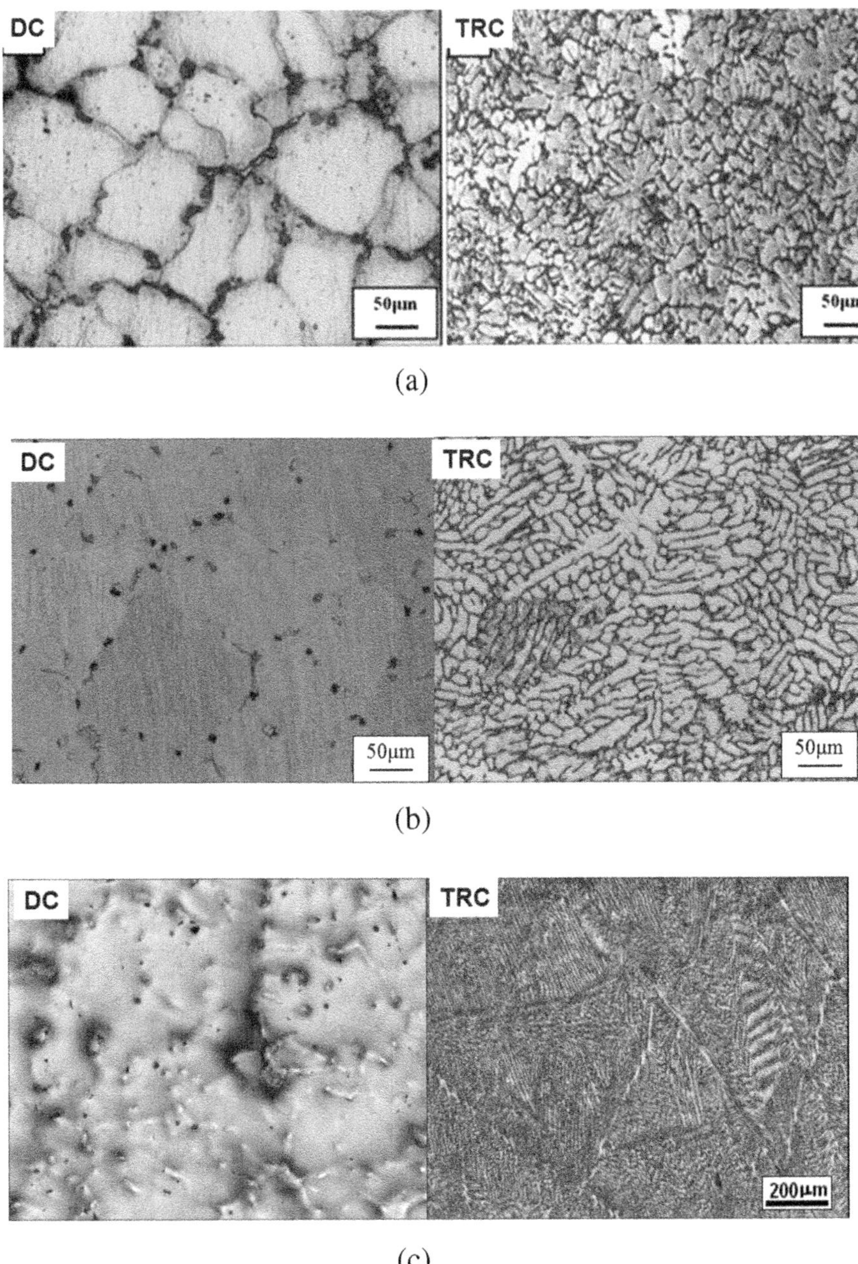

FIGURE 6.18 Comparison of optical micrographs between DC and TRC sheets. (a) ZK60, (b) AZ41, and (c) AM31.

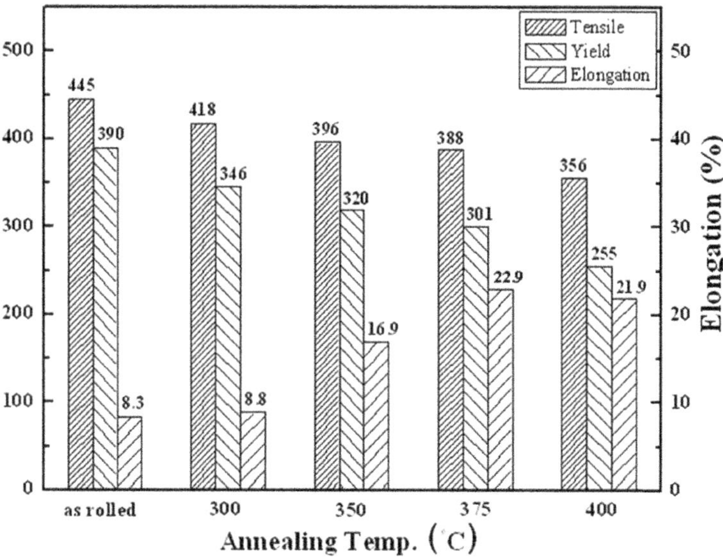

FIGURE 6.19 Variation of mechanical properties of ZK60 sheets fabricated by the TRC process with annealing temperature.

fabricated by TRC were mainly reviewed. The velocity of the top roll, Vtop, was held constant at 10 mpm (meter per minute), and that of the bottom roll, Vbot, was varied intentionally at 10, 12, and 15 mpm.

Figure 6.20 shows the variation of the strain rate and reduction area during rolling. At 448 K (175°C), asymmetric rolling processes have higher strain rates than conventional rolling. The strain rates increased with the roll-speed ratio. Solution heat-treated (T4) samples possess higher strain rates than aged (T6) samples at the same roll-speed ratio. At higher temperature of 498 K (225°C), the variational trends of strain rates were similar to those at 448 K (175°C). The strain rate increased with the roll speed. Strength usually decreases with temperature, and thus strain rates increase with temperature, assuming the same rolling conditions. The overall strain rates at 498 K (225°C) were greater than those at 448 K (175°C).

Texture variation in the surface regions was examined using XRD. The orientation distribution function (ODF) and complete pole figures were computed from incomplete pole figures that were experimentally measured from the top surface and the center of the sheets. Based on ODF, the volume fraction of the specific texture components can be computed, and it can provide easy quantitative understanding of the texture evolution. The volume fractions of texture components were obtained using an appropriate approach [37]. Basal textures of magnesium alloys consisted of two fibers (i.e., Type A, 0002 112¯0, and Type B, 0002 101¯0).

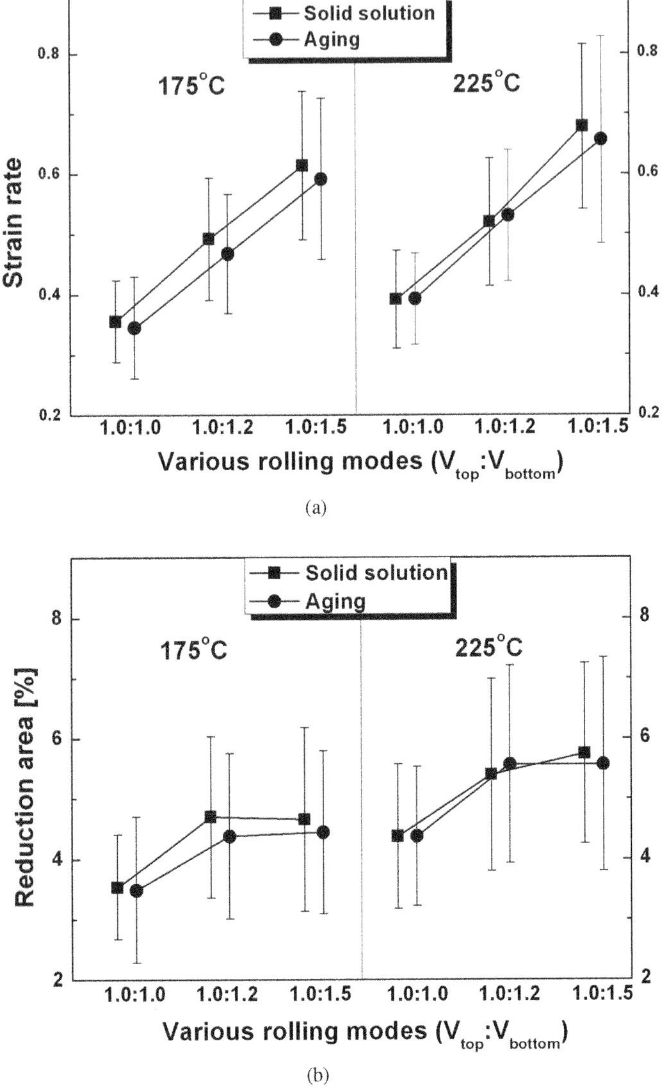

FIGURE 6.20 Variations of (a) Strain rate and (b) Reduction area during warm rolling of ZK60. (From Cho, J.-H. et al., *Metall. Mater. Trans. A*, 41, 2575–2583, 2010.)

Figure 6.21 shows the volume fraction of basal fibers of the rolled sheets. As-received sheets have basal textures in the (0002) pole figures in the surface region. Basal intensities increased during both asymmetric and conventional rolling. Asymmetric rolling usually lowers the basal textures more effectively than conventional rolling because of shear deformation. At 448 K (175°C), basal intensities decrease with an increase in the roll-speed ratio. Effect of asymmetric rolling was clearer at a speed ratio of 1.5, than at other ratios. At 498 K (225°C), basal intensities decreased with roll-speed ratios of 1.0 and 1.2, and then, they increase again

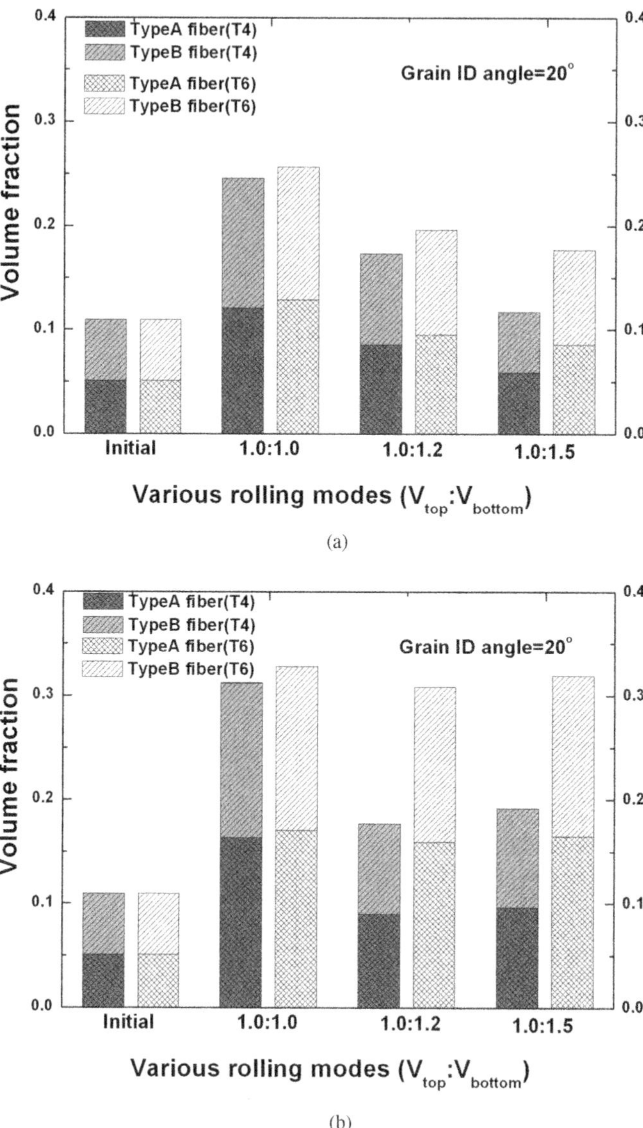

FIGURE 6.21 Volume fractions (top surface regions of the TRC sheets) after rolling at temperatures of (a) 448 K (175°C) and (b) 498 K (225°C). The Type A fiber is given by {0002}(11$\bar{2}$0), and the Type B fiber is given by {0002}(10$\bar{1}$0). (From Cho, J.-H. et al., *Metall. Mater. Trans. A*, 41, 2575–2583, 2010.)

at a ratio of 1.5, as shown in Figure 6.21b. There may exist some optimized rolling conditions that produce an appropriate shear deformation, or maximize the effect of asymmetric rolling [38]. Both solution heat-treated (T4) and artificially aged (T6) samples exhibited an increase in the number of basal fibers during rolling. Aged (T6) samples have higher intensities than solution (T4) samples in the surface regions.

The working temperature could affect texture variation during rolling. This is because deformation mechanisms change with temperature and DRX easily occurs at elevated temperatures. The pole figures of the samples rolled at 498 K (225°C) have higher basal intensities than those of the samples rolled at 448 K (175°C). Shear band and twinning usually accommodate external loading at lower temperatures than at elevated temperatures, and this results in weakening of basal textures. Slip deformation, rather than twinning and other inhomogeneities, appeared to be more active at 498 K (225°C) than at 448 K (175°C), and thus the texture further strengthened at 498 K (225°C).

Texture variation in the center regions was also examined, as shown in Figure 6.22. The initial texture at the center exhibited lower intensity than that at the surface. Basal intensities decreased with roll-speed ratios of 1 and 1.2. Overall, solution heat-treated (T4) samples exhibited larger basal intensities than did aged (T6) samples. There existed some difference between texture evolution at the surface and at the center. After 50% reduction, the basal texture at the surface was still higher than that at the center. Plane strain compression in the center region usually results in stronger rolling texture. In these experiments, however, the basal fibers were smaller in the center region than at the surface. The initial texture at the surface had stronger intensity than in the center. It seems that the initial intensities affected deformation texture during rolling. Volume fractions of basal textures for solution heat-treated (T4) samples had a lesser gradient along the thickness direction than those for aged (T6) samples, considering textures at both the surface and the center. It should also be noted that both the T4 and the T6 samples had lower and more widely spread basal intensities in the center region, than in the surface region.

The twin-roll casting process involved both fast solidification and subsequent hot rolling. There also exists a strong temperature gradient across the thickness of the layer. In conventional rolling, the surface region usually has less plane strain compression and more shearing. This results in weakened rolling textures, or basal textures, in magnesium alloys. From the viewpoint of texturing during TRC, basal textures at the center had lower intensities than those at the surface. In addition, a split in the texture was found at the center, along the RD. The texture at the surface, however, consisted of strong basal fibers concentrated at the center of the pole figures. The texture split in the central region was related to the fabrication process of the strips. Studies on plane strain compression carried out at various temperatures and strain rates revealed that the distribution of basal textures was affected by the deformation conditions [39]. In particular, deformation at high temperature and low strain rate prevented basal fibers from forming and developing. The basal poles of magnesium alloys during plane strain compression at 673 K (400°C) were split along the RD. During TRC, the center region of the strips experienced plane strain compression greater than the shearing. The deformation temperature was also higher at the center than at the surface. These deformation conditions may affect texture evolution. As a result, basal textures were split along the RD and had low intensities. An intensity split in the basal fibers was also reported during warm rolling of Mg alloys [40–42]. The split in the basal poles during rolling decreased with increasing temperature. The increase in the split along the RD was attributed

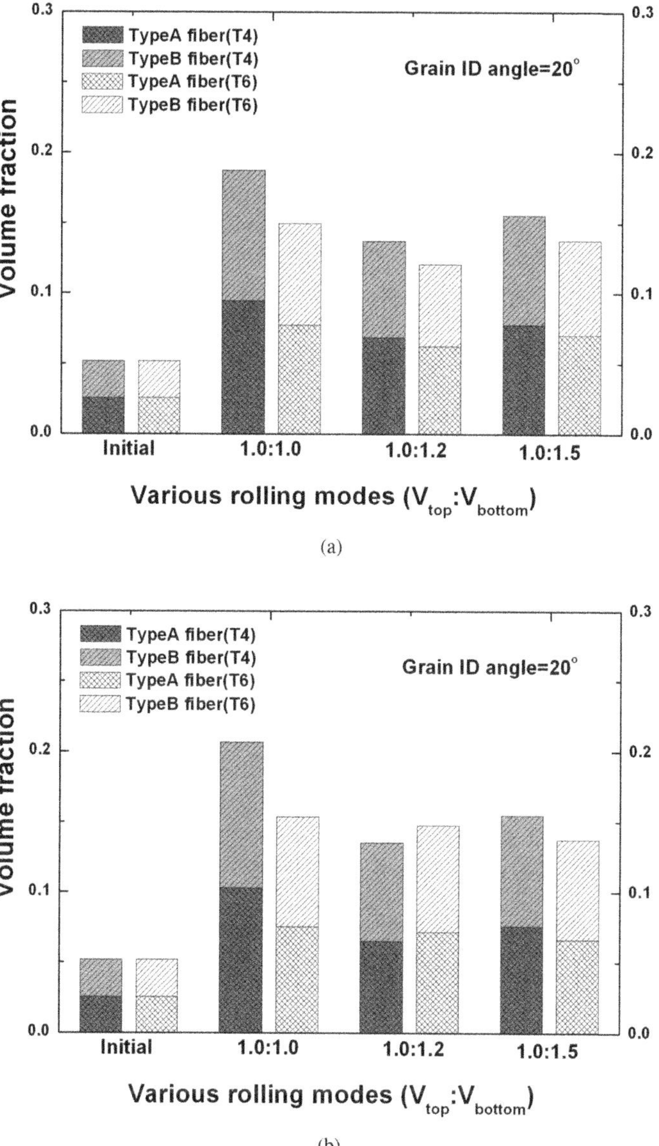

FIGURE 6.22 Volume fractions (center regions of the TRC sheets) after rolling at temperatures of (a) 448 K (175°C) and (b) 498 K (225°C). (From Cho, J.-H. et al., *Metall. Mater. Trans. A*, 41, 2575–2583, 2010.)

to double twinning, and other deformation heterogeneities also occurred at low temperatures. Peak broadening was also observed and it was attributed to single (tensile) twinning [42] or the activation of a prismatic a slip [41]. It appears that, depending on the rolling temperatures and alloying, the balance of deformation mechanisms changes, and texture evolution is also affected.

6.6 SUMMARY FOR THE FUTURE STUDIES

Various Mg alloys have the potential to replace aluminum or steel alloys because of its light weight. However, further attempts to improve their strength and elongation are still required. Here we mainly discussed mechanical properties and microstructure of Mg-Zn-Zr alloys. Zn addition provides precipitate hardening, and Zr contributes to strength by grain refinement. As a casting method, a twin-roll casting was introduced. The twin-roll casting process of Mg-Zn-Zr alloys has been successfully carried out, and high strength and elongation of the system was obtained, compared to ingot casting process. Although the twin-roll casting is an effective fabrication method as mentioned before, its working condition is usually tight. Depending on alloying composition, various working conditions should be carefully determined.

In addition to thermo-mechanical processing, alloying design is also crucial to obtain improved mechanical properties. Various alloying approaches have been carried out on Mg-Zn systems. Cu [43], Co [44], and Ba [45] were used for macro-alloying additions to Mg-Zn based alloys. Temperature of solution heat treatment for the ternary systems of Mg-Zn-X (X = Cu, Ba, and Co) can be increased. The increase in the temperature of heat treatment is related to an increased eutectic temperature in the Mg-Zn-X alloys. The addition of Co to a Mg-Zn system allows a much higher solution treatment temperature that can lead to higher concentrations of Zn atoms and vacancies in Mg grains of the solution-treated samples. This causes a significant increase in number density of precipitates and greater age-hardening response [44]. Alloying with Ba in Mg-Zn alloys results in an increased number density of the strengthening precipitates [45]. The hardness produced by artificial aging increased twice. A trace amount of Cr or V is soluble in the Mg lattice in the presence of Zn [46,47]. The presence of Cr or V significantly improves the age hardening response of Mg – Zn alloy by increasing the number density of the precipitates. The kinetics of precipitation is accelerated and the highest level of hardening in MgZnCr (or V) alloys is produced by aging at intermediate temperatures.

The addition of Ag, or the combined addition of Ag and Ca to a Mg-Zn alloys [48,49] also can result in a significant enhancement in age-hardening response and tensile yield strength. The improved age-hardening response is associated with a refined distribution of rod-shaped precipitates. Alloys of nominal composition Mg-3Zn-xCa (x = 0, 0.5, 1.0) wt.% were studied in [50]. Ca addition causes the increase in hardness, and much refined and more homogeneous distribution of the precipitates. The age-hardening of the ternary alloy is attributed to the fine disc-shape plates lying on the basal plane of the matrix. The additions of Y to Mg-Zn alloys led to the formation of relatively large particles of a quasicrystalline phase [51,52].

A lot of efforts have been made to improve mechanical properties of Mg-Zn based alloys, including alloy design, casting, and heat treatments. In addition, there are also many micromechanical issues to overcome. For the age-hardening Mg-Zn based alloys, it is necessary to comprehensively understand various factors to control the nucleation and growth of precipitates. Their orientation relationship with the matrix, shape and morphology, and interactions between the precipitates and dislocations and twins during deformation also should be investigated.

REFERENCES

1. J. Nie, *Metall. Mater. Trans. A* 43 (11) (2012) 3891–3939.
2. L. Y. Wei, G. L. Dunlop, H. Westengen, *Metall. Mater. Trans. A* 26 (1995) 1947–1955.
3. L. Y. Wei, G. L. Dunlop, H. Westengen, *Metall. Mater. Trans. A* 26 (1995) 1705–1716.
4. J. Nie, *Scripta. Mater.* 48 (2003) 1009–1015.
5. X. Gao, J. F. Nie, *Scr. Mater.* 56 (8) (2007) 645–648.
6. A. Nayeb-Hashemi, J. Clark, *Phase Diagrams of Binary Magnesium Alloys*, ASM International, Materials Park, OH, 1988.
7. S. Wasiur-Rahman, M. Medraj, *Intermetallics* 17 (10) (2009) 847–864.
8. M. Mezbahul-Islam, A. OmarMostafa, M. Medraj, *J. Mater.* (704283) (2014) 1–33.
9. D. Y. Maeng, T. S. Kim, J. H. Lee, S. J. Hong, S. K. Seo, B. S. Chun, *Scr. Mater.* 43 (2000) 385–389.
10. S. R. Agnew, M. H. Yoo, C. N. Tome, *Acta Mater.* 49 (2001) 4277–4289.
11. S. R. Agnew, O. Duygulu, *Int. J. Plast.* 21 (2005) 1161–1193.
12. S. H. Choi, E. J. Shin, B. S. Seong, *Acta Mater.* 55 (2007) 4181–4192.
13. J. H. Cho, Y. M. Jin, H. W. Kim, S. B. Kang, *Mater. Sci. Forum* 558–559 (2007) 159–164.
14. H. Chen, S. B. Kang, H. Yua, J. H. Cho, H. W. Kim, G. Mina, *J. Alloys Compd.* 476 (2009) 324–328.
15. A. Galiyev, R. Kaibyshev, G. Gottstein, *Acta Mater.* 49 (2001) 1199–1207.
16. C. Y. Wang, X. J. Wang, H. Chang, K. Wu, M. Y. Zheng, *Mater. Sci. Eng. A* 464 (2007) 52–58.
17. J.-H. Cho, S.-H. Han, H.-T. Jeong, S.-H. Choi, *J. Alloys Compd.* 743 (2018) 553–563.
18. J.-H. Cho, S. S. Jeong S.-B. Kang, *Mater. Des.* 110 (2016) 214–224.
19. H. Chen, S. B. Kan, H. Yua, H. W. Kim, G. Mina, *Mater. Sci. Eng. A* 492 (2008) 317–327.
20. S. Wang, S. B. Kang, J.-H. Cho, *J. Mater. Sci* 44 (2009) 5475–5484.
21. X. Gong, S. B. Kang, L. Saiyi, J.-H. Cho, *Mater. Des* 30 (2009) 3345–3350.
22. H. Chen, Y. Huashun, S. B. Kang, J. H. Cho, G. Min, *Mater. Sci. Engr. A* 527 (2010) 1236–1242.
23. S. Wang, S. B. K. M. Wang, J.-H. Cho, *Trans. Nonferrous Met. Soc. China* 20 (2010) 763–768.
24. X. Gong, S. B. Kang, J.-H. Cho, L. Saiyi, *Mater. Des.* 97 (2014) 183–188.
25. C. S. Roberts, *Magnesium and Its Alloys*, John Wiley & Sons, New York, 1960.
26. J. Robson, N. Standford, M. Barnett, *Acta. Mater.* 59 (2011) 1945–1956.
27. G. Lorimer, L. Mackenzie, *Microsc. Microanal.* 11 (Suppl 2) (2005) 192–193.
28. J.-H. Cho, S.-H. Kim, S.-H. Han, S.-B. Kang, *Microsc. Microanal.* 19(S5) (2013) 8–12.
29. J. Cho, S. S. Jeong, H. Kim, S. Kang, *Mater. Sci. Engr. A* 566 (2013) 40–46.
30. L. Remy, *Metall. Mater. Trans. A* 12 (1981) 387.
31. G. E. Dieter, *Mechanical Metallurgy*, Mc Graw Hill, McGraw-Hill Book Company Limited, 1988.
32. A. Datta, U. V. Waghmare, U. Ramamurty, *Acta Mater.* 56 (2008) 2531–2539.
33. S. Sandlobes, S. Zaefferer, I. Schestakow, S. Yi, R. Gonzalez-Martinez, *Acta Mater.* 59 (2011) 429–439.
34. J. Bae, C. Kang, S. Kang, *J. Mater. Process. Technol.* 191 (2007) 251–255.
35. J. Zeng, R. Koitzsch, H. Pfeifer, B. Friedrich, *J. Mater. Process. Technol.* 209 (2009) 2321–2328.
36. Y. S. Lee, H. W. Kim, J. H. Cho, *J. Mater. Process. Technol.* 218 (2015) 48–56.
37. J. H. Cho, A. D. Rollett, K. H. Oh, *Metall. Mater. Trans. A* 35 (2004) 1075–1086.
38. J. Cho, H. Kim, S. Kang, T. S. Han, *Acta Mater.* 59 (2011) 5638–5651.
39. T. Al-Samman, G. Gottstein, *Mater. Sci. Forum* 539–543 (2007) 3401–3406.

40. M. J. Philippe, *Mater. Sci. Forum* 157–162 (1994) 1337–1350.
41. L. Mackenzie, M. Pekguleryuz, *Mater. Sci. Eng. A* 480 (2008) 189–197.
42. K.-H. Kim, B.-C. Suh, J. H. Bae, M.-S. Shim, S. Kim, N. J. Kim, *Scr. Mater.* 63 (7) (2010) 716–720.
43. G. Lorimer, *Proceedings of the London Conference on Magnesium Technology*, C. Baker, G.W. Lorimer, and W. Unsworth, Eds. (1987) 47–53.
44. J. Geng, X. Gao, X. Fang, J. Nie, *Scripta Mater.* 64 (2011) 506–509.
45. J. Buha, *Mater. Sci. Eng. A* 491 (2008) 70–79.
46. J. Buha, *Mater. Sci. Eng. A* 492 (2008) 293–299.
47. J. Buha, *Acta Mater.* 56 (2008) 3533–3542.
48. C. Mendis, K. Oh-ishi, Y. Kawamura, T. Honma, S. Kamado, K. Hono, *Acta Mater.* 57 (2009) 749760.
49. K. Oh-ishi, R. Watanabe, C. Mendis, K. Hono, *Mater. Sci. Eng. A* 526 (2009) 177184.
50. A. W. Katarzyna K., Lidia L., Joanna W, *Mater. Sci. Forum* 765 (2013) 481485.
51. I. Kim, D. Bae, D. Kim, *Mater. Sci. Eng. A* 359 (2003) 313318.
52. A. Singh, M. Nakamura, M. Watanabe, A. Kato, A. Tai, *Scripta Mater.* 49 (2003) 417422.
53. J.-H. Cho, H.-M. Chen, S.-H. Choi, H.-W. Kim, S.-B. Kang, *Metall. Mater. Trans. A* 41 (2010) 2575–2583.

7 Biodegradable Magnesium Alloys with Aluminum, Lithium and Rare Earth Additions

Filip Pastorek

CONTENTS

7.1 INTRODUCTION

Significant progress in the field of various implant type development (pins, rods, screws, sutures for fixation) appeared started in nineteenth century. The most researches are focused on three groups of possible biodegradable materials, including metals, ceramics and polymers. Among them, magnesium and its alloys are under massive investigation in last two centuries. Already in mentioned nineteenth century, magnesium wires were used as ligature for bleeding vessels. Instead of the problem with rapid magnesium implant corrosion causing subcutaneous gas cavities surrounding the implant, other complications such as infections or pain were observed by the patients with magnesium implants in a minimal level during the postoperative follow-up. The main clinical groups interested in the application of biodegradable magnesium alloys are musculoskeletal, cardiovascular and general surgery. The purpose of the usage of biodegradable implant materials such as magnesium alloys is the reduction of traditional implant materials (titanium alloys, stainless steel, etc.) problems including stress shielding effect, further removal and refitting of the new implant connected with the additional expense of the procedures and pain of the patients.

7.2 BIODEGRADABLE MAGNESIUM ALLOYS

Biodegradable magnesium-based metals are potential to be used as a new class of biodegradable medical implant materials since they possess many advantages over the current applied and developed biomaterials such as good biological behaviors (Mg is an essential element for human body that is able to absorb and excrete direct Mg corrosion products in urines [1–4]), good mechanical properties (more suitable strengths for load bearing applications than currently developed biodegradable polymers, ceramics and bioactive glasses), closer modulus and density with bones compared with titanium alloys, stainless steels and cobalt based alloys [3,5–7]. However, a critical problem to be settled is how to well control the degradation rates of magnesium-based metals [7]. HP (high purity) Mg reaches very positive biodegradation rate (measured as hydrogen evolution rate) of 0.008 $mL/cm^2/day$ while acceptable level for human body was set to 0.01 $mL/cm^2/day$. Unfortunately, HP magnesium has yield strength (cast Mg typically <55 MPa) lower than natural bone (e.g., femur ~110 MPa). Hence, it should be specially cast or mechanically processed to improve its yield strength for bioapplication [8]. Another option is a suitable alloying of Mg, thus creating magnesium alloy with multiple positive properties. Here, however, it must be taken into consideration that such elements simultaneously modify the material's corrosion behavior (rate and type of corrosion) [9]. As a rule, the effects which the alloying elements produce in this respect are deliberate and intentionally obtained [10]. Most used chemical elements in magnesium implant materials are Al, Mn, Zn, Ca, Li, Zr and Y [11].

These alloying elements improve the mechanical and physical properties of Mg alloys for orthopaedic applications by: (a) optimizing grain size, (b) improving corrosion resistance, (c) providing mechanical strength by the formation of intermetallic states and (d) ease the manufacture process of Mg alloys [12,13]. As these materials are used in the body, care must be taken to choose alloying elements that are non-toxic [3]. Moreover, elements such as iron, nickel, and copper, which are recurrently found in magnesium alloys are frequently unintentional and due to the manufacturing process [10]. In this respect, iron in particular plays a significant role owing to its frequently inevitable contact with magnesium alloys during their manufacture. During the melting and stirring of the melt in the steel crucible, which occasionally contains nickel, the iron and nickel contents can precipitate out of the crucible's wall and thus end up in the magnesium's melt in which it represents just one constituent of the alloy [10,14]. Contaminants containing copper mainly arise by employing "impure" aluminum alloying elements [10,15]. Even in small amounts, iron, copper, and nickel as constituents in pure magnesium lead to an increase in the corrosion rate [10]. Typically, Cu is limited to 100–300 ppm, Ni should not exceed 20–50 ppm and Fe and Be are limited to 35–50 ppm and 5 ppm, respectively [13].

7.3 Mg-Al ALLOYS

Aluminum (Al) is the most commonly used alloying element for Mg alloys with a maximum solubility of 12.7 wt.% in Mg. Several Mg–Al based alloy systems have been developed for industrial applications, such as AZ and AM alloy systems [16].

Addition of Al (1%–5%) leads to the reduction of grain size, while content above 5% does not further affect the grain size [13,17]. Generally, Al dissolves partly in Mg solid solutions and precipitate as $Mg_{17}Al_{12}$ secondary phases along the grain boundaries [13,18]. The $Mg_{17}Al_{12}$ phases increase with increasing Al content, and show a net-shaped distribution when the content of Al is above 3%, resulting an increase in the yield strength of the Mg alloy. During a tensile test, the $Mg_{17}Al_{12}$ phase will be broken before any plastic deformation can occur [19]. The as-cast Mg-Al alloys show α-Mg matrix and β-phases mainly consisting of $Mg_{17}Al_{12}$ phases. In the presence of electrolytes, these phases show different electrode potentials. The $Mg_{17}Al_{12}$ phase exhibits a passive behavior, acting as a cathode with respect to the α-phase of Mg matrix, thereby accelerating the corrosion of the alloy [13,19]. However, the inert nature of $Mg_{17}Al_{12}$ phase can itself act as a corrosion barrier, thereby reducing the corrosion in AZ91D alloys [13,20]. If β-phase volume fraction is higher and distributed along the grain boundary, it might also act as a corrosion barrier surrounding the α-Mg matrix [9,13]. Once the network of the β phases breaks down after a deformation process, or is destroyed and distributed discontinuously in the Mg matrix, the action as a barrier is undermined, resulting in accelerated corrosion [19–21].

Alloying by Al provide both solid solution strengthening and precipitation strengthening. Additionally the increase of the content of Al lowers the temperature of liquidus and solidus lines and enhances the castability of alloys with high Al contents [11]. Increasing amounts of Al up to 6 wt.% improves the UTS of binary Mg–Al alloy but any higher amount reduces the alloy strength. The alloy elongation increases with increasing Al content [16].

As a potential biodegradable material several magnesium alloys with Al as a main alloying element were studied [19] including AE21 [22–25], AZ21 [18,26], AZ31 [27–30], AZ31B [31], AZ91 and AZ91D [27,32–34]. Mochizuky and Kaneda [35] studied the biodegradation properties and blood compatibility of AZ series of Mg alloy containing 3% to 9% of Al and 1% of Zn. In terms of the biodegradation property, it was found that the elution of Mg^{2+} and Zn^{2+} ions into the blood and the associated change in the pH of the blood were suppressed with an increase in aluminum content in the alloy. This indicated that the degradation resistance in blood is markedly improved with an increase in the aluminum content of the alloy. Thus, the AZ series Mg alloy can be regarded as a potential biomaterial for use in medical devices that are in contact with the blood [35]. Wen et al. [36] indicated that the degradation rate of AZ31, AZ61, and AZ91D alloys is relatively high (3–8 mm/yr^{-1}) in the first 24 h immersion period in m-SBF and slows down with a prolonged immersion period up to 24 d in an order of AZ91D (1.23 $mm.yr^{-1}$) < AZ61 (1.32 mm/yr^{-1}) < AZ31 (2 $mm.yr^{-1}$). In this in vitro experiment, the AZ91D alloy exhibited relatively uniform corrosion morphology with a few shallow pits, while obvious corrosion pits were seen on the surface of AZ31 and AZ61 [16,36]. Song et al. [8], compared biodegradable properties of AZ91 with commercial pure (CP) and ultra-pure (UP) magnesium through hydrogen evolution measurements. AZ91D reached better corrosion resistance in terms of lower hydrogen evolution compared to CP magnesium but higher hydrogen evolution than UP magnesium. The slower hydrogen evolution rates of AZ91 than that of CP-Mg (26 $mL.cm^{-2}.day^{-1}$) verify that alloying can retard the biodegradation process. The estimated hydrogen volume was consistent very well with the subcutaneous gas bubble generated by the AZ91D nail in a guinea pig femora for 4 weeks after implantation, where the subcutaneous bubble was 1 cm in diameter, i.e. 4 mL in volume. It was reported that the subcutaneous bubbles in guinea pigs appeared within 1 week after surgery and disappeared after 2–3 weeks [27]. This phenomenon implies that by a certain mechanism, guinea pigs can gradually release, absorb or consume the generated bubbles at a rate slightly lower than the hydrogen evolution rate of AZ91 in them. In other words, the measured hydrogen evolution rate of the AZ91D nail must be very close to the tolerable level in a guinea pig. It signifies that a hydrogen evolution rate lower than that of AZ91D may not lead to a subcutaneous bubble or can only lead to an insignificant volume of subcutaneous bubble in a guinea pig [8]. Therefore, Song et al. [8] set a hydrogen evolution rate 0.01 $mL.cm^{-2}.day^{-1}$, six times lower than that of AZ91D as a tolerated level. The tolerated rate can be used to screen out candidate biodegradable magnesium alloys for further tests in the human body [8]. Witte et al. [37] reported that the implantation of AZ91 magnesium alloy showed no significant harm to its neighboring tissues and also exhibited good biocompatibility. Recent investigations into the application of coronary stents for commercial magnesium alloys including AE21 [22] and AZ31B [38], have also opened up new opportunities for biodegradable magnesium alloys [39].

Although Al is the most widely used element for Mg alloys such as AZ21, AZ91D and AZ31, due to its excellent effects on the refining of the microstructure and enhancement of the corrosion resistance, medical research has found that an accumulation of Al in the brain may harm intelligence and cause neuropathologically relevant issues [19,40]. Released aluminum ions could easily combine with inorganic phosphates, leading to a lack of phosphate in the human body, and an increased concentration of Al ions in the brain. This seems to be associated with Alzheimer's disease [39]. Furthermore, Al has a significant impact on immunology, and vaccines containing Al may lead to lymphocytes and inconspicuous muscle fiber damage [19,41]. The total body burden of Al in healthy adults is 30–50 mg and the safe dose of Al containing medications can take a much larger amount of Al than in the diet, possibly as high as 12–71 $mg.kg^{-1}.day^{-1}$. Adverse effects may be seen if the dose is exceeded in humans [19,42].

7.4 Mg-RE ALLOYS

The E in the designation of a number of magnesium alloys represents rare earth (RE) elements in general (yttrium has its own designation letter, W) [43]. RE elements are a group of 17 elements, including 15 lanthanides, scandium (Sc) and yttrium (Y). They are normally added to Mg alloys by master alloys such as mischmetal (typically 50% cerium (Ce), 45% lanthanum (La), small amounts of neodymium (Nd) and praseodymium (Pr)), Y-, Ce- or Nd-rich hardeners [11,13,44]. These master alloys or hardeners contain one or two RE elements in larger quantity and almost any other RE elements in different amounts [11]. However, this mischmetal can vary in its composition depending on its source. For biomedical applications a more defined approach is desirable, since reproducibility is a major requirement for medical devices [45]. The RE elements can be classified into two groups (i) High solubility group (yttrium (Y), gadolinium (Gd), terbium (Tb), dysprosium (Dy), holmium (Ho), erbium (Er), thulium (Tm), ytterbium (Yb), and lutetium(Lu)) and (ii) limited solubility group (neodymium (Nd), lanthanum (La), cerium (Ce), praseodymium (Pr), samarium (Sm), europium (Eu)) [11,13]. RE elements are widely used in different commercially available alloys such as AE, QE, WE or ZE series alloys [43,45,46].

Suitable additions of RE elements in magnesium alloys can significantly improve many of their properties. Some amount of the RE elements is kept in solid solution and therefore RE elements can strengthen the material by solid solution strengthening. Additionally, all RE elements can form complex intermetallic phases with Al or Mg. These intermetallic phases act as obstacles for the dislocation movement at elevated temperatures and cause precipitation strengthening. The RE elements with limited solubility forms intermetallic phases early during solidification. Thus, RE elements can arrest grain boundaries at elevated temperatures and contribute to strength mainly by precipitation strengthening. This mechanism increases the service temperature of Mg alloys and improves creep resistance as well as corrosion resistance [11].

Rare-earth-containing magnesium alloys are among the prime candidates for the successful application of magnesium in orthopedic and trauma surgery [46,47] although with respect to the biological effects of rare earth elements, the situation is not so clear. In vitro, rare earth oxide additives enhanced osteoblastic differentiation and collagen production of human mesenchymal stem cells and the effects of RE elements ions on the bone resorbing function of osteoclasts depend on the concentration [45,48,49]. In a feeding experiment in ovariectomized rats in vivo, RE elements showed antiresorptive effects, an activation of osteoblasts and positive effects on bone density [45]. Additionally, there is increasing evidence that many RE elements exhibit anticarcinogenic properties, which could lead to multifunctionality of the designated alloys [43,50–53]. Feyerabend et al. [46] have comprehensively investigated the short-term effects on various cells of some RE elements, including Y, Nd, Dy, Pr, Gd, La, Ce and Eu. La and Ce showed the worst biocompatibility with the highest cytotoxicity on cells, whereas the highly soluble Dy and Gd seem to be more suitable. Nakamura et al. [54] suggested that RE elements can be chemically classified into three groups on the principle of their ionic radii: (i) light RE elements: La, Ce and Pr; (ii) medium RE elements: Nd, Pm, Sm, Eu and Gd; and (iii) heavy RE elements: Tb, Dy, Ho, Er, Tm, and Yb. The light RE elements, Ce and Pr, usually induce severe hepatotoxicity, including symptoms of fatty liver and jaundice; medium RE elements are mainly distributed into the spleen and lungs [19,54]. The absorption of RE elements is dependent not only on the concentration but also the size of the elements [19].

Several RE elements doped Mg alloys have been investigated as a potential biodegradable materials [13,55]. The most researches were focused on RE elements such as Gd, Dy, Nd, Y, and additional RE elements. Gd; can be used to adjust mechanical properties with a wide range of alloy compositions and heat treatments due to its large solubility of 23.49 wt.% at the eutectic temperature and the formation of intermetallic phases like Mg_5Gd [43]. As a single alloying element, Gd is present in solid solution, and can be used in a concentration-dependent manner to contribute to precipitation strengthening. Although many authors state that gadolinium is highly toxic, the acute toxicity is only moderate. The intraperitoneal LD50 dose of $GdCl_3$ was 550 $mg.kg^{-1}$ in mice, while $GdNO_3$ induced acute toxicity at a concentration of 300 $mg.kg^{-1}$ in mice and 230 mg.kg^{-1} in rats, respectively [56,57]. Tests regarding the cytotoxicity of Gd in osteoblastlike cells showed that it could be a suitable element with which to design Mg–Gd-based implant materials for medical applications [43,46].

Potential candidate for biomedical use is magnesium alloy with Dy (Mg–10Dy) providing a low corrosion rate, uniform corrosion behavior and good cytocompatibility [58,59]. However, the mechanical properties of Mg–10Dy alloy are not satisfactory. Moreover, Mg–10Dy alloy shows no age-hardening response during aging, which limits the potential for tailoring mechanical properties by heat treatment [60,61].

Better results were reached on Mg–Nd–Zn–Zr alloy (2.0–4.0 wt.% Nd, 0.1–0.5 wt.% Zn, and 0.3–0.6 wt.% Zr) designed for medical use, which exhibited good mechanical properties after extrusion and the subsequent cyclic extrusion and compression (approximately 300 MPa in UTS and 30% in elongation) [16,62]. Zhao et al. [63] reported that the corrosion resistance of Nd was inferior to Mg and Nd would produce more stable $Nd(OH)_3$ in neutral 0.1 M NaCl electrolytes than Mg. The enrichment of surface film with Nd significantly reduced the film hydration [64] and hindered the diffusion of chloride ions [65], and thus increased the protective nature of the surface film. Due to the limited solid solubility of Nd in Mg (3.6 wt.%), the precipitation of the Nd intermetallic compounds occurs, which is more stable than the Mg matrix. The incorporation of Nd in the surface film increases its protectiveness and thereby decreases the corrosion rate. In a comparison with Mg–La and Mg–Ce alloys, Mg-Nd alloy exhibited lower corrosion rate both in in vivo and in vitro studies [45].

Family of Mg–Y–Zn alloys has been developed with the aim of achieving the mechanical, electrochemical and biological characteristics suitable for medical applications [66,67]. These Mg–Y–Zn alloys (denoted as ZW21 and WZ21) are assumed to meet the degradation demands. In addition, these alloys also exhibit high ductility (uniform elongation of 17%–20%) and appropriate strength (ultimate tensile strength: 250–270 MPa) [66]. The high ductility is ascribed both to the fine-grained microstructure (enabling complementary deformation modes) and to the weak texture of the Mg–Y–Zn alloys. This weak texture, in turn, generates a very low tension–compression asymmetry (ratio of yield stress in tension and yield stress in compression 1), which is advantageous with respect to the fatigue behavior of the alloys, where the lower of the tension or compression yield stresses determines the fatigue strength. For stent applications the fatigue strength is of great importance because the stent device is subject to a permanent cyclic load (heartbeat) in the blood vessel [67].

Since the second phases in the Mg–Y–Zn alloys developed are also very stable [66], their growth-inhibiting potential can be deployed even at elevated temperatures. This allows the application of various forming procedures and heat treatments. Because of the stability of the second phases and the thus-constricted grain growth, the mechanical properties can be preserved even at elevated temperatures or restored after recrystallization processes. This is of particular advantage for the stent fabrication process, where the material is repeatedly subjected to thermo-mechanical processing steps, i.e. tube drawing and stress-relief annealing [67].

Biocompatibilities of an absorbable Mg stent with Y and some RE elements additives were studied in vivo and in vitro, and it was indicated that Mg alloys without Al but containing small amounts of Y and RE elements would be appropriate for biomedical applications. Y is a particularly disputed alloying element, and it is essential to further investigate the effect of an addition of Y in Mg alloys on biocompatibility [19,68]. German company Biotronik adopted WE43 alloy, a rare earth (RE) element-strengthened magnesium alloy, as a biodegradable magnesium alloy for the development of bio-absorbable coronary stents [39]. In 2005, the implantation of a 3 mm diameter stent was performed in a hybrid procedure on a baby weighing 1.7 kg [69]. Despite their small size, the artery vessels of the baby tolerated the degradation process. The mechanical and degradation characteristics of the magnesium alloy stent proved adequate to secure the reperfusion of the previously occluded left pulmonary artery [39,69]. Actually, a drug-eluting magnesium stent and a magnesium screw consisting of MgYREZr (a material similar to WE43) that contains >90 wt.% magnesium are already in clinical use [45,70,71]. The implantation of WE43 alloy (containing RE elements mischmetal based on neodymium, cerium and lanthanum [72]) into the femora of guinea pigs showed good biocompatibility with the corrosion layer in direct contact with the surrounding bone [27]. Stents made of WE43 showed antiproliferative properties and a drastically reduced restenosis rate compared to conventional stainless-steel stents [45,73].

7.5 Mg-Li ALLOYS

A major problem facing hexagonal close packed (hcp) metals, such as magnesium, is limited ductility and low-temperature formability [74]. Lithium is the only element known that is able to change the lattice structure from hexagonal close-packed (h.c.p.) to body-centered cubic (b.c.c.) crystal structure in magnesium alloys [11,13]. Therefore, it can be used to enhance ductility and formability of magnesium alloys but unfortunately it has a negative effect on strength [11]. Li is more reactive than Mg and has a pronounced effect on the corrosion behavior of Mg alloys. Li content below 9% in pure Mg is beneficent to corrosion resistance, whereas excess addition of Li is detrimental to the corrosion resistance [13,75,76]. Li can improve the corrosion resistance of a magnesium alloy through a mechanism that corroding Li shifts the pH value of the solution to a pH > 11.5, which stabilizes the $Mg(OH)_2$ film on the alloy surface. However, the in vivo alkalization is a hazard to the human body [8]. Suitable option for improving the corrosion behavior of magnesium alloys with lithium is additional alloying by Al [77] and RE elements [78].

The usage of lithium in the field of biomedicine is debatable. Since Li was discovered, it has attracted a great deal of attention, due to its potential toxicity [19,79]. Li has numerous effects in humans and in other organisms as it inhibits the functioning of multiple enzymes in the body [80]. Li is a teratogenic hazard to the cardiovascular system of the human body, as when Li is given to mice and rats they could produce skeletal and craniofacial defects. Doses of Li (10 mg. L^{-1} in serum) in humans induce bipolar disorder, and at 20 mg. L^{-1} Li in serum there is a risk of death. Li has specific toxicity presenting with several features: acute abnormalities from Li poisoning and chronic changes such as nephrogenic diabetes insipidus, epithelial cell disease, and chronic kidney disease [19,80,81]. However lithium is medically used to treat neurological disorders such as depression or schizophrenia [82].

Mg–Li alloys are among the lowest density metallic materials. The density of Mg alloy can be decreased from 1.8 to 1.35 g.cm^{-3} with a little loss of elastic modulus. As Li content is less than 5 mass%, the alloy is composed of α phase, which is a solid solution of Li in Mg. The alloy with Li content higher than 11 mass% is composed of β phase, a solid solution of Mg in Li. As Li content is higher than 5 mass% and lower than 11 mass%, the alloy is composed of α phase and β phase [83]. These alloys have high stiffness ratio, good machining property, good magnetic screen and shock resistance ability. They are generally used widely in the fields of electricity, weapon industry and spaceflight, etc. [84–87].

Several binary Mg-Li alloys were studied including Mg–3.3Li [88], Mg–4Li [89], Mg–5Li [90] and Mg–8.8Li [75]. However, the mechanical strengths of these alloys are compromised, and the loss of strength of Mg–Li binary alloys can hardly be regained through a heat treatment procedure [91]. To avoid this problem, ternary and even multi-component alloy systems were recently developed by means of adding alloying elements such as aluminum, zinc, silicon and RE elements to Mg–Li binary alloys [91–96].

A number of evaluations of Mg–Li-based alloys for biomedical applications was performed since 2004. Thin AL36 alloy wires were developed as resorbable sutures [97]. The mechanical properties of AL36 magnesium wires were found to be sufficient to meet the requirements of sutures, although they were lower than those of commercially available polymeric sutures. Long-term degradation behaviors of three Mg–Li–Al– (RE elements) alloys with different concentrations of aluminum and rare earth elements were evaluated as candidate materials for potential cardiovascular stent application. The results indicated that the alloy LA92 degraded even more slowly than the WE-type alloy after immersion in Hanks' balanced salt solution for 94 days and displayed a steady hydrogen evolution rate over the whole period of immersion tests [91,98]. Moreover, short-term evaluation to determine the effect of lithium on primary cells and cell lines demonstrated that lithium did not have a negative impact on cell viability [46]. It is therefore of great interest to develop Mg–Li-based alloys further toward cardiovascular stent application [91].

Lithium is the main alloying element used in one of the most popular magnesium alloy under research as a biodegradable material (LAE442) [27,32,91,99,100]. This alloy containing 4 wt% lithium, 4 wt% aluminum and 2 wt% rare earth exhibits reduced density, improved ductility and enhanced corrosion resistance [91]. Different implant designs have been tested in different animal models, all indicating a good biocompatibility and a slow homogeneous degradation of this alloy [82]. Ullman et al. [101] studied the influence of the grain size on the in vivo degradation behavior of the magnesium alloy LAE442. It was showed that implants having finer grains reached, on average, a slower corrosion rate and a better clinical tolerance than implants possessing coarse grains. Defects on the implant's surface lead to an initially accelerated corrosion in comparison to implants without defects [101]. Witte et al. [27] compared the in vivo degradation of LAE442 alloy with other potential biodegradable magnesium alloys (AZ31, AZ91, WE43) in femurs of guinea pigs. The slowest rate of corrosion was noted for LAE442, while AZ31, AZ91, and WE43 degraded at similar rates.

7.6 DEGRADATION MECHANISM OF MAGNESIUM-BASED METALS IN HUMAN BODY FLUIDS

Magnesium-based metals are generally known to be corroded in an aqueous environment via an electrochemical reaction, which produces magnesium hydroxide and hydrogen gas. The overall corrosion reaction of magnesium in an aqueous environment is given below [7]:

$$Mg(s) + 2H_2O(aq) = Mg(OH)_2(s) + H_2(g)$$

This overall reaction may include the following partial reactions [11]:

$$Mg(s) \leftrightarrow Mg^{2+}(aq) + 2e^- \text{ (anodic reaction)}$$

$$2H_2O(aq) + 2e^- \leftrightarrow H_2(g) + 2OH^-(aq) \text{ (cathodic reaction)}$$

$$Mg^{2+}(aq) + 2OH^-(aq) \leftrightarrow Mg(OH)_2(s) \text{ (product formation)}$$

It is well known that Cl^- ions easily induce pitting corrosions to happen on magnesium alloys. When the chloride concentration in the corrosive environment rises above 30 mmol.l^{-1}, formed magnesium hydroxide will continue to react with magnesium to form a highly soluble magnesium chloride and thus the degradation rate is increased [7]. Therefore, severe pitting corrosion can be observed on magnesium alloys in vivo where the chloride content of the body fluid is about 150 mmol.l^{-1} [11,27,32,102]. Proteins such as albumin have been demonstrated to form a corrosion blocking layer on the magnesium alloys in the in vitro experiment. This layer can be enriched with calcium phosphates that concomitantly participate in the corrosion protection. However, organic compounds, such as amino acids, promote the dissolution of magnesium [7,103].

A very important parameter in assessing the biodegradability of magnesium and its alloys is the hydrogen evolution rate because: (1) hydrogen evolution is a detrimental process impeding the development and application of biodegradable magnesium alloys; (2) 1 mL of H_2 evolved corresponds to 1 mg of Mg dissolved, and hence measuring hydrogen evolution rate is equivalent to measuring the biodegradation or corrosion rate of a magnesium alloy; (3) 1 mole H_2 evolved is equivalent to 2 moles OH generated in solution, so the hydrogen evolution rate to some extent indicates the degree of alkalization of solution; and (4) unanodized and anodized magnesium alloys usually suffer from localized corrosion and hydrogen mainly comes from the local corroding areas, which implies that the hydrogen evolution rate can reliably reflect the local degradation rate of magnesium alloys [8,104]. The evolution of H_2 gas after adding Mg and its alloys to aqueous solutions has been extensively observed [20,104], as has the formation of gas cavities in vivo [105,106]. Very intensive biodegradation of the magnesium implant causes the rapid increase of created hydrogen amount and the painful growth of the cavities as the body is not able to immediately absorb all the hydrogen [107]. In this situation, the only way is to withdrawn the hydrogen by aspiration or incision. The amount of locally evolving gas depends on the corrosion rate of the implant, the size of the implant, the acidity of the tissue environment and the extent to which the metal is exposed to the attack from blood serum. If the Mg oxidation is slow, as it is when the metal is well buried in cortical bone, the gas absorbs about as rapidly as it forms [108].

Another mentioned concern is that the local pH increases at the implant due to the hydroxyl groups. High pH inhibits cell proliferation and tissue formation. In a static solution, pH can increase up to 10 [104]. These problems may be more severe in a laboratory bench top experiment than in the body. What is really happening in vivo in a cardiovascular situation can be more accurately simulated in vitro by using an electrolyte having appropriate chloride, phosphate, and protein concentrations, by applying a mechanical stimulus, and using a flow-cell environment with convection and diffusion [109]. The increase of pH on the corroding surface can be neutralized in the body's buffered solutions and through shear flow systemically. But increasing the flow rate increases the corrosion rate of biodegradable implants and the mechanical stimulus creates stress-corrosion cracking, pit-growth, and fatigue corrosion [110].

It is also important to note that magnesium alloys are potentially susceptible to stress corrosion cracking (SCC) in environments containing chloride. Some commercial magnesium alloys such as AZ91, AZ31 and AM30 are susceptible to SCC even in mild environment like distilled water, suggesting that the SCC threshold could also be easily reached in an environment like the body. For magnesium-based implants in a loading-bearing application – such as coronary stents under the loading of blood vessel and blood flow, plates and screws for orthopedic fixation under the loading of body weight and movement—SSC should therefore be a major consideration [39,111,112].

7.7 SUMMARY

Although the quantum of various studies on different magnesium alloys systems was performed in the last centuries, searching for the best possible biodegradable magnesium alloy with none or minimal weak sites is still not finished. The more complex alloy is used, the more impact has to be focused on its multispectral mechanical, corrosion, chemical and medical analysis taking into consideration many factors like the size of the implant, the loads, situation of the implant, human body response, etc. Human body is an extremely sophisticated system, hence the effects of the biodegradable implant material have to be studied in a very complex manner. Actually, among the evaluated magnesium alloys with Al, Li, or RE elements additions, the most appropriate alternatives for biomedical use appear to be multicomponent alloys (combining many positive features of individual alloying elements) such as WE43 and LAE442. However, these alloys have still many limitations and their use for more complicated applications in human body is limited. Hence, further research in the field of suitable micro alloying, mechanical and surface treatment of the biodegradable implants composed of magnesium needs to be performed, so that the usage of standard implant materials can be minimized.

ACKNOWLEDGMENTS

This research was supported by European project Research centre of the University of Žilina—Second Phase: ITMS 313011D011 and by Science Grant Agency of the Slovak Republic through project No. 1/0045/17. Authors are grateful to the Slovak Research and Development Agency for support in experimental works by the project No. APVV-16-0276.

REFERENCES

1. Mhaede, M., Pastorek, F., Hadzima, B. Influence of shot peening on corrosion properties of biocompatible magnesium alloy AZ31 coated by dicalcium phosphate dihydrate (DCPD). *Materials Science and Engineering: C* 2014, 39, 330–335.
2. Song, G., Song, S. A possible biodegradable magnesium implant material. *Advanced Engineering Materials* 2007, 9(4), 298–302.
3. Staiger, M.P., Pietak, A.M., Huadmai, J., Dias, G. Magnesium and its alloys as orthopedic biomaterials: A review. *Biomaterials* 2006, 27(9), 1728–1734.
4. Wan, Y., Xiong, G., Luo, H., He, F., Huang, Y., Zhou, X. Preparation and characterization of a new biomedical magnesium–calcium alloy. *Materials & Design* 2008, 29(10), 2034–2037.
5. Zhou, Y.L., Luo, D.M., Hu, W.Y., Li, Y.C., Hodgson, P.D., Wen, C.E. Compressive properties of hot-rolled Mg-Zr-Ca alloys for biomedical applications. *Advanced Materials Research* 2011, 197, 56–59.
6. Zhang, W., Li, M., Chen, Q., Hu, W., Zhang, W., Xin, W. Effects of Sr and Sn on microstructure and corrosion resistance of Mg–Zr–Ca magnesium alloy for biomedical applications. *Materials & Design* 2012, 39, 379–383.
7. Tan, L., Yu, X., Wan, P., Yang, K. Biodegradable materials for bone repairs: A review. *Journal of Materials Science & Technology* 2013, 29(6), 503–513.
8. Song, G. Control of biodegradation of biocompatible magnesium alloys. *Corrosion Science* 2007, 49(4), 1696–1701.
9. Song, G.L., Atrens, A. Corrosion mechanisms of magnesium alloys. *Advanced Engineering Materials* 1999, 1(1), 11–33.
10. Waizy, H., Seitz, J.M., Reifenrath, J., Weizbauer, A., Bach, F.W., Meyer-Lindenberg, A., Denkena, B., Windhagen, H. Biodegradable magnesium implants for orthopedic applications. *Journal of Materials Science* 2013, 48(1), 39–50.
11. Witte, F., Hort, N., Vogt, C., Cohen, S., Kainer, K.U., Willumeit, R., Feyerabend, F. Degradable biomaterials based on magnesium corrosion. *Current Opinion in Solid State and Materials Science* 2008, 12(5), 63–72.
12. Poinern, G.E.J., Brundavanam, S., Fawcett, D. Biomedical magnesium alloys: A review of material properties, surface modifications and potential as a biodegradable orthopaedic implant. *American Journal of Biomedical Engineering* 2012, 2(6), 218–240.
13. Agarwal, S., Curtin, J., Duffy, B., Jaiswal, S. Biodegradable magnesium alloys for orthopaedic applications: A review on corrosion, biocompatibility and surface modifications. *Materials Science and Engineering: C* 2016, 68, 948–963.
14. Makar, G.L., Kruger, J. Corrosion of magnesium. *International Materials Reviews* 1993, 38(3), 138–153.
15. Bühler, K. Korrosion und Korrosionsschutz von Magnesiumlegierungen. *Metall* 1990, 44(8), 748–53.
16. Zheng, Y.F., Gu, X.N., Witte, F. Biodegradable metals. *Materials Science and Engineering: R: Reports* 2014, 77, 1–34.

17. Lee, Y.C, Dahle, A.K, StJohn, D.H. The role of solute in grain refinement of magnesium. *Metallurgical and Materials Transactions A* 2000, 31(11), 2895–906.
18. Song, G., Atrens, A., Wu, X., Zhang, B. Corrosion behaviour of AZ21, AZ501 and AZ91 in sodium chloride. *Corrosion Science* 1998, 40(10), 1769–1791.
19. Ding, Y., Wen, C., Hodgson, P., Li, Y. Effects of alloying elements on the corrosion behavior and biocompatibility of biodegradable magnesium alloys: A review. *Journal of Materials Chemistry B* 2014, 2(14), 1912–1933.
20. Song, D., Ma, A.B., Jiang, J.H., Lin, P.H., Yang, D.H., Fan, J.F. Corrosion behaviour of bulk ultra-fine grained AZ91D magnesium alloy fabricated by equal-channel angular pressing. *Corrosion Science* 2011, 53(1), 362–373.
21. Zhang, T., Li, Y., Wang, F. Roles of β phase in the corrosion process of AZ91D magnesium alloy. *Corrosion Science* 2006, 48(5), 1249–1264.
22. Heublein, B., Rohde, R., Kaese, V., Niemeyer, M., Hartung, W., Haverich, A. Biocorrosion of magnesium alloys: A new principle in cardiovascular implant technology? *Heart* 2003, 89(6), 651–656.
23. Minárik, P., Král, R., Janeček, M. Effect of ECAP processing on corrosion resistance of AE21 and AE42 magnesium alloys. *Applied Surface Science* 2013, 281, 44–48.
24. Hadzima, B., Bukovina, M., Doležal, P. Shot peening influence on corrosion resistance of AE21 magnesium alloy. *Materials Engineering* 2010, 14(4), 14.
25. Hadzima, B., Bukovinová, L. Electrochemical characteristics of shot-peened and phospatized AE21 magnesium alloy. *Acta Metallurgica Slovaca* 2011, 17(4), 228.
26. Pietak, A., Mahoney, P., Dias, G.J., Staiger, M.P. Bone-like matrix formation on magnesium and magnesium alloys. *Journal of Materials Science: Materials in Medicine* 2008, 19(1), 407–415.
27. Witte, F., Kaese, V., Haferkamp, H., Switzer, E., Meyer-Lindenberg, A., Wirth, C.J., Windhagen, H. In vivo corrosion of four magnesium alloys and the associated bone response. *Biomaterials* 2005, 26(17), 3557–3563.
28. Alvarez-Lopez, M., Pereda, M.D., Del Valle, J.A., Fernandez-Lorenzo, M., Garcia-Alonso, M.C., Ruano, O.A., Escudero, M.L. Corrosion behaviour of AZ31 magnesium alloy with different grain sizes in simulated biological fluids. *Acta Biomaterialia* 2010, 6(5), 1763–1771.
29. Song, Y., Shan, D., Chen, R., Zhang, F., Han, E.H. Biodegradable behaviors of AZ31 magnesium alloy in simulated body fluid. *Materials Science and Engineering: C* 2009, 29(3), 1039–1045.
30. Müller, W.D., Nascimento, M.L., Zeddies, M., Córsico, M., Gassa, L.M., Mele, M.A.F.L.D. Magnesium and its alloys as degradable biomaterials: Corrosion studies using potentiodynamic and EIS electrochemical techniques. *Materials Research* 2007, 10(1), 5–10.
31. Ren, Y., Huang, J., Zhang, B., Yang, K. Preliminary study of biodegradation of AZ31B magnesium alloy. *Frontiers of Materials Science in China* 2007, 1(4), 401–404.
32. Witte, F., Fischer, J., Nellesen, J., Crostack, H.A., Kaese, V., Pisch, A., Beckmann, F., Windhagen, H. In vitro and in vivo corrosion measurements of magnesium alloys. *Biomaterials* 2006, 27(7), 1013–1018.
33. Wang, Y.M., Wang, F.H., Xu, M.J., Zhao, B., Guo, L.X., Ouyang, J.H. Microstructure and corrosion behavior of coated AZ91 alloy by microarc oxidation for biomedical application. *Applied Surface Science* 2009, 255(22), 9124–9131.
34. Abidin, N.I.Z., Atrens, A.D., Martin, D., Atrens, A. Corrosion of high purity Mg, Mg2Zn0. 2Mn, ZE41 and AZ91 in Hank's solution at 37°C. *Corrosion Science* 2011, 53(11), 3542–3556.
35. Mochizuki, A., Kaneda, H. Study on the blood compatibility and biodegradation properties of magnesium alloys. *Materials Science and Engineering: C* 2015, 47, 204–210.

36. Wen, Z., Wu, C., Dai, C., Yang, F. Corrosion behaviors of Mg and its alloys with different Al contents in a modified simulated body fluid. *Journal of Alloys and Compounds* 2009, 488(1), 392–399.
37. Witte, F., Ulrich, H., Palm, C., Willbold, E. Biodegradable magnesium scaffolds: Part II: Peri-implant bone remodeling. *Journal of Biomedical Materials Research Part A* 2007, 81(3), 757–765.
38. Li, H., Zhong, H., Xu, K., Yang, K., Liu, J., Zhang, B., Zheng, F., Xia, Y., Tan, L., Hong, D. Enhanced efficacy of sirolimus-eluting bioabsorbable magnesium alloy stents in the prevention of restenosis. *Journal of Endovascular Therapy* 2011, 18(3), 407–415.
39. Yang, K., Tan, L. Control of biodegradation of magnesium (Mg) alloys for medical applications. In *Corrosion Prevention of Magnesium Alloys*. Woodhead Publishing, Cambridge, UK, 2013, 509–543.
40. Krewski, D., Yokel, R.A., Nieboer, E., Borchelt, D., Cohen, J., Harry, J., Kacew, S., Lindsay, J., Mahfouz, A.M., Rondeau, V. Human health risk assessment for aluminium, aluminium oxide, and aluminium hydroxide. *Journal of Toxicology and Environmental Health, Part B* 2007, 10(S1), 1–269.
41. Shingde, M., Hughes, J., Boadle, R., Wills, E.J., Pamphlett, R. Macrophagic myofasciitis associated with vaccine-derived aluminium. *The Medical Journal of Australia* 2005, 183(3), 145–146.
42. Verstraeten, S.V., Aimo, L., Oteiza, P.I. Aluminium and lead: Molecular mechanisms of brain toxicity. *Archives of Toxicology* 2008, 82(11), 789–802.
43. Hort, N., Huang, Y., Fechner, D., Störmer, M., Blawert, C., Witte, F., Vogt, C. et al., Magnesium alloys as implant materials–principles of property design for Mg–RE alloys. *Acta Biomaterialia* 2010, 6(5), 1714–1725.
44. Friedrich, H.E., Mordike, B.L. *Magnesium Technology*. Berlin, Germany: Springer-Verlag, 2006, 219–430.
45. Willbold, E., Gu, X., Albert, D., Kalla, K., Bobe, K., Brauneis, M., Janning, C. et al., Witte, F. Effect of the addition of low rare earth elements (lanthanum, neodymium, cerium) on the biodegradation and biocompatibility of magnesium. *Acta Biomaterialia* 2015, 11, 554–562.
46. Feyerabend, F., Fischer, J., Holtz, J., Witte, F., Willumeit, R., Drücker, H., Vogt, C., Hort, N. Evaluation of short-term effects of rare earth and other elements used in magnesium alloys on primary cells and cell lines. *Acta Biomaterialia* 2010, 6(5), 1834–1842.
47. Imwinkelried, T., Beck, S., Iizuka, T., Schaller, B. Effect of a plasmaelectrolytic coating on the strength retention of in vivo and in vitro degraded magnesium implants. *Acta Biomaterialia* 2013, 9(10), 8643–8649.
48. Karakoti, A.S., Tsigkou, O., Yue, S., Lee, P.D., Stevens, M.M., Jones, J.R., Seal, S. Rare earth oxides as nanoadditives in 3-D nanocomposite scaffolds for bone regeneration. *Journal of Materials Chemistry* 2010, 20(40), 8912–8919.
49. Zhang, J., Xu, S., Wang, K., Yu, S. Effects of the rare earth ions on bone resorbing function of rabbit mature osteoclasts in vitro. *Chinese Science Bulletin* 2003, 48(20), 2170–2175.
50. Dai, Y., Li, J., Li, J., Yu, L., Dai, G., Hu, A., Yuan, L., Wen, Z. Effects of rare earth compounds on growth and apoptosis of leukemic cell lines. *In Vitro Cellular & Developmental Biology-Animal* 2002, 38(7), 373–375.
51. Ji, Y.J., Xiao, B., Wang, Z.H., Cui, M.Z., Lu, Y.Y. The suppression effect of light rare earth elements on proliferation of two cancer cell lines. *Biomedical and Environmental Sciences: BES* 2000, 13(4), 287–292.
52. Kostova, I., Momekov, G., Stancheva, P. New samarium (III), gadolinium (III), and dysprosium (III) complexes of coumarin-3-carboxylic acid as antiproliferative agents. *Metal-Based Drugs* 2007, 2007, 8.
53. Magda, D., Miller, R.A. Motexafin gadolinium: A novel redox active drug for cancer therapy. *Seminars in Cancer Biology* 2006, 16 (6), 466–476.

54. Nakamura, Y., Tsumura, Y., Tonogai, Y., Shibata, T., Ito, Y. Differences in behavior among the chlorides of seven rare earth elements administered intravenously to rats. *Toxicological Sciences* 1997, 37(2), 106–116.
55. Xin, Y., Hu, T., Chu, P.K. In vitro studies of biomedical magnesium alloys in a simulated physiological environment: A review. *Acta biomaterialia* 2011, 7(4), 1452–1459.
56. Haley, T.J., Raymond, K., Komesu, N., Upham, H.C. Toxicological and pharmacological effects of gadolinium and samarium chlorides. *British Journal of Pharmacology and Chemotherapy* 1961, 17(3), 526–532.
57. Bruce, D.W., Hietbrink, B.E., DuBois, K.P. The acute mammalian toxicity of rare earth nitrates and oxides. *Toxicology and Applied Pharmacology* 1963, 5(6), 750–759.
58. Yang, L., Huang, Y., Peng, Q., Feyerabend, F., Kainer, K.U., Willumeit, R., Hort, N. Mechanical and corrosion properties of binary Mg–Dy alloys for medical applications. *Materials Science and Engineering: B* 2011, 176(20), 1827–1834.
59. Yang, L., Hort, N., Laipple, D., Höche, D., Huang, Y., Kainer, K.U., Willumeit, R., Feyerabend, F. Element distribution in the corrosion layer and cytotoxicity of alloy Mg–10Dy during in vitro biodegradation. *Acta Biomaterialia* 2013, 9(10), 8475–8487.
60. Yang, L., Huang, Y., Feyerabend, F., Willumeit, R., Kainer, K.U., Hort, N. Influence of ageing treatment on microstructure, mechanical and bio-corrosion properties of Mg–Dy alloys. *Journal of the Mechanical Behavior of Biomedical Materials* 2012, 13, 36–44.
61. Yang, L., Huang, Y., Feyerabend, F., Willumeit, R., Mendis, C., Kainer, K.U., Hort, N. Microstructure, mechanical and corrosion properties of Mg–Dy–Gd–Zr alloys for medical applications. *Acta Biomaterialia* 2013, 9(10), 8499–8508.
62. Zhang, X., Yuan, G., Wang, Z. Mechanical properties and biocorrosion resistance of Mg-Nd-Zn-Zr alloy improved by cyclic extrusion and compression. *Materials Letters* 2012, 74, 128–131.
63. Zhao, X., Shi, L.L., Xu, J. A comparison of corrosion behavior in saline environment: Rare earth metals (Y, Nd, Gd, Dy) for alloying of biodegradable magnesium alloys. *Journal of Materials Science & Technology* 2013, 29(9), 781–787.
64. Nordlien, J.H., Nisancioglu, K., Ono, S., Masuko, N. Morphology and structure of water-formed oxides on ternary MgAl alloys. *Journal of the Electrochemical Society* 1997, 144(2), 461–466.
65. Kubásek, J., Vojtěch, D., Lipov, J., Ruml, T. Structure, mechanical properties, corrosion behavior and cytotoxicity of biodegradable Mg–X (X = Sn, Ga, In) alloys. *Materials Science and Engineering: C* 2013, 33(4), 2421–2432.
66. Hänzi, A.C., Sologubenko, A.S., Uggowitzer, P.J. Design strategy for new biodegradable Mg–Y–Zn alloys for medical applications. *International Journal of Materials Research* 2009, 100(8), 1127–1136.
67. Hänzi, A.C., Gerber, I., Schinhammer, M., Löffler, J.F., Uggowitzer, P.J. On the in vitro and in vivo degradation performance and biological response of new biodegradable Mg–Y–Zn alloys. *Acta Biomaterialia* 2010, 6(5), 1824–1833.
68. Loos, A., Rohde, R., Haverich, A., Barlach, S. In vitro and in vivo biocompatibility testing of absorbable metal stents. *Macromolecular Symposia* 2007, 253(1), 103–108.
69. Zartner, P., Cesnjevar, R., Singer, H., Weyand, M. First successful implantation of a biodegradable metal stent into the left pulmonary artery of a preterm baby. *Catheterization and Cardiovascular Interventions* 2005, 66(4), 590–594.
70. Campos, C.M., Muramatsu, T., Iqbal, J., Zhang, Y.J., Onuma, Y., Garcia-Garcia, H.M., Haude, M., Lemos, P.A., Warnack, B., Serruys, P.W. Bioresorbable drug-eluting magnesium-alloy scaffold for treatment of coronary artery disease. *International Journal of Molecular Sciences* 2013, 14(12), 24492–24500.

71. Windhagen, H., Radtke, K., Weizbauer, A., Diekmann, J., Noll, Y., Kreimeyer, U., Schavan, R., Stukenborg-Colsman, C., Waizy, H. Biodegradable magnesium-based screw clinically equivalent to titanium screw in hallux valgus surgery: Short term results of the first prospective, randomized, controlled clinical pilot study. *Biomedical Engineering Online* 2013, 12(1), 62.
72. Takenaka, T., Narazaki, Y., Uesaka, N., Kawakami, M. Improvement of corrosion resistance of magnesium alloys by surface film with rare earth element. *Materials Transactions* 2008, 49(5), 1071–1076.
73. Kirkland, N.T., Birbilis, N., Walker, J., Woodfield, T., Dias, G.J., Staiger, M.P. In-vitro dissolution of magnesium–calcium binary alloys: Clarifying the unique role of calcium additions in bioresorbable magnesium implant alloys. *Journal of Biomedical Materials Research Part B: Applied Biomaterials* 2010, 95(1), 91–100.
74. Bach, F.W., Schaper, M., Jaschik, C. Influence of lithium on hcp magnesium alloys. *Materials Science Forum* 2003, 419, 1037–1042.
75. Song, Y., Shan, D., Chen, R., Han, E.H. Corrosion characterization of Mg–8Li alloy in NaCl solution. *Corrosion Science* 2009, 51(5), 1087–1094.
76. Zeng, R.C., Sun, L., Zheng, Y.F., Cui, H.Z., Han, E.H. Corrosion and characterisation of dual phase Mg–Li–Ca alloy in Hank's solution: The influence of microstructural features. *Corrosion Science* 2014, 79, 69–82.
77. Mueller, W.D., Nascimento, M.L., De Mele, M.F.L. Critical discussion of the results from different corrosion studies of Mg and Mg alloys for biomaterial applications. *Acta Biomaterialia* 2010, 6(5), 1749–1755.
78. Witte, F., Fischer, J., Nellesen, J., Vogt, C., Vogt, J., Donath, T., Beckmann, F. In vivo corrosion and corrosion protection of magnesium alloy LAE442. *Acta Biomaterialia* 2010, 6(5), 1792–1799.
79. Alexander, M.P., Farag, Y.M.K., Mittal, B.V., Rennke, H.G., Singh, A.K. Lithium toxicity: A double-edged sword. *Kidney International* 73(2), 233–237.
80. Aral, H., Vecchio-Sadus, A. Toxicity of lithium to humans and the environment—A literature review. *Ecotoxicology and Environmental Safety* 2008, 70(3), 349–356.
81. Giles, J.J., Bannigan, J.G. Teratogenic and developmental effects of lithium. *Current Pharmaceutical Design* 2006, 12(12), 1531–1541.
82. Angrisani, N., Reifenrath, J., Zimmermann, F., Eifler, R., Meyer-Lindenberg, A., Vano-Herrera, K., Vogt, C. Biocompatibility and degradation of LAE442-based magnesium alloys after implantation of up to 3.5 years in a rabbit model. *Acta Biomaterialia* 2016, 44, 355–365.
83. Li, J.F., Zheng, Z.Q., Li, S.C., Ren, W.D., Zhang, Z. Preparation and galvanic anodizing of a Mg–Li alloy. *Materials Science and Engineering: A* 2006, 433(1), 233–240.
84. Haferkamp, H., Niemeyer, M., Boehm, R., Holzkamp, U., Jaschik, C., Kaese, V. Development, processing and applications range of magnesium lithium alloys. *Materials Science Forum* 2000, 350, 31–42.
85. Crawford, P., Barrosa, R., Mendez, J., Foyos, J., Es-Said, O.S. On the transformation characteristics of LA141A (Mg-Li-Al) alloy. *Journal of Materials Processing Technology* 1996, 56(1–4), 108–118.
86. Watanabe, H., Tsutsui, H., Mukai, T., Kohzu, M., Tanabe, S., Higashi, K. Deformation mechanism in a coarse-grained Mg–Al–Zn alloy at elevated temperatures. *International Journal of Plasticity* 2001, 17(3), 387–397.
87. Wang, T., Zhang, M., Wu, R. Microstructure and properties of Mg–8Li–1Al–1Ce alloy. *Materials Letters* 2008, 62(12), 1846–1848.
88. Liu, T., Wang, Y.D., Wu, S.D., Peng, R.L., Huang, C.X., Jiang, C.B., Li, S.X. Textures and mechanical behavior of Mg–3.3% Li alloy after ECAP. *Scripta Materialia* 2004, 51(11), 1057–1061.

89. Trojanova, Z., Drozd, Z., Lukáč, P., Chmelik, F. Deformation behaviour of Mg–Li alloys at elevated temperatures. *Materials Science and Engineering: A* 2005, 410, 148–151.
90. Shao, Y., Huang, H., Zhang, T., Meng, G., Wang, F. Corrosion protection of Mg–5Li alloy with epoxy coatings containing polyaniline. *Corrosion Science* 2009, 51(12), 2906–2915.
91. Zhou, W. R., Zheng, Y.F., Leeflang, M.A., Zhou, J. Mechanical property, biocorrosion and in vitro biocompatibility evaluations of Mg–Li–(Al)–(RE) alloys for future cardiovascular stent application. *Acta Biomaterialia* 2013, 9(10), 8488–8498.
92. Liu, T., Wu, S.D., Li, S.X., Li, P.J. Microstructure evolution of Mg–14% Li–1% Al alloy during the process of equal channel angular pressing. *Materials Science and Engineering: A* 2007, 460, 499–503.
93. Kim, W.J., Kim, M.J., Wang, J.Y. Ultrafine-grained Mg–9Li–1Zn alloy sheets exhibiting low temperature superplasticity. *Materials Science and Engineering: A* 2009, 516(1), 17–22.
94. Zhang, M. L., Wu, R.Z., Wang, T., Liu, B., Niu, Z.Y. Microstructure and mechanical properties of Mg-xLi-3Al-1Ce alloys. *Transactions of Nonferrous Metals Society of China* 2007, 17(1A), 381–384.
95. Wu, R., Qu, Z., Zhang, M. Effects of the addition of Y in Mg–8Li–(1, 3) Al alloy. *Materials Science and Engineering: A* 2009, 516(1), 96–99.
96. Drozd, Z., Trojanová, Z., Kúdela, S. 2004. Deformation behaviour of Mg–Li–Al alloys. *Journal of Alloys and Compounds* 2004, 378(1), 192–195.
97. Seitz, J.M., Wulf, E., Freytag, P., Bormann, D., Bach, F.W. The manufacture of resorbable suture material from magnesium. *Advanced Engineering Materials* 2010, 12(11), 1099–1105.
98. Leeflang, M.A., Dzwonczyk, J.S., Zhou, J., Duszczyk, J. Long-term biodegradation and associated hydrogen evolution of duplex-structured Mg–Li–Al–(RE) alloys and their mechanical properties. *Materials Science and Engineering: B* 2011, 176(20), 1741–1745.
99. Witte, F., Abeln, I., Switzer, E., Kaese, V., Meyer-Lindenberg, A., Windhagen, H. Evaluation of the skin sensitizing potential of biodegradable magnesium alloys. *Journal of Biomedical Materials Research Part A* 2008, 86(4), 1041–1047.
100. Krause, A., Von der Höh, N., Bormann, D., Krause, C., Bach, F.W., Windhagen, H., Meyer-Lindenberg, A. Degradation behaviour and mechanical properties of magnesium implants in rabbit tibiae. *Journal of Materials Science* 2010, 45(3), 624.
101. Ullmann, B., Reifenrath, J., Seitz, J.M., Bormann, D., Meyer-Lindenberg, A. Influence of the grain size on the in vivo degradation behaviour of the magnesium alloy LAE442. *Proceedings of the Institution of Mechanical Engineers, Part H: Journal of Engineering in Medicine* 2013, 227(3), 317–326.
102. Xu, L., Yu, G., Zhang, E., Pan, F., Yang, K. In vivo corrosion behavior of Mg-Mn-Zn alloy for bone implant application. *Journal of Biomedical Materials Research Part A* 2007, 83(3), 703–711.
103. Yamamoto, A., Hiromoto, S. Effect of inorganic salts, amino acids and proteins on the degradation of pure magnesium in vitro. *Materials Science and Engineering: C* 2009, 29(5), 1559–1568.
104. Song, G. Recent progress in corrosion and protection of magnesium alloys. *Advanced Engineering Materials* 2005, 7(7), 563–586.
105. Kaya, R.A., Çavuşoğlu, H., Tanik, C., Kaya, A.A., Duygulu, Ö., Mutlu, Z., Zengin, E., Aydin, Y. The effects of magnesium particles in posterolateral spinal fusion: An experimental in vivo study in a sheep model. *Journal of Neurosurgery: Spine* 2007, 6(2), 141–149.

106. Höh, N.V.D., Bormann, D., Lucas, A., Denkena, B., Hackenbroich, C., Meyer-Lindenberg, A. Influence of different surface machining treatments of magnesium-based resorbable implants on the degradation behavior in rabbits. *Advanced Engineering Materials* 2009, 11(5), B47–B54.
107. Kuhlmann, J., Bartsch, I., Willbold, E., Schuchardt, S., Holz, O., Hort, N., Höche, D., Heineman, W.R., Witte, F. Fast escape of hydrogen from gas cavities around corroding magnesium implants. *Acta Biomaterialia* 2013, 9(10), 8714–8721.
108. Witte, F. The history of biodegradable magnesium implants: A review. *Acta Biomaterialia* 2010, 6(5), 1680–1692.
109. Hiromoto, S., Yamamoto, A., Maruyama, N., Somekawa, H., Mukai, T. Influence of pH and flow on the polarisation behaviour of pure magnesium in borate buffer solutions. *Corrosion Science* 2008, 50(12), 3561–3568.
110. Yun, Y., Dong, Z., Lee, N., Liu, Y., Xue, D., Guo, X., Kuhlmann, J. et al., Revolutionizing biodegradable metals. *Materials Today* 2009, 12(10), 22–32.
111. Kannan, M.B., Dietzel, W., Raman, R.K.S., Lyon, P. Hydrogen-inducedcracking in magnesium alloy under cathodic polarization. *Scripta Materialia* 2007, 57, 579–581.
112. Winzer, N., Atrens, A., Dietzel, W., Song, G., Kainer, K.U. Magnesium stress corrosion cracking. *Transactions of Nonferrous Metals Society of China* 2007, 17(1), 150–155.

8 Effect of Heat Treatment Parameters on the Microstructure of Mg-9Al Magnesium Alloy

Andrzej Kiełbus, Janusz Adamiec, and Robert Jarosz

CONTENTS

8.1 INTRODUCTION

Most commercial magnesium alloys are based on the Mg-Al system. These alloys are relatively cheap compared with other magnesium alloys available. They are often used for many technical applications, including the aerospace, automobile and motor vehicle, metallurgical, chemical and electronical industries [1,2].

The basic Mg-Al magnesium alloys contain aluminium, zinc and manganese which allow obtaining suitable properties. Aluminium enhances both tensile strength and hardness, and improves casting properties of an alloy. The best ratio of mechanical to plastic properties is obtained with a 6% Al content. Zinc is added to improve the room temperature strength and fluidity. An addition of zinc in combination with Al aims at improving tensile strength at a room temperature, however 1% of Zn with a $7 \div 10\%$ Al content in an alloy enhances hot cracking. Manganese does not have much effect on tensile strength, however, it does increase yield strength slightly. Addition of manganese is required to control the corrosion behaviour. Mg-Al magnesium alloys with manganese are commonly used for components where good ductility and impact strength are required. Its most important function is to improve the saltwater resistance of Mg-Al alloys by removing iron and other heavy metal elements into relatively harmless intermetallic compounds. The quantity of manganese in magnesium alloys is limited by its relatively low solubility in magnesium. Manganese content in alloys with an Al addition does not exceed 0.3% and 1.5% in alloys without Al addition [3,4].

Magnesium alloys are subjected to heat treatment mostly for the purpose of improvement of their mechanical properties or as an intermediary operation, to prepare the alloy to other specific treatment processes. The type of heat treatment depends on the chemical composition of the alloy, its form (casting or after plastic working) and on the anticipated service parameters.

Solution heat treatment of magnesium alloys enhances their strength, with maximum ductility and resistance to dynamic loads. Aging of solution heat treated alloys allows obtaining maximum hardness and yield point, with a decrease of ductility, whereas aging treatment without prior solution heat treatment, as well as annealing, result in a decrease of casting stresses and partial increase of mechanical properties when under tension. Annealing significantly decreases the mechanical properties and causes improvement of plastic properties, thus facilitating further treatment. A change of the heat treatment basic parameters has an influence on a change of the properties [5–7].

After solution treatment a solid solution of aluminum is present in magnesium (α-Mg) and possibly, some areas of undissolved precipitates of $Mg_{17}Al_{12}$ phase [8]. After aging, two types of $Mg_{17}Al_{12}$ precipitates occur in Mg-Al alloy: continuous and discontinuous. In most cases, the precipitates occur simultaneously. The continuous precipitation is a result of nucleation and growth of individual $Mg_{17}Al_{12}$ phase particles, which leads to changes in the matrix composition. Whereas discontinuous precipitations nucleate on the boundaries of the solid solution grains and when growing, they take the form resembling nodules [9]. Mg-Al alloys containing $5 \div 10$ wt.% of Al are dominated by the continuous precipitations of $Mg_{17}Al_{12}$ phase. However, it has been found that the morphology and quantity of precipitates of the $Mg_{17}Al_{12}$ phase in Mg-Al alloys depends on the chemical composition (Al content) and heat treatment parameters.

8.2 MATERIAL AND RESEARCH METHODOLOGY

8.2.1 Material for Research

The material for the research was a sand casting Mg-9Al magnesium alloy. Sand casting was carried out at about 700°C. The chemical composition of the analysed alloy is shown in Table 8.1.

TABLE 8.1
Chemical Composition of the Mg-9Al Magnesium Alloy in wt. %

Al	Zn	Mn	Si	Fe	Ni	Mg
8,7	0,8	0,25	0,03	0,002	0,001	balance

8.2.2 Research Methodology

The study was conducted on Mg-9Al magnesium alloy in as-cast condition and after heat treatment.

8.2.2.1 Heat Treatment Parameters

The solution and aging treatment parameters are presented in Table 8.2 and on Figures 8.1 and 8.2.

8.2.2.2 Microstructural Investigations

Attempts to reveal the Mg-9Al microstructure after heat treatment were made on the surface of the metallographic specimen. The specimens' rough cut was made with Phoenix cut-off machine. The precise cutting was performed with a Beuhler Isomet 5000 cut-off machine. The specimens were first ground and then polished according to the recommendations of Buehler company. For the microstructure observation

TABLE 8.2
Heat Treatment Parameters

	Solution Treatment				Aging Treatment			
Designation	Heating [°C/h]	Temperature [°C]	Time [h]	Cooling	Heating [°C/h]	Temperature [°C]	Time [h]	Cooling
N	As-cast							
N1	25	360	3	water	–	–	–	–
N2		N1 + 415	24	water	–	–	–	–
N3		N2			50	170	8	air
N4	25	360	3	–	–	–	–	–
		415	24	water				
N5		N4			50	170	8	air

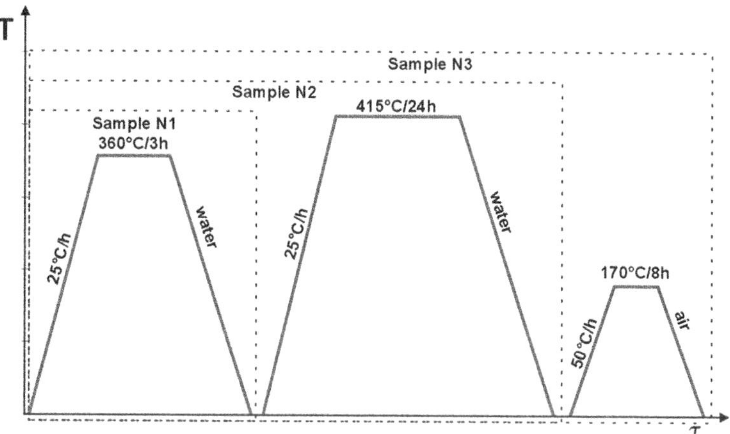

FIGURE 8.1 Heat treatment parameters of N1 ÷ N3 samples.

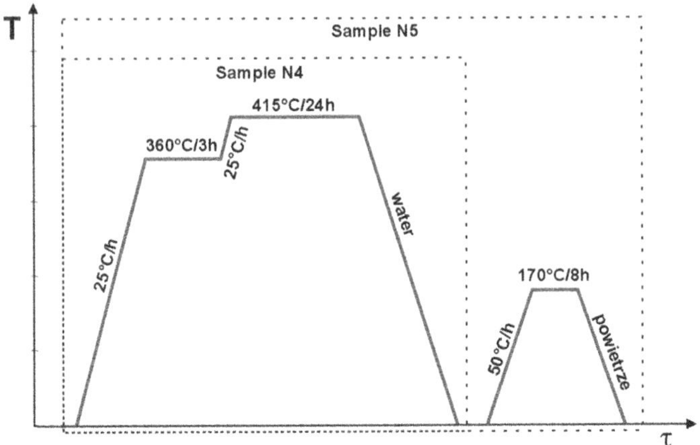

FIGURE 8.2 Heat treatment parameters of N4 ÷ N5 samples.

an OLYMPUS GX71 metallographic microscope and a HITACHI S-3400N scanning electron microscope with a Thermo Noran EDS spectrometer equipped with SYSTEM SIX were used. The best images of the alloy microstructure after solution heat treatment and after both solution heat treatment and aging are presented on Figure 8.3.

It has been found that:

- The best etching reagent for the alloy after solution heat treatment is reagent: 10 ml HF + 96 ml H_2O. Etching with this reagent allows for a clear identification of phases in the structure and for its qualitative description (Figure 8.3a);
- In order to evaluate the grain size in the alloy after solution treatment, the etching reagent is: 20 ml acetic acid + 80 ml H_2O + 5g $NaNO_2$;

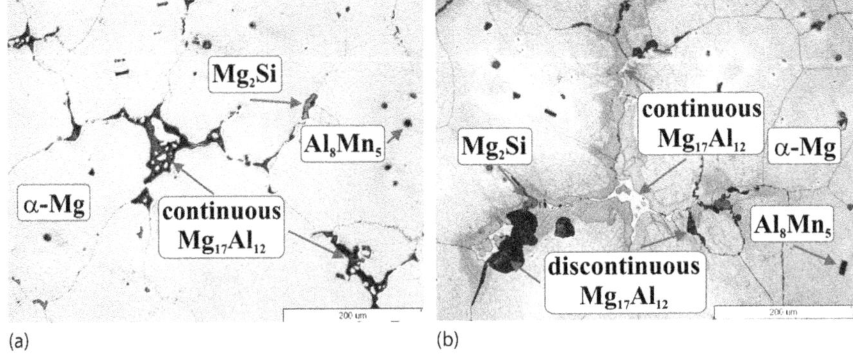

FIGURE 8.3 Microstructure of Mg-9Al alloy: (a) after solution heat treatment – specimen N2, (b) after solution heat treatment and aging – specimen N3.

- The best reagent for the alloy after aging to be used for a qualitative and quantitative evaluation is: 5–20 ml acetic acid, 80–95 ml H_2O, this reagent also allows the determination of the grain size in the structure. A good etching reagent is also: 19 ml H_2O + 60 ml ethylene glycol + 20 ml acetic acid + 1 ml HNO_3, which allows individual phases in the structure to be detected (Figure 8.3b);
- It is recommended that in all cases, observations on a light microscope in bright field at a 500× magnification should be carried out for a qualitative evaluation, whereas the 100× to 200× magnification should be used for a quantitative evaluation.

8.2.2.3 Quantitative Evaluation of Phases

The quantitative evaluation of phases detected in the Mg-9Al magnesium alloy in as-cast state and after heat treatment was performed using a light microscope OLYMPUS GX-71, equipped with an automatic table for image stitching in XYZ axes and AnalySIS Pro® software as well as MetIlo® software. It was assumed that, based on the metallographic investigations, the surface fractions of the following phases would be analyzed: continuous $Mg_{17}Al_{12}$ phase, discontinuous $Mg_{17}Al_{12}$ phase, regions enriched with Al, Mg_2Si phase and the area occupied by solid solution. The detection of phases is shown on Figure 8.4.

The following parameters were measured to characterize microstructure:

- Volume fraction of intermetallic phases throughout the sample section, V_V [%],
- Mean area of solid solution grain plain section.

The procedure of converting the input image into a binary image to be used for grain size measurement is presented in Table 8.3 and for intermetallic phase precipitates (on the example of continuous $Mg_{17}Al_{12}$ phase) in Table 8.4.

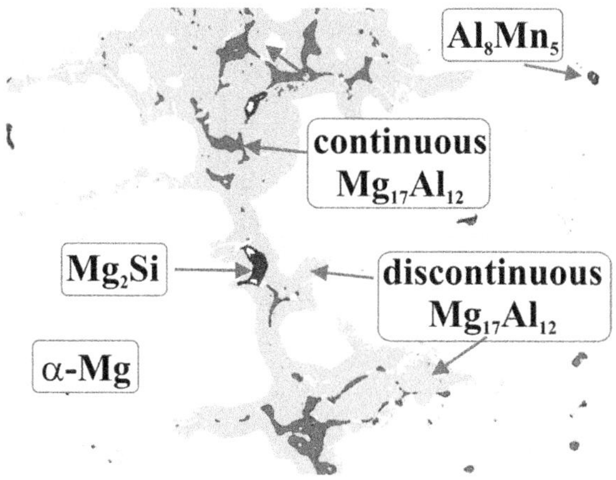

FIGURE 8.4 The detection of phases occurring in the Mg-9Al alloy structure.

TABLE 8.3
The Procedure of a Quantitative Evaluation of Solid Solution Grains in Mg-9Al Alloy, LM

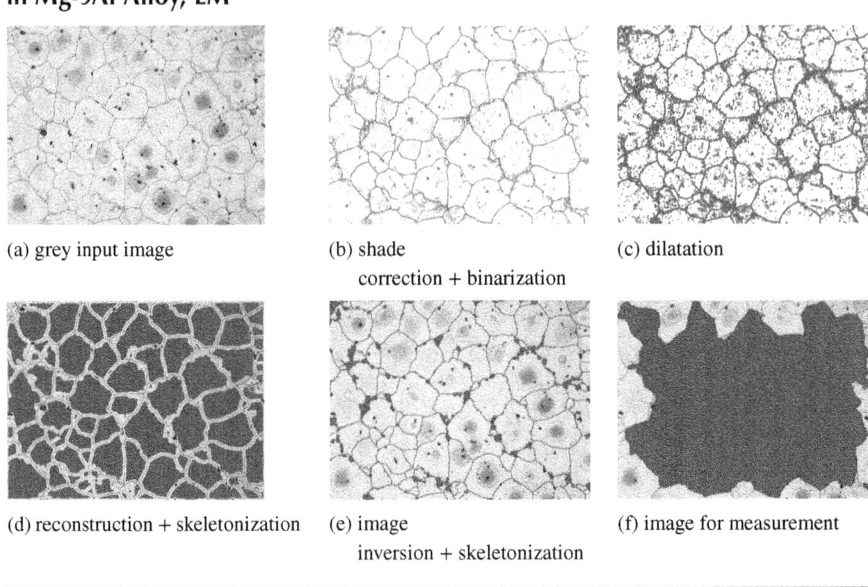

(a) grey input image (b) shade correction + binarization (c) dilatation

(d) reconstruction + skeletonization (e) image inversion + skeletonization (f) image for measurement

TABLE 8.4
The Procedure of a Quantitative Evaluation of Discontinuous $Mg_{17}Al_{12}$ Phase in Mg-9Al Alloy

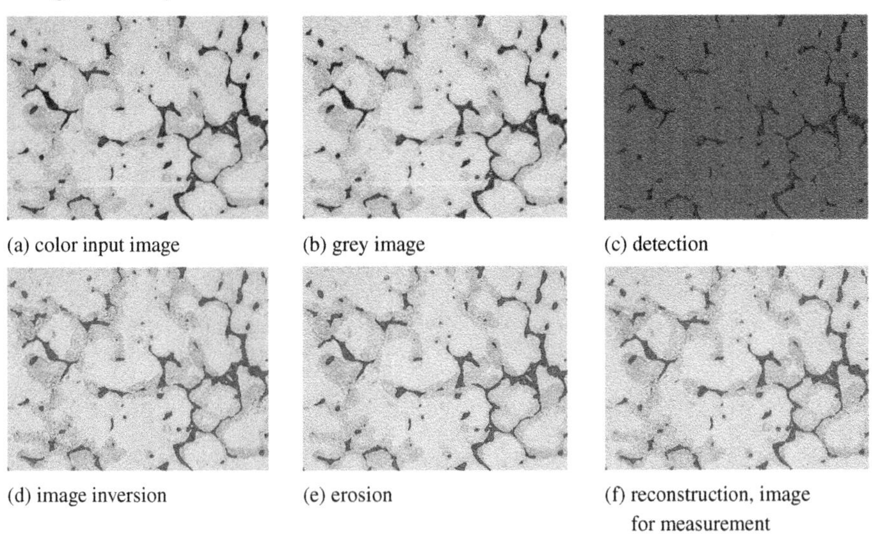

(a) color input image (b) grey image (c) detection

(d) image inversion (e) erosion (f) reconstruction, image for measurement

8.3 MICROSTRUCTURE OF THE Mg-9Al MAGNESIUM ALLOY IN AS-CAST CONDITION

The Mg-9Al magnesium alloy in as-cast condition is characterized by a solid solution structure α-Mg with discontinuous and continuous precipitates of $Mg_{17}Al_{12}$ phase at grain boundaries. Moreover, the occurrence of Laves' phase in the form of Mg_2Si and precipitates of Mn_5Al_8 phase, has been provided (Figure 8.5). A quantitative evaluation of as-cast structure has shown that the mean plane section area of the α-Mg solid solution grains equals $\bar{A} = 5442\ \mu m^2$. The average area fraction of continuous $Mg_{17}Al_{12}$ phase was $A_A = 6.66\%$ and discontinuous (α-Mg + $Mg_{17}Al_{12}$ areas) was $A_A = 19.92\%$. The area fraction of Mg_2Si phase was equal $A_A = 0.15\%$ (Table 8.5).

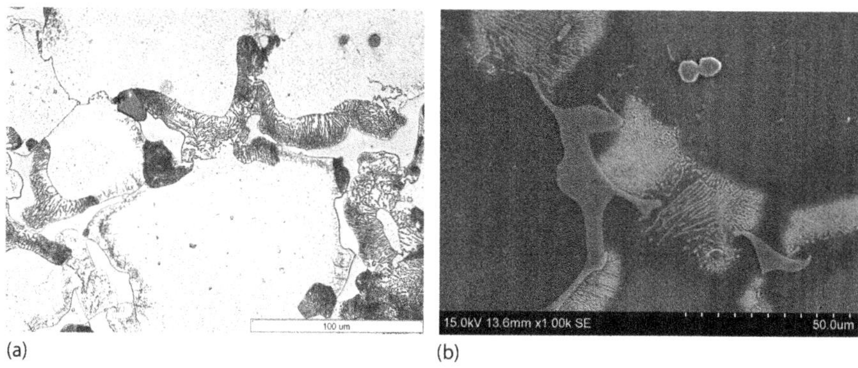

FIGURE 8.5 Microstructure of the as-cast Mg-9Al alloy (sample N), (a) LM image, (b) SEM image.

TABLE 8.5
Assessment of Phases Area Fractions in the Mg-9Al Alloy Microstructure in as Cast Condition (Specimen N)

Phase	Area Fraction A_A [%]	Standard Deviation σ [%]	Variability Index V [%]
$Mg_{17}Al_{12}$ (continuous)	6.66	1.57	23.57
$Mg_{17}Al_{12}$ (discontinuous)	19.92	5.49	27.54
Mg_2Si	0.15	0.07	47.14
α-Mg	73.26	4.8	6.56

8.4 MICROSTRUCTURE OF THE Mg-9Al MAGNESIUM ALLOY AFTER SOLUTION TREATMENT

After solution treatment 360°C/3h/water (sample N1) the $Mg_{17}Al_{12}$ intermetallic phase dissolves in the matrix (Figure 8.6). The area fraction of continuous $Mg_{17}Al_{12}$ phase decrease (~2 times) from A_A = 6.66% after casting to A_A = 3.41% and discontinuous $Mg_{17}Al_{12}$ phase decrease (~4 times) from A_A = 19.92% after casting to A_A = 5.37%. The area fraction of Mg_2Si phase was equal A_A = 0.39% (Table 8.6). The mean plane section area of α-Mg grains equals $\bar{A}$ = 6798 μm^2 and is higher (about 1300 μm^2) compared to the as cast state (Table 8.7).

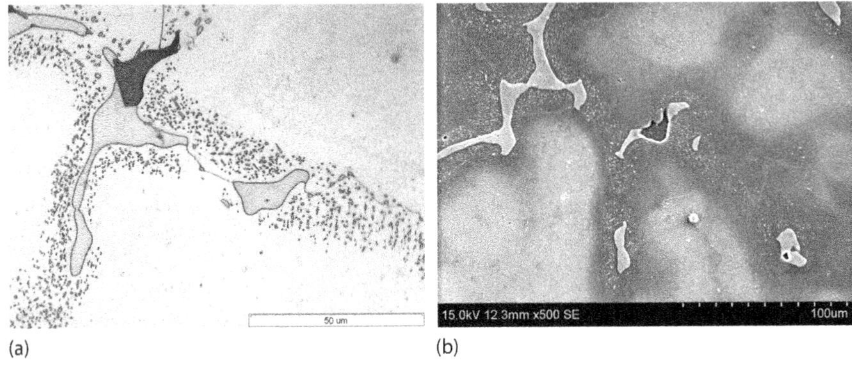

FIGURE 8.6 Microstructure of the Mg-9Al alloy after solution treatment 360°C/3h/air (sample N1), (a) LM image, (b) SEM image.

TABLE 8.6
Assessment of Phases Area Fractions in the Mg-9Al Alloy Microstructure After Solution Treatment

Sample	Phase	Area Fraction A_A [%]	Standard Deviation σ [%]	Variability Index $\nu(A_A)$ [%]
N1	$Mg_{17}Al_{12}$ (continuous)	3,41	1,29	38
	$Mg_{17}Al_{12}$ (discontinuous)	5,37	0,28	5
	Mg_2Si	0,39	0,06	24
	Solution α	88,27	3,51	15
N2	$Mg_{17}Al_{12}$ (continuous)	1,19	0,28	24
	$Mg_{17}Al_{12}$ (discontinuous)	0	0,00	0
	Mg_2Si	0,34	0,07	21
	Solution α	97,39	1,12	1
N4	$Mg_{17}Al_{12}$ (continuous)	0,7	0,3	43
	$Mg_{17}Al_{12}$ (discontinuous)	0,81	0,15	18
	Mg_2Si	0,1	0,04	42
	Solution α	96,50	1,49	2

TABLE 8.7
Influence of the Solution Treatment Parameters on the Mean Plane Section Area of α-Mg Grains

Sample	Mean Plane Section Area Ā [μm]	Variability Index V [%]
N	5442	58
N1	6798	65
N2	7555	71
N4	7544	71

Solution treatment 415°C/24h/water after treatment 360°C/3h/water (sample N2) caused a considerable (4-times) decrease of the $Mg_{17}Al_{12}$ phase quantity compared to the state after treatment at 360°C/3h/water (Figure 8.7). The area fraction of continuous $Mg_{17}Al_{12}$ phase decrease to $A_A = 1.19\%$. The precipitates of discontinuous $Mg_{17}Al_{12}$ phase weren't observed (Table 8.6). The area fraction of Mg_2Si phase was equal $A_A = 0.34\%$. The mean plane section area of the solid solution α-Mg grain increased to $\bar{A} = 7555\ \mu m^2$ and was higher (about 750 μm^2) compared to the state after treatment 360°C/3h/water (Table 8.7).

Solution treatment at 360°C/3h/without cooling + 415/24/water (sample N4) also caused a reduction of the number of $Mg_{17}Al_{12}$ phase precipitates (Figure 8.8). A quantitative evaluation of Mg-9Al alloy after this treatment has shown that area fraction of continuous $Mg_{17}Al_{12}$ phase was equal $A_A = 0.7\%$, discontinuous $Mg_{17}Al_{12}$ phase was equal $A_A = 0{,}81\%$ and of Mg_2Si phase was equal $A_A = 0.1\%$ (Table 8.4). This treatment led to reduction of discontinuous β phase (Table 8.6). The mean plane section area of solid solution α grain was $\bar{A} = 7544\ \mu m^2$ and was identical as after solution treatment 360°C/3h/water + 415/24/water (sample N2) (Table 8.7).

FIGURE 8.7 Microstructure of the Mg-9Al alloy after solution treatment 360°C/3h/air + 415/24/air (sample N2), LM image.

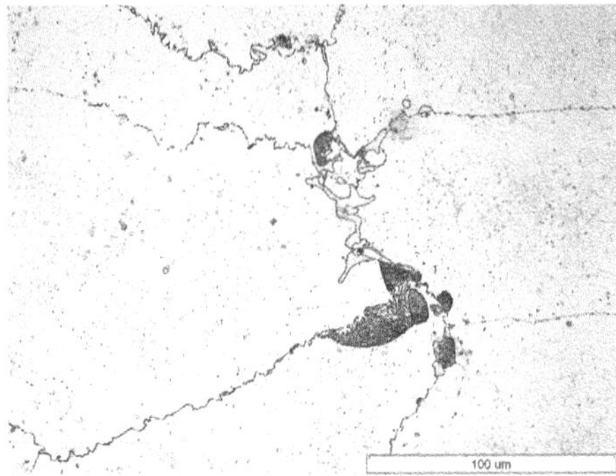

FIGURE 8.8 Microstructure of the Mg-9Al alloy after solution treatment 360°C/3h + 415/24air (sample N4), LM image.

8.5 MICROSTRUCTURE OF THE Mg-9Al MAGNESIUM ALLOY AFTER AGING TREATMENT

The microstructure of Mg-9Al alloy after aging treatment 170°C/8h/air applied after solution treatment 360°C/3h/water + 415/24/water (sample N3) is shown on Figure 8.9. This treatment caused precipitation of discontinuous $Mg_{17}Al_{12}$ phase. The area fraction of this phase increase $A_A = 10.56\%$. The significant influence of this treatment was not affirmed on the quantity of continuous $Mg_{17}Al_{12}$ phase precipitates ($A_A = 1.1\%$) compared to the state after solution treatment. The area fraction of Mg_2Si phase was equal $A_A = 0.37\%$ (Table 8.8).

The aging treatment applied after solution treatment 360°C/3h/without cooling + 415/24/water also caused increase of discontinuous $Mg_{17}Al_{12}$ phase area fraction ($A_A = 2.81\%$), but this increase was very small (Figure 8.10). The area fraction of

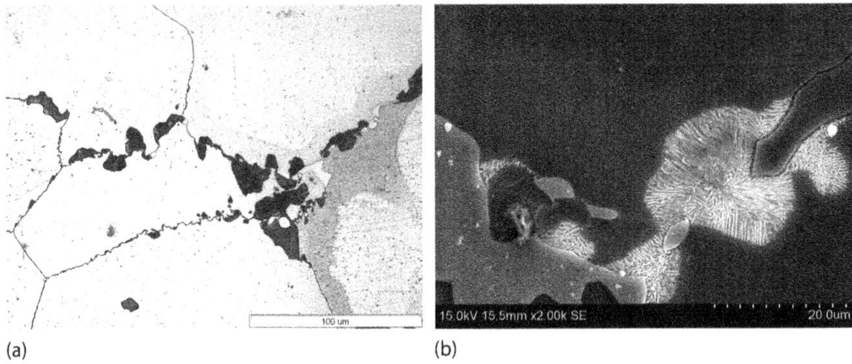

FIGURE 8.9 Microstructure of the Mg-9Al alloy after solutioning 360°C/3h/air + 415/24/air and aging 170°C/8h/air (sample N3), (a) LM image, (b) SEM image.

TABLE 8.8
Assessment of Phases Area Fractions in the Mg-9Al Alloy Microstructure after Aging Treatment

Sample	Phase	Area Fraction A_A [%]	Standard Deviation σ [%]	Variability Index $\nu(A_A)$ [%]
N3	$Mg_{17}Al_{12}$ (continuous)	1.1	0.47	42
	$Mg_{17}Al_{12}$ (discontinuous)	10.56	2.82	27
	Mg_2Si	0.37	0.17	46
	Solution α	82.29	10.93	13
N5	$Mg_{17}Al_{12}$ (continuous)	1.57	0.7	44
	$Mg_{17}Al_{12}$ (discontinuous)	2.80	1.86	66
	Mg_2Si	0.15	0.06	41
	Solution α	91.95	7.67	8

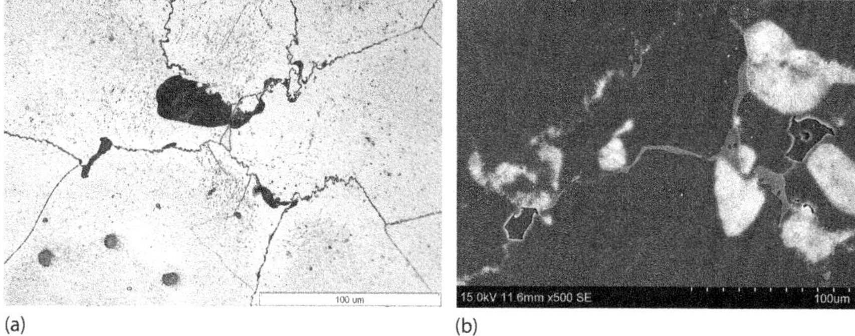

FIGURE 8.10 Microstructure of the Mg-9Al alloy after solutioning 360°C/3h + 415/24/air and aging 170°C/8h/air (sample N5), (a) LM image, (b) SEM image.

continuous $Mg_{17}Al_{12}$ phase was slightly higher ($A_A = 1.57\%$) compared to the state after solution treatment. The area fraction of Mg_2Si phase was equal $A_A = 0.15\%$ (Table 8.8). The mean plane section area of α-Mg grains didn't change after aging treatment.

8.6 SUMMARY

The subject of the research carried out was an evaluation of the effect of heat treatment parameters on the microstructure of the Mg-9Al magnesium alloy. It was found that in as cast condition, the Mg-9Al alloy had a solid solution α-Mg structure with discontinuous and continuous precipitates of $Mg_{17}Al_{12}$ phase at grain boundaries. Moreover, the occurrence of Mg_2Si and Mn_5Al_8 phases has been provided.

Application of solutioning has led to dissolving of continuous and discontinuous precipitates of $Mg_{17}Al_{12}$ phase regardless of parameters used. In case of the first solutioning variant i.e. 360°C/3h with water cooling (sample N1) an incomplete supersaturation of solid solution is being acquired.

The area fraction of the continuous precipitates of $Mg_{17}Al_{12}$ phase is being reduced almost two times to the value $A_A = 3.41\%$, while the area fraction of discontinuous precipitates is being reduced almost four times to the value $A_A = 5.37\%$ (Figure 8.11). Renewed solutioning from temperature 415°C/24h/water (sample N2) leads to reduction of area fraction of the continuous precipitates of $Mg_{17}Al_{12}$ phase to value $A_A = 1.19\%$ and to complete dissolving of discontinuous precipitates of $Mg_{17}Al_{12}$ phase in α-Mg solid solution. Regardless of heat treatment parameters used any significant changes in area fraction of Mg_2Si phase weren't detected. In all cases content of this phase was about $A_A = 0.1 \div 0.4\%$.

In case of the second treatment variant i.e. solutioning 360°C/3h/ without cooling with following heating up to temperature 415°C and holding the temperature for 24 h (sample N4) the reduction of area fraction of the continuous precipitates $Mg_{17}Al_{12}$ phase to value $A_A = 0.45\%$ and absence of complete dissolving of discontinuous precipitates $Mg_{17}Al_{12}$ phase in α-Mg solid solution are being acquired. The area fraction of this phase equals $A_A = 0.81\%$ (Figure 8.11).

Solutioning leads to increase of the mean plane section area of solid solution α-Mg grain along with increase of temperature and heating time (Figure 8.12). In the initial state the mean plane section area of solid solution α grain was equal $\bar{A} = 5442\ \mu m^2$. After solution treatments designated N2 and N4 the mean plane section area slightly increase to $\bar{A} = 7555\ \mu m^2$ and $\bar{A} = 7544\ \mu m^2$, respectively.

Application of aging 170°C/8h with cooling in air has led to significant changes in area fraction of discontinuous precipitates of $Mg_{17}Al_{12}$ phase. After full solutioning according to first variant (sample N2) with next aging the area fraction of discontinuous precipitates of $Mg_{17}Al_{12}$ phase increase from $A_A = 0\%$ to $A_A = 10.56\%$, however

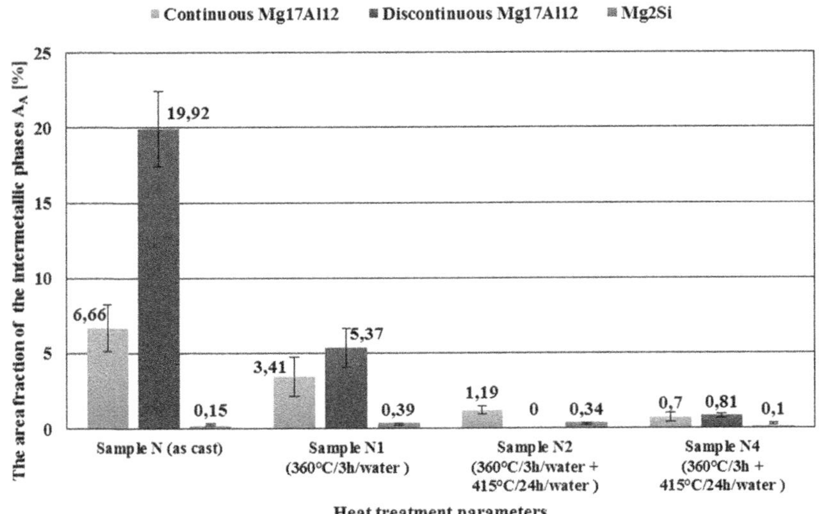

FIGURE 8.11 The influence of solution treatment parameters on the intermetallic phases quantity in Mg-9Al alloy.

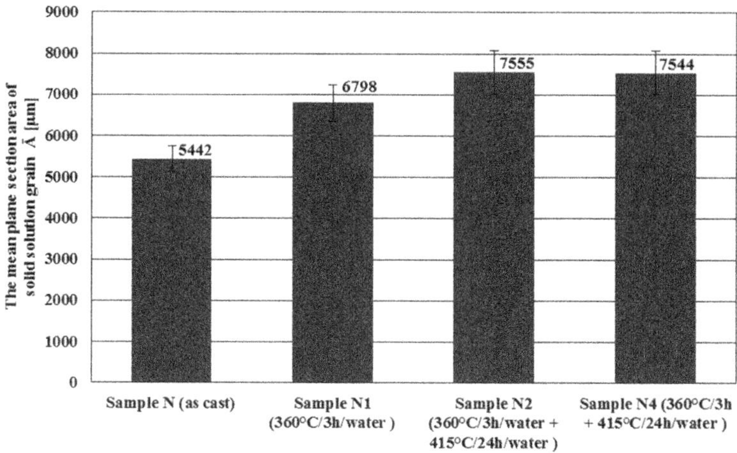

FIGURE 8.12 The influence of solution treatment parameters on the mean plane section area of α-Mg grain in Mg-9Al alloy.

the area fraction of continuous precipitates of $Mg_{17}Al_{12}$ phase doesn't change and equals $A_A = 1.1\%$ (Figure 8.13). Solutioning and aging according to second heat treatment variant (sample N5) has led to precipitation of the considerably smaller quantity of discontinuous precipitates of $Mg_{17}Al_{12}$ phase ($A_A = 2.81\%$). The smaller quantity of this phase is caused by incomplete supersaturation of the alloy Mg-9Al after applying the second variant of solutioning (sample N4) (Figure 8.13).

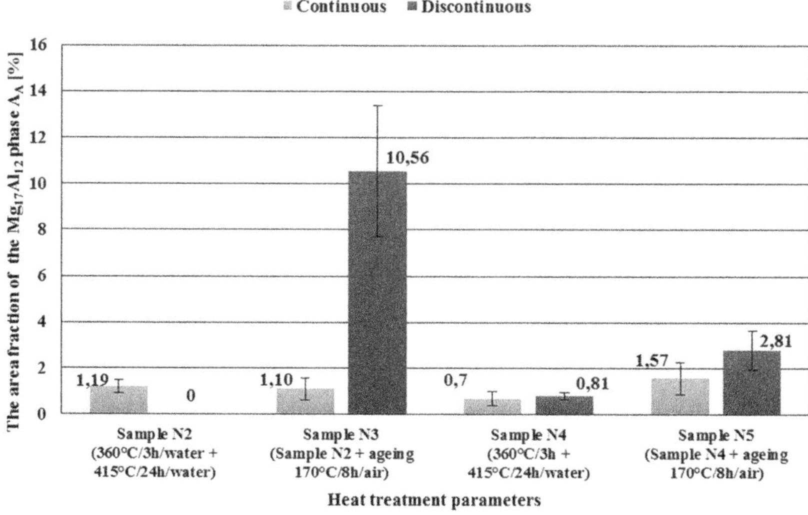

FIGURE 8.13 The influence of aging treatment parameters on the $Mg_{17}Al_{12}$ phases quantity in Mg-9Al alloy.

8.7 CONCLUSIONS

Based on the research results obtained, it has been found that:

1. The Mg-9Al magnesium alloy in as cast condition is characterized by a solid solution structure α-Mg with discontinuous and continuous precipitates of $Mg_{17}Al_{12}$ phase at grain boundaries. Moreover, the occurrence of Mg_2Si and Mn_5Al_8 phases, has been provided. The mean plane section area of solid solution α-Mg grain equals $\bar{A} = 5442\ \mu m^2$. The average area fraction of continuous $Mg_{17}Al_{12}$ phase was $A_A = 6.66\%$ and discontinuous (α-Mg + $Mg_{17}Al_{12}$ areas) was $A_A = 19.92\%$.
2. Independently from solution treatment parameters a reduction of the quantity of continuous and discontinuous $Mg_{17}Al_{12}$ phase precipitates is observed. The mean plane section area of α-Mg grains is higher (about 2000 μm^2) compared to the as cast state.
3. The aging treatment applied after solution treatment caused increase of discontinuous $Mg_{17}Al_{12}$ phase area fraction mainly along the solid solution grain boundaries.
4. Aging treatment at after solution treatment 360°C/3h/without air cooling + 415°C/24/water (sample N3) cause more significant (~5 times) increase of discontinuous $Mg_{17}Al_{12}$ phase area fraction comparing to solution treatment with water cooling after heating at 360°C.

ACKNOWLEDGMENTS

This work was supported by the Polish Ministry of Education and Science under the research project No. 6 ZR7 2005 C/06609.

REFERENCES

1. Friedrich, H., Schumann, S. Research for a "new age of magnesium" in the automotive industry. *Journal of Material Processing Technology* 2001, 117 (3), 276–281.
2. Aghion, E., Bronfin, B., Eliezer D. The role of magnesium industry in protecting the environment. *Journal of Material Processing Technology* 2001, 117 (3), 381–385.
3. Luo, A.A. Magnesium: Current and potential automotive applications. *JOM* 2002, 54 (1), 42–48.
4. Dahle, A., Lee, Y., Nave, M., Schaffer, P., StJohn, D. Development of the as-cast microstructure in magnesium-aluminium alloys. *Journal of Light Metals* 2001, 1 (1), 61–72.
5. Caceres, C.H., Davidson, C.J., Griffiths, J.R., Newton, C.L. Effects of solidification rate and ageing on the microstructure and mechanical properties of AZ91 alloy. *Materials Science and Engineering: A* 2002, 325 (1–2), 344–355.
6. Stratton, P., Chang, E. Protective atmospheres for the heat treatment of magnesium alloys. *In Magnesium Technology 2000.* 1st Ed. Kaplan H. I., Hryn J. N., Clow B. B., Eds., John Wiley & Sons, Hoboken, NJ, 2000, 71–76.

7. Cerri, E., Cabibbo, M., Evangelista, E. Microstructural evolution during high-temperature exposure in a thixocast magnesium alloy. *Materials Science and Engineering: A* 2002, 333 (1–2), 208–217.
8. Zhang, M., Kelly, P.M. Crystallography of $Mg_{17}Al_{12}$ precipitates in AZ91D alloy. *Scripta Materialia* 2003, 48 (5), 647–652.
9. Duly, D., Simon, J.P., Brechet, Y. On the competition between continuous and discontinuous precipitations in binary Mg-Al alloys. *Acta Metallurgica et Materialia* 1995, 43 (1), 101–106.

9 Composition, Structure, and Protective Properties of Air-Formed Oxide Films on Magnesium Alloys

Sebastián Feliu Jr.

CONTENTS

9.1 INTRODUCTION

Normally, metals and alloys spontaneously form an oxide or hydroxide surface layer when in contact with the air. The surface of a metal or alloy is immediately covered in less than a millisecond by a thin (1–4 nm) oxide nanolayer. The film is transparent and invisible to the naked eye. In certain cases (for example, for stainless steels, titanium and aluminum alloys), these layers are continuous and uniform films that greatly retard the corrosion of the metal substrate. Magnesium and its alloys also spontaneously form thin layered structures consisting of oxide, hydroxide and carbonate layers on the surface upon exposure to the laboratory atmosphere, as Figure 9.1 shows, which affect their propensity for corrosion in a natural environment.

In contrast to the protective and stable oxide/hydroxide surface film that forms on other metals and alloys, the air-formed film on a magnesium alloy is considered to be discontinuous, provides considerably less protection and is consequently a primary factor that explains the poor corrosion resistance of this material. However, the literature indicates that very low corrosion rates have been observed for some combinations of magnesium alloys and surface conditions during the first stages of exposure to a 0.6 M NaCl solution, which agrees with the notable differences observed for the oxide films formed on the surface prior to the immersion test. Additionally, the literature notes the presence of non-corroded areas on the exposed surface of some magnesium alloys after prolonged immersion. This entry addresses the nature of the air-formed films on commercial magnesium alloys, and the changes in their chemical composition induced by exposure to dry oxygen at room temperature, to the ambient

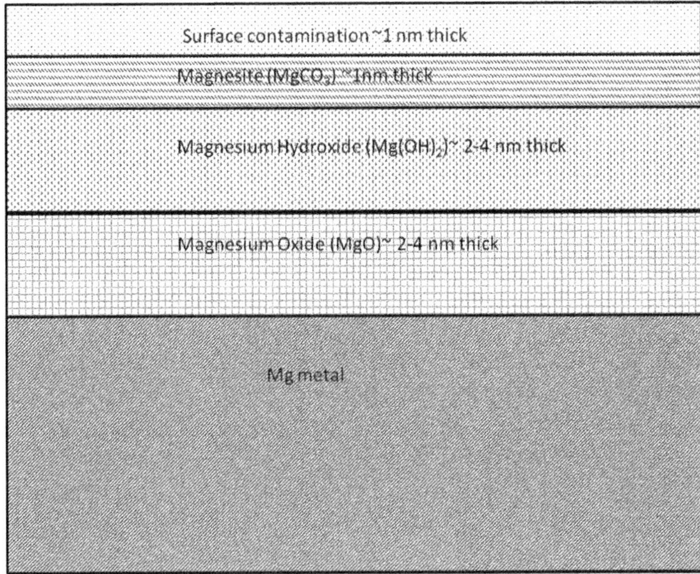

FIGURE 9.1 Schematic diagram of proposed multi-layer air formed oxide films structure on Mg surface in contact with laboratory atmosphere.

atmosphere or to a highly humid atmosphere, as well as to an aqueous solution, with the aim of elucidating the protective properties of these films.

9.2 RELEVANCE OF SURFACE FILMS ON THE CORROSION RESISTANCE OF MAGNESIUM ALLOYS

Magnesium alloys have many applications in aerospace construction, aircraft, guided weapons, laptop computers, the manufacture of mobile phones, cameras, electronic components, sports equipment, and in the automotive industry because of their light weight, high specific strength and stiffness, good electro-magnetic shielding capacity, high-damping capacity, heat dissipation, good machinability, excellent castability and excellent recyclability.[1] Magnesium is the lightest of the structural metals, with a weight density of 1.74 g cm^{-3}, which is approximately 2/3 that of aluminum and 1/4 that of iron.[2] It is expected that magnesium and its alloys will have significant applications in the transportation industry, where weight is important, to lower fuel consumption and CO_2 emissions.[3] The biocompatibility and biodegradability of magnesium alloys is interesting for medical implants and their hydrogen storage capability must also be mentioned.[4] However, the engineering applications of magnesium alloys are primarily limited by their low corrosion resistance, especially in a chloride-containing aqueous environment. Electrochemically, magnesium is the most active metal of the various structural materials.[5] Therefore, the majority of its applications are restricted to mild environments, and exposure to environments containing H_2O and Cl^- must be avoided.[6] The further expansion of their applications appears to be unlikely until an effective solution to the corrosion of Mg alloys becomes available. A better understanding of the corrosion mechanism and of the factors that affect the corrosion is expected to lead to more corrosion-resistant Mg alloys.[7]

The corrosion of Mg alloys generally begins on a portion of the surface, then expands over the entire surface due to the breakdown of the surface film; therefore, the corrosion resistance of the metal in a corrosive environment is initially associated with the protective properties and the chemistry of the surface oxide film.[8] Any attempt to improve the corrosion resistance of Mg alloys requires a profound understanding of the relationship between the composition, structure and the protective properties of the surface films on unalloyed (i.e., commercially pure) Mg and how these relationships are modified in alloys[3] or under changing exposure conditions. The structure and morphology of an oxide surface film formed naturally on magnesium exposed to air has been studied to a limited extent.[9] Unfortunately, information regarding the chemistry and the degree of protection provided by a natural oxide surface film is scarce, and its effect on the corrosion resistance of magnesium alloys is still not comprehensively understood.[10] Additionally, most of the naturally formed surface films on Mg alloys are not protective.[11]

One of the primary obstacles for characterizing the air-formed films on magnesium alloys is their thickness, which does not exceed several nanometers.[12] Such a thin oxide film does not produce a sufficient signal for conventional characterization techniques (i.e., optical microscopy, scanning electron microscopy (SEM)/energy dispersive X-ray microanalysis (EDX) or X-ray diffraction (XRD)).[13] The use of the X-ray photoelectron spectroscopy (XPS) and its capability of analyzing a film only

3 nm thick has been demonstrated to be particularly effective for the surface characterization of materials.[14,15] The strength of XPS lies in its ability to quantify the composition of an extremely thin oxide film and to distinguish the chemical state of the elements detected. In addition, using intensity ratio of the oxidic to the metallic components, it is possible to measure the thickness of a thin oxide film that typically forms on a metal surface in contact with the air.[8,15] Consequently, a large number of XPS studies have recently been conducted on the surface characteristics of thin oxide films formed on magnesium exposed to diverse conditions.[14] In this entry, information on the formation, the composition, the structure and the thickness of the surface oxide films formed on magnesium alloys has been primarily obtained from XPS studies.

9.3 OXIDE FILMS FORMED ON THE MAGNESIUM SURFACE AND MAGNESIUM ALLOYS AFTER EXPOSURE TO OXYGEN AT ROOM TEMPERATURE

After oxygen exposure at room temperature, magnesia (MgO) is spontaneously formed on the surface of an Mg alloy according to the reaction[16–18]:

$$2Mg + O_2 = 2MgO \tag{9.1}$$

Magnesium has a very high affinity for oxygen, and the standard Gibbs free energy of the oxidation reaction (Eq. 9.1) ($\Delta G_f^0 = -1138$ KJ mol O_2 at 25°C) is one of the most highly negative for any metal[18], and it ensures that a surface oxide layer forms rapidly.[19] MgO is a highly efficient insulator, the only oxide assumed to grow on pure magnesium[12] and the predominant form produced on magnesium alloys after room temperature exposure to an oxygen containing environment.[17] MgO has a face-centered cubic (fcc) crystal halite structure with a lattice parameter of 0.42 nm[16–18,20] and cubic close-packing of O ions, with Mg ions occupying all of the octahedral sites.[16–18] For Mg-Al alloys, containing more than 4% Al oxidized at 31°C, a substantial enrichment in Al in both the oxide film and the subsurface region of the alloy substrate[21] is present because the Al has a greater affinity for oxygen than it does for magnesium.[16,22]

It is generally accepted that the initial Mg oxidation kinetics progress in three phases: oxygen chemisorption below the topmost magnesium layer, nucleation-lateral growth, and an increase in the thickness.[18,23,24] In the first stage, a monolayer of oxygen is chemisorbed on a clean metal surface.[18] An adsorbed species may act by reordering the valence electrons of the metal, which favors the chemical bond formation.[18,23] In the second stage, oxide islands nucleate from the adsorbed oxygen and grow laterally across the surface. Place exchange and surface diffusion are the important factors during this stage.[18,23] The growth of the islands is the initial formation stage of a three-dimensional oxide film and has been reported to be rapid and to have a linear dependence on duration of the oxygen exposure.[18,23] The oxide thickening stage following the coalescence of the oxide is slow and driven by diffusion.[23] This final stage has some of the characteristics of a steady-state, and the growth of the oxide film approaches a limiting value.[25] According to the Cabrera-Mott theory on the growth of very thin oxide films,[26] the diffusion of oxygen or metal ions through the oxide film proceeds by migration under a high electrostatic field associated with the presence of negatively charged oxygen species adsorbed on the outer surface of the metal.[21,27] The electric field in a very thin film is enormous at first, but the thickening of the film weakens the field strength, and the oxide growth virtually ceases after few minutes.[28]

In an early study in 1923, Pilling and Bedworth[29] suggested that the protective nature of the oxide film that naturally forms on a metallic material can be indicated by the Pilling–Bedworth ratio (P-B ratio), which is defined as the ratio of the molecular volume of the oxide to the atomic volume of the corresponding metal from which the oxide is formed. Due to the large difference in the density of the oxide and metal, as indicated by the MgO to Mg volume ratio of 0.81, the oxide scale is subject to increased tension, discontinuous, porous and not compact, and may not protect the substrate from oxidation.[16,18–20,23,30] A P-B ratio of 1–2 is typically considered a minimum but is not a sufficient prerequisite for protective film formation.[31]

Based on the possible similarity to the passive film formed on stainless steel, Tan et al.[16] considered likely that the MgO formed at a low temperature is crystalline. Additionally, Reichel et al.[32] have recently used thermodynamic model calculations to demonstrate that the initial development of an amorphous oxide phase is unlikely on a Mg{0 0 0 1} metal surface. However, Czwerwinsky[17,24] thought that amorphous MgO forms on Mg alloys at room or at low temperature. Nordlien et al.[9,22,33] visualized cross sections of the thin oxide film by transmission electron microscopy (TEM), and used an ultramicrotome analysis and electron diffraction to reveal that the sections have an amorphous structure.[18] An initially amorphous oxide overgrowth has been postulated by Kurth et al.[34] Because a MgO film formed on an Mg-based alloy as a result of exposure to oxygen at ambient temperature may be amorphous, continuous, insulating, dense, compact and have a low density of defects,[35] it should offer better protection than a crystalline structure.[17] While experimental evidence exists about the growth of a stable and amorphous aluminum oxide upon the initial oxidation of a bare surface of pure Al metal at room temperature, experimental evidence about the structure of the initial MgO overgrowth on a bare Mg surface is still lacking.[21]

The oxide formed on pure Mg during oxidation in dry oxygen is very thin. After 15 min of exposure at room temperature, the thickness of the MgO film was 1.5 nm, and it was 2.6 nm after the same exposure duration at 300°C, which was approximately the same thickness as an Al_2O_3 layer formed on pure Al.[16,35] Our thermogravimetric results are shown in Figure 9.2, and imply that a minimal formation of magnesium oxide occurs on a freshly polished AZ31 (nominally 3 wt% Al—1 wt% Zn) magnesium alloy tested at 200°C for as long as 60 minutes.[30] The lack of an easy path for rapid Mg^{2+} transport via solid-state diffusion through adherent oxide areas could be a possible explanation for this highly protective behavior.[17,25] Figure 9.3 shows the details at a nanometer scale of the typical surface roughness of test specimens with both as-received and polished surfaces. As it is shown in Figure 9.2, a considerably higher weight gain is observed in the alloy AZ31 in the as-received condition than for the same alloy when polished, which is probably associated with the formation of an additional oxide layer comprising a mixture of spinel ($MgAl_2O_4$) and MgO,[27] which seems to reduce the protective properties of the homogeneous, passive, uniform thin layer on a polished surface (Figure 9.3a and c). An increase of the temperature above 400°C[16] might cause the oxide to change structurally from an amorphous

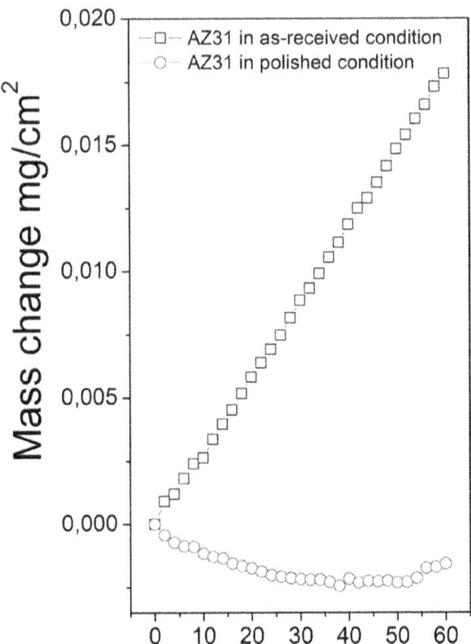

FIGURE 9.2 Evolution of mass change values obtained in the AZ31 alloy in the as-received and polished surface condition as a function of the time of heating at 200°C in an air environment.

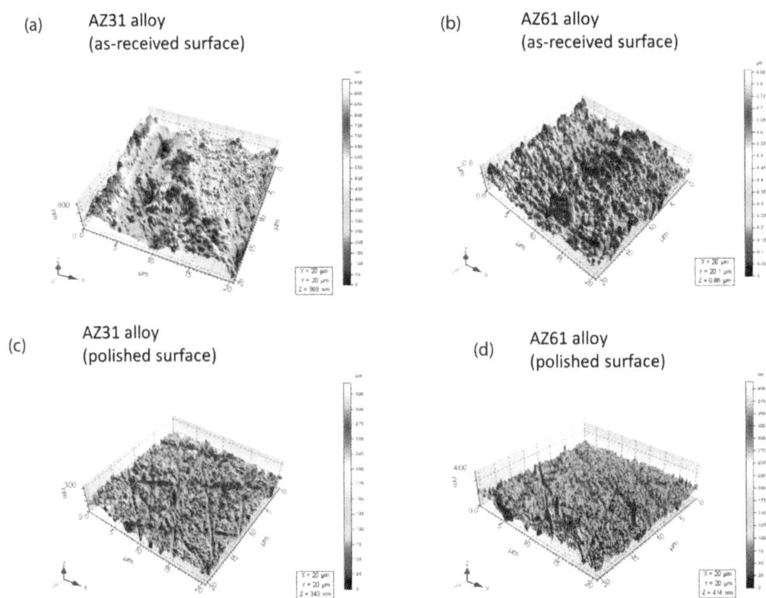

FIGURE 9.3 Atomic-force microscope (AFM) images of the surfaces in the as-received (a,b) and polished surface conditions (c,d) for AZ31 alloy (a,c) and AZ61 alloy (b,d).

to a crystalline phase.[12] A growth of thick (i.e., in the micrometer range[21]) MgO oxide scales with a normal cubic lattice forms under tension and tends to be non-protective; therefore, oxide nodules form and coalesce into a loose fine-grained structure.[18,23,24]

9.4 AIR-FORMED OXIDE FILMS SPONTANEOUSLY FORMED ON MAGNESIUM ALLOYS IN CONTACT WITH A LABORATORY ATMOSPHERE

9.4.1 Chemical Composition, Structure and Thickness

When exposed to air, the fresh surface of pure Mg is covered by gray oxide film[36] visible to the eye[23] and is composed of oxides, hydroxides or other oxidized products.[37] The reaction of Mg with O_2 to form MgO is well-known, but the reaction of water with Mg at room temperature is less well understood.[38] The direct reaction of Mg and water may be generally described by the following reaction schemes of the dissociation of adsorbed H_2O molecules (Eqs. 9.2 and 9.3) on the surface, which lead to the respective formation of MgO, and $Mg(OH)_2$[37,38]:

$$\text{Complete dissociation: } H_2O(ad) \rightarrow O(ad) + 2H(ad) \quad (9.2)$$

$$\text{Partial dissociation: } H_2O(ad) \rightarrow OH(ad) + H(ad) \quad (9.3)$$

XPS studies of an *in situ* oxide growth on pure Mg and Mg-Al alloys performed in an ultra-high vacuum containing trace amounts of heavy water indicated that the initial interaction with water vapor resulted in a three step oxidation process similar to that observed for oxygen interactions with magnesium.[39,40] It was also reported that the interaction of water and humid air with the Mg surface first results in the growth of a partially hydrated and defective thin film of crystalline MgO (c-MgO), which is formed by the combination of completely dissociated H_2O molecules with Mg atoms (Eq. 9.2).[37,39] However, the reaction rate was much slower with water vapor.[41] Splinter and McIntyre[40] suggest that the rate of oxide nucleation and growth is enhanced on a Mg-Al surface compared to a pure Mg surface. Al^{3+} cations are incorporated into the growing oxide film, especially at a higher Al content and a longer exposure duration.

The formation of $Mg(OH)_2$ is also expected due to the chemical hydration of MgO.[37] The adsorbed water molecules dissociate (i.e., undergo proteolysis) to form OH^- and H^+, which results in the hydroxylation of the MgO surface[20]:

$$O^{2-}\text{ (surface)} + H_2O(g) \rightarrow 2OH^-\text{ (surface)} \quad (9.4)$$

Equivalently:

$$MgO\text{(surface)} + H_2O(g) \rightarrow Mg(OH)_2\text{(surface)} \quad (9.5)$$

The hydroxylation of the MgO film (Eqs. 9.4 and 9.5) is the primary reaction that corrodes Mg in the presence of water vapor in air or aqueous immersion.[20] It may be noted that the compounds' thermodynamic equilibrium constants indicate that

$Mg(OH)_2$ is more chemically stable than MgO[42] such that the precipitation of the hydroxide predominates over that of the oxide even at low water vapor concentrations (>1 ppm H_2O).[20] Accordingly, the hydration of MgO in the surface layer of the oxide film by the water vapor in the laboratory results in a layered surface film consisting of an ultrathin $Mg(OH)_2$ layer on top of a relatively thicker MgO inner layer adjacent to the Mg metal.[4,20]

At low RH (relative humidity) or in a dry cycle, the amount of liquid water on the magnesium is very limited (i.e., an adsorbed water layer less than 1 μm thick). In this case, the degradation of magnesium occurs predominantly via oxidation and hydration.[43] The brucite, $Mg(OH)_2$ first forms on the surface of a magnesium alloy (Eq. 9.5) but is only stable at a low CO_2 pressure.[43] Atmospheric CO_2 reacts with $Mg(OH)_2$ under dry conditions at ambient temperature to form magnesite ($MgCO_3$), which is the thermodynamically favored corrosion product, as indicated by Eq. 9.6:

$$Mg(OH)_2 + CO_2 \rightarrow MgCO_3 + H_2O \tag{9.6}$$

XPS studies consistently detect $Mg(OH)_2$ in air-formed films on pure Mg in contact with a laboratory atmosphere as a single layer mixed with MgO or in the outer layer of a duplex structure.[3,4,8,15,35,37–41,44–46] Chen et al.[8] suggested that the interaction of water and humid air with the Mg surface results in the conversion of the MgO layer to $Mg(OH)_2$. McIntyre and Chen[44] observed that during the first few weeks, the O/Mg ratio slowly increased from approximately 1.1 to 1.8, which may be associated with hydroxide formation. Asami and Ono[15] reported that the initial surface film formed on a scratched Mg electrode in air consists of mixed oxy-hydroxide (hydrated) layers measuring 2.7 nm in thickness. Yao et al.[38] concluded that the film formed on a dry-ground, melt-spun, pure Mg ribbon oxidized in air was a mixture of MgO and $Mg(OH)_2$, with $Mg(OH)_2$ predominantly formed on the outermost surface. Liu et al.[4] found that the concentration of $Mg(OH)_2$ in the surface layer formed on a mechanically ground Mg surface was significantly higher than that of MgO. Also, Santamaria et al.[45] used XPS to show that the film formed on an Mg electrode after mechanical polishing consisted of an inner thin oxide layer of MgO 2.5 nm thick, and a 2.2 nm thick outer layer of $Mg(OH)_2$. Additionally, Fotea et al.[46] reported significant amounts of bidentate-bound carboxylate salts on the Mg surface as a result of the reaction/adsorption of volatile organic acids and the water present in the atmosphere.

Alloying elements affect the composition, thickness and structure of air-formed films on the surface of various magnesium alloys.[47] Makar and Kruger[48] showed that the air-formed oxide on the Mg-Al alloys has a layered structure measuring 5–15 nm in thickness and is composed of a MgO/Mg-Al-oxide/substrate, and the Mg-rich oxide becomes thinner as the Al content increases. Song et al.[42] found that the inner layers of MgO formed on Mg-Zn alloys are enriched in ZnO. Song et al.[47] also observed four layers on the surface of Mg–Li alloys in contact with the ambient atmosphere. The top layer was a mixture of $Mg(OH)_2$ and Li_2O, the second layer was a mixture of $Mg(OH)_2$, Li_2O and MgO, the third layer was a mixture of $Mg(OH)_2$, MgO, LiOH, Li_2O and Mg, and the bottom layer was a mixture of MgO, Li_2O, Li and Mg. Nouri el al.[49] demonstrated that the addition of a small amount of yttrium reduced the OH^-/O ratio in the surface oxide film formed on the magnesium alloy AZ31, rendering it stronger and more stable.

We used XPS in a previous study[50] to determine the respective thicknesses of native air-formed oxide films of 2.6, 2.8, 4.4 and 6.3 nm for pure magnesium, AZ31, AZ91D (nominally 9 wt% Al—1 wt% Zn) and AZ80 (nominally 8 wt% Al—1 wt% Zn) magnesium alloys. A direct relationship seems to exist between the air-formed oxide film thickness and the degree of microstructural complexity of the surface on which it forms. In other studies[27,51], XPS analysis indicated notable differences in the surface air-formed oxide films on the AZ31 and AZ61 (nominally 6 wt% Al—1 wt% Zn) magnesium alloys under different surface conditions. Attention was drawn to the higher thickness of the air-formed oxide film on a polished AZ61 alloy compared to AZ31 (Figure 9.4a and b). Additionally, Song et al.[42] observed that a higher Zn content results in a thicker air-formed oxide film on Mg-Zn alloys. We concluded that the very significant presence of the second phase β ($Mg_{17}Al_{12}$) on the grain boundary of the AZ61 alloy (Figure 9.5b) compared to its absence on AZ31 (Figure 9.5a) makes the outer surface more favorable for the movement of the reaction products, favoring the nucleation and growth of a thicker air-formed oxide film. Additionally, the O/(Al + Mg) ratios of 1.2–1.5 determined by XPS on the surface of the AZ31 and AZ61 alloys in as-received condition suggested that an additional layer of a spinel and MgO mixture formed as a result of the manufacturing process (Figure 9.4c and d).[51]

Several XPS studies on Mg and Mg alloys have reported that it is likely that the surface oxide film formed after exposure to the ambient atmosphere also comprised a

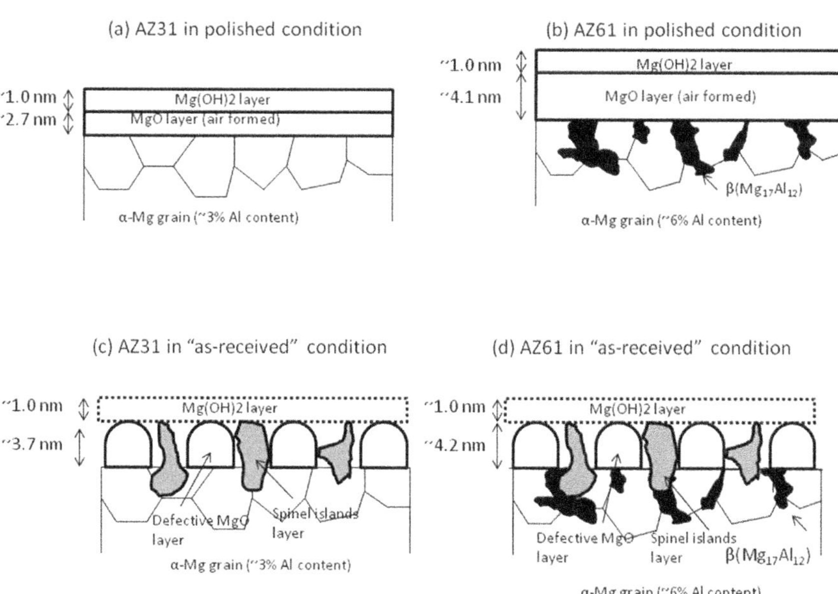

FIGURE 9.4 Schematic model of the chemical composition, structure and thickness of the air formed film on the surface of freshly polished (a,b) and as received condition (c,d) commercial AZ31 (a,c) and AZ61 alloys (b,d).

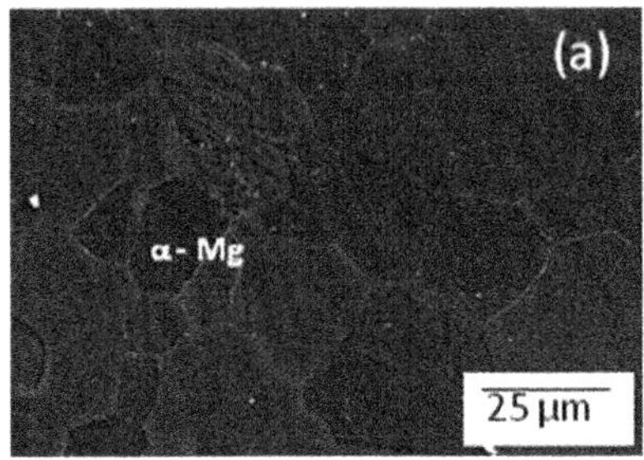

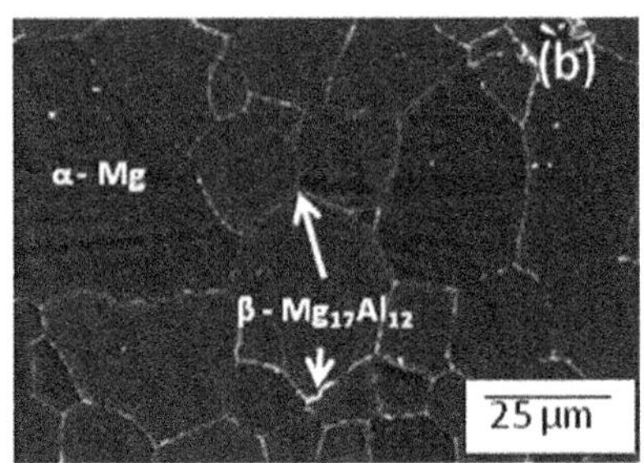

FIGURE 9.5 BSE images showing the microstructure for AZ31 (a) and AZ61 magnesium alloys (b).

magnesium carbonate-rich layer at the uppermost part of the material surface, above the $MgO/Mg(OH)_2$ layer.[3,4,13,35,37,38,42,45,46,51,52] Fournier et al.[35] observed that a small amount of water vapor and CO_2 in a high vacuum furnace can cause the formation of hydroxide and carbonate at room temperature. Fotea et al.[46] concluded from XPS data that a layer of $MgCO_3$ rapidly forms on the surface of $Mg(OH)_2$ upon exposure to ambient air, and the total amount decreases with time. Song et al.[42] found that the outer layer of an air-formed film on Mg-Zn alloys consisted of basic magnesium carbonate and MgO.

Our XPS results for magnesium alloys stored at room temperature for one year in the laboratory indicated that the greatest amount of magnesium carbonate was present in the air-formed film on the alloy AZ80 than on the alloys AZ91 and AZ31.[13] In accordance with the thickness differences (on a nm scale) in the air formed layer commented above, we attributed the increased amount of magnesium carbonate to an increased fraction of the second phase β on the surface of the magnesium alloy and the higher porosity of the oxide/hydroxide film formed on the surface of the magnesium–aluminum alloy, which may facilitate the higher diffusion of CO_2.[13]

In all of these studies, no direct imaging of the surface film in cross-section was performed.[3] Nordlien et al.[9] used an ultramicrotomed cross-section and TEM/EDX and limited plan view sections to discuss the morphology, composition and structure of the surface film formed just after exposing a fresh surface by scratching in low-humidity (35%–55%) air for 0.25–1 h. The air-formed film was found to be thin (20–50 nm), dense and to have an amorphous structure consisting of a mixture of MgO and $Mg(OH)_2$. The amount of magnesium hydroxide was estimated by XPS to be in the range from 50 to 60 wt%.[33] No further increase in the film thickness is observed in a low humidity atmosphere. The air-formed film on a Mg-Al alloy has an amorphous structure of mixed MgO and Al_2O_3[18,22] which is greatly enriched in Al due to its greater oxygen affinity, especially in alloys containing more than 4% Al that produce an oxide containing 35% Al in the innermost layer. The thickness of the oxide film of a Mg–Al alloy is 20–40 nm and it decreases with an increase of the Al concentration.[18,22]

9.4.2 Protective Capacity in Laboratory Atmosphere

Air-formed films on Mg alloys can provide good atmospheric corrosion resistance in dry air (i.e., <80% RH) at a normal ambient temperature[37] that is typically found in the interior of vehicles or buildings.[20] Atmospheric corrosion is only superficial and typically slower than the corrosion of mild steel, and better than that of some aluminum alloys because of the presence of an air-formed film on the magnesium surface.

Even if the air contains corrosive chlorides, its corrosivity is not too high if it is sufficiently dry.[19] The thickness of the air-formed films on a magnesium specimen left in laboratory air for a few months changed very little (3.09–3.13); however, surface hydration caused a significant increase in the quantity of water molecules detected (from 1 to 3).[15] The oxide product was a 5–6 nm thick MgO film that formed independently of the exposure time between 7 days and 7 years, and a surface film formed in air after 7 years largely consisted of more ionized components such as carbonate and hydroxide.[38] McIntyre and Chen[44] have suggested that an oxide approximately 4.5 nm thick was formed when the ambient exposure time reached 10 months. The rate of attack is negligible at low humidity.[46]

9.5 EFFECT OF HIGH RELATIVE HUMIDITY ON THE SURFACE CHARACTERISTICS AND CORROSION BEHAVIOUR OF AIR-FORMED OXIDE FILMS ON MAGNESIUM ALLOYS

9.5.1 Chemical Composition and Structure

Under dry atmospheric conditions at normal ambient temperature, as discussed in previous section, a thin protective oxide film mainly comprising a relatively thick MgO inner layer forms on the Mg surface and the rate of attack is negligible.[53,54] Pure Mg exposed to the atmosphere corrodes only if the humidity is sufficiently high for an aqueous solution to form on the sample surface.[54,55] When the RH approaches 100%, multiple layers of adsorbed water exist on the magnesium surface, with properties resembling those of bulk water.[6,20] An electrochemical reaction occurs on the surface under these conditions, is closely associated with the corrosion of the magnesium in an aqueous solution,[55] and is based on three reactions.[43]

The cathodic reaction of a magnesium alloy in a humid atmospheric environment (Eqs. 9.7 and 9.8) includes the reduction of water and oxygen:[10,43]

$$2H_2O + 2e^- \rightarrow H_2 + 2OH^- \text{ (cathodic reaction)} \quad (9.7)$$

$$O_2 + 2H_2O + 4e^- \rightarrow 4OH^- \text{ (cathodic reaction)} \quad (9.8)$$

Simultaneously, the anodic reaction (Eq. 9.9)

$$Mg \rightarrow Mg^{2+} + 2e^- \text{ (anodic reaction)} \tag{9.9}$$

As the Mg^{2+} ion concentration near a corroded site exceeds its solubility, a layer of crystalline magnesium hydroxide precipitates as shown by Eq. 9.10:[6,43,56]

$$Mg^{2+}(aq) + 2OH^-(aq) \rightarrow Mg(OH)_2(s) \tag{9.10}$$

In the absence of CO_2, high humidity promotes the formation of $Mg(OH)_2$ on the magnesium surface.[57]

At an RH up to 90%, the film formed in humid air is amorphous $Mg(OH)_2$ and nearly invisible.[53,54] As the humidity increases further, the initially clear structure of the film becomes a heavier tarnish, and the principal corrosion product is crystalline $Mg(OH)_2$. In humid air or in an aqueous environment, the initial air-formed oxide film is thinner, and may be hydrolyzed and dissolved by liquid water and undermined by the formation of a less stable, hydrated oxide.[54]

Exposure to a humid atmosphere produced a much thicker film (e.g., 100–150 nm after 4 days exposure to 65% relative humidity) than was produced in dry air, and it had a duplex structure.[9,33,54] The 20–40 nm thick outer oxide layer is similar to the complete oxide film formed in air: although it has a compact, dense, amorphous morphology, acts as a membrane permeable to the ingress of water and the egress of soluble Mg species and becomes separated from the substrate.[9,23,33,54] The inner oxide layer is a cellular-like hydrated layer, which results from water molecules that penetrate the initial air-formed film to react with the Mg substrate.[9,33,54]

As previously mentioned for low humidity, the presence of atmospheric CO_2 also has a major effect on the chemistry and properties of an air-formed film in high humidity due to the incorporation of magnesium carbonates.[58] The CO_2 in the atmosphere dissolves in the thin layer of electrolytes present on the metal surface (Eq. 9.11) to form carbonic acid (H_2CO_3 (aq))(Eq. 9.12)[20,36] which is in a dissociation equilibrium with the H^+, HCO_3^-, and CO_3^{2-} ions in solution according to the Eqs. 9.13 and 9.14[20,36,59]:

$$CO_2(g) \leftrightarrow CO_2(aq) \tag{9.11}$$

$$CO_2(aq) + H_2O(l) \leftrightarrow H_2CO_3(aq) \tag{9.12}$$

$$H_2CO_3(aq) + OH^-(aq) \leftrightarrow HCO_3^-(aq) + H_2O(l) \tag{9.13}$$

$$HCO_3^-(aq) + OH^-(aq) \leftrightarrow CO_3^{2-}(aq) + H_2O(l) \tag{9.14}$$

CO_2 also reacts with the primary corrosion product brucite, and forms the more chemically stable magnesium hydroxycarbonates by precipitation from an aqueous solution (Eq. 9.15) and by the direct reaction of CO_2 with $Mg(OH)_2$ (see, e.g., Eq. 9.16):

$$5Mg^{2+}(aq) + 4CO_3^{2-}(aq) + 2OH^-(aq) + 4H_2O(l) \rightarrow Mg_5(CO_3)_4(OH)_2 \cdot 4H_2O \quad (9.15)$$

$$5Mg(OH)_2 + 4CO_2 \rightarrow Mg_5(CO_3)(OH)_2 \cdot 4H_2O \quad (9.16)$$

Laboratory studies[60] reveal that magnesite ($MgCO_3$) is formed on the alloy AZ91D in humid air (i.e., 95% RH) as indicated by Eq. 9.6 and then transforms into nesquehonite ($MgCO_3 \cdot 3H_2O$) from the direct hydration of magnesite (Eq. 9.17) after 2–3 days. This reaction does not involve CO_2:

$$MgCO_3(s) + 3H_2O \rightarrow MgCO_3 \cdot 3H_2O(s) \quad (9.17)$$

Long-term exposure leads to the formation of hydromagnesite ($Mg_5(CO_3)_4(OH)_2 \cdot 4H_2O$) via the dehydration of nesquehonite as indicated by Eq. 9.18:

$$4MgCO_3 \cdot 3H_2O \rightarrow 3MgCO_3 \cdot Mg(OH)_2 \cdot 3H_2O + CO_2 + 8H_2O \quad (9.18)$$

9.5.2 Protective Capacity of the Air-Formed Films at High Relative Humidity

9.5.2.1 Corrosion Resistance of the $Mg(OH)_2$ Products Formed

Atmospheric corrosion resistance deteriorates considerably at >90% relative humidity.[49] Corrosion damage can be severe if the Mg alloy surface is sprayed with atmospheric aerosols (e.g., a marine environment) or exposed to NaCl on winter roads where deicing salts are used.[1,61,62] Chloride ions are reported to enhance the dissolution of the passive film and the anodic dissolution of metal and increases the amount of electrolyte on the metal surface and thus the surface conductivity.[55,63]

Several key factors affect the comparatively poor corrosion resistance of the $Mg(OH)_2$ formed in high humidity, including (i) the P-B ratios of the oxide and hydroxide (0.81 and 1.77, respectively), that cause the hydrolysis of the MgO to $Mg(OH)_2$ to result in local volumetric changes, which likely promotes the formation of cracks and spalls inside the oxide film and provides little resistance to the transport of water vapor or oxygen[23,49,57] (ii) the intense hydrogen evolution reaction which accompanies the formation of $Mg(OH)_2$, results in cracking of the film[64] (iii) an unstable $Mg(OH)_2$ layer that forms on the Mg, which is soluble over a wide pH range from approximately 2–10.5 is susceptible to dissolution in most aqueous or humid environments.[6,20,64] ($Mg(OH)_2$ is protective and stable only under highly alkaline conditions at pH values greater than 11[10,19]) (iv) brucite is a hexagonal crystalline structure in which the layers are held together by hydrogen bonding, which facilitates basal cleavage.[57,64]

9.5.2.2 Corrosion Resistance of the Formed Magnesium Carbonate Products

In the presence of ambient concentrations of CO_2, the corrosion resistance of Mg alloys in high humidity is improved due to the formation of carbonate-containing films in the outer layer of the air-formed film that is more protective than the hydroxide film that forms in the absence of CO_2.[6,10,20,43,55,57,58,60,65] Lindstrom et al.[6] proposed that magnesium hydroxyl carbonate products, which are thicker than the magnesium hydroxide film, were non-conducting and may thus retard corrosion by physically blocking both the anodic and cathodic sites. Also the solubility of magnesium hydroxyl carbonate in the presence of ambient levels of CO_2 is lower than that of brucite at neutral pH.[55] In addition, it is generally accepted that OH^- consumption by HCO_3^- and the carbonate buffering effect counteracts the development of a surface pH gradient, impeding the development of macroscopic corrosion cells.[6,66] Hallam et al.[57] proposed that the magnesite structures held together by ionic bonds resist the compressive stress arising from corrosion and can form a fully protective layer, unlike the weaker brucite structure. For the Mg-Al alloys, apart from the beneficial effect of CO_2 attributed to the formation of a magnesium hydroxyl carbonate, it was suggested that neutralization of the cathodic area by CO_2 resulted in the decreased solubility of aluminum and the stabilization of the Al-containing surface film.[55,65] Lindstrom et al.[6] suggested that carbonates can increase the stability of the relatively uniform corrosion layer and inhibit the pitting corrosion.

In a previous study,[13] we observed a strong effect of the presence of carbonates on an air-formed surface film on the atmospheric corrosion of Mg-Al alloy. A linear relationship was found between the amount of magnesium carbonate detected by XPS on the surface of magnesium alloys of an alloy stored at room temperature for one year in a laboratory atmosphere and the subsequent corrosion in a humid environment. The air-formed oxide film on the surface of the alloy AZ80 had a higher content of $MgCO_3$ than the alloys AZ31 and AZ91 and commercially pure Mg and had the highest atmospheric corrosion resistance in humid air.

In another study,[67] mass gain vs. time curves for pure Mg and the alloy AZ31 exposed to 98% RH at 50°C were carried out and indicated that corrosion rate steadily decreased with the exposure time. Also, the mass gain values behaved according to an approximately parabolic law (Figure 9.6) typical of a process that is controlled by diffusion, possibly associated with the accumulation of

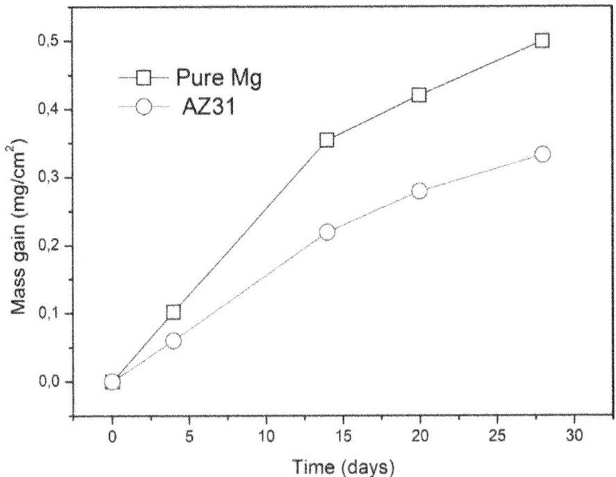

FIGURE 9.6 Mass gain values for samples of pure Mg and AZ31 alloys exposed for 28 days at 98% RH and 50°C.

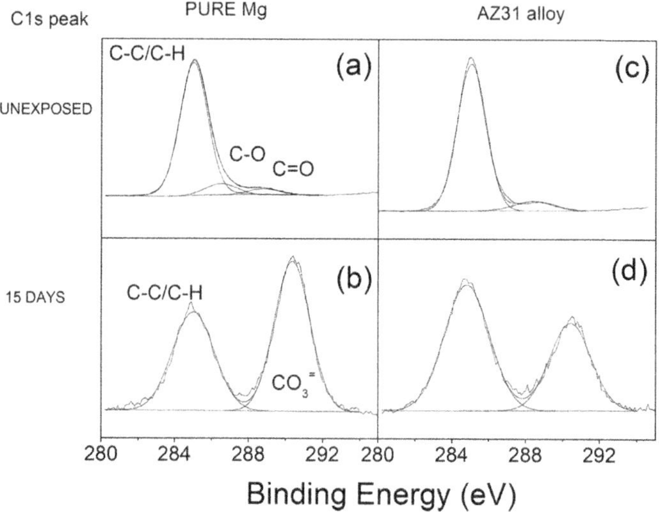

FIGURE 9.7 High-resolution Cls spectra obtained on the unexposed pure Mg (a) and AZ31 alloy (c) and their change after 15 days of exposure at 98% RH and 50°C (b,d).

protective corrosion products on the metal surface. After 15 days exposure to a high relative humidity, XPS analysis revealed a substantial enrichment of magnesium carbonate on the surface of both pure Mg and the alloy AZ31 (Figure 9.7), which is commonly associated with the diffusion of CO_2 from the environment that reacts with the air-formed film, and which seems to reduce the susceptibility to localized corrosion of magnesium exposed to humid air.

9.6 EFFECT OF AQUEOUS IMMERSION ON THE SURFACE CHARACTERISTICS AND CORROSION BEHAVIOUR OF AN AIR-FORMED OXIDE FILM ON MAGNESIUM ALLOYS

9.6.1 Chemical Composition and Structure

The composition of surface films formed under aqueous conditions is similar to that previously discussed for a high humidity atmosphere, and comprises a thin MgO inner layer and a thicker $Mg(OH)_2$ outer layer. The outer layer is thickened by hydration of the inner layer.[57] Because $Mg(OH)_2$ is a thermodynamically stable solid phase in a Mg corrosion product in an aqueous solution, the general belief is that a MgO-rich inner layer forms during the preparation of a Mg sample when it is in contact with the laboratory air.[4,9,33,45,68]

Nordlien et al.[9,33] used TEM and EDX on ultramicrotomed cross sections of Mg surface films formed by immersion in distilled water after 48 hours and observed that the oxide film had a three-layered structure with an overall plate-like morphology. The outer layer had a thick platelet-like morphology measuring 0.4–0.6 μm thick, consisting of a porous structure rich in magnesium hydroxide. Below the "platelets," a dense thin intermediate magnesium oxide (MgO)-rich layer of 20–40 nm was observed. A 0.4–0.6 μm hydrated inner cellular layer of Mg–$Mg(OH)_2$ is located below the MgO layer. The inner and the intermediate layers have a structure similar to the air-formed film[9,33] and act as a membrane permeable to the egress of soluble Mg cations. The outer platelet-like, porous $Mg(OH)_2$ layer forms by a dissolution-precipitation reaction.[9,33]

Taheri et al.[3,68] used TEM and focused ion beam (FIB) to investigate Mg specimens after 48 h of water exposure, and found that a duplex surface film was formed, consisting of a thinner, more-porous nanocrystalline MgO inner layer approximately 50–90 nm thick, and a thicker, less-porous $Mg(OH)_2$-rich outer platelet layer approximately 500 nm thick. The inner MgO region may be associated with the initial air-formed native film layer. The outer $Mg(OH)_2$ layer thickens as the dissolution proceeds by the hydration of the bulk inner MgO layer.[3,68]

XPS analysis of the composition and structure of the native surface film formed on Mg after a short exposure to H_2O indicated a duplex film consisting of an outer $Mg(OH)_2$-rich layer on top of an inner MgO-rich layer.[4,38,45] Yao et al.[38] observed that the film formed on melt-spun Mg in distilled water after 60 s was mainly a mixture of $Mg(OH)_2$ and MgO. Depth profiling revealed that $Mg(OH)_2$ was predominant in the top layer and decreased gradually with depth, while MgO had the opposite behavior. According to XPS data reported by Santamaria et al.,[45] the initial film formed on Mg after dry-polishing and immersion in ultrapure water for 0–300 s showed a bi-layer structure, consisting of an ultra-thin MgO layer at the metal interface (~2.5 nm) and a $Mg(OH)_2$ external layer. The thickness of the $Mg(OH)_2$ layer increased with increasing immersion time and became the predominant component at exposure times greater than 30 s, but the composition and thickness of the inner (MgO) layer was unchanged. Similar results were produced in a study by Liu et al.[4] and Song et al.[69] reported a three layer structure for the surface film formed on the magnesium-aluminum-zinc alloy AZ21 (nominally 2 wt% Al—1 wt% Zn) and the AZ91 alloy immersed in an alkaline NaCl solution (pH 11) with a porous outer layer of $Mg(OH)_2$, a dense, middle layer that consisted mainly of MgO and an inner, Al_2O_3 rich barrier layer.

Baril et al.[70] used electrochemical impedance spectroscopy (EIS) to characterize the electrical properties of a film formed on the surface of pure magnesium in a sodium sulfate solution. The advantage of EIS to estimate surface film thickness is that the measurement is performed *in situ* and the specimen is in contact with the electrolyte. In contrast, an XPS measurement is made *ex situ*, in an ultrahigh vacuum, and the possibility that the film composition will dry and change exists. They proposed the formation of a double layer corrosion film comprising a thin inner layer of MgO in contact with the metal substrate, and a relatively thick outer porous layer of $Mg(OH)_2$.[70] They calculated that the MgO inner layer thickness was approximately 1 nm[70] which was consistent with the *ex situ* XPS values.[4,45,61] Some correspondence was also observed in a previous study[27] between the higher initial capacitance obtained by EIS for the alloy AZ31 compared with those for the alloy AZ61 and the presence of thinner air-formed films on the alloy AZ31 indicated by XPS (Figure 9.4a and b).

Taheri et al.[3,68,71] proposed that the complete formation of a $Mg(OH)_2$-rich outer layer in pure H_2O involves the stress-rupture (hydration) of the dense air-formed layer as a precursor to corrosion. The breakdown of the air-formed MgO film to a reasonable extent is due to the energetically favorable hydration of MgO to $Mg(OH)_2$,[3] because the cubic MgO is converted to hexagonal $Mg(OH)_2$ with a volume twice that of the oxide. The resultant disruption of the film allows the H_2O electrolyte to directly contact the fresh metal surface and enhances the egress of Mg cations that are produced from the corrosion of the Mg alloy, to the interface between the surface film and H_2O.[38] After the Mg^{2+} ions exceed their limited solubility, $Mg(OH)_2$ could form by precipitation on the outer surface layer. Therefore, the more-hydrated outer layer continues to grow largely at the expense of the Mg substrate.

9.6.2 Corrosion Resistance of an Air-Formed Film in an Aqueous Solution

Atrens et al.[61] concluded that the surface film formed during specimen preparation plays a key role in the corrosion of Mg in an aqueous solution during the early stages of the immersion test until steady state corrosion occurs, and it is not uncommon that parts of the surface remain un-attacked after prolonged immersion in NaCl. The initial surface film formed during specimen preparation provided temporary protection, and the partial corrosion reactions were primarily limited to imperfections in the protective film or at locations where no surface film was present.[61] The corrosion rate increases until the film formed during preparation has broken down, then the corrosion expands throughout the entire surface and a steady state is achieved.[53,61,72] Additionally, the existence of a relatively resistant oxide surface film is supported by the filiform-like corrosion morphology that has been regularly observed across the Mg alloy surface.[61,73]

Nordlien et al.[22] have suggested a significant improvement in the corrosion resistance of Mg-Al alloys in an aqueous environment as the Al concentration increases to approximately 4% which can be explained by the formation of a continuous passivation network with a skeletal structure of an Al_2O_3 component in an amorphous matrix that acts as an effective barrier impeding the diffusion of corrosive species to the substrate and reduces the degree of hydration.[4] As previously mentioned, the Al concentration in the oxide film formed in an Al-containing Mg alloy exposed to oxygen reaches a terminal value of approximately 35% when the Al content of the alloy exceeds 4%.[22] These researchers also suggested that the film properties become predominantly determined by the presence of Al in the oxide with a corrosion resistance that is regarded as superior to that of the $Mg(OH)_2$ or the MgO layer.[22,45]

Laboratory tests have demonstrated that the corrosion rate of both the AM50 (nominally 4.5–5.3 wt% Al—0.28–0.5 wt% Mn) and AZ91D magnesium alloys was strongly affected by the surface state before exposure, and the average corrosion rate of an as-received surface resulting from the casting process was approximately twice that of a polished surface, mainly because the as-received specimens had an active, larger surface area.[74] We also observed in previous studies[27,51,73] that the chemistry and structure of an air-formed oxide surface film on the commercial AZ31 and AZ61

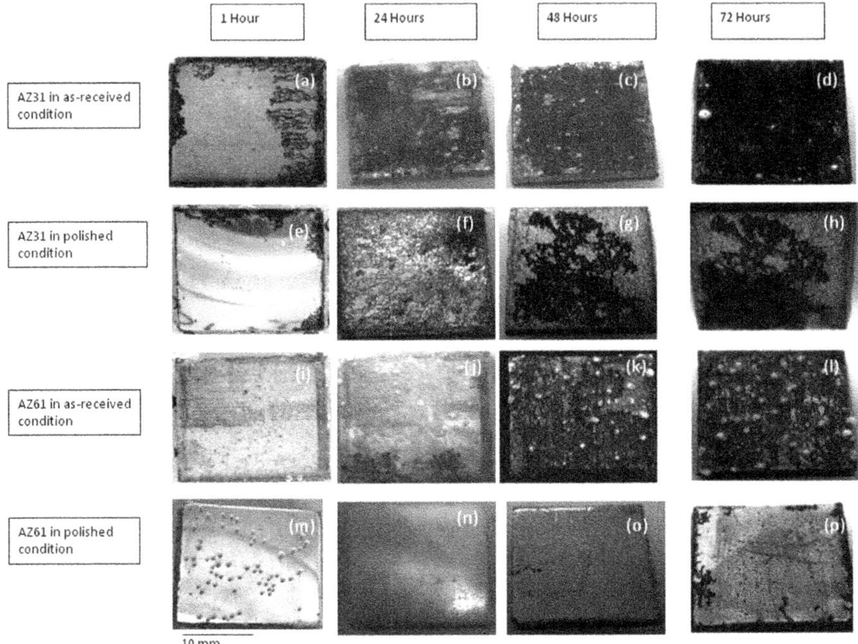

FIGURE 9.8 Photographic images of the evolution of corrosion morphology with immersion time in a 0.6 M NaCl solution for alloys AZ31 and AZ61 in the as-received and polished surface conditions.

alloys with different surface finishes affect the corrosion results during the early stage of exposure (i.e., before the formation of visible black spots on the surface), in immersion tests in 0.6 M NaCl solution. Figure 9.8 shows examples of the initiation and evolution of the corrosion morphology of the AZ31 and AZ61 alloys as a function of surface conditions and immersion time. Corrosion progressively darkened the surface of the AZ61 and AZ31 specimens in as-received condition and the polished AZ31 specimen from the typical shiny reflective metallic appearance of un-attacked areas, which was attributed to the formation and growth of filiform corrosion (Figure 9.8a–l), whereas the surface of polished AZ61 specimens remained almost unchanged even after 72 h immersion (Figure 9.8m–p). In these studies,[27,51,73] some correspondence was observed between the better corrosion resistance in a NaCl aqueous environment and the thicker MgO layer measured by XPS that was spontaneously formed on the AZ61 polished surface in contact with the laboratory atmosphere (Figure 9.4).

Figure 9.9 compares backscattered electron (BSE) images of the surface of the AZ31 and AZ61 alloys after 2 and 4 days immersion in a 0.6 M NaCl solution. After 2 days, the surface of the AZ31 specimen was completely corroded, and the corrosion layer was thick and loose with many pores (Figure 9.9a). After 4 days of immersion, the surface of the AZ31 specimen (Figure 9.9b) showed a dense corrosion layer

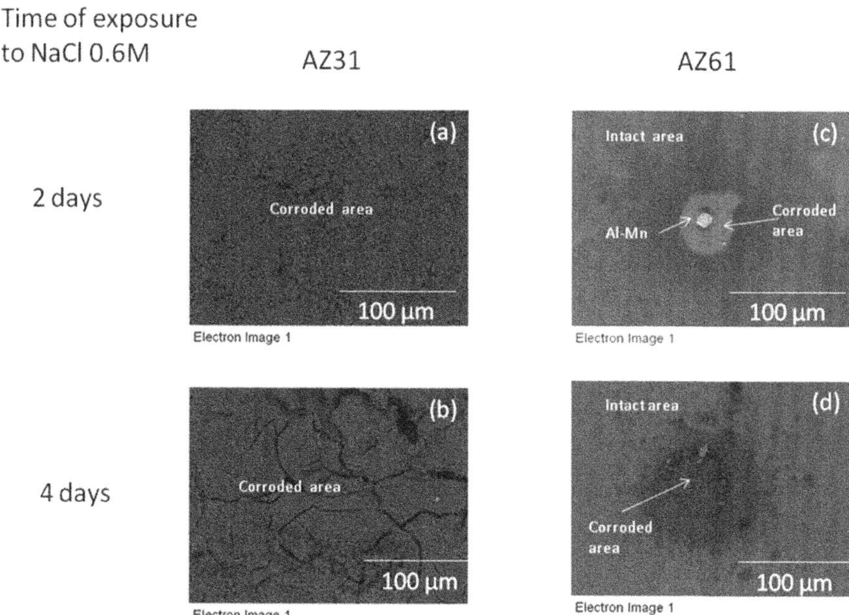

FIGURE 9.9 BSE images showing the surface of the alloys after immersion in a 0.6 M NaCl solution for 2 and 4 days. AZ31 (a,b) and AZ61 alloys (c,d).

with numerous wide cracks and discontinuities.[75] In contrast, a significant fraction of the surface area of AZ61 specimens after immersion for 2 (Figure 9.9c) and 4 days (Figure 9.9d) was intact, and the corrosion was localized around Al-Mn inclusions which were not covered by a protective air-formed film. We speculated that these differences in corrosion resistance may be associated with a thicker air-formed oxide film on the polished AZ61 alloy (Figure 9.4b) than on the AZ31 alloy (Figure 9.4a) prior to the immersion test.

Electrochemical impedance results have provided additional quantitative information on the differences in the protective properties of air-formed films on the surface of magnesium alloys. The impedance values in Figure 9.10 show a clear trend of values three or four times higher for the AZ61 alloy during the first 2 days of immersion than for the alloy AZ31 for the same immersion time. Additionally, it is interesting to note that these values do not noticeably decrease during the initial period (Figure 9.10b). This higher resistance is in agreement with our previous results[27,53,71] which attributed it to a particularly protective effect of an air-formed oxide film on the polished surface of the AZ61 alloy. It is noteworthy that a marked tendency exists for impedance values of the AZ61 alloy to decrease after 3 days of immersion (Figure 9.10b), which suggests that the protective action of the film covering the metallic surface becomes weaker, probably due to the progressive deterioration of the air-formed film in contact with the corrosive 0.6 M NaCl solution.

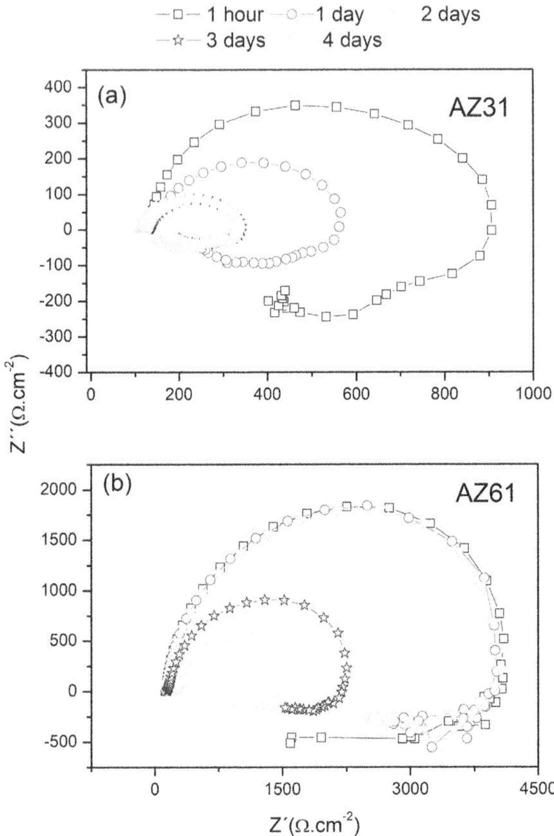

FIGURE 9.10 Variation in Nyquist plots for AZ31 (a) and AZ61 (b) samples with immersion time in 0.6 M NaCl solution.

9.7 CONCLUSIONS

1. The air-formed surface film on magnesium alloys may be sufficiently protective in mild or weakly corrosive environments. In 0.6 M NaCl, the surface films are protective only during the early stages of immersion.
2. The surface film in dry air mainly comprises MgO, and its thickness increases with exposure time to a limiting value of approximately 2 to 4 nm. This oxide film is highly protective against oxidation below 400°C. The oxidation of magnesium alloys (e.g., AZ31) is strongly affected by the surface state.
3. The dissolution reaction of a MgO layer in contact with water vapor in a laboratory atmosphere resulted in the formation of an ultrathin $Mg(OH)_2$ layer on top of a thicker MgO inner layer. The air-formed film on magnesium alloys provides good resistance at low humidity Additionally, atmospheric CO_2 reacts with $Mg(OH)_2$ to form magnesium carbonate ($MgCO_3$).

The carbonate enrichment in the air-formed oxide film upon exposure to laboratory atmosphere observed by XPS reduces the corrosion of magnesium alloys in a humid environment.

4. In humid air, the electrochemical reaction of Mg corrosion caused a significant increase of the film thickness and the principal product was $Mg(OH)_2$ which tended to be non-protective. The corrosion resistance of Mg alloys is highly improved in the presence of atmospheric CO_2 and high humidity, due to the formation of a thick, protective carbonate deposit in the outer layer.
5. The protective properties of the MgO layer formed in contact with air plays a key role in the corrosion soon after immersion in an aqueous solution. During the initial immersion period (up to 3 days), a strong link is observed between corrosion resistance and the thickness and uniformity of an air-formed film on magnesium alloys.

ACKNOWLEDGMENTS

The author gratefully acknowledges the financial support for this project from the Spanish Ministry of Economy and Competitiveness (MAT2015-65445-C2-1-R).

REFERENCES

1. Xu, R.Z., Yang, X.B., Li, P.H., Suen, K.W., Wu, S., Chu, P.K. Electrochemical properties and corrosion resistance of carbon-ion-implanted magnesium. *Corros. Sci.* **2014**, *82*, 173–179.
2. Mori, Y., Koshi, A., Liao, J., Asoh, H., Ono, S. Characteristics and corrosion resistance of plasma electrolytic oxidation coatings on AZ31B Mg alloy formed in phosphate—Silicate mixture electrolytes. *Corros. Sci.* **2014**, *88*, 254–262.
3. Taheri, M., Phillips, R.C., Kish, J.R., Botton, G.A. Analysis of the surface film formed on Mg by exposure to water using a FIB cross-section and STEM–EDS. *Corros. Sci.* **2012**, *59*, 222–228.
4. Liu, M., Zanna, S., Ardelean, H., Frateur, I., Schmutz, P., Song, G.L., Atrens, A., Marcus, P. A first quantitative XPS study of the surface films formed, by exposure to water, on Mg and on the Mg–Al intermetallics: Al_3Mg_2 and $Mg_{17}Al_{12}$. *Corros. Sci.* **2009**, *51*(5), 1115–1127.
5. Lu, X.Y., Zuo, Y., Zhao, X.H., Tang, Y.M. The improved performance of a Mg-rich epoxy coating on AZ91D magnesium alloy by silane pretreatment. *Corros. Sci.* **2012**, *60*, 165–172.
6. Lindstrom, R., Johansson, L.G., Thompson, G.E., Skeldon, P., Svensson, J.E. Corrosion of magnesium in humid air. *Corros. Sci.* **2004**, *46*(5), 1141–1158.
7. Cao, F.Y., Shi, Z.M., Song, G.L., Liu, M., Dargusch, M.S., Atrens, A. Influence of casting porosity on the corrosion behaviour of Mg0.1Si. *Corros. Sci.* **2015**, *94*, 255–269.
8. Chen, C., Splinter, S.J., Do, T., McIntyre, N.S. Measurement of oxide film growth on Mg and Al surfaces over extended periods using XPS. *Surf. Sci.* **1997**, *382*(1–3), L652–L657.
9. Nordlien, J.H., Ono, S., Masuko, N., Nisancioglu, K. Morphology and structure of oxide-films formed on magnesium by exposure to air water. *J. Electrochem. Soc.* **1995**, *142*(10), 3320–3322.
10. Liao, J.S., Hotta, M. Corrosion products of field-exposed Mg-Al series magnesium alloys. *Corros. Sci.* **2016**, *112*, 276–288.

11. Liu, Y.F., Yang, W., Qin, Q.L., Wu, Y.C., Wen, W., Zhai, T., Yu, B., Li, D.Y., Luo, A., Song, G.L. Microstructure and corrosion behavior of die-cast AM60B magnesium alloys in a complex salt solution: A slow positron beam study. *Corros. Sci.* **2014**, *81*, 65–74.
12. Kurth, M., Graat, P.C.J., Carstanjen, H.D., Mittemeijer, E.J. The initial oxidation of magnesium: An in situ study with XPS, HERDA and ellipsometry. *Surf. Interface Anal.* **2006**, *38*(5), 931–940.
13. Feliu, S. Jr., Pardo, A., Merino, M.C., Coy, A.E., Viejo, F., Arrabal, R. Correlation between the surface chemistry and the atmospheric corrosion of AZ31, AZ80 and AZ91D magnesium alloys. *Appl. Surf. Sci.* **2009**, *255*(7), 4102–4108.
14. Saleh, H., Weling, T., Seidel, J., Schmidtchen, M., Kawalla, R., Mertens, F.O.R.L., Vogt, H.P. An XPS study of native oxide and isothermal oxidation kinetics at 300°C of AZ31 twin roll cast magnesium alloy. *Oxid. Met.* **2013**, *81*(5–6), 529–548.
15. Asami, K., Ono, S. Quantitative X-ray photoelectron spectroscopy characterization of magnesium oxidized in air. *J. Electrochem. Soc.* **2000**, *147*(4), 1408–1413.
16. Tan, Q.Y., Atrens, A., Mo, N., Zhang, M.X. Oxidation of magnesium alloys at elevated temperatures in air: A review. *Corros. Sci.* **2016**, *112*, 734–759.
17. Czerwinski, F. Oxidation characteristics of magnesium alloys. *JOM*, **2012**, *64*(12), 1477–1483.
18. Eliezer, D., Alves, H. Corrosion and oxidation of magnesium alloys. In *Handbook of Materials Selection*; Kutz, M., Ed; John Wiley & Sons: New York, 2002; 267–291.
19. Gusieva, K., Davies, C.H.J., Scully, J.R., Birbilis, N. Corrosion of magnesium alloys: The role of alloying. *Int. Mater. Rev.* **2015**, *60*(3), 169–194.
20. Esmaily, M., Svensson, J.E., Fajardo, S., Birbilis, N., Frankel, G.S.,Virtanen, S., Arrabal, R., Thomas, S., Johansson, L.G. Fundamentals and advances in magnesium alloy corrosion. *Prog. Mater Sci.* **2017**, *89*, 92–193.
21. Jeurgens, L.P.H., Vinodh, M.S., Mittemeijer, E.J. Initial oxide-film growth on Mg-based MgAl alloys at room temperature. *Acta Mater.* **2008**, *56*(17), 4621–4634.
22. Nordlien, J.H., Nisancioglu, K., Ono, S., Masuko, N. Morphology and structure of oxide films formed on MgAl alloys by exposure to air and water. *J. Electrochem. Soc.* **1996**, *143*(8), 2564–2572.
23. Ghali, E. *Corrosion Resistance of Aluminum and Magnesium Alloys: Understanding, Performance, and Testing*; John Wiley & Sons: Hoboken, NJ, 2010.
24. Czerwinski, F. The oxidation behaviour of an AZ91D magnesium alloy at high temperatures. *Acta Mater.* **2002**, *50*, 2639–2654.
25. Liu, J.W., Li, Y., Wang, F.H. The high temperature oxidation behavior of Mg–Gd–Y–Zr alloy. *Oxid. Met.* **2009**, *71*(10), 319–334.
26. Cabrera, N., Mott, N.F. Theory of the oxidation of metals, *Rep. Prog. Phys.* **1948**, *12*, 163–184.
27. Feliu, S. Jr., Maffiotte, C., Samaniego, A., Galvan, J.C., Barranco, V. Effect of naturally formed oxide films and other variables in the early stages of Mg-alloy corrosion in NaCl solution. *Electrochim. Acta.* **2011**, *56*(12), 4554–4565.
28. Feliu, S. Jr., Bartolomé, M.J., González, J.A., López, V., Feliu, S. Passivating oxide film and growing characteristics of anodic coatings on aluminium alloys. *Appl. Surf. Sci.* **2008**, *254*(9), 2755–2762.
29. Pilling, N.B., Bedworth, R.E. The oxidation of metals at high temperatures. *J. Inst. Met.* **1923**, *29*, 529–582.
30. Feliu, S. Jr., Samaniego, A., Barranco, V., El-Hadad, A.A., Llorente, I., Serra, C., Galvan, J.C. A study on the relationships between corrosion properties and chemistry of thermally oxidised surface films formed on polished commercial magnesium alloys AZ31 and AZ61. *Appl. Surf. Sci.* **2014**, *295*, 219–230.

31. Unocic, K.A., Elsentriecy, H.H., Brady, M.P., Meyer, H.M., Song, G.L., Fayek, M., Meisner, R.A., Davis, B. Transmission electron microscopy study of aqueous film formation and evolution on magnesium alloys. *J. Electrochem. Soc.* **2014**, *161*(6), C302–C311.
32. Reichel, F, Jeurgens, L.P.H., Mittemeijer, E.J. The thermodynamic stability of amorphous oxide overgrowths on metals. *Acta Mater.* **2008**, *56*(19), 659–674.
33. Nordlien, J.H., Ono, S., Masuko, N., Nisancioglu, K. A TEM investigation of naturally formed oxide films on pure magnesium. *Corros. Sci.* **1997**, *39*(8), 1397–1414.
34. Kurth, M., Graat, P.C.J., Mittemeijer, E.J. The oxidation kinetics of magnesium at low temperatures and low oxygen partial pressures. *Thin Solid Films* **2006**, *500*(1–2), 61–69.
35. Fournier, V., Marcus, P., Olefjord, I. Oxidation of magnesium. *Surf. Interface Anal.* **2002**, *34*(1), 494–497.
36. Esmaily, M., Shahabi-Navid, M., Svensson, J.E., Halvarsson, M., Nyborg, L., Cao, Y., Johansson, L.G. Influence of temperature on the atmospheric corrosion of the Mg–Al alloy AM50. *Corros. Sci.* **2015**, *90*, 420–433.
37. Taheri, M. Influence of the surface film stability on the corrosion resistance of Mg in aqueous solutions. Ph. D. Thesis; McMaster University: Ontario, Canada, 2013.
38. Yao, H.B., Li, Y., Wee, A.T.S. An XPS investigation of the oxidation/corrosion of melt-spun Mg. *Appl. Surf. Sci.* **2000**, *158*(1–2), 112–119.
39. Splinter, S.J., McIntyre, N.S., Lennard, W.N., Griffiths, K., Palumbo, G. An AES and XPS study of the initial oxidation of polycrystalline magnesium with water vapour at room temperature. *Surf. Sci.* **1993**, *292*(1–2), 130–144.
40. Splinter, S.J., McIntyre, N.S. The initial interaction of water vapour with Mg-Al alloy surfaces at room temperature. *Surf. Sci.* **1994**, *314*(2), 157–171.
41. Fuggle, J.C., Watson, L.M., Fabian, D.J., Affrossman, S. X-ray photoelectron studies of the reaction of clean metals (Mg, Al, Cr, Mn) with oxygen and water vapour. *Surf. Sci.* **1975**, *49*(1), 61–76.
42. Song, Y.W., Han, E.H., Dong, K.H., Shan, D.Y., Yim, C.D., You, B.S. Microstructure and protection characteristics of the naturally formed oxide films on Mg-xZn alloys. *Corros. Sci.* **2013**, *72*, 133–143.
43. Cui, Z.Y., Li, X.G., Xiao, K., Dong, C.F. Atmospheric corrosion of field-exposed AZ31 magnesium in a tropical marine environment. *Corros. Sci.* **2013**, *76*, 243–256.
44. McIntyre, N.S., Chen, C. Role of impurities on Mg surfaces under ambient exposure conditions. *Corros. Sci.* **1998**, *40*(10), 1697–1709.
45. Santamaria, M., Di Quarto, F., Zanna, S., Marcus, P. Initial surface film on magnesium metal: A characterization by X-ray photoelectron spectroscopy (XPS) and photocurrent spectroscopy (PCS). *Electrochim. Acta* **2007**, *53*(3), 1314–1324.
46. Fotea, C., Callaway, J., Alexander, M.R. Characterisation of the surface chemistry of magnesium exposed to the ambient atmosphere. *Surf. Interface Anal.* **2006**, *38*(10), 1363–1371.
47. Song, Y.W., Shan, D.Y., Chen, R.S., Han, E.H. Investigation of surface oxide film on magnesium lithium alloy. *J. Alloys Compd.* **2009**, *484*(1–2), 585–590.
48. Makar, G.L., Kruger, J. Corrosion studies of rapidly solidified magnesium alloys. *J. Electrochem. Soc.* **1990**, *137*(2), 414–421.
49. Nouri, M., Liu, Z.R., Li, D.Y., Yan, X.G., Tahreen, N., Chen, D.L. The role of minor yttrium in tailoring the failure resistance of surface oxide film formed on Mg alloys. *Thin Solid Films* **2016**, *615*, 29–37.
50. Feliu, S. Jr., Galván, J.C., Pardo, A, Merino, M.C., Arrabal, R. Native air-formed oxide film and its effect on magnesium alloys corrosion. *Open Corros. J.* **2010**, *3*, 80–91.

51. Feliu, S. Jr., Maffiotte, C., Samaniego, A., Galvan, J.C., Barranco, V. Effect of the chemistry and structure of the native oxide surface film on the corrosion properties of commercial AZ31 and AZ61 alloys. *Appl. Surf. Sci.* **2011**, *257*(20), 8558–8568.
52. Abreu, J.B., Soto, J.E., Ashley-Facey, A., Soriaga, M.P., Garst, J.F., Stickney, J.L. The interfacial chemistry of the Grignard reaction: The composition of the film formed on air-exposed magnesium. *J. Colloid Interface Sci.* **1998**, *206*(1), 247–251.
53. Song, G.L., Atrens, A. Corrosion Mechanisms of Magnesium Alloy. *Adv. Eng. Mater.* **1999**, *1*(1),11–33.
54. Alves, H., Koster, U., Aghion, E., Eliezer, D. Environmental behavior of magnesium and magnesium alloys. *Mater. Technol.* **2001**, *16*(2), 110–126.
55. Shahabi-Navid, M., Esmaily, M., Svensson, J.E., Halvarsson, M., Nyborg, L., Cao, Y., Johansson, L.G. NaCl-induced atmospheric corrosion of the MgAl alloy AM50-the influence of CO_2. *J. Electrochem. Soc.* **2014**, *161*(6), C277–C287.
56. Yang, W., Zhu, Z.J., Wang, J.J., Wu, Y.C., Zhai, T., Song, G.L. Slow positron beam study of corrosion behavior of AM60B magnesium alloy in NaCl solution. *Corros. Sci.* **2016**, *106*, 271–280.
57. Hallam, K.R., Minshall, P.C., Heard, P.J., Flewitt, P.E.J. Corrosion of the alloys Magnox AL80, Magnox ZR55 and pure magnesium in air containing water vapour. *Corros Sci.* **2016**, *112*, 347–363.
58. Arrabal, R., Mingo, B., Pardo, A., Matykina, E., Mohedano, M., Merino, M.C., Rivas, A., Maroto, A. Role of alloyed Nd in the microstructure and atmospheric corrosion of as-cast magnesium alloy AZ91. *Corros. Sci.* **2015**, *97*, 38–48.
59. Feliu, S. Jr., Llorente, I. Corrosion product layers on magnesium alloys AZ31 and AZ61: Surface chemistry and protective ability. *Appl. Surf. Sci.* **2015**, *347*, 736–746.
60. Jonsson, M., Persson, D., Thierry, D. Corrosion product formation during NaCl induced atmospheric corrosion of magnesium alloy AZ91D. *Corros. Sci.* **2007**, *49*(3), 1540–1558.
61. Atrens, A, Song, G.L., Liu, M., Shi, Z.M., Cao, F.Y., Dargusch, M.S. Review of recent developments in the field of magnesium corrosion. *Adv. Energy Mater.* **2015**, *17*(4), 400–453.
62. Liu, W.J., Cao, F.H., Xia, Y., Chang, L.R., Zhang, J.Q. Localized corrosion of magnesium alloys in NaCl solutions explored by scanning electrochemical microscopy in feedback model. *Electrochim. Acta* **2014**, *132*, 377–388.
63. Song, G., Atrens, A., St John, D., Wu, X., Nairn, J. The anodic dissolution of magnesium in chloride and sulphate solutions. *Corros. Sci.* **1997**, *39*(10–11), 1981–2004.
64. Song, Y.W., Han, E.H., Dong, K.H., Shan, D.Y., Yim, C.D., You, B.S. Study of the corrosion product films formed on the surface of Mg–xZn alloys in NaCl solution. *Corros. Sci.* **2014**, *88*, 215–225.
65. Lindstrom, R., Svensson, J.E., Johansson, L.G. The influence of carbon dioxide on the atmospheric corrosion of some magnesium alloys in the presence of NaCl. *J. Electrochem. Soc.* **2002**, *149*(4), B103–B107.
66. Lindstrom, R., Johansson, L.G., Svensson, J.E. The influence of NaCl and CO_2 on the atmospheric corrosion of magnesium alloy AZ91. *Mater. Corr.* **2003**, *54*(8), 587–594.
67. Feliu, S. Jr., Merino, M.C., Arrabal, R., Coy, A.E., Matykina, E. XPS study of the effect of aluminium on the atmospheric corrosion of the AZ31 magnesium alloy. *Surf. Interface Anal.* **2009**, *41*(3), 143–150.
68. Taheri, M., Danaie, M., Kish, J.R. TEM examination of the film formed on corroding Mg prior to breakdown. *J. Electrochem. Soc.* **2014**, *161*(3), C89–C94.
69. Song, G.L., Atrens, A., Wu, X.L., Zhang, B. Corrosion behaviour of AZ21, AZ501 and AZ91 in sodium chloride. *Corros. Sci.* **1998**, *40*(10), 1769–1791.

70. Baril, G., Galicia, G., Deslouis, C., Pebere, N., Tribollet, B., Vivier, V. An impedance investigation of the mechanism of pure magnesium corrosion in sodium sulfate solutions. *J. Electrochem. Soc.* **2007**, *154*(2), C108–C113.
71. Taheri, M., Kish, J.R. Nature of Surface Film Formed on Mg Exposed to 1 M NaOH. *J. Electrochem. Soc.* **2013**, *160*(1), C36–C41.
72. Song, G.L., Atrens, A. Understanding magnesium corrosion—A framework for improved alloy performance. *Adv. Energy Mater.* **2003**, *5*(12), 837–858.
73. Samaniego, A., Llorente, I., Feliu, S. Jr. Combined effect of composition and surface condition on corrosion behaviour of magnesium alloys AZ31 and AZ61. *Corros. Sci.* **2013**, *68*, 66–71.
74. LeBozec, N., Jonsson, M., Thierry, D. Atmospheric corrosion of magnesium alloys: Influence of temperature, relative humidity, and chloride deposition. *Corrosion* **2004**, *60*(4), 356–361.
75. Delgado, M.C., Garcia-Galvan, F.R., Llorente, I., Perez, P., Adeva, P., Feliu, S. Jr. Influence of aluminium enrichment in the near-surface region of commercial twin-roll cast AZ31 alloys on their corrosion behaviour, *Corros. Sci.* **2017**, *123*, 182–196.

10 Importance of Cleanliness for Magnesium Alloys

Shae K. Kim

CONTENTS

10.1 INTRODUCTION

The importance of Mg alloys can be described by the never ending request of weight reduction not only for environment as global effort but also for performance as consumer expectation. From the point of environment protection, the transportation industry should weigh down vehicles to decrease fuel consumption and emission. Consumer expects high fuel efficiency at current high oil price and always high performance, for which the weight reduction plays important and main role. Beyond the lightest structural metallic material, Mg alloys are also important in terms of designer's choice due to the unique properties not found in other materials.

To make the best use of Mg alloys as industry-friendly material, the inherent SHE (Safety, Health, Environment) obstacles that have been taken for granted have to get solved. Mg alloys burn so easily during process and may burn in service due to the poor oxidation and ignition resistances (Safety). To prevent the oxidation and ignition during process, toxic SO_2 gas (Health) or super global warming SF_6 gas (Environment) can and have been generally used with auxiliary toxic beryllium alloying (Health). It is obvious that SHE obstacles are mainly attributed to the non-protective porous surface oxide nature of Mg alloys. The most serious and important problem that have been ignored is that this instability causes the surface oxides to be incorporated into molten Mg alloys. This incorporation again makes the melts and then the products unclean and unsuitable, which causes in turn the uncontrolled and deteriorated process-ability and property. This is the maxim of this article to emphasize the importance of melt and product cleanliness and related SHE obstacles in connection with the non-protective porous surface oxide nature of Mg alloys.

The environmental benefit and better performance provided by Mg alloys with lightweight, unique properties, recyclability and limitless resource can be enhanced if Mg industry is working together to make the solution for SHE obstacles. Moreover, getting back the cleanliness can get back the ideal process-abilities and properties of Mg alloys by minimizing the gap between real and ideal values. This is the well-known advantage of defect control. It can be to generate the protective stable surface oxide nature in Mg alloys. The challenge and significant task is that the solution should at least maintain the real process-ability of fluidity and mechanical property of ductility of Mg alloys. This is because a solution to generate the stable surface and then solve SHE obstacles should be a solution at the same time for the cleanliness. It must not damage the cleanliness and moreover decrease the real values of process-ability and property. The real values of Mg alloys are not the ideal but the already deteriorated values due to the already existing incorporated oxides in common Mg alloys. More than that, the solution should not be different for each Mg alloy but have the expandability to be applied in the same way to all types of Mg alloys.

As one possible solution to satisfy all the issues above, ECO-Mg alloys are addressed, which can be simply made by adding CaO powders into current all types of Mg alloys and even into new Mg alloys. The stable oxide nature on all ECO-Mg alloys can ensure no burning during process and application (Safety), non-SO_2 processing and Be-free alloying (Health), and non-SF_6 processing (Environment), which in turn ensure the melt cleanliness, and therefore, can get back, at least, the ideal process-abilities and properties. The thermodynamic consideration of CaO and the difference between Ca alloying and CaO addition in terms of process and property will be addressed in detail.

The material choice in the transportation industry depends on the balance of weight, quality and cost (QCD). If the cost plays the main role, Mg alloys are hard to keep and/or win back their territory. This situation should be understood differently for casting and wrought Mg alloys, because the QCD competitiveness differs, in comparison with Al alloys, as indicated in Table 10.1. Casting Mg alloys have better castability due to low density and high fluidity with the little higher material cost than casting Al alloys, while there are the big cost gap and formability difference between wrought Mg alloys of HCP structure and Al alloys of FCC structure. The strategic point of casting ECO-Mg alloys for industry is to improve further the castability for thinner and more complicated die casting parts with low defect rate and high productivity. Assuming no friction-related surface oxidation and ignition, unconventional relative higher temperature forming could be applied for wrought ECO-Mg alloys, which can make the use of high forming velocity, high productivity and large reduction ratio, reduced forming step. This is the tailored targets to be achieved by casting and wrought ECO-Mg alloys.

TABLE 10.1
Performance and Cost between Mg and Al Alloys for Casting and Wrought Processes

		Magnesium	Aluminum
Casting	• Castability	<0.5 t	>0.8 t
	• Property (Elongation)	6%–8%	1%–2%
	• Alloy	less than 10	more than 50
	• Market share (Alloy price)	over 96% [3.5 ~ 25$/kg]	30% [2 ~ 10$/kg]
Wrought	• Formability	HCP	FCC
	• Property (Elongation)	20%	30%
	• Alloy	less than 10	more than 50
	• Market share (Alloy price) (extrusion/rolling)	3% [15 ~ 40]/1% [20 ~ 50] (bar, tube/plate, sheet)	30% [3 ~ 10]/40% [3 ~ 15] (bar, tube/plate, sheet)

10.2 ECO-Mg ALLOYS

The way to produce ECO-Mg alloys is simple—adding CaO powders into molten Mg alloys.[1] The way has been improved to meet the efficient mass production and control the CaO content exactly. ECO-Mg master alloys with high weight percentage of CaO are prepared in advance, and then mixed and diluted to produce ECO-Mg alloys in the CaO range of 0.2% and 1.5%. The content of CaO and the base alloy compositions should be tailored according to the purpose. There are various ECO-Mg alloys, which are mass produced based on pure Mg-, AZ-, AM-, AS-, AE-, AEJ-, ZE-, ZW-, and WZ-basis, etc. The reduction of CaO in molten Mg and Mg alloys is obvious thermodynamically.[2] It is the spontaneous reaction by the reduction to form Mg_2Ca in pure Mg and Mg alloys without Al and to form Al_2Ca and/or $(Mg, Al)_2Ca$ in Al-bearing Mg alloys. The reaction based on the Ellingham diagram is $Mg_{(l)} + CaO_{(s)} = MgO_{(s)} + Ca_{(l)}$, $\Delta G^o_{(l)} = 37.84$ kJ at 973 K, which tells that CaO is more stable than MgO, assuming Ca and Mg in the equation as pure liquids. If we consider Ca and Mg as the liquid solution, the thermodynamics tells CaO should be reduced not only in Mg alloys but also in pure Mg. The understanding of the thermodynamic reduction of CaO may raise the additional question about the difference between Ca alloying and CaO addition. In terms of microstructure, the exceptional difference is Ca content in α-phase (solid solution phase). Although the strong segregation tendency, Ca alloyed pure Mg solidifies first as α-phase with Ca solid solution and then as divorce eutectic phase of Mg and Mg_2Ca, according to Ca content and process parameters. On the contrary, CaO added pure Mg solidifies first as α-phase with almost no Ca solid solution, regardless of CaO content and process parameters.[3] Two things should be noted clearly. The content of CaO is not the weight fraction of added CaO powders but the content of Ca in the form of Mg_2Ca or Al_2Ca and/or $(Mg, Al)_2Ca$, the thermodynamically stable phases, which are formed by the reduction from CaO during ECO-Mg manufacturing. The CaO on the top of and in the surface oxide layer is not the added CaO to make ECO-Mg alloys but the re-oxidized CaO from initially reduced Mg_2Ca or Al_2Ca.

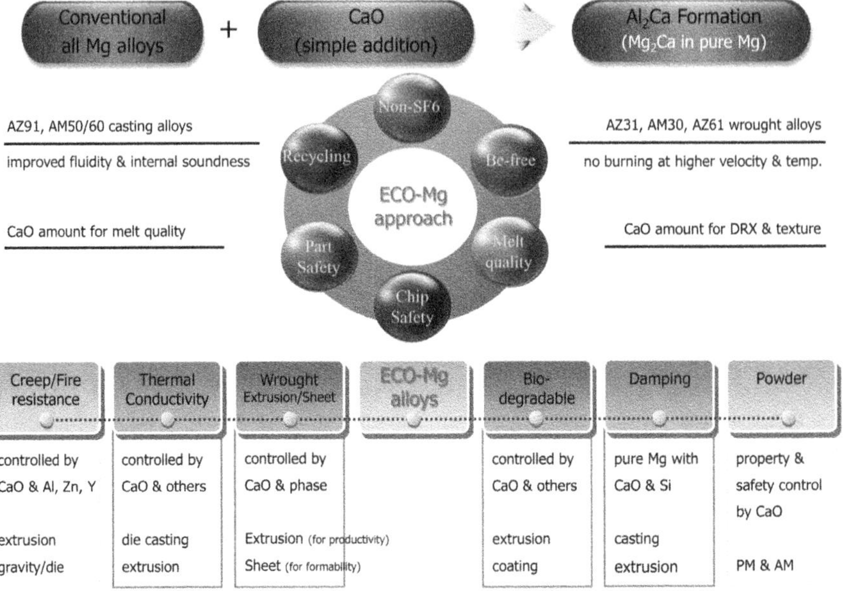

FIGURE 10.1 Introduction of and R&D scopes of ECO-Mg alloys.

As noted earlier and discussed later, ECO-Mg alloys can provide the stable protective surface oxide nature in all types of Mg alloys as well as maintain, at least, the real process-abilities and properties. The stable oxide nature in all ECO-Mg alloys can ensure no burning during process and application (Safety), non-SO_2 processing and Be-free alloying (Health), and non-SF_6 processing (Environment), which in turn ensure the melt cleanliness, and therefore, can get back, at least, the ideal process-abilities and properties. By adding CaO powders in the range of 0.2% and 0.7% into current casting and wrought Mg alloys, ECO-Mg alloys can ensure (1) non-SF_6 or non-SO_2 melting and casting, (2) Be-free alloying, (3) improved melt cleanliness, (4) improved process-abilities for casting, extrusion, rolling, forming, welding as well as surface treatment, (5) improved mechanical properties, (6) no burning during processing and application and even of machined chips, and (7) improved recyclability. CaO of over 0.7% can be introduced for special (1) creep-resistant, (2) ignition-resistant and fire-proof, (3) high-strength, (4) damping, (5) bio-degradable, and (6) desulphurizing ECO-Mg alloys, which is summarized in Figure 10.1.

10.3 IMPORTANCE AND CONTROL OF MELT CLEANLINESS IN Mg ALLOYS

Mg even as an alloying element for Al alloys greatly decreases the oxidation resistance.[4] The oxide formed on the as-cast surface in Al alloys is the usually filmlike Al_2O_3, while the oxide in Al-Mg alloys is particulate spinel, $MgAl_2O_4$ due to the break-away oxidation after the first-formed crystalline MgO. Mg alloys have poor oxidation and ignition resistances, because of the non-protective porous surface oxide nature. Czerwinski represents well the schematic surface oxide development of Mg alloys at various oxidation stages.[5] The poor oxidation resistance of Mg alloys, however, is not the serious problem by itself. The serious problem is that the burning follows after the oxidation and moreover extinguishing the Mg burning is difficult, in part because the flame temperature can reach 3370K (3100°C, 5610°F). The most serious and important problem, the main purpose of this article, is that this instability causes the surface oxides to be incorporated into molten Mg alloys and this incorporation again makes the melts and then the products unclean, defect-full, unsuitable, which in turn causes the uncontrolled and deteriorated process-ability and property.

Table 10.2 compares Mg and Al alloys in terms of surface oxide stability, surface tension, density ratio, and separation (refining) way, based on which we can understand why there are so much oxide inclusions and why it is difficult to prevent and control them in Mg alloys compared with Al alloys. The surface oxide of Al alloys can be represented as the protective, stable, filmlike surface oxide, while Mg alloys have the non-protective, porous, and particulate surface oxide. When the stable oxide film forms on Al alloys, it can keep its filmlike morphology and passivation. However, the surface oxide keeps growing in Mg alloys due to its non-protective and porous nature. This difference can be explained by Pilling-Bedworth ratio (PBR), the ratio of the volume of a metal oxide to the volume of the corresponding metal.[6] PBR of less than 1 in Mg alloys means that the oxide layer is likely broken and non-protective and grows continuously, while PBR of over 1 in Al alloys means that the oxide film is passivating and protective against further oxidation. According to the surface tension and the density ratio of a metal oxide to the corresponding metal, we can estimate the possibility and easiness of particulate surface oxides to be incorporated into the molten metal. Regardless of the different oxide morphologies: particulate of Mg alloys and filmlike of Al alloys, the lower surface tension and higher density ratio give us the understanding of why there are so much particulate oxides, oxide inclusions, present inside molten Mg alloys. It is worth noting that Mg is intentionally alloyed to make particulate-reinforced Al composites through molten metal processes, because Mg alloying can make the reinforcement to be incorporated and wetted well. The oxide separating (refining) direction in connection with the surface tension and density ratio explains how to remove and

TABLE 10.2
Comparisons of Mg and Al Alloys in Terms of Oxide Stability, Surface Tension, Density Difference between a Metal and Its Corresponding Oxide, and Separation (Refining) Way

	Magnesium	Aluminum
Oxide stability, Pilling-Bedworth Ratio	MgO/Mg = 0.81	Al_2O_3/Al = 1.28
Surface Tension, mN/m	577-0.26(T-923K)	1024-0.27(T-923K)
Density (Density Ratio)	MgO 3.58, Mg 1.8 (1.99)	Al_2O_3 3.97, Al 2.7 (1.47)
Oxide Separation (Refining) Direction	Settling down	Floating

the difficulty of removing incorporated particulate oxides. For Al alloys, argon gas bubbling is effective to float and remove the oxides. However, there is no practical way to settle down oxide inclusions inside molten Mg alloys except flux refining, which can't be applied for in-house melting and recycling. Unlike Al alloys, it is very important to remove repeatedly the dross on the top of molten Mg alloys, because the dross grows continuously and is easy to be incorporated during melting and holding. The preheating is also prerequisite to minimize the amount of surface oxides on Mg ingots. Based on the facts discussed above, it becomes clear that the degree of uncleanliness in Mg alloys is too severe to be compared with other metallic materials. We can turn this problem as an opportunity to improve the processability and property of Mg alloys simply by controlling oxide inclusions, that is, improving melt cleanliness.

Figure 10.2 compares the spiral fluidities of AZ21, AZ31 and AZ61 Mg alloys of different Chinese ingot makers, together with those of Ca alloyed AZ31 Mg alloys. Although the spiral fluidity test is not the quantitative method, the results demonstrate well that increasing Al increases the spiral fluidities of various Mg alloys and moreover its comparative reliability. Increasing Al content from 3% to 6% increases the fluidities from 28 cm to 34 cm (6 cm increase) in the ingot H and from 35 cm to 41 cm (6 cm increase) in the ingot R, respectively. Based on the results, what should be explained is the 7 cm difference between the ingots H and R, in AZ31 and AZ61 Mg alloys. It should also be answered that AZ61 in the ingot H has the lower fluidity length than that of AZ31 in the ingot R. To fully understand the fluidity, two different sides should be considered. One is the theoretical side, which consists of alloy factors, mold factors, and test variables. Viscosity, freezing range, surface tension and latent heat belong to the alloy factors.[7] In this experiment, there is no

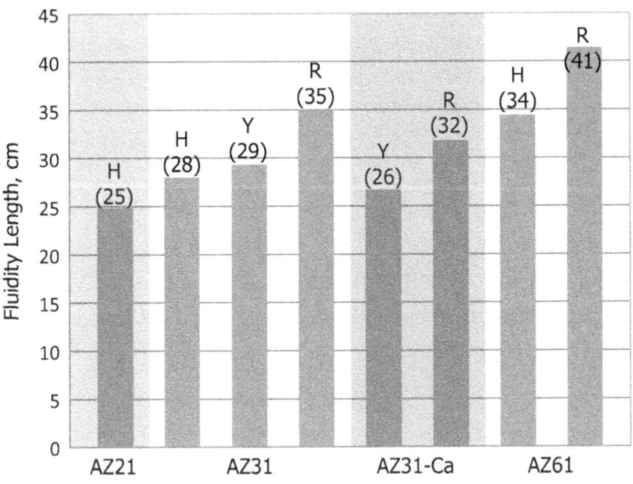

FIGURE 10.2 Spiral fluidity results of AZ21, AZ31, AZ61 and AZ31 + Ca Mg alloys manufactured by different Chinese ingot makers.

difference in composition: the same amount of Al and the almost same Fe, Cu and Ni contents. The same spiral fluidity mold and the same melting and casting variables were utilized under the same protective atmosphere of SF_6 gas. Therefore, we can ignore the first side to understand the results about the spiral fluidity differences. The other is the practical side, which represents the amount of inclusions and surface oxide characteristics. The copied expressions in the ref[7] are "viscosity is not strongly influential," "the true effect of surface tension is overpowered by the influence of surface oxide film characteristics," and "inclusions dramatically reduce the fluidity." It becomes possible to understand the fluidity difference based on the amount of oxide inclusions in molten Mg alloys, because, except that, there is no difference at all. The results of the fluidity difference can be used as a guideline to select and buy appropriate Mg alloy ingots and can be the clear evidence about the influence and the adverse effect of oxide inclusions on the fluidity in Mg alloys.

Figure 10.3 shows the comparative results of the spiral fluidity about Ca alloyed and CaO added AZ31 Mg alloys with respect to Ca and CaO contents. Once again, the content of CaO is the content of Ca in the form of Al_2Ca, which is formed by the reduction from CaO. It is general that increasing Ca decreases the castability of Mg alloys,[8] which is matched with and confirmed by the spiral fluidity results about Ca alloyed AZ31 Mg alloys in Figure 10.3 and also the results in Figure 10.2. Although it is well known, the adverse effect of Ca on the fluidity of Mg alloys can't be clearly understood yet. In this case, both the theoretical and practical sides should be considered together, because Ca alloying changes the viscosity and freezing range as

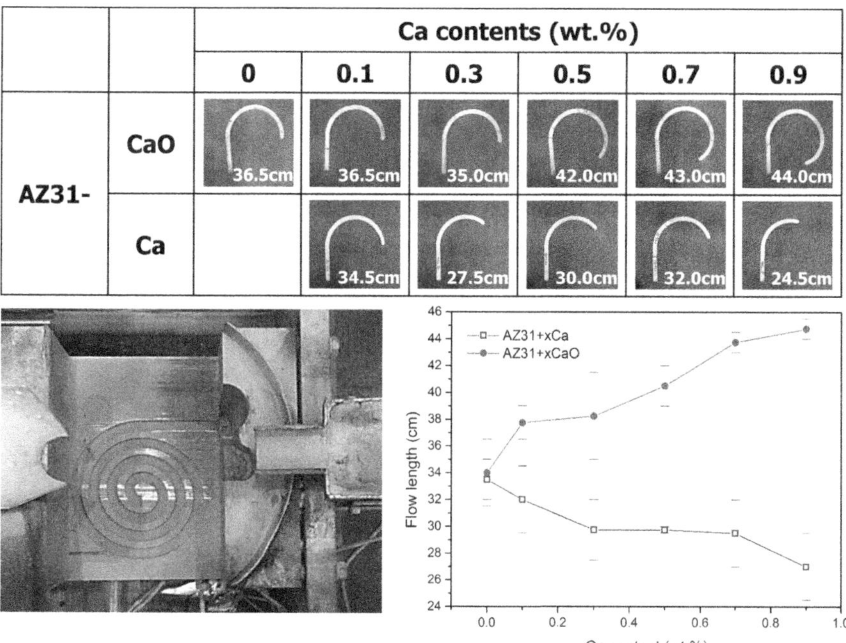

FIGURE 10.3 Spiral fluidity results of Ca alloyed and CaO added AZ31 Mg alloys.

well as increases the amount of oxide inclusions in the melt due to its high oxidation tendency. It is worth noting that Ca is used as a viscosity intensifier to make Al foam metals and the degree of viscosity increase can't be explained only by the alloy factors in the theoretical side. Unlike Ca alloyed AZ31 Mg alloys, it is interesting that increasing CaO does, at least, maintain or increase the real fluidity of the alloys. There is one-to-one correspondence between CaO content and the fluidity length, which has been verified by more than 5-year mass production in worldwide Mg industry. More quantitative and detailed research are being carried out to study the difference between Ca alloying and CaO addition in terms of the fluidity. The current understanding is based on the different amount of oxide inclusions present in Ca alloyed and CaO added Mg alloys.

Former Hydro Magnesium investigated the adverse effect of oxides on the tensile properties for as-die cast specimens of AM50 Mg alloy, redrawn in Figure 10.4.[9]

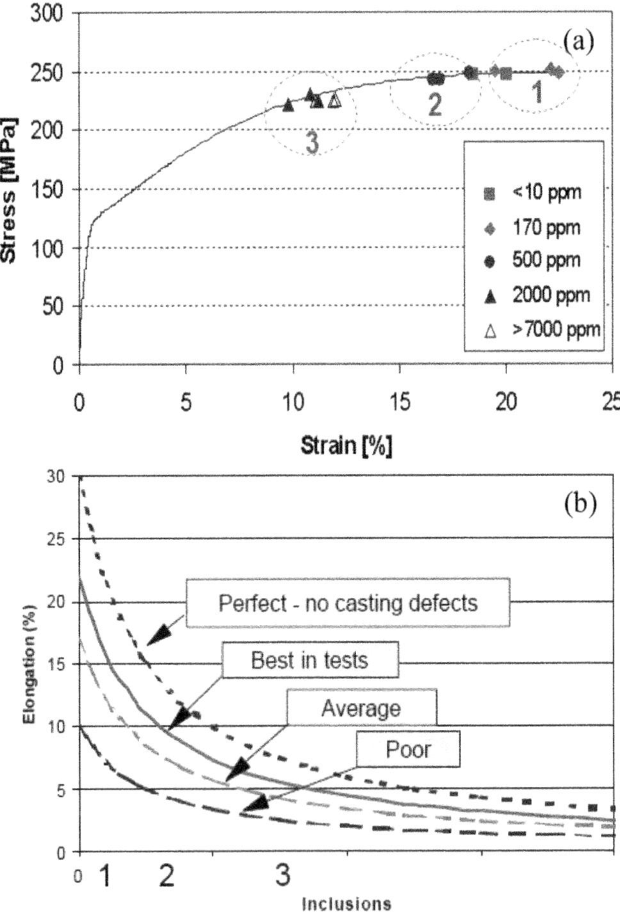

FIGURE 10.4 Metal cleanliness vs. mechanical properties of Mg alloys.

The typical yield strength, ultimate tensile strength and elongation values of AM50 Mg alloy are 120MPa, 200MPa and 8%, respectively. The average elongation value at the region 2 in Figure 10.4b represents the typical elongation value of AM50 Mg alloy. The elongation value over 15% in as-die cast condition has great industrial meaning by itself for automotive industry. What should be demonstrated is that the 2 fold improvement of the elongation value can be made by simple defect control, that is, by melt cleanliness.

Figure 10.5 shows the results of the non-SF_6 die cast 0.25% ~ 0.65% CaO added AM60 ECO-Mg alloys by using a Buhler 1,450-ton cold chamber machine. The molten AM60 Mg alloy was prepared in advance by using a Rauch 1-ton melting and dozing furnace and then 5 kg ingot of 1.2% CaO added AM60 ECO-Mg alloy was added repeatedly into the furnace to investigate the influence of CaO content. The samples for chemical analysis were picked up at every 30 minutes and analyzed by ICP-mass spectrometer. The 2-cavity of the tensile test specimen was added at the end of oil pan cavity around the overflow sections in the die casting mold. Tensile test specimens with the gauge length of 30 mm and the diameter of 6 mm were directly die cast. The typical values of yield strength, ultimate tensile strength and elongation are 130MPa, 220MPa and 6% for AM60 Mg alloy, respectively. In 0.25% ~ 0.65% CaO added AM60 ECO-Mg alloys, the

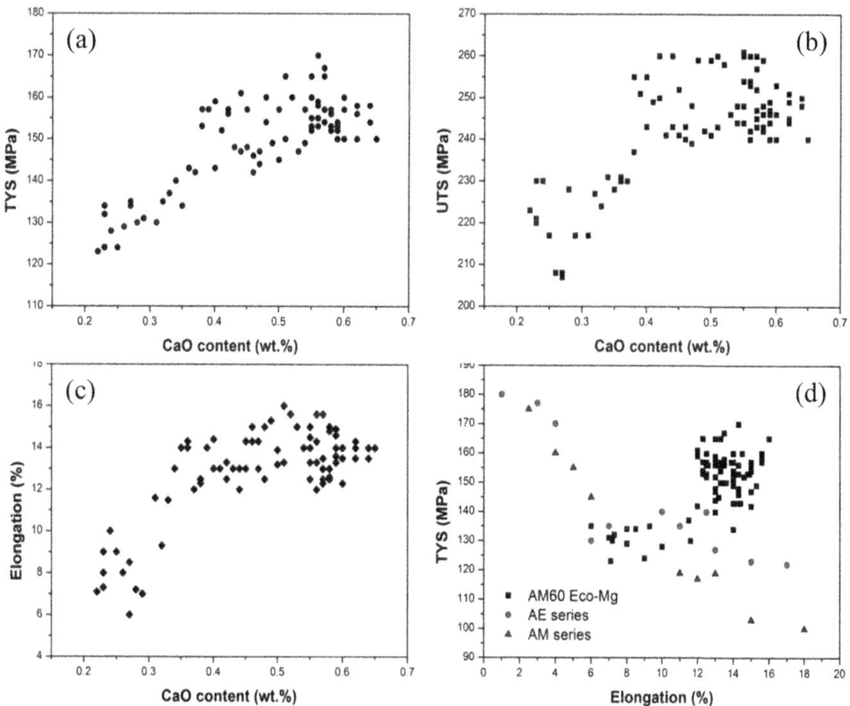

FIGURE 10.5 Effect of CaO on (a) yield strength, (b) ultimate tensile strength, (c) elongation, and (d) the balanced yield strength and elongation of CaO added AM60 ECO-Mg alloys.

yield strength, ultimate tensile strength and elongation values increase linearly with CaO increasing up to 0.5% CaO and then saturate. The maximum values are 160 MPa, 260 MPa and 15%, respectively. The increase of yield strength can be understood by the combined effect of α-phase grain refinement and the Al_2Ca phase in eutectic region. The point of the results is the influence of oxide inclusions on the improved elongation value, which is well matched with the explanation for Figure 10.4. The trade-off of strength and ductility is normal in metallic materials, as shown also for normal Mg alloys in Figure 10.5d. What should be emphasized is that the strength and ductility can be together improved by controlling oxide inclusions or melt cleanliness for Mg alloys. The comprehensive research is necessary to understand clearly the difference of Ca alloyed and CaO added Mg alloys in terms of ductility. The results of Figures 10.2 through 10.5 demonstrate the importance of melt and product cleanliness and the difference of the amount of oxide inclusions between Mg and ECO-Mg alloys as a possible solution to control the cleanliness practically.

10.4 NON-SF_6 PROCESSING AND BE-FREE ALLOYING

It is needless to say the importance of non-SF_6 processing in terms of environmental protection. SF_6 gas has the 23,900 times global warming potential of CO_2 gas. The global warming potential of 1 bombe of SF_6 gas equals the amount of CO_2 gas emitted in a year from 240 cars. Moreover, every amount of SF_6 gas used for melting and casting Mg alloys is discharged atmospherically with no possible system and way to collect it. There are two approaches for non-SF_6 processing: developing alternative gases with lower environmental impacts or alloys that can be processed without protective gas. There are several alternative gases developed but they are insufficient and cause side effect to substitute SF_6 entirely. Therefore, old-fashioned SO_2 gas has been used in Europe as the replacement of SF_6, although it is toxic for human. New Mg alloys with the stability of surface oxide can be the best, industry-friendly way under auxiliary dry-air, argon or reduced SO_2 atmosphere. This is because a protective gas can't ensure the melt cleanliness and solve the safety issue of Mg machined chips and products.

Researches have been carried out to evaluate the effect of Ca alloying on non-SF_6 processing of Mg alloys, which revealed it was useful only for non-SF_6 processing.[10] As noted, there occur the side effects of deteriorated castability and ductility due to Ca alloying.[8] Therefore, Ca alloying can't be applied generally to all types of Mg alloys but has been used limitedly for creep and ignition resistant alloys.

Non-SF_6 processing is also possible by ECO-Mg alloys, the point of which can at least maintain the real fluidity and ductility and can be applied in the same way to all types of Mg alloys. Figure 10.6 shows non-SF_6 casting for (a) pure Mg, (b) pure Mg + 0.3% CaO, (c) AM60, (d) AM60 + 0.3CaO, (e) AZ91, and (f) AZ91 + 0.3% CaO Mg and ECO-Mg alloys, indicating the possibility of the small of amount CaO on non-SF_6 processing. In pure Mg, the severe oxidation begins with tarnished surface just after casting, and the burning starts in 10 seconds and continues with white flame. Almost similar phenomena occur in AM60 and

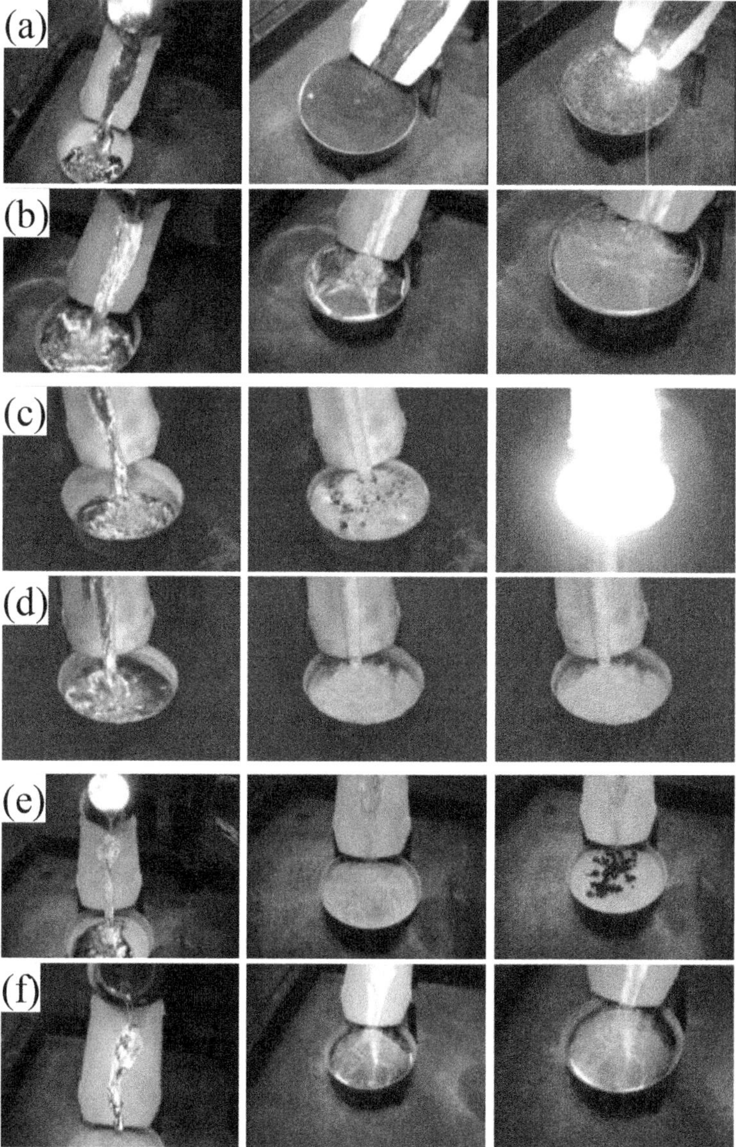

FIGURE 10.6 Non-SF_6 casting for (a) pure Mg, (b) pure Mg + 0.3% CaO, (c) AM60, (d) AM60 + 0.3CaO, (e) AZ91, and (f) AZ91 + 0.3% CaO Mg and ECO-Mg alloys.

AZ91 Mg alloys. The results of non-SF_6 processing are well agreed with the general facts that the continuous oxidation and resultant burning occur at the temperature over 500°C due to the porous surface nature. In addition, the results, obtained during melting and casting, tell that the oxidation and ignition resistances increase with Al increasing. However, it is totally in conflict with the adverse effect of Al on the oxidation resistance during solid oxidation at 500°C. It was revealed that the oxidation resistance continuously diminished with Al increasing and was significantly cut back at 9%Al. Even in pure Mg based ECO-Mg alloy only with 0.3% CaO, no burning occurs even with the yellow surface due to the little oxidation. No burning occurs and shiny surfaces can be maintained both in AM60 + 0.3% CaO and AZ91 + 0.3% CaO. For practical melting and longer holding, the optimum content of CaO is being investigated as the functions of alloy composition and ambient atmosphere under auxiliary dry air or argon, some results of which are given in Figure 10.7, together with the issue of beryllium-free alloying.

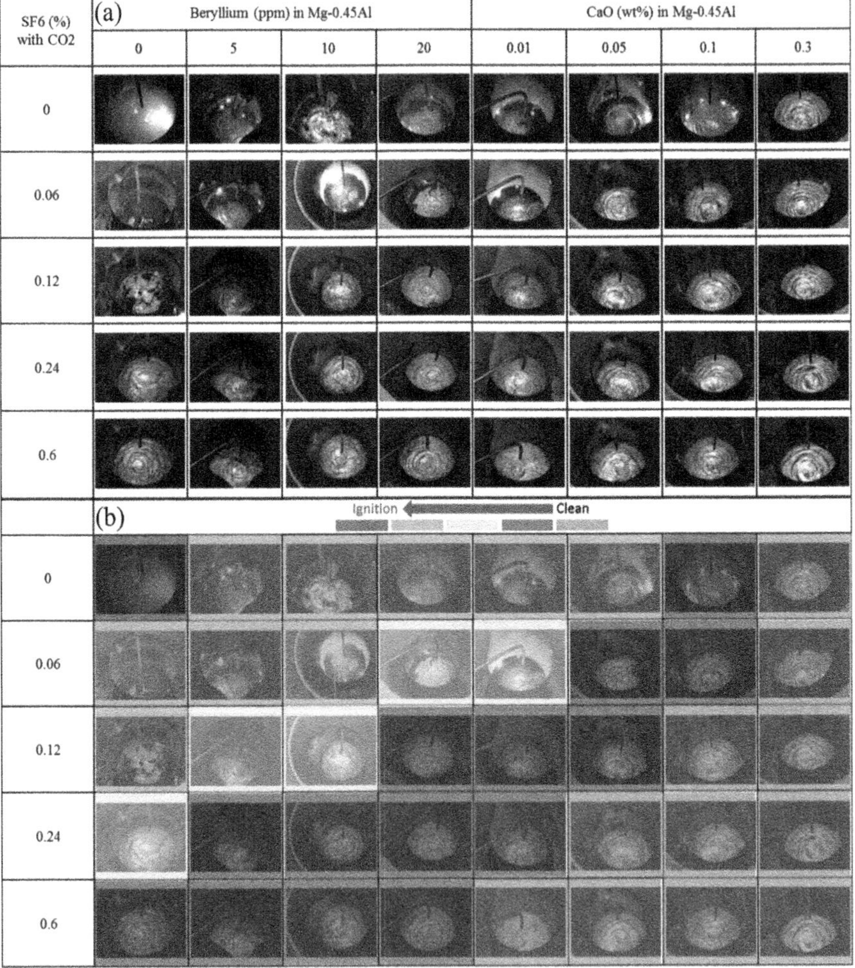

FIGURE 10.7 Mg melting phenomena under (a) CaO and SF_6 without beryllium, compared with those under beryllium and SF_6 and (b) the arbitrary cleanliness ranking based on (a).

Every Mg alloy contains beryllium of 10 ~ 20 ppm. During producing primary alloys and recycling scraps, the ingot surface changes into black without beryllium, even under SF_6. Just like SO_2, however, the use of beryllium should be prevented for human being. Figure 10.7 shows the combined effects of CaO and SF_6 without beryllium, compared with those of beryllium and SF_6. Figure 10.7b is the arbitrary cleanliness ranking based on Figure 10.7a. Mg alloys with containing much more beryllium, not in ppm order, can show much better oxidation and ignition resistances. The current level of beryllium, 10 ~ 20 ppm, has been set up as the optimum condition by considering both oxidation resistance and toxicity, because beryllium has to play auxiliary role for SF_6 or SO_2 gas. It is clearly seen that the small amount of CaO addition can protect melt oxidation and ignition and reveal the silver-like bright surface with possible melt cleanliness. Once again, it should be noted that it is vital to prevent the use of SF_6, SO_2 and beryllium. However, it should be done without damaging the real fluidity and ductility and, more than that, should ensure much better melt cleanliness to reduce down the gap to the ideal process-ability and property of Mg alloys.

The issue of non-SF_6 processing and Be-free alloying is of great significance in welding processes. To improve the performance of welding and the welded properties, it is recommended to use SF_6 or SO_2 but it is not easy to change the system to mix a general shielding gas for welding with SF_6 gas. Therefore, Mg alloys with self-passivating characteristics are important not only for Mg product to be welded and but also for Mg welding wire for arc welding. FSW (Friction Stir Welding) may be an alternative for fusion welding processes but it is much better to take advantage of every welding process for Mg alloys. The issue of Be-free alloying is always important but is of great interest for welding because of the fume of beryllium generated. The photos taken during 2 welding processes for Mg alloys are given in Figure 10.8: Figure 10.8a and b are for laser welding AZ31 Mg and AZ31 + 1.5% CaO ECO-Mg plates and Figure 10.8c and d are for arc welding by AZ31 Mg and AZ31 + 1.0% CaO ECO-Mg welding wires of 1.6 mm in diameter. The fume generated by laser welding may cause the change of alloy composition and deteriorate the welded properties. It is metallic fume with beryllium in Mg alloys. AZ31 Mg alloy with beryllium generates much bigger fume than AZ31 + 1.5% CaO ECO-Mg alloys without beryllium. The Mg and ECO-Mg alloys welding wires also show the different fume generation during arc welding. Friction stir welding was also applied to and could show the better performance with 0.5% and 1.0% added AZ31 ECO-Mg alloys.[11] The mechanism that operates for friction stir welding AZ31 ECO-Mg alloys is not related to the self-passivating surface nature but the dynamic recrystallization due to Al_2Ca. The dynamic recrystallization can cause the significant grain refinement in the stir zones and the presence of Al_2Ca can maintain the same fine grains even after annealing at 400°C in AZ31 ECO-Mg alloys. No grain refinement was obtained in the stir zone and the significant grain growth occurred after annealing 400°C in AZ31 Mg alloys.

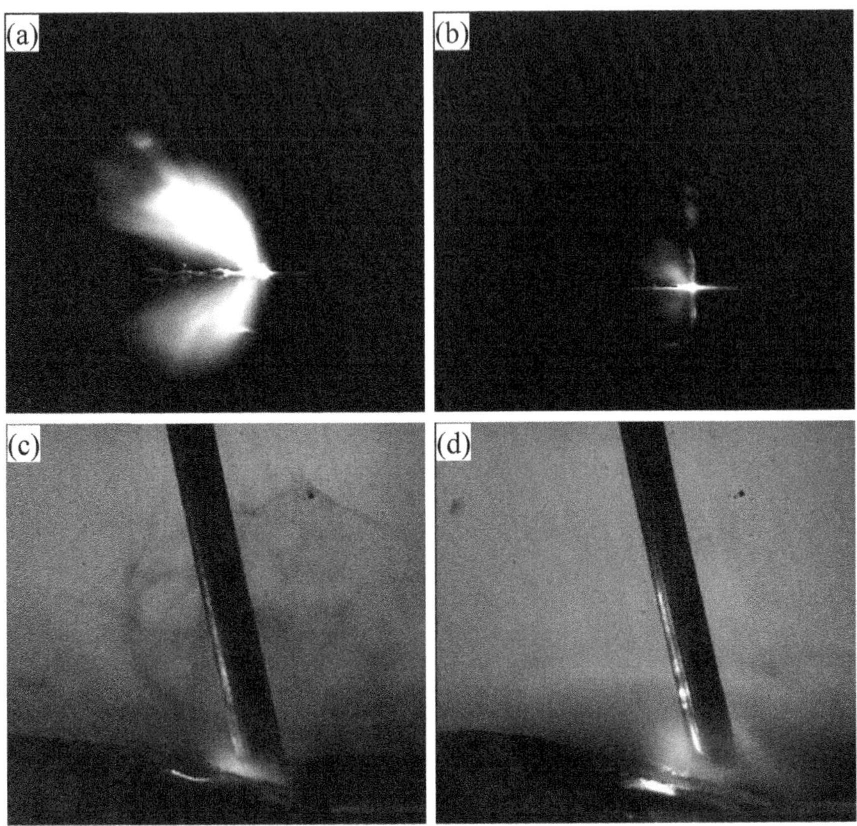

FIGURE 10.8 Fume generation during Mg welding processes: laser welding (a) AZ31 Mg and (b) AZ31 + 1.5% CaO ECO-Mg plates; and arc welding by (c) AZ31 Mg and (d) AZ31 + 1.0% CaO ECO-Mg welding wires.

10.5 IGNITION RESISTANT Mg ALLOYS FOR SAFETY

Although global warming potential and toxicity, SF_6 gas or SO_2 gas can prevent Mg alloys from burning during processing with auxiliary beryllium alloying. During application, however, there is no way to prevent the burning of Mg products, and moreover water and common extinguisher can't be used to extinguish the burning. There has been no restriction nor regulation for the burning of Mg alloys in automotive and train. FAA had prevented the use of Mg alloys in commercial aircrafts and has established the regulation and requirement for ignition resistant Mg alloys. The time to ignite and the time to self-extinguish, if ignition occurs, are the two key factors and the third factor for the weight loss less than 10% is added in the final version, as summarized in Table 10.3.[12] The change of requirements has been made in the best way to reflect the regulation of aircraft materials. From the side of Mg alloys, the change should be welcome because the little bit lightened requirements can open up the possibility to utilize other benefits of various Mg alloys, on the basis

TABLE 10.3
Change of Requirements for FAA Ignition Test

	1st Version	2nd Version	3rd Version	Final Version
Time to ignition, min	2 + 8	2 + 3	2 + 2	2 + 2
Time to burner off	Ignition + 30s	Ignition + 30s	Ignition	Ignition
Delta t, sec	15	15	120	180

of satisfying the required ignition resistance. To combine two key factors is of great importance because the stability of surface oxides in Mg alloys is not permanent but relatively valid. The time to self-extinguish is important to evaluate its passivating characteristics, because there is a continuous chance for a Mg alloy with the ignition resistance to ignite. To satisfy the two key factors together with the third factor, the alloy design should consider altogether the alloying elements (1) having the lower oxide formation energy than MgO, (2) having the stability of surface oxides, (3) having higher melting temperatures of each oxide and the oxide mixtures, and (4) having the higher surface segregated tendency than Mg. It should be noted that the one alloying element, or more than one alloying elements, that can give enough oxidation and ignition resistances for Mg alloys, can satisfy at least 3 points out of 4. The alloying elements are Ca, Sc and Y, which are appropriate for (1) the lower oxide formation energies than MgO. One method to access the stability of surface oxides for the point (2) is to select alloying elements that have the average PBR of over 1. The average PRB of binary oxides in Mg alloys is easily calculated by doing the sum of PBR of MgO and PRB of the oxide of an alloying element and then dividing the sum by 2. The PBRs of each oxide are 1.89 (SiO_2), 1.59 (ZnO), 1.29 (Al_2O_3), 1.13 (Y_2O_3), 1.07 (CeO_2), 0.66 (SrO), 0.65 (CaO), and 0.81 (MgO). The average PBRs can be over 1 with Si (1.35), Zn (1.2) and Al (1.05) and just below 1 with Y (0.97) and Ce (0.94). Although the relative oxide stability can be explained based on PBR previously between Mg and Al alloys, it is well known that the oxidation resistances and behaviors of Mg alloys can't be explained entirely by PBR alone. As an alloying element in Mg alloys, Si with the average PBR of 1.35 causes severe oxidation and burning and increasing Al also degrades the oxidation resistance, while Ca and Sr with the average PBRs of 0.73, on the contrary, have revealed to improve the oxidation resistance and generate the stable surface oxides by experiments. Although there has been no theory to predict the oxidation resistance and the mechanism is not clear, Ca, Sr, Y and rare earth are most promising in terms of the stability of surface oxides in Mg alloys. For the point (3), the oxides of the corresponding alloying elements in Mg alloys that have the higher melting temperatures are CaO (2,572°C), Y_2O_3 (2,425°C), and Al_2O_3 (2,054°C). Although there is not full data about the eutectic temperatures of the binary oxides, the highest eutectic temperature is 2,374°C between MgO (2,825°C) and CaO (2,572°C) and the lowest is 1,465°C between MgO and SiO_2 (1,723°C). Surface segregation is the enrichment of one element of an alloy. As well known, it is important because surface segregation changes surface composition and the change in turn changes oxidation resistance. The surface segregation is also of great importance in terms of self-extinguishment to recover its passivating characteristics. Various models have been proposed to understand surface segregation. Recently, Aydin, D. S. and Bayindir, Z. et al.[13] can summarize the surface segregation tendency for Mg alloys based on Wigner-Seitz radii, which indicates that an alloying element with the larger Wigner-Sietz radius (lower average electron density) segregates to the surface. Based on it, alloying elements having the surface segregation tendency with the larger radius over Mg (0.18) are Ca (0.21), Te (0.21), Ce (0.20), Nd (0.20), Sn (0.19), and Sr (0.19). What becomes clear as the summary is that only Ca can satisfy all 4 points; and Y and Sr are promising candidates. CaO added ECO-Mg alloys can also satisfy all 4 points.

To meet and pass the current FAA requirements, the alloy design should optimize the enough, not excessive, CaO, Ca, Sr or Y and the mixture of them, and consider additional alloying elements to optimize the process-abilities and properties of Mg alloys. Figure 10.9 shows the preliminary burning test results of 0.5%, 1.0%, and 1.5% CaO added AZ31 ECO-Mg alloys in the form of hollow tube by FAA. Even 0.5% CaO added AZ31 ECO-Mg alloy with having the lowest oxidation and ignition resistances among three can pass the 4-minute exposure and maintain the clean surface. Figure 10.10 shows the recent burning test results of 1.0% CaO added AZ31 and AZ61 wrought ECO-Mg alloys and AM60 and AZ91 casting ECO-Mg alloys by a FAA approved burning test lab., according to the requirements and the requested test and report procedures. The reported data include the times to start to warp, sag, melt and ignite; the weight loss percentage; and the total burn durations of bar and residue. Ban-lifted are only five parts in aircraft seat—leg, spreader, cross tube, seat-back frame, and lower baggage. Although the ban-lifted five parts are made by wrought processes, fire-proof ECO-Mg alloys plus die casting process can give the additional weight reduction by part integration and thin wall thickness and the benefit of cost effectiveness.

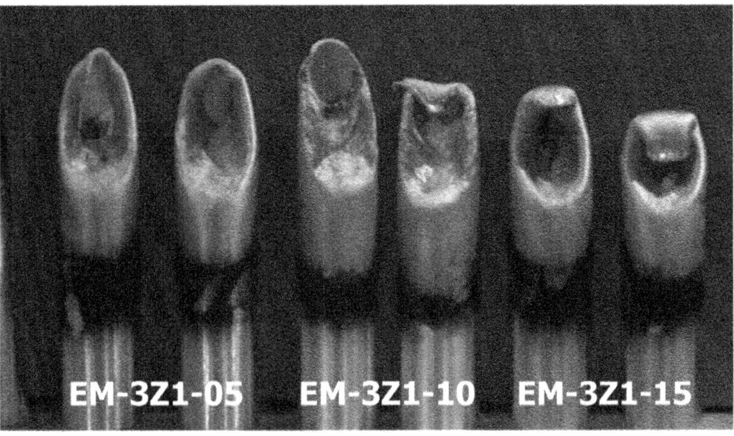

FIGURE 10.9 Ignition test results of 0.5%, 1.0%, and 1.5% CaO added AZ31 ECO-Mg alloys in the form of hollow tube by FAA.

Test No.	Run	Date	Alloy	Time Warp	Time Sag (Sec)	Melt (Sec)= ANY Bar Separation	Ignition (Sec)= first sustained bar bum	Burner Off (Sec)	Bar Out (Sec)	Residue Out (Sec)	Initial Weight (lbs)	Final Bar Wt.	Weight Loss	Weight Loss (%)	Total Bar Burn Duration	Total Residue Burn Duration	Post Flame Bum
Pass-Criteria							< 120 sec							< 10%	< 180 sec		
10014	1	20.03.2014	EM3Z1-10	105	190	196	N/A	240	N/A	N/A	0,479	0,4785	0,000	0,00%	N/A	N/A	N/A
10014	2	20.03.2014	EM3Z1-10	120	159	164	N/A	240	N/A	N/A	0,480	0,48	0,000	0,00%	N/A	N/A	N/A
10014	3	20.03.2014	EM3Z1-10	86	113	127	N/A	240	N/A	N/A	0,479	0,479	0,000	0,00%	N/A	N/A	N/A
10114	1	20.03.2014	EM9Z1-10	128	169	174	N/A	240	N/A	N/A	0,4865	0,486	0,0005	0,10%	N/A	N/A	N/A
10114	2	20.03.2014	EM9Z1-10	136	152	184	N/A	240	N/A	N/A	0,494	0,494	0,000	0,00%	N/A	N/A	N/A
10114	3	20.03.2014	EM9Z1-10	120	150	180	N/A	240	N/A	N/A	0,492	0,492	0,000	0,00%	N/A	N/A	N/A
10214	1	20.03.2014	EM6Z1-10	86	118	180	N/A	240	N/A	N/A	0,488	0,488	0,000	0,00%	N/A	N/A	N/A
10214	2	20.03.2014	EM6Z1-10	94	125	168	N/A	240	N/A	N/A	0,489	0,4885	0,000	0,00%	N/A	N/A	N/A
10214	3	20.03.2014	EM6Z1-10	90	111	178	N/A	240	N/A	N/A	0,489	0,489	0,000	0,00%	N/A	N/A	N/A
10314	1	20.03.2014	EM6M0-10	119	160	202	N/A	240	N/A	N/A	0,484	0,484	0,000	0,00%	N/A	N/A	N/A
10314	2	20.03.2014	EM6M0-10	124	0	143	N/A	240	N/A	N/A	0,4855	0,485	0,001	0,10%	N/A	N/A	N/A
10314	3	20.03.2014	EM6M0-10	115	0	138	N/A	240	N/A	N/A	0,490	0,49	0,000	0,00%	N/A	N/A	N/A

FIGURE 10.10 Ignition test results of 1.0% CaO added AZ31 and AZ61 wrought ECO-Mg alloys and AM60 and AZ91 casting ECO-Mg alloys by a FAA approved burning test lab, according to the requirements and the requested test and report procedures.

Automotive and railway industries may make their own regulation and requirement for Mg products. For R&D stage to develop new ignition resistant Mg alloys, it is needless to say the necessity of a simple and quantitative method to evaluate the ignition resistance by utilizing small sample. Some researchers still use a furnace chip test even for comparison and qualification. As illustrated in Figure 10.11, the furnace chip test shows over 200 k deviation of the ignition temperature for the same alloy, according to the way of chip preparation and test parameters such as the type and size of a machining tool and its sharpness; the shape, size and thickness of machined chips; the degree of chip compaction; the material and shape of a container for chip; and furnace temperature. Mg alloys with no ignition under the furnace temperature of 800°C readily ignite under the furnace temperature of 1,000°C. The same chips prepared by the same tool show the big deviation because the thickness of the chips grow thicker as the sharpness of the tool becomes dulled. To minimize efforts to examine every parameter involved and to obtain reliable and quantitative data, it is recommended to use the existing, already accepted method, DTA (Differential Thermal Analysis) to evaluate the ignition temperatures of Mg alloys. What we have to fix are only the shape and weight of specimen, heating rate and atmosphere of dry air or argon—the other parameters are automatically fixed by DTA.

As shown in Figure 10.12a, the effect of Al content on the ignition temperature was investigated to verify DTA as a reliable and quantitative method for a spherical specimen under dry air atmosphere. The ignition temperatures of AZ31, AM60, and

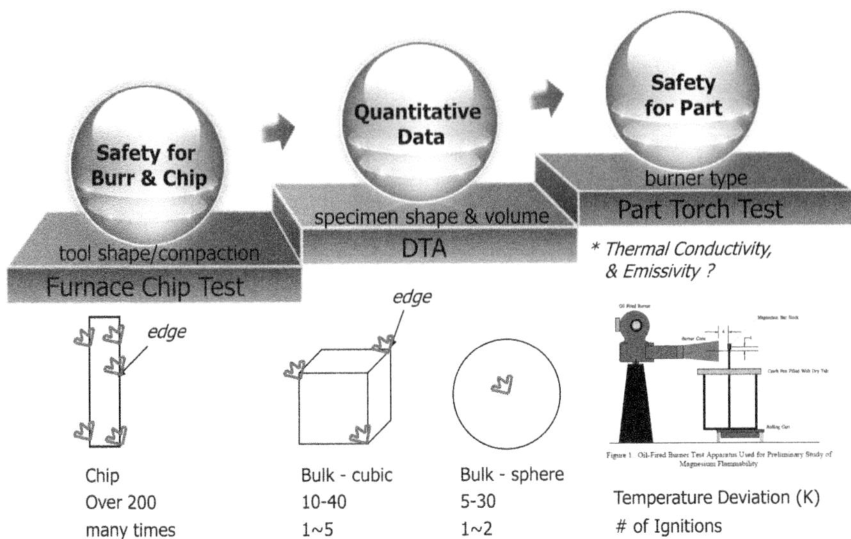

FIGURE 10.11 Advantages and limitations of each burning test method: furnace chip test, DTA, and part torch test.

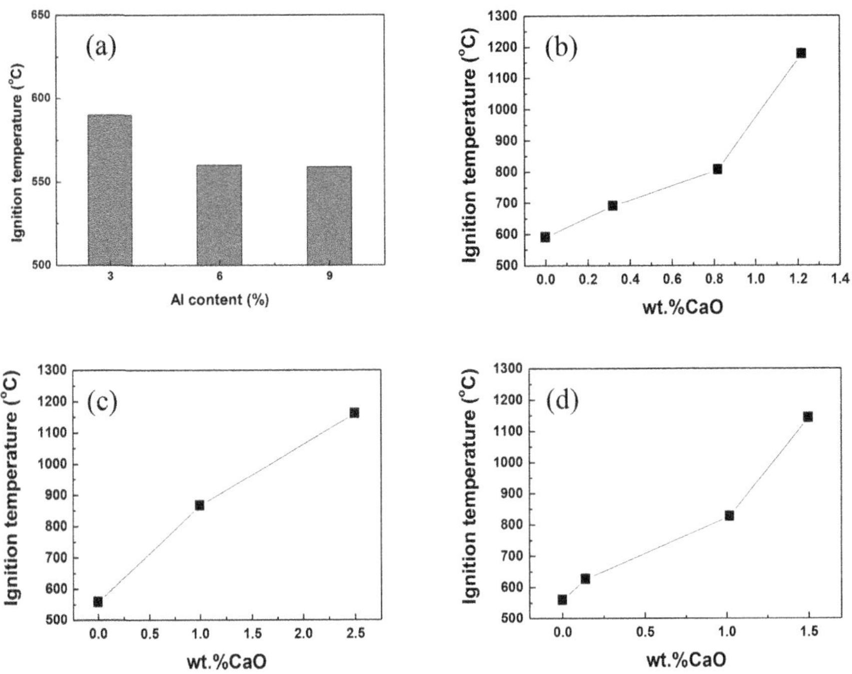

FIGURE 10.12 DTA results of (a) AZ31, AM60, and AZ91 Mg alloys; (b) AZ31, (c) AM60, and (d) AZ91 ECO-Mg alloys as a function of CaO content.

AZ91 Mg alloys are 590°C, 560°C, and 559°C, respectively. Figure 10.12 also shows the DTA results of AZ31, AM60, and AZ91 Mg alloys as a function of CaO content. The ignition temperatures of Mg alloys increase almost linearly with increasing CaO content. The ignition temperatures of AZ31, and 0.32%, 0.82%, and 1.22% CaO added AZ31 ECO-Mg alloys are 590°C, 691°C, 806°C, and 1,177°C, respectively. DTA is much easier than the furnace chip test and the reliable and quantitative method. It should be cautious to fix the weight of a spherical specimen and do the final polish by SiC emery paper 800 grit to remove surface fouling and keep the same surface roughness.

The issue of severe oxidation and ignition is of interest for wrought process. Forming Mg alloys at higher velocity can be assessed at relative high temperature forming due to more activated slip systems. One obstacle for it is the friction-enhanced oxidation and ignition, for which the use of SF_6 is difficult to be applied in similar situation for Mg products in service. Alloy development to improve the formability of Mg alloys is the common way to be taken. To improve the formability by relative higher temperature forming can be an alternative in connection with ignition-proof Mg alloys. As a part of evidence, Figure 10.13 demonstrates the possibility of relative higher temperature extrusion at the higher forming velocity of 3 m/min without the friction-enhanced surface oxidation and burning in the case of 0.3% CaO added AZ31 ECO-Mg alloy, compared with the case of AZ31 Mg alloy with the black surface.

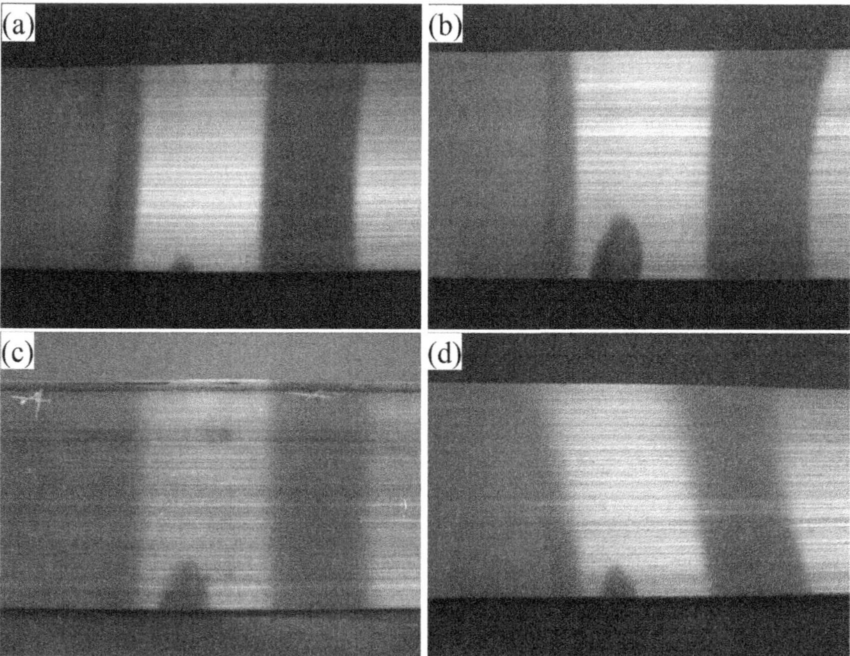

FIGURE 10.13 Effect of high velocity forming on the friction-enhanced surface oxidation and ignition of (a) AZ31 and (b) AZ31 + 0.3% CaO at 1m/min; and (c) AZ31 and (d) AZ31 + 0.3% CaO at 3 m/min.

10.6 CONCLUSION

The importance of and the possibility to solve (1) the melt cleanliness and (2) the inherent SHE (Safety, Health, Environment) obstacles of Mg alloys are addressed, with an emphasis on (3) the fact that two issues are mainly attributed to the non-protective porous surface oxide nature. If the melt cleanliness of Mg alloys was achieved by generating the stable self-passivating surface oxide nature, SHE obstacles could be solved simultaneously and the gap between real and ideal values could be reduced to be able to improve the fluidity and ductility of Mg alloy. ECO-Mg alloys are demonstrated as one of possible ways to satisfy all the issues together, even without damaging real mechanical property of ductility and process-abilities of castability and formability and with the expandability to be applied to all types of Mg alloys in the same way. The melt cleanliness of Mg alloys without SHE problems can make the best use of the environmental benefits provided by Mg alloys with lightweight, unique properties, recyclability and limitless resource.

ACKNOWLEDGMENTS

ECO-Mg alloys have been developed under the financial supports of Ministry of Trade, Industry and Energy (MOTIE), Ministry of Science, ICT and Future Planning (MSIP), Korea Institute of Industrial Technology (KITECH), and the companies of EMK, NICE and Boeing. The author thanks Dr. Jin-Kyu Lee of EMK for experimental work. The author also thanks Dr. Hyun-Kyu Lim of KITECH, Prof. In-Ho Jung of McGill University, and Dr. Donald Shih of Boeing for valuable discussion.

REFERENCES

1. Kim, S. K. Design and development of high-performance ECO-Mg alloys. In *Magnesium Alloys—Design, Processing and Properties*; Czerwinski, F., Ed.; InTech: Rijeka, Croatia, 2011; 431–468.
2. Jeong, J. W., Im, J. S., Song, K., Kwon, M. H., Kim, S. K., Kang, Y. B., Oh, S. H. Transmission electron microscopy and thermodynamic studies of CaO-added AZ31 Mg alloys. *Acta. Mater.* 2013, 61, 3267–3277.
3. Ha, S. H., Lee, J. K., Kim, S. K. Effect of CaO on oxidation resistance and microstructure of pure Mg. *Mater. Trans.* 2008, 49 (5), 1081–1083.
4. Habashi, F. *Aluminum Alloys: Preparation, Properties, Applications*; Wiley-VCH: New York, 1998.
5. Czerwinski, F. The early stage oxidation and evaporation of Mg-9%Al-1%Zn. *Corros. Sci.* 2004, 46 (2), 377–386.
6. Pilling, N. B., Bedworth, R. E. The oxidation of metals in high temperature. *J. Inst Metals* 1923, 29, 529–582.
7. Kearney, A. L., Elwin, L. R. Aluminum foundry products. In *Properties and Selection: Nonferrous Alloys and Special-Purpose Materials*, Metal Handbook, 10th ed.; ASM International: Cleveland, OH, 1990; Vol. 2, 123–151.
8. Aghion, E., Bronfin, B., Katzir, M., Bar-Yosef, O., Shmelkin, E., Lautzker, M., Lere, E., Eliezer, D., Moscovitch, N. Development of new magnesium casting alloys. In *Magnesium Alloys Science, Technology and Applications*; Aghion, E., Eliezer, D., Eds.; Israeli Consortium for the Development of Magnesium Technologies: Israel, 2004; 137–158.

9. Hydro Magnesium. Quality assurance indicators of magnesium alloy ingots, *IMA Educational Seminar*, Chenguo Tian, China, October 23, 2002.
10. Sakamoto, M., Akiyama, S. Suppression of ignition and burning of molten Mg alloys by Ca bearing stable oxide film. *J. Mater. Sci. Lett.* 1997, 16, 1048–1050.
11. Choi, D. H., Kim, S. K., Jung, S. B. The microstructures and mechanical properties of friction stir welded AZ31 with CaO Mg alloys. *J. Alloys Comp.* 2013, 554, 162–168.
12. Marker, T. R. *Development of a Laboratory-Scale Flammability Test for Magnesium Alloys used in Aircraft Seat Construction*, DOT/FAA/TC-13/52; U.S. Department of Transportation: NJ, 2014.
13. Aydin, D. S., Bayindir, Z., Pekguleryuz, M. O. The effect of strontium (Sr) on the ignition temperature of magnesium (Mg): A look at the pre-ignition stage of Mg-6wt%Sr. *J. Mater. Sci.* 2013, 48, 8117–8132.

11 Application of Computational Thermodynamics for Magnesium Alloys Development

Yu Zhong and Zi-Kui Liu

CONTENTS

11.1 INTRODUCTION

With a density being two-thirds that of aluminum and one-quarter of steel, magnesium (Mg) alloys are of growing importance. The development of Mg-based light alloys for vehicle structures has been promoted in light of global climate change due to greenhouse gases [1]. Computational thermodynamics, based on the CALPHAD (CALculation of PHAse Diagram) approach [2–4], is a key enabling technology to help accelerate the pace of materials research and development through building phase relations that can reduce the design-to-alloy production cycle time [5,6]. This technology has the capability to solve a vast number of materials-related problems related to phase equilibria, phase stability, and phase transformations by using well-developed CALPHAD software packages such as Thermo-Calc [7], FactSage [8], and PANDAT [9], where a modeled thermodynamic database is prerequisite for any thermodynamic analyses [1,10]. However, measured thermodynamic data from experiments, thermochemical data in particular, are scarce, and that can result in uncertainties in thermodynamic modeling, particularly the distribution of Gibbs energy into enthalpy and entropy contributions in some cases. Fortunately, the advanced computational tools available today, for instance first-principles calculations based on the density functional theory [11], can provide considerable insight into these basic materials properties. First-principles calculations have now become an important part of computational thermodynamics. This chapter will include the CALPHAD modeling and the new first-principles calculations of Mg-based alloys and compounds.

The remainder of this chapter is organized as follows. In Section 2, the background of computational thermodynamics is presented, including the fundamentals of thermodynamics; experimental measurements and estimations; CALPHAD modeling; and fundamentals of first-principles calculations. Section 3 presents the applications of computational thermodynamics, including multi-component thermodynamic databases, its equilibrium and non-equilibrium applications, and first-principles calculations. Finally Section 4 shows future trends of computational thermodynamics.

11.2 BACKGROUND OF COMPUTATIONAL THERMODYNAMICS

11.2.1 Fundamentals of Thermodynamics

Based on the combined first and second law of thermodynamics, the change of internal energy, U, of a system is given by [12]

$$dU = TdS - PdV + \sum \mu_i dN_i - Dd\xi \tag{11.1}$$

The internal energy is thus a function of the natural variables S (entropy), V (volume), N_i (amount of independent component i), and ξ (internal process) of the system, i.e., $U(S,V,N_i,\xi)$. The other variables T (temperature), P (pressure), μ_i (chemical potential), and D (driving force) are dependent variables and can be represented by partial derivatives of the internal energy with respect to their corresponding natural variables while keeping other natural variables constant. Noting that one major focus of this chapter is the thermochemical properties of individual phases in a closed system without mass change (i.e., $dN_i = 0$) and at the equilibrium condition (i.e., $Dd\xi = 0$); therefore the variables N_i and ξ are ignored herein unless otherwise mentioned.

In experiments, the variables of temperature T and pressure P are typically controlled. While in theoretical calculations, it is more convenient to control temperature T and volume V. For the convenience of both experiments and calculations, thermodynamic variables of enthalpy H, Helmholtz energy F, and Gibbs energy G are defined as follows:

$$H = U + PV \tag{11.2}$$

$$F = U - TS \tag{11.3}$$

$$G = U - TS + PV = H - TS = F + PV \tag{11.4}$$

Other commonly used thermodynamic variables can be obtained with the following derivatives:

$$\alpha_V = \left(\frac{\partial \ln V}{\partial T}\right)_P = \frac{1}{V}\left(\frac{\partial V}{\partial T}\right)_P = \frac{1}{V}\frac{\partial^2 G}{\partial P \partial T} = \frac{1}{B_T}\left(\frac{\partial P}{\partial T}\right)_V = \frac{1}{B_T}\left(\frac{\partial S}{\partial V}\right)_T = 3\alpha_L \tag{11.5}$$

$$B_T = -V\left(\frac{\partial P}{\partial V}\right)_T = V\left(\frac{\partial^2 F}{\partial V^2}\right)_T \tag{11.6}$$

$$B_S = B_T C_P / C_V = B_T + (\alpha_V B_T)^2 TV / C_V \tag{11.7}$$

$$C_V = \left(\frac{\partial U}{\partial T}\right)_V = T\left(\frac{\partial S}{\partial T}\right)_V = -T\left(\frac{\partial^2 F}{\partial T^2}\right)_V \tag{11.8}$$

$$C_P = \left(\frac{\partial H}{\partial T}\right)_P = T\left(\frac{\partial S}{\partial T}\right)_P = -T\left(\frac{\partial^2 G}{\partial T^2}\right)_P = C_V + \alpha_V{}^2 B_T TV \tag{11.9}$$

where α_V is volume thermal expansion, α_L the linear volume thermal expansion, B_T the isothermal bulk modulus, B_s the isentropic (adiabatic) bulk modulus, C_V the heat capacity at constant volume, and C_P the heat capacity at constant pressure.

The third law of thermodynamics states that entropy is zero at zero Kelvin for an ordered phase, i.e., $S_0 = 0$. However, the absolute value of internal energy U is unknown in the framework of materials thermodynamics that primarily concerns chemical reactions, and we thus must select a reference state. The same is true for H, F, and G. A widely used reference state in the CALPHAD community for thermodynamic modeling is to set $H_{298.15} = 0$ for pure elements at their respective stable structures at room temperature (298.15 K) and ambient pressure. This is known as the stable element reference (SER). From the viewpoint of theoretical predictions, e.g., first-principles calculations, thermodynamic properties can be easily obtained using the above relations once F or G is known.

11.2.2 Experimental Measurements and Empirical Estimations

Computational thermodynamics greatly rely on the phase equilibria data and thermochemical data from experimental measurements and empirical estimations before the adoption of the first principles calculations based on the density functional theory [13]. Details regarding methods for measuring equilibrium data (i.e., phase diagram data) between phases and thermochemical data for individual phases can be found in [14–18].

11.2.2.1 Phase Equilibria Data from Experiments

The most common method to determine phase equilibria is to use equilibrated materials. It typically involves material preparation through high temperature melting or powder metallurgy, homogenization heat treatment, isothermal or cooling/heating procedure, and identifications of crystal structures and phase compositions. The phase diagram data can be determined via diffusion couples or multiples, differential thermal analysis (DTA) and differential scanning calorimetry (DSC). Both attempt to measure the difference in temperature with the same amount of power supplied between a sample and an inert standard during heating or cooling. Extreme care is needed in interpreting the temperature determination and the amount of heat associated with the DTA/DSC curves as discussed in detail in the reference [17].

The major challenge in the equilibrated material approach is to ensure that the whole sample reaches equilibrium. On the other hand, diffusion couples and multiples do not require the whole sample to be in equilibrium and are based on the assumption that any two phases in contact are in equilibrium with each other at the phase interface, and the phase compositions can be obtained by extrapolation of concentration profiles in the two phases to the phase interface.

Besides the above two approaches, additional methods such as electrical resistivity measurement, magnetic transition measurement can also provide phase equilibria data.

11.2.2.2 Thermochemical Data from Experiments

Thermochemical properties can be determined by electromotive forces (EMF), calorimetric methods including DTA and DSC, and vapor pressure method. The calorimetric method is divided into solution, combustion, direct reaction, and heat capacity calorimetry, respectively. Performing and evaluating such experiments requires scrutiny, detailed in the special publication from NIST [19].

The solution calorimetry was detailed in the book edited by Marsh and O'Hare [20]. In one experiment, the enthalpy of solution of a single phase is measured in a particular solvent. In another experiment, the enthalpy of solution of constitutive pure elements or compounds is measured in the solvent as identical as possible to that used in the first experiment. The difference of the two enthalpies of solution thus gives the enthalpy of formation of the single phase from its constitutive elements or compounds at the temperature of the samples before they are dropped into the solvent, usually at room temperature.

In combustion calorimetry, the sample is ignited and reacts with reactive gases like oxygen or fluorine. To accurately calculate the enthalpy of formation from the enthalpy of combustion, the reliable characterization of reactants and reaction products is critical, such as incomplete combustion, impurities in the reactants which are often ill-defined, and more than one reaction gaseous species and condensed phases. Combustion calorimeters are usually of isoperibol type around room temperature in a water bath.

In the Knudsen effusion method (vapor pressure method) [21], a small amount of volatile species in the gas phase effuses through a small orifice of 0.1–1 mm with negligible influence of the equilibrium in the Knudsen cell. The vapor is ionized and analyzed with a mass spectrometer. The partial pressure of a species can be calculated from its ionization area and intensity through a calibration factor determined by a reference material with known partial pressure.

11.2.2.3 Thermochemical Data from Empirical Estimation

The thermochemical data can also be predicted using empirical approaches, i.e., the enthalpy of formation estimated by Miedema's method [22], Le Van's method, or Slobodin's method, see [14] for more details. The thermochemical properties can be predicted by first-principles calculations in the case that experimental measurements are not available. The accuracy from first principles calculations are still not comparable with the direct experimental measurements in most cases. However, with the significant progress in last decade, the accuracy has been greatly improved.

11.2.3 Fundamentals of CALPHAD Modeling

One common expression for Gibbs energy of pure elements and stoichiometric compounds at finite temperatures and ambient pressure is as follows:

$$G = a + bT + cT\ln T + dT^2 + eT^{-1} + fT^3 \tag{11.10}$$

where a, b, c, d, e, and f are model parameters. The expressions for other thermodynamic properties can be obtained from G, for example (see Eq. 11.9):

$$C_P = -c - 2dT - 2eT^{-2} - 6fT^2. \tag{11.11}$$

For solution phases such as the substitutional solutions of liquid, bcc, fcc, and hcp, the Gibbs energy is written as follows [2,3]:

$$G = \sum_i x_i G_i^0 + RT\sum_i x_i \ln x_i + G^{ex} \tag{11.12}$$

where G_i^0 is the Gibbs energy of the pure element i usually expressed by Eq. 11.10 with functions from the SGTE database [23] commonly used, x_i the mole fraction of i, and R the gas constant. G^{ex} is the excess Gibbs energy, expressed using the Redlich-Kister polynomial [24]

$$G^{ex} = \sum_i \sum_{<j} x_i x_j \sum_{n=0} L_{ij}^n (x_i - x_j)^n + \sum_i \sum_{<j} \sum_{<k} x_i x_j x_k L_{ijk} + \dots \tag{11.13}$$

where L_{ij}^{n} is the nth binary interaction parameter between elements i and j. L_{ijk} represents the ternary interaction parameter among elements i, j, and k, i.e., $L_{ijk} = x_i L_{ijk}^0 + x_j L_{ijk}^1 + x_k L_{ijk}^2$. These L parameters may depend on temperature

$$L = A + BT + CT \ln T \tag{11.14}$$

where A, B, and C are the modeling parameters to be evaluated.

For the convenience of CALPHAD modeling, the Gibbs energy of a stoichiometric compound can be expressed by

$$G = \sum_i x_i G_i^0 + \Delta G_f \tag{11.15}$$

where G_i^0 is the Gibbs energy of pure element i. The Gibbs energy of formation ΔG_f is temperature dependent and follows the same format as Eq. 11.10

$$\Delta G_f = a + bT + cT \ln T + dT^2 + eT^{-1} + fT^3 \tag{11.16}$$

For non-stoichiometric compounds, one typically adopts the sublattice model, for details see [2,3]. The method developed by Hillert and Jarl [25] is commonly used by CALPHAD community to incorporate magnetic contributions into thermodynamic descriptions of phases, and was recently modified by Palumbo et al. [26].

11.2.4 Fundamental of First-Principles Calculations

First-principles calculations, such as the one based on density functional theory (DFT) [11], only require the knowledge of atomic species and crystal structure to define the energetics of the structure and hence are predictive in nature. In DFT the total energy of a many-electron system in an external potential is expressed by a unique functional of the electron density $\rho(\vec{r})$, and the total energy has its minimum at the ground-state electron density [27]. The Kohn-Sham one-electron equation for DFT is [13]

$$\hat{H}_{KS}\psi_i = \varepsilon_i \psi_i \tag{11.17}$$

where ψ_i are wave functions of single particles and ε_i the single-particle energies. The Kohn-Sham Hamiltonian is

$$\hat{H}_{KS} = -\frac{\hbar^2}{2m_e}\vec{\nabla}_i^2 + \frac{e^2}{4\pi\varepsilon_0}\int \frac{\rho(\vec{r}')}{|\vec{r}-\vec{r}'|} d\vec{r}' + V_{xc} + V_{ext} \tag{11.18}$$

where the first term is the kinetic energy operator for the electrons with mass m_e at position $\vec{r}$, the second term describes the Coulomb interactions between electrons, V_{xc} is the exchange-correlation operator, and the system-specific information of atomic species and crystal structure is contained within the external potential V_{ext}, i.e., the electron-ion interaction. Unlike the Hartree-Fock method, the Kohn-Sham Hamiltonian treats only the exchange exactly and does not include the correlation.

In DFT, both exchange and correction are treated approximately. To solve Eq. 11.17, we first need to express the wave function ψ_i by a given basis set, such as the plane wave basis set used in the pseudopotential method of the VASP code [28,29]. Secondly, we need a method to account for the electron-ion interactions, such as the ultrasoft pseudopotentials (USPP) or the projector augmented wave method (PAW) [29,30] used in the VASP code. Finally, we need a method to describe the exchange-correlation functional. The two widely used approximations herein include (i) the local density approximation (LDA) [31] and (ii) the improved LDA by adding the gradient of the density, i.e., the generalized gradient approximation (GGA) [32,33]. By construction, LDA is expected to adequately describe a slowly varying density, but it works well for most cases, especially in materials with band gaps (e.g., insulators) and for surface properties. GGA is typically superior to LDA for metallic systems. For example the ground-state of bcc Fe can be described correctly by GGA but not by LDA [34]. Generally speaking, GGA overestimates the lattice parameters while LDA underestimates them, and both GGA and LDA underestimate the band gap widths. In order to overcome the limitations of GGA and LDA, several improved exchange-correlation functionals have been developed, such as the PBEsol GGA [35] which uses a reduced gradient dependence for better description of surface and solid properties. Hybrid functionals, which mix the (partial) exact exchange energy with a DFT exchange functional, offer a better description of band gaps and other properties. Examples of hybrid functionals include PBE0 [36], HSE06 [37], and HSEsol [38], etc.

Two parameters are extremely important to the practice of first-principles calculations: the *k*-point sampling mesh and the energy cutoff for the wave functions. "Bigger is better" for both of these parameters and will yield results with improved convergence; however, an increase in convergence requires more computational time. For more details of DFT-based first-principles calculations, see [27,34,39,40].

11.3 APPLICATION OF COMPUTATIONAL THERMODYNAMICS

11.3.1 Multi-component Mg Alloy Thermodynamic Databases from Various Sources

The thermodynamic databases available for magnesium alloys together with the elements included are listed in Table 11.1, two of which (TCMG3 and PanMagnesium) are commercial databases from Thermo-Calc [41] and CompuTherm [42], respectively. Besides the commercial ones, magnesium alloy databases developed by groups of Du [43] and Liu [44] are listed in Table 11.1. In addition, a 19-element magnesium database has been developed by the group in Penn State based on available publications and the critical thermodynamic modeling of binary and ternary systems [45–47], which contains two different features as described below. First, the predicted first-principles energetics have been used in evaluation of model parameters for compounds and solid solutions when the energetic data are inaccurate or absent in the literature in some key systems. For instance, the thermodynamic modeling of the Al-Mg system used the data of first-principles enthalpies of formation for ε-$Al_{30}Mg_{23}$, γ-$Al_{12}Mg_{17}$ and three Laves phases and the enthalpies of mixing of fcc and hcp solution phases predicted by first-principles

TABLE 11.1
Elements Included in the Available Thermodynamic Database for Mg Alloys

Source	Elements																									
TC[41]	Mg	Ag	Al		Ca	Ce	Cu		Fe	Gd	K	La	Li	Mn	Na	Nd	Ni	Pr	Sc		Si	Sn	Sr	Th	Y	Zn	Zr
Pan[42]	Mg	Ag	Al	C	Ca	Ce	Cu	Dy	Fe	Gd		La	Li	Mn		Nd	Ni	Pr	Sc	Se	Si	Sn	Sr		Y	Zn	Zr
Du[43]	Mg		Al						Fe					Mn							Si						
Liu[44]	Mg		Al			Ce																			Y	Zn	
PSU	Mg		Al			Ce	Cu		Fe		K	La	Li	Mn	Na	Nd					Si	Sn	Sr		Y	Zn	Zr

Source: Thermocalc Software, Stockholm, Sweden, http://www.Thermocalc.com; CompuTherm LLC, Madison, Wisconsin, http://www.computherm.com; Du, Y., *Z. Metallkd*, 96, 1351–1362, 2005; Liu, X. et al., *Rare Metals*, 25, 441–447, 2006.

special quasirandom structures [46,48]. Second, the alkali elements sodium and potassium, together with iron, are involved in order to analyze the influence of impurities in Al-Mg alloys such as the high temperature embrittlement (HTE) as discussed below [47,49].

11.3.2 Equilibrium Calculations with Thermodynamic Databases

With the Gibbs energy description of individual phases, the calculation of phase equilibria and thermochemical properties can be carried out by the minimization of Gibbs energies, which are the two types of equilibrium calculations using thermodynamic databases.

11.3.2.1 Phase Equilibria Prediction

For binary systems, the binary phase diagram can be plotted out as shown in Figure 11.1 for Mg-Sr system [50] in comparison with the experimental liquidus measurements. However, some of the thermochemical properties are still assigned arbitrary values due to the lack of experimental data. Table 11.2 summarizes the modeled L parameters (see Eq. 11.12 through Eq. 11.14) for the liquid and hcp phases of the Mg-X binary systems from a thermodynamic database developed for Mg-based alloys [1], where X is one of the alloying elements Al, Ca, Ce, Cu, Fe, K, La, Li, Mn, Na, Nd, Pr, Si, Sn, Sr, Y, Zn or Zr. Two or three L parameters are typically used to describe the liquid and hcp phases, and one to three fitting parameters are employed for each L parameter.

For multi-component systems, calculations such as isothermal section, isopleth section, liquidus and solidus prediction can be carried out as shown in Figure 11.2 for the Ca-Mg-Sr ternary system [50]. The thermodynamic databases for multi-component systems were used to compare with the experimental observations.

The following three examples are taken from works validating the database PanMagnesium [51,52]. The invariant temperatures from 11 ternary systems were

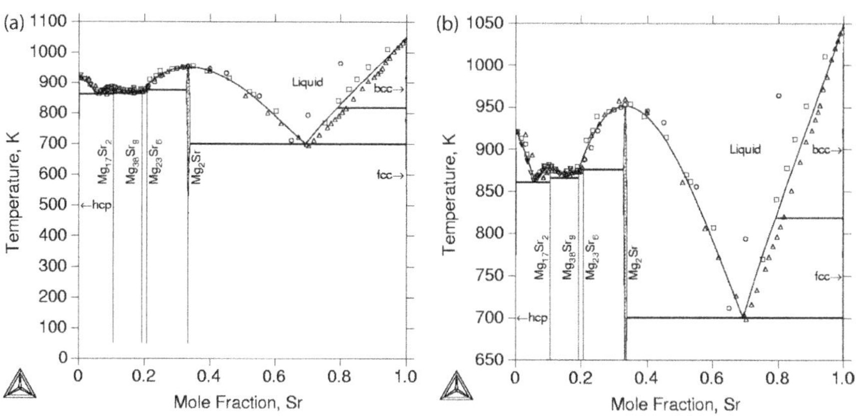

FIGURE 11.1 Comparison between the calculated Mg-Sr (a) phase diagram and (b) the experimental data. (From Zhong, Y. et al., *J. Alloy. Compd.*, 421, 172–178, 2006.)

TABLE 11.2
Modeled *L* Parameters of Liquid and hcp Phases for Mg-X Binary Systems

System and References	Phase	L^0	L^1	L^2
Mg-Al [49]	liquid	−9031 + 4.855T	−891 + 1.137T	434
	hcp	4272–2.190T	−1.077–1.015T	−965
Mg-Ca [90]	liquid	−32322.4 + 16.721T	60.3 + 6.549T	−5742.3 + 2.760T
	hcp	−9183.2 + 16.981T		
Mg-Ce [91]	liquid	−36703 + 13.831T	30962–17.297T	−15090
	hcp	−94338 + 79.952T		
Mg-Cu [92]	liquid	−36962.7 + 4.744T	−8182.19	
	hcp	22500-3T		
Mg-Fe [93]	liquid	61343 + 1.5T	−2700	
	hcp	92400		
Mg-K [47]	liquid	37272.2		
	hcp	−1072.6		
Mg-La [94]	liquid	−32472.5 + 8.367T	35610.1–24.012T	−13162.4
	hcp	−25000		
Mg-Li [93]	liquid	−14935 + 10.371T	−1789 + 1.143T	6533–6.6915T
	hcp	−6856	4000	4000
Mg-Mn [95]	liquid	+25922.4 + 9.036T	−3470.8	
	hcp	+37148–1.810T		
Mg-Na [47]	liquid	26025.779	4509.96384	
	hcp	75698.31		
Mg-Nd [96]	liquid	−43547.1 + 19.415T	−30060.5 + 10.775T	−26879.5 + 22.230T
	hcp	−13200		
Mg-Pr [97]	liquid	−41498.5 + 13.863T	−38739.5 + 27.386T	
	hcp	−10000		
Mg-Si [93][a]	liquid	−83864.3 + 32.4T	18027.4–19.612T	2486.67–0.311T
	hcp	−7148.79 + 0.894T		
Mg-Sr [50]	liquid	−18647.1 + 9.070T	−12831.6 + 7.509T	
	hcp	10000		
Mg-Y [98]	liquid	−41165.3 + 17.564T	−15727.0 + 4.705T	
	hcp	−26612.8 + 13.946T	−2836.2	
Mg-Zn [99]	liquid	−81439.7 + 518.25T–64.71Tln(T)	2627.5 + 2.931T	−1673.3
	hcp	−1600.8 + 7.624T	−3823.0 + 8.026T	
Mg-Zr [100]	liquid	4961.4 + 38.180T		
	hcp	30384.0 + 13.723T	18588.9–25.175T	

Note: See Eq. 11.13 and Eq. 11.14.

[a] For liquid phase L^3 = 18541.2–2.318T; L^4 = −12338.8 + 1.542T

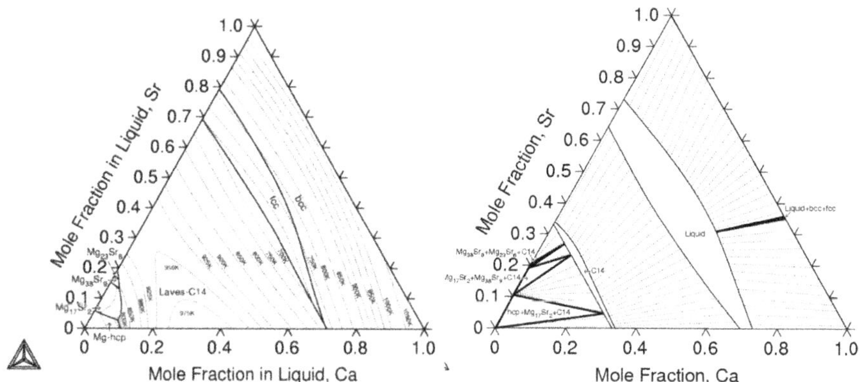

FIGURE 11.2 Liquidus projection to the composition triangle in the Ca-Mg-Sr system with isotherms (dotted lines) superimposed (left), Calculated isothermal section of the Ca-Mg-Sr system at 750 K (right). (From Zhong, Y. et al., *J. Alloy. Compd.*, 421, 172–178, 2006.)

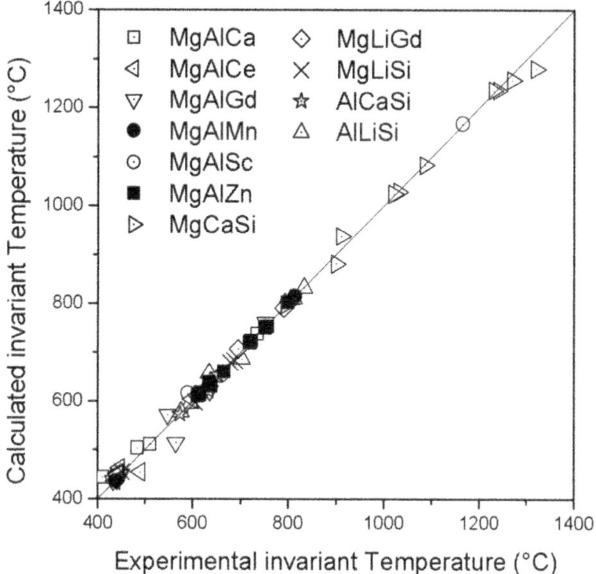

FIGURE 11.3 Invariant temperatures for various ternary alloys. (From Schmid-Fetzer, R. et al., *Essential Readings in Magnesium Technology*, 451–456, 2007.) comparison between calculated and experimental data.

plotted in comparison with experimental data (Figure 11.3 [51]), which shows very good agreement. The liquidus projection and isothermal sections for Mg-Al-Zn ternary system were plotted (Figure 11.4 [51]), which shows reasonable agreement with experimental data with the primary solidifying phases. The isopleth section of $Mg_{65}Al_{35}$ to Sr corner was plotted (Figure 11.5 [52]), which shows good agreement with experimental DSC analysis for the composition range close to $Mg_{65}Al_{35}$. The disagreement was

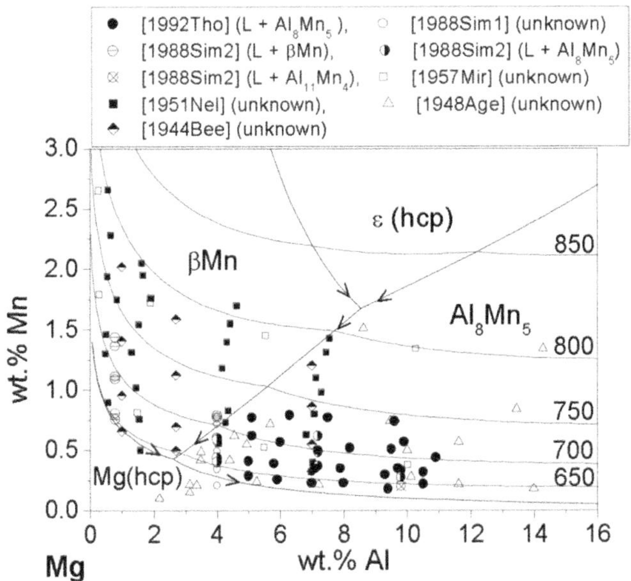

FIGURE 11.4 Calculated partial liquidus surface with the compositions of experimental alloys superimposed. The primary solid phase is specified in some experimental data. (From Schmid-Fetzer, R. et al., *Essential Readings in Magnesium Technology*, 451–456, 2007.)

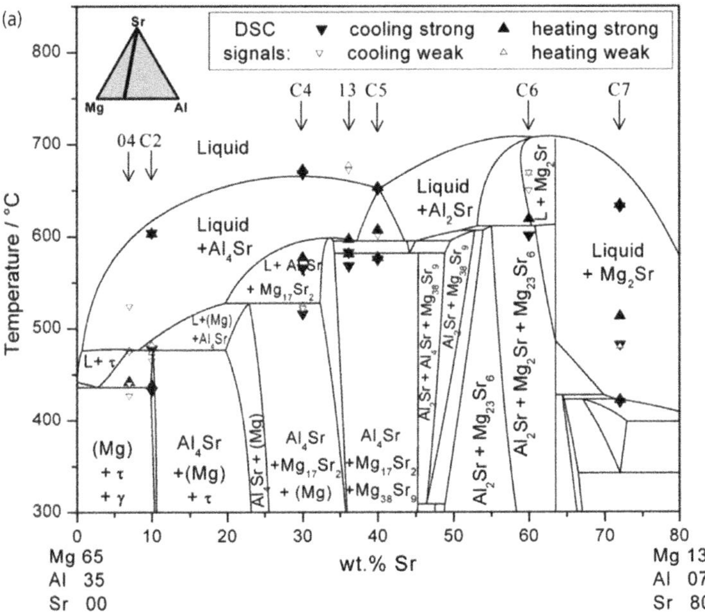

FIGURE 11.5 Calculated isopleth in the Mg-Al-Sr system from $Mg_{65}Al_{35}$ to the Sr-corner. (From Janz, A. et al., *Intermetallics*, 15, 506–519, 2007.)

observed in the composition range higher than 70wt%Sr, which indicated the possible existence of a new ternary Sr rich phase. This can be solved with the inclusion of the new ternary phase in the thermodynamic database in the future [51].

11.3.2.2 Calculations of Thermochemical Properties

With the thermodynamic database, the thermochemical properties such as the enthalpy of mixing, enthalpy of formation, and activity can be calculated. For each compound presented in the Mg-X binary systems from the aforementioned databases [41–44] in Table 11.1, the enthalpy and entropy of formation have been modeled with the CALPHAD method and compared with first principles (at 0 K). For most cases, the ΔH values modeled with the CALPHAD method agree reasonably well with those of first-principles predictions, indicating the capabilities of first-principles methodology. For more details regarding these comparisons, see [53].

11.3.3 Non-equilibrium Applications of Thermodynamic Databases

Besides the equilibrium calculations listed in 11.3.2, there are several non-equilibrium applications.

11.3.3.1 Solidification Simulation

The equilibrium calculations and Scheil simulations were used to simulate cooling processes. In equilibrium calculations, the global equilibrium is assumed at each temperature. In the Scheil simulations, it is assumed that (1) the liquid phase is homogenous all the time; (2) there is no diffusion in the solid; and (3) Local equilibrium exists at the liquid/solid interface. Scheil simulation is a reasonable approximation because diffusion in the solid is slow, whereas the composition of the liquid should be almost homogeneous due to rapid diffusion and convection. The equilibrium calculations and the Scheil simulations represent two extreme situations, while the practical casting sample should be between these two simulations. One example is the phase evolution in AZ61 (Mg-6Al-1Zn in wt.%) alloy. Figure 11.6 shows the distributions of alloying elements aluminum and zinc in the solid (hcp), liquid, and phases in AZ61 alloy as a function of temperature predicted by equilibrium calculation and Scheil model [1]. However, it has been experimentally demonstrated that for typical AZ alloys the solidification follows more closely the Scheil conditions even for cooling rates as low as 1 K/min [54].

11.3.3.2 High Temperature Embrittlement

The thermodynamic database for aluminum/magnesium alloys can also used to understand the sodium-induced high-temperature embrittlement (HTE) in Al alloys, where sodium is an undesirable impurity introduced in the normal manufacturing process [47,49]. Although sodium is present only in a very small amount in Al-Mg alloys, it causes HTE due to the formation of an intergranular sodium-rich liquid phase, weakening significantly the strength of grain boundaries. To understand the mechanism of sodium-induced HTE in Al + 5wt.% magnesium alloys with variable Na contents, for example, the (Al + 5Mg)-Na phase diagram is calculated and shown in Figure 11.7 [47,49]. The HTE is closely related to the formation of the

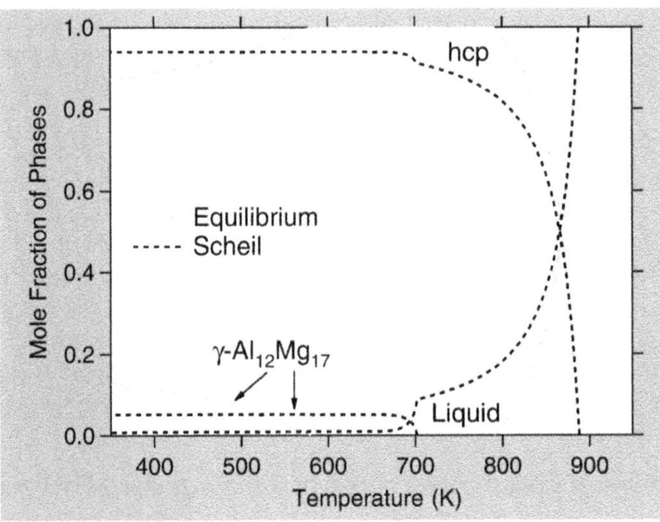

FIGURE 11.6 Predicted fractions of solid (hcp), liquid, and phases in AZ61 alloy by equilibrium and Scheil models. (From Shang, S. et al., *JOM*, 60, 45–47, 2008.)

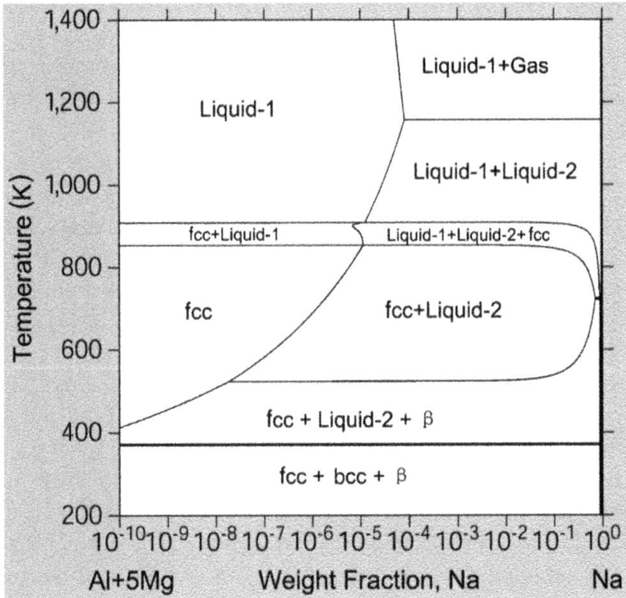

FIGURE 11.7 Calculated isopleth of the Al (+ 5 wt.% Mg)- Na system. (From Zhang, S., Thermodynamic Investigation of Alkali-Metal Impurities on the Processing of Al and Mg Alloys, Ph. D. Thesis, Pennsylvania State University, 2006; Zhang, S. et al., *Philos. Mag.*, 87, 147–157, 2007.)

liquid-2 phase of almost pure Na (most likely at the grain boundaries) [47,49], and therefore the hot-rolling safe zone is located in the single-phase fcc region as shown in Figure 11.7. High-temperature embrittlement can be suppressed by avoiding the formation of liquid-2 phase by either lowering the Na content or quenching from temperatures where the liquid-2 phase is not stable. The phase boundary between fcc and "fcc + liquid-2" thus separates the hot-rolling safe zone and the sensitive zone, respectively.

11.3.3.3 Prediction of New Bulk Metallic Glass

Bulk metallic glass (BMG) can be easily formed around a eutectic reaction region. The liquidus of the Mg-Ca-Zn ternary system was plotted as shown in Figure 11.8 in comparison with the experimental compositions of BMG, from the work of Kyeong et al. [55]. Four blue squares correspond to the eutectic reaction compositions, while red, green and blue cycles correspond to the BMG compositions reported in the literature. Based on the calculations of the liquidus projection and invariant reactions, the new Zn-based BMG was successfully made [55]. Similarly, the W-Fe-Hf-Pd-Ta and W-Fe-Si-C based thermodynamic database was developed to search for potentially W-based metallic glass with high glass forming ability as shown in Figure 11.9 [56]. Furthermore, by combining the CALPHAD method with kinetic approaches, the glass forming ability (GFA) can be evaluated in terms of critical cooling rate and fragility. The driving forces for crystallization from the undercooled

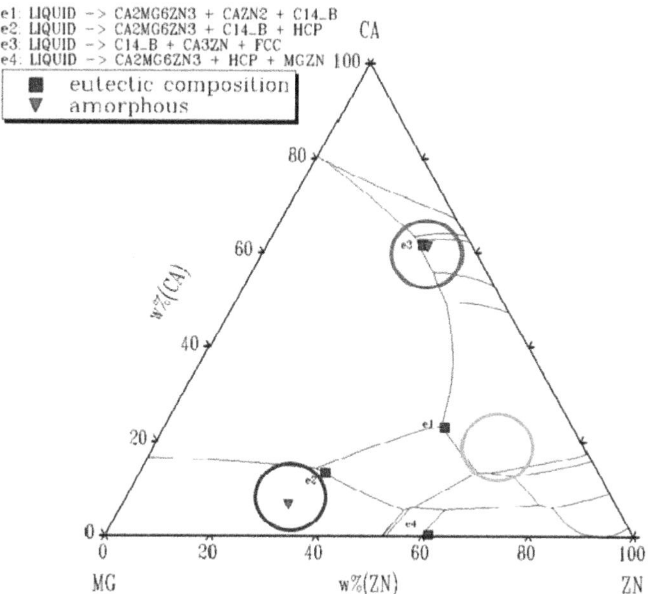

FIGURE 11.8 Bulk metallic glass formation regions on the calculated liquid projection and 4 eutectic reaction in the Mg-Ca-Zn ternary system. (From Kyeong, J.S. et al., Thermodynamic assessment and its application of ternary Mg-Zn-Ca system using CALPHAD method, in TMS The Minerals, Metals & Materials Society, *Magnes. Technol.*, 167–172, 2009.)

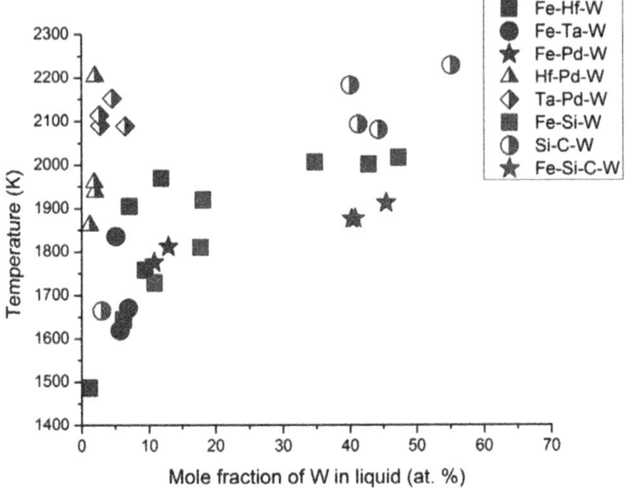

FIGURE 11.9 Summary of invariant reactions with W compositions larger than 1 at.% and reaction temperatures lower than 2300 K. (From Hu, Y.J. et al., *Intermetallics*, 48, 79–85, 2014.)

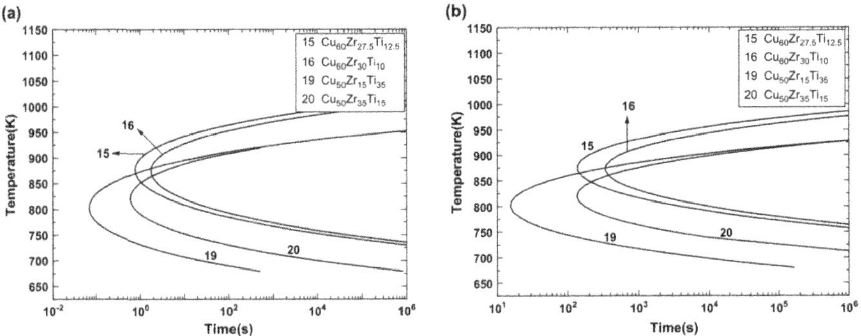

FIGURE 11.10 Calculated TTT curve of four Cu-Zr-Ti alloys by (a) Turnbull model and (b) TS model. (From Ge, L. et al., *Acta. Phys-Chim. Sin.*, 23, 895–899, 2007.)

liquid alloys can be calculated by using Turnbull and Thompson-Spaepen (TS) Gibbs free energy approximate equations, respectively. Time-temperature-transformation (TTT) curves of these alloys can be obtained with Davies-Uhlmann kinetic equations based on classical nucleation theory. Such approach has been successfully applied to Cu-based and W-based BMG as shown in Figure 11.10 [57–59], which can be used to explore other BMG-forming alloys as well.

11.3.3.4 Linkage with Mechanical Properties

Some mechanical properties such as creep resistance can be linked with the computational thermodynamics. One example is the creep resistance improvement in the Mg-Sc-Mn system. Sc is added into Mg because of several reasons: the melting point of the Mg-hcp solid solution increases with the addition of Sc; Sc has higher melting temperature comparing with other rare earth elements indicating a lower diffusivity in magnesium; the density of Sc (3 g/cm^3) is lower than alternative alloying elements. In addition, Mn was added to reduce the amount of Sc needed based on the prediction from computational thermodynamics. Figure 11.11a shows the calculated isothermal section [60]. It shows with the addition of Mn, the high thermal stability Mn_2Sc phase will precipitate out, which will increase the creep resistance of Mg-Sc alloys. Figure 11.11b shows creep curves of various Mg-Sc-based alloys together

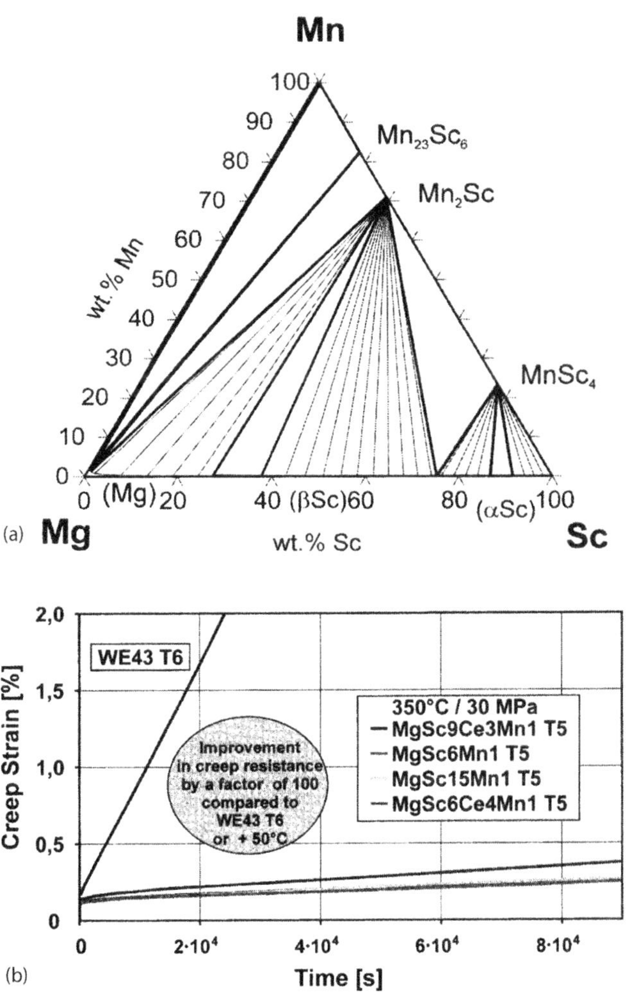

FIGURE 11.11 (a) Calculated isothermal section of the Mg-Sc-Mn system. (From Pisch, A. et al., in *Magnesium Alloys and Their Applications*, Mordike, B.L. and Kainer, K.U. (Eds.), pp. 181–186, 1998.) (b) Creep curves of new developed Mg alloys. (From Buch von, F., Entwicklung hochkriechbeständiger Magnesiumlegierungen des Typs Mg–Sc(-X-Y), Mg–Gd, undMg–Tb Dissertation, Technische Universität Clausthal, 1999.)

with that for WE43 T6 [61]. These curves were measured at 350°C and 30 MPa. It can be seen that the creep rate of Mg-Sc-Mn alloys is two orders of magnitude lower than that of WE43.

Similar work has been carried out to explain the creep resistance improvement with the Ca addition to the Mg-Al based Mg alloys [46]. In 2000s, the work by General Motors and verified that the addition of Ca on the improvement of the creep resistance of Mg-Al based alloys [62], owing to the replacement of the γ-$Al_{12}Mg_{17}$ phase by more stable Laves phases [63,64].

Figure 11.12a [46] shows the simulation results for the Mg-4.5wt.% Al alloy. In the equilibrium calculation, the liquid phase solidifies completely into Mg-hcp phase at 850 K and the γ phase will precipitate from Mg-hcp matrix when the temperature reaches 525 K. It will form 9at.% γ. However, in Scheil simulation, the liquid phase will transform into 97at.% hcp and 3at.% γ, where the γ phase forms at 711 K. To improve the creep resistance of AM50 alloys at elevated temperatures, two criteria were used to measure the effect of alloying elements, i.e., decreasing the stability of γ-$Al_{12}Mg_{17}$ phase and increasing the solidification temperature [65].

The addition of calcium to Mg-Al based alloys was reported to improve the creep resistance in 1960 [66]. It claimed the beneficial effect of calcium content of 0.5~3wt.%. Later work in 1970s claimed a creep improvement with addition about 1wt.%Ca [67]. In 1996, Mg-5wt.%Al-0.8wt.%Ca alloy (AX51) was developed which provided similar creep resistance to AE42 [68]. Figure 11.12b [46] shows the simulation results for Mg-4.5wt.%Al-1wt.%Ca alloy. In the equilibrium calculation, the liquid phase transformed into Mg-hcp, 7at.% γ, and 1.8at.% C15. In the Scheil simulation, the liquid phase transformed into Mg-hcp, 1.8at.% C15, and 1at.% γ. The Scheil solidification temperature is 709 K. As the amount of the γ phase is much lower than that in the Mg-4.5wt.%Al alloy, its creep resistance should be better than Mg-4.5wt.%Al alloy, as shown by experimental observation [68]. Its creep resistance

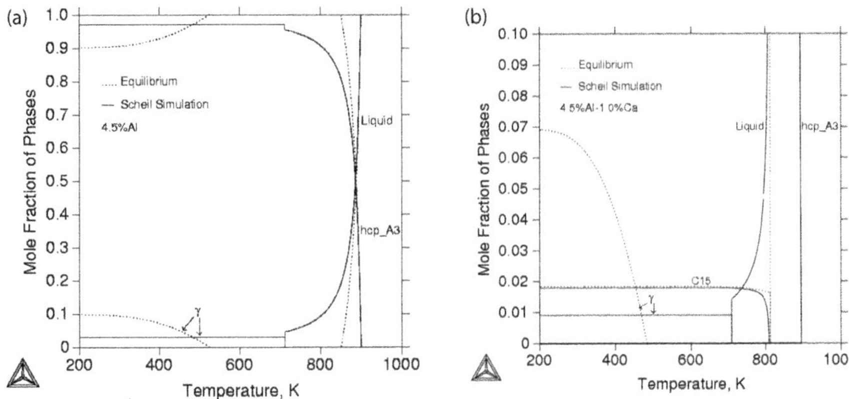

FIGURE 11.12 Calculate phase fractions as a function of temperature for the Mg-4.5wt.%Al-1.0wt.%Ca alloy. (From Zhong, Y., Investigation in Mg-Al-Ca-Sr-Zn System by Computational Thermodynamics Approach Coupled with First-Principles Energetics and Experiments, Ph. D. Thesis, The Pennsylvania State University, 2005.)

can be further improved if the γ phase is completely substituted by the C15 phase and its solidification temperature increased to much higher than Mg-4.5wt.%Al alloy, which can be achieved by adjusting alloy compositions [46].

11.3.4 Applications of First-Principles Calculations

First-principles calculations can be used to predict the existence of new phases in multi-component systems. For example, there were two ternary phases predicted by the first-principles calculations and then verified by experiments in the Mg-Al-Ca ternary system. It was first reported by Luo et al. that (Al, Mg)$_2$Ca exists in Mg-Al-Ca alloy [62], however, it was only identified it is not C15 phase and has hexagonal structure. Systematic first principles calculations then have been carried out for the lave phase structures in the composition range of $(Al_xMg_{1-x})_2Ca$ in Mg-Al-Ca system. It indicated the existence of a possible C36 lave phase with the composition of (Al, Mg)$_2$Ca. The FP prediction was reported in 2004 TMS conference [69,70] and was later experimentally verified by detail TEM analysis in [71–73].

Similarly the total energies and the enthalpies of formation of the three Laves crystal structures, i.e., C14, C15, and C36, at Al_2Mg and Al_2Ca compositions were calculated as reported in Table 11.3. The enthalpies of formation of the three Laves phase structures of Al_2Mg vary from −2.301 kJ/mol-atom to −2.802 kJ/mol-atom, while those of Al_2Ca vary from −32.722 kJ/mol-atom to −33.981 kJ/mol-atom. The C15 structure, which has the lowest enthalpy of formation at the Al_2Ca composition, is the stable phase in the Al-Ca binary system. The C36 structure has the lowest enthalpy of formation at the Al_2Mg composition among the three Laves phase structures, while none of them is stable in the Al-Mg system. However, the enthalpy of formation of C14 and C36 is only 0.312 kJ/mol-atom and 0.223 kJ/mol-atom higher than the stable convex hull according to the latest thermodynamic database of the Al-Mg binary system [49].

TABLE 11.3
Structure Properties and Enthalpies of Formation for the Al-Mg Binary System in a Variety of Order Structures

	Lattice Constants (Å)	Total Energy (eV/atom)	Enthalpies of Formation (kJ/mol-atom)
hcp-Mg	a = b = 3.177, c = 5.172	−1.4852	/
fcc-Al	a = b = c = 4.041	−3.6892	/
fcc-Ca	a = b = c = 5.486	−1.9477	/
C14-Al_2Ca	a = b = 5.670, c = 9.231	−3.4478	−32.722
C15-Al_2Ca	a = b = c = 8.005	−3.4609	−33.981
C36-Al_2Ca	a = b = 5.686, c = 18.347	−3.4571	−33.617
C14-Al_2Mg	a = b = 5.448, c = 8.742	−2.9827	−2.713
C15-Al_2Mg	a = b = c = 7.667	−2.9784	−2.301
C36-Al_2Mg	a = b = 5.450, c = 17.514	−2.9836	−2.802

Source: Zhong, Y. et al., *Scripta. Mater.*, 55, 573–576, 2006.

The calculations show that there is a very small difference in enthalpies of formation among these three Laves phase structures. The maximum difference is only 0.501 kJ/mol-atom for Al_2Mg and 1.259 kJ/mol-atom for Al_2Ca. These differences are small considering the uncertainty of first-principles calculations, viz., 1 kJ/mol-atom, and indicate that a ternary Laves phase Al_2(Mg, Ca) may exist in the Al_2Ca-Al_2Mg section. Simple experiments with a diffusion couple between Mg-30 wt.% Ca alloy and pure Al verified the existence of the Al_2(Mg, Ca) Laves phase [74].

11.4 FUTURE TRENDS

Computational thermodynamics has made significant progress in the past four decades because of the advance in methodologies and computer recourses. In addition to data from experiments and estimations, thermodynamic data from first-principles calculations are playing more and more important roles. Furthermore, CALPHAD modeling of kinetic properties broadens the applications of computational thermodynamics to meso-scale kinetic simulations and macro-scale kinetic simulations. In particular, the ongoing and future research trends include the following:

1. With the advance of first-principles calculations, more properties will be predicted directly theoretically, such as Helmholtz energy for non-stable phases; defect energies, which include vacancy formation energy, stacking fault energy, the generalized stacking fault energy, anti-phase boundary energy, and surface and interfacial energies [75–78]; diffusion coefficient of solid phases [79–83]; creep rate [84]; tensile and shear strength [85,86]; solute strengthening [87,88].
2. With support from first-principles calculations, the thermodynamic properties can be predicted efficiently. The existing thermodynamic database of many assessed systems need to be revisited [48]. Any changes on binary systems will have a snowball effect on the higher order systems. It is extremely difficult to make such modifications manually especially for multi-component thermodynamic databases. The automation of multi-component thermodynamic database development is necessary such as ESPEI approach [89]. The advantage of this approach is the database has the self-improvement capability when new information becomes available.
3. Kinetic simulations incorporated with computational thermodynamics will become more and more common. There can be three major categories of the treatment:
   - The state reactor modeling in which the overall reaction is separated into several stages, and the local equilibrium is assumed in each stage.
   - The meso-scale kinetics such as popular phase-field approaches in which thermodynamic driving force and kinetic parameters are directly taken from CALPHAD databases.
   - The macro-scale kinetics in successful industrial CFD (computational fluid dynamics) software such as Fluent and Aspen in which simulations can be further improved by incorporating the thermodynamic databases.

ACKNOWLEDGMENTS

This work has been funded by National Science Foundation through grants of 9983532, 0510180, and 1006557.

REFERENCES

1. Shang, S., Zhang, H., Ganeshan, S., and Liu, Z.K., The development and application of a thermodynamic database for magnesium alloys, *JOM*, 2008. 60(12): 45–47.
2. Saunders, N. and Miodownik, A.P., *CALPHAD (Calculation of Phase Diagrams): A Comprehensive Guide*. 1998, Pergamon, Oxford.
3. Lukas, H.L., Fries, S.G., and Sundman, B., *Computational Thermodynamics: The Calphad Method*. 2007, Cambridge University Press, Cambridge, UK.
4. Liu, Z.K., First-principles calculations and CALPHAD modeling of thermodynamics, *J. Phase. Equilib. Diff.*, 2009. 30(5): 517–534.
5. Olson, G.B., Computational design of hierarchically structured materials, *Science*, 1997. 277: 1237–1242.
6. Allison, J., Liu, B., Boyle, K., Beals, R., and Hector, L., Integrated computational materials engineering (ICME) for magnesium: An international pilot project, in *Magnesium Technology*. 2008. The Minerals, Metals and Materials Society, Warrendale, PA, 14–18.
7. Andersson, J.O., Helander, T., Hoglund, L.H., Shi, P.F., and Sundman, B., Thermo-Calc and Dictra, computational tools for materials science, *Calphad*, 2002. 26(2): 273–312.
8. Bale, C.W., Belisle, E., Chartrand, P., Decterov, S.A., Eriksson, G., Hack, K., Jung, I.H. et al., Fact Sage thermochemical software and databases: Recent developments, *Calphad*, 2009. 33(2): 295–311.
9. Cao, W., Chen, S.L., Zhang, F., Wu, K., Yang, Y., Chang, Y.A., Schmid-Fetzer, R., and Oates, W.A., PANDAT software with PanEngine, PanOptimizer and PanPrecipitation for multi-component phase diagram calculation and materials property simulation, *Calphad*, 2009. 33(2): 328–342.
10. Shang, S.L., Wang, Y., and Liu, Z.K., ESPEI: Extensible, self-optimizing phase equilibrium infrastructure for magnesium alloys, in *Magnesium Technology 2010*. 2010. The Minerals, Metals, and Materials Society, Warrendale, PA, 617–622.
11. Hohenberg, P. and Kohn, W., Inhomogeneous electron gas, *Phys. Rev.*, 1964. 136(3B): B864–B871.
12. Hillert, M., *Phase Equilibria, Phase Diagrams, and Phase Transformations: Their Thermodynamic Basis*, Second ed. 2008, Cambridge University Press, Cambridge, UK.
13. Kohn, W. and Sham, L.J., self-consistent equations including exchange and correlation effects, *Phys. Rev.*, 1965. 140(4A): A1133–A1138.
14. Kubaschewski, O., Alcock, C.B., and Spencer, P.J., *Materials Thermochemistry*, 6th ed. 1993, Pergamon press, New York.
15. Goodwin, A.R.H., Marsh, K.N., and Wakeham, W.A., eds. *Measurement of The Thermodynamic Properties of Single Phases*. 2003, Elsevier, Amsterdam, the Netherlands.
16. Weir, R.D. and Loos, T.W.D., eds. *Measurement of the Thermodynamic Properties of Multiple Phases*. 2005, Elsevier, Amsterdam, the Netherlands.
17. Zhao, J.C., *Methods for Phase Diagram Determination*. 2007, Elsevier Science, Amsterdam, the Netherlands.
18. Sarge, S.M., Höhne, G.W.H., Hemminger, W. *Calorimetry: Fundamentals, Instrumentation and Applications*. 2014: WILEY-VCH, Weinheim, Germany.

19. Boettinger, W.J., U.R, K., Moon, K.-W., and Perepezko, J.H., DTA and Heat-flux DSC Measurements of Alloy Melting and Freezing. Vol. NIST Special Publication 960–15. 2006: U.S. Department of Commerce, Washington, DC, pp. 90.
20. Marsh, K.N. and O'Hare, P.A.G., eds. *Solution Calorimetry*. 1994, Blackwell Scientific Publications.
21. Copland, E.H. and Jacobson, N.S., Measuring thermodynamic properties of metals and alloys with Knudsen effusion mass spectrometry, 2010.
22. Miedema, A.R., Dechatel, P.F., and Deboer, F.R., Cohesion in alloys: Fundamentals of a semi-empirical model, *Physica B & C*, 1980. 100(1): 1–28.
23. Dinsdale, A.T., SGTE data for pure elements, *Calphad*, 1991. 15: 317–425.
24. Redlich, O. and Kister, A.T., Algebraic representation of thermodynamic properties and the classification of solutions, *Ind. Eng. Chem. Res.*, 1948. 40(2): 345–348.
25. Hillert, M. and Jarl, M., Model for alloying effects in ferromagnetic metals, *Calphad*, 1978. 2(3): 227–238.
26. Palumbo, M., Burton, B., Costae Silva, A., Fultz, B., Grabowski, B., Grimvall, G., Hallstedt, B. et al., Thermodynamic modelling of crystalline unary phases, *Physica Status Solidi (b)*, 2014. 251(1): 14–32.
27. Hafner, J., Atomic-scale computational materials science, *Acta Materialia*, 2000. 48(1): 71–92.
28. Kresse, G. and Furthmuller, J., Efficient iterative schemes for ab initio total-energy calculations using a plane-wave basis set, *Phys. Rev. B*, 1996. 54(16): 11169–11186.
29. Kresse, G. and Joubert, D., From ultrasoft pseudopotentials to the projector augmented-wave method, *Phys. Rev. B*, 1999. 59(3): 1758–1775.
30. Blochl, P.E., Projector augmented-wave method, *Phys. Rev. B*, 1994. 50(24): 17953–17979.
31. Perdew, J.P. and Zunger, A., Self-interaction correction to density-functional approximations for many-electron systems, *Phys. Rev. B*, 1981. 23(10): 5048–5079.
32. Perdew, J.P. and Wang, Y., Accurate and simple analytic representation of the electron-gas correlation-energy, *Phys. Rev. B*, 1992. 45(23): 13244–13249.
33. Perdew, J.P., Burke, K., and Ernzerhof, M., Generalized gradient approximation made simple, *Phys. Rev. Lett.*, 1996. 77(18): 3865–3868.
34. Jones, R.O. and Gunnarsson, O., The density functional formalism, its applications and prospects, *Rev. Mod. Phys.*, 1989. 61(3): 689–746.
35. Perdew, J.P., Ruzsinszky, A., Csonka, G.I., Vydrov, O.A., Scuseria, G.E., Constantin, L.A., Zhou, X.L., and Burke, K., Restoring the density-gradient expansion for exchange in solids and surfaces, *Phys. Rev. Lett.*, 2008. 100(13): 136406.
36. Perdew, J.P., Emzerhof, M., and Burke, K., Rationale for mixing exact exchange with density functional approximations, *J. Phys. Chem.*, 1996. 105(22): 9982–9985.
37. Krukau, A.V., Vydrov, O.A., Izmaylov, A.F., and Scuseria, G.E., Influence of the exchange screening parameter on the performance of screened hybrid functionals, *J. Phys. Chem.*, 2006. 125(22): 224106.
38. Schimka, L., Harl, J., and Kresse, G., Improved hybrid functional for solids: The HSEsol functional, *J. Phys. Chem.*, 2011. 134(2): 024116.
39. Cottenier, S., *Density Functional Theory and The Family of (L)APW-Methods: A Step-by-Step Introduction*. 2002, KU Leuven, Brussels, Belgium.
40. Martin, R.M., *Electronic Structure: Basic Theory and Practical Methods*. 2004, Cambridge University Press, Cambridge.
41. Thermocalc Software, Stockholm, Sweden, http://www.Thermocalc.com.
42. CompuTherm LLC, Madison, Wisconsin, http://www.computherm.com.
43. Du, Y., Thermodynamic description of the Al-Fe-Mg-Mn-Si system and investigation of microstructure and microsegregation during directional solidification of an Al-Fe-Mg-Mn-Si alloy, *Z. Metallkd.*, 2005. 96: 1351–1362.

44. Liu, X., Wang, C., Wen, M., Chen, X., and Pan, F., Thermodynamic database of the phase diagrams in the Mg-Al-Zn-Y-Ce system, *Rare Metals*, 2006. 25(5): 441–447.
45. Ozturk, K., Investigation in Mg-Al-Ca-Sr system by Computational Thermodynamics Approach Coupled with First-principles Energetics and Experiments, Ph. D. Thesis, Pennsylvania State University, 2003.
46. Zhong, Y., Investigation in Mg-Al-Ca-Sr-Zn System by Computational Thermodynamics Approach Coupled with First-Principles Energetics and Experiments, Ph. D. Thesis, The Pennsylvania State University, 2005.
47. Zhang, S., Thermodynamic Investigation of Alkali-Metal Impurities on the Processing of Al and Mg Alloys, Ph. D. Thesis, Pennsylvania State University, 2006.
48. Zhong, Y., Yang, M., and Liu, Z.-K., Contribution of first-principles energetics to Al–Mg thermodynamic modeling, *Calphad*, 2005. 29(4): 303–311.
49. Zhang, S., Han, Q., and Liu, Z.K., Fundamental understanding of Na-induced high temperature embrittlement in Al–Mg alloys, *Philosophical Magazine*, 2007. 87(1): 147–157.
50. Zhong, Y., Sofo, J.O., Luo, A.A., and Liu, Z.-K., Thermodynamics modeling of the Mg–Sr and Ca–Mg–Sr systems, *J. Alloy. Compd.*, 2006. 421(1–2): 172–178.
51. Schmid-Fetzer, R., Grobner, J., Mirkovic, D., Janz, A., and Kozlov, A., *Essential Readings in Magnesium Technology*. 2007. 451–456.
52. Janz, A., Gröbner, J., Mirković, D., Medraj, M., Zhu, J., Chang, Y.A., and Schmid-Fetzer, R., Experimental study and thermodynamic calculation of Al–Mg–Sr phase equilibria, *Intermetallics*, 2007. 15(4): 506–519.
53. Zhang, H., Shang, S.L., Saal, J.E., Saengdeejing, A., Wang, Y., Chen, L.Q., and Liu, Z.K., Enthalpies of formation of magnesium compounds from first-principles calculations, *Intermetallics*, 2009. 17(11): 878–885.
54. Mirković, D. and Schmid-Fetzer, R., Solidification curves for commercial Mg alloys determined from differential scanning calorimetry with improved heat-transfer modeling, *Metall. Mater. Trans. A*, 2007. 38(10): 2575–2592.
55. Kyeong, J.S., Lim, H.K., Lee, H.D., Kim, W.T., Kim, D.H., and Lee, B.J., Thermodynamic assessment and its application of ternary Mg-Zn-Ca system using CALPHAD method, in TMS (The Minerals, Metals & Materials Society), *Magnes. Technol.*, 2009. 167–172.
56. Hu, Y.J., Lieser, A.C., Saengdeejing, A., Liu, Z.K., and Kecskes, L.J., Glass formability of W-based alloys through thermodynamic modeling: W-Fe-Hf-Pd-Ta and W-Fe-Si-C, *Intermetallics*, 2014. 48: 79–85.
57. Ge, L., Hui, X.D., Chen, G.L., and Liu, Z.K., Prediction of the glass-forming ability of Cu-Zr binary alloys, *Acta Phys-Chim Sin.*, 2007. 23(6): 895–899.
58. Ge, L., Hui, X.D., Chen, G.L., and Liu, Z.K., Prediction of glass forming ability in Cu-Zr-Ti ternary amorphous alloys, *Rare Metal Mat. Eng.*, 2008. 37(4): 589–593.
59. Ge, L., Hui, X., Wang, E.R., Chen, G.L., Arroyave, R., and Liu, Z.K., Prediction of the glass forming ability in Cu-Zr binary and Cu-Zr-Ti ternary alloys, *Intermetallics*, 2008. 16(1): 27–33.
60. Pisch, A., Schmid-Fetzer, R., Buch, F.V., Mordike, B.L., Juch- mann, P., Bach, F.-W., and Haferkamp, H., Computer aided design of novel Mg-alloys, in *Magnesium Alloys and Their Applications*, B.L. Mordike and K.U. Kainer, Editors. 1998. Frankfurt, Germany, 181–186.
61. Mordike, B.L. and Ebert, T., Magnesium properties-applications-potential, *Mater. Sci. Eng. A*, 2001. 302: 37–45.
62. Luo, A.A., Balogh, M.P., and Powell, B.R., Creep and microstructure of magnesium-aluminum-calcium based alloys, *Metall. Mater. Trans. A*, 2002. 33(3): 567–574.
63. Pekguleryuz, M.O. and Baril, E., Development of creep resistant Mg-Al-Sr alloys, in *Minerals, Metals and Materials Society/AIME, Magnesium Technology 2001 (USA)*. 2001. Minerals, Metals and Materials Society/AIME, New Orleans, LA, 184 Thorn Hill Road, Warrendale, PA 15086–7528, 119–125.

64. Luo, A., Balogh, M.P., and Powell, B.R., Tensile creep and microstructure of magnesium-aluminum-calcium alloys for powertrain applications: Part 2 of 2, SAE Technical Paper 2001-01-0423, 2001. doi:10.4271/2001-01-0423.
65. Luo, A.A., Recent magnesium alloy development for elevated temperature applications, *Int. Mater. Rev.*, 2004. 49(1): 13–30.
66. Fuchs, H.J., Patent, GB 847992, Magnesium Alloys Having a High Resistance to Permanent Creep Deformation at Elevated Temperatures, 1960; UK.
67. Hollrigl-Rosta, F., Just, E., Kohler, J., and Melzer, H.J., Magnesium in the volkswagen, *Light Met. Age*, 1980. 38(7–8): 22–23, 26–29.
68. Pekguleryuz, M.O. and Luo, A., Creep-resistant magnesium alloy die castings, Patent, WO96/25529, 1996.
69. Zhong, Y., Luo, A.A., Sofo, J., and Liu, Z.-K., First-principles investigation of laves phases in Mg-Al-Ca system, in *Materials Science Forum*, W. Ke, E.H. Han, Y.F. Han, K. Kainer and A.A. Luo, Editors. 2004. Charlotte, NC.
70. Zhong, Y., Luo, A.A., Sofo, J.O., and Liu, Z.K., First-principles investigation of laves phases in Mg-Al-Ca system, *Mater. Sci. Forum*, 2005. 488: 169–176.
71. Suzuki, A., Saddock, N.D., Jones, J.W., and Pollock, T.M., Structure and transition of eutectic (Mg, Al) 2Ca Laves phase in a die-cast Mg–Al–Ca base alloy, *Scripta Mater.*, 2004. 51(10): 1005–1010.
72. Suzuki, A., Saddock, N.D., Jones, J.W., and Pollock, T.M., Solidification paths and eutectic intermetallic phases in Mg-Al-Ca ternary alloys, *Acta Materialia*, 2005. 53(9): 2823–2834.
73. Suzuki, A., Saddock, N.D., Jones, J.W., and Pollock, T.M., Phase equilibria in the Mg-Al-Ca ternary system at 773 and 673 K, *Metal. Mater. Trans. A*, 2006. 37A(3A): 975–983.
74. Zhong, Y., Liu, J., Witt, R.A., Sohn, Y.-h., and Liu, Z.-K., Al2(Mg, Ca) phases in Mg–Al–Ca ternary system: First-principles prediction and experimental identification, *Scripta. Mater.*, 2006. 55(6): 573–576.
75. Wang, Y., Liu, Z.K., Chen, L.Q., and Wolverton, C., First-principles calculations of beta"-Mg5Si6/alpha-Al interfaces, *Acta Mater.*, 2007. 55(17): 5934–5947.
76. Wang, Y., Chen, L.Q., Liu, Z.K., and Mathaudhu, S.N., First-principles calculations of twin-boundary and stacking-fault energies in magnesium, *Scripta Mater.*, 2010. 62(9): 646–649.
77. Manga, V.R., Saal, J.E., Wang, Y., Crespi, V.H., and Liu, Z.K., Magnetic perturbation and associated energies of the antiphase boundaries in ordered Ni3Al, *J Appl Phys.*, 2010. 108(10): 103509.
78. Han, J., Su, X.M., Jin, Z.H., and Zhu, Y.T., Basal-plane stacking-fault energies of Mg: A first-principles study of Li- and Al-alloying effects, *Scripta Mater.*, 2011. 64(8): 693–696.
79. Mantina, M., Wang, Y., Arroyave, R., Chen, L.Q., Liu, Z.K., and Wolverton, C., First-principles calculation of self-diffusion coefficients, *Phys. Rev. Lett.,*, 2008. 100(21): 215901.
80. Mantina, M., Shang, S.L., Wang, Y., Chen, L.Q., and Liu, Z.K., 3d transition metal impurities in aluminum: A first-principles study, *Phys. Rev. B*, 2009. 80(18): 184111.
81. Ganeshan, S., Hector, L.G., and Liu, Z.K., First-principles study of self-diffusion in hcp Mg and Zn, *Comput. Mater. Sci.*, 2010. 50(2): 301–307.
82. Ganeshan, S., Hector, L.G., and Liu, Z.K., First-principles calculations of impurity diffusion coefficients in dilute Mg alloys using the 8-frequency model, *Acta Mater.*, 2011. 59(8): 3214–3228.
83. Shang, S.L., Hector, L.G., Jr., Wang, Y., and Liu, Z.K., Anomalous energy pathway of vacancy migration and self-diffusion in hcp Ti, *Phys. Rev. B*, 2011. 83(22): 224104.

84. Somekawa, H., Hirai, K., Watanabe, H., Takigawa, Y., and Higashi, K., Dislocation creep behavior in Mg-Al-Zn alloys, *Mater. Sci. Eng. A*, 2005. 407(1–2): 53–61.
85. Ogata, S., Li, J., and Yip, S., Ideal pure shear strength of aluminum and copper, *Science*, 2002. 298(5594): 807–811.
86. Clatterbuck, D.M., Chrzan, D.C., and Morris, J.W., The ideal strength of iron in tension and shear, *Acta Mater.*, 2003. 51(8): 2271–2283.
87. Leyson, G.P.M., Curtin, W.A., Hector, L.G., and Woodward, C.F., Quantitative prediction of solute strengthening in aluminium alloys, *Nature Mater.*, 2010. 9(9): 750–755.
88. Yasi, J.A., Hector, L.G., and Trinkle, D.R., First-principles data for solid-solution strengthening of magnesium: From geometry and chemistry to properties, *Acta Mater.*, 2010. 58(17): 5704–5713.
89. Campbell, C.E., Kattner, U.R., and Liu, Z.-K., The development of phase-based property data using the CALPHAD method and infrastructure needs, *Integr. Mater. Manuf. Innov.*, 2014. 3(1): 12.
90. Zhong, Y., Ozturk, K., Sofo, J.O., and Liu, Z.K., Contribution of first-principles energetics to the Ca-Mg thermodynamic modeling, *J. Alloy. Compd.*, 2006. 420(1–2): 98–106.
91. Zhang, H., Wang, Y., Shang, S.L., Chen, L.Q., and Liu, Z.K., Thermodynamic modeling of Mg-Ca-Ce system by combining first-principles and CALPHAD method, *J. Alloy. Compd.*, 2008. 463(1–2): 294–301.
92. Zuo, Y. and Chang, Y.A., Thermodynamic calculation of Mg-Cu phase-diagram, *Z. Metallkd.*, 1993. 84(10): 662–667.
93. Ansara, I., Dinsdale, A.T., and Rand, M.H., *COST 507: Thermochemical Database For Light Metal Alloys*. Vol. 2. 1998, European Commission, Luxembourg.
94. Guo, C.P. and Du, Z.M., Thermodynamic assessment of the La-Mg system, *J. Alloy. Compd.*, 2004. 385(1–2): 109–113.
95. Grobner, J., Mirkovic, D., Ohno, M., and Schmid-Fetzer, R., Experimental investigation and thermodynamic calculation of binary Mg-Mn phase equilibria, *J. Phase. Equilib. Diff.*, 2005. 26(3): 234–239.
96. Meng, F.G., Liu, H.S., Liu, L.B., and Jin, Z.P., Thermodynamic optimization of Mg-Nd system, *T. Nonferr. Metal. Soc.*, 2007. 17(1): 77–81.
97. Guo, C.P. and Du, Z.M., Thermodynamic assessment of the Mg-Pr system, *J. Alloy Compd.*, 2005. 399(1–2): 183–188.
98. Fabrichnaya, O.B., Lukas, H.L., Effenberg, G., and Aldinger, F., Thermodynamic optimization in the Mg-Y system, *Intermetallics*, 2003. 11(11–12): 1183–1188.
99. Agarwal, R., Fries, S.G., Lukas, H.L., Petzow, G., Sommer, F., Chart, T.G., and Effenberg, G., Assessment of the Mg-Zn System, *Z. Metallkd.*, 1992. 83(4): 216–223.
100. Arroyave, R., Shin, D., and Liu, Z.K., Modification of the thermodynamic model for the Mg-Zr system, *Calphad*, 2005. 29(3): 230–238.

12 Surprises and Pitfalls in the Development of Magnesium Powder Metallurgy Alloys

Paul Burke, Yiannis G. Kipouros, William D. Judge, and Georges J. Kipouros

CONTENTS

12.1 INTRODUCTION

Recent trends in the transportation sector have been pushing for lighter weight materials to replace traditional steel structural parts. The weight loss due to replacing traditional materials with lighter materials can improve fuel economy and performance, or allow the installation of additional equipment and features without a weight penalty over the original design.

The use of aluminum alloys has been a major step forward in this goal, and research into new alloys and processing methods has been fervent in the past quarter century. Aluminum now enjoys a catalog of thousands of alloys and many easily achieved forming methods. A powder metallurgy aluminum method was developed in this laboratory about 25 years ago. It was based primarily on the hypothesis that the addition of one percent magnesium into aluminum will help to eliminate the passive layer formed on the surface of the powder allowing sintering to occur [1,2].

Magnesium, which enjoys an even lower density than aluminum, is the next step in this evolution. While die cast magnesium alloy parts are increasing in use, the overall scale of magnesium is very small compared to aluminum. The primary issues hindering the use of magnesium are difficulty in cold and hot forming, the small number of developed alloys and high corrosion rates.

Powder metallurgy (P/M) is a relatively new processing method, having only been in use for 20 years for light metals. P/M began with ferrous alloys, and the industry is still dominated by steel and stainless steel parts. Copper P/M alloys have also been in use for many years and have a large market share. Aluminum is quickly becoming a large part of the P/M industry, and the majority of research in the field over the past 15 years has been directed toward utilizing aluminum. The key benefits of P/M processing are the production of net and near-net shape parts, and the ability to realize new novel alloys through rapid solidification of powders. Net shape parts are desirable to reduce secondary operations necessary to put the part into service, reducing cost and increasing precision. Rapid solidification of powders can lead to the development of novel alloys because of the increased solubility of alloying elements which are trapped by the very fast cooling rate. Additive manufacturing processes initially developed for rapid prototyping are now becoming a major competitor to traditional manufacturing techniques particularly for complex near net shape parts. The initial material for additive manufacturing was metal powders but, presently, metal wires are becoming more attractive as feed materials.

Magnesium P/M is an excellent method for achieving the goal of increased magnesium usage by addressing its three primary issues. Poor formability is circumvented because parts are net or near-net shaped and require only slight machining to achieve high tolerance. New alloy development can be accelerated by the use of rapidly solidified powders of novel compositions, and new alloys that address corrosion issues can be explored.

However, magnesium P/M is not without its challenges. Sourcing high purity magnesium powders is difficult because of lack of commercial interest thus far. The primary use for particulate magnesium is currently as a de-oxidizer in steel making, which does not require high-grade powder. Because of this, the majority of available magnesium powders are produced by mechanical grinding, which results in high oxygen contamination due to the long exposure to atmosphere during grinding. Ground magnesium powders are unsuitable for P/M use because the surface layer formed during exposure to atmosphere is then a barrier to the metal diffusion that is necessary during sintering to produce bonds and structurally sound parts.

Recently magnesium powder produced by centrifugal atomization has become commercially available [3]. Powder produced by this method has less impurities due to the inert argon atmosphere in which it is made; therefore the surface layer is much thinner than in the case of the ground powder, but it is still present and acting as a barrier to metal diffusion during sintering. Other metal powders are affected by surface layers as well, but methods have been discovered to reduce, disrupt, or otherwise remove the barrier to allow successful sintering as was the case of aluminum where magnesium played a principal role.

The difficulty in removing the layer as a barrier is amplified in the case of reactive metals, where any oxide formed is very stable and challenging to remove. Because magnesium is one of the most reactive structural metals, it is important for the future of magnesium P/M to identify a method to remove the surface layer barrier. While some methods do exist, they are expensive and would be difficult to achieve on a large industrial scale. For magnesium P/M to compete on a level with aluminum P/M, simple, industrially relevant processing using existing P/M equipment is necessary. Because the economics of P/M require high volume, a robust process utilizing uni-axial compaction and mesh belt furnaces is needed to compete with established magnesium production methods like die casting [4–7].

12.2 THERMOCHEMICAL STUDIES

Two of the most fundamental and powerful tools for studying sintering are kinetics and thermodynamics, which study the changes with time and temperature, respectively. The main process variables responsible for the success of sintering are time, temperature, heating schedule, and sintering atmosphere.

The powder metallurgy process begins with raw magnesium powders of a 75 μm typical size that are blended with alloying elements and compacted into the desired geometry in a die at high pressure. The main variables in the compaction process are the applied pressure, the powder size, and the fluidity of the powder, the latter being influenced by the size distribution. The "green" compacts are then sintered in a furnace at high temperature to metallurgically bond the powders [5].

Sintering is the step in the P/M process where the powder compacts are heated in a controlled atmosphere to a temperature lower than the melting temperature, but high enough to allow rapid diffusion between particles. This diffusion between particles bonds them together and results in a solid part [5]. By retaining the parts below their melting temperature, it is possible to realize net shape products that, unlike casting methods, require little to no secondary processing. The main process variables responsible for the success of sintering are time, temperature, heating schedule, and sintering atmosphere.

The effect of time and temperature on the properties of the sintered parts has been tested for many light metals and alloys, including numerous aluminum alloys along with pure magnesium and magnesium alloys [6–8]. It had been determined in these previous studies that temperature was the most important factor in magnesium P/M processing [7]. Two of the most fundamental and powerful tools for studying sintering are kinetics and thermodynamics, which study the changes with time and temperature, respectively. Because temperature was previously determined to be of greater importance to the success of Mg sintering, this review will focus on thermodynamics. The thermodynamics of sintering have been studied in ceramics [8–19], where typical sintering times are in the order of hours. These systems have been studied with various approaches, including verifying the fit of published Gibbs free energy values determined using bulk samples to sintering experiments, application of pressure-composition isotherms and Van't Hoff diagrams, determination of sintering force using surface and grain boundary tension equations, and the calculation of arbitrary equilibrium constants.

There is very little in the literature about thermodynamics applied to the sintering of metals. The reason for this is the primary assumption required when discussing thermodynamics that the system is at equilibrium, and the very short sintering times for many alloy systems, on the order of 20 minutes, negates this assumption. However, in the case of magnesium the assumption is valid because it requires a longer than typical sintering time of 40–60 minutes. Also, the previous findings that temperature has a greater effect on sintering of magnesium than time points to a thermodynamically controlled process. The process is not perfectly at equilibrium, however, so the term quasi-equilibrium is applied.

Sintering is fundamentally an irreversible process and inhibited by the formation of a surface film which could consist of oxides, hydroxides, carbonates, and hydrates. In aluminum powder metallurgy, a small addition of magnesium is always present to reduce the surface film. In magnesium sintering, calcium has been suggested as an addition to remove the surface film. In P/M of metals and alloys, two-dimensional change measurements do not always reflect the degree of sintering, therefore the present study considers changes in density and its comparison to the theoretical density. Density measurements are standardized for P/M components, are more precise, and reflect a clearer indication of the degree of sintering.

The thermodynamic and kinetic study which constitutes the basis of the thermochemical analysis was based on the sintering of pure aluminum, pure magnesium, with no alloying additions, sintering aides or pre-treatments applied, to gain knowledge of the fundamental characteristics of the surface layer which inhibits sintering. The procedure was repeated with the addition of appropriate metal that can help sintering by reducing the surface film. With the advent of nano-powders, kinetic treatments like the one presented in this paper may be very useful.

The experimental thermochemical analysis was performed in this laboratory with the following methodology, fully described elsewhere [22,23]. The commercially available aluminum powders were obtained from Ecka Granules while Arcos Organics supplied the calcium granules of approximately 2 mm average size. Tangshan Weihao Magnesium Powders supplied the argon protected centrifugally atomized pure magnesium powder, which had an average particle size of 38 μm. Powders were characterized by laser Malvern size analyzer in water to determine the size distribution. All powders were available from previous work in the laboratory [24–29].

12.2.1 Compaction

To minimize exposure of the samples to atmosphere uniform compaction of the atomized magnesium powders was performed using a Loomis cold isostatic press (CIP) on samples prepared in a glove box under argon. Samples measured approximately 15 mm in diameter and 30 mm length when compacted at 25 ksi (173 MPa), which was the maximum capacity of the available CIP equipment.

12.2.2 Sintering

Sintering took place in a horizontal tube furnace, which was first evacuated and then purged several times under a flowing argon atmosphere. The samples were heated to the corresponding sintering temperature, which ranged from 500°C to 625°C in 25°C increments and kept at the sintering temperature for 60 minutes. The sintering times were established by conducting sintering experiments at constant temperatures and recording the achieved density. The optimum time was determined as such that prolonged time will not further increase density. After sintering, the furnace was turned off and the samples were quenched by moving the samples into a water-cooled jacket and cooled to room temperature.

12.2.3 Density Measurements

The density of the material was measured on the sintered products and was determined using the MPIF Standard 42 "Density of Compacted or Sintered Powder Metallurgy Products" [30]. This method involved weighing the sample in air according to Equation (12.1),

$$\rho_{green} = \frac{m_{air} \cdot \rho_{water}}{m_{air} - m_{water}} \tag{12.1}$$

where ρ_{green} is the green density, the weight of the compact in air is m_{air}, the weight suspended in water is m_{water}, and the density of water is ρ_{water}. After sintering, samples were impregnated with a suitable oil (Nuto H 46) by vacuum impregnation in a vacuum chamber. The samples were placed in the oil bath and a vacuum was drawn on the chamber. After the impregnation process, the samples were re-weighed in a water bath containing the appropriate amount of PhotoFlo as a wetting agent. The weights of the dried, oil impregnated and suspended in water samples were recorded and used to calculate sintered density according to Equation (12.2):

$$\rho_{\text{sintered}} = \frac{m_a \cdot \rho_{water}}{m_{a,o} - m_{a,ow}} \tag{12.2}$$

where the sintered density is ρ_{sintered}, the weight of the oil filled compact in air is $m_{a,o}$, and the weight suspended in water is $m_{a,ow}$. Porosity was determined by subtracting the resulting densities from the theoretical density of magnesium, 1.74 g/cm^3 and aluminum 2.70 g/cm^3, respectively.

12.2.4 Sintering

Figure 12.1 shows an SEM micrograph of the magnesium powder before compaction. The powder size from the SEM micrograph agreed well with the size determined by the laser size analysis supplied by the producer. The overwhelming majority of the powder is of a spherical nature, a shape expected for centrifugally atomized powders.

The compaction curve of pure magnesium powder under uniaxial die compaction (UDC) pressing in an Instron frame is shown in Figure 12.2. It is clear that the density approaches theoretical values in pressures at 300 MPa. Compacting in a cold isostatic press (CIP) offers the additional advantage of avoiding exposing the samples which are prepared in a glove box to atmospheric conditions. Furthermore, it was found that of the two techniques, uniaxial die compaction

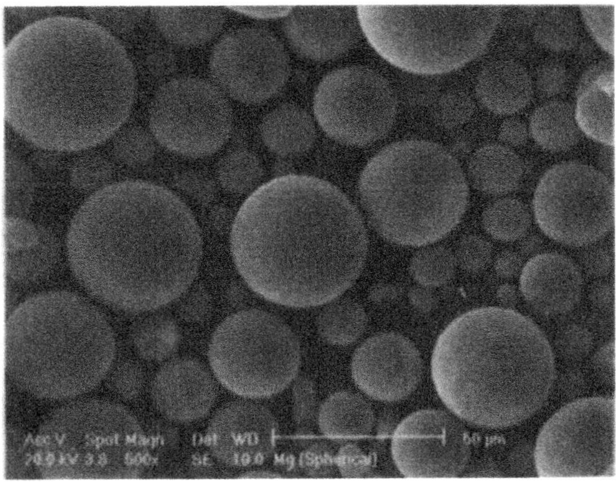

FIGURE 12.1 SEM micrograph of pure Mg powder before compaction. (From Burke, P., Investigation of the sintering fundamentals of magnesium powders, PhD Thesis, Dalhousie University, Halifax, Canada, 2011.)

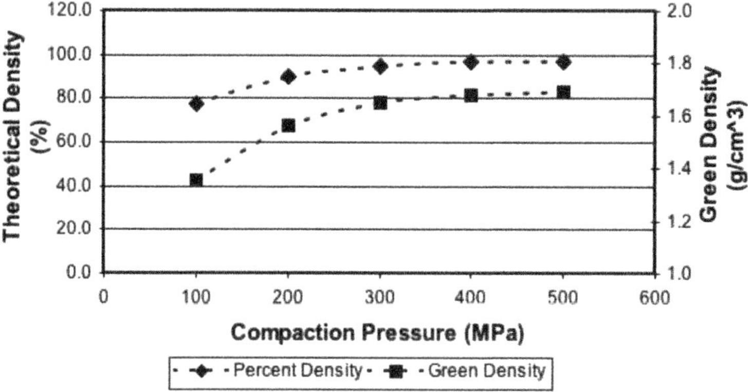

FIGURE 12.2 Uniaxial die compaction curve of pure Mg powder. (From Burke, P., Investigation of the sintering fundamentals of magnesium powders, PhD Thesis, Dalhousie University, Halifax, Canada, 2011.)

(UDC) and cold isostatic pressing (CIP), the latter produced better results for the sintering of pure magnesium powders.

An optical micrograph of a sample sintered at 600°C is shown in Figure 12.3, and a focused ion beam micrograph of a single Mg powder particle, highlighting the surface layer, is shown in Figure 12.4.

Figure 12.5 shows the effect of sintering time on the density and hardness of a cold isostatically pressed atomised magnesium samples sintered at 600°C [3]. It can be seen that in general, the higher sintering temperatures result in a higher final density in the compacts.

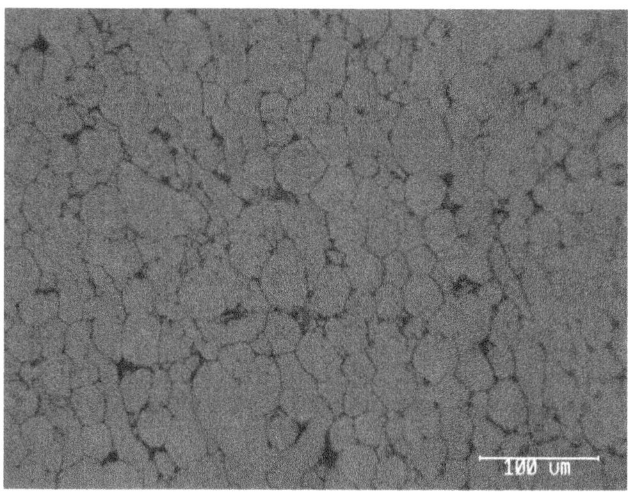

FIGURE 12.3 Optical micrograph of a pure Mg compact sintered at 600°C for 60 minutes. (From Burke, P. et al., *Can. Metall. Quart.*, 48, 123, 2009.)

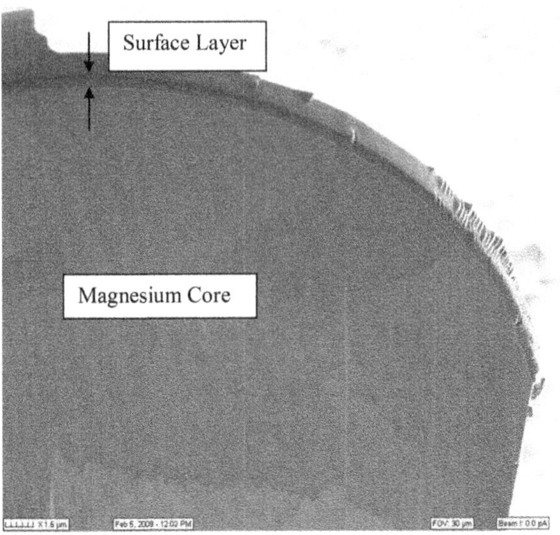

FIGURE 12.4 Focused ion beam micrograph of a single Mg particle, showing the pure Mg core and the surface layer enveloping the perimeter.

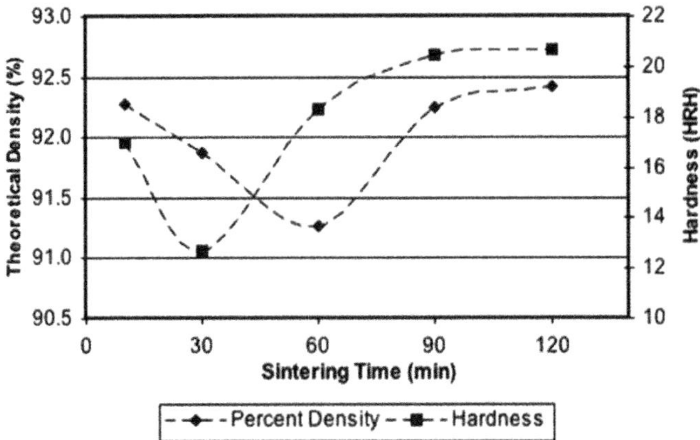

FIGURE 12.5 The effect of sintering time on the density and hardness of a cold isostatically pressed atomized magnesium samples sintered at 600°C. (From Burke, P., Investigation of the sintering fundamentals of magnesium powders, PhD Thesis, Dalhousie University, Halifax, Canada, 2011.)

12.2.5 Time Dependence Kinetics

For an nth order of reaction/phenomenon and a rate constant, K, the kinetic equation may be written as Equation (12.3)

$$\int_{\rho \text{ at } t=t_0 \text{ point 5}}^{\rho \text{ at } t=t} \rho^{-n} d\rho = \mathrm{K} \int_{t=t_0 \text{ point 5}}^{t=t} dt \tag{12.3}$$

where ρ is the density, t is the time, and K is the rate constant. Upon integration Equation (12.3) gives Equation (12.4):

$$(-n+1)\,\mathrm{Ln}\frac{(\rho)_{at\ t=t}}{(\rho)_{\rho\ \mathrm{at}\ t=t_0\ \mathrm{point}\ 5}} = \mathrm{Ln}(-n+1)\,K + \mathrm{Ln}t \tag{12.4}$$

Equation (12.4) can be used in its empirical form [6] shown by Equation (12.5)

$$\rho = \rho^0 + K\mathrm{Ln}(t/t^0) \tag{12.5}$$

where all the symbols are in accordance to equation (12.3).

This treatment is consistent with the quasi-thermodynamic approach which was introduced by L. Tisza [31] and allows for the temperature dependence of the kinetic transformations. Porosity is selected as the variable for sintering and it is calculated from the measured densities of the sample in comparison to the theoretical density of pure magnesium or aluminum. The rate constant, K, was then calculated by Equation (12.6):

$$K_a = (P_i - P)/(P) \tag{12.6}$$

where P_i is the initial and greatest specific porosity of the sample sintered at the lowest temperature, and P is the porosity at a given temperature. K_a was calculated for each subsequent temperature and a plot of $\ln K_a$ vs. 1/T gives the slope of $-\Delta H^0/R$ from which the apparent activation enthalpy, ΔH^0 can be calculated. The intercept of the plot is given by Equation (12.7)

$$A = R/Nh\ \exp(\Delta S^0/R) \tag{12.7}$$

where R = gas constant, N = Avogadro number, h = Planck constant. Equation (12.7) may reveal the apparent entropy, ΔS^0, of sintering. Finally, the apparent free energy of sintering, ΔG, is found by Equation (12.8).

$$\Delta G^0 = \Delta H^0 - T\Delta S^0 \tag{12.8}$$

Tables 12.1 through 12.4 show the calculation of $\ln K_a$ from experimental data for pure magnesium, pure aluminum, aluminum-1wt% Mg, and magnesium-1wt% Ca, respectively.

For each table describing the results of the experiments, a Van't Hoff plot of the rate constant as a function of the reciprocal sintering temperature was constructed. Figure 12.6 for example shows the Van't Hoff plot of the rate constant as a function of the reciprocal sintering temperature for the sintering of pure magnesium powder.

Similar plots were constructed for Tables 12.2 through 12.4 and the resulting plots are shown in Figures 12.7 through 12.9, respectively.

TABLE 12.1
Sintered Density of Mg Compacts Sintered from 500°C to 625°C for 60 Minutes

T (°C)	T (K)	1/T (K) 10^{-3}	ρ (g/cm³)	P	K_a	Ln K_a
500	773	1.29	1.547	0.193		
525	798	1.25	1.551	0.189	0.022	−3.82
550	823	1.22	1.557	0.183	0.057	−2.86
575	848	1.18	1.564	0.177	0.093	−2.38
600	873	1.15	1.577	0.163	0.183	−1.70
625	898	1.11	1.605	0.135	0.426	−0.85

Note: T is sintering temperature, ρ is sintered density, P is porosity, K_a is the rate constant.

TABLE 12.2
Sintered Density of Al Compacts Sintered from 500°C to 625°C for 60 Minutes

T (°C)	T (K)	1/T (K) × 10^{-3}—	ρ (g/cm³)	P	K_a	Ln K_a
500	773	1.29	2.456	0.244		
525	798	1.25	2.486	0.214	0.139	−1.973
550	823	1.22	2.604	0.96	1.5	−0.405
575	848	1.18	2.506	0.194	0.25	−1.386
600	873	1.15	2.512	0.188	0.286	−1.252
625	898	1.11	2.525	0.175	0.385	−0.955

Note: T is sintering temperature, ρ is sintered density, P is porosity, K_a is the rate constant.

TABLE 12.3
Sintered Density of Al with 1 wt% Magnesium Compacts Sintered from 500°C to 625°C for 60 Minutes

T (°C)	T (K)	1/T (K) × 10^{-3}	ρ (g/cm³)	P	K_a	Ln K_a
500	773	1.29	2.451	0.339		
525	798	1.25	2.494	0.296	0.146	−1.973
550	823	1.22	2.313	0.477	0.290	−1.252
575	848	1.18	2.496	0.294	0.153	−1.876
600	873	1.15	2.501	0.289	0.172	−1.761
625	898	1.11	2.451	0.339	0.001	−1.773

Note: T is sintering temperature, ρ is sintered density, P is porosity, K_a is the rate constant.

TABLE 12.4
Sintered Density of Mg with 1 wt% Calcium Compacts Sintered from 500°C to 625°C for 60 Minutes

T (°C)	T (K)	1/T (K) × 10^{-3}	ρ (g/cm³)	P	K_a	Ln K_a
500	773	1.29	1.5965	0.144		
525	798	1.25	1.6102	0.130	0.108	−2.233
550	823	1.22	1.6069	0.133	0.079	−2.538
575	848	1.18	1.5791	0.131	0.1095	−2.211
600	873	1.15	1.5897	0.138	0.0459	−3.081

Note: T is sintering temperature, ρ is sintered density, P is porosity, K_a is the rate constant.

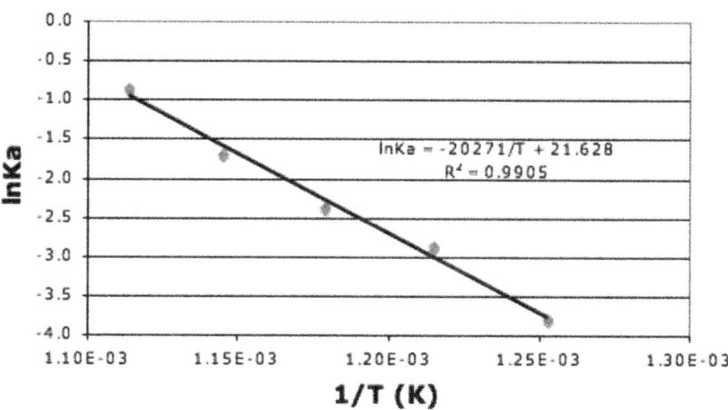

FIGURE 12.6 Van't Hoff plot of the rate constant as a function of the reciprocal sintering temperature for the sintering of pure magnesium powder.

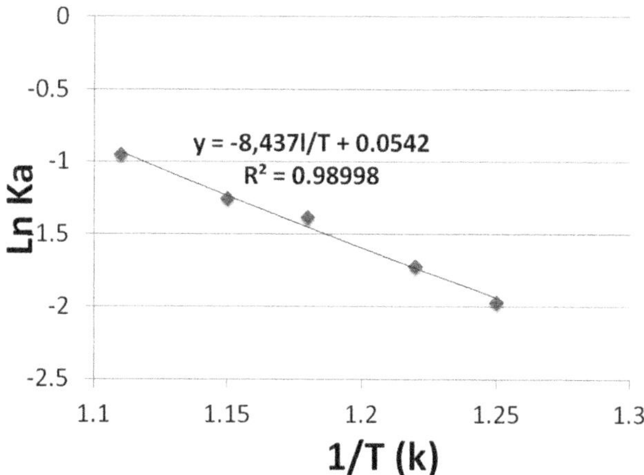

FIGURE 12.7 Van't Hoff plot of the rate constant as a function of the reciprocal sintering temperature for the sintering of pure aluminum powder.

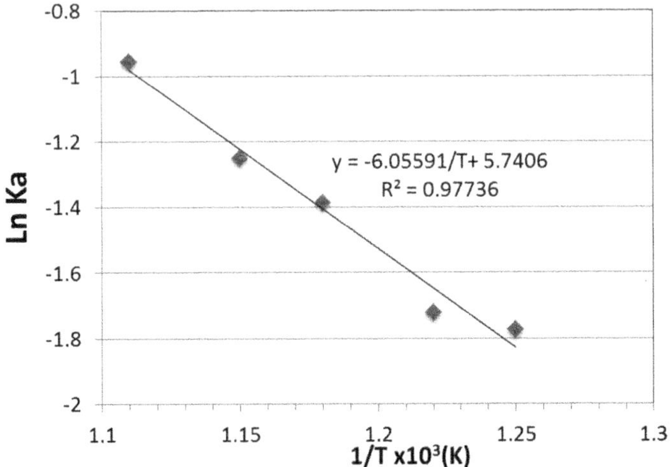

FIGURE 12.8 Van't Hoff plot of the rate constant as a function of the reciprocal sintering temperature for the sintering of aluminum with 1 wt% magnesium powder.

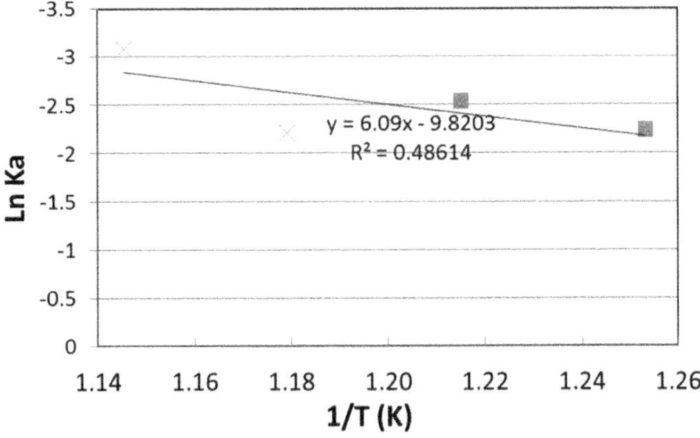

FIGURE 12.9 Van't Hoff plot of the rate constant as a function of the reciprocal sintering temperature for the sintering of magnesium sintered together with 1 wt% calcium powder.

All the Van't Hoff plots generated from the data presented in Tables 12.1 through 12.4 are presented in Figure 12.10. It is important to point that the plot of Figure 12.9, which is also included in Figure 12.10, is of questionable statistic value because of the low regression coefficient and demonstrates the difference between an alloy which was originally produced by powder atomization (Figure 12.8) and the alloy which produced by mixing and sintering the powders of the components.

From these plots the corresponding apparent thermodynamic data of the sintering process for the relevant cases can be calculated and they are shown in Table 12.5.

These results indicate that sintering of pure magnesium in the presence of the surface layer is endothermic in nature and shows an increase in entropy. Sintering will not occur spontaneously, even up to the melting temperature of the metal (650°C).

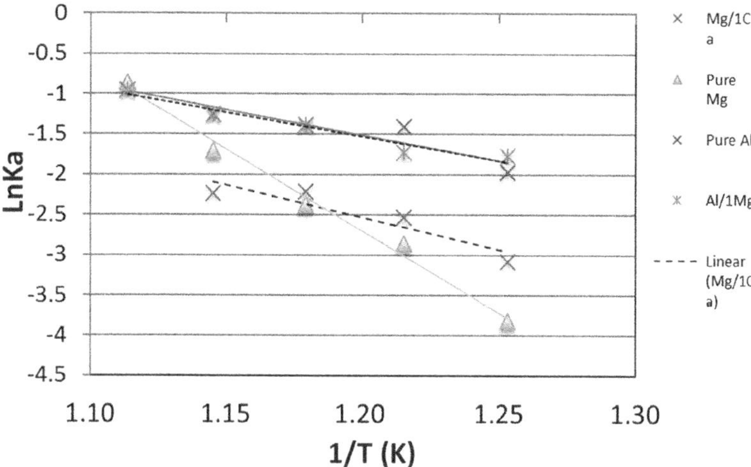

FIGURE 12.10 Plots of $\ln K_a$ vs. 1/T for the data presented in Tables 12.1 through 12.4.

TABLE 12.5
Apparent Activation Thermodynamic Quantities ΔH°, ΔS°, ΔG°, for Sintering Pure Aluminum, Pure Magnesium, Aluminum with 1 wt% Magnesium, and Magnesium with 1wt% Calcium Compacts

	Apparent Activation Thermodynamic Quantities for Sintering		
	Enthalpy, ΔH ($J\ mol^{-1}$)	Entropy, ΔS ($J\ K^{-1}\ mol^{-1}$)	Free Energy, ΔG ($J\ mol^{-1}$)
Pure magnesium	168,909	180.3	ΔG° = 168,909–180.3 T
Pure aluminum	70,111	27.07	ΔG° = 70,111—27.07 T
Aluminum with 1wt% magnesium	50,341	47.85	ΔG° = 50,341—47.85 T
Magnesium with 1wt% calcium	51,156	82.49	ΔG° = 51,156—82.49 T

According to corresponding free energy equation, a sintering temperature of ~665°C, would be required to successfully sinter pure magnesium.

Sintering too close to the melting temperature of the metal may cause distortions in the part, and the net-shape benefit of P/M is lost. Because of this, successful sintering of pure magnesium with a surface layer present in the absence of a reducing atmosphere will not be possible. This explains the difficulty found in sintering pure magnesium compacts, due to the highly stable surface layer which inhibits diffusion between adjacent particles during sintering.

The success of the sintering is determined by comparing the density of the sintered part to the theoretical density. The apparent thermodynamic data are: for sintering pure magnesium, Gibbs free energy (ΔG° = 168, 909–80.3 T J mol^{-1}) in the presence of the surface layer. For the sintering pure aluminum the corresponding Gibbs free energy (ΔG° = 70, 111–7.07 T J mol^{-1}). For sintering aluminum containing 1 wt% magnesium the corresponding Gibbs free energy ΔG° = 50,341–47.85 T

J mol^{-1}). In the case of sintering magnesium containing 1 wt% calcium, the corresponding Gibbs free energy ($\Delta G° = 51{,}156 - 82.49\ T\ J\ mol^{-1}$). These results represent quantifiable measurements that can be used to compare the effectiveness of sintering.

The sintering of pure magnesium will not occur spontaneously at any temperature below its melting point. This is attributed to the surface layer found on the magnesium powder, which is an effective barrier to diffusion, and therefore complicates sintering. The results obtained by the thermochemical experiments were obtained without any assumption for the composition of the surface layer. It is commonly referred to as an oxide. However, given the fact that aluminum passivates while magnesium corrodes it was important to investigate further the nature of the surface layer.

To allow for successful sintering, alloying additions, sintering aids, or powder pre-treatments need to be utilized to decrease the enthalpy and/or increase the entropy. From previous studies [14], it is known that these techniques lead to successful results. Calcium and yttrium additions for the sintering of magnesium have been tested and the results are presented in reference [15].

12.3 SURFACE LAYER

Most metal powders exposed to air are covered by a surface layer. The layer is typically composed of oxides, but may also contain hydroxides, carbonates, or other compounds formed by reactions with the atmosphere. This layer prevents solid-state sintering in low melting point metals. Most commercial magnesium powders are produced by mechanical grinding. The low cost and less-restrictive requirements for its main intended use as a reactant make grinding attractive. For P/M applications the angular morphology of the powder gives good green strength because of mechanical interlocking, but the powder particles are typically covered by a thick surface layer due to long exposure to air during the grinding process. Recently, magnesium powder was also produced commercially by centrifugal atomization under an inert atmosphere and it was shipped in sealed plastic envelopes covered by a fine layer of an organic substance. The black envelopes protected the contents from the effect of daylight. The product has very little surface oxidation due to the inert production conditions, and vacuum for transportation and spherical morphology [3].

There are several possible reactions between magnesium and atmospheric gases that form compounds that may be found in the surface layer covering magnesium powders. The plausible reactions are:

$$Mg + 1/2O_2 = MgO \tag{12.9}$$

$$Mg + 1/2O_2 + CO_2 = MgCO_3 \tag{12.10}$$

$$Mg + 1/2O_2 + H_2O = Mg(OH)_2 \tag{12.11}$$

The presence of these compounds has been confirmed by a combination of differential scanning calorimetry, x-ray photoelectron spectroscopy, and transmission electron microscopy [3].

FIGURE 12.11 Focused ion beam image of the cross section of a single Mg powder particle. Note the surface layer (thin white layer) enveloping the Mg core. (From Burke, P.J. et al., *Appl. Spectrosc.*, 66, 510, 2012.)

TABLE 12.6
Thickness of Surface Layer Formed on Pure Mg Particle Due to Exposure to Atmosphere for Times

	As Received	24 hour	72 hour
Thickness (nm)	14.6	30.3	85.3

A focused ion beam (FIB) microscope was utilized to prepare thin foils of pure magnesium powder that were examined in the TEM to visually determine the thickness of the surface layer formed on powders exposed to atmosphere for various lengths of time [3]. Figure 12.11 shows a FIB image of a single particle of atomized magnesium [3]. Table 12.6 shows the thickness of the as received powder, and of samples after 24 and 72 hours exposure [3]. The surface layer of magnesium is not stable, and the thickness increases with increased exposure time. Therefore, it is important to limit open air processing and work under controlled atmosphere as much as possible.

12.3.1 Compaction Effects

One way to bypass the surface layer is by creating cracks and pores through it. During compaction, the force exerted on the powder is enough to break the

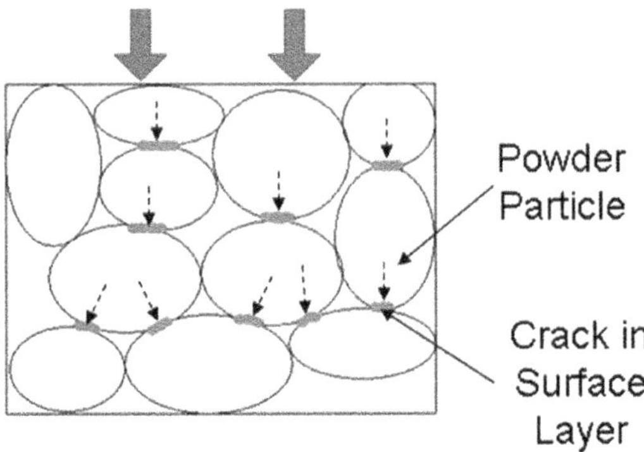

FIGURE 12.12 Force of compaction creates cracks in surface layer.

surface layer, as shown in Figure 12.12, creating short circuit pathways for unimpeded sintering. Also, during heating of the powder, any hydroxide or carbonate that is part of the layer may decompose, which will create pores in the layer as gas is released. If there are adjacent cracks or pores in two particles, rapid sintering will take place at that local area. However, if a point contact between particles does not have a break in the layer, diffusion is still impeded.

12.3.2 Temperature Effects

Of the compounds found in the layer, $MgCO_3$ and $Mg(OH)_2$ undergo decomposition reactions that will occur spontaneously below the melting point of magnesium. Thermodynamic software was utilized to estimate the temperature at which decomposition initiates, and was confirmed experimentally by differential scanning calorimeter [3]:

$$Mg(OH)_2 = MgO + H_2O \sim 280°C \tag{12.12}$$

$$MgCO_3 = MgO + CO_2 \sim 520°C \tag{12.13}$$

To expand on the results presented in Ref. [3], Figure 12.13 shows the change in mass of the pure magnesium sample exposed to air as it is heated to 600°C under nitrogen flowing gas in a thermal differential device. A small weight loss is apparent beginning at 280°C which corresponds to the loss of water vapor from the hydroxide decomposition. After the small initial loss, and the exposure of pure magnesium as the surface layer is disrupted the mass increases as magnesium nitride forms. Figure 12.13 is scaled to show the small mass loss at 280°C and the total gain is not shown, but the maximum mass attained at 600°C was 223 mg.

The presence of calcium, which is a more reactive metal than magnesium, disrupts the magnesium surface layer through the following reactions:

$$MgO + Ca = Mg + CaO \tag{12.14}$$

$$Mg(OH)_2 + Ca = Mg + Ca(OH)_2 \tag{12.15}$$

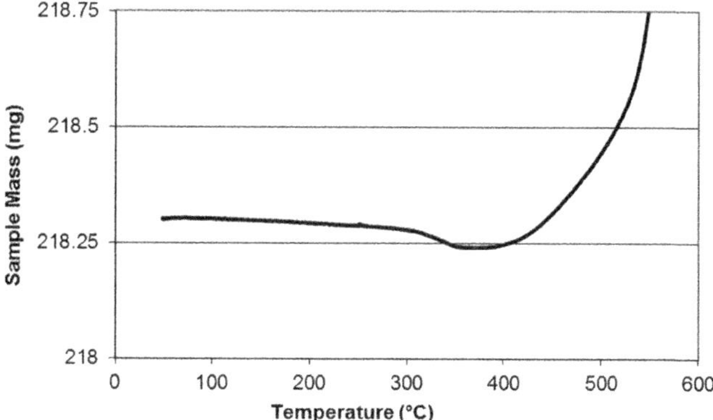

FIGURE 12.13 Mass change of pure magnesium sample while heating to 600°C. Maximum mass was 223 mg at 600°C. (From Burke, P. et al., *Can. Metall. Quart.*, 55, 45, 2016.)

$$MgCO_3 + Ca = Mg + CaCO_3 \tag{12.16}$$

Continuing from previous DSC work [15], the TGA data was collected for calcium additions in magnesium. Figure 12.14 shows the mass change while heating, and the result is similar to the pure magnesium case. The maximum mass attained was less however, at 217 mg. This may be due to liquid phases shielding the pure magnesium from contact with the nitrogen atmosphere.

Yttrium is another metal that is more reactive than magnesium and has the ability to disrupt the surface layer. The Mg-O_2-H_2O-CO_2-Y system is different from the calcium system because yttrium is more reactive than calcium, and also it has the ability to decompose only the oxide of magnesium, but not the hydroxide or carbonate. Yttrium also forms a nitride that is more stable than magnesium nitride.

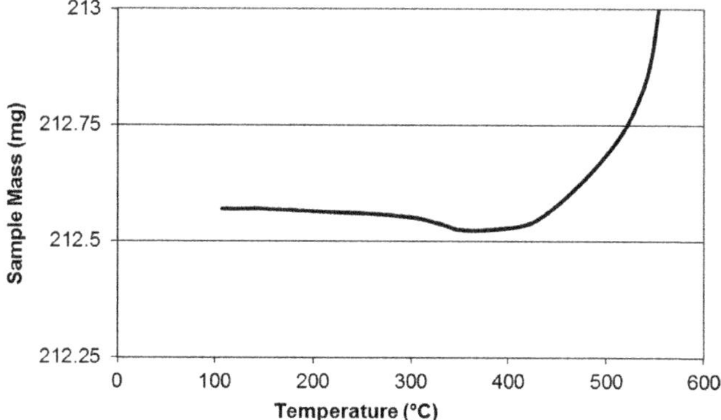

FIGURE 12.14 Mass change of Mg-1Ca sample while heating to 600°C. Maximum mass was 217 mg at 600°C. (From Burke, P. et al., *Can. Metall. Quart.*, 55, 45, 2016.)

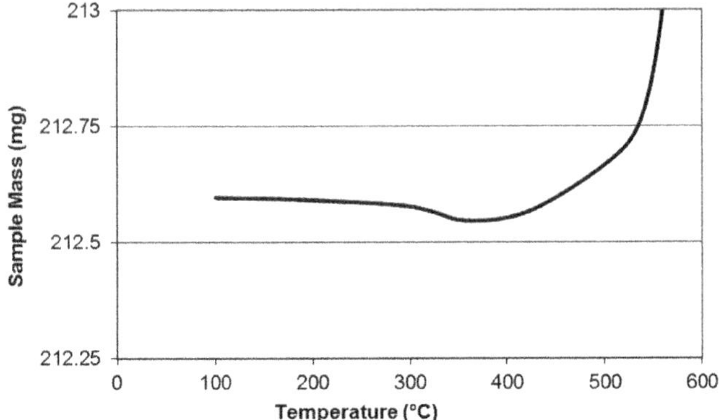

FIGURE 12.15 Mass change of Mg-2Y sample while heating to 600°C. Maximum mass was 226 mg at 600°C. (From Burke, P. et al., *Can. Metall. Quart.*, 55, 45, 2016.)

Continuing from previous DSC work [15], the TGA data was collected for yttrium additions in magnesium. Figure 12.15 shows the change is mass on heating and is similar to the pure magnesium and Mg-1Ca case with the exception of the nitride formation beginning at a lower temperature. The maximum mass gained was also higher than previous samples, at 226 mg, indicating more mass was gained due to the formation of YN along with Mg_3N_2.

Reduction to elemental magnesium would be more beneficial but is not possible below the melting point of magnesium. However, the disruption to the surface layer caused by the decomposition may well expose areas of magnesium below the layer, allowing for better diffusion between adjacent particles during sintering of P/M parts.

12.3.3 Methods for Layer Disruption

If compaction and decomposition do not sufficiently disrupt the layer, the layer needs to be reduced or otherwise removed to enable effective sintering. This can be accomplished with a thermodynamic unstable oxide, where some combination of temperature, pressure, atmosphere, or the addition of a more reactive metal will cause the oxide layer to become unstable [20]. The oxide is then reduced, and the sintering kinetics are similar to the pure metal. In the case of magnesium, the oxide is especially stable and the only option to disrupt the surface layer is the addition of a metal with a more stable oxide.

If the oxide layer is unstable with respect to dissolution in the parent metal at high temperature, this can be utilized for sintering. The sintering process is preceded by an incubation period dictated by the kinetics of dissolution and diffusion of oxygen and the parent metal. At sintering temperature, the oxide layer then dissolves into the parent metal. The incubation period for iron is very short at approximately 10 seconds. Therefore, iron can be readily sintered even when covered by an oxide layer. However, the incubation period for low melting

TABLE 12.7
Diffusion Rates of a Metal through Itself and through Its Oxide

	D_M m^2 s^{-1}	D_{Ox} m^2 s^{-1}
Cu	5.65×10^{-13}	6.65×10^{-12}
Al	1.84×10^{-12}	5.51×10^{-30}
Mg	3.01×10^{-12}	5.25×10^{-24}

Source: Kubota, K. et al., *J. Mater. Sci.*, 34, 2255–2262, 1999.

temperature metals such as magnesium have incubation periods upwards of one hundred days. This explains why ferrous alloys dominate the P/M industry.

Another mechanism to remove an oxide layer can be used with metals that have a low oxygen solubility and high oxide stability. In this case metal atoms will diffuse through the oxide layer and the neck will form on the surface. If the rate of diffusion through the metal oxide is similar to the metal diffusing through itself, then sintering is not retarded by this process. As shown in Table 12.7 [4], the rate of diffusion of copper through its oxide is actually faster than through itself, so sintering is not hindered by the presence of a surface layer. This leads to copper P/M being second only to ferrous P/M. In the case of aluminum and magnesium, the diffusion rate through its oxide is orders of magnitude slower than though itself. However, sintering is still possible, but at an increased duration.

The use of a liquid phase is an alternative to solid-state sintering and may reduce the problem of an oxide layer on powder particles by further exploiting breaks in the surface layer. As the liquid spreads between the particles, any cracks or pores become short circuit pathways, regardless of whether they are directly adjacent to another crack in a neighbouring particle. This allows for reduced sintering duration.

Another method for disrupting the surface layer is to add a small amount of a more reactive metal. During sintering, the additive reacts with the surface of the surrounding base powder, reducing the oxides of the base to pure metal and forming its own oxide. While the total amount of oxide is still the same, the cohesive layer that was the barrier to diffusion has been disrupted, providing pathways for unimpeded diffusion. This technique is used successfully in aluminum P/M [32,33], where a small amount of magnesium results in an increase in density and mechanical properties.

In the case of magnesium, the oxide is especially stable and the only option to disrupt the surface layer is the addition of a metal with a more stable oxide. Metals that have an oxide with a lower free energy than MgO and therefore have the ability to disrupt the surface oxide of magnesium are shown in Figure 12.16 [3]. Additions of these metals will disrupt the cohesive MgO on the surface of the power as the oxide of the other metal is formed, allowing unimpeded diffusion and increased interparticle bonding.

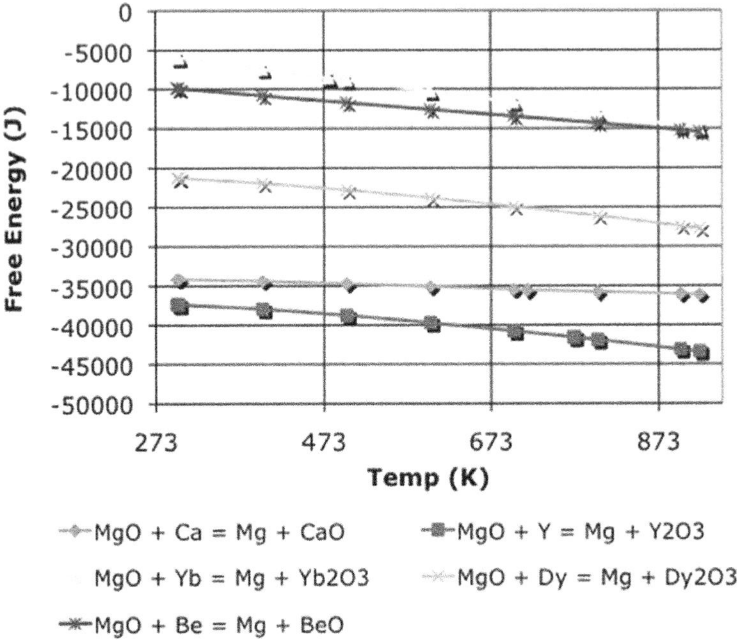

FIGURE 12.16 Free energy—temperature diagram showing metals that may reduce MgO. (From Burke, P., Investigation of the sintering fundamentals of magnesium powders, PhD Thesis, Dalhousie University, Halifax, Canada, 2011.)

12.4 P/M ALLOYING THEORY

In a powder metallurgy product, where the starting material is fine powder and is never subjected to temperatures above melting, grain refinement is not possible or necessary. The grain size largely established by the original size of powder, and sintering kinetics determine final grain size. Therefore, traditional alloying additions used for grain refinement are redundant for P/M alloys. However, grain refinement of wrought and cast alloys takes place through surface modification of the base magnesium. In P/M alloys where grain refinement is not possible, this surface modification may still be beneficial to sintering.

The group of alloying additions that are used to change crystal structure are also not necessary for a P/M product. One of the key advantages of powder metallurgy is the net and near-net shape processing, where no secondary forming is required to produce the final product. Because of this feature, the difficulty in forming that is inherent with hexagonal close packed (HCP) crystal structure is circumvented, and changing to a cubic structure is unnecessary.

The design of alloys for powder metallurgy is mainly concerned with improving strength, ductility, and corrosion resistance. The common approach for new P/M alloys is to mimic wrought and cast alloys. While this is useful in comparing the properties of P/M to cast and wrought, it makes use of alloying additions that may not be useful for P/M processing and does not address challenges that are P/M specific. New P/M specific alloying additions and alloys will need to be designed to take advantage of the benefits, and avoid the problems associated with powder metallurgy.

To increase the sintering response of a P/M alloy, a low melting point metal is often added to provide more area for diffusion to occur. Additions of tin or lithium, which are not traditional magnesium alloying elements, will form a liquid phase at low temperature. Also, traditional strengthening element zinc forms a similar liquid phase. Another consideration is low melting point eutectics that form between magnesium and the alloying addition, or between two alloying additions. Because of the local composition effect in P/M, even very small additions that would form solid solutions in wrought and cast alloys may briefly form a liquid phase. The traditional alloying addition of aluminum has two possible eutectic liquids with magnesium, and alloying elements added for surface layer disruption, such as calcium and yttrium, also may form liquids.

Because the alloy is never fully molten during processing, the surface layer on the original powder is of great concern. Any impurities on the surface may become a barrier to diffusion and densification. Also, while the bulk of the part does not melt, there may be certain constituents of the alloy that form a liquid during sintering. The fraction of liquid must be controlled to prevent excessive dimensional change during sintering, and the behaviour of the liquid, either transient or persistent, will affect sintering performance. If a metal with a more stable oxide to disrupt the surface layer, or a liquid phase forming metal is added, there is the possibility that some of the addition will remain unreacted and go into solid solution or form an intermetallic compound. Therefore, it is important to consider the effect on other properties when adding alloying elements for P/M specific purposes.

12.4.1 Phase Diagrams of Interest for Mg P/M

The following alloy systems are of interest to research in magnesium powder metallurgy either as traditional additions to improve mechanical properties or corrosion resistance, or as powder metallurgy specific additions to form liquid phases or to disrupt surface layers on the magnesium base powder.

Aluminum is a traditional alloying addition for wrought and cast magnesium alloys as it strengthens through solid solution with magnesium, and also improves the corrosion resistance. The phase diagram suggests that precipitation hardening is possible, but the precipitates are bulky and incoherent to the matrix, and little strengthening is realized.

The Al–Mg phase diagram [32] has two low melting temperature eutectics at 450°C and 437°C that will form liquid phases during sintering. However, the solubility of aluminum in magnesium is high, and the liquid is quickly absorbed into the matrix at high sintering temperatures or long sintering times. It is also possible that one of the bulky, incoherent intermetallics may form to the detriment of physical properties.

Calcium has been used in some wrought alloys to improve strength, and the intermetallic, Mg_2Ca, has a higher melting point than magnesium, which if precipitated, may increase high temperature strength. Calcium is of interest to magnesium P/M primarily because it is one of the few elements that has a higher affinity for oxygen, and has the ability to disrupt the surface layer on magnesium particles that interfere with sintering.

The low melting temperature eutectics at 516°C and 445°C allow liquid formation during sintering, which may help quickly draw calcium through the compact by capillary action to disrupt a larger area of magnesium surface layer. The low solubility of calcium in magnesium will impede the diffusion of calcium through magnesium particles, however.

Lithium has been used in wrought alloys to transform the hexagonal crystal structure of magnesium to a cubic structure. An addition of 12 wt% or more lithium is required, and improved formability is realized without severely affecting the advantageous properties of magnesium.

In the case of powder metallurgy, lithium is of interest because it is the closest of all the elements to forming an ideal liquid phase with magnesium. Lithium has a low melting point of 180°C, magnesium is highly soluble in lithium for fast diffusion, and no intermetallic compounds are formed. However, the solubility of lithium in magnesium is relatively high, and will form a transient liquid phase in low concentrations.

Tin has been used as an alloying addition to wrought magnesium with the purpose of strengthening through solid solution and precipitation hardening. The intermetallic, Mg_2Sn, has the unique property of a higher melting point than either magnesium or tin, and high temperature properties are improved. Tin is a possible liquid phase forming addition to magnesium P/M, but the only feature it shares with an ideal liquid phase is a low melting temperature.

Yttrium is used in some high-temperature wrought alloys, where mechanical properties are secondary to creep resistance. Yttrium is of use to magnesium P/M because, like calcium, it will disrupt the oxide of magnesium to improve sintering response. Above 584°C, a eutectic liquid forms which may help distribute yttrium to have a greater layer disruption effect. However, the large number of possible phases and intermetallics make it difficult to predict the behaviour of yttrium during sintering. Also, yttrium has such a strong affinity for oxygen that it may form its own surface layer, hindering any beneficial reactions with magnesium.

Zinc is commonly used in cast and wrought alloys as a solid solution strengthener. Zinc is almost always used along with aluminum additions because it has the ability to refine the bulky Al-Mg precipitates and increases their hardening effect. In the case of P/M, the low melting point of zinc allows for liquid phase formation, but it is highly transient. There are also a number of intermetallics that may form and reduce the amount of pure zinc available to contribute to the liquid phase.

12.4.2 Metal Matrix Composites

Magnesium metal matrix composites (MMC) require a different approach to alloying. MMCs have been studied [33–35,39], where P/M can alleviate some of the matrix/reinforcement interface issues when the matrix metal is in the molten state. The P/M method also allows a highly homogenous mixture of the reinforcement particles within the matrix, resulting in consistent mechanical properties in all directions. Difficulty to achieve successful sintering has brought into question the possibility of P/M to produce magnesium matrix composites [34]. The state in which the surface of the magnesium powder is before the P/M processing controls the sintering process. Obviously the methods of production of magnesium powder are critical in achieving the properties required for a successful sintering. It appears that no attention has been paid into the characterization of the surface properties of the magnesium powders.

12.4.3 Superplastic Alloys

Superplastic alloys are another major research focus that requires different alloying techniques. Superplastic P/M alloys use nano-sized base powders to produce a very fine grain structure in the final sintered product. With the greatly reduced grain size as compared to parts produced through ingot metallurgy, the strength and superplastic properties of the metal are proportionally increased [20]. Powders of alloys that contain rare earth elements and other additions are produced by atomization, where rapid solidification allows extended solubility not possible with ingot metallurgy. Alloying additions are chosen to improve strength, creep resistance and superplasticity [21].

12.5 MAGNESIUM P/M PROCESSING ROUTES

Limited research has been done on the use of powder metallurgy routes for the production of magnesium products. Traditional processing by use of press and sinter methods has the advantage of net and near net shape products, but the disadvantage of difficulty with the stable surface layer of magnesium. The majority of the research that has involved magnesium powder mainly uses P/M as a means to produce alloys that cannot be produced with traditional ladle metallurgy but does not take advantage of other P/M specific properties such as net shape processing. Other research uses novel techniques that overcome the surface layer in different ways, but sacrifice some of the ease of the press and sinter method.

12.5.1 Traditional Press and Sinter

12.5.1.1 Pure Magnesium

The traditional press and sinter methods are difficult to achieve for pure reactive metals. Typically, some additions are required to disrupt the layer and allow unimpeded sintering. However, a complete study of sintering pure magnesium has been completed [26]. In this study the method of powder production, initial powder size, method of compaction, compaction pressure, sintering temperature, and sintering time were addressed. Sample preparation and compaction were done in open air conditions to simulate current industrial processes, and sintering was accomplished under flowing argon. Conclusions were drawn from pre and post sintering density, hardness testing and examination of microstructures.

Powders produced by mechanical grinding were compared to atomized powders and it was found that atomized powders were superior due to lower impurity levels.

Powders of both types were also sieved into four size categories ranging from 38 μm to 75 μm and it was found that the smallest size powders were superior due to the strong dependence on small final grain size in HCP metals due to the Hall-Petch effect. Compaction by uni-axial pressing and cold isostatic pressing (CIP) were tested and it was found that CIP was a more reliable method, and also the only method suitable for the atomized powder. This was believed to be due to the increased shear forces of isostatic compaction that more effectively disrupted the surface layer. In the case of the atomized powder, the increased disruption of the layer leads to more magnesium-to-magnesium contact between particles and the cold welding necessary to have bonding in the compacted state. Higher compaction pressures for both compaction methods were superior due to the same phenomenon.

Samples were sintered for 30 minutes at temperatures of 400°C, 500°C and 600°C and it was found that higher sintering temperatures were beneficial. Although the mechanism was not well understood at the time of publishing, it was shown in later work [3] that the higher temperature allows for more complete decomposition of the $Mg(OH)_2$ and $MgCO_3$ in the surface layer. This creates more pathways for unimpeded diffusion and results in better sintering. Sintering times of 10, 30, 60, 90, 120 and 360 minutes were tested at a sintering temperature of 600°C. The properties increased with sintering time, with 360 minutes resulting in a density of 95% of theoretical, a Rockwell H hardness of 65, a tensile strength of 72 MPa and an elongation of 1.5%.

The increase in properties is therefore related to the longer time required to diffuse through the small area where there are adjacent cracks in the surface layer of two particles. Because all the necking can only take place at these areas, a long time is needed for necks to grow large enough to bond other particles that do not have adjacent cracks. The properties of the samples sintered for 360 minutes are quite comparable to pure sand cast magnesium, and represent a successful method to produce pure magnesium P/M components in the traditional open air uni-axial frame press and controlled atmosphere furnace.

Sintering of pure magnesium has also been achieved by M. Wolff et al. [35]. The study focused on methods of protecting powders from further oxidation by conducting sample preparation and pressing under inert conditions in a glove box, and sintering with the presence of a getter material. The getter material consisted of pure magnesium powder that completely surrounded a crucible that contained the samples to be sintered. Powders of 45 μm size were compacted uni-axially to 100 MPa and sintered under argon for 64 hours at 630°C. Resulting properties were a density of 85% theoretical, with a compressive strength of 167 MPa and a fracture strain of 26%.

The study concludes that sintering of pure magnesium is highly dependent on the purity of sintering conditions so that the powder is not further oxidized. Extensive protection of the samples by a volume of getter magnesium much greater than the sample volume is required to achieve even a low density of 85%. The authors believed that the existing oxide layer was destabilized at the sintering temperature and sintering was accomplished because further oxidation does not take place when surrounded by getter material. However, these conclusions do not fit the behavior noted previously [25].

In the 2009 study [26], it was found that compaction pressure played a larger role in successful sintering due to the disruption of the surface layer. The 100 MPa pressure used in the 2010 study [35] may be insufficient to allow practical sintering, even after the long 64-hour duration. The authors' conclusion that the existing oxide layer is not the issue, but that oxidation during sintering is the problem may stem from the especially long 64-hour time at temperature. At more industrially relevant sintering times, as used in the 2009 study [26], oxidation during sintering was shown to have little effect, and that disrupting the existing stable surface layer was the key factor.

12.5.1.2 Magnesium Alloys

Because pure magnesium does not have sufficient mechanical properties for structural use, alloys are required. Early work was based on replicating common wrought alloys [27], where alloy AZ31 (3% aluminum and 1% zinc, by weight) was tested. Samples were prepared with ground magnesium powder and compacted uni-axially similarly to [26]. Test variables included compaction pressures of 300, 400, or 500 MPa, sintering temperatures of 500°C, 550°C, or 600°C, sintering times of 20, 40, or 60 minutes and post sinter quenching at 375°C or 450°C. The highest properties achieved were a theoretical density of 94% and a Rockwell H hardness of 78 when compacted at 500 MPa, sintered at 500°C for 60 minutes and quenched at 450°C. The mechanical properties were poor however, with a tensile strength of only 31 MPa at and elongation of 0.12%.

The study concluded that the transient liquid phase formed from the low melting point eutectics in the Mg-Al-Zn system are the main barrier to successful sintering, and that a balance in sintering time and temperature must be struck to keep the liquid from absorbing too fast into the magnesium matrix, where it can no longer act as a diffusion bridge, and remaining at the grain boundaries as brittle frozen intermetallics. From further tensile data, it was shown that the intermediate sintering time of 40 minutes resulted in the best mechanical properties, even though the density and hardness were slightly lower than the 60-minute samples. The study also made note that while the system could be optimized by balancing the transient liquid phase behavior, there was an underlying issue that resulted in poor mechanical properties.

It is now known that the surface layer, especially on the ground magnesium powder, plays the most significant role in successful sintering. In light of this new information, select conditions were tested again using atomized magnesium powder, with all other factors remaining the same. The samples compacted at 500 MPa, sintered at 500°C for 40 minutes and quenched at 450°C had the best properties, with a density of 95.1%, a Rockwell H hardness of 58, and a tensile strength of 48 MPa at 0.136% elongation. The resulting microstructure is shown in Figure 12.17.

Therefore the reduction of the existing surface layer by utilizing the atomized powder indeed resulted in superior properties in all tested aspects. The thinner surface layer that surrounds the atomized powder is easier to break during compaction than the thicker layer found on the ground powder. Because diffusion through cracks in the surface layer formed during compaction is the only mechanism for bonding in this case, the atomized powder results in improved properties. However, the continued presence of the surface layer, even though it is less thick, is still a sufficient barrier to diffusion.

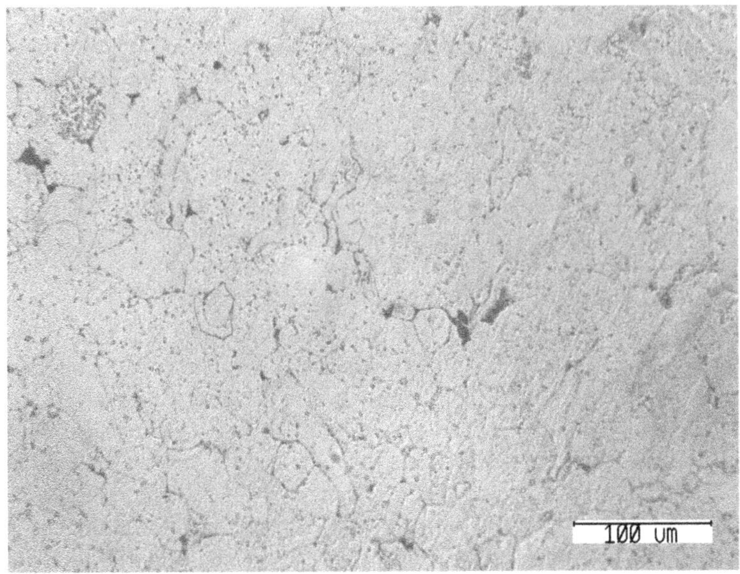

FIGURE 12.17 Microstructure of AZ31 alloy using atomized magnesium powder. Compacted at 500 MPa, sintered at 500°C for 40 minutes.

12.5.2 Novel Compaction Methods

12.5.2.1 Canned Powder Hot Extrusion

Canned powder hot extrusion [21] is the P/M method utilized to produce the majority of samples in both the MMC and grain refinement research primarily due to safety considerations. In this process, the base powders are blended and loaded into a thin-walled cylindrical vessel made from a highly formable metal such as aluminum or copper. The powders are vacuum de-gassed at an elevated temperature to remove air trapped between particles. Following de-gassing, the can is sealed. Extrusion takes place at sintering temperature and the reduction is typically between 10:1 and 20:1, allowing the powders to be compacted and sintered in one step. This process produces parts with very high densities, in the range of 98%+, but it is mainly a laboratory batch type procedure. To implement such a process on an industrial scale would be both costly and time consuming. It also forfeits near-net-shape processing, one of P/Ms key advantages. However, these types of experiments indicate that if surface conditions are good sintering of magnesium can be successful.

12.5.2.2 Equal Channel Angular Pressing

Recently, new sintering and compacting strategies have been identified that may better address issues unique to magnesium. Equal channel angular pressing (ECAP) [35] is a method to provide extremely high strain rates, equivalent to extruding at a 30:1 ratio, without a change in cross sectional dimensions. The high strain breaks any surface layer on the particles and plastically deforms the powders to ensure good interparticle bonding and provides adequate pathways for diffusion. While cross sectional dimensions are retained, it is still only possible to produce rods of material, and not near-net-shape parts.

12.5.3 Novel Surface Layer Disrupting Additions

12.5.3.1 Calcium

Calcium has the ability to naturally reduce MgO, the main component of the surface layer, which disrupts the layer. Early experiments show that small calcium additions do improve sinterablity [28]. In that work, a large granule of calcium was compacted into one end of a cold isostatically pressed magnesium cylinder, which was then sintered under argon for 40 minutes at 600°C. The result was first a calcium/magnesium eutectic liquid forms, which then travels through the grain boundaries of the magnesium matrix by capillary action. The liquid readily wets the magnesium grains and disrupts the layer, allowing more sites for necking to take place. The liquid also acts as a traditional liquid phase sintering addition, creating a diffusion bridge for mass flow between particles that are not in direct point contact.

The study concluded from visual examination of the microstructure that sintering was complete and porosity was very low in an area close to the calcium granule. Examining the amount of calcium present in this well-sintered portion of the cylinder found that an amount of 1% by weight of calcium would be sufficient to disrupt the layer. Areas farther from the calcium granule where the eutectic liquid did not penetrate were not well sintered and were typical in microstructure to pure magnesium.

Following that study, Wolff et al. performed experiments with 1% calcium additions, prepared by blending magnesium/calcium alloys with pure magnesium [26]. Sintering conditions were similar to the pure magnesium study included in the same publication, where samples were contained in a covered crucible surrounded by getter material and sintered at 610°C for 64 hours. Best results were found using Mg-7Ca master alloy for the 1% addition of calcium, where density was 98% of theoretical and compressive strength was 255 MPa at 25% elongation. Reducing the calcium content to 0.6% increases the compressive strength to 284 MPa, which the authors contribute to less Mg_2Ca intermetallic formation.

The study concludes that the liquid phase formed is the main contributing factor to the success of the calcium containing samples versus pure magnesium. The interaction of the calcium with the oxide is mentioned but not elaborated upon, and oxygen content of the matrix or calcium rich grain boundaries is not reported.

In light of the success of the calcium additions from the previous studies, Mg-1Ca samples were prepared using traditional press and sinter P/M methods. Pure calcium granules were ground in a glove box and sieved to <100 μm size and blended with pure magnesium. The homogenous mixture was compacted open to air in a uni-axial press. Sintering was performed under argon at a temperature of 600°C for

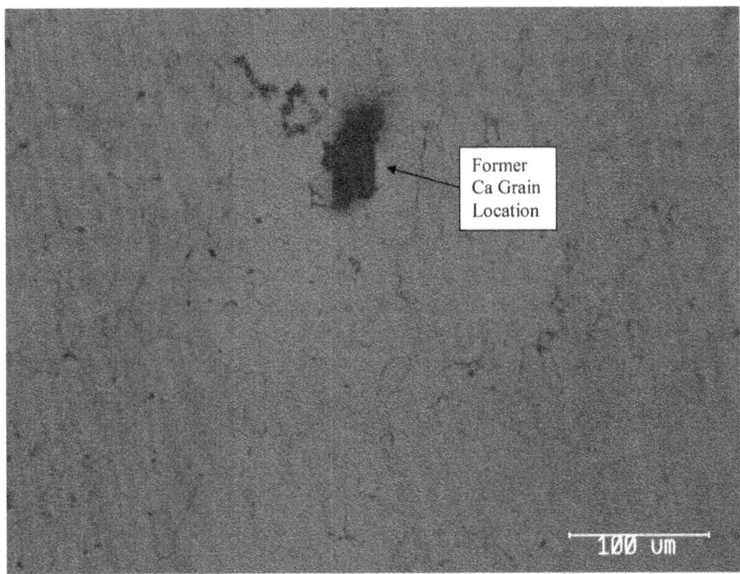

FIGURE 12.18 Mg-1Ca alloy sintered at 600°C for 40 minutes.

40 minutes. Figure 12.18 shows the resulting microstructure. Sintering is complete and grains are well bonded, and the majority of pores are small and round, located at grain boundaries. The exception is the large pore in which a calcium powder particle formerly occupied, as detailed in Figure 12.19, which shows an SEM image of a similar pore, along with the EDS calcium map. It is clear the pure calcium particle has reacted with magnesium to form the eutectic liquid, which has then spread across the matrix grains disrupting the oxide layer. Any remaining unreacted calcium is then drawn into the matrix grains as a solid solution. The EDS map indicates that the calcium is homogenously distributed in the magnesium matrix.

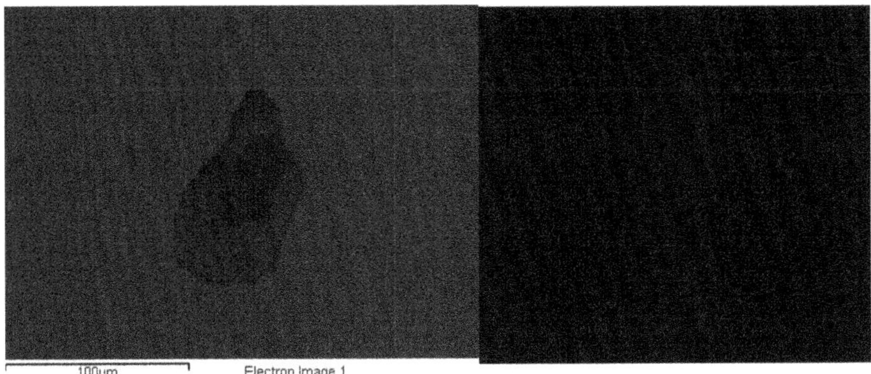

FIGURE 12.19 SEM image of large pore in Mg-1Ca alloy. Ca EDS map showing homogenous distribution in surrounding matrix, New AZ31 1%Ca.

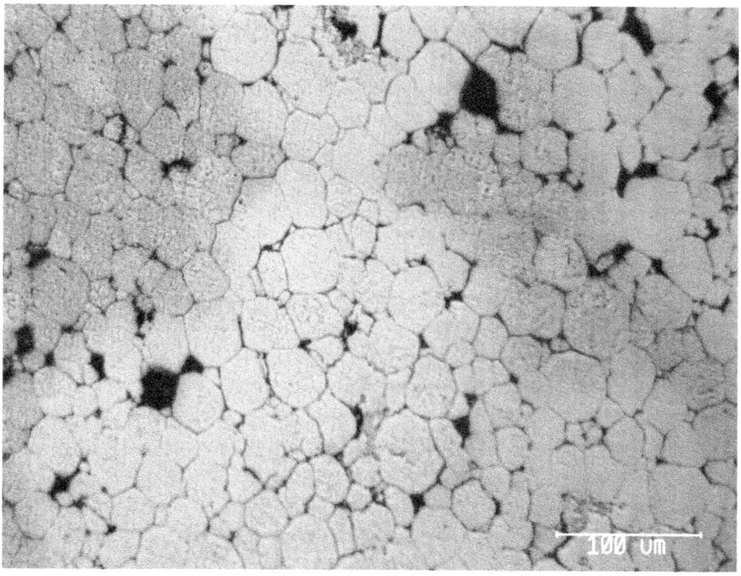

FIGURE 12.20 AZ31 with 1 wt% Ca addition. Sintered at of 600°C for 40 minutes.

The resulting properties of the Mg-1Ca samples were a density of 91% of theoretical, a Rockwell H hardness of 55, and a shrinkage of approximately 2%. The density is low because of the macro pores created when calcium reacts with magnesium to form the eutectic liquid. This leads to the conclusion that calcium additions should be as fine as possible, so they may fit in the interstitial space between magnesium particles. This way when they react to form liquid, the resulting pore will be no larger than the natural pore space between matrix particles.

Calcium additions have also been tested with alloy AZ31, building on work shown in a previous section on AZ31 P/M alloys produced with atomized magnesium powder. Processing is similar save for the addition of 1 wt% elemental calcium ground and sieved to <100 μm in a glove box. Samples were sintered at 600°C for 40 minutes. The properties achieved were a theoretical density of 94.1% with a tensile strength of 85 MPa at an elongation of 1.2%. The microstructure is shown in Figure 12.20. Porosity is fairly high due to the many low melting point eutectics that form in the Mg-Al-Zn-Ca system. Areas formerly occupied by aluminum, zinc, and calcium become pores as the metals react and liquefy. Bonding is strong between particles however, as shown by the relatively high tensile strength even with the large amount of porosity.

12.5.3.2 Yttrium

Yttrium is another metal which is more reactive than magnesium and therefore has the capacity to disrupt the magnesium surface layer. Yttrium was added to pure magnesium in the amount of 1 wt%, compacted uni-axially to 500 MPa and sintered for 40 minutes at 600°C. The resulting properties were a theoretical density of 96%, a Rockwell H hardness of 58 and shrinkage of approximately 1.5%. The microstructure

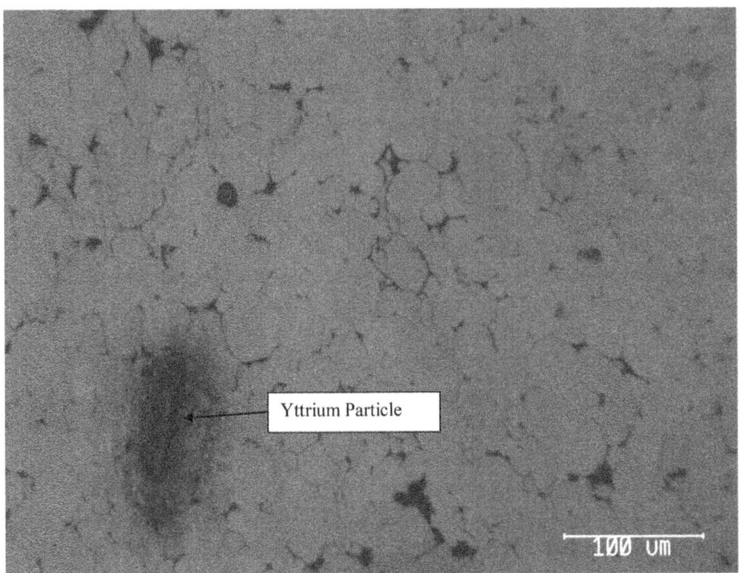

FIGURE 12.21 Mg-1Y alloy sintered at 600°C for 40 minutes.

is shown in Figure 12.21 where an yttrium particle that has not fully reacted is visible. Figure 12.22 shows a SEM image and corresponding EDS of an area with a similar yttrium particle, and it can be seen that while the particle is not fully reacted, the surrounding area has a homogenous distribution of yttrium in magnesium. Similar to the calcium case, it would be beneficial to have the yttrium addition as small as possible to avoid any large incoherent particles in the matrix.

Yttrium was also used as an addition to Mg-8Al alloy in the amount of 2 wt% [29]. Samples were pressed uni-axially at 500 MPa and sintered for 40 minutes at 600°C. The resulting properties were 95% theoretical density, a Rockwell H hardness of 81, and a dimensional change of 2%. Further testing not completed in time to be included

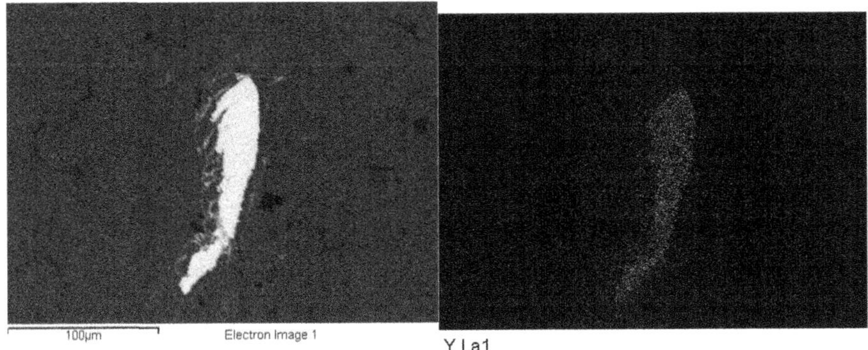

FIGURE 12.22 SEM image of yttrium particle in Mg-1Y alloy. Yttrium EDS map showing homogenous distribution in surrounding matrix.

in the original paper showed the Mg-8Al-2Y alloy had a tensile strength of 128 MPa at an elongation of 3.2%. These results are very promising, with a large increase in mechanical properties compared to other alloys tested.

12.5.4 Novel Sintering Methods

12.5.4.1 Spray Forming

Spray forming has been used to produce magnesium alloy billets [36]. Spray forming allows the direct rapid solidification and deposition of high density from a liquid alloy. The process has the benefit that contamination is minimized because powder production, compaction and consolidation are completed in one step. The process is somewhat net-shaped, as the spray can be manipulated as the layers are deposited, so simple two-dimensional parts are possible. However, to realize full strength and density, the parts were first extruded.

12.5.4.2 Spark Plasma Sintering

Spark plasma sintering (SPS) has been used to produce alloys [37] and MMCs [38]. SPS enjoys very short sintering times, and the high energy pulse has been attributed to cleaning contaminates from powder surfaces in-situ. This can allow sintering of pure metals or alloy with no addition of sintering aids. Dies for the process are either graphite or metal and allow for near-net-shape production. However, the process is completed under vacuum and under pressure, so the equipment is complicated and batch in operation.

12.5.4.3 Microwave

Microwave sintering of magnesium was demonstrated by Gupta et al. [40]. Compacts were placed inside a conventional microwave oven that had been specially converted to achieve a controlled atmosphere. A susceptor material surrounded the bottom and sides of the samples so that a two directional heating was achieved. Microwave heating from the inside out, and microwave coupled external heating from the outside in allows rapid sintering. After sintering samples were hot extruded to minimize porosity and improve mechanical properties. In the study, microwave sintering was compared to conventional sintering, and it was found that a microwave sintering length of 25 minutes gave improved results to two hours of conventional sintering. It was concluded that the rapid microwave sintering minimized the extent of high temperature exposure. Although the authors did not mention, it is possible that any further oxidation that takes place at temperature is reduced due to the reduced time at high temperature.

12.5.5 Post Sintering Operations

12.5.5.1 Hot Work

Many of the studies that have been completed on magnesium P/M use some high sheer stress process after sintering to bring the density of the samples near theoretical [20,21,24,26,34,36]. Obvious benefits to mechanical properties are realized with the elimination of porosity. Most studies do not report properties prior to the application of the high sheer stress hot work, with the exception of Ref. [37]. In the study a Mg-8Al-2Y alloy was hot rolled after sintering at 350°C in 0.127 mm increments until signs of cracking were apparent.

The hot rolled samples attaining a near-theoretical density of 99% from 95%, which was accomplished in a reduction of 26%. The effect of increasing the density was most apparent in hardness, where it had increased by a large factor in all samples successfully rolled, climbing an entire scale in the Rockwell test from 81 H to 81 E. While there are no wrought magnesium alloys with exact compositions, a similar alloy, AZ91 (Mg-9Al-1Zn), has a hardness of 75 HRE.

This demonstrates the large effect that secondary high sheer stress hot working has on the samples, and why most researchers choose to complete this step without regard for properties previous to hot working. However, to take advantage of the near-net shape benefits of P/M, secondary hot work is not advisable. It is for this reason that the current research trend is toward alloys and additions that make near full density possible without the need for secondary hot working [29,36–38].

12.5.6 Additive Manufacturing of Magnesium

With additive manufacturing, it is possible to directly consolidate magnesium or magnesium alloy powders to net-shape components by selective laser melting. Additive manufacturing is well-established for some engineering materials based on iron, nickel, or titanium alloys [41]. However, additive manufacturing of magnesium components is still in the developmental stages where research remains focused on developing the processing rather than finished properties [42]. There is a strong interest in developing additively manufactured magnesium alloys for automotive, aerospace, and biomedical applications owing to magnesium's high stiffness, low density, biocompatibility, and biodegradability.

Despite the many challenges remaining in processing magnesium, some generalizations for additive manufacturing may be drawn from investigations on selectively laser melting pure magnesium [43–45], Mg-9Al [46], AZ91 (Mg-9Al-1Zn) [47–49], AZ31 (Mg-3Al-1Zn) [50], WE43 (Mg-4Y-3Rare Earth) [51,52], and ZK60 (Mg-6Zn-0.5Zr) [53,54] alloys. Gas or centrifugally atomized, pre-alloyed magnesium alloy powder in the size range 40–100 μm tends to produce the highest quality parts. If the powder is too fine, the surface area to volume ratio increases and, therefore, so does the content of contaminants from the surface layer. Pre-alloyed powder helps ensure the small melting zone (beam focal point of ~100 μm) contains the appropriate alloy additions to disrupt the surface layer (i.e., Ca or Y). Under carefully chosen processing conditions, it is possible to manufacture magnesium components with densities routinely exceeding 95% theoretical, even up to 99.5% theoretical, and good mechanical properties. Tensile and yield strength are usually comparable to, or exceed, the values for corresponding cast magnesium alloys, however their elongation remains inferior [47,55].

Most of the major challenges in additively manufacturing magnesium components arise from the reactivity and physical properties of magnesium metal. At its melting point, liquid magnesium metal is thermodynamically favoured to form MgO under partial pressures of oxygen greater than 10^{-56} atm. Therefore, it is practically impossible to prevent the oxidation of magnesium during manufacturing as such a low partial pressure of oxygen is impossible to maintain in the build chamber. In practice, even when a flowing atmosphere of high-purity argon or helium is maintained in the build chamber, the final oxygen content of magnesium components becomes significant (up to 6–10 at.% at the surface [43,44]), especially if the powder bed is pre-heated [56].

In cast magnesium operations, it is well known that atmospheres containing sulfur hexafluoride (SF_6) or carbon dioxide help minimize oxidation [57], but this has not been explored in additive manufacturing of magnesium. In the build chamber, there is essentially a new surface layer formed at each step which must be disrupted through alloy additions (i.e., Ca or Y). Most studies on additive manufacturing have not considered the role of surface layer disrupting additions. In the unused powder bed, there is a high powder attrition rate because the growth of surface layers reduces powder recyclability, especially for longer build times.

Certain physical properties of elemental magnesium also pose a challenge in its additive manufacturing. Magnesium's high vapor pressure (~4×10^{-3} atm at the melting point), relatively small difference between melting and boiling points (650°C and 1091°C, respectively), and low enthalpy of vaporization require precise control of the laser energy input to prevent vaporization of magnesium [48]. Zinc is also relatively volatile and may selectively vaporize from certain magnesium alloys. Vaporization is somewhat reduced by operating the laser in a pulsed, rather than continuous, mode [44] or by pressurizing the build chamber (to 1.3 atm absolute pressure) to increase the boiling point of magnesium [58]. Another challenge is the relatively low viscosity, low surface tension, and high solidification shrinkage of liquid magnesium which affects the stability of the melt pool and can increase porosity [42].

Even with the difficulties encountered, the outlook for additively manufactured magnesium components is promising. There is a high potential in biomedical applications such as bone implants [59] or biodegradable porous scaffolds [52]. New alloys are being developed specifically for selectively laser melting [60] and new applications are being discovered such as battery components [61]. Another promising technique to overcome the oxidation challenges is to develop wire-fed systems heated by an electron beam, laser, plasma, or electric arc. A wire-fed electric arc system was able to manufacture components with up to 25% elongation [62].

Another possibility is electrically enhancing the manufacturing process by high frequency AC or pulse DC methods, similar to welding technologies for magnesium.

12.6 CONCLUSIONS

Thermochemical results indicate that the sintering of pure magnesium is difficult in the solid state. Disruptive additions lead to the formation of alloys that seem to reduce the surface film and may allow the sintering and consolidation bellow the melting temperature. Surface analysis indicate that the surface film on magnesium consists of oxide, hydroxide and carbonate. Hydroxides and carbonate decompose at relatively low temperatures allowing significant improvement in solid consolidation. Additions of calcium and rare earth elements have the tendency to disrupt the surface film to make sintering possible. Further consolidation of the magnesium alloys may be achieved by additive manufacturing techniques.

ACKNOWLEDGMENTS

The authors wish to acknowledge the financial support of the Natural Sciences and Engineering Research Council (NSERC) of Canada. The Minerals Engineering Centre (MEC) and MATNET of Dalhousie University for allowing the use of the characterization equipment. Also acknowledged are the efforts by Mr. Piotr Cebula to verify some of the results.

REFERENCES

1. D.P. Bishop, G.J. Kipouros and W.F. Caley. "Diffusion-based microalloying via reaction sintering." *Journal of Materials Science*, 32, 2353–2358 (1997).
2. D.P. Bishop, J.R. Cahoon, M.C. Chaturvedi, G.J. Kipouros and W.F. Caley. "Diffusion-based microalloying of aluminum alloys by powder metallurgy and reaction sintering." *Journal of Materials Science*, 33, 3927–3934, (1998).
3. P. Burke. "Investigation of the sintering fundamentals of magnesium powders." PhD Thesis, Dalhousie University, Halifax, Canada, January 2011.
4. H.E. Friedrich and B.L. Mordike. *Magnesium Technology – Metallurgy, Design Data, Applications*, Springer–Verlag, Berlin, Germany, 2006.
5. R.M. German. "Thermodynamics of sintering." In *Sintering of Advanced Materials: Fundamentals and Processes*, Ed. Z.Z. Fang, Woodhead Publishing, Oxford, UK, pp. 3–32, 2010.
6. M.N. Rahaman. "Kinetics and mechanisms of densification." In *Sintering of Advanced Materials: Fundamentals and Processes*, Ed. Z.Z. Fang, Woodhead Publishing, Oxford, UK, pp. 33–64, 2010.
7. P. Burke, D. Fancelli and G.J. Kipouros. "Investigation of the sintering fundamentals of magnesium powders." Ed. M.O. Pekguleryuz, *Light Metals in Transport Applications, COM 2007*, Toronto, Ontario, pp. 183–195, 2007.
8. G.M. Gross, H.J. Seifert and F. Aldinger. "Thermodynamic assessment and experimental check of fluoride sintering aids for AlN." *Journal of the European Ceramic Society*, 18 (7), 871–877, 1998.
9. R. Zhang, L. Gao and J. Guo. "Thermodynamic behavior of copper-coated silicon carbide particles during conventional heating and spark plasma sintering." *Journal of the American Ceramic Society*, 86 (8), 1446–1448, 2003.
10. M. Medraj, Y. Baik, W.T. Thompson and R.A.L. Drew. "Understanding AlN sintering through computational thermodynamics combined with experimental investigation." *Journal of Materials Processing Technology*, 161 (3), 415–422, 2005.
11. F. Wakai, D. Gomez-Garcia and A. Dominguez-Rodriguez. "Pore channel closure in sintering of a ring of three spheres." *Journal of the European Ceramic Society*, 27 (11), 3365–3370, 2007.
12. I. Fernandez, G. Gonzalez, F.C. Gennari and G.O. Meyer. "Influence of sintering parameters on formation of Mg-Co hydrides based on their thermodynamic characterization." *Journal of Alloys and Compounds*, 462 (1–2), 119–124, 2008.
13. J.A. Puszkiel, P. Arneodo Larochette and F.C. Gennari. "Thermodynamic and kinetic studies of Mg-Fe-H after mechanical milling followed by sintering." *Journal of Alloys and Compounds*, 463 (1–2), 134–142, 2008.
14. Y. Sarikaya, K. Ada and M. Onal. "A model for initial-stage sintering thermodynamics of an alumina powder." *Powder Technology*, 188 (1), 9–12, 2008.
15. D. Tan, H. Ma, G. Li, H. Liu and D. Zou. "Sintering reaction of pseudoleucite syenite: Thermodynamic analysis and process evaluation." *Earth Science Frontiers*, 16 (4), 269–276, 2009.

16. D. Li, S. Chen, D. Wang, Y. Li, W. Shao, Y. Long, Z. Liu and S.P. Ringer. "Thermo-analysis of nanocrystalline TiO2 ceramics during the whole sintering process using differential scanning calorimetry." *Ceramics International*, 36 (2), 827–829, 2010.
17. J.R Shallenberger, D.A. Cole, S.W. Novak, R.L. Moore, M.J. Edgell, S.P. Smith and C.J. Hitzman, et al. "Oxide thickness determination by XPS, AES, SIMS, RBS and TEM." *Ion Implantation Technology Proceedings*, 1998 International Conference on, vol. 1, pp. 79–82, 1999.
18. R.N. Lumley, T.B. Sercombe and G.B. Schaffer. "Surface oxide and the role of magnesium during the sintering of aluminium." *Metallurgical and Materials Transactions A*, 30 (2), 457–463, 1999.
19. B.J. Kellet and F.F. Lange. "Thermodynamics of densification: I, sintering of simple particle arrays, equilibrium configurations, pore stability, and shrinkage." *Journal of the American Ceramic Society*, 72 (5), 725–734, 1989.
20. K. Kubota, M. Mabuchi and K. Higashi. "Review processing and mechanical properties of fine-grained magnesium alloys." *Journal of Materials Science*, 34, 2255–2262, 1999.
21. Y. Kawamura and A. Inoue. "Development of high strength magnesium alloys by rapid solidification." *Materials Science Forum*, 419–422, 709–714, 2003.
22. Y.G. Kipouros, P. Burke and G.J. Kipouros. "A quasi-thermodynamic approach to the sintering of pure magnesium powder." *Thermodynamics 2011*, Athens, Greece, 31 August–3 September 2011 Ed. G. Jackson & I.G. Economou, pp. 283–284 (2011).
23. Y.G. Kipouros. "Kinetic considerations of the sintering process of aluminum and magnesium powders, B.Eng." Thesis, Dalhousie University, 30 November, 2012.
24. J. Li, G.J. Kipouros, P. Burke and C. Bibby. "Investigation of surface film Formed on Fine Mg Particles." *Microscopy & Microanalysis*, 15, 362–363, 2009.
25. P. Burke, J. Li and G.J. Kipouros. "DSC and FIB/TEM investigation of calcium and yttrium additions in the sintering of magnesium powder." *Canadian Metallurgical Quarterly*, 55, 1–12, 2016.
26. P. Burke, D. Fancelli. V. Laverdiere and G.J. Kipouros. "Sintering fundamentals of magnesium powders." *Canadian Metallurgical Quarterly*, 48 (2), 123–132, 2009.
27. P. Burke and G.J. Kipouros. "Magnesium powder metallurgy AZ31 alloy using commercially available powders." *High Temperature Materials and Processes*, 30 (1–2), 51–61, 2011.
28. P. Burke and G.J. Kipouros. "Powder metallurgy of magnesium: Is it feasible?" Ed. S. R. Agnew, *Magnesium Technology 2010, TMS 2010*, Seattle, WA, pp. 115–120, 2010.
29. P. Burke, C. Petit, V. Vuaroqueaux and G.J. Kipouros. "The effect of processing sintering parameters and post-sintering operations in the production of magnesium powder metallurgy parts." *COM 2010, Aerospace Materials and Manufacturing: Advances in Materials, Processes and Repair Technologies*, pp. 225–234, 2010.
30. Metal Powders Industries Federation, *Standard Test Methods for Metal Powders and Powder Metallurgy Products*, Princeton, NJ, 1999 Edition.
31. L. Tisza. *Generalized Thermodynamics*, 42, MIT Press, Cambridge, MA, 1966.
32. *ASM Handbook of Binary Phase Diagrams.*
33. D.J. Loyd. "Particle reinforced aluminum and magnesium matrix composites." *International Materials Review*, 39, 1–23, 1994.
34. W. Xie, Y. Lui, D.S. Li, J. Zhang, Z.W. Zhang and J. Bi. "Influence of sintering routes to the mechanical properties of magnesium alloy and its composites produced by PM technique." *Journal of Alloys and Compounds*, 431, 162–166, 2007.
35. German Sintering Magnesium, 2010.
36. C.J. Bettles, M.H. Moss and R. Lapovok. "A Mg-Al-Nd alloy produced via a powder metallurgical route." *Materials Science and Engineering: A*, 515 (1–2), 26–31, 2009.

37. Y. Li, Y. Chen, H. Cui, B. Xiong and J. Zhang. "Microstructure and mechanical properties of spray-formed AZ91 magnesium alloy." *Materials Characterization*, 60 (3), 240–245, 2009.
38. H.T. Son, J.M. Hong, I.H. Oh, J.S. Lee, T.S. Kim and K. Maruyama. "Microstructure and mechanical properties of Mg-Zn-Y alloys fabricated by rapid solidification and spark plasma sintering processes." *Solid State Phenomena: B*, 124–126 (2), 1517–1520, 2007.
39. J. Umeda, K. Kondoh and H. Imai. "Friction and wear behavior of sintered magnesium composite reinforced with CNT-Mg2Si/MgO." *Materials Science and Engineering: A*, 504 (1–2), 157–162, 2009.
40. M. Gupta and W.L.E. Wong. Enhancing overall mechanical performance of metallic materials using two-directional microwave assisted rapid sintering, *Scripta Materialia*, 52 (6), 479–483, 2005.
41. T. DebRoy, H.L. Wei, J.S. Zuback, T. Mukherjee, J.W. Elmer, J.O. Milewski, A.M. Beese, A. Wilson-Heid, A. De and W. Zhang. "Additive manufacturing of metallic components–Process, structure and properties." *Progress in Materials Science*, 92, 112–224, 2018.
42. V. Manakari, G. Parande and M. Gupta. "Selective laser melting of mangesium and magnesium alloy powders: A review." *Metals*, 7 (1), 2, 2017.
43. C.C. Ng, M.M. Savalani, H.C. Man and I. Gibson. "Layer manufacturing of magnesium and its alloy structures for future applications." *Virtual and Physical Prototyping*, 5 (1), 13–19, 2010.
44. C.C. Ng, M. Savalani and H.C. Man. "Fabrication of magnesium using selective laser melting technique." *Rapid Prototyping Journal*, 17 (6), 479–490, 2011.
45. D. Hu, Y. Wang, D. Zhang, L. Hao, J. Jiang, Z. Li and Y. Chen. "Experimental investigation on selective laser melting of bulk net-shape pure magnesium." *Materials and Manufacturing Processes*, 30 (11), 1298–1304, 2015.
46. B. Zhang, H. Liao and C. Coddet. "Effects of processing parameters on properties of selective laser melting Mg–9%Al powder mixture." *Materials and Design*, 34, 753–758, 2012.
47. K. Wei, M. Gao, Z. Wang and X. Zeng. "Effect of energy input on formability, microstructure and mechanical properties of selective laser melted AZ91D magnesium alloy." *Materials Science and Engineering A*, 611, 212–222, 2014.
48. G. Yingchun, Z. Wei, Z. Hongyu, L. Zhongli, S.H. Leng and H. Minghui. "Analysis of selective vaporization behavior in laser melting of magnesium alloy by plume deposition." *Laser and Particle Beams*, 32 (1), 49–54, 2014.
49. D. Schmid, J. Renza, M.F. Zaeh and J. Glasschroeder. "Process influences on laser-beam melting of the magnesium alloy AZ91." *Physics Procedia*, 83, 927–936, 2016.
50. A. Pawlak, M. Rosienkiewicz and E. Chlebus. "Design of experiments approach in AZ31 powder selective laser melting process optimization." *Archives of Civil and Mechanical Engineering*, 17 (1), 9–18, 2017.
51. R. Tandon, T. Wilks, M. Gieseke, C. Noelke, S. Kaierle and T. Palmer. "Additive manufacturing of elektron® 43 alloy using laser powder bed and directed energy deposition." In *Euro PM2015. European Congress and Exhibition on Powder Metallurgy. AM—Special Processes and Materials*, October 2015, pp. 1–7.
52. Y. Li, J. Zhou, P. Pavanram, M.A. Leeflang, L.I. Fockaert, B. Pouran, N. Tümer, K. Schröder et al. "Additively manufactured biodegradable porous magnesium." *Acta Biomaterialia*, 67, 378–392, 2018.
53. K. Wei, Z. Wang and X. Zeng. "Influence of element vaporization on formability, composition, microstructure, and mechanical performance of the selective laser melted Mg–Zn–Zr components." *Materials Letters*, 156, 187–190, 2015.

54. C. Shuai, Y. Yang, P. Wu, X. Lin, Y. Liu, Y. Zhou, P. Feng, X. Liu and S. Peng. "Laser rapid solidification improves corrosion behavior of mg-Zn-Zr alloy." *Journal of Alloys and Compounds*, 691, 961–969, 2017.
55. C.C. Ng, M.M. Savalani, M.L. Lau and H.C. Man. "Microstructure and mechanical properties of selective laser melted magnesium." *Applied Surface Science*, 257, 7447–7454, 2011.
56. M.M. Savalani and J.M. Pizarro. "Effect of preheat and layer thickness on selective laser melting (SLM) of magnesium." *Rapid Prototyping Journal*, 22, 1, 115–122, 2016.
57. K. Aarstad. Protective Films on Molten Magnesium. PhD thesis, Norwegian University of Science and Technology, 2004. http://hdl.handle.net/11250/248760.
58. M. Gieseke, C. Noelke, S. Kaierle, V. Wesling and H. Haferkamp. "Selective laser melting of magnesium and magnesium alloys." In *Magnesium Technology 2013*, TMS, pp. 65–68, 2013.
59. J. Matena, S. Petersen, M. Gieseke, M. Teske, M. Beyerbach, A. Kampmann, H. Escobar, N. Gellrich, H. Haferkamp and I. Nolte. "Comparison of selective laser melted titanium and magnesium implants coated with PCL." *International Journal of Molecular Sciences*, 16 (6), 13287–13301, 2015.
60. T. Long, X. Zhang, Q. Huang, L. Liu, Y. Liu, J. Ren, Y. Yin, D. Wu and H. Wu. "Novel mg-based alloys by selective laser melting for biomedical applications: Microstructure evolution, microhardness and in vitro degradation behaviour." *Virtual and Physical Prototyping*, 13 (2), 71–81, 2018.
61. Y.T. Chen, F.Y. Hung, T.S. Lui and J.Z. Hong. "Microstructures and charge-discharging properties of selective laser sintering applied to the anode of magnesium matrix." *Materials Transactions*, 58 (4), 525–529, 2017.
62. J. Guo, Y. Zhou, C. Liu, Q. Wu, X. Chen and J. Lu. "Wire arc additive manufacturing of AZ31 magnesium alloy: Grain refinement by adjusting pulse frequency." *Materials*, 9 (10), 823, 2016.
63. P.J. Burke, Z. Bayindir and G.J. Kipouros. "X-ray photoelectron spectroscopy (XPS) investigation of the surface film on magnesium powders." *Applied Spectroscopy*, 66, 510, 2012.

13 Metallography of Magnesium Alloys

S.V.S. Narayana Murty, Sushant K. Manwatkar, V.S.K. Chakravadhanula, and P. Ramesh Narayanan

CONTENTS

13.1 INTRODUCTION TO MAGNESIUM ALLOYS

Magnesium alloys are the lightest (density 1.8 g/cc) structural metallic materials with attractive combination of properties like good strength, stiffness, high damping capacity, high thermal diffusivity and constitute a special class of alloys for light weight applications [1,2]. Processing difficulties arising from high reactivity and relatively low workability as well as low corrosion resistance are few limitations of these alloys to be overcome for more extensive use of magnesium alloys. Pure magnesium has hexagonal closed packed (HCP) structure, shows very poor workability and deforms through basal slip at room temperature. As the temperature of deformation is increased, slip occurs on additional non-close packed planes viz. $(10\bar{1}0)$, (1120), (1121) and thus it fulfills the requirement of five independent slip systems for compatible deformation. Alloying with aluminium and zinc improves its workability by promoting slip on non-basal planes. Minor additions of zirconium to Mg-Zn and Mg-RE alloy systems results in refinement of crystal grains thereby enhancing the workability of the alloy. Some alloying elements like lithium and indium improve workability remarkably by changing the crystal structure from HCP to body centered cubic (BCC) [3]. Accordingly, there are three ways for improving the workability of magnesium viz. by alloying, by grain refinement and by changing crystal structure. All the above three methods are widely used for the development of magnesium alloys for various applications.

Microstructural characterization of magnesium alloys is important for understanding the various constituent phases present and to relate it to the mechanical properties. Varieties of microstructures are developed in these alloys depending on alloying/trace element additions, method of manufacture, thermomechanical processing performed and temper condition imparted to the final product. The solidification microstructure in the as-cast condition reveals the cast grain size/cell size and interdendritic arm spacing, constituent particles, cast defects like porosities due to gas or shrinkage and cracks. These features can be readily observed in an optical microscope after polishing and etching of specimens with a suitable reagent. The grain structure and orientation can be observed in thermomechanically processed materials like plates/sheets/extrusions using an optical microscope. If higher resolution of these microscopic features is desired or chemical composition of these microscopic features is to be obtained, a scanning electron microscope (SEM) fitted with an energy dispersive spectroscope (EDS) can be used. The microstructural changes that take place during the age hardening can however be observed only in a transmission electron microscope (TEM). This includes formation of fine strengthening precipitates during ageing and observation of dislocation structures in thermomechanically processed materials. A TEM equipped with an EDS or an electron energy loss spectroscopy (EELS) would provide the chemical composition of these phases.

For complete understanding of the microstructure-mechanical property correlation in magnesium alloys, it is essential to understand the microstructures and/or compositions at different length scales. With the challenges imposed by structural designers for meeting various requirements, physical metallurgists are synthesizing tailor made alloys for specific applications. This is in strong contrast to the much simpler binary and ternary alloy systems. These tailor-made alloys are microstructurally far more complex and their understanding is essential for their usage in critical applications. Therefore, the purpose of this chapter is to provide a comprehensive understanding on the metallographic observation of magnesium alloys. This includes preparation of the specimens for optical microscopic examination, optical photomicrographs of different grades of commercial and experimental magnesium alloys, characterization of phases by SEM and finally characterization of precipitates and nanoscale substructure features through (S)-TEM.

13.2 METALLOGRAPHY OF MAGNESIUM ALLOYS

Magnesium and its alloys are among the most difficult metallic materials to prepare for metallographic examination [4]. Pure magnesium is very soft with low hardness and strength and is unsuitable for structural applications. Therefore, it is always alloyed with other elements to impart higher strength, improved workability, better corrosion resistance and improved creep resistance. These alloying additions will result in increasing the hardness of the matrix through solid solution strengthening and result in the formation of intermetallic phases. These intermetallics or precipitates that form as a result of age hardening are much harder than the magnesium matrix. This poses difficulty in eliminating scratches and in controlling relief (height difference between matrix and second phases) in the polished specimen. The high reactivity of magnesium also poses problems with usage of water during the specimen preparation, especially during the final polishing steps. Therefore, considerable care must be taken to prepare metallographic specimens of magnesium alloys for microstructural examination. The following sections present the various steps involved in the preparation of magnesium alloy specimens for microscopic examination.

13.3 OPTICAL MICROSCOPY OF MAGNESIUM ALLOYS

Optical microscopy is the most common and vital tool for microstructural examination of magnesium and its alloys. Despite having lower resolution, it is used up to magnification of 2500X with features as small as 0.1 μm resolved clearly. Optical microscopic examination of freshly polished surface provides useful information on size, shape and distribution of second phase particles, presence of casting/welding defects such as porosity, cracks; eutectic melting due to overheating and presence of extraneous inclusions. After etching with a suitable reagent, the microstructure reveals the shape and size of the (sub)grains, phases present, features resulting from thermomechanical processing such as grain shape and their orientation. However, it must be noted that optical microscopy cannot reveal the precipitates formed during age hardening or other sub-structure information such as the presence of dislocations, for which transmission electron microscopy is required. The identification of intermetallic phases present is important because, their shape, size, volume fraction and distribution can affect the mechanical properties. These intermetallic phases are the result of equilibrium/non-equilibrium reactions occurring in the alloy during solidification and their fragmentation and distribution during subsequent thermomechanical processing.

Metallographic examination of magnesium alloys is usually conducted on polished and etched specimens for research and development, quality control and failure analysis investigations. This a well executed series of steps progressively consisting of specimen sectioning, mounting, grinding, polishing and etching. Sectioning is essential in most cases to obtain a small piece for metallographic examination. This step is very critical for magnesium alloys as will be discussed in the following sections. During sectioning of a sample from bulk, care should be taken to ensure that the sample is a true representation of the bulk material. Mounting of the sectioned specimen in a thermoplastic resin or thermosetting compound is carried out for ease of handling or where microstructural features at the edge are required. However, mounting is not essential where specimen size is big enough to polish comfortably.

It is a good practice to microscopically examine the as polished surface before etching which provides useful information about the presence of inclusions, porosities, eutectic melting, excessive heating, presence of weld defects like porosities and cracks. Etching of the specimens with a suitable reagent is performed to reveal the (sub)grain structure, grain size, constituent phases, detect segregation, orientation of grains and effects of cold work. The following are the procedures followed for microstructural evaluation of magnesium alloys.

13.3.1 Sectioning

Magnesium alloys are relatively soft and can be easily sectioned by any standard cutting technique. However, care should be taken to reduce the amount and depth of damage produced at the sectioned surface. Sectioning can introduce considerable damage which can affect the quality of final polishing. An abrasive cut-off saw produces relatively smooth surfaces with little damage by suitably selecting the wheel, coolant, feed rate and load. A precision diamond saw produces little surface damage and gives excellent surface finish for subsequent grinding/polishing. Magnesium dust is a fire hazard and therefore dry cutting of magnesium specimens should be completely avoided. A coolant for cutting may be used to avoid rise in temperature and microstructure modification in the specimen. Fracture surfaces must be carefully preserved against contamination. If the metallographic examination is for the purpose of quality control, the specimen should represent the lot. The sectioned specimen should be properly engraved with the relevant details for proper identification.

13.3.2 Mounting

Mounting is an essential step for obtaining a perfectly flat surface after grinding/polishing suitable for microscopic examination. Even though magnesium alloys have relatively low melting point, low solution annealing/aging temperatures, hot compression mounting can be used for good edge retention. Care should be taken to select the pressure during mounting so that it will not affect the specimen microstructure, viz. high purity magnesium can twin under load. However, if the specimens are in solution treated condition, epoxy resins should be preferred as they produce very little heat during polymerization. If multiple specimens are being mounted together, dissimilar materials/grades or materials under different heat treatment conditions should be avoided in a single mount as the polishing and etching characteristics are different. The intermetallics from harder specimen can scratch a softer specimen and it will result in poor surface finish for metallographic examination. It is a good practice to engrave or mark the specimens on the back of the mount for proper identification of specimens during microstructural examination.

13.3.3 Grinding

Grinding can be carried out either manually or using automatic grinding wheels. Silicon carbide is the usual choice for grinding for magnesium and its alloys. Specimens can be ground using a series of silicon carbide (SiC) abrasive sheets of increasingly finer sizes. If sectioning of the metallographic specimen has been done with minimal damage, grinding can start with a relatively finer grit size. Mechanical grinding is normally carried out in successive steps using SiC abrasive papers of different grit sizes 180, 220, 320, 400, 600, 800, 1000, 1200 and 2400. With each progressive grinding step, the specimen should be rotated by 90 degrees to remove the scratches from the previous step. The starting grit depends on the method used for sectioning. Even though wax coated silicon carbide papers have been suggested for grinding magnesium alloys to avoid getting the SiC particles getting embedded in the specimens, practically, it is not essential. Water can be used to flush the grinding debris from the paper surface as well as for cooling the specimen. At the end of the final grinding step, the specimen is ready for subsequent polishing.

Use of excessive force can change the surface microstructure and the force for grinding should be kept very low to avoid deep deformation. When polishing softer tempers of magnesium alloys, grinding with heavy load for longer time can result in embedment of abrasive particles in to the matrix which may mislead the interpretation of results obtained. Caution must be exercised while grinding specimens with defects such as pores and cracks as the abrasive left in the cavities can mislead interpretations of the results in microscopy/chemical analysis in scanning electron microscope fitted with energy dispersive spectroscope (SEM-EDS). In such cases, focused ion beam (FIB) fitted to an SEM is very useful.

13.3.4 Polishing

The surface after grinding is not ready for microscopic examination as it still contains many scratch marks on the surface and needs to be polished further. Initial rough polishing is performed using 3 µm and 1 µm diamond paste on a short nap cloth followed by final polishing with a 0.25 µm diamond paste. Cleaning can be done with water in between the polishing steps, but generally water is avoided after final polishing step. Instead, ethyl/methyl alcohol can be used for cleaning. It is important to wash the sample after final step in an ultrasonic bath to remove the abrasive particles. In cases where polishing with 0.25 µm diamond paste does not produce a mirror like, scratch-free surface, the final polishing can be carried out using colloidal silica or alumina suspensions. Alumina has been used as an aqueous suspension and is available in sizes down to 0.05 microns. Another simple and easy way of obtaining scratch free surfaces is through electrolytic polishing. This technique is ideally suited for polishing pure magnesium and magnesium alloys in soft tempers which are difficult to polish mechanically. This can be used for polishing large specimens.

13.3.5 Etching

Etching of the freshly polished specimen reveals the microstructure and brings in contrast for the microscopic examination. Etching is a controlled corrosion process arising out of potential different between the constituents, thereby revealing the microstructure. Potential differences exist between differently oriented grains, between grain boundaries and grain interiors, between impurity phases and matrix or between two phases in multi-phase alloys. Numerous etchants have been developed for revealing the microstructure of aluminum and its alloys. Table 13.1 gives the list of selected etchants for magnesium alloys. The selection of etchants and etching times for microstructural examination depends on the alloy chemistry and heat treatment condition. In solution treated condition, the etching time can be longer (up to one minute) whereas in the as-cast condition or in aged condition, the times can be 5–10 seconds. Etching can be carried out by swabbing or by immersion. Etching should be stopped when

TABLE 13.1
Selected Etchants for Macroscopic and Microscopic Examination of Magnesium Alloys

Etchant	Etching Practice	Features Revealed
Nital 1–5 mL concentrated HNO_3 in 100 mL of ethanol or methanol	Swab or immerse the specimen for few seconds up to a minute. Wash the etched specimen in water followed by alcohol and dry.	General microstructure
Glycol 1 mL concentrated HNO_3 in 24 mL water and 75 mL ethylene glycol	Swab or immersion etching for few seconds to a minute. Wash the etched specimen in water followed by alcohol and dry.	General microstructure. Reveals constituents in magnesium-rare earth/magnesium-thorium alloys
Acetic glycol 20 mL acetic acid, 1 mL concentrated HNO_3, 60 mL ethylene glycol, 20 mL water	Immersion etch with gentle agitation for 1–10 seconds. Wash the etched specimen in water followed by alcohol and dry.	General microstructure. Reveals grain boundaries in magnesium-rare earth/ magnesium-thorium alloys
Phospho-picral 0.7 mL H_3PO_4, 5 g picric acid, 100 mL ethanol	Immersion etch for 10–20 seconds until polished surface fades. Wash in alcohol and dry.	Estimating the amount of massive phase
Acetic-picral 5 mL acetic acid, 6 g picric acid, 100 mL ethanol	Immersion etch with gentle agitation until surface fades. Wash in alcohol and dry.	Reveals grain boundaries in most alloys and tempers, cold work and twinning

sufficient surface dullness is achieved. Immediately after etching, the specimen should be washed with running water, rinsed with ethanol and dried. If multiple etchants are to be used to observe different microstructural features, the previous etched layer should be completely removed prior to subsequent etching. Multiple etch-polish steps are occasionally used to improve the quality of surface finish.

13.4 MICROSTRUCTURAL EVOLUTION IN MAGNESIUM ALLOYS

The atlas of microstructures at the end of this chapter provides the optical, scanning electron microscopic and (scanning)-transmission electron microscopic observations of different grades of magnesium alloys obtained in the authors laboratory. The procedures followed in the preceding sections were followed for obtaining the metallographic specimens and etchants used for optical microscopic examination were given in the caption of the photomicrograph. Optical photomicrographs were recorded in an inverted optical metallurgical microscope on the freshly polished and etched specimens. The optical microstructures obtained at different magnifications are presented in the atlas to highlight the microstructural features (Figures 13.1 through 13.185). Optical microstructures of specimens in the as-cast condition and after different heat treatment conditions have been presented to show the microstructural features under different conditions. In order to further study the intermetallic phases, their morphology and chemical composition, the freshly polished and etched specimens have been observed under SEM-EDS. The microstructures recorded in SEM are presented and the constituent intermetallics and phases are analyzed using EDS (Figures 13.186 through 13.259). The sub-microscopic examination of the phases present in the material and compositional analysis were studied in a transmission electron microscopic (TEM) examination (Figures 13.260 through 13.360). The (S)TEM studies of the samples studied here were prepared by cutting, polishing the bulk sample, followed by disc punching the specimen to 3.05 mm disc. Further the 3.05 mm disc was lapped to achieve a thickness below 50 μm, followed by dimpling to achieve a thickness at the center of the disc between 10 and 20 μm. Then the sample is subjected to ion milling to achieve an ultrathin area at the center of the specimen to achieve electron transparency. For magnesium, extra care has been taken to avoid ion milling induced damage to the specimen by carrying out the ion beam thinning at −30°C.

FIGURE 13.1 Optical microstructure of 30 mm thick AZ31 magnesium alloy casting made in cast iron mould. The etchant is Acetic Picral.

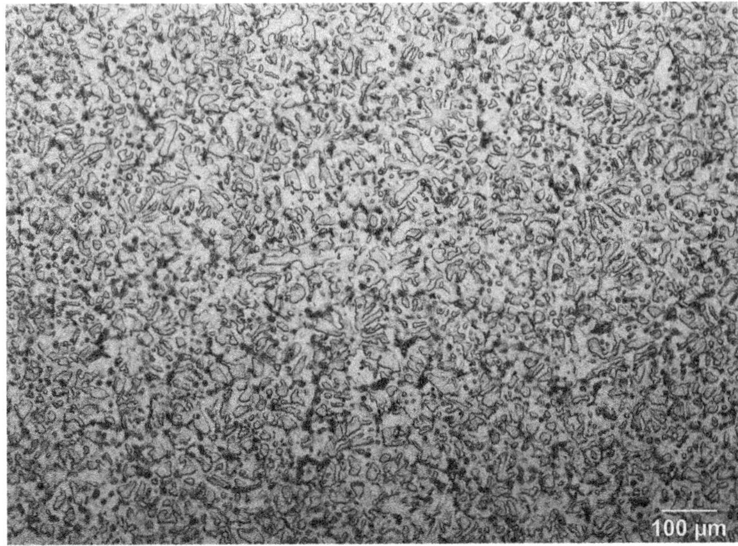

FIGURE 13.2 Optical microstructure of 30 mm thick AZ31 magnesium alloy casting made in cast iron mould taken at high magnification. Fine flowery type dendritic structure can be seen. The etchant is Acetic Picral.

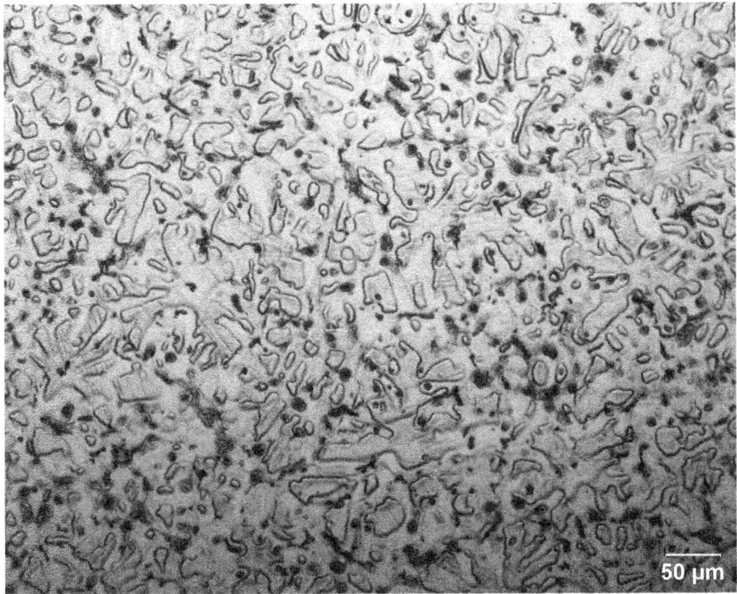

FIGURE 13.3 Optical microstructure of 30 mm thick AZ31 magnesium alloy casting made in cast iron mould taken at high magnification. Fine flowery type dendritic structure along with black $Mg_{17}Al_{12}$ phase can be seen. The etchant is Acetic Picral.

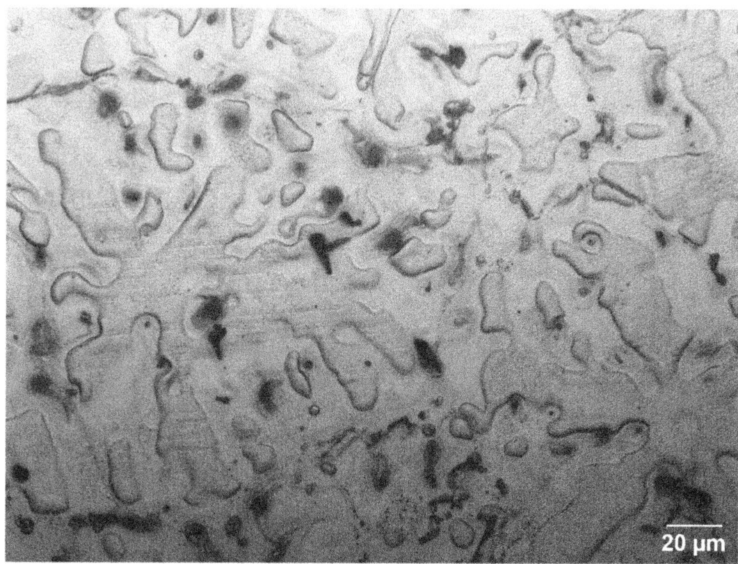

FIGURE 13.4 Optical microstructure of 30 mm thick AZ31 magnesium alloy casting made in cast iron mould taken at high magnification. Fine flowery type dendritic structure along with black $Mg_{17}Al_{12}$ phase can be seen. The etchant is Acetic Picral.

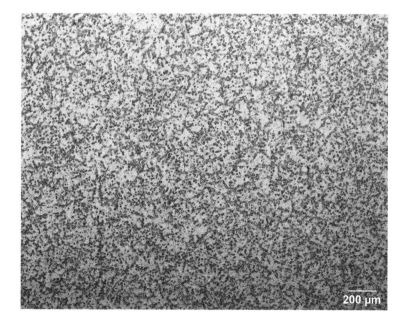

FIGURE 13.5 Optical microstructure of 30 mm thick AZ31 magnesium alloy casting made in cast iron mould. $Mg_{17}Al_{12}$ phase can be clearly seen. The etchant is Acetic Glycol.

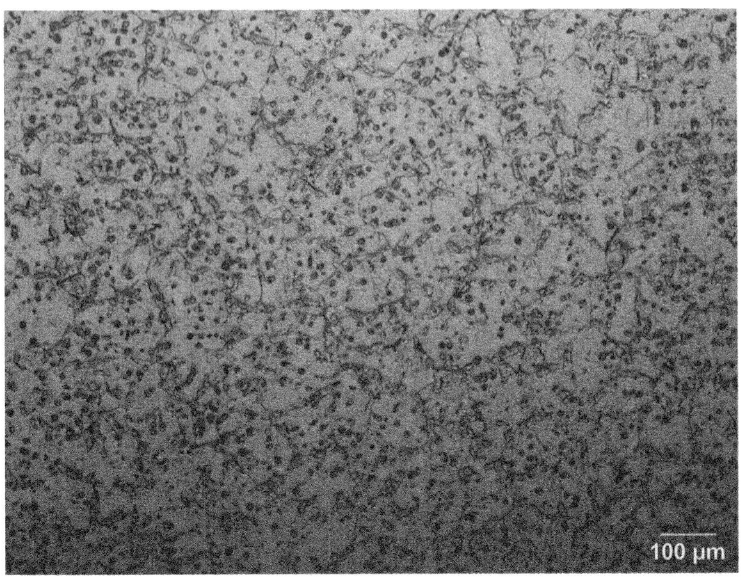

FIGURE 13.6 Optical microstructure of 30 mm thick AZ31 magnesium alloy casting made in cast iron mould. $Mg_{17}Al_{12}$ phase can be clearly seen. The etchant is Acetic Glycol.

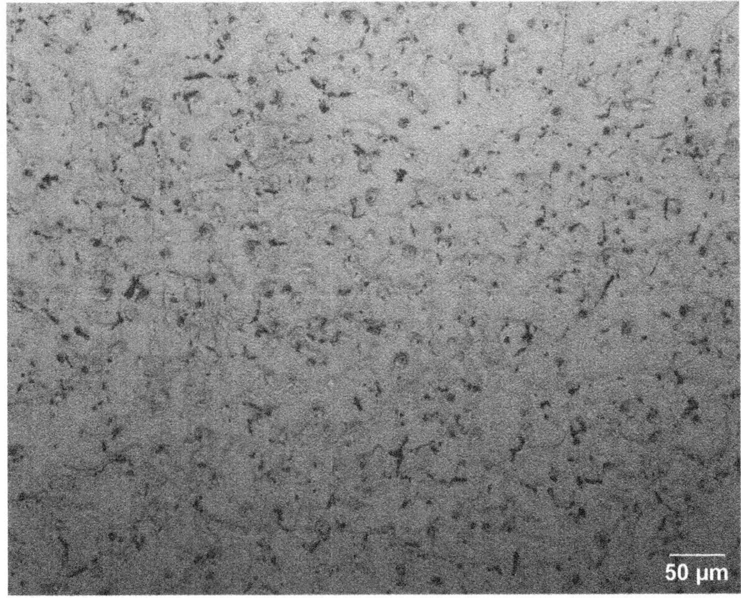

FIGURE 13.7 Optical microstructure of 30 mm thick AZ31 magnesium alloy casting made in cast iron mould. $Mg_{17}Al_{12}$ phase can be clearly seen. The etchant is Acetic Glycol.

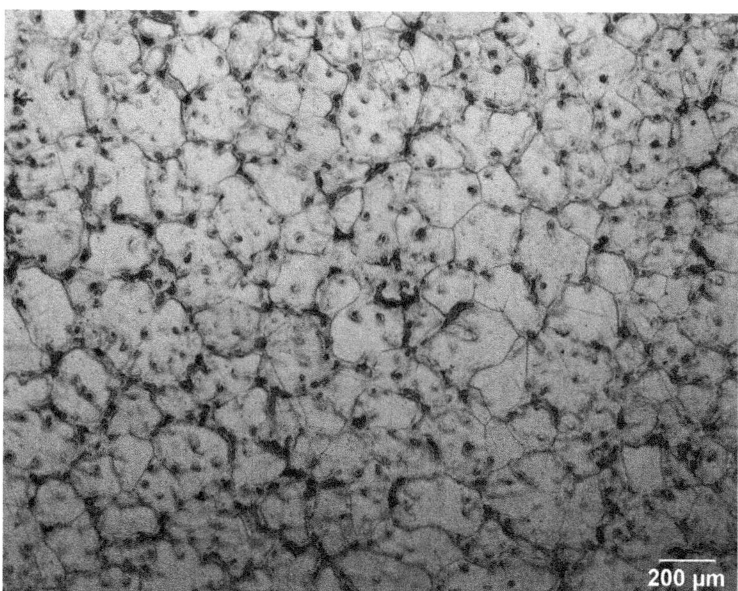

FIGURE 13.8 Optical microstructure of as-cast magnesium alloy AZ31. The microstructure reveals grain structure with eutectics at the grain boundaries. The etchant is Acetic Picral.

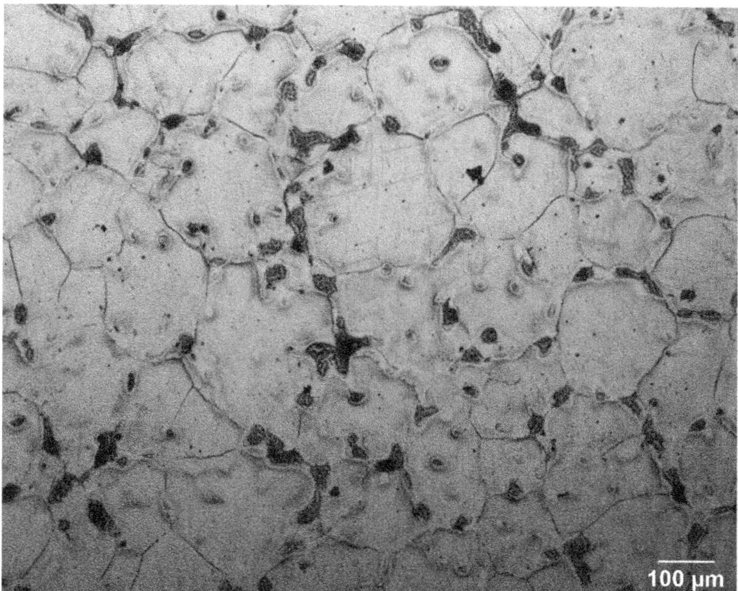

FIGURE 13.9 Optical microstructure of as-cast magnesium alloy AZ31 at high magnification. The microstructure reveals grain structure with eutectics at the grain boundaries. The etchant is Acetic Picral.

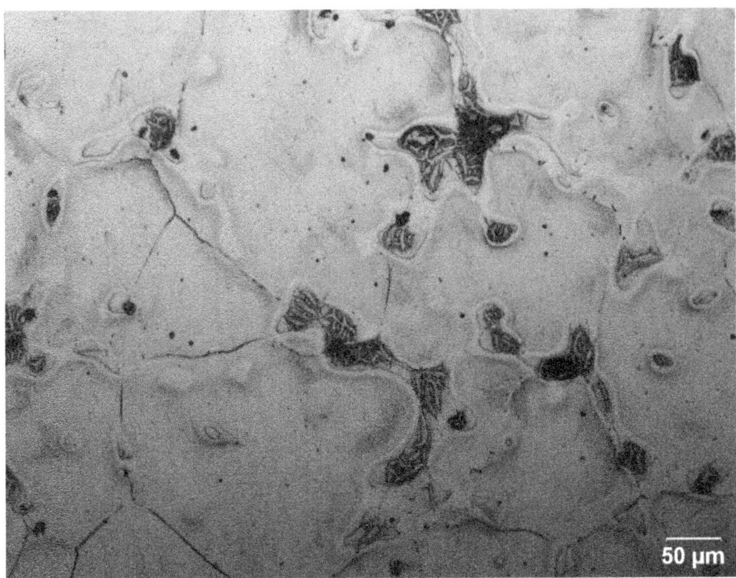

FIGURE 13.10 Optical microstructure of as-cast magnesium alloy AZ31 at high magnification. The microstructure reveals grain structure with lamellar eutectic of $Mg_{17}Al_{12}$ at the grain boundaries. The etchant is Acetic Picral.

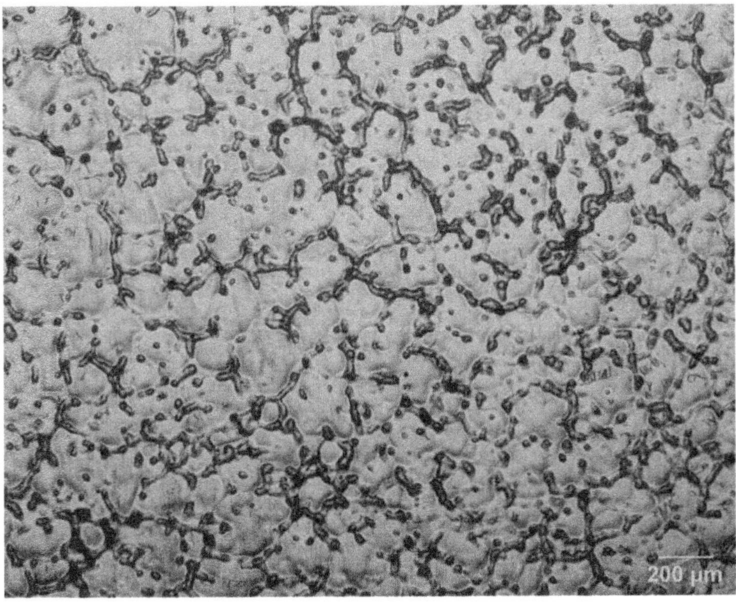

FIGURE 13.11 Optical microstructure of as-cast magnesium alloy AZ31. The microstructure reveals grain structure with clear eutectics at the grain boundaries. The etchant is Acetic Glycol.

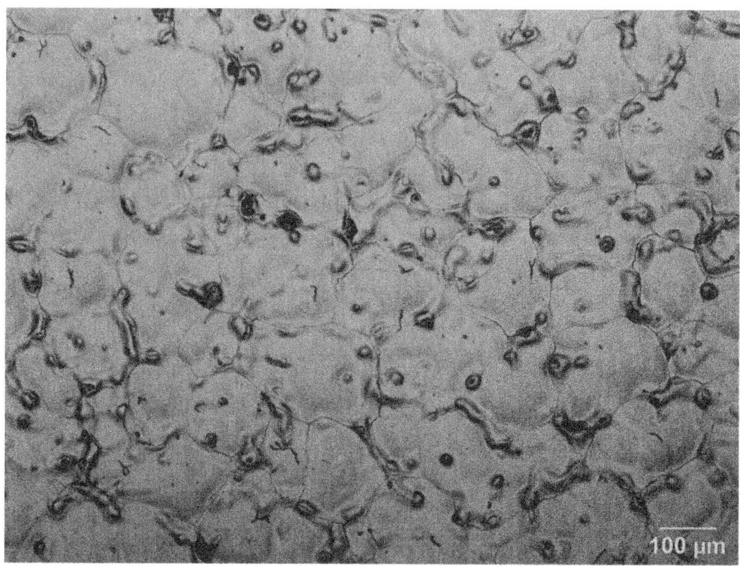

FIGURE 13.12 Optical microstructure of as-cast magnesium alloy AZ31 taken at high magnification. The microstructure reveals grain structure with clear eutectics at the grain boundaries. The etchant is Acetic Glycol.

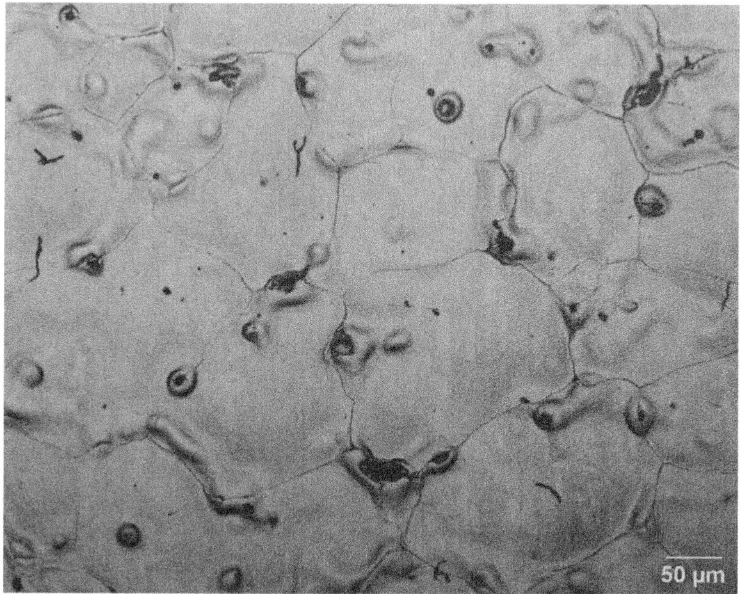

FIGURE 13.13 Optical microstructure of as-cast magnesium alloy AZ31 taken at high magnification. The microstructure reveals grain structure with clear eutectics at the grain boundaries. The etchant is Acetic Glycol.

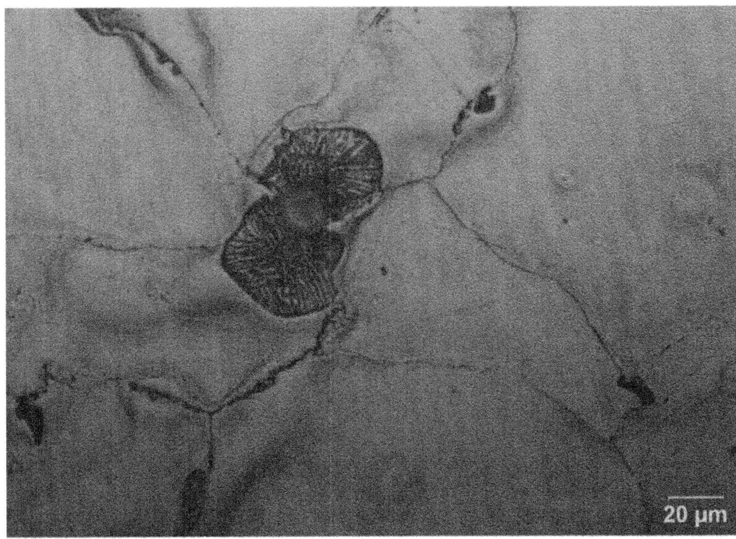

FIGURE 13.14 Optical microstructure of as-cast magnesium alloy AZ31 taken at high magnification. The microstructure reveals grain structure with clear lamellar eutectic of $Mg_{17}Al_{12}$ at the grain boundaries. The etchant is Acetic Glycol.

FIGURE 13.15 Optical microstructure of AZ31 magnesium alloy plate in longitudinal direction. Fine equiaxed grain structure along with the fragmented particles of $Mg_{17}Al_{12}$ can be seen. The etchant is Acetic Picral.

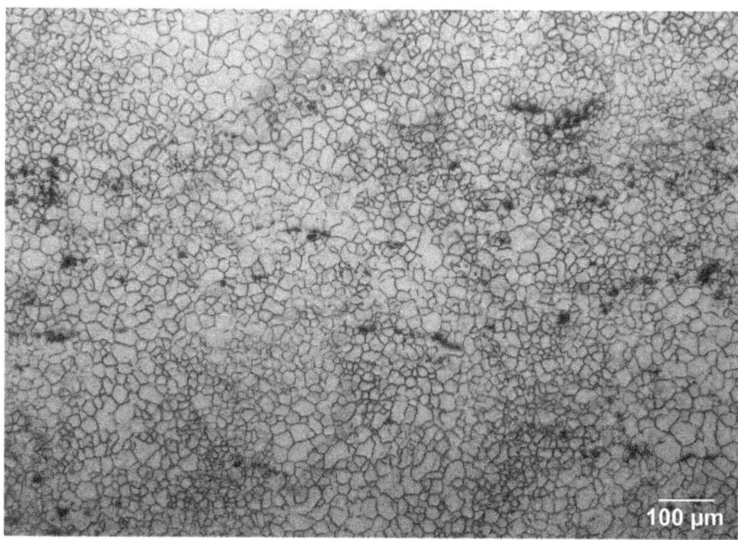

FIGURE 13.16 Optical microstructure of AZ31 magnesium alloy plate in longitudinal direction taken at high magnification. Fine equiaxed grain structure along with the fragmented particles of $Mg_{17}Al_{12}$ can be seen. The etchant is Acetic Picral.

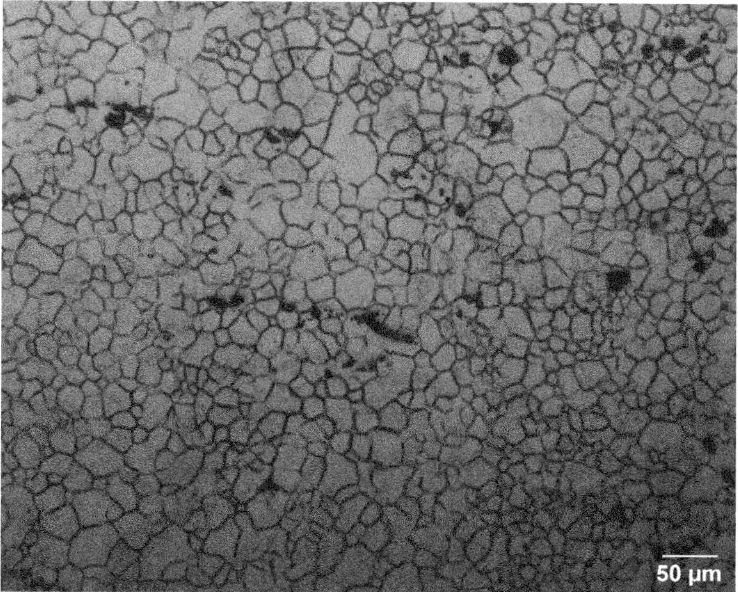

FIGURE 13.17 Optical microstructure of AZ31 magnesium alloy plate in longitudinal direction taken at high magnification. Fine equiaxed grain structure along with the fragmented particles of $Mg_{17}Al_{12}$ can be seen. The etchant is Acetic Picral.

FIGURE 13.18 Optical microstructure of AZ31 magnesium alloy plate in long transverse direction showing fine equiaxed grains with flow lines along rolling direction. The etchant is Acetic Picral.

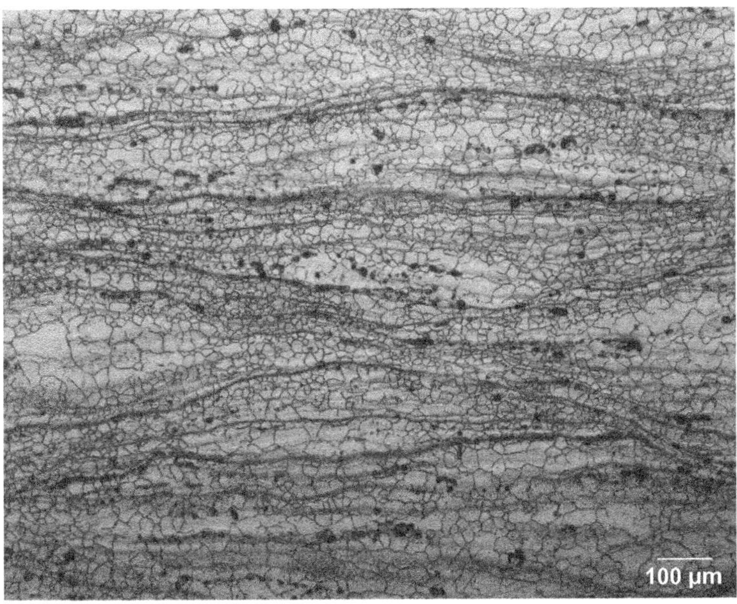

FIGURE 13.19 Optical microstructure of AZ31 magnesium alloy plate in long transverse direction taken at high magnification. A mixture of fine and coarse grains can be seen along with fragmented particles of $Mg_{17}Al_{12}$. The etchant is Acetic Picral.

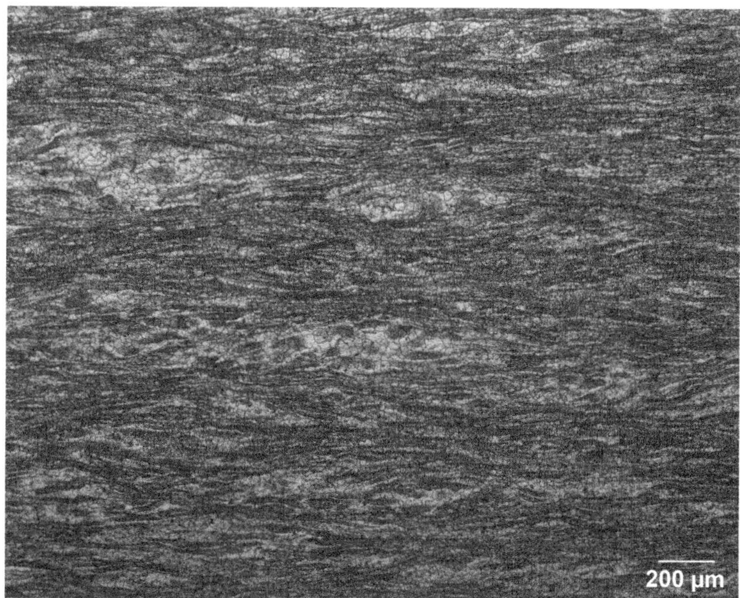

FIGURE 13.20 Optical microstructure of AZ31 magnesium alloy plate in short transverse direction showing fine equiaxed grains with flow lines. The etchant is Acetic Picral.

FIGURE 13.21 Optical microstructure of AZ31 magnesium alloy plate in short transverse direction taken at high magnification showing fine equiaxed grains with bimodal grain size distribution and flow lines. The etchant is Acetic Picral.

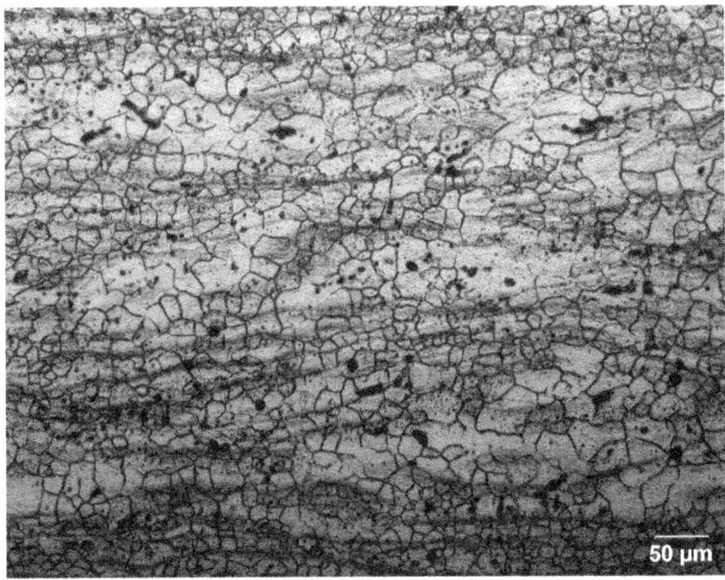

FIGURE 13.22 Optical microstructure of AZ31 magnesium alloy plate in short transverse direction taken at high magnification showing fine equiaxed grains with bimodal grain size distribution and flow lines. The etchant is Acetic Picral.

FIGURE 13.23 Optical microstructure of AZ31 magnesium alloy bar rolled at 250°C. The microstructure is taken on the cross section. Fine equiaxed grains can be seen. The etchant is Acetic Picral.

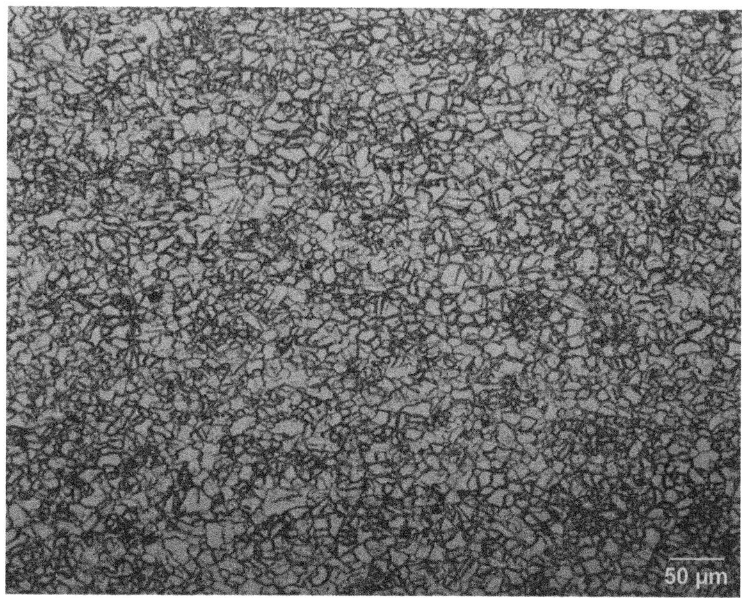

FIGURE 13.24 Optical microstructure of AZ31 magnesium alloy bar rolled at 250°C taken at high magnification. The microstructure is taken on the cross section. Fine equiaxed grains can be seen. The etchant is Acetic Picral.

FIGURE 13.25 Optical microstructure of AZ31 magnesium alloy rolled at 300°C. The microstructure is taken on the cross section. Fine equiaxed grains can be seen. The etchant is Acetic Picral.

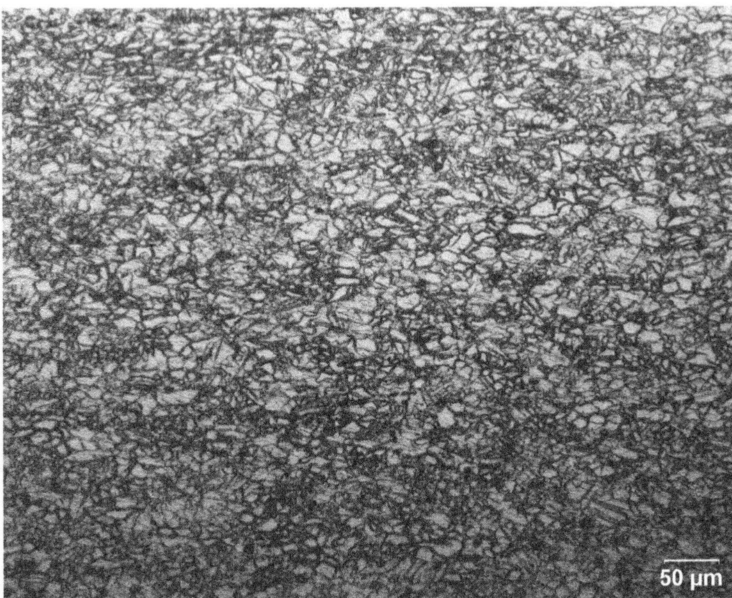

FIGURE 13.26 Optical microstructure of AZ31 magnesium alloy rolled at 300°C taken at high magnification. The microstructure is taken on the cross section. The specimen was etched with Acetic Picral.

FIGURE 13.27 Optical microstructure of AZ31 magnesium alloy rolled at 350°C. The microstructure is taken on the cross section. Fine equiaxed grains can be seen. The etchant is Acetic Picral.

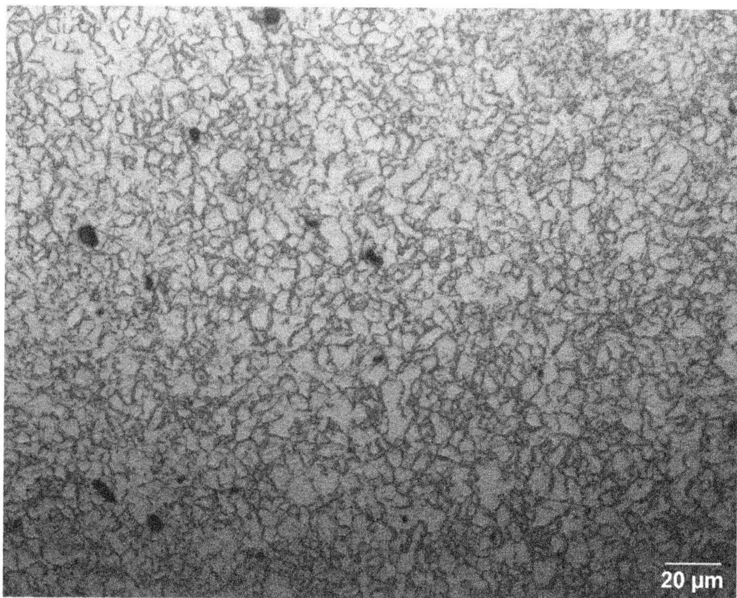

FIGURE 13.28 Optical microstructure of AZ31 magnesium alloy rolled at 350°C taken at high magnification. The microstructure is taken on the cross section. The specimen was etched with Acetic Picral.

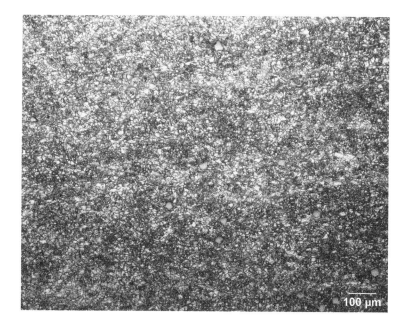

FIGURE 13.29 Optical microstructure of AZ31 magnesium alloy rolled at 400°C. The microstructure is taken on the cross section. Fine equiaxed grains can be seen. The etchant is Acetic Picral.

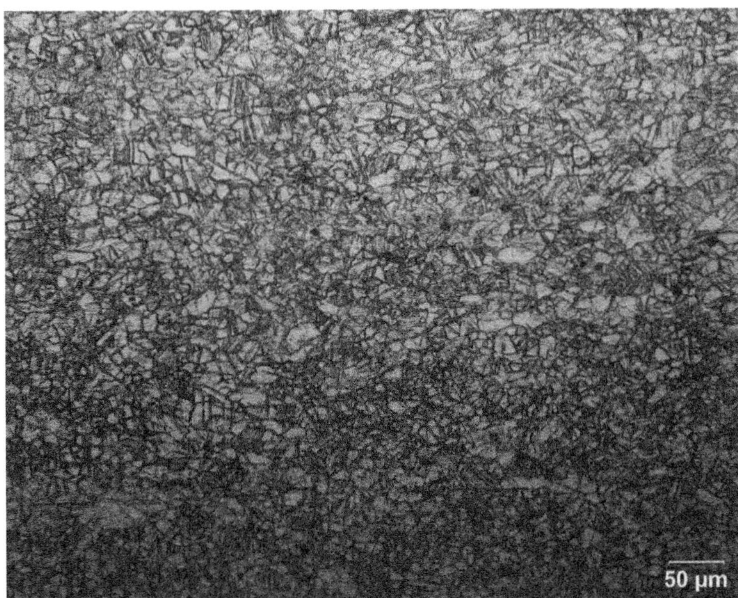

FIGURE 13.30 Optical microstructure of AZ31 magnesium alloy rolled at 400°C taken at high magnification. The microstructure is taken on the cross section. The specimen was etched with Acetic Picral.

FIGURE 13.31 Optical microstructure of AZ31 magnesium alloy rolled at 450°C. The microstructure is taken on the cross section. Fine equiaxed grains can be seen. The etchant is Acetic Picral.

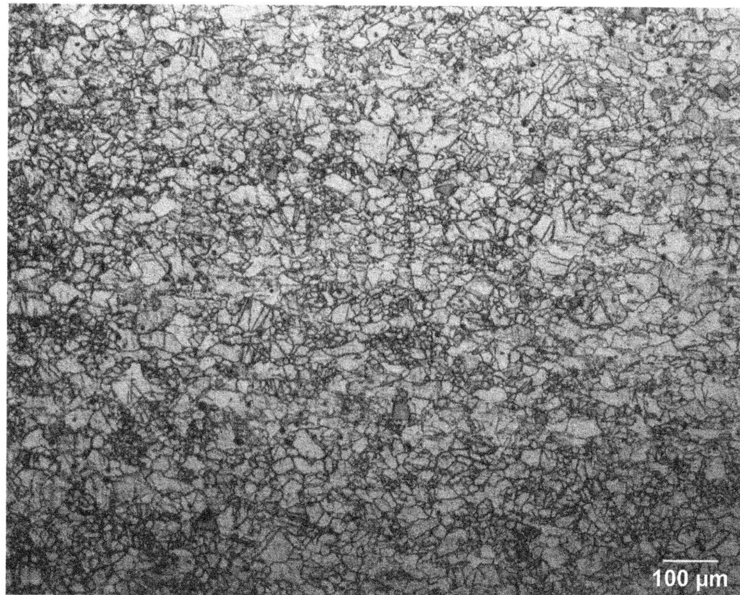

FIGURE 13.32 Optical microstructure of AZ31 magnesium alloy rolled at 450°C taken at high magnification. The microstructure is taken on the cross section. The etchant is Acetic Picral.

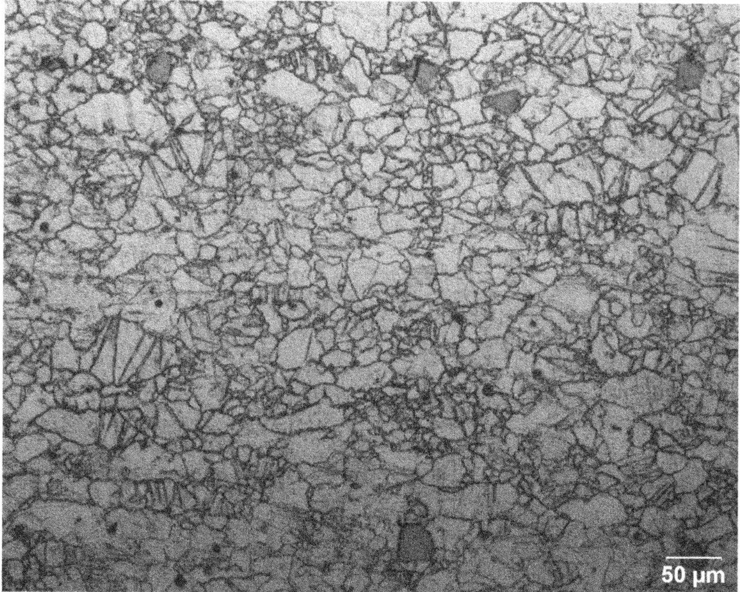

FIGURE 13.33 Optical microstructure of AZ31 magnesium alloy rolled at 450°C taken at high magnification. The microstructure is taken on the cross section. The etchant is Acetic Picral.

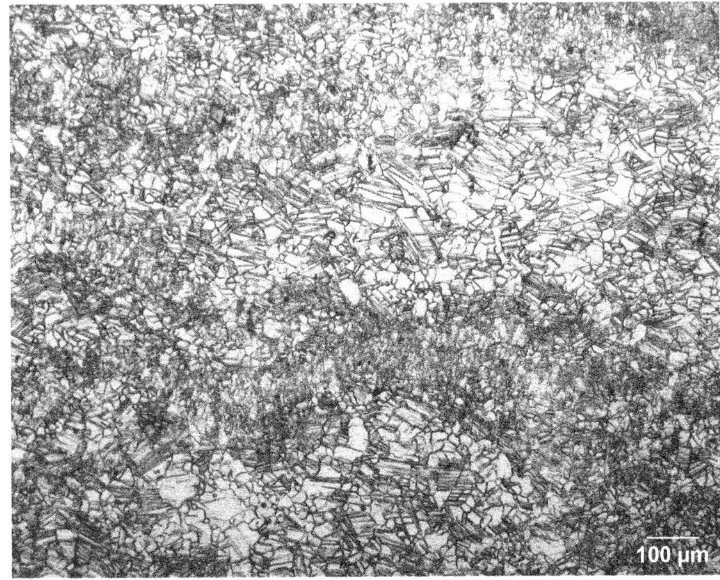

FIGURE 13.34 Optical microstructure of AZ31 magnesium alloy rolled at 300°C and annealed at 300°C for 8 hours. The microstructure is taken on the cross section. Annealing twins can be seen in the microstructure. The etchant is Acetic Picral.

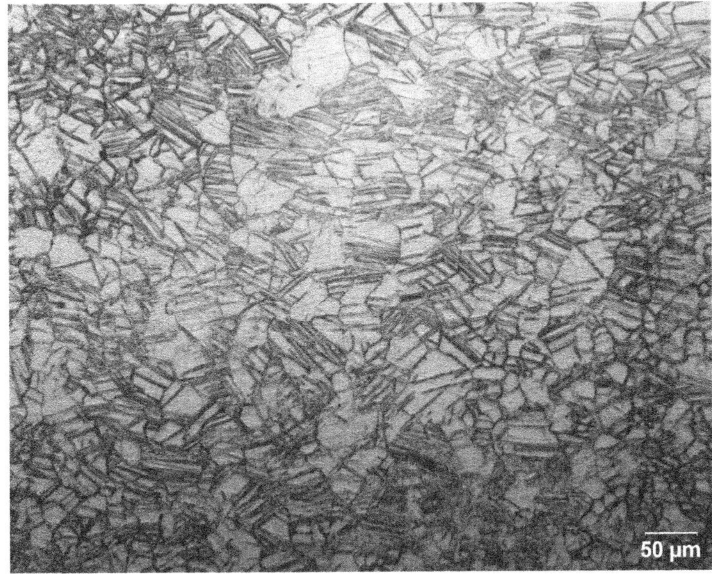

FIGURE 13.35 Optical microstructure of AZ31 magnesium alloy rolled at 300°C and annealed at 300°C for 8 hours taken at high magnification. The microstructure is taken on the cross section. Annealing twins can be seen in the microstructure. The etchant is Acetic Picral.

FIGURE 13.36 Optical microstructure of AZ31 magnesium alloy hot rolled plate of 45 mm thickness. The microstructure was recorded at the center of the specimen thickness cross section. Coarse elongated grains can be seen along with fine recrystallized equiaxed grains. The etchant is Acetic Picral.

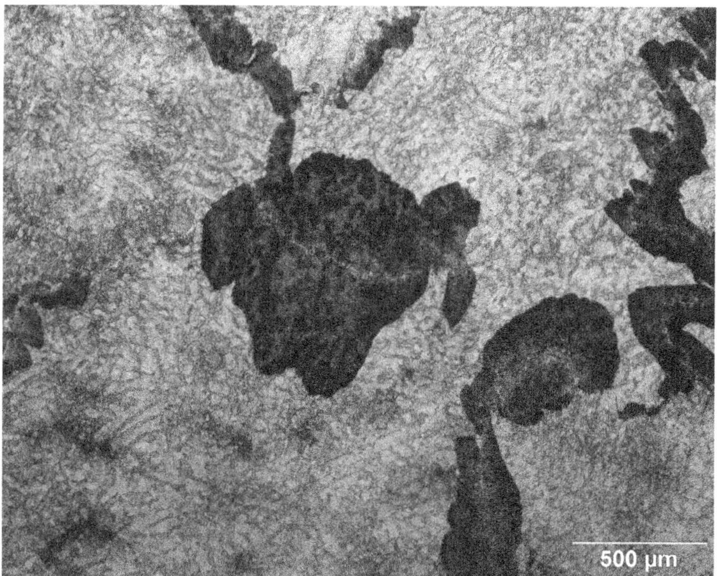

FIGURE 13.37 Optical microstructure of AZ80 magnesium alloy forging heat treated to T5 condition taken along longitudinal direction. Massive lamellar eutectics of $Mg_{17}Al_{12}$ can be seen. The etchant is Acetic Picral.

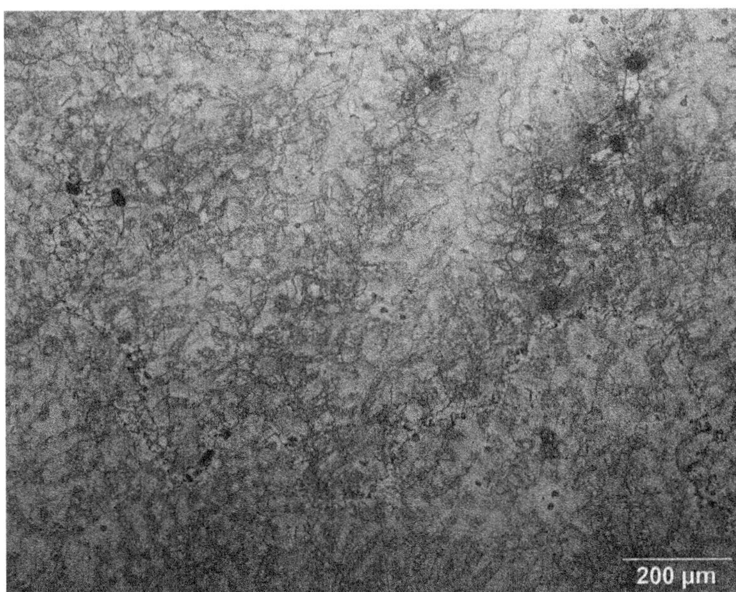

FIGURE 13.38 Optical microstructure of AZ80 magnesium alloy forging heat treated to T5 condition taken along longitudinal direction at high magnification. The etchant is Acetic Picral.

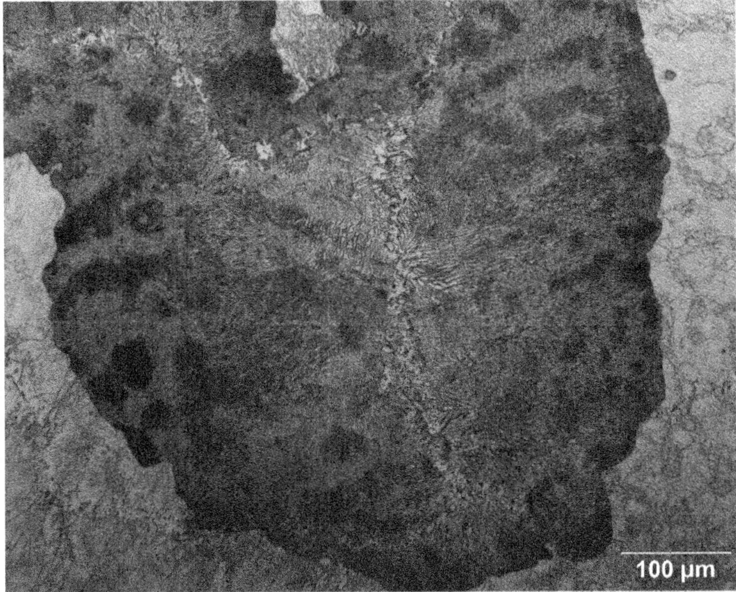

FIGURE 13.39 Optical microstructure of AZ80 magnesium alloy forging heat treated to T5 condition taken along longitudinal direction showing massive lamellar eutectic of $Mg_{17}Al_{12}$. The etchant is Acetic Picral.

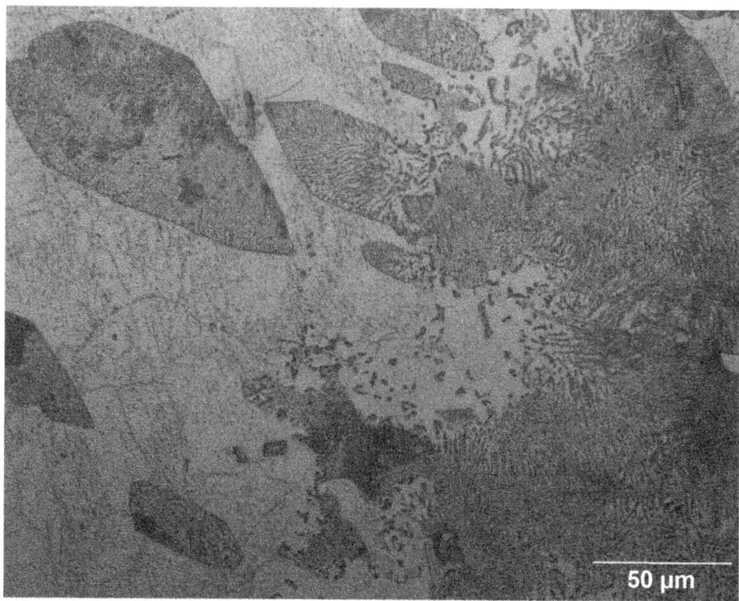

FIGURE 13.40 Optical microstructure of AZ80 magnesium alloy forging heat treated to T5 condition taken along longitudinal direction showing massive lamellar eutectic of $Mg_{17}Al_{12}$ at high magnification. The etchant is Acetic Picral.

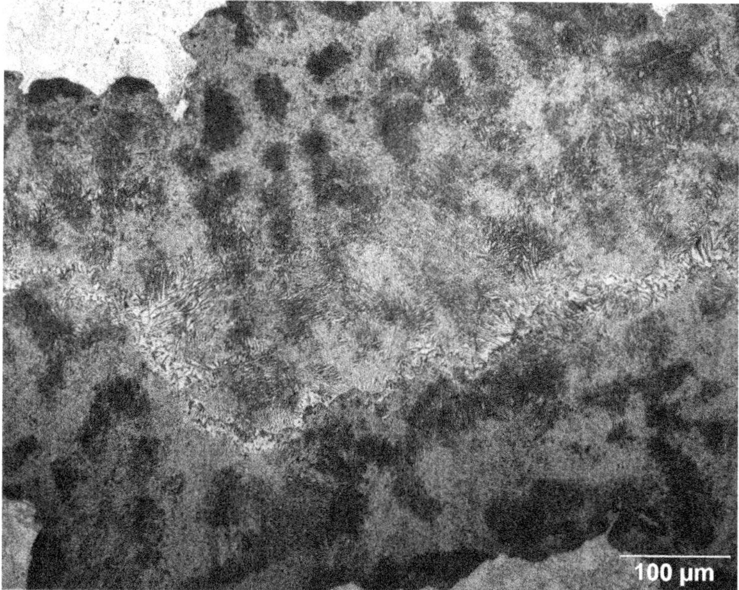

FIGURE 13.41 Optical microstructure of AZ80 magnesium alloy forging heat treated to T5 condition taken along long transverse direction showing massive lamellar eutectic of $Mg_{17}Al_{12}$ at high magnification. The etchant is Acetic Picral.

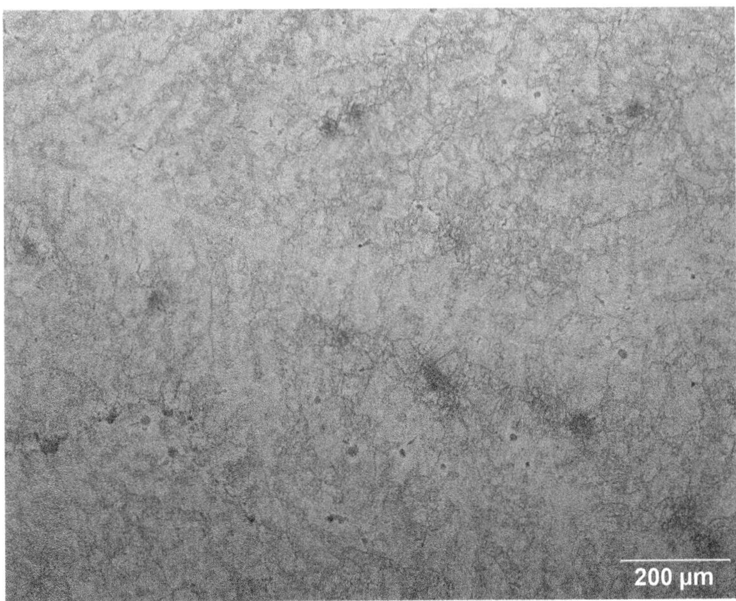

FIGURE 13.42 Optical microstructure of AZ80 magnesium alloy forging heat treated to T5 condition taken along long transverse direction. Thermomechanically processed grain structure can be seen. The etchant is Acetic Picral.

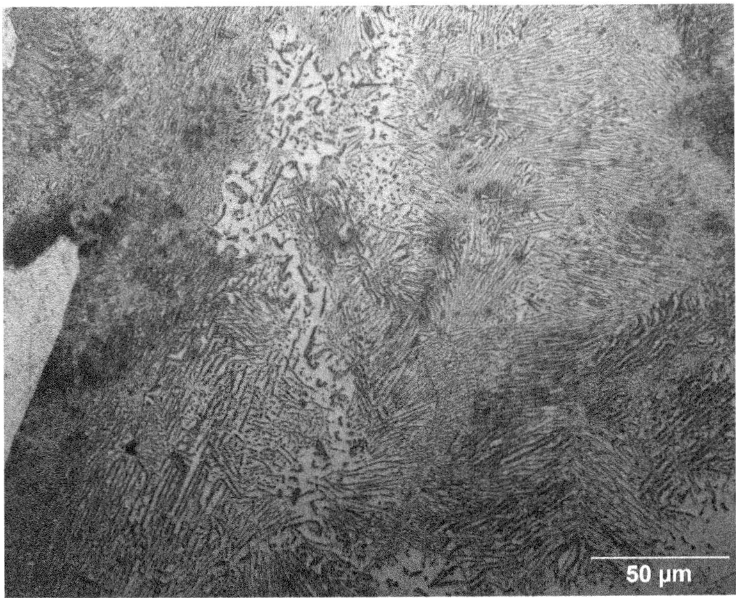

FIGURE 13.43 Optical microstructure of AZ80 magnesium alloy forging heat treated to T5 condition taken along long transverse direction showing massive lamellar eutectic of $Mg_{17}Al_{12}$ at high magnification. The etchant is Acetic Picral.

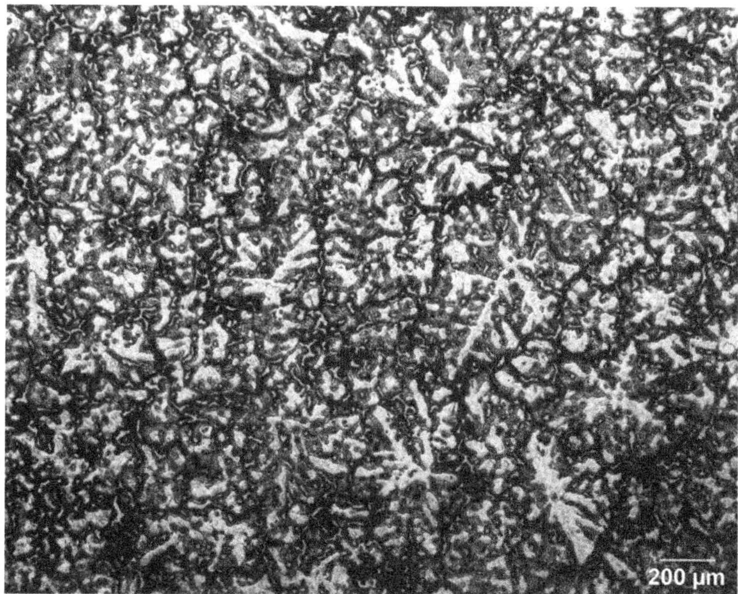

FIGURE 13.44 Optical microstructure of AZ91 magnesium alloy in as-cast condition showing clear dendritic structure. The etchant is Acetic Picral.

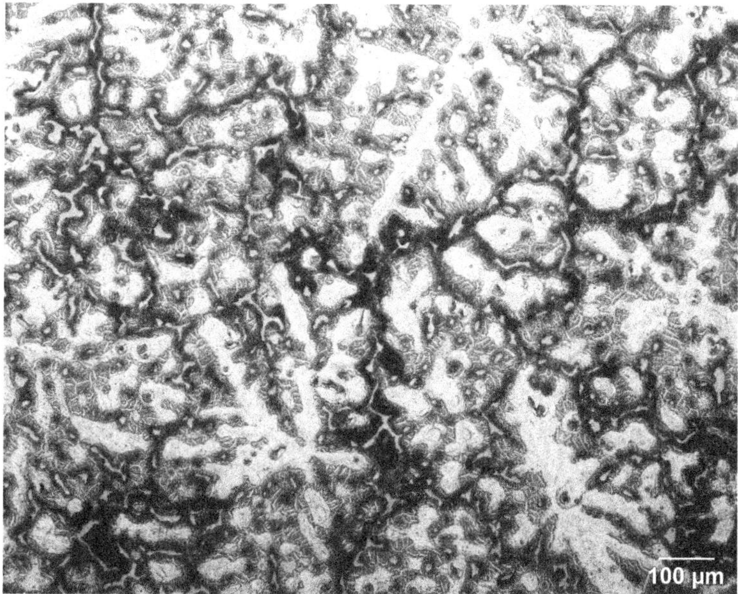

FIGURE 13.45 Optical microstructure of AZ91 magnesium alloy in as-cast condition showing clear dendritic structure taken at high magnification. The etchant is Acetic Picral.

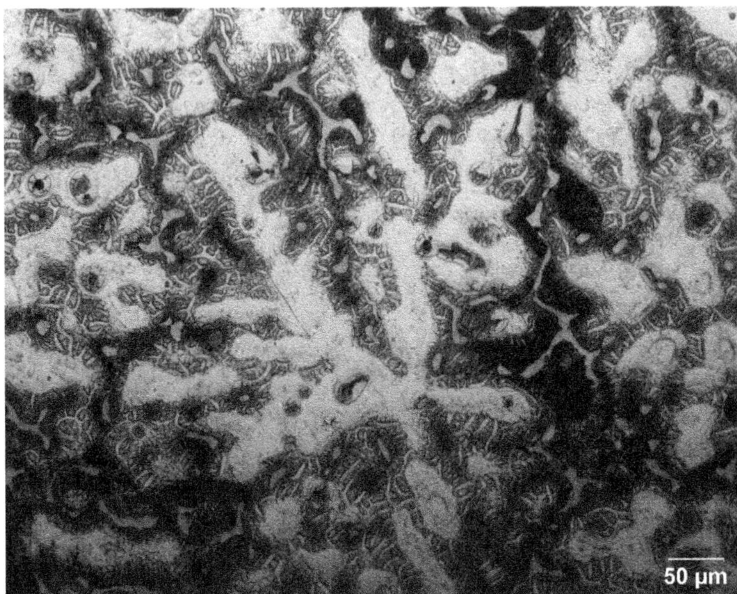

FIGURE 13.46 Optical microstructure of AZ91 magnesium alloy in as-cast condition showing clear dendritic structure taken at high magnification. The etchant is Acetic Picral.

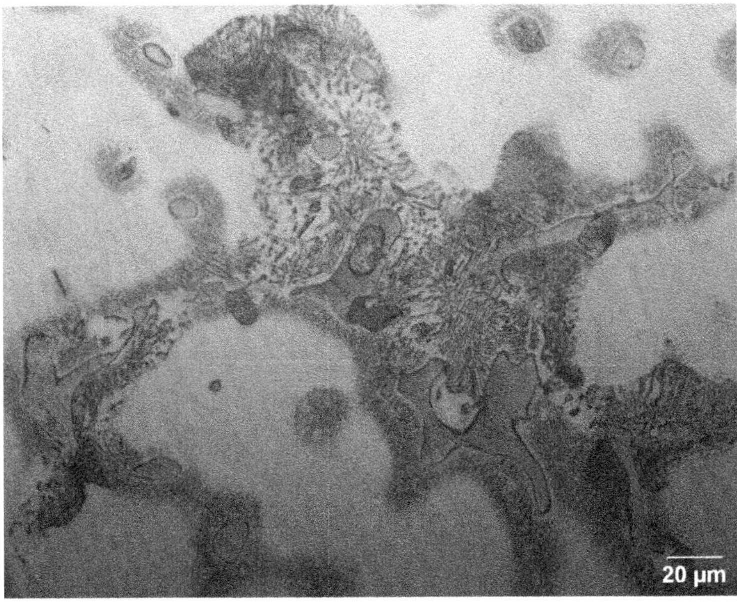

FIGURE 13.47 Optical microstructure of AZ91 magnesium alloy in as-cast condition showing clear dendritic structure taken at high magnification. Massive $Mg_{17}(Al,Zn)_{12}$ grain boundary phase along with fine lamellar eutectic of $Mg_{17}Al_{12}$ is clearly seen. The etchant is Acetic Picral.

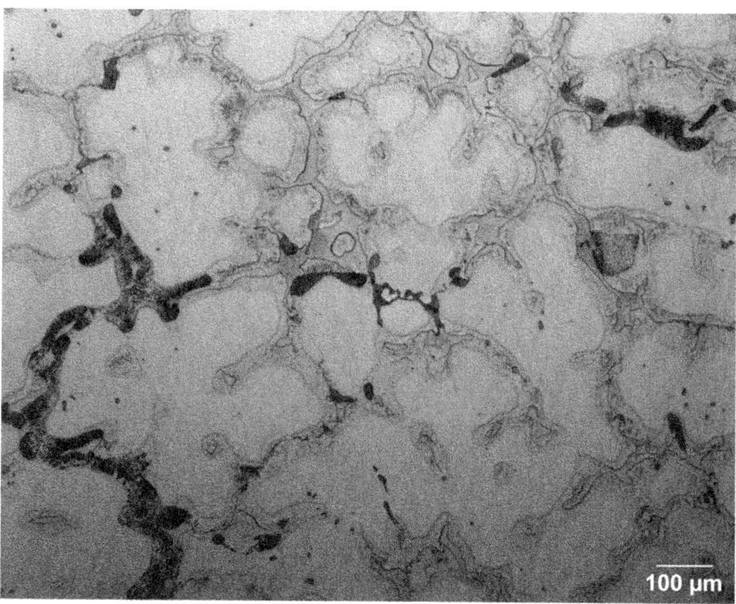

FIGURE 13.48 Optical microstructure of AZ91 magnesium alloy in as forged condition. Massive Mg_{17} $(Al,Zn)_{12}$ grain boundary phase along with fine lamellar eutectic of $Mg_{17}Al_{12}$ is clearly seen. The etchant is Acetic Picral.

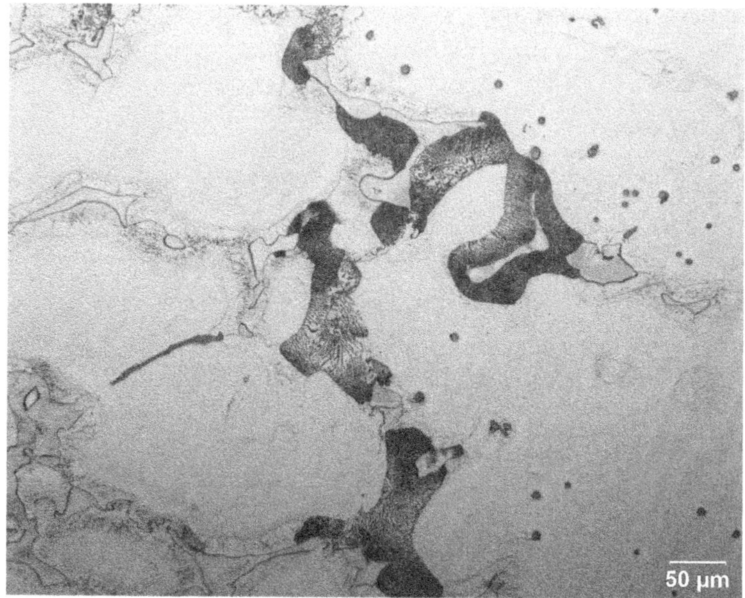

FIGURE 13.49 Optical microstructure of AZ91 magnesium alloy in as forged condition taken at high magnification. Massive Mg_{17} $(Al,Zn)_{12}$ grain boundary phase along with fine lamellar eutectic of $Mg_{17}Al_{12}$ is clearly seen. The etchant is Acetic Picral.

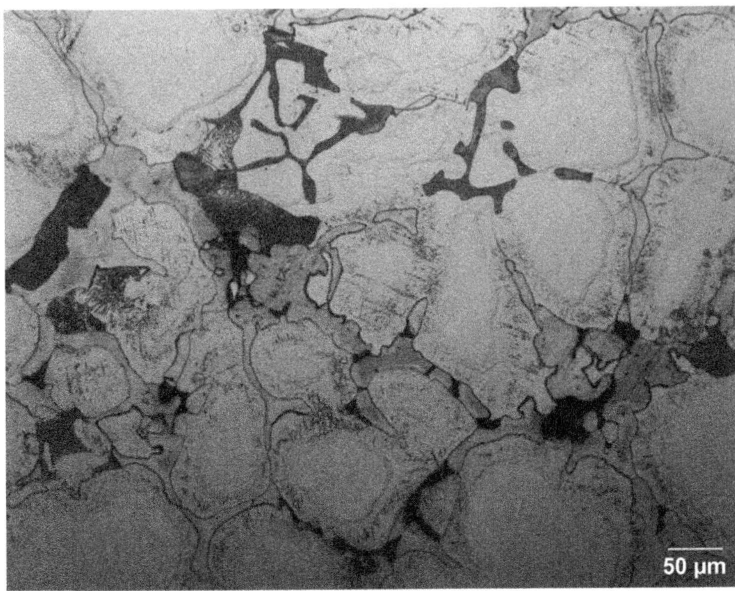

FIGURE 13.50 Optical microstructure of AZ91 magnesium alloy in as forged condition taken at high magnification. Massive Mg_{17} $(Al,Zn)_{12}$ grain boundary phase along with fine lamellar eutectic of $Mg_{17}Al_{12}$ is clearly seen. The etchant is Acetic Picral.

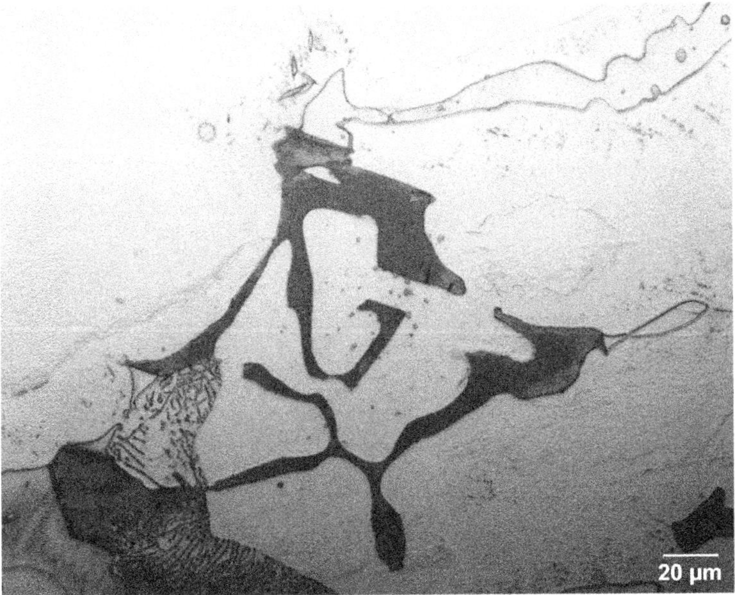

FIGURE 13.51 Optical microstructure of AZ91 magnesium alloy in as forged condition taken at high magnification. Massive Mg_{17} $(Al,Zn)_{12}$ grain boundary phase along with fine lamellar eutectic of $Mg_{17}Al_{12}$ is clearly seen. The etchant is Acetic Picral.

FIGURE 13.52 Optical microstructure of AZ91 magnesium alloy in as forged and heat treated to T4 condition. Massive Mg_{17} $(Al,Zn)_{12}$ grain boundary phase along with fine lamellar eutectic of $Mg_{17}Al_{12}$ is clearly seen. The etchant is Acetic Picral.

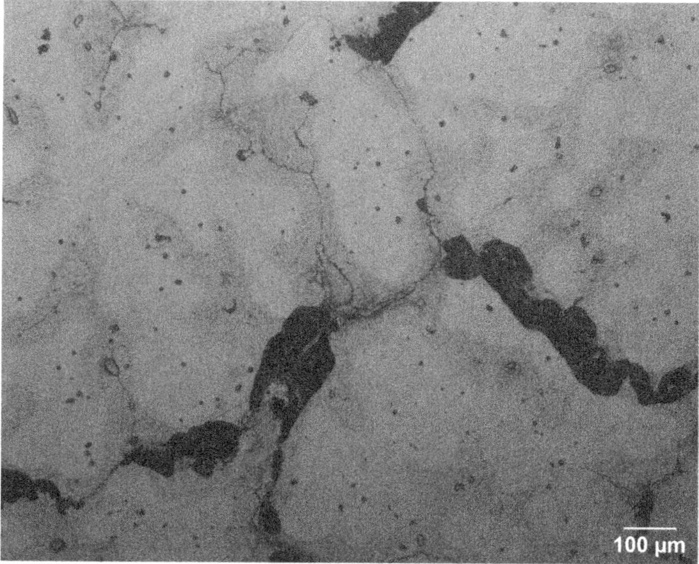

FIGURE 13.53 Optical microstructure of AZ91 magnesium alloy in as forged and heat treated to T4 condition taken at high magnification. Massive Mg_{17} $(Al,Zn)_{12}$ grain boundary phase along with fine lamellar eutectic of $Mg_{17}Al_{12}$ is clearly seen. The etchant is Acetic Picral.

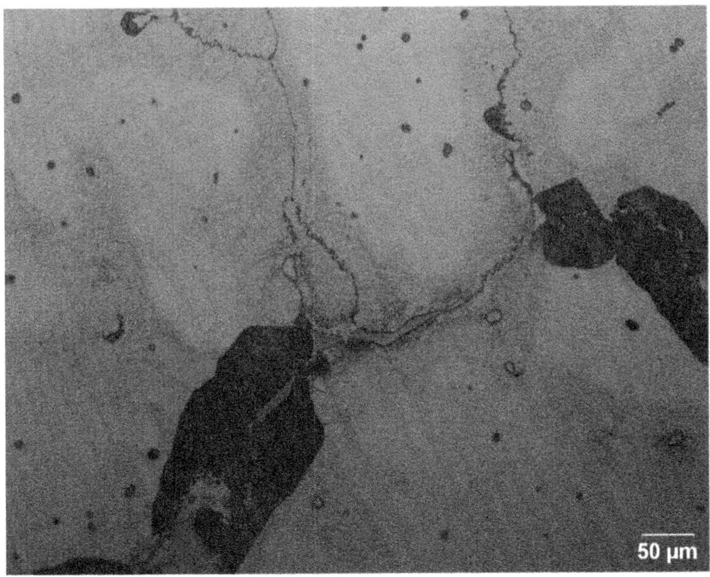

FIGURE 13.54 Optical microstructure of AZ91 magnesium alloy in as forged and heat treated to T4 condition taken at high magnification. Massive Mg_{17} $(Al,Zn)_{12}$ grain boundary phase along with fine lamellar eutectic of $Mg_{17}Al_{12}$ is clearly seen. Corrugated grain boundaries with fine particles of Al_4Mn can be seen inside the grains. The etchant is Acetic Picral.

FIGURE 13.55 Optical microstructure of AZ91 magnesium alloy in as forged and heat treated to T4 condition showing fine lamellar eutectic of $Mg_{17}Al_{12}$. The etchant is Acetic Picral.

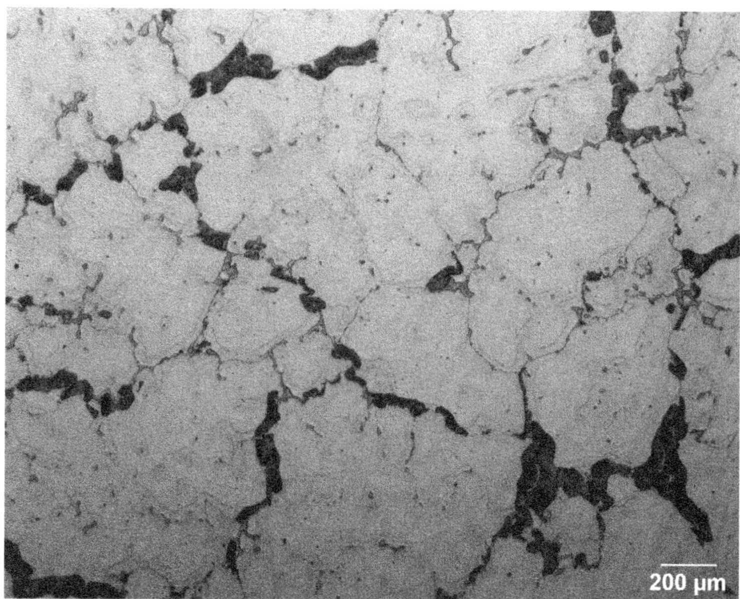

FIGURE 13.56 Optical microstructure of AZ91 magnesium alloy in as forged and heat treated to T6 condition. $Mg_{17}(Al,Zn)_{12}$ grain boundary phase along with fine lamellar eutectic of $Mg_{17}Al_{12}$ is clearly seen. The etchant is Acetic Picral.

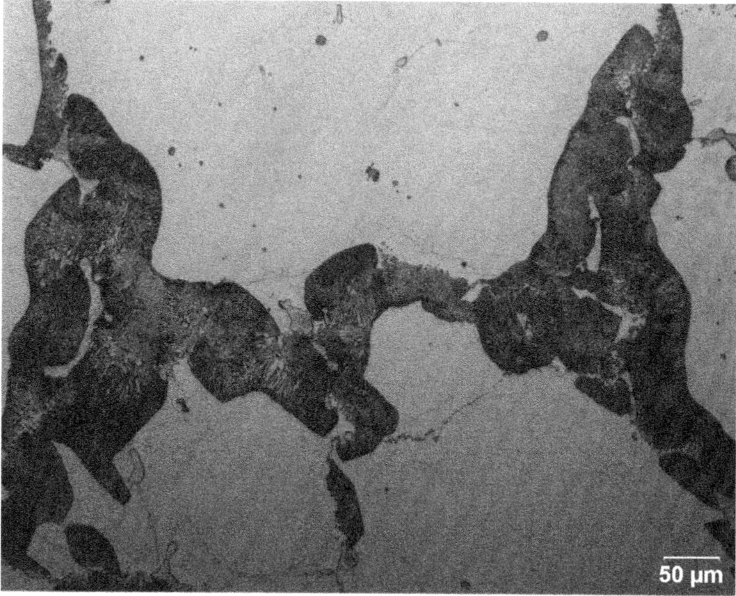

FIGURE 13.57 Optical microstructure of AZ91 magnesium alloy in as forged and heat treated to T6 condition showing fine lamellar eutectic of $Mg_{17}Al_{12}$. The etchant is Acetic Picral.

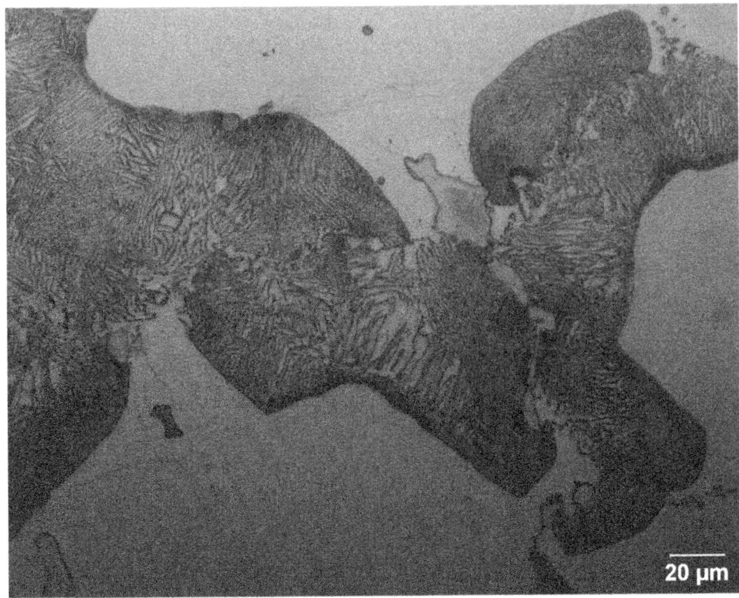

FIGURE 13.58 Optical microstructure of AZ91 magnesium alloy in as forged and heat treated to T6 condition taken at high magnification showing fine lamellar eutectic of $Mg_{17}Al_{12}$. The etchant is Acetic Picral.

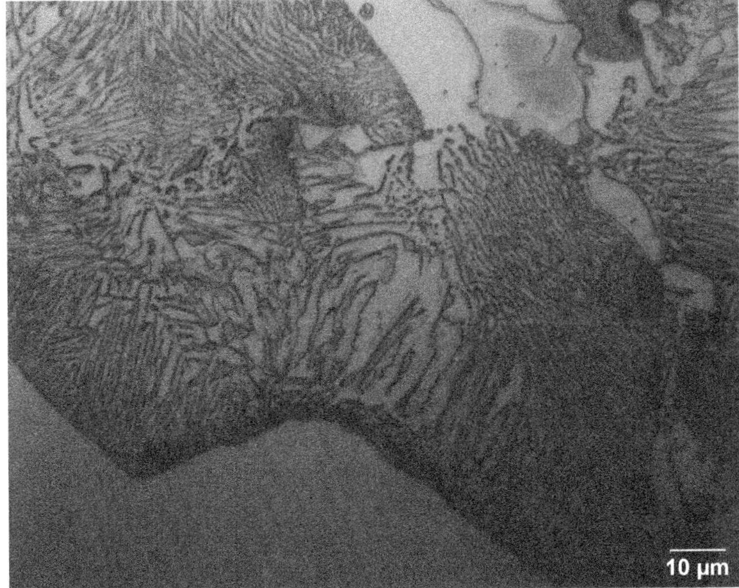

FIGURE 13.59 Optical microstructure of AZ91 magnesium alloy in as forged and heat treated to T6 condition taken at high magnification showing fine lamellar eutectic of $Mg_{17}Al_{12}$ is clearly seen. The etchant is Acetic Picral.

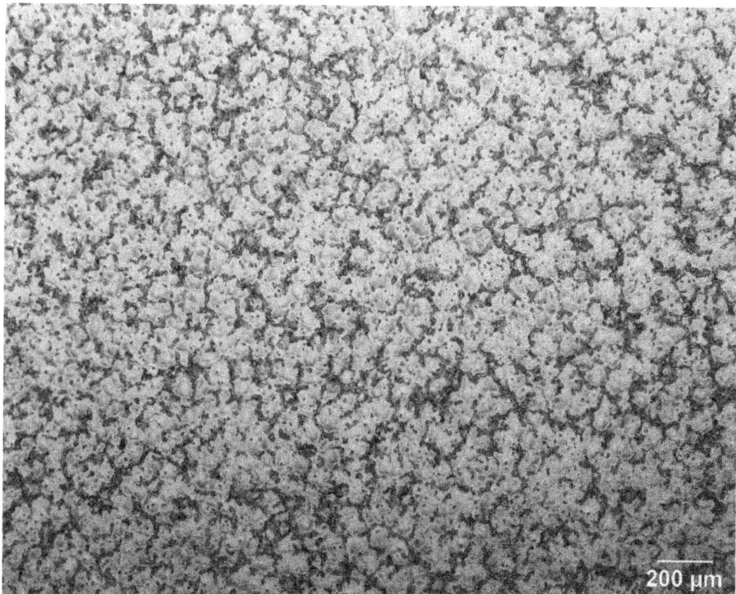

FIGURE 13.60 Optical microstructure of 30 mm thick AZ91 magnesium alloy casting made in cast iron mould. The etchant is Acetic Picral.

FIGURE 13.61 Optical microstructure of 30 mm thick AZ91 magnesium alloy casting made in cast iron mould taken at high magnification. The etchant is Acetic Picral.

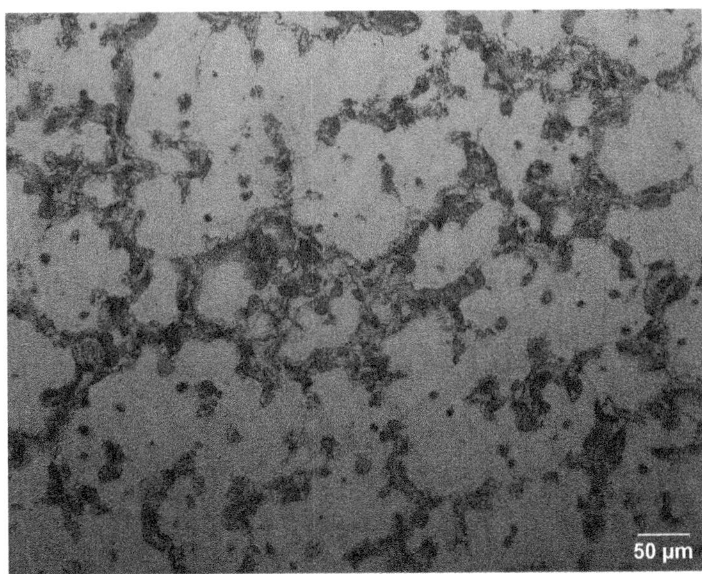

FIGURE 13.62 Optical microstructure of 30 mm thick AZ91 magnesium alloy casting made in cast iron mould taken at high magnification. The microstructure reveals grain boundary lamellar eutectic phase of $Mg_{17}Al_{12}$ phase along with small amount of Mg_{17} $(Al,Zn)_{12}$ phase. The etchant is Acetic Picral.

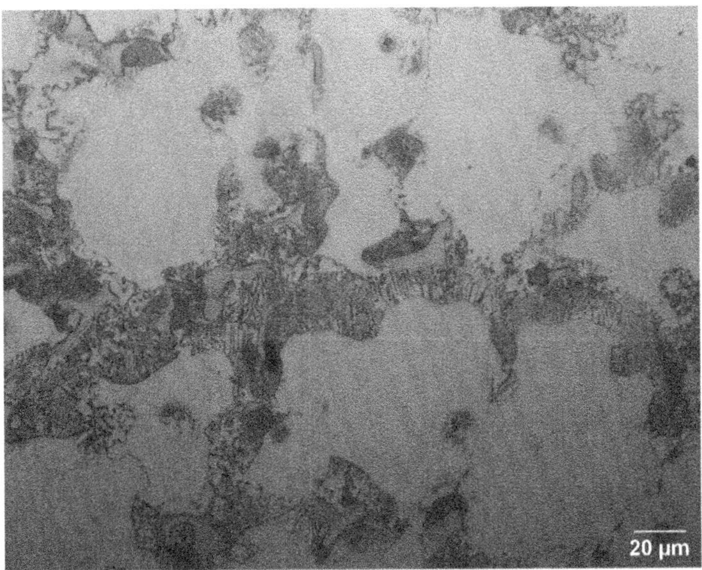

FIGURE 13.63 Optical microstructure of 30 mm thick AZ91 magnesium alloy casting made in cast iron mould taken at high magnification. The microstructure reveals grain boundary lamellar eutectic phase of $Mg_{17}Al_{12}$ phase along with small amount of Mg_{17} $(Al,Zn)_{12}$ phase. The etchant is Acetic Picral.

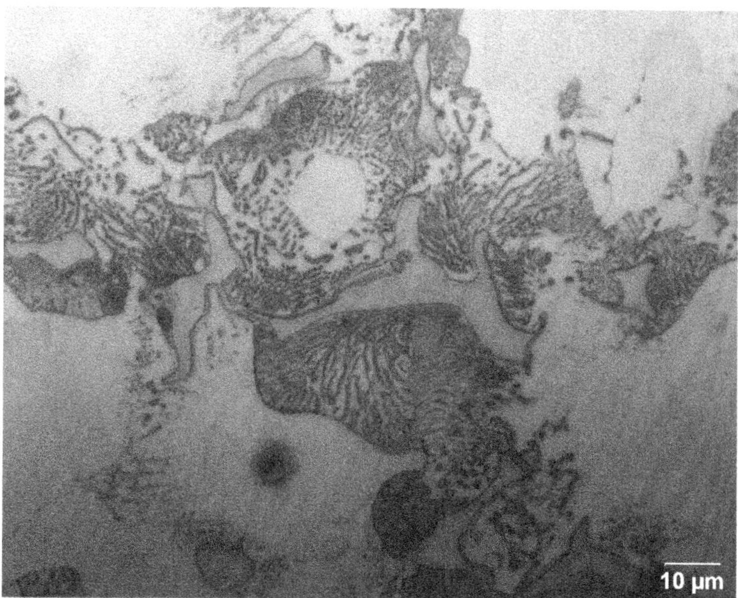

FIGURE 13.64 Optical microstructure of 30 mm thick AZ91 magnesium alloy casting made in cast iron mould taken at high magnification. The microstructure reveals grain boundary lamellar eutectic phase of $Mg_{17}Al_{12}$ phase along with small amount of Mg_{17} $(Al,Zn)_{12}$ phase. The etchant is Acetic Picral.

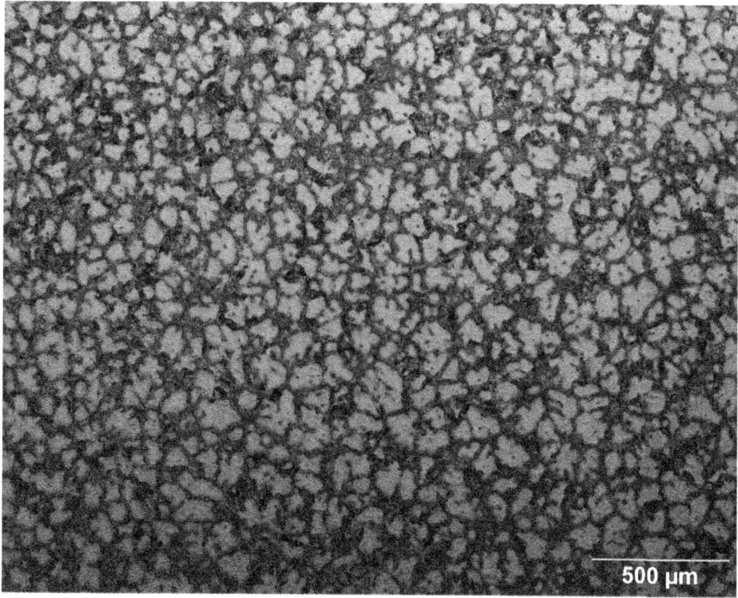

FIGURE 13.65 Optical microstructure of AZ91 magnesium alloy cast plate in as-cast condition. Fine dendritic structure can be seen. The etchant is Acetic Picral.

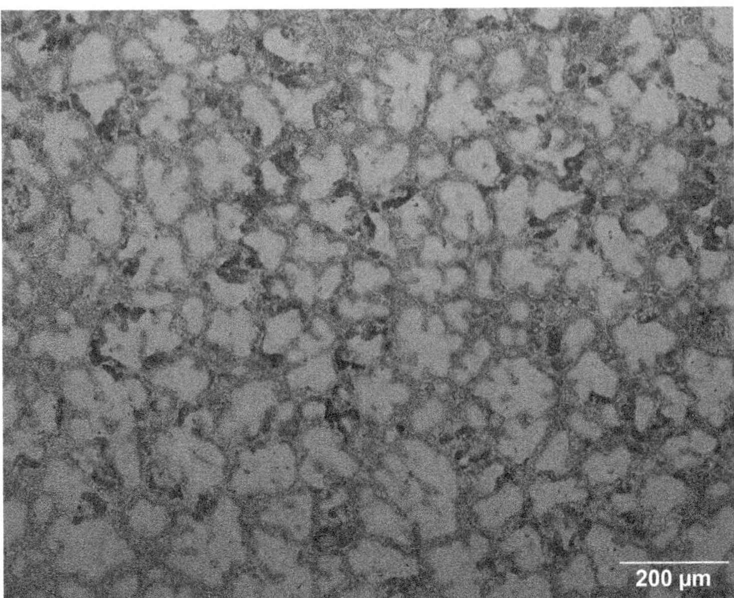

FIGURE 13.66 Optical microstructure of AZ91 magnesium alloy cast plate in as-cast condition taken at high magnification. Massive Mg_{17} $(Al,Zn)_{12}$ grain boundary phase along with fine lamellar eutectic of $Mg_{17}Al_{12}$ is clearly seen. The etchant is Acetic Picral.

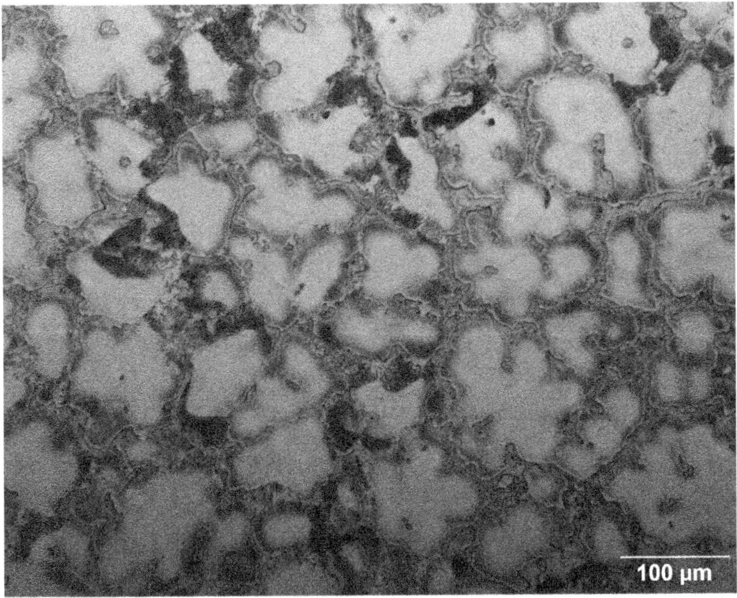

FIGURE 13.67 Optical microstructure of AZ91 magnesium alloy cast plate in as-cast condition taken at high magnification. Massive Mg_{17} $(Al,Zn)_{12}$ grain boundary phase along with fine lamellar eutectic of $Mg_{17}Al_{12}$ is clearly seen. The etchant is Acetic Picral.

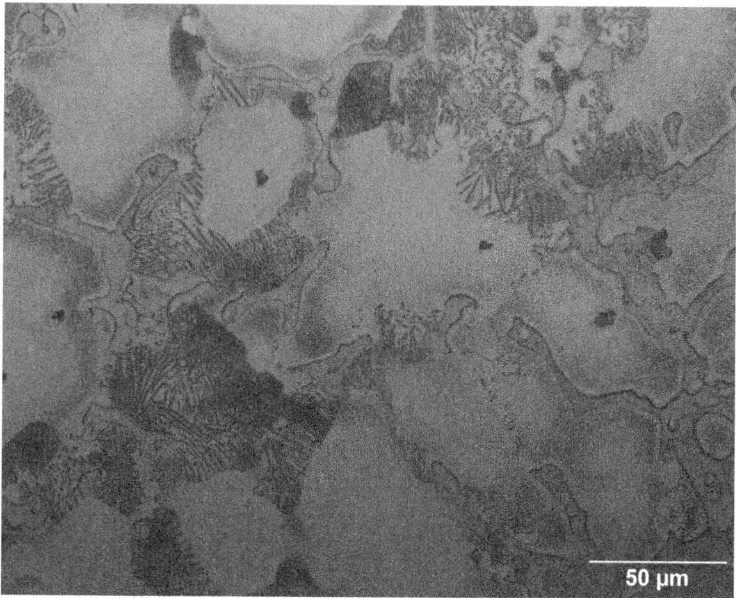

FIGURE 13.68 Optical microstructure of AZ91 magnesium alloy cast plate in as-cast condition taken at high magnification. Massive Mg_{17} $(Al,Zn)_{12}$ grain boundary phase along with fine lamellar eutectic of $Mg_{17}Al_{12}$ is clearly seen. The etchant is Acetic Picral.

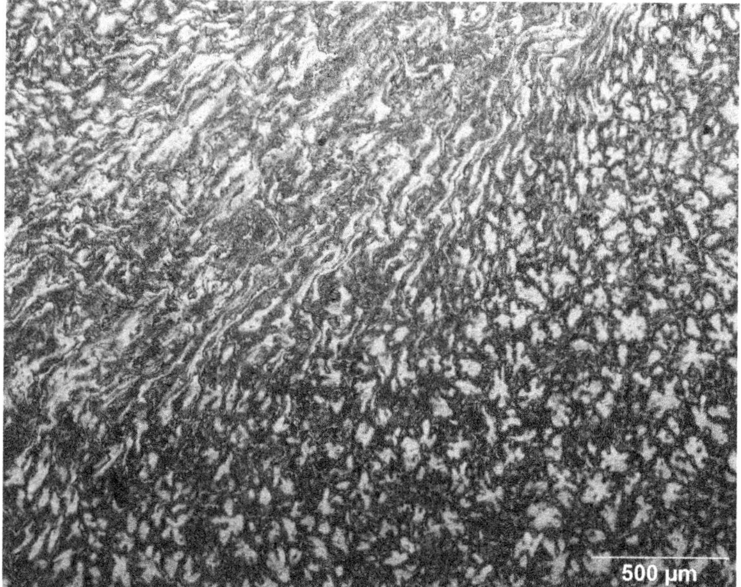

FIGURE 13.69 Optical microstructure of AZ91 magnesium alloy cast plate that has been subjected to friction stir processing. The microstructure was obtained on the surface showing the interface of processed zone (left) and parent material (right). The etchant is Acetic Picral.

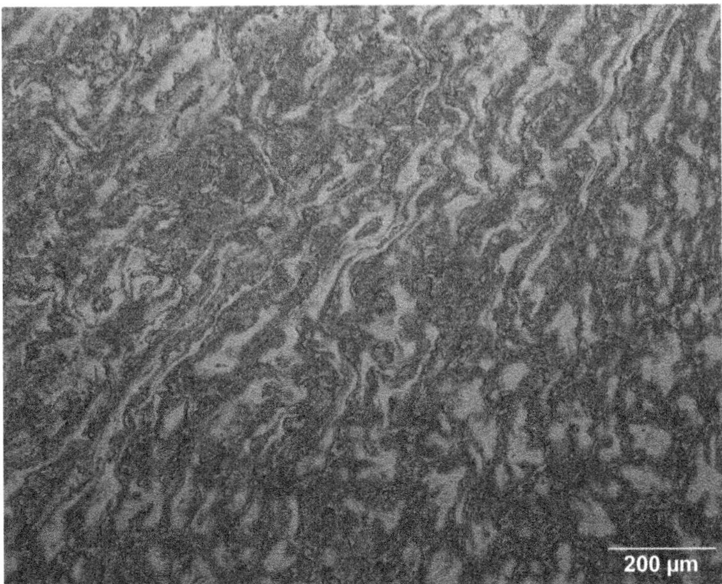

FIGURE 13.70 Optical microstructure of AZ91 magnesium alloy cast plate that has been subjected to friction stir processing. The microstructure was obtained on the surface in the stirred zone. The etchant is Acetic Picral.

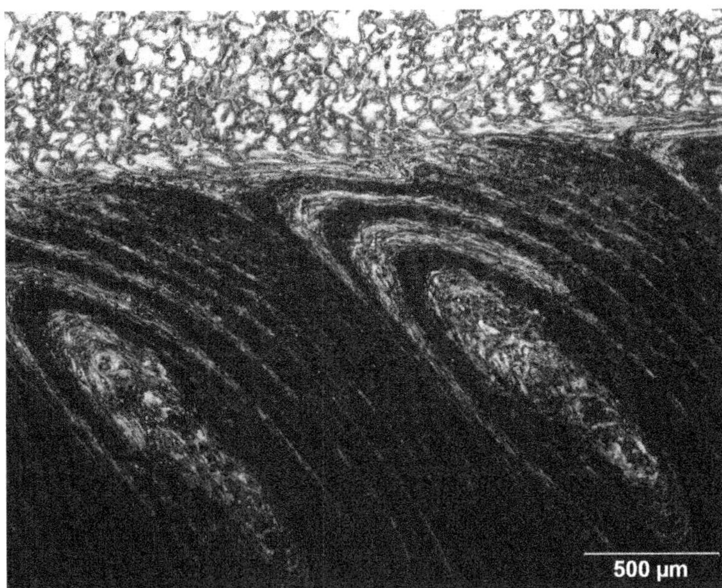

FIGURE 13.71 Optical microstructure of AZ91 magnesium alloy cast plate that has been subjected to friction stir processing showing 'onion ring' like circular grain flow. The microstructure was obtained on the cross section at the interface between the parent material and stirred zone. The etchant is Acetic Picral.

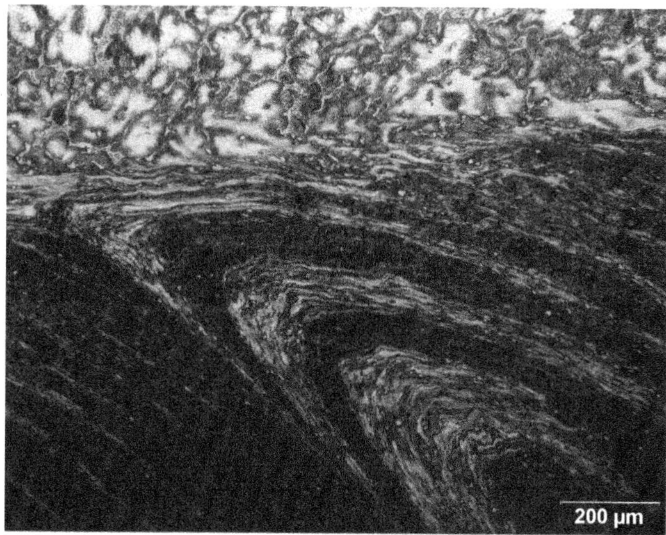

FIGURE 13.72 Optical microstructure of AZ91 magnesium alloy cast plate that has been subjected to friction stir processing showing 'onion ring' like circular grain flow at high magnification. The microstructure was obtained on the cross section at the interface between the parent material and stirred zone. Fragmented particles along the grain flow can be seen. The etchant is Acetic Picral.

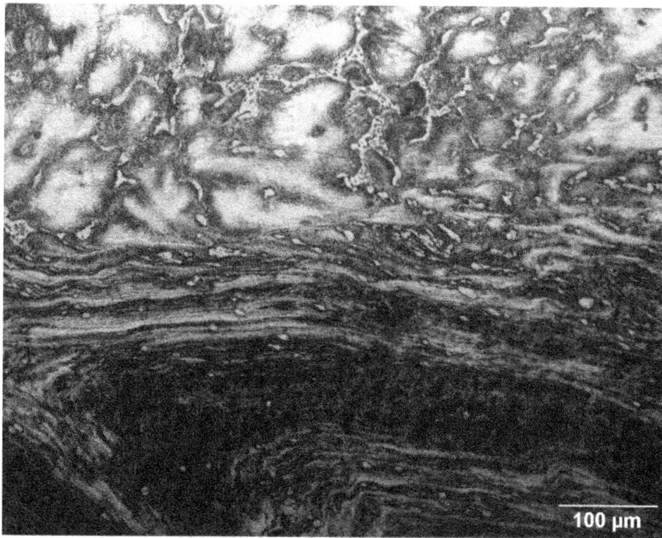

FIGURE 13.73 Optical microstructure of AZ91 magnesium alloy cast plate that has been subjected to friction stir processing showing onion-ring-like circular grain flow at high magnification. The microstructure was obtained on the cross section at the interface between the parent material and stirred zone. Fragmented particles along the grain flow can be seen. The etchant is Acetic Picral.

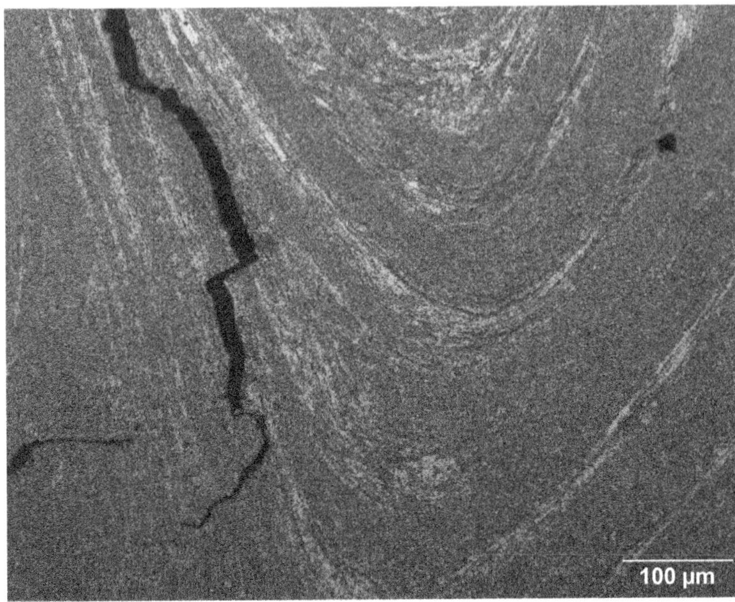

FIGURE 13.74 Optical microstructure of AZ91 magnesium alloy cast plate that has been subjected to friction stir processing. The microstructure was obtained on the cross section it the stirred zone. Weld defects in the form of cracks in the weld nugget can be seen. The etchant is Acetic Picral.

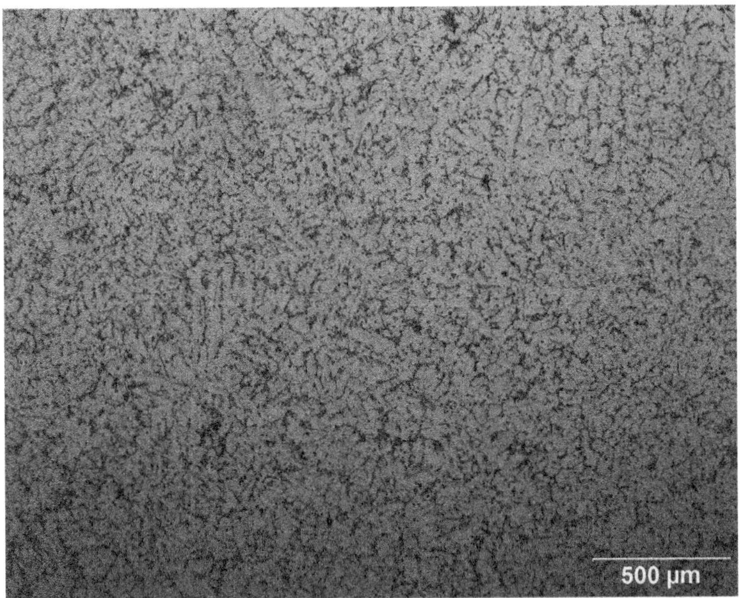

FIGURE 13.75 Optical microstructure of as-cast magnesium alloy AE42 taken at high magnification. Fine dendritic structure can be seen. The etchant is Acetic Picral.

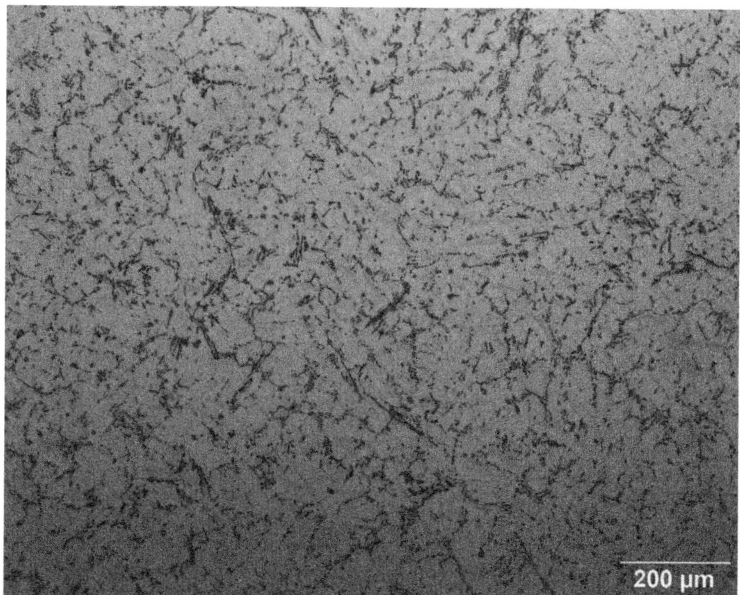

FIGURE 13.76 Optical microstructure of as-cast magnesium alloy AE42 taken at high magnification. Fine dendritic structure can be seen. The etchant is Acetic Picral.

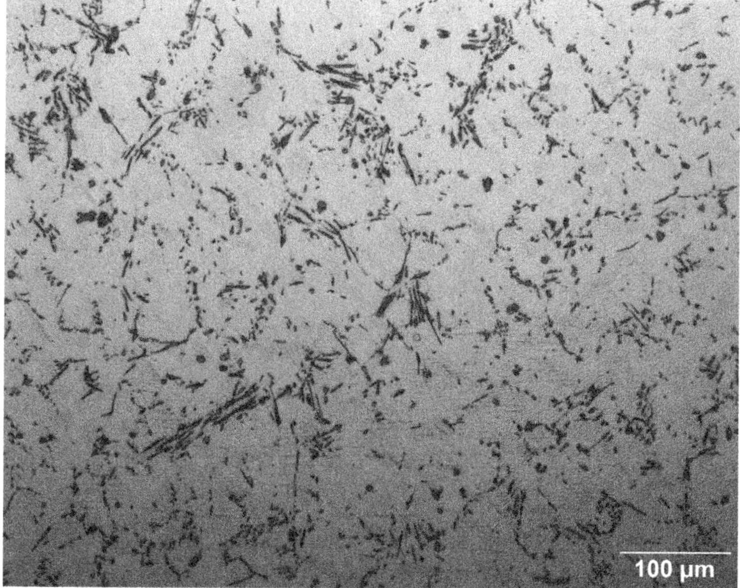

FIGURE 13.77 Optical microstructure of as-cast magnesium alloy AE42 taken at high magnification. Fine dendritic structure can be seen. Grain boundary $Al_{11}RE_3$ phase along with $Al_{10}Ce_2Mn_7$ particles can be seen. The etchant is Acetic Picral.

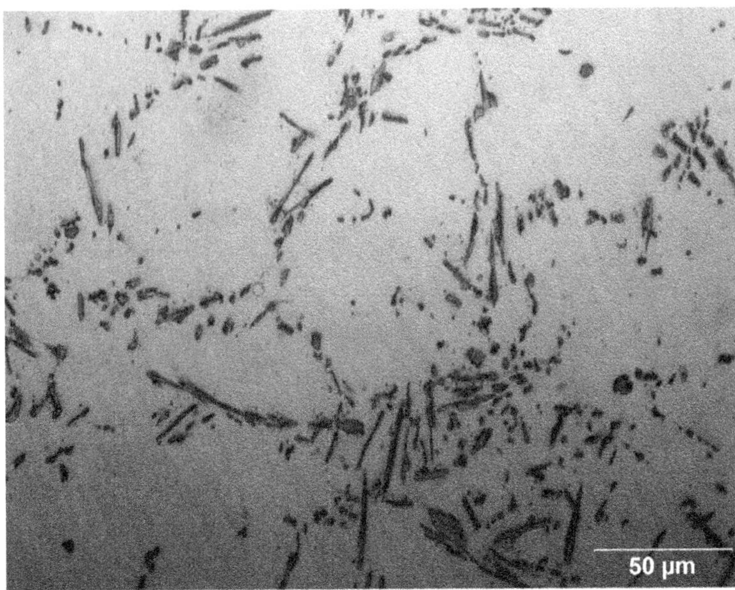

FIGURE 13.78 Optical microstructure of as-cast magnesium alloy AE42 taken at high magnification. Fine dendritic structure can be seen. Grain boundary $Al_{11}RE_3$ phase along with $Al_{10}Ce_2Mn_7$ particles can be seen. The etchant is Acetic Picral.

FIGURE 13.79 Optical microstructure of as-cast and extruded magnesium alloy AE42 taken along the longitudinal direction. Flow lines can be seen. The etchant is Acetic Picral.

FIGURE 13.80 Optical microstructure of as-cast and extruded magnesium alloy AE42 taken along the longitudinal direction. Flow lines can be seen along with fragmented second phase particles. The etchant is Acetic Picral.

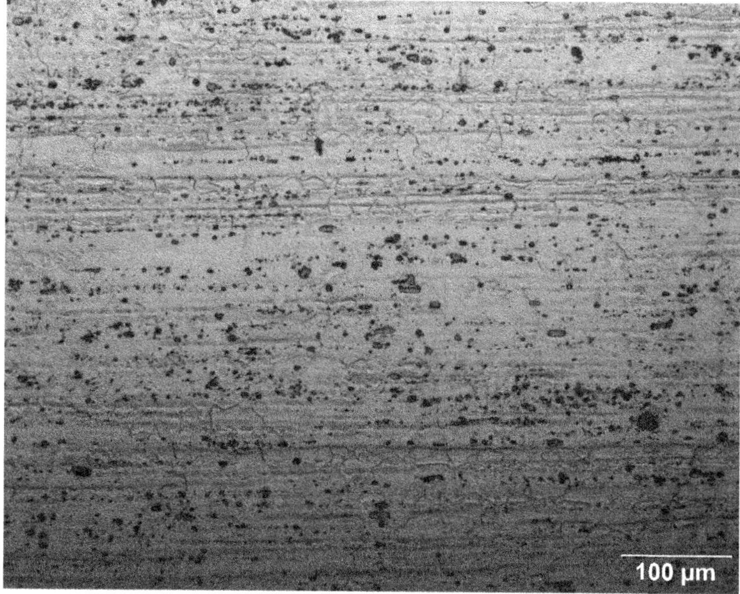

FIGURE 13.81 Optical microstructure of as-cast and extruded magnesium alloy AE42 taken along the longitudinal direction taken at high magnification. Flow lines can be seen along with fragmented second phase particles. The etchant is Acetic Picral.

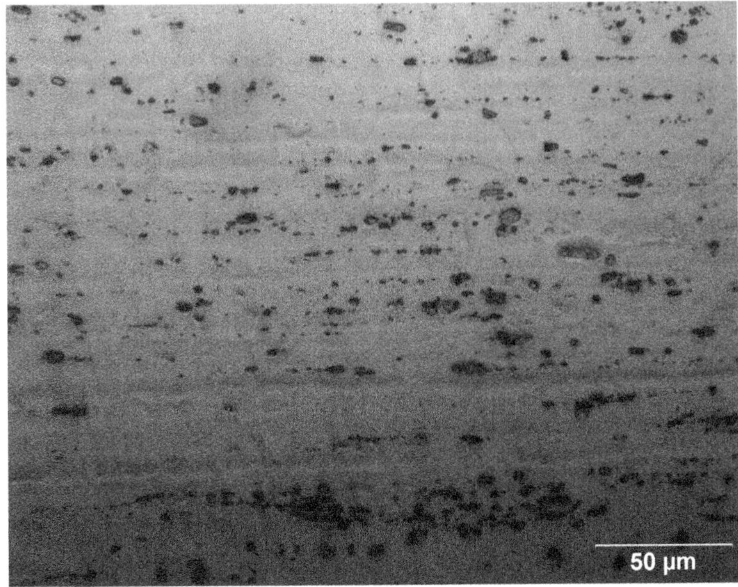

FIGURE 13.82 Optical microstructure of as-cast and extruded magnesium alloy AE42 taken along the longitudinal direction taken at high magnification. Flow lines can be seen along with fragmented second phase particles. The etchant is Acetic Picral.

FIGURE 13.83 Optical microstructure of as-cast and extruded magnesium alloy AE42 taken on the cross section. The etchant is Acetic Picral.

FIGURE 13.84 Optical microstructure of as-cast and extruded magnesium alloy AE42 taken on the cross section taken at high magnification. The etchant is Acetic Picral.

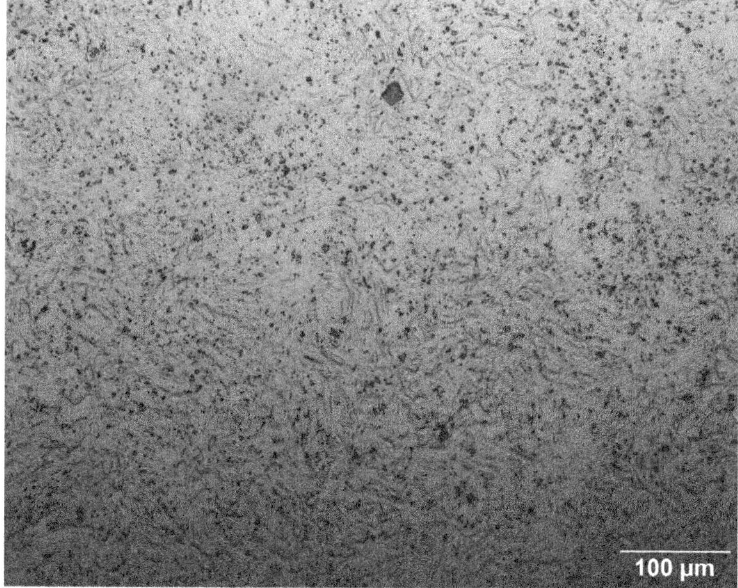

FIGURE 13.85 Optical microstructure of as-cast and extruded magnesium alloy AE42 taken on the cross section taken at high magnification. Fragmented second phase particles can be seen. The etchant is Acetic Picral.

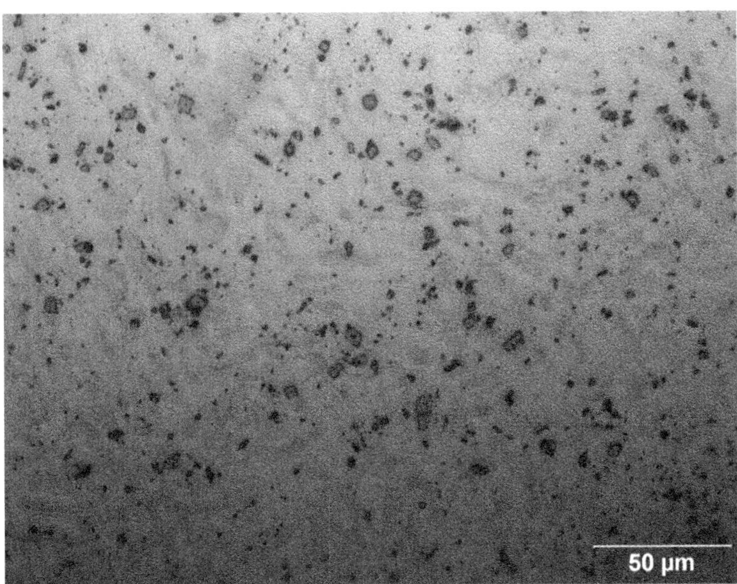

FIGURE 13.86 Optical microstructure of as-cast and extruded magnesium alloy AE42 taken on the cross section taken at high magnification. Fragmented second phase particles can be seen. The etchant is Acetic Picral.

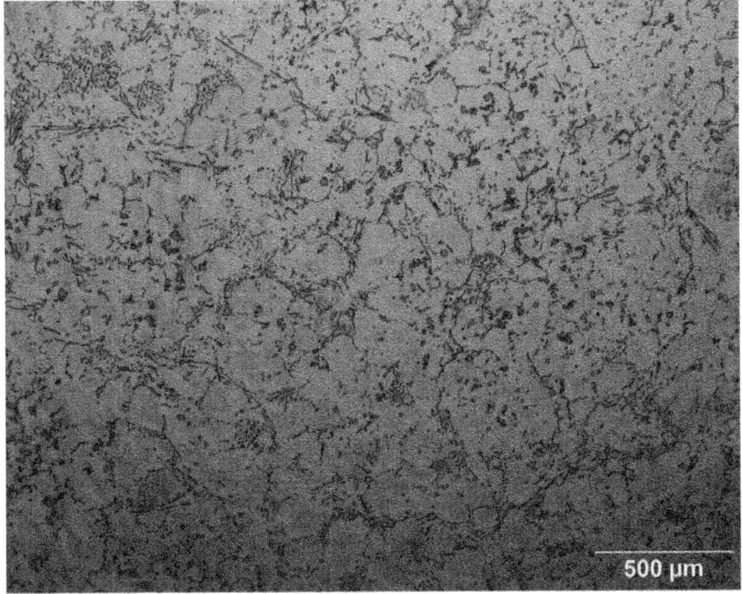

FIGURE 13.87 Optical microstructure of as-cast and extruded magnesium alloy AE44 taken on the cross section. The etchant is Acetic Picral.

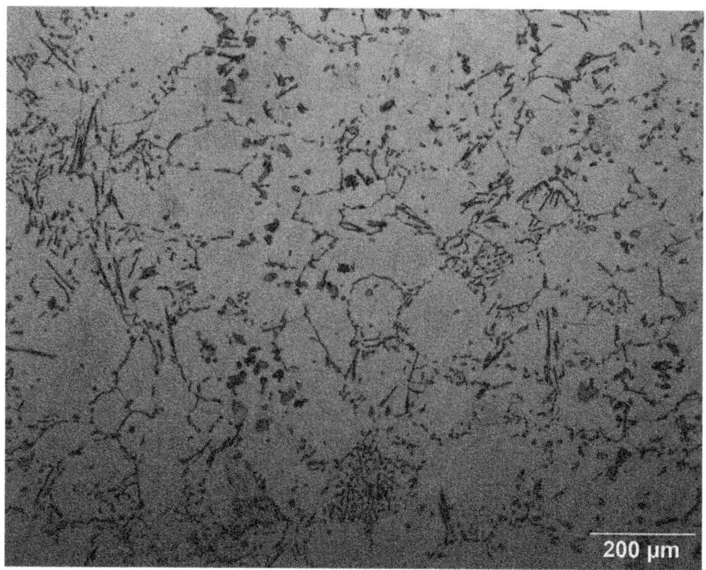

FIGURE 13.88 Optical microstructure of as-cast and extruded magnesium alloy AE44 taken on the cross section taken at high magnification. The microstructure consists of grain boundary phase $Al_{11}(La,Ce,Nd)_3$, needles of $Al_{11}Ce_3$ and blocky particles of $Al_{10}(Ce,Nd)_2Mn_7$. The etchant is Acetic Picral.

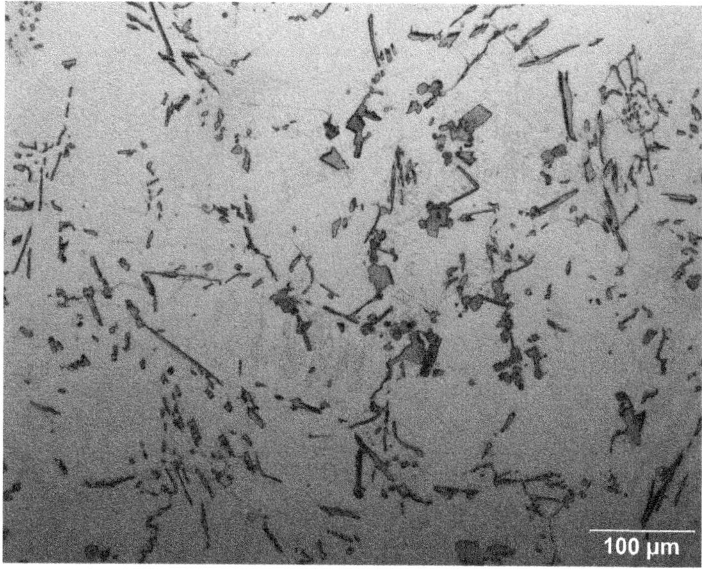

FIGURE 13.89 Optical microstructure of as-cast and extruded magnesium alloy AE44 taken on the cross section taken at high magnification. The microstructure consists of grain boundary phase $Al_{11}(La,Ce,Nd)_3$, needles of $Al_{11}Ce_3$ and blocky particles of $Al_{10}(Ce,Nd)_2Mn_7$. The etchant is Acetic Picral.

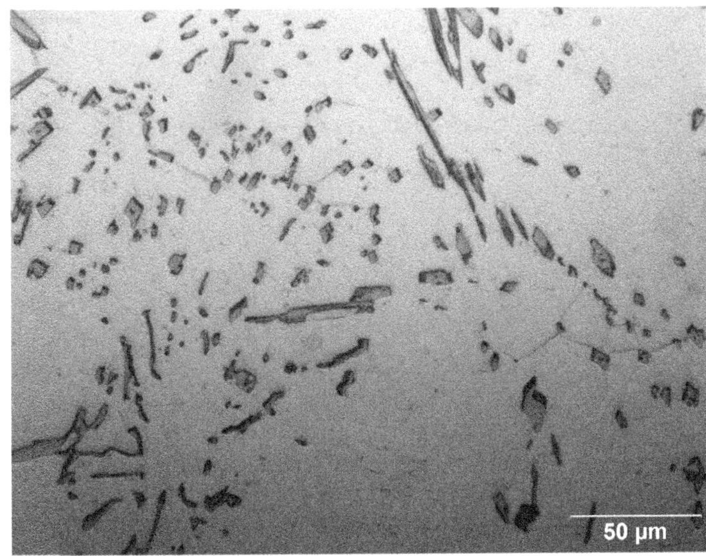

FIGURE 13.90 Optical microstructure of as-cast and extruded magnesium alloy AE44 taken on the cross section taken at high magnification. The microstructure consists of grain boundary phase $Al_{11}(La,Ce,Nd)_3$, needles of $Al_{11}Ce_3$ and blocky particles of $Al_{10}(Ce,Nd)_2Mn_7$. The etchant is Acetic Picral.

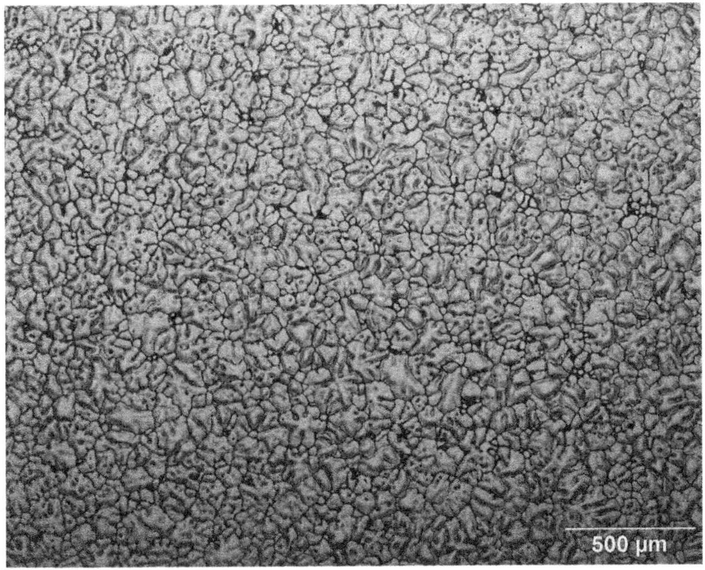

FIGURE 13.91 Optical microstructure of as-cast magnesium alloy ZC63. Fine dendritic structure with continuous network of eutectic along the grain boundaries can be seen. The etchant is Acetic Picral.

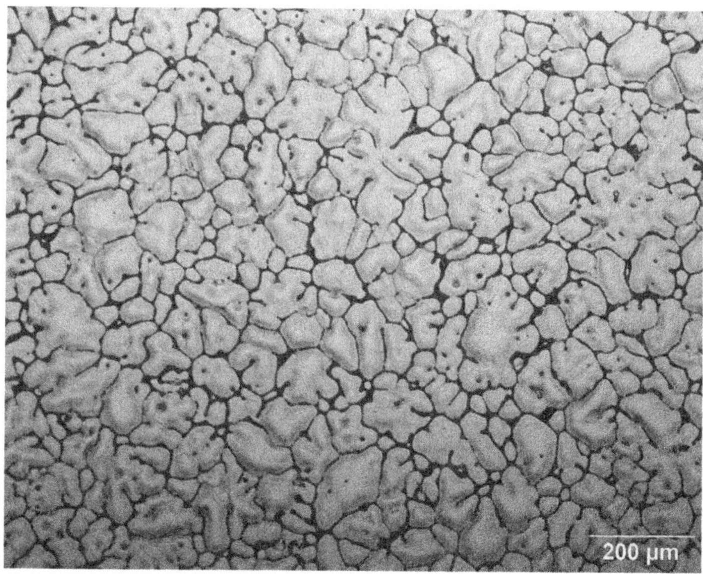

FIGURE 13.92 Optical microstructure of as-cast magnesium alloy ZC63. Optical microstructure of as-cast magnesium alloy ZC63 taken at high magnification. Fine dendritic structure with continuous network of eutectic along the grain boundaries can be seen. The etchant is Acetic Picral.

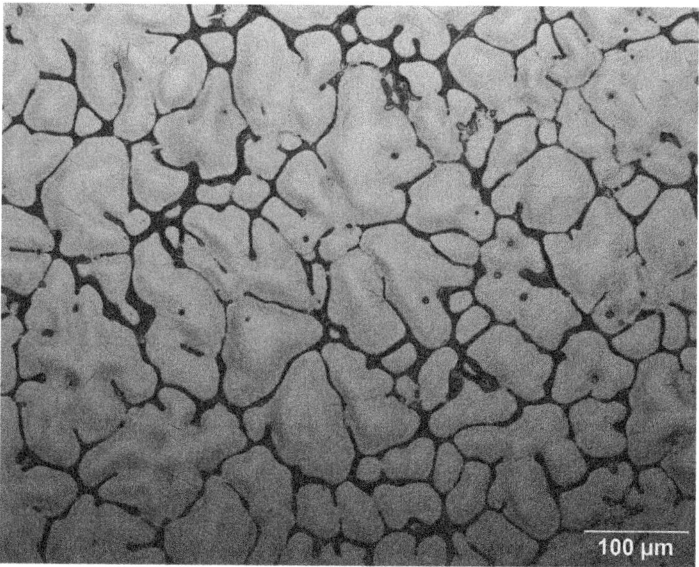

FIGURE 13.93 Optical microstructure of as-cast magnesium alloy ZC63. Optical microstructure of as-cast magnesium alloy ZC63 taken at high magnification. Fine dendritic structure with continuous network of eutectic along the grain boundaries can be seen. The etchant is Acetic Picral.

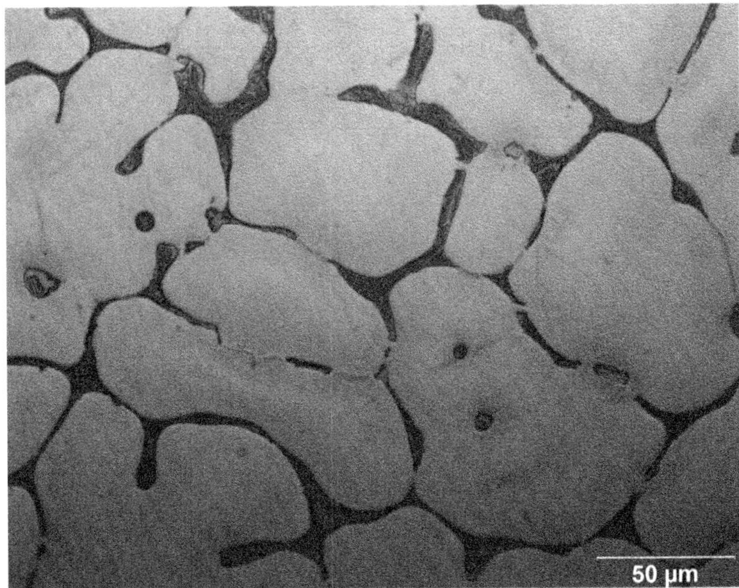

FIGURE 13.94 Optical microstructure of as-cast magnesium alloy ZC63. Optical microstructure of as-cast magnesium alloy ZC63 taken at high magnification. Fine lamellar eutectic of $Mg(Zn,Cu)_2$ can be seen at the grain boundaries along with the particles of Mg (Zn,Cu) inside the grains. The etchant is Acetic Picral.

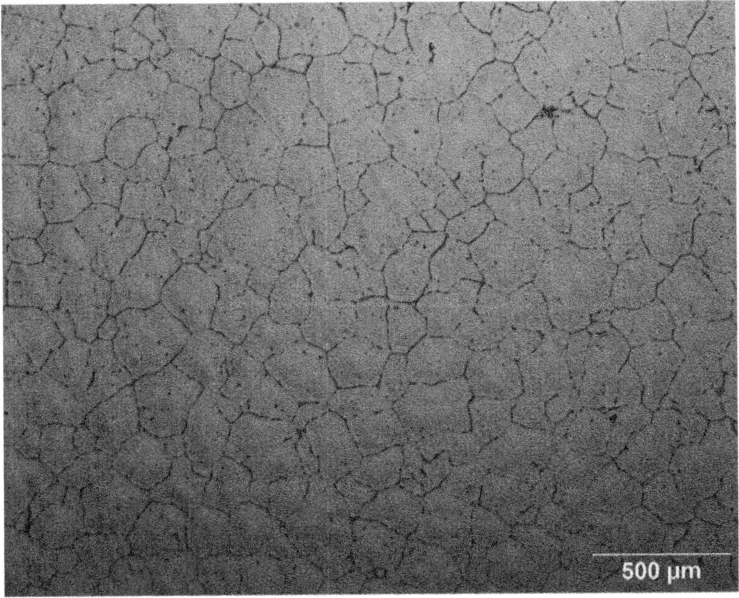

FIGURE 13.95 Optical microstructure of as-cast magnesium alloy ZE10. Fine equiaxed grain structure along with grain boundary phase can be seen. The etchant is Acetic Picral.

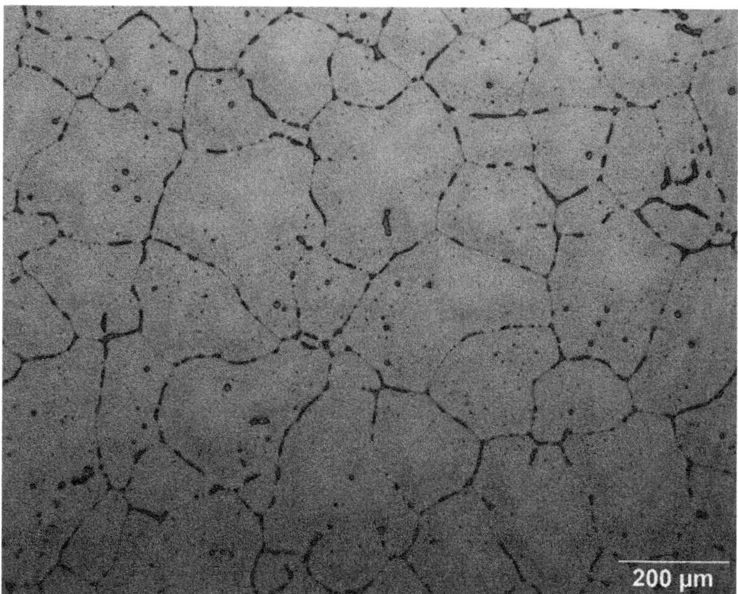

FIGURE 13.96 Optical microstructure of as-cast magnesium alloy ZE10 taken at high magnification. Equiaxed grain structure along with grain boundary phase Mg_7Zn_3RE can be seen. The specimen was etched with Acetic Picral.

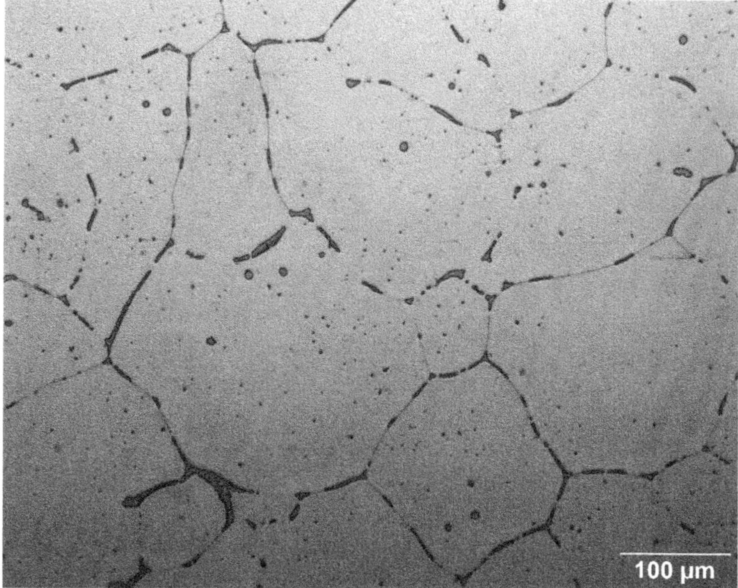

FIGURE 13.97 Optical microstructure of as-cast magnesium alloy ZE10 taken at high magnification. Equiaxed grain structure along with grain boundary phase Mg_7Zn_3RE can be seen. The etchant is Acetic Picral.

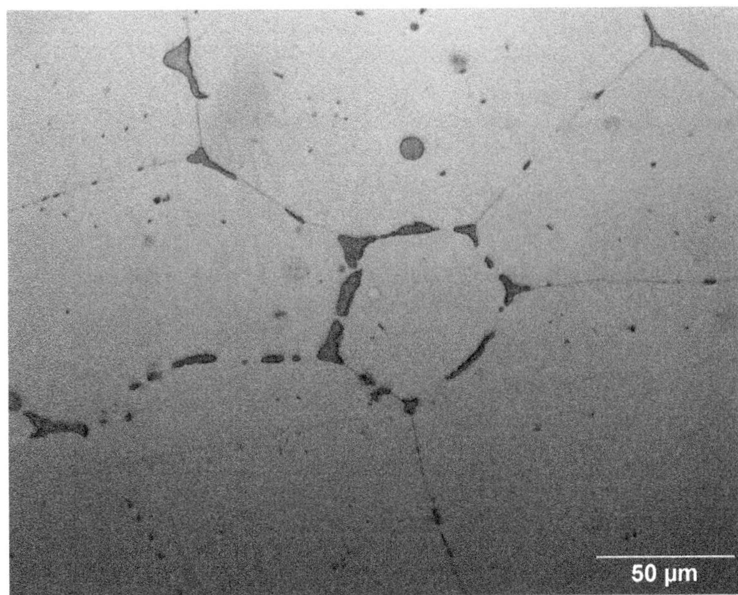

FIGURE 13.98 Optical microstructure of as-cast magnesium alloy ZE10 taken at high magnification. Equiaxed grain structure along with grain boundary phase Mg_7Zn_3RE can be seen. The etchant is Acetic Picral.

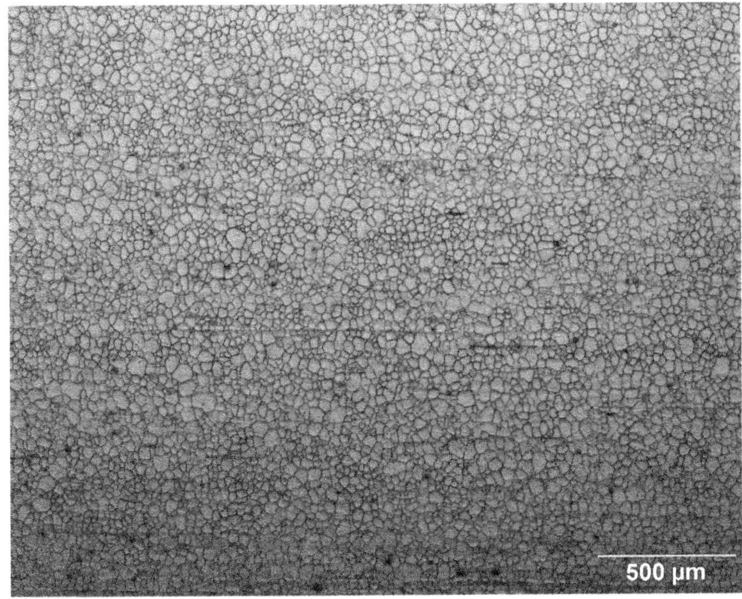

FIGURE 13.99 Optical microstructure of as-cast and extruded magnesium alloy ZE10 taken along the longitudinal direction. Fine equiaxed recrystallized grains along with the fragmented particles of Mg_7Zn_3RE can be seen. The etchant is Acetic Picral.

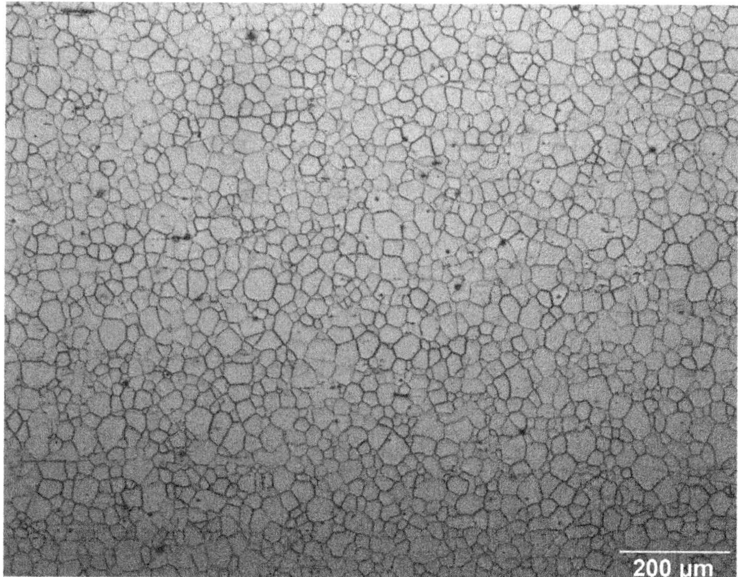

FIGURE 13.100 Optical microstructure of as-cast and extruded magnesium alloy ZE10 taken along the longitudinal direction. Fine equiaxed recrystallized grains along with the fragmented particles of Mg_7Zn_3RE can be seen. The etchant is Acetic Picral.

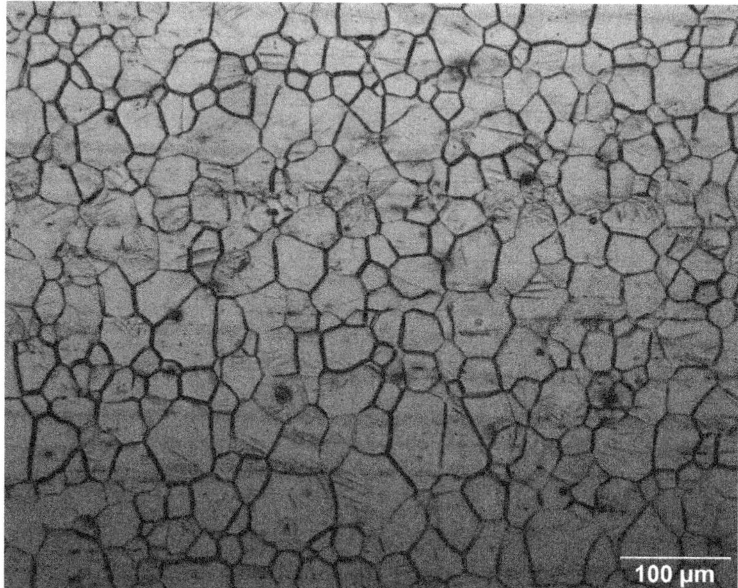

FIGURE 13.101 Optical microstructure of as-cast and extruded magnesium alloy ZE10 taken along the longitudinal direction taken at high magnification. The etchant is Acetic Picral.

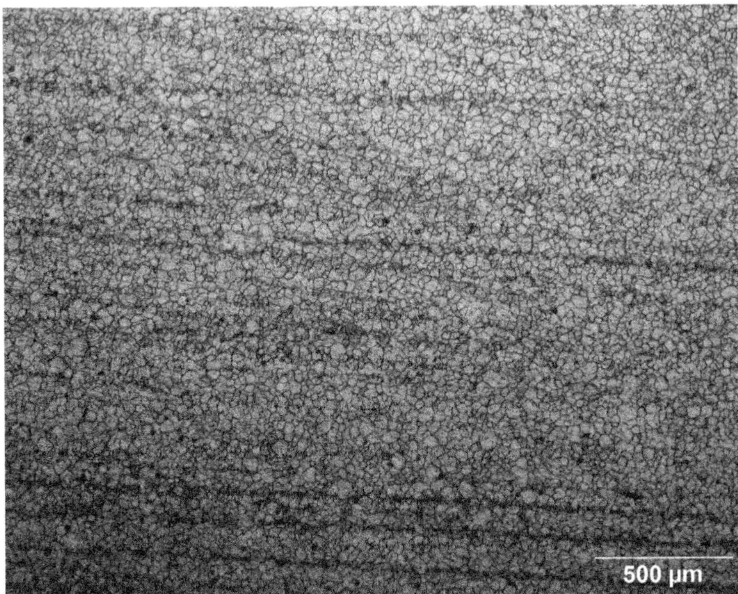

FIGURE 13.102 Optical microstructure of as-cast and extruded magnesium alloy ZE10 taken on the cross section. Fine equiaxed recrystallized grains along with the fragmented particles of Mg_7Zn_3RE can be seen. The etchant is Acetic Picral.

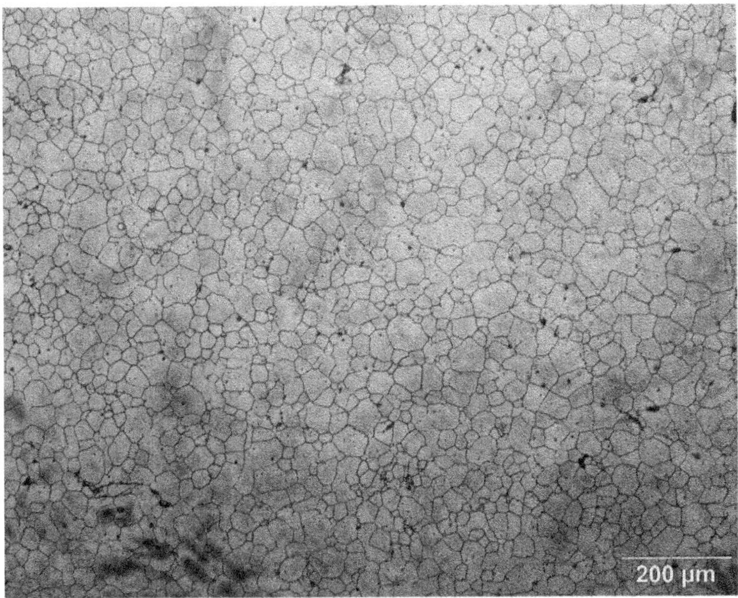

FIGURE 13.103 Optical microstructure of as-cast and extruded magnesium alloy ZE10 taken on the cross section taken at high magnification. Fine equiaxed recrystallized grains along with the fragmented particles of Mg_7Zn_3RE can be seen. The etchant is Acetic Picral.

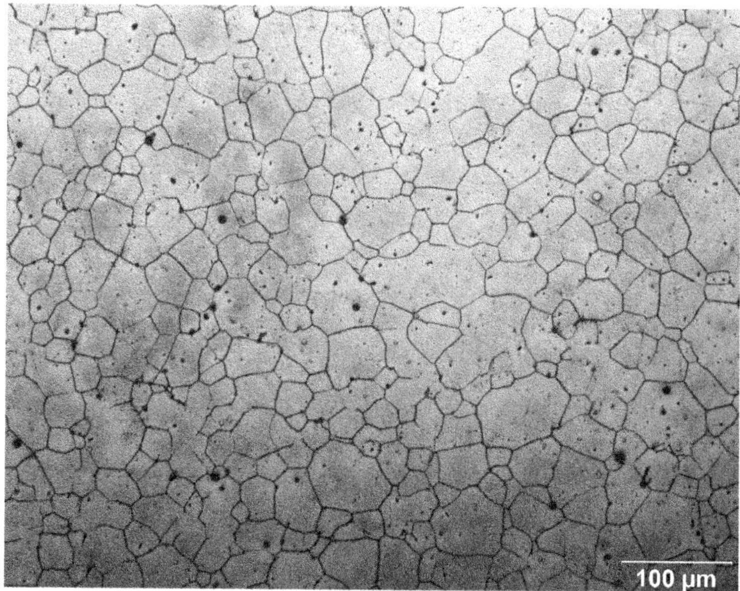

FIGURE 13.104 Optical microstructure of as-cast and extruded magnesium alloy ZE10 taken on the cross section taken at high magnification. Fine equiaxed recrystallized grains along with the fragmented particles of Mg_7Zn_3RE can be seen. The etchant is Acetic Picral.

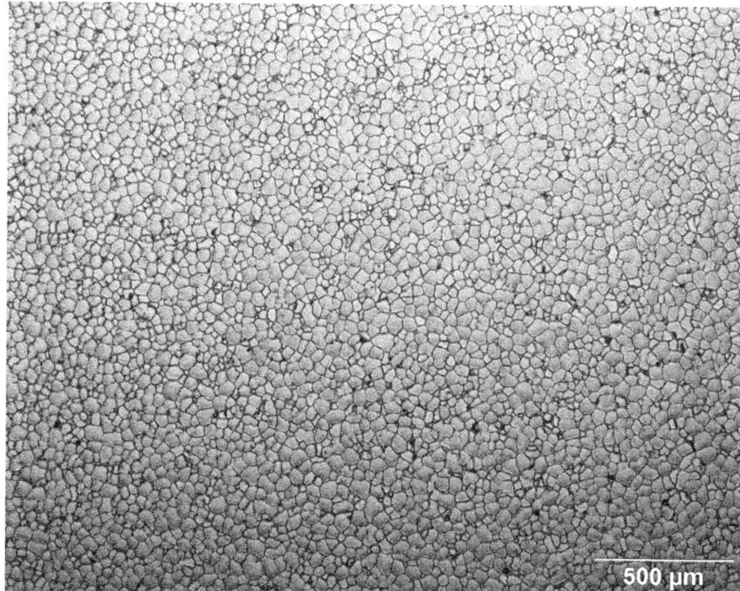

FIGURE 13.105 Optical microstructure of as-cast magnesium alloy ZE41. Fine equiaxed grain structure can be seen with Mg_7Zn_3RE eutectic at grain boundaries. The etchant is Acetic Picral.

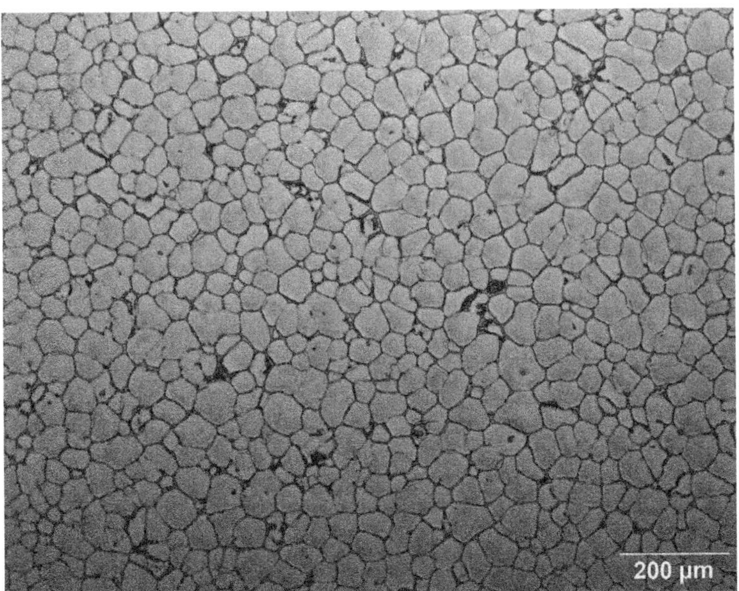

FIGURE 13.106 Optical microstructure of as-cast magnesium alloy ZE41 taken at high magnification. Fine equiaxed grain structure can be seen with Mg_7Zn_3RE eutectic at grain boundaries. The etchant is Acetic Picral.

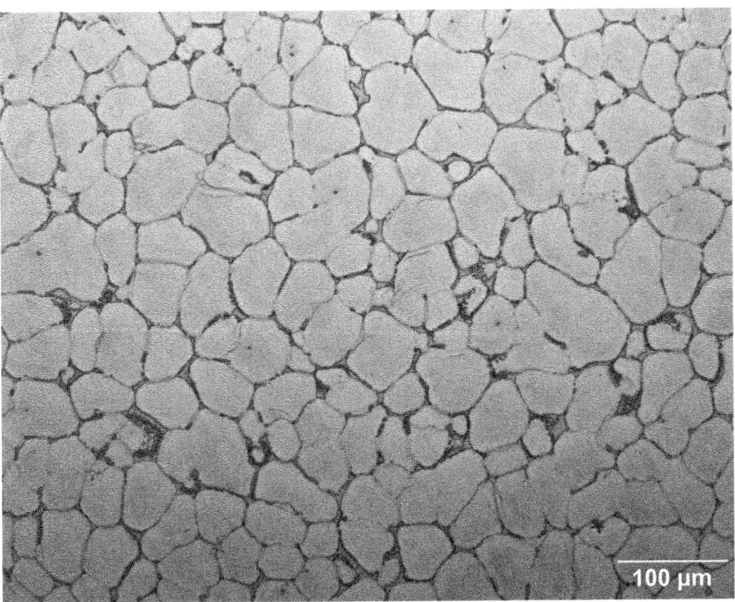

FIGURE 13.107 Optical microstructure of as-cast magnesium alloy ZE41 taken at high magnification. Fine equiaxed grain structure can be seen with Mg_7Zn_3RE eutectic at grain boundaries. The etchant is Acetic Picral.

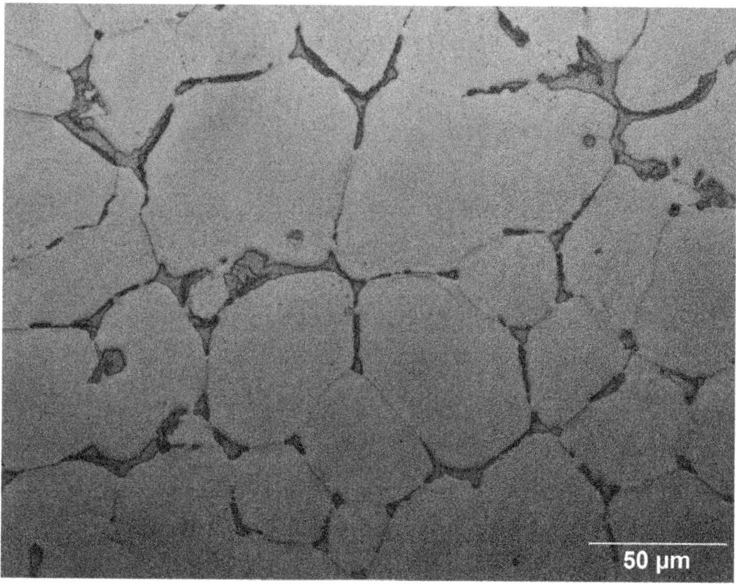

FIGURE 13.108 Optical microstructure of as-cast magnesium alloy ZE41 taken at high magnification. Fine equiaxed grain structure can be seen with Mg_7Zn_3RE eutectic at grain boundaries. The etchant is Acetic Picral.

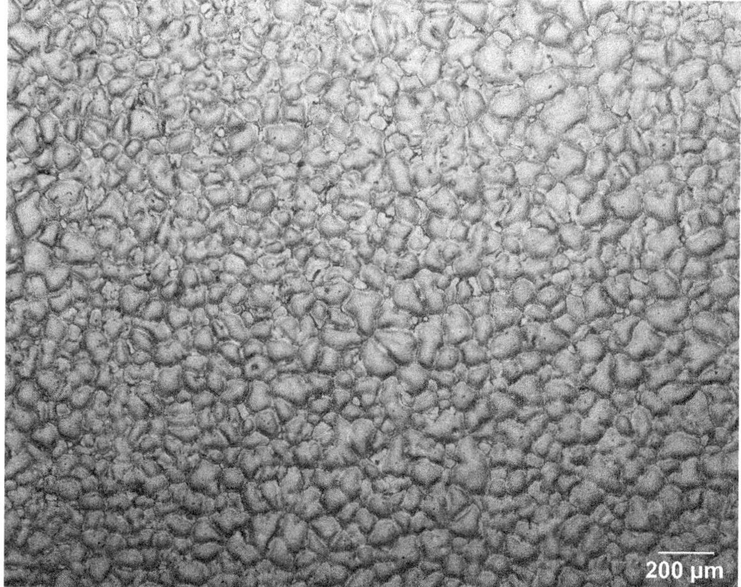

FIGURE 13.109 Optical microstructure of as-cast magnesium alloy ZE41 (R25). Dendritic structure with flowery pattern can be seen. The etchant is Acetic Picral.

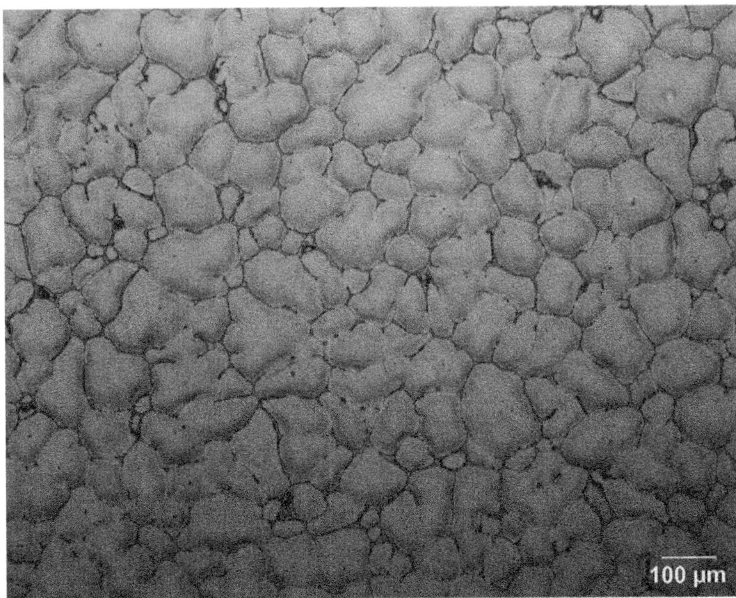

FIGURE 13.110 Optical microstructure of as-cast magnesium alloy ZE41 (R25) taken at high magnification. Dendritic structure with flowery pattern can be seen. Mg_7Zn_3RE eutectic at grain boundaries can be seen. The etchant is Acetic Picral.

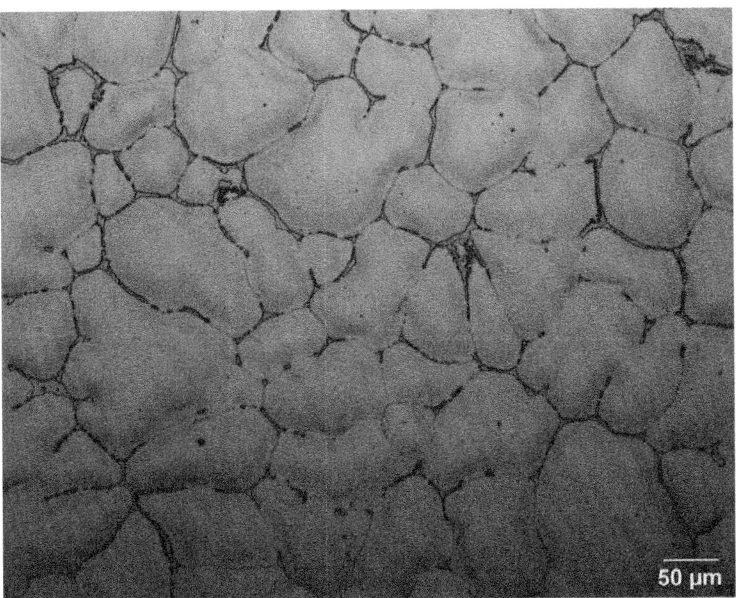

FIGURE 13.111 Optical microstructure of as-cast magnesium alloy ZE41 (R25) taken at high magnification. Dendritic structure with flowery pattern can be seen. Mg_7Zn_3RE eutectic at grain boundaries can be seen. The etchant is Acetic Picral.

FIGURE 13.112 Optical microstructure of as-cast magnesium alloy ZK60. Dendritic structure with fine particles of Mg_2Zn_3 can be seen. The etchant is Acetic Picral.

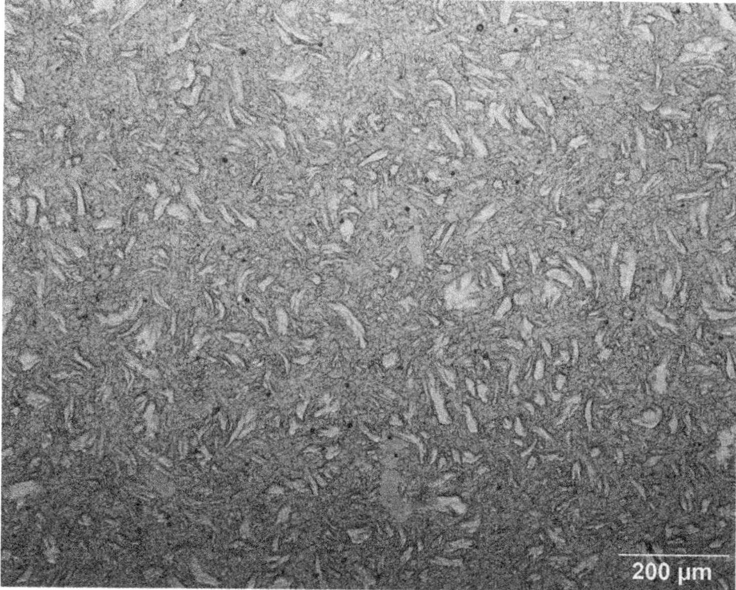

FIGURE 13.113 Optical microstructure of as-cast magnesium alloy ZK60 taken at high magnification. Dendritic structure with fine particles of Mg_2Zn_3 can be seen. The etchant is Acetic Picral.

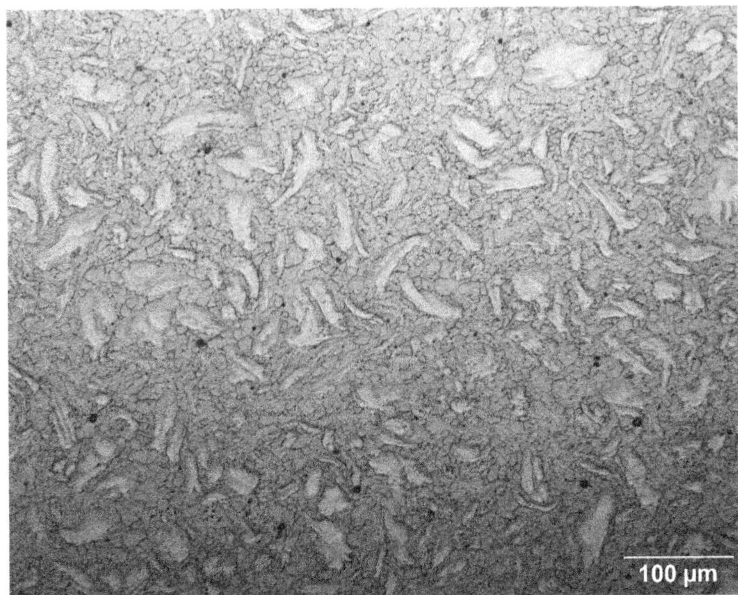

FIGURE 13.114 Optical microstructure of as-cast magnesium alloy ZK60 taken at high magnification. Dendritic structure with fine particles of Mg_2Zn_3 can be seen. The etchant is Acetic Picral.

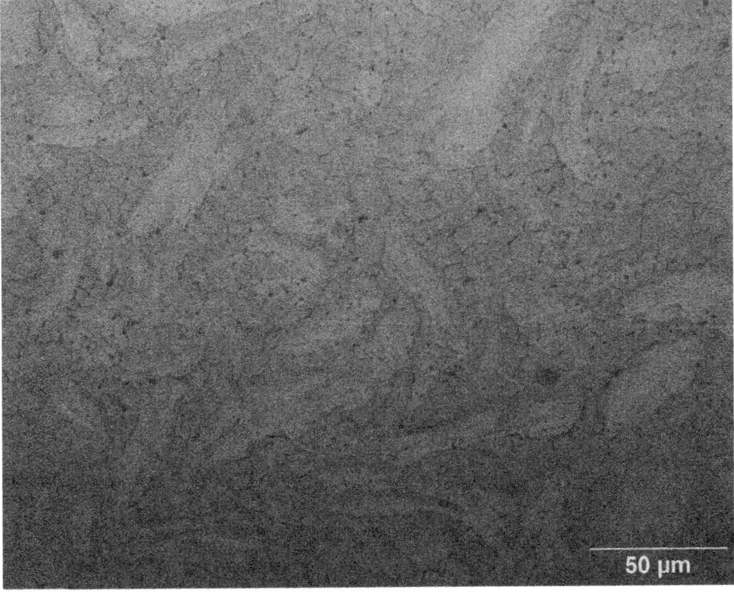

FIGURE 13.115 Optical microstructure of as-cast magnesium alloy ZK60 taken at high magnification. Dendritic structure with fine particles of Mg_2Zn_3 can be seen. Bigger particles of Mg(Zn,Zr) are also seen. The etchant is Acetic Picral.

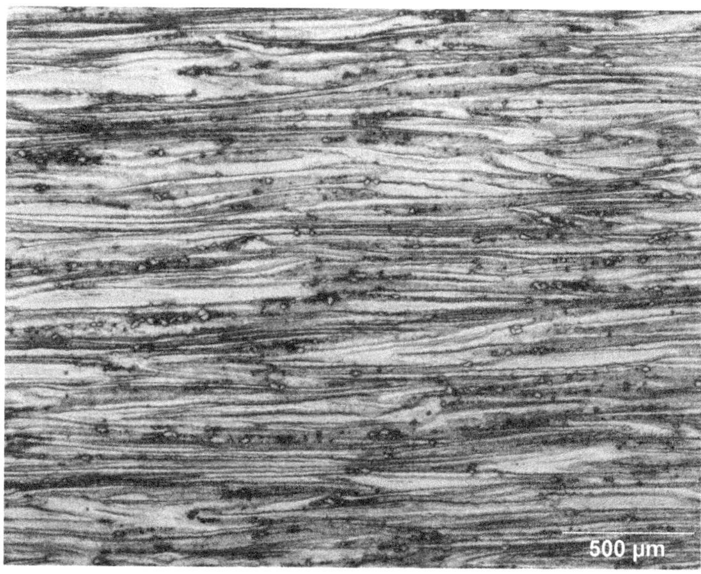

FIGURE 13.116 Optical microstructure of as-cast and extruded magnesium alloy ZK60 taken along the longitudinal direction. The microstructure shows flow lines along the rolling direction with fragmentation and distribution of second phase particles. The etchant is Acetic Picral.

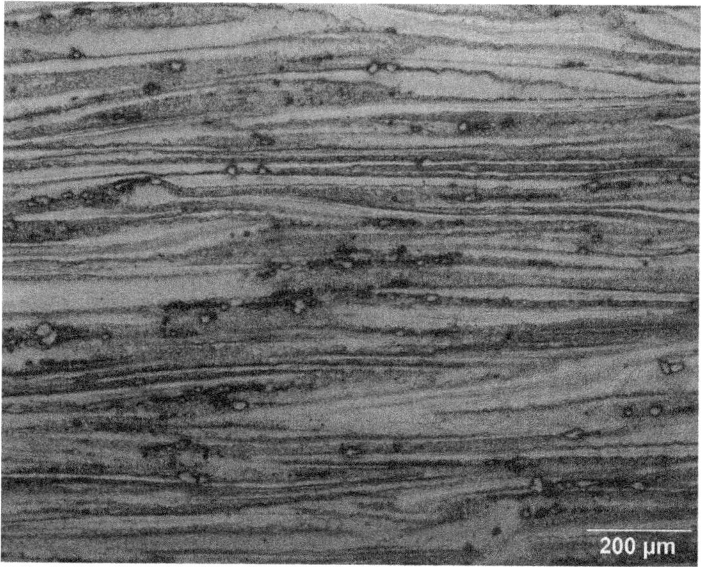

FIGURE 13.117 Optical microstructure of as-cast and extruded magnesium alloy ZK60 taken along the longitudinal direction taken at high magnification. The microstructure shows flow lines with fine recrystallized grains in the rolling direction with fragmentation and distribution of second phase particles. The etchant is Acetic Picral.

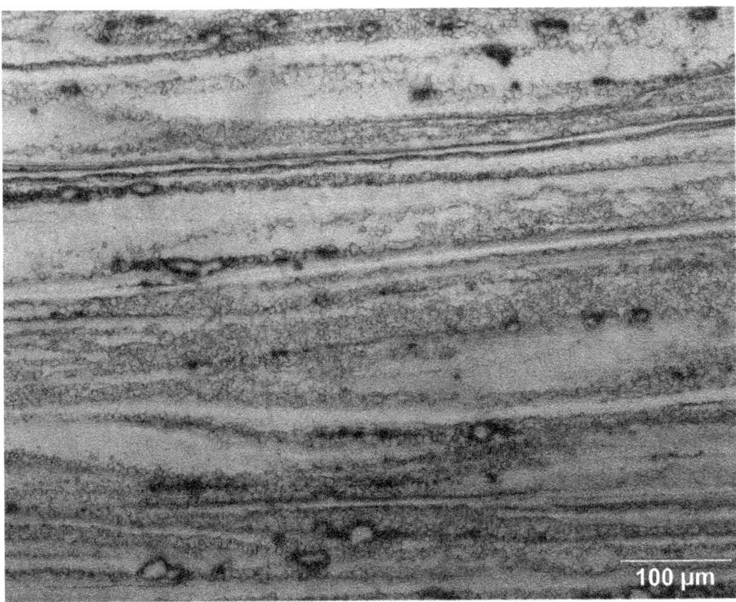

FIGURE 13.118 Optical microstructure of as-cast and extruded magnesium alloy ZK60 taken along the longitudinal direction taken at high magnification. The microstructure shows flow lines with fine recrystallized grains in the rolling direction with fragmentation and distribution of second phase particles. The etchant is Acetic Picral.

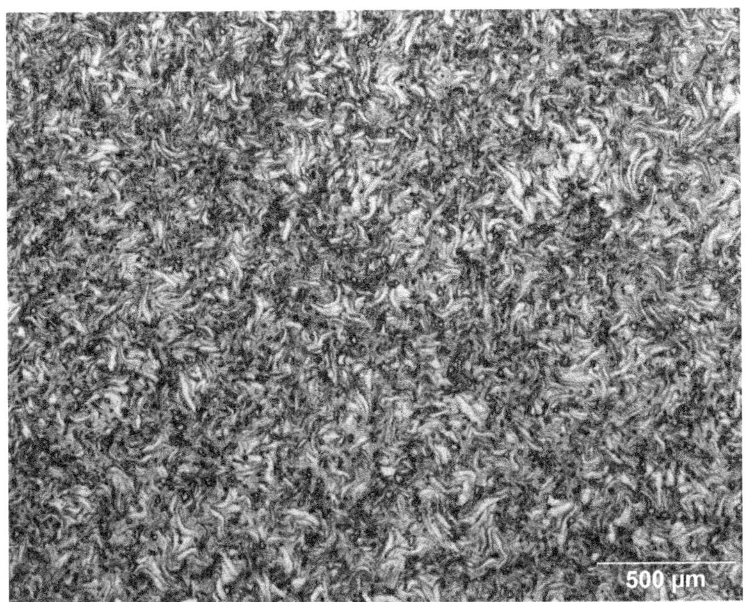

FIGURE 13.119 Optical microstructure of as-cast and extruded magnesium alloy ZK60 taken at the cross section. The etchant is Acetic Picral.

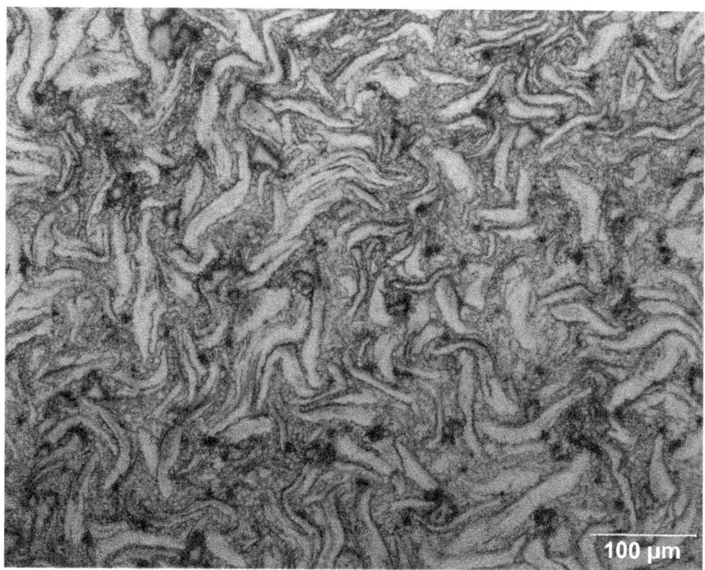

FIGURE 13.120 Optical microstructure of as-cast and extruded magnesium alloy ZK60 taken at the cross section. The microstructure shows pattern of metal flow with fine recrystallized grains with fragmentation and distribution of second phase particles. The etchant is Acetic Picral.

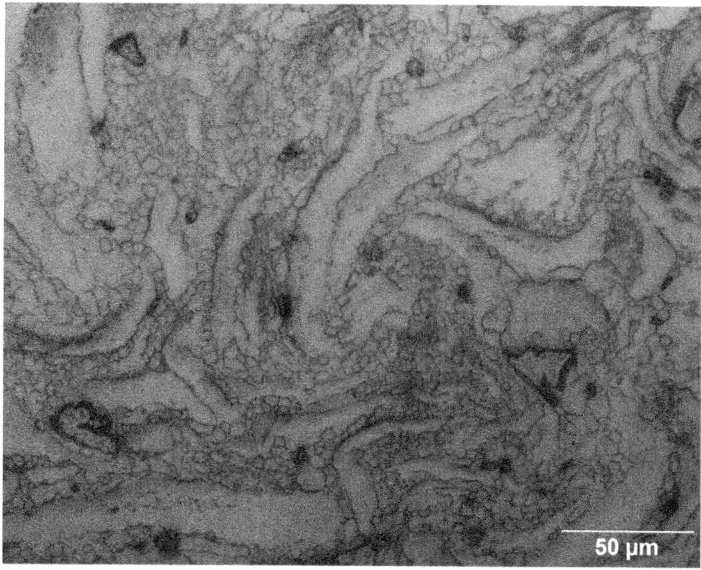

FIGURE 13.121 Optical microstructure of as-cast and extruded magnesium alloy ZK60 taken at the cross section. The microstructure shows pattern of metal flow with fine recrystallized grains with fragmentation and distribution of second phase particles. The etchant is Acetic Picral.

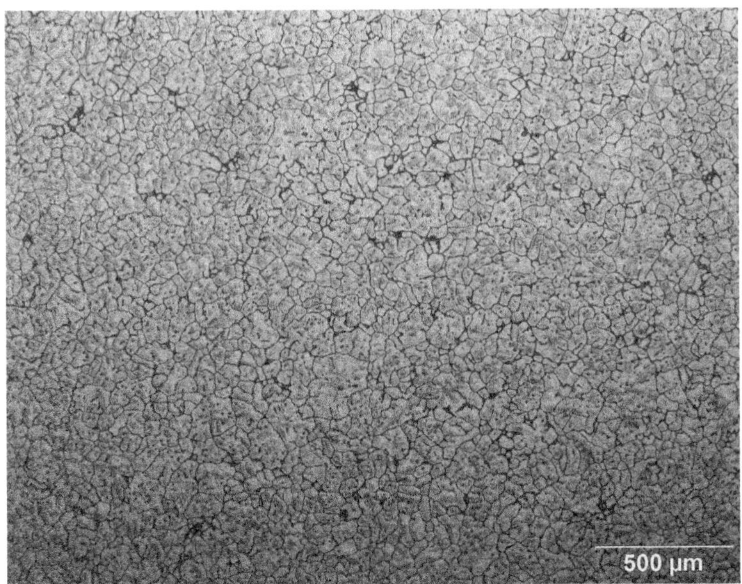

FIGURE 13.122 Optical microstructure of as-cast magnesium alloy QE22. Fine equiaxed grain structure can be seen. The etchant is Acetic Picral.

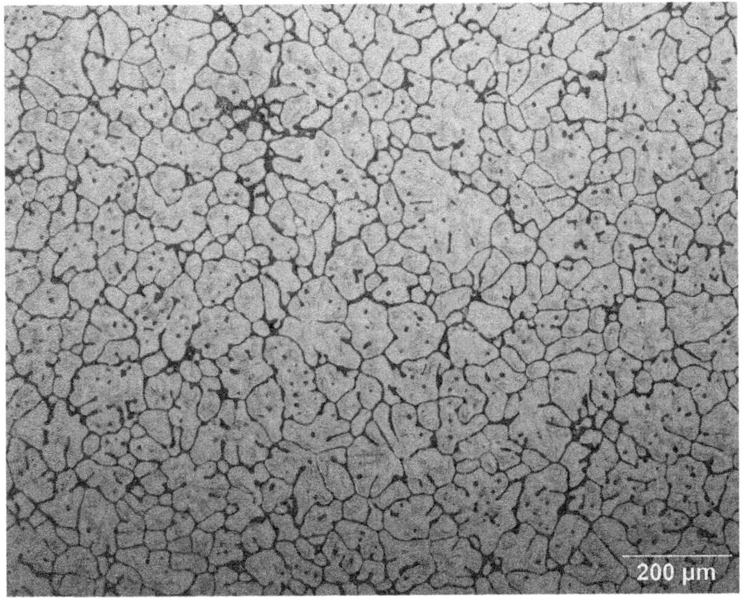

FIGURE 13.123 Optical microstructure of as-cast magnesium alloy QE22. Dendritic solidification pattern with $Mg_{12}Nd_2Ag$ phase at the grain boundaries can be seen. The etchant is Acetic Picral.

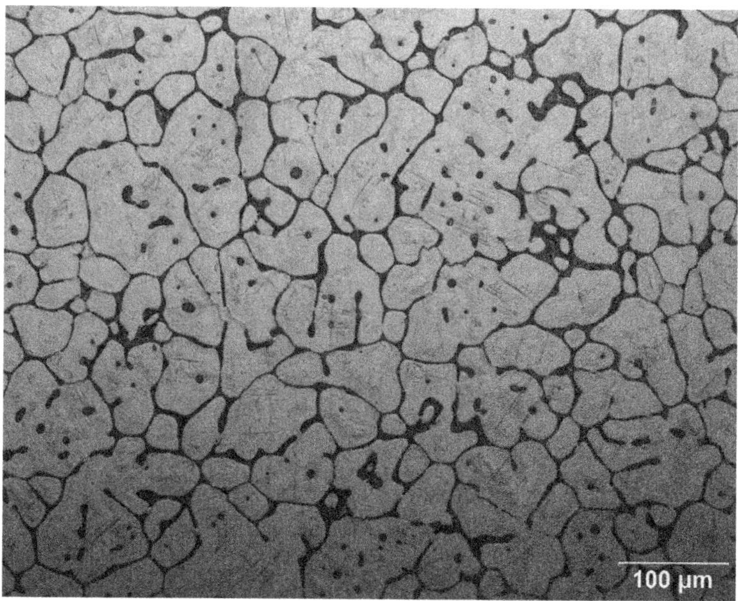

FIGURE 13.124 Optical microstructure of as-cast magnesium alloy QE22 taken at high magnification. Dendritic solidification pattern with $Mg_{12}Nd_2Ag$ phase at the grain boundaries can be seen. The etchant is Acetic Picral.

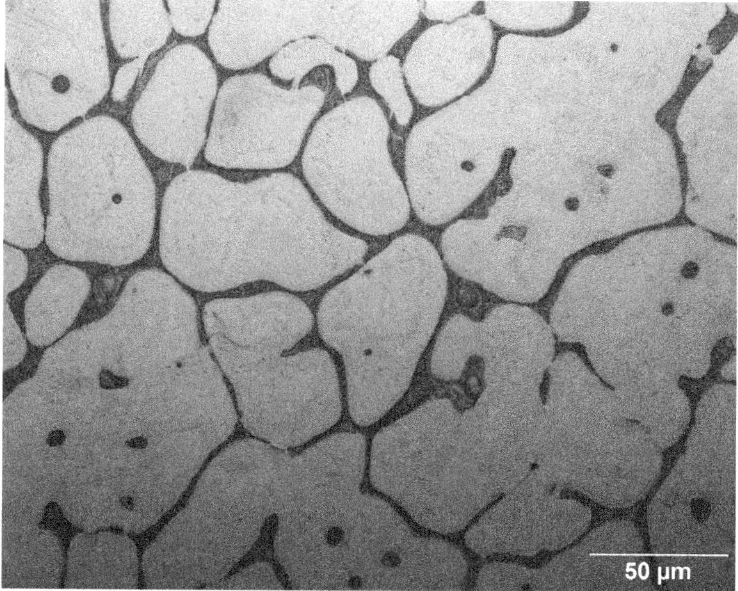

FIGURE 13.125 Optical microstructure of as-cast magnesium alloy QE22 taken at high magnification. Dendritic solidification pattern with $Mg_{12}Nd_2Ag$ phase at the grain boundaries can be seen. The etchant is Acetic Picral.

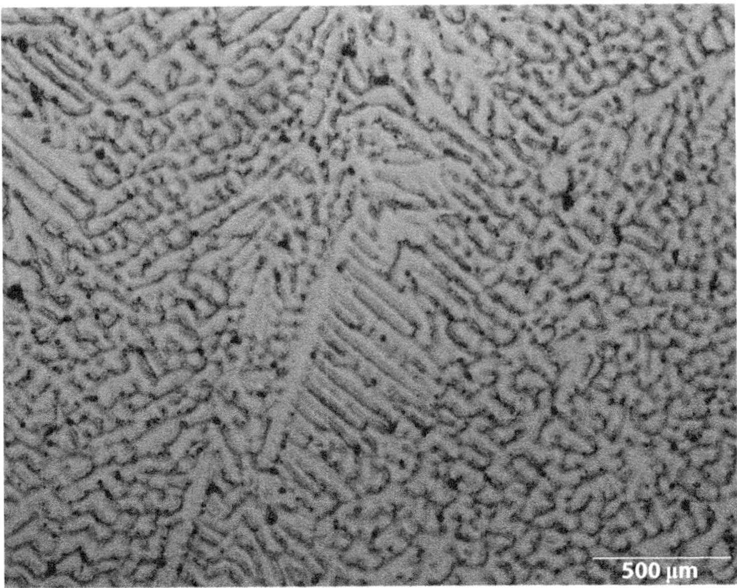

FIGURE 13.126 Optical microstructure of as-cast magnesium alloy WE54. Dendritic solidification pattern can be seen. The etchant is Acetic Picral.

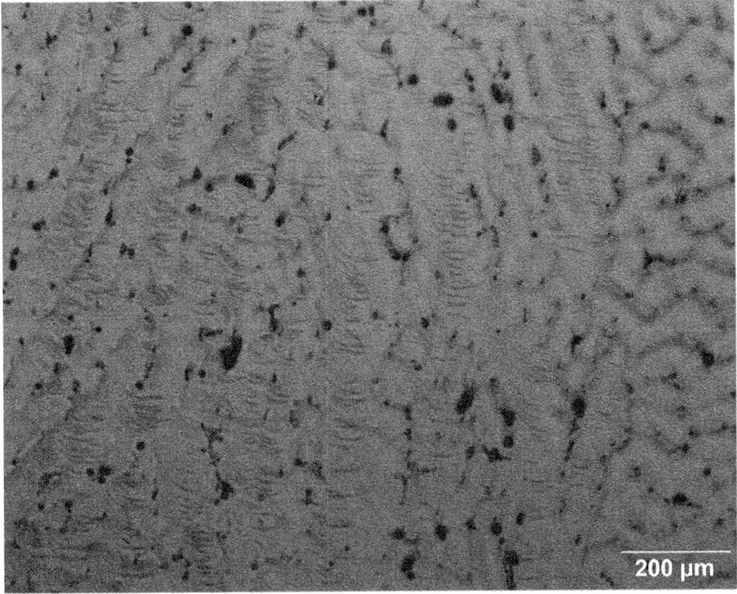

FIGURE 13.127 Optical microstructure of as-cast magnesium alloy WE54 taken at high magnification. Dendritic solidification pattern can be seen. $Mg_{12}NdY$ phase and $Mg_{24}Y_5$ particles can be seen. The etchant is Acetic Picral.

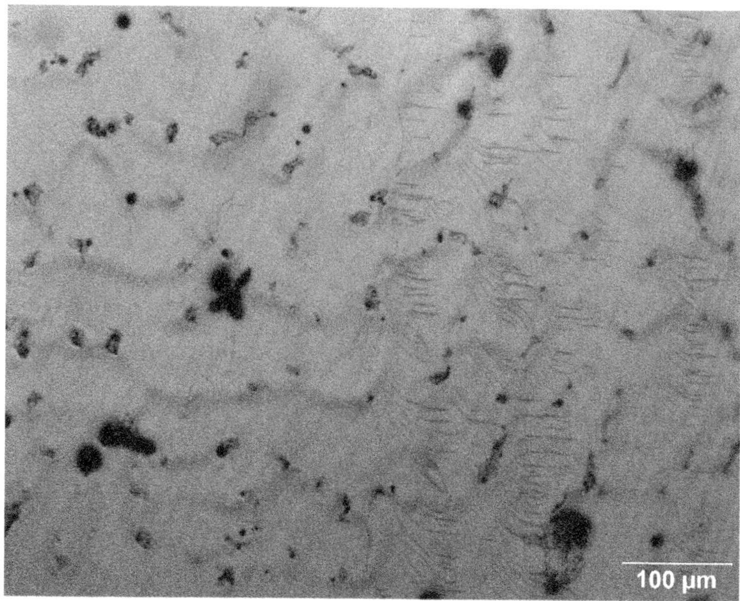

FIGURE 13.128 Optical microstructure of as-cast magnesium alloy WE54 taken at high magnification. Dendritic solidification pattern can be seen. $Mg_{12}NdY$ phase and $Mg_{24}Y_5$ particles can be seen. The etchant is Acetic Picral.

FIGURE 13.129 Optical microstructure of as-cast and extruded magnesium alloy (6 mm diameter) WE54 taken along the longitudinal direction. Fragmented particles of $Mg_{12}NdY$ and $Mg_{24}Y_5$ can be seen distributed in the microstructure. The etchant is Acetic Picral.

FIGURE 13.130 Optical microstructure of as-cast and extruded magnesium alloy (6 mm diameter) WE54 taken along the longitudinal direction taken at high magnification. Equiaxed recrystallized grain structure with fragmented particles of $Mg_{12}NdY$ and $Mg_{24}Y_5$ can be seen distributed in the microstructure. The etchant is Acetic Picral.

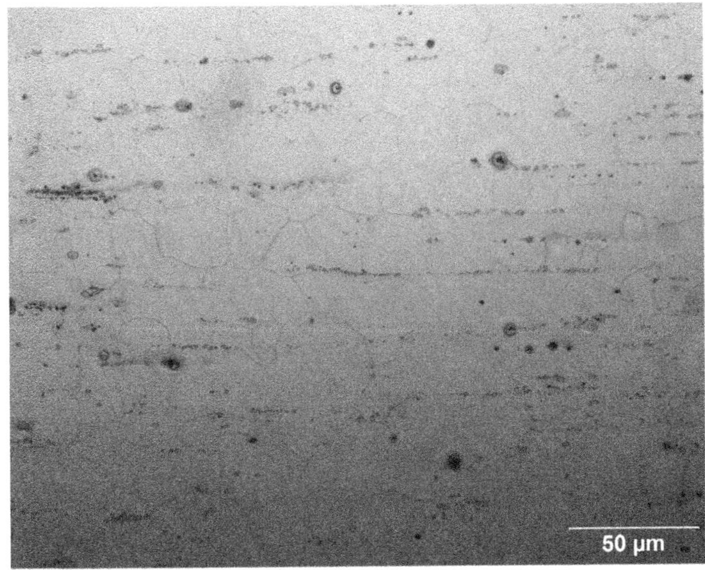

FIGURE 13.131 Optical microstructure of as-cast and extruded magnesium alloy (6 mm diameter) WE54 taken along the longitudinal direction taken at high magnification. Equiaxed recrystallized grain structure with fragmented particles of $Mg_{12}NdY$ and $Mg_{24}Y_5$ can be seen distributed in the microstructure. The etchant is Acetic Picral.

FIGURE 13.132 Optical microstructure of as-cast and extruded magnesium alloy (6 mm diameter) WE54 taken at the cross section. Fine equiaxed recrystallized grain can be seen. The etchant is Acetic Picral.

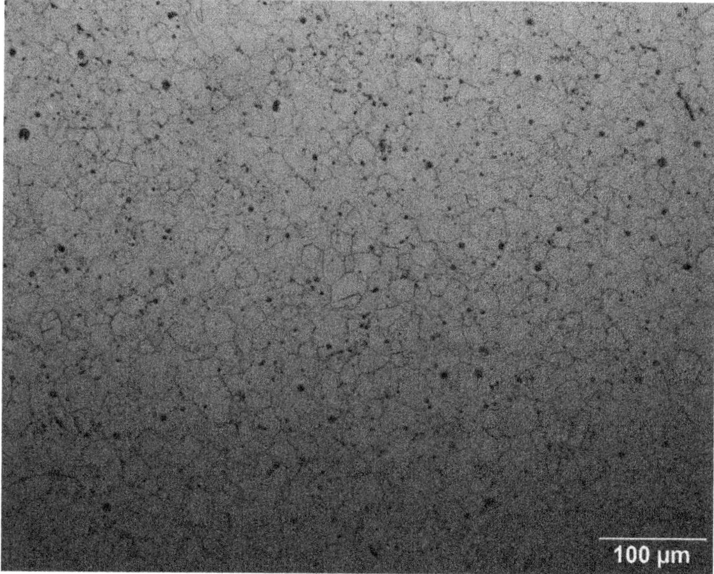

FIGURE 13.133 Optical microstructure of as-cast and extruded magnesium alloy (6 mm diameter) WE54 taken at the cross section taken at high magnification. Fine equiaxed recrystallized grains with fragmented particles of $Mg_{12}NdY$ and $Mg_{24}Y_5$ can be seen distributed in the microstructure. The etchant is Acetic Picral.

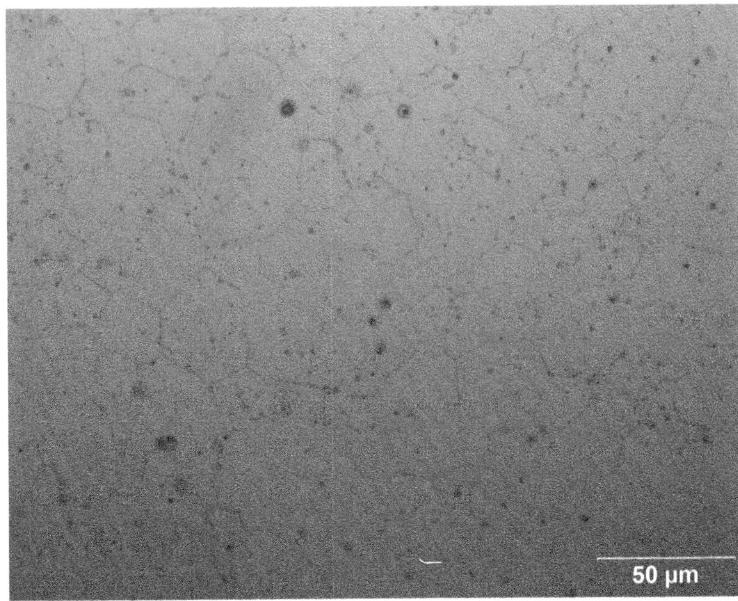

FIGURE 13.134 Optical microstructure of as-cast and extruded magnesium alloy (6 mm diameter) WE54 taken at the cross section taken at high magnification. Fine equiaxed recrystallized grains with fragmented particles of $Mg_{12}NdY$ and $Mg_{24}Y_5$ can be seen distributed in the microstructure. The etchant is Acetic Picral.

FIGURE 13.135 Optical microstructure of as-cast and extruded magnesium alloy (3 mm diameter) WE54 taken along the longitudinal direction. The etchant is Acetic Picral.

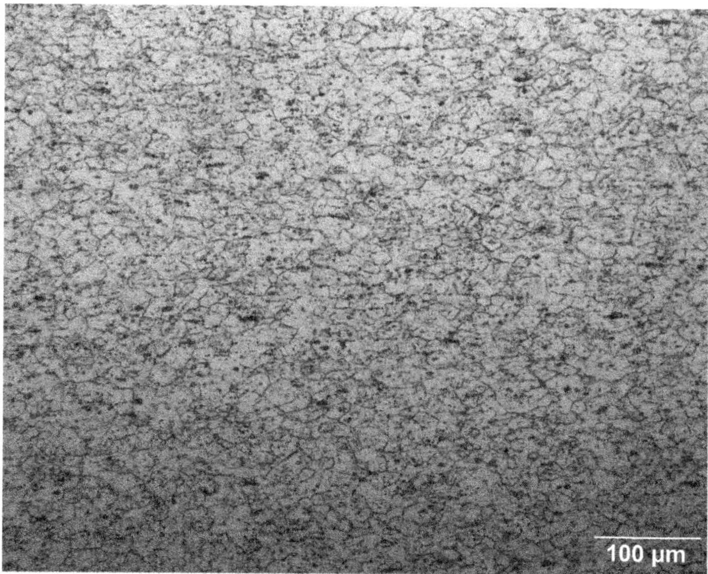

FIGURE 13.136 Optical microstructure of as-cast and extruded magnesium alloy (3 mm diameter) WE54 taken along the longitudinal direction taken at high magnification. The etchant is Acetic Picral.

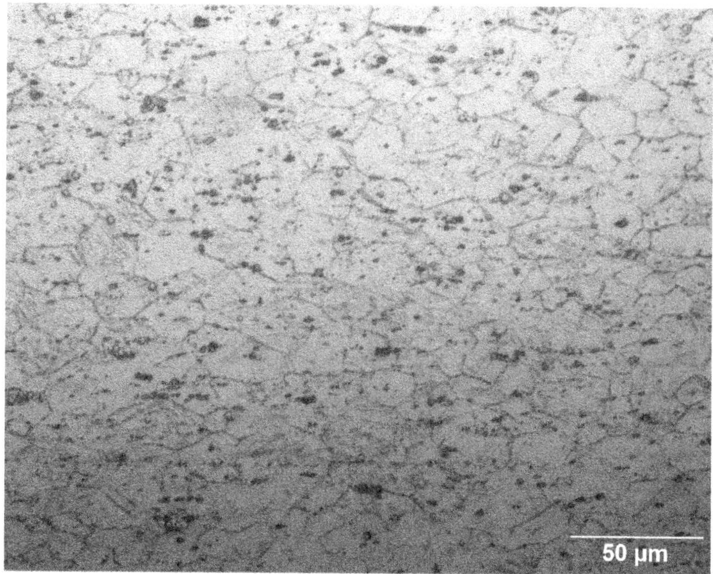

FIGURE 13.137 Optical microstructure of as-cast and extruded magnesium alloy (6 mm diameter) WE54 taken at the cross section taken at high magnification. Fine equiaxed recrystallized grains with fragmented particles of $Mg_{12}NdY$ and $Mg_{24}Y_5$ can be seen distributed in the microstructure. The etchant is Acetic Picral.

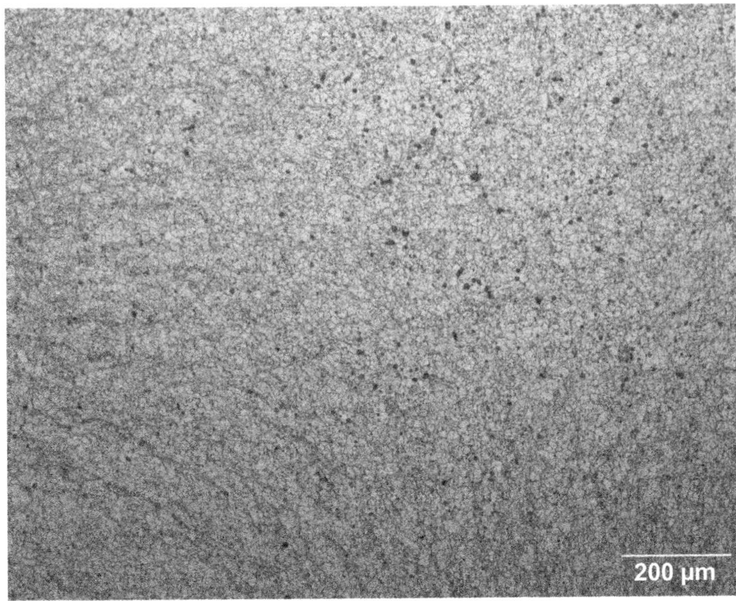

FIGURE 13.138 Optical microstructure of as-cast and extruded magnesium alloy (3 mm diameter) WE54 taken at the cross section. The etchant is Acetic Picral.

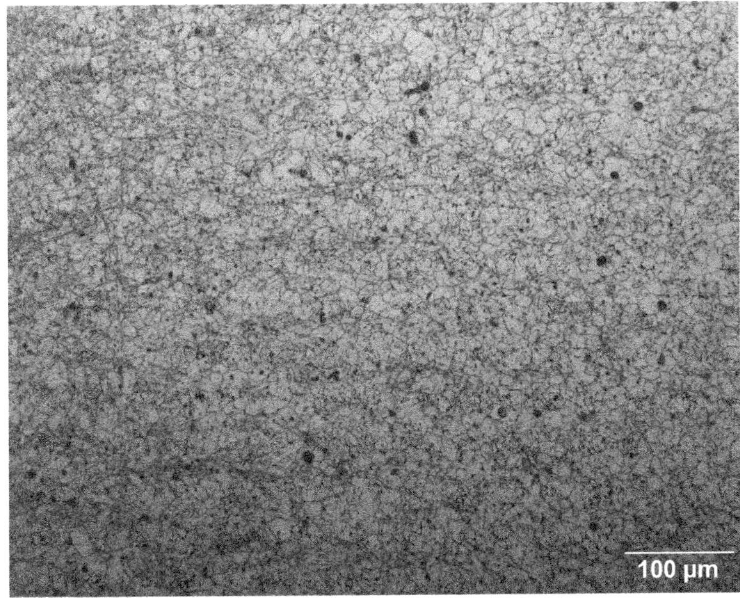

FIGURE 13.139 Optical microstructure of as-cast and extruded magnesium alloy (3 mm diameter) WE54 taken at the cross section. The etchant is Acetic Picral.

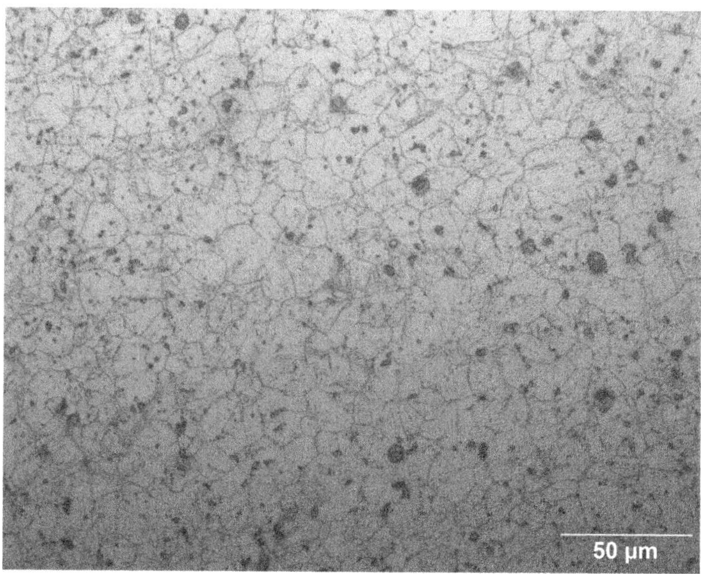

FIGURE 13.140 Optical microstructure of as-cast and extruded magnesium alloy (3 mm diameter) WE54 taken at the cross section. Fine equiaxed recrystallized grains with fragmented particles of $Mg_{12}NdY$ and $Mg_{24}Y_5$ can be seen distributed in the microstructure. The etchant is Acetic Picral.

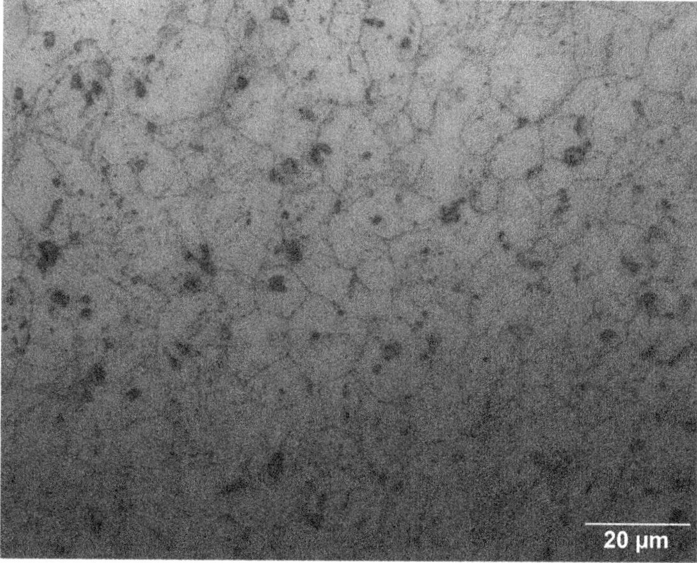

FIGURE 13.141 Optical microstructure of as-cast and extruded magnesium alloy (3 mm diameter) WE54 taken at the cross section. Fine equiaxed recrystallized grains with fragmented particles of $Mg_{12}NdY$ and $Mg_{24}Y_5$ can be seen distributed in the microstructure. The etchant is Acetic Picral.

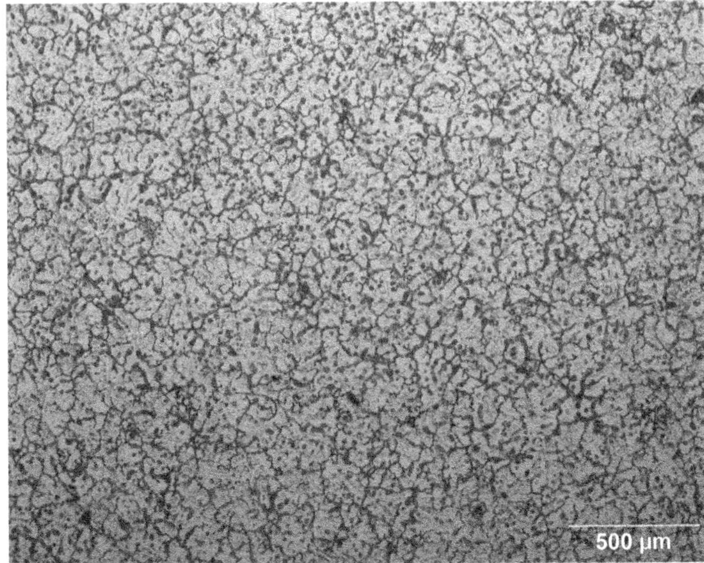

FIGURE 13.142 Optical microstructure of as-cast magnesium alloy MRI153M. As-cast dendritic structure with grain boundary eutectic of $(Mg,Al)_2Ca$ phase can be seen. The etchant is Acetic Picral.

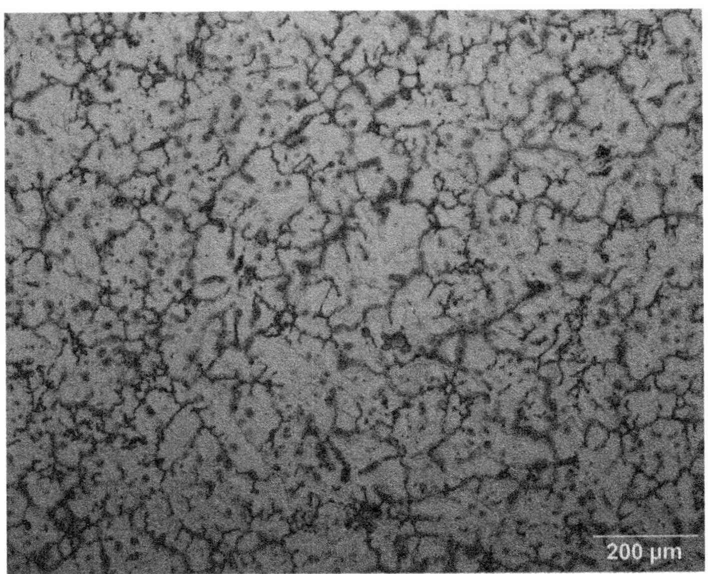

FIGURE 13.143 Optical microstructure of as-cast magnesium alloy MRI153M taken at high magnification. As-cast dendritic structure with grain boundary eutectic of $(Mg,Al)_2Ca$ phase can be seen. The etchant is Acetic Picral.

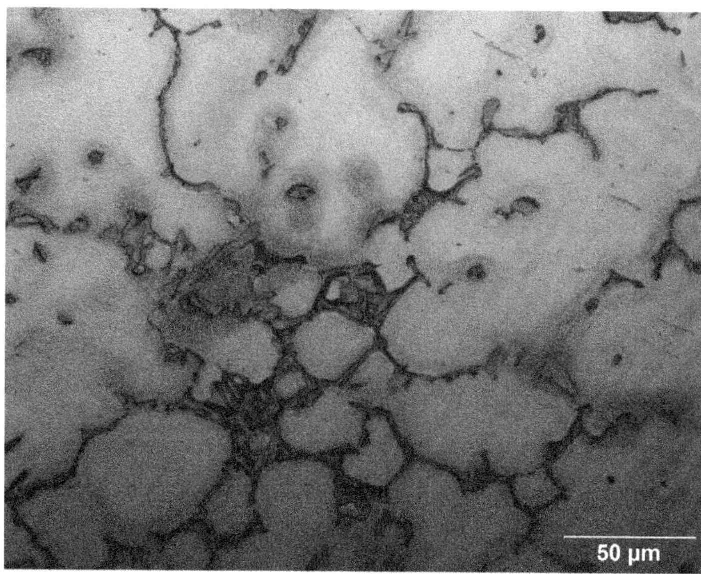

FIGURE 13.144 Optical microstructure of as-cast magnesium alloy MRI153M taken at high magnification. As-cast dendritic structure with grain boundary eutectic of $(Mg,Al)_2Ca$ phase can be seen. Particles of AlMn can also be seen. The etchant is Acetic Picral.

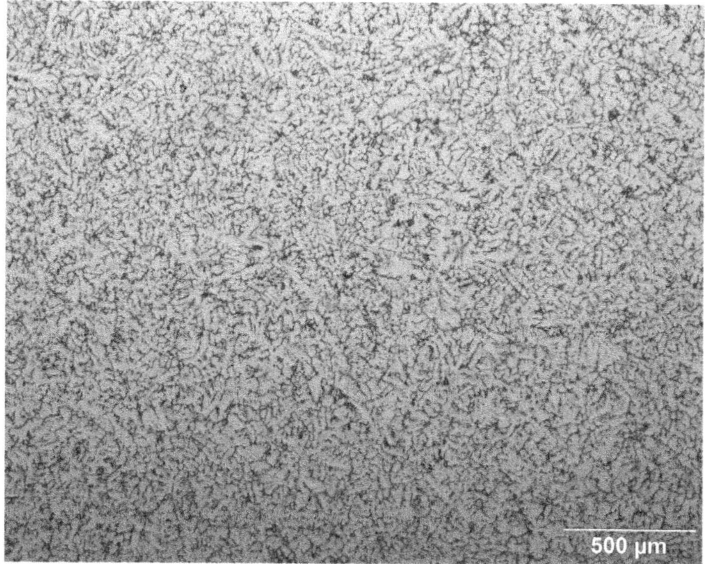

FIGURE 13.145 Optical microstructure of as-cast magnesium alloy MRI230D. As-cast dendritic structure with grain boundary eutectic of $(Mg,Al)_2Ca$ phase can be seen. The etchant is Acetic Picral.

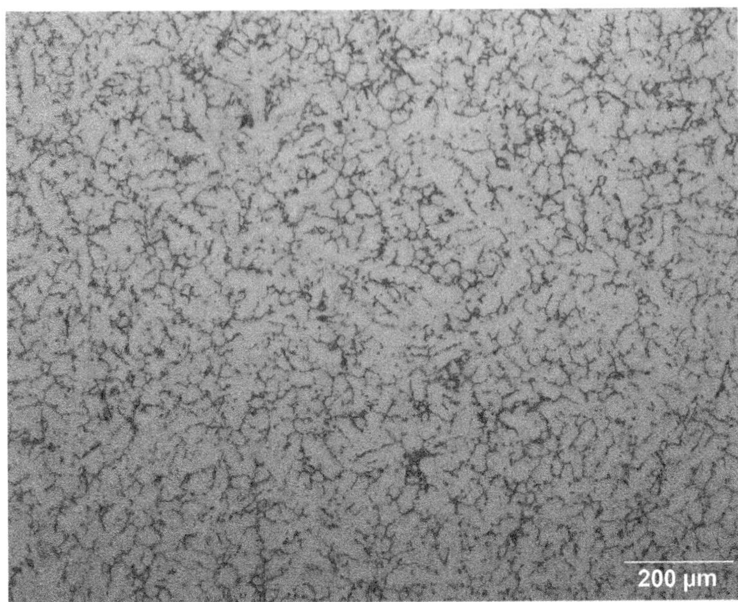

FIGURE 13.146 Optical microstructure of as-cast magnesium alloy MRI230D taken at high magnification. As-cast dendritic structure with grain boundary eutectic of $(Mg,Al)_2Ca$ phase can be seen. Small particles of CaMgSn can also be seen. The etchant is Acetic Picral.

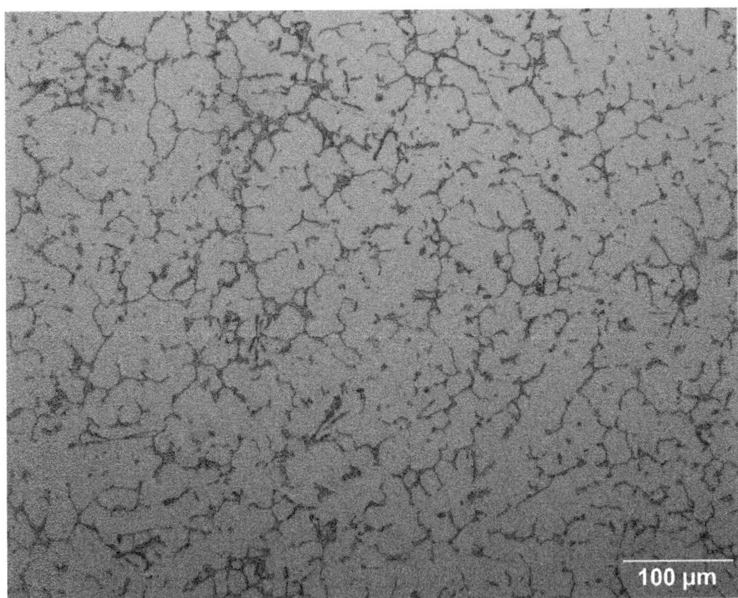

FIGURE 13.147 Optical microstructure of as-cast magnesium alloy MRI230D taken at high magnification. As-cast dendritic structure with grain boundary eutectic of $(Mg,Al)_2Ca$ phase can be seen. Small particles of CaMgSn can also be seen. The etchant is Acetic Picral.

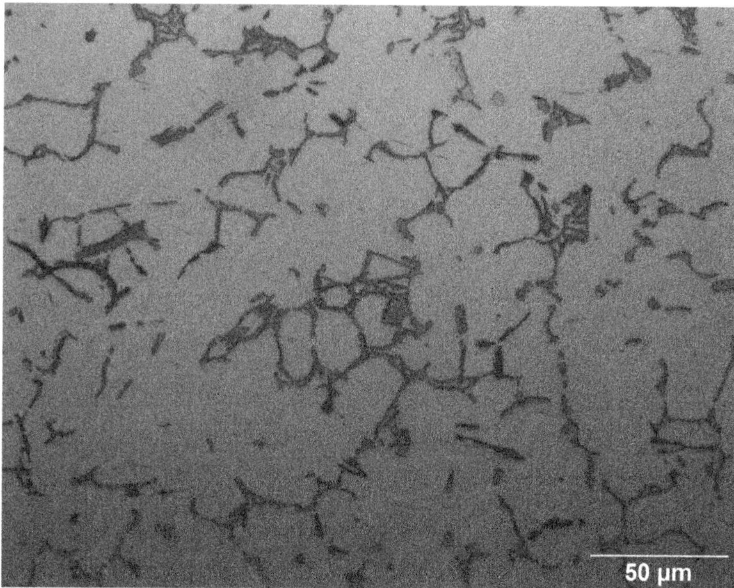

FIGURE 13.148 Optical microstructure of as-cast magnesium alloy MRI230D taken at high magnification. As-cast dendritic structure with grain boundary eutectic of $(Mg,Al)_2Ca$ phase can be seen. Small particles of CaMgSn can also be seen. The etchant is Acetic Picral.

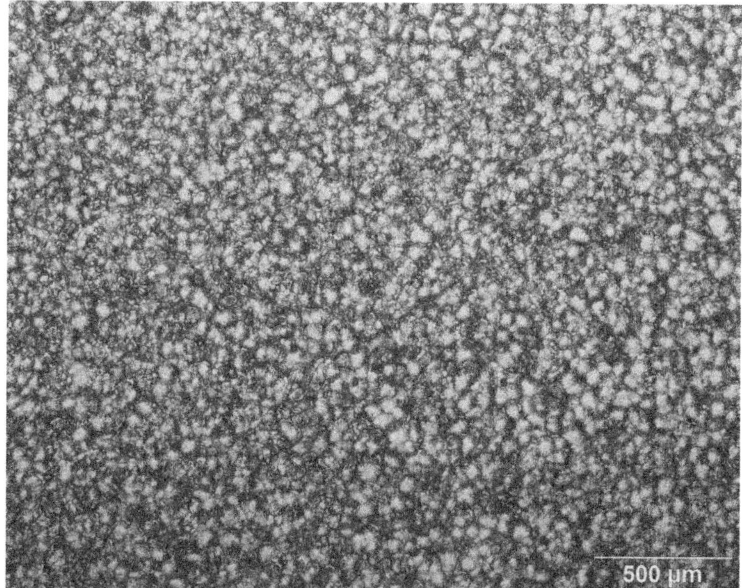

FIGURE 13.149 Optical microstructure of as-cast magnesium alloy Elektron 21. Dendritic solidification with fine equiaxed grains with $Mg_{12}(Nd,Gd)$ grain boundary phase can be seen. The etchant is Acetic Picral.

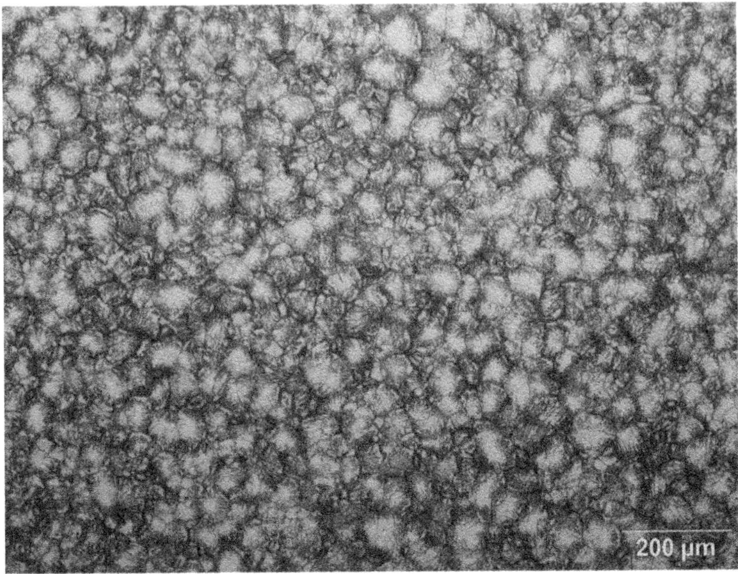

FIGURE 13.150 Optical microstructure of as-cast magnesium alloy Elektron 21 taken at high magnification. Dendritic solidification with fine equiaxed grains with $Mg_{12}(Nd,Gd)$ grain boundary phase can be seen. The etchant is Acetic Picral.

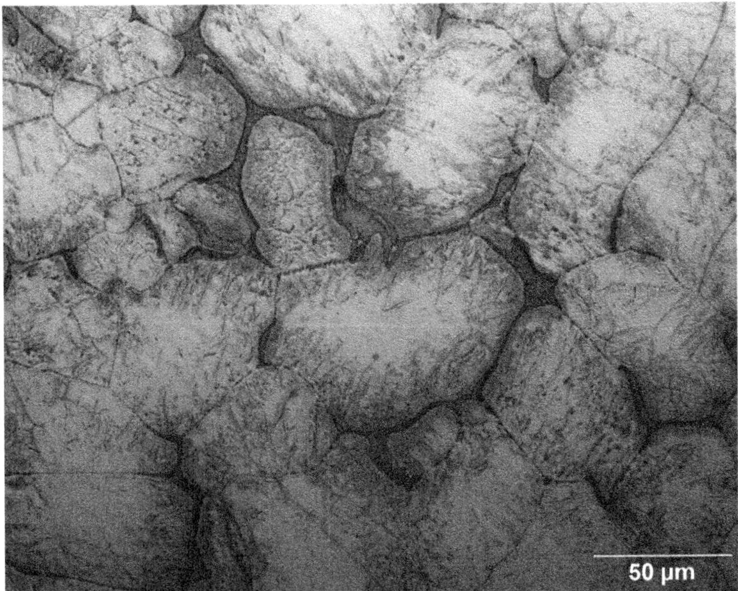

FIGURE 13.151 Optical microstructure of as-cast magnesium alloy Elektron 21 taken at high magnification. Dendritic solidification with fine equiaxed grains with $Mg_{12}(Nd,Gd)$ grain boundary phase can be seen. The etchant is Acetic Picral.

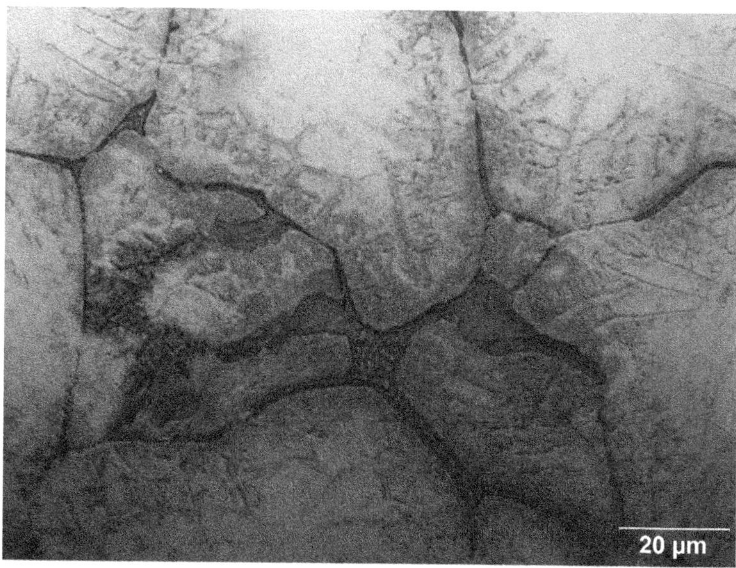

FIGURE 13.152 Optical microstructure of as-cast magnesium alloy Elektron 21 taken at high magnification. Dendritic solidification with fine equiaxed grains with $Mg_{12}(Nd,Gd)$ grain boundary phase can be seen. The etchant is Acetic Picral.

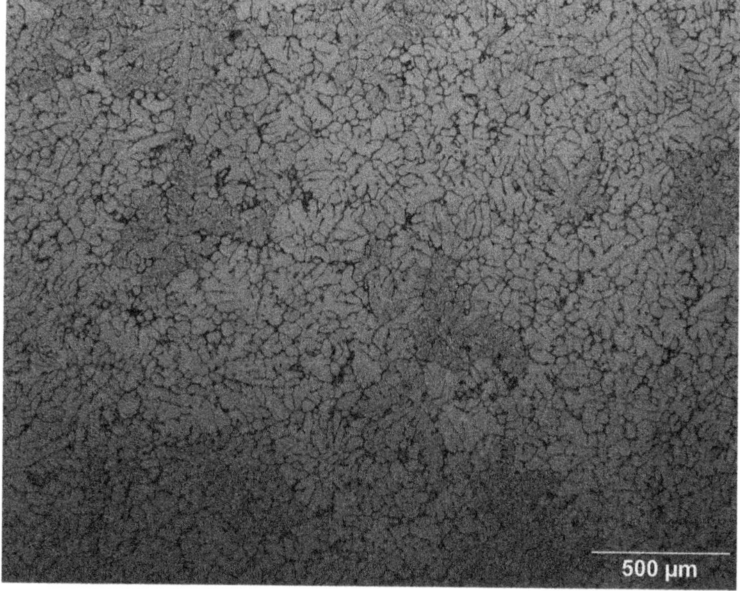

FIGURE 13.153 Optical microstructure of as-cast magnesium alloy Diemag 422. Dendritic solidification pattern can be seen. The etchant is Acetic Picral.

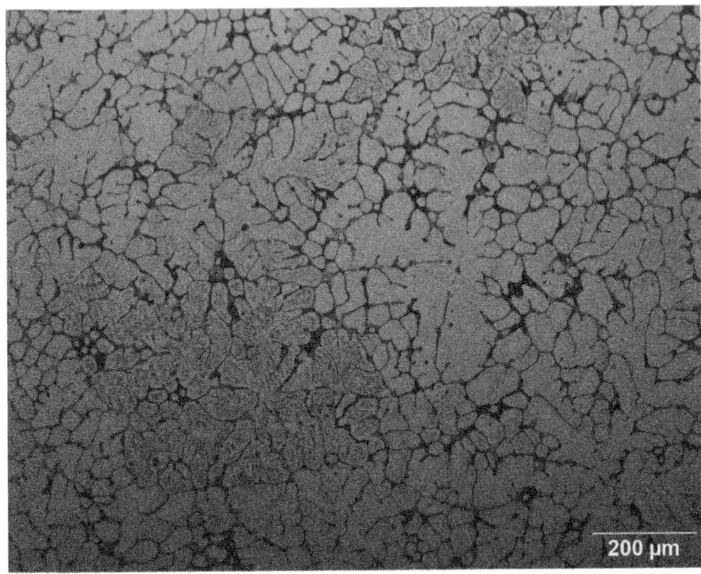

FIGURE 13.154 Optical microstructure of as-cast magnesium alloy Diemag 422 taken at high magnification. Compact Ba-rich phase $Mg_{21}Al_3Ba_2$ and lamellar Al_2Ca phase at grain boundaries can be seen. Dendritic solidification pattern can be seen. The etchant is Acetic Picral.

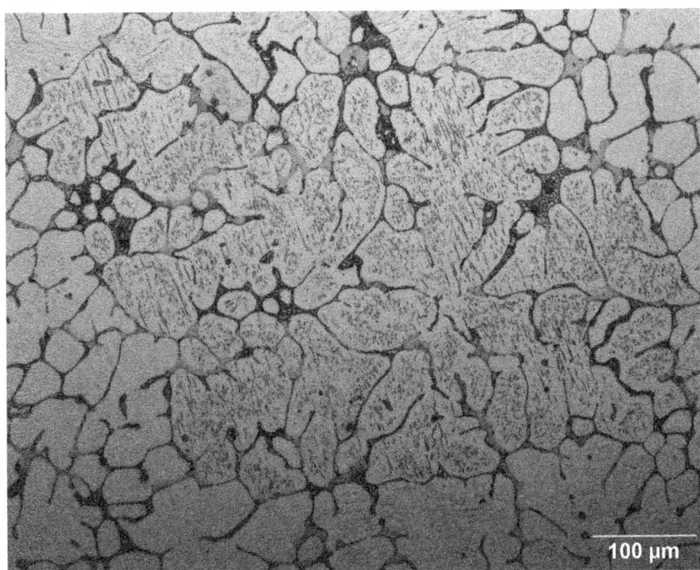

FIGURE 13.155 Optical microstructure of as-cast magnesium alloy Diemag 422 taken at high magnification. Compact Ba-rich phase $Mg_{21}Al_3Ba_2$ and lamellar Al_2Ca phase at grain boundaries can be seen. Dendritic solidification pattern can be seen. The etchant is Acetic Picral.

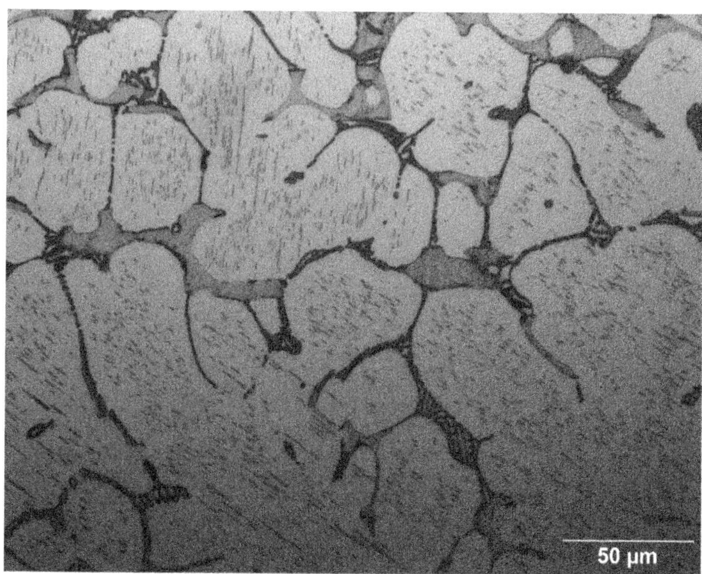

FIGURE 13.156 Optical microstructure of as-cast magnesium alloy Diemag 422 taken at high magnification. Compact Ba-rich phase $Mg_{21}Al_3Ba_2$ and lamellar Al_2Ca phase at grain boundaries can be seen. Dendritic solidification pattern can be seen. The etchant is Acetic Picral.

FIGURE 13.157 Optical microstructure of as-cast and extruded magnesium alloy Diemag 422 taken along the longitudinal direction. Fragmented particles of $Mg_{21}Al_3Ba_2$ and Al_2Ca can be seen distributed in the microstructure. The etchant is Acetic Picral.

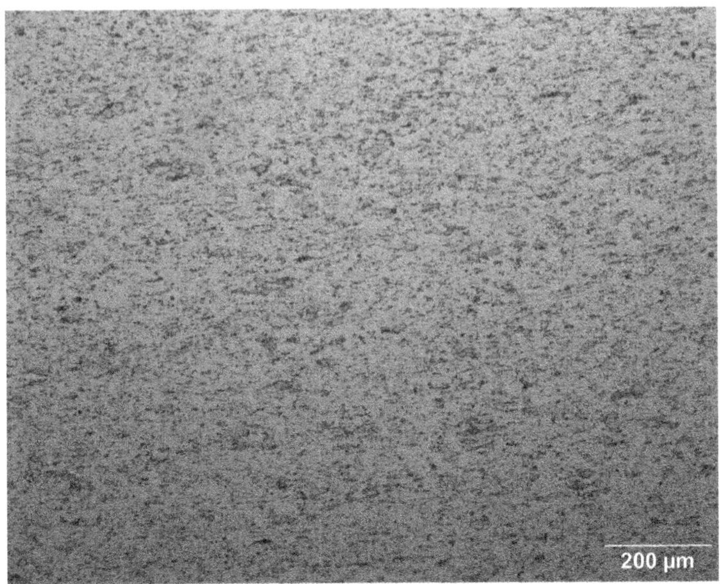

FIGURE 13.158 Optical microstructure of as-cast and extruded magnesium alloy Diemag 422 taken along the longitudinal direction taken at high magnification. Fragmented particles of $Mg_{21}Al_3Ba_2$ and Al_2Ca can be seen distributed in the microstructure. The etchant is Acetic Picral.

FIGURE 13.159 Optical microstructure of as-cast and extruded magnesium alloy DieMag 422 taken along the longitudinal direction taken at high magnification. Fragmented particles of $Mg_{21}Al_3Ba_2$ and Al_2Ca can be seen distributed in the microstructure. The etchant is Acetic Picral.

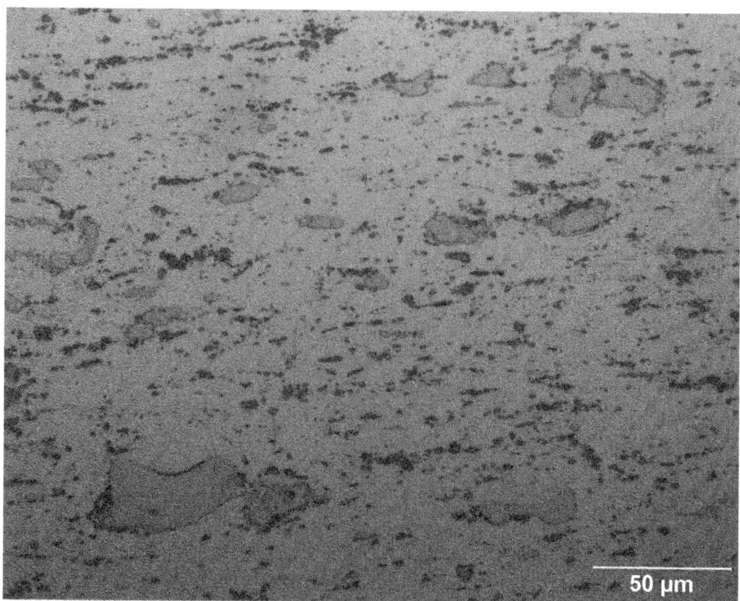

FIGURE 13.160 Optical microstructure of as-cast and extruded magnesium alloy DieMag 422 taken along the longitudinal direction taken at high magnification. Fragmented particles of $Mg_{21}Al_3Ba_2$ and Al_2Ca can be seen distributed in the microstructure. The etchant is Acetic Picral.

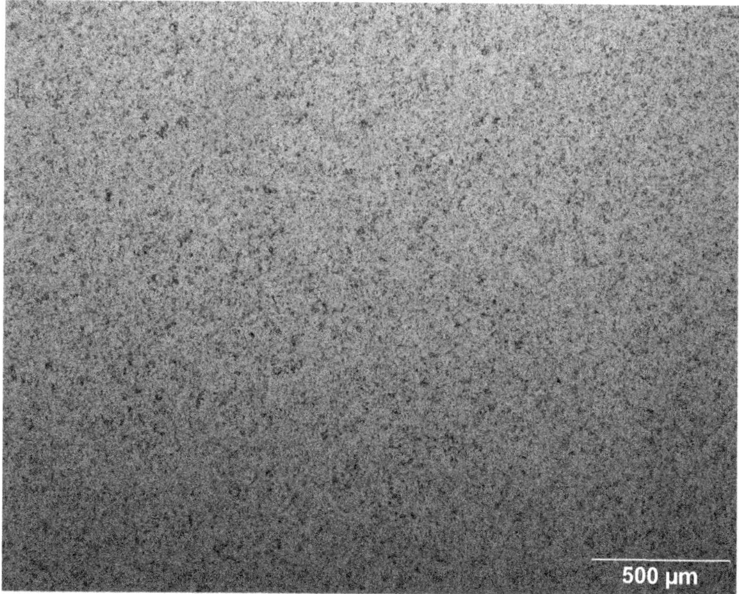

FIGURE 13.161 Optical microstructure of as-cast and extruded magnesium alloy DieMag 422 taken at the cross section. The specimen was etched with Acetic Picral.

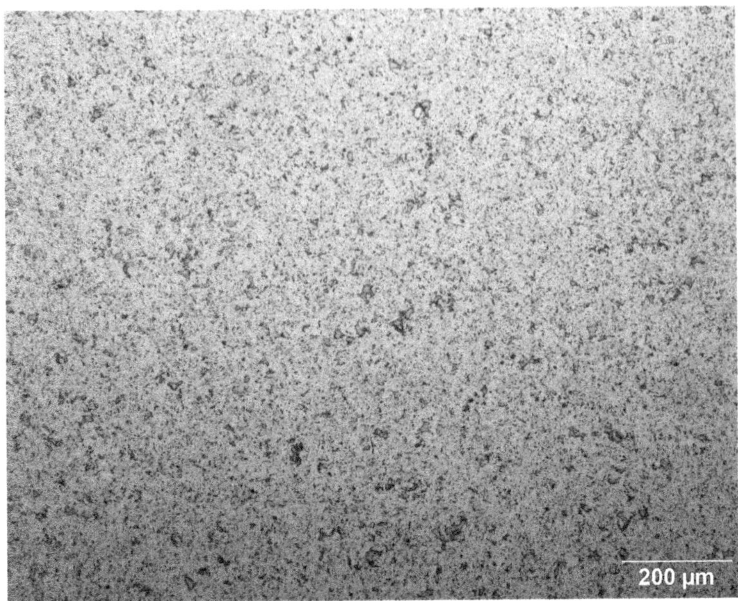

FIGURE 13.162 Optical microstructure of as-cast and extruded magnesium alloy DieMag 422 taken at the cross section at high magnification. Fragmented particles of $Mg_{21}Al_3Ba_2$ and Al_2Ca can be seen distributed in the microstructure. The etchant is Acetic Picral.

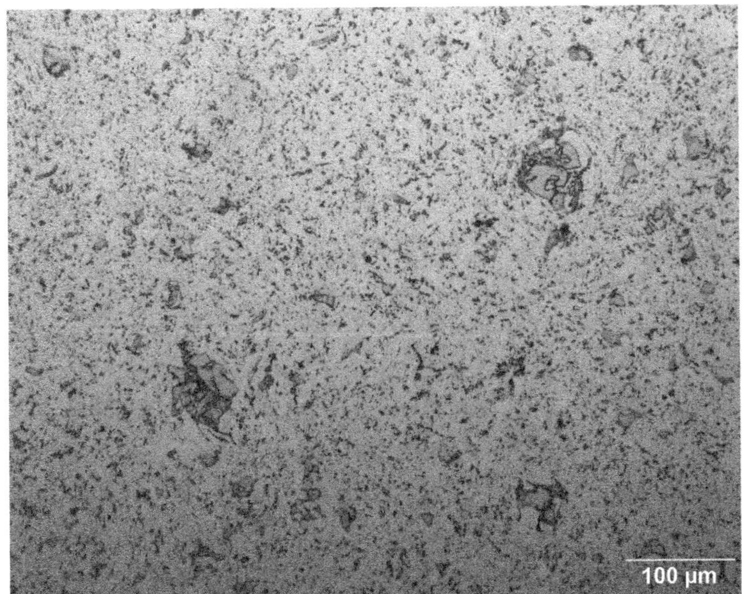

FIGURE 13.163 Optical microstructure of as-cast and extruded magnesium alloy DieMag 422 taken at the cross-section at high magnification. Fragmented particles of $Mg_{21}Al_3Ba_2$ and Al_2Ca can be seen distributed in the microstructure. The etchant is Acetic Picral.

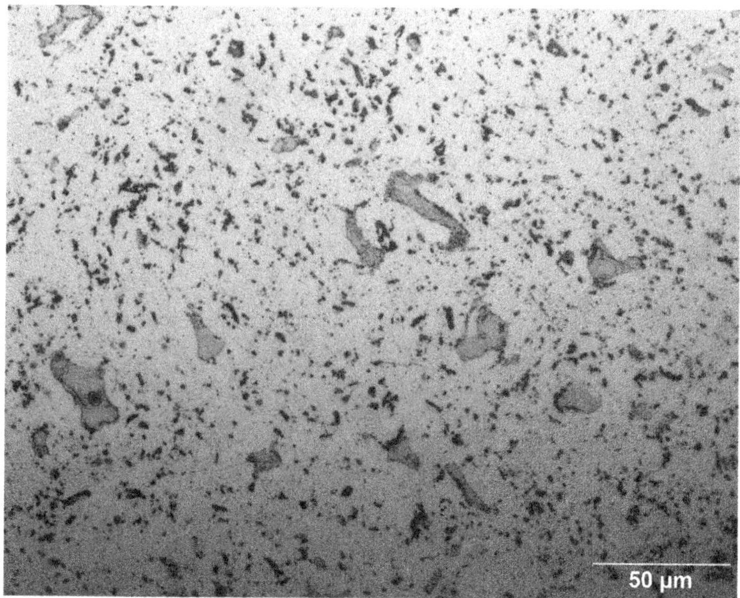

FIGURE 13.164 Optical microstructure of as-cast and extruded magnesium alloy DieMag 422 taken at the cross-section at high magnification. Fragmented particles of $Mg_{21}Al_3Ba_2$ and Al_2Ca can be seen distributed in the microstructure. The etchant is Acetic Picral.

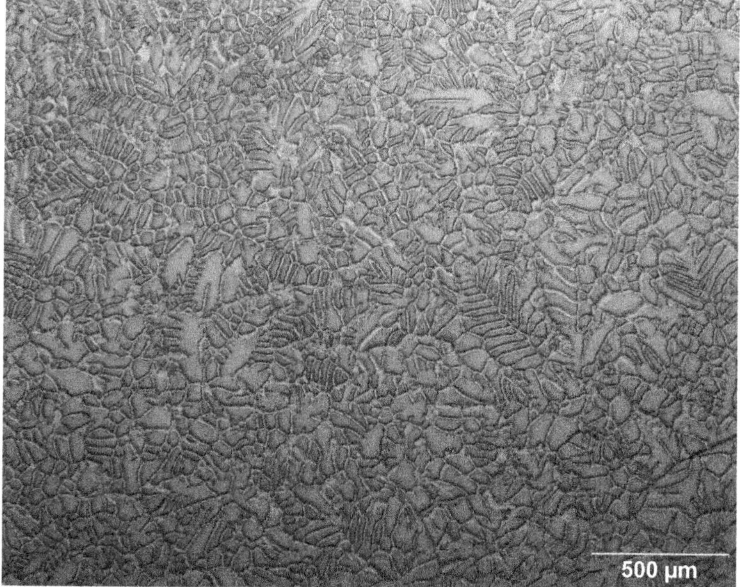

FIGURE 13.165 Optical microstructure of as-cast Mg-4Ag alloy. As-cast dendritic structure can be seen. The etchant is Acetic Picral.

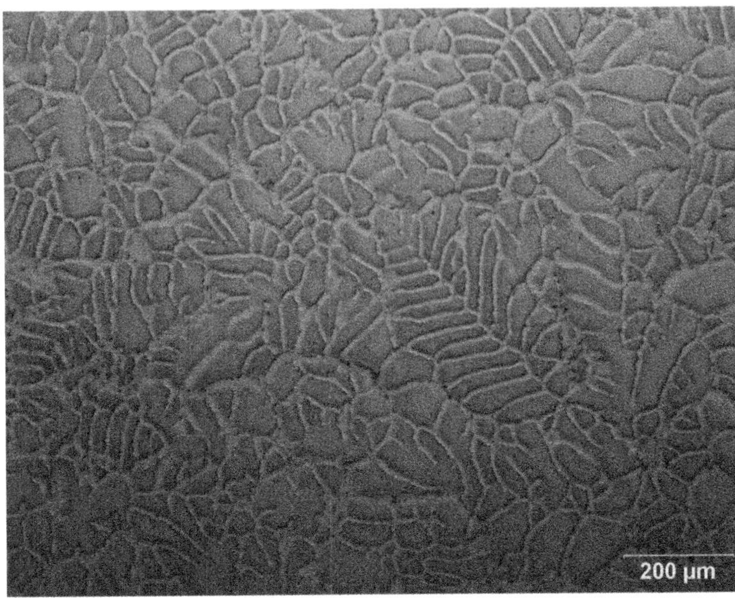

FIGURE 13.166 Optical microstructure of as-cast Mg-4Ag alloy at high magnification. As-cast dendritic structure can be seen. The etchant is Acetic Picral.

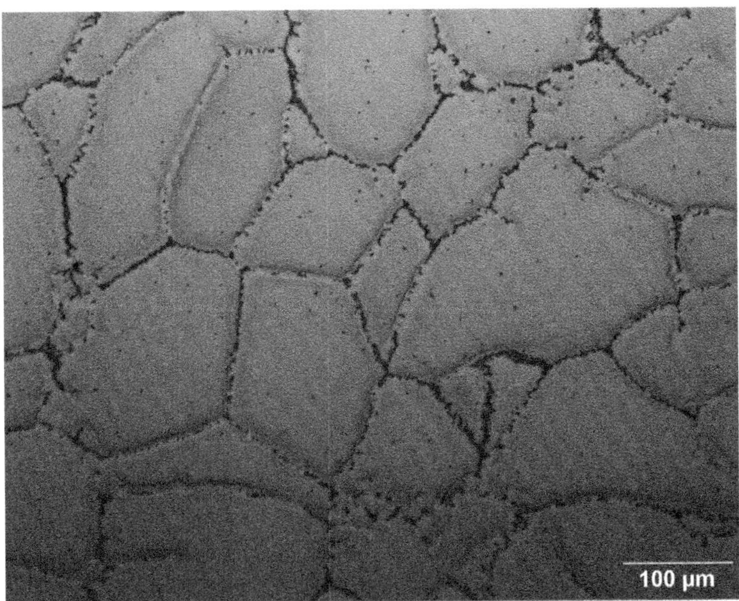

FIGURE 13.167 Optical microstructure of as-cast Mg-4Ag alloy at high magnification. Precipitation of Mg_4Ag at the grain boundaries can be seen. The etchant is Acetic Picral.

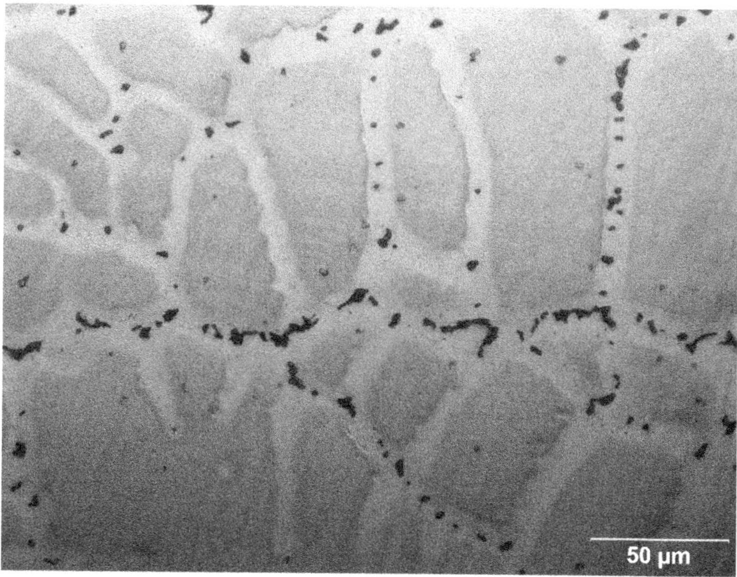

FIGURE 13.168 Optical microstructure of as-cast Mg-4Ag alloy at high magnification. Precipitation of Mg_4Ag at the grain boundaries and silver rich light regions along the grain boundaries can be seen. The etchant is Acetic Picral.

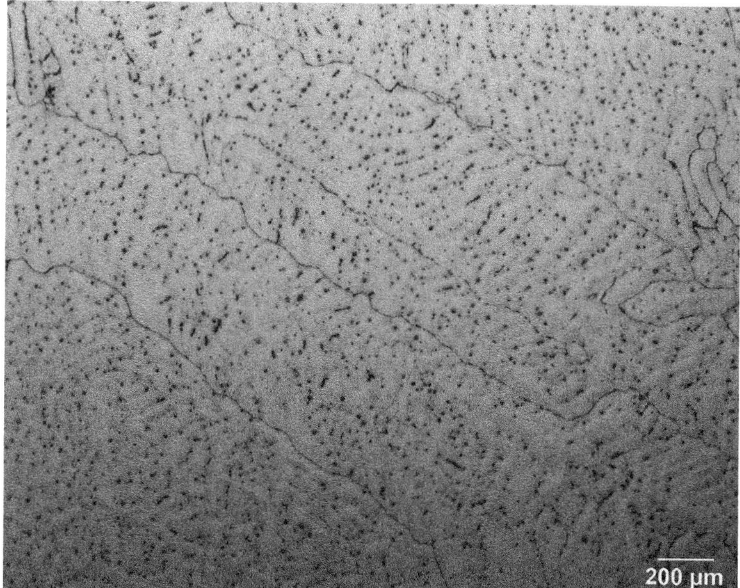

FIGURE 13.169 Optical microstructure of as-cast Mg-2Gd-2Zn alloy. As-cast dendritic structure can be seen. The etchant is Acetic Picral.

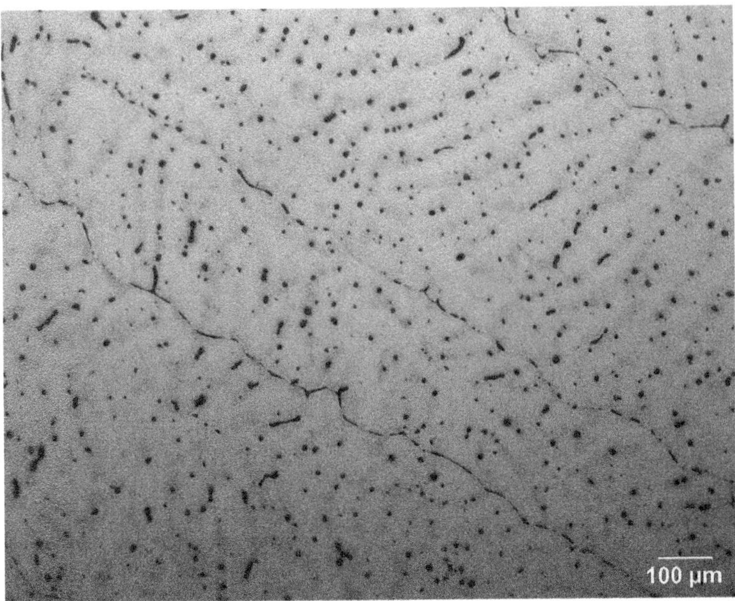

FIGURE 13.170 Optical microstructure of as-cast Mg-2Gd-2Zn alloy at high magnification. As-cast dendritic structure and $(Mg,Zn)_3Gd$ at the grain boundaries and grain interiors can be seen. The etchant is Acetic Picral.

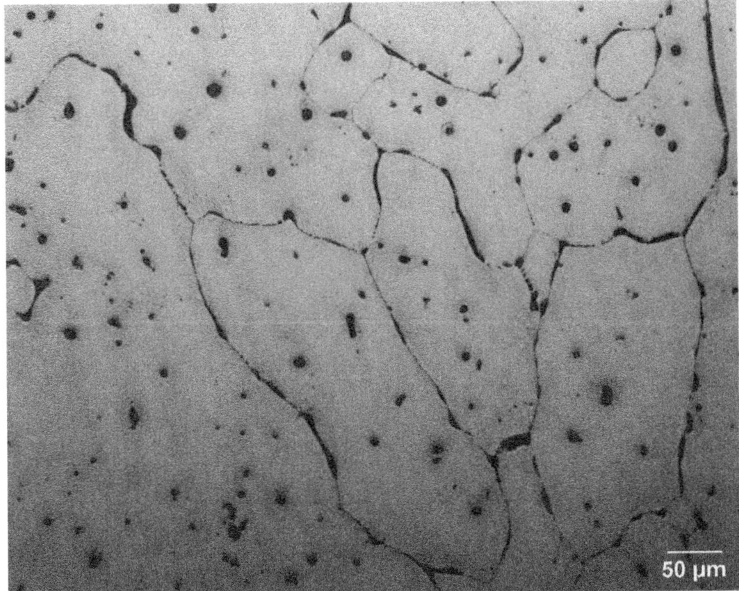

FIGURE 13.171 Optical microstructure of as-cast Mg-2Gd-2Zn alloy at high magnification. As-cast dendritic structure and $(Mg,Zn)_3Gd$ at the grain boundaries and grain interiors can be seen. The etchant is Acetic Picral.

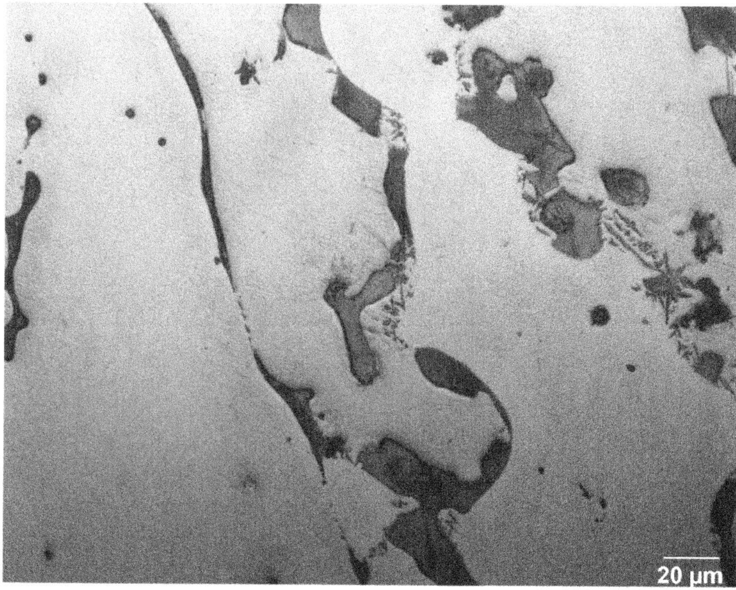

FIGURE 13.172 Optical microstructure of as-cast Mg-2Gd-2Zn alloy at high magnification. Lamellar eutectics of $(Mg,Zn)_3Gd$ at the grain boundaries and grain interiors can be seen. The etchant is Acetic Picral.

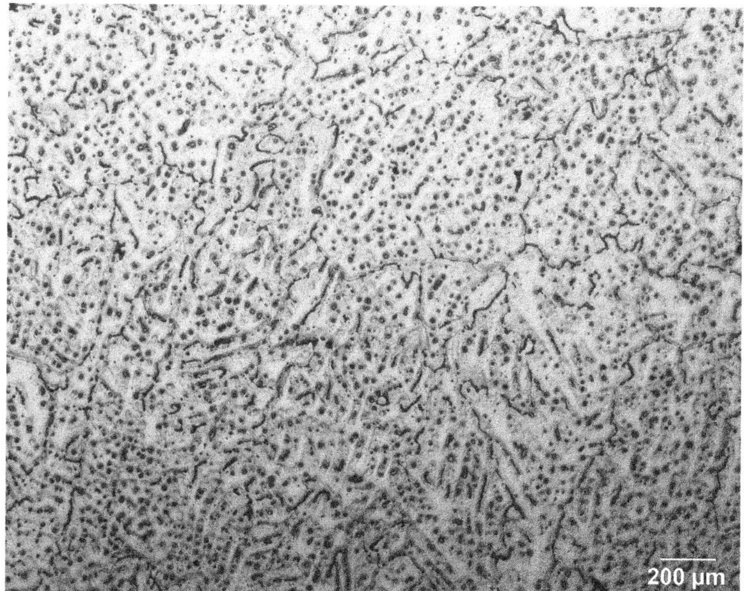

FIGURE 13.173 Optical microstructure of as-cast Mg-2Gd-6Zn alloy. As-cast dendritic structure can be seen. The etchant is Acetic Picral.

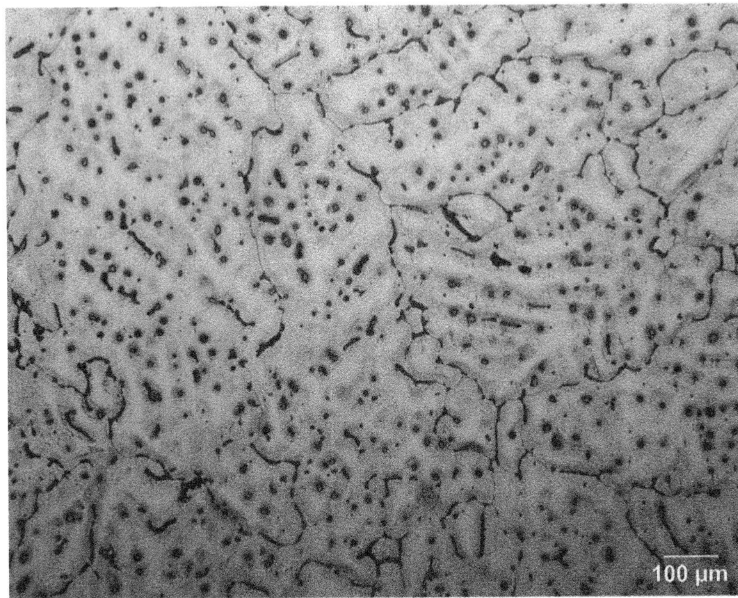

FIGURE 13.174 Optical microstructure of as-cast Mg-2Gd-6Zn alloy at high magnification. As-cast dendritic structure and $(Mg,Zn)_3Gd$ at the grain boundaries and grain interiors can be seen. The etchant is Acetic Picral.

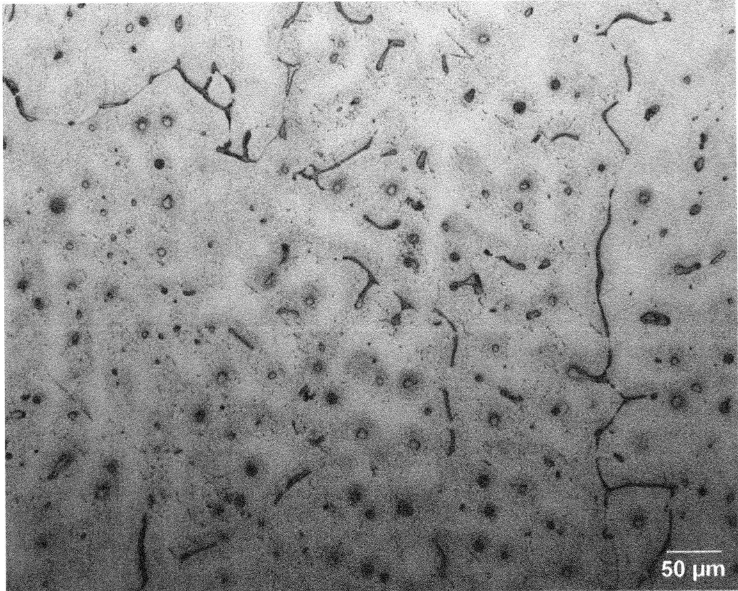

FIGURE 13.175 Optical microstructure of as-cast Mg-2Gd-6Zn alloy at high magnification. $(Mg,Zn)_3Gd$ at the grain boundaries and grain interiors can be seen. The etchant is Acetic Picral.

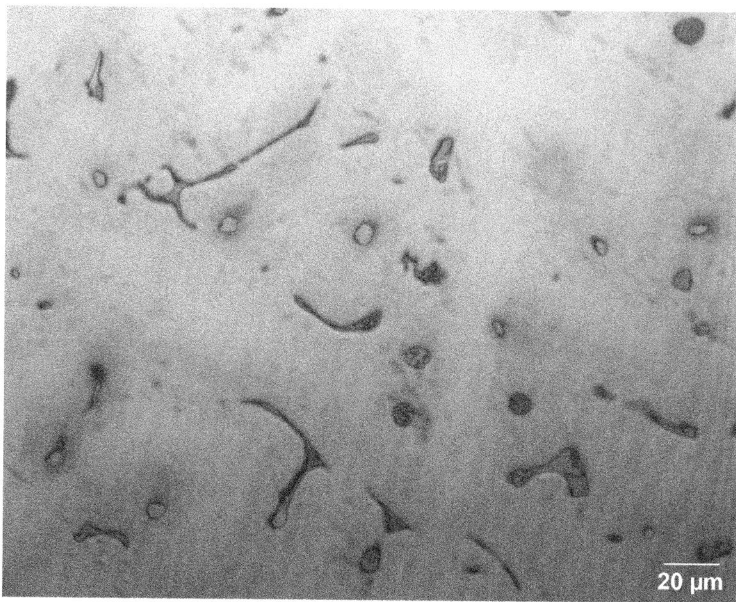

FIGURE 13.176 Optical microstructure of as-cast Mg-2Gd-6Zn alloy at high magnification $(Mg,Zn)_3Gd$ at the grain boundaries and grain interiors can be seen. The etchant is Acetic Picral.

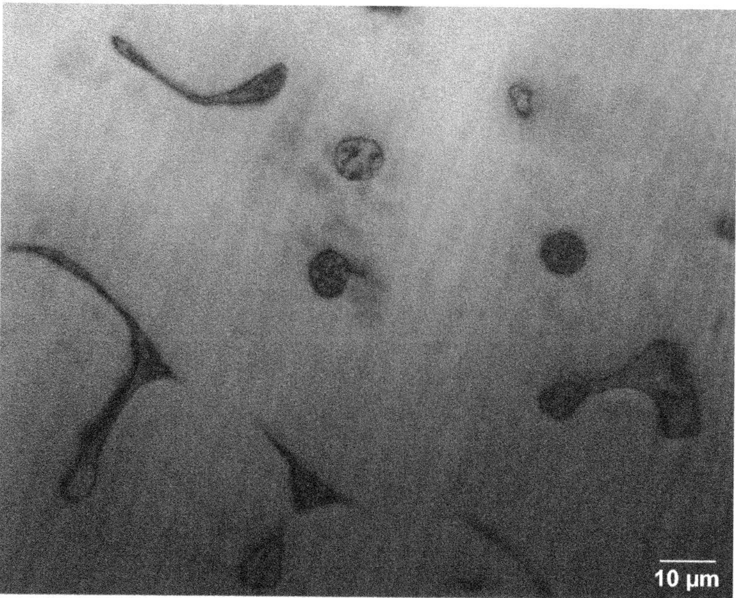

FIGURE 13.177 Optical microstructure of as-cast Mg-2Gd-6Zn alloy at high magnification. Lamellar eutectics of $(Mg,Zn)_3Gd$ at the grain boundaries and grain interiors can be seen. The etchant is Acetic Picral.

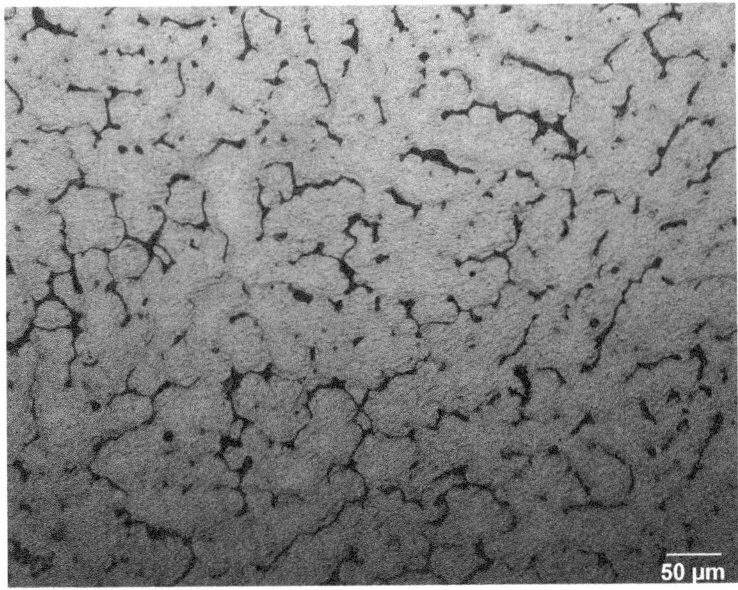

FIGURE 13.178 Optical microstructure of as-cast Mg-10Gd-2Zn alloy. As-cast dendritic structure and network of $(Mg,Zn)_3Gd$ at the grain boundaries and grain interiors can be seen. The etchant is Acetic Picral.

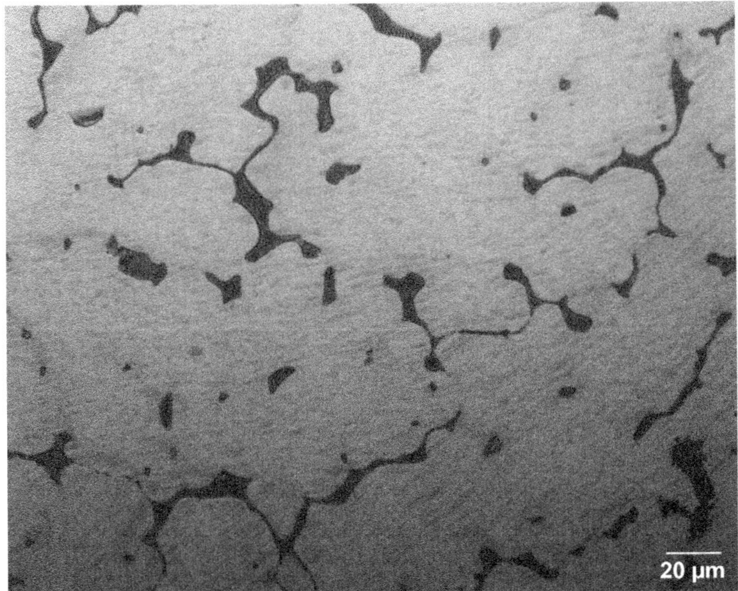

FIGURE 13.179 Optical microstructure of as-cast Mg-10Gd-2Zn alloy at high magnification. Lamellar eutectics of $(Mg,Zn)_3Gd$ at the grain boundaries and grain interiors can be seen. The etchant is Acetic Picral.

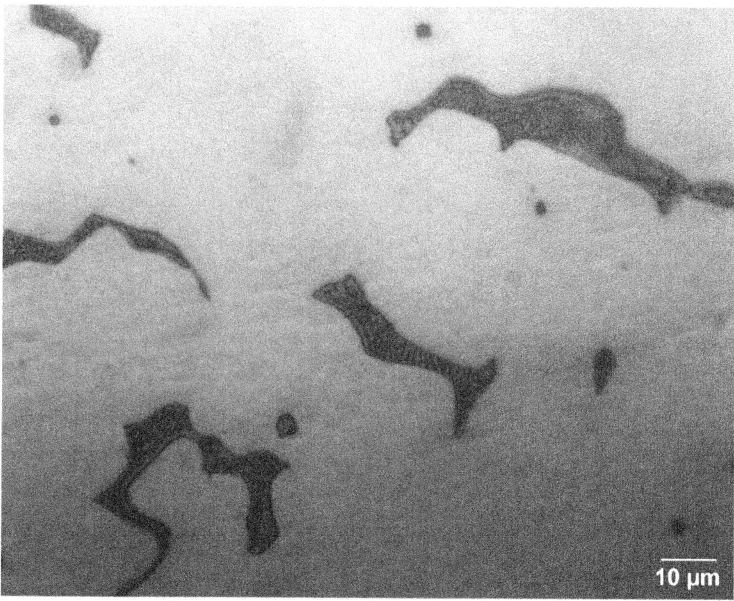

FIGURE 13.180 Optical microstructure of as-cast Mg-10Gd-2Zn alloy at high magnification. Lamellar eutectics of $(Mg,Zn)_3Gd$ at the grain boundaries and grain interiors can be seen. The etchant is Acetic Picral.

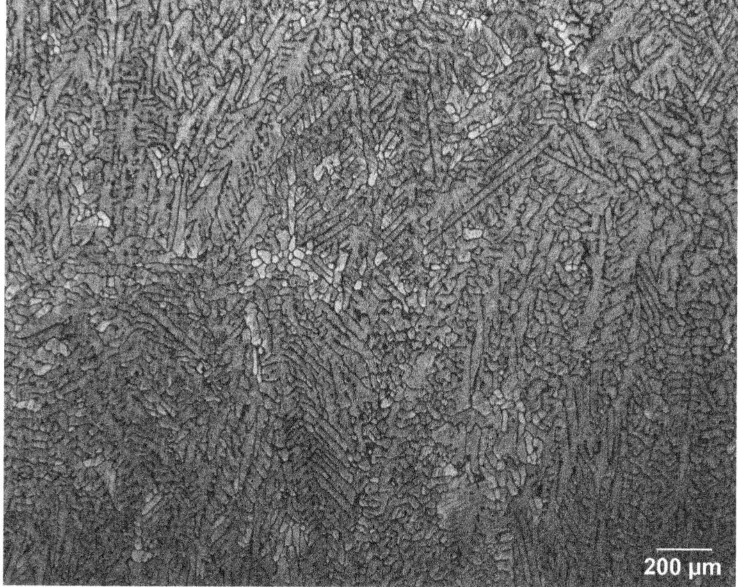

FIGURE 13.181 Optical microstructure of as-cast Mg-10Gd-6Zn alloy. As-cast dendritic structure can be seen. The etchant is Acetic Picral.

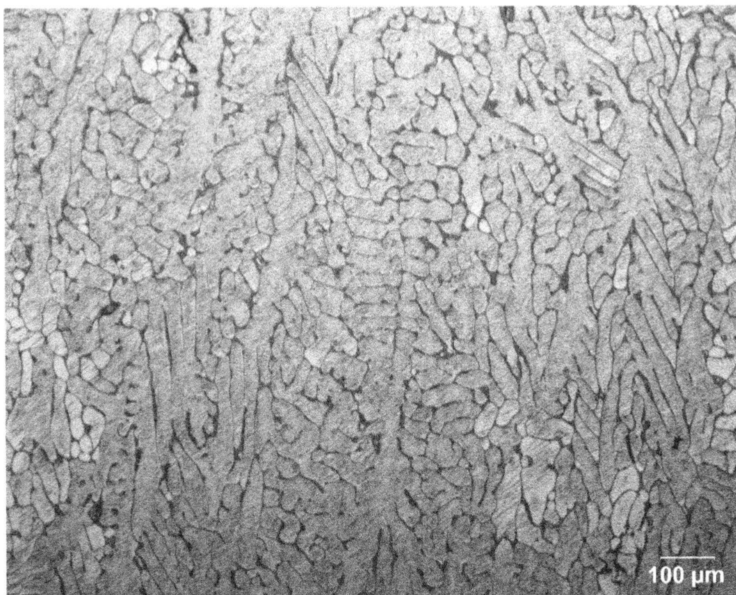

FIGURE 13.182 Optical microstructure of as-cast Mg-10Gd-6Zn alloy at high magnification. As-cast dendritic structure can be seen. The etchant is Acetic Picral.

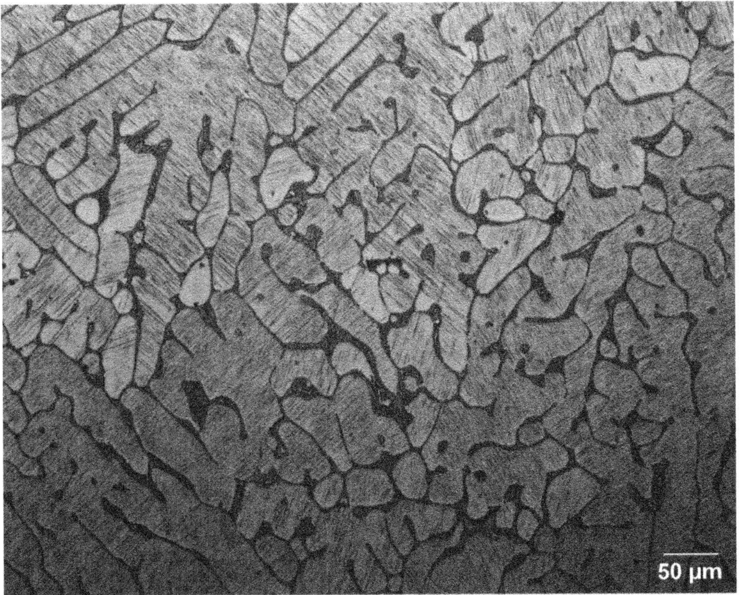

FIGURE 13.183 Optical microstructure of as-cast Mg-10Gd-6Zn alloy at high magnification. Lamellar eutectics of $(Mg,Zn)_3Gd$ at the grain boundaries and grain interiors can be seen. The etchant is Acetic Picral.

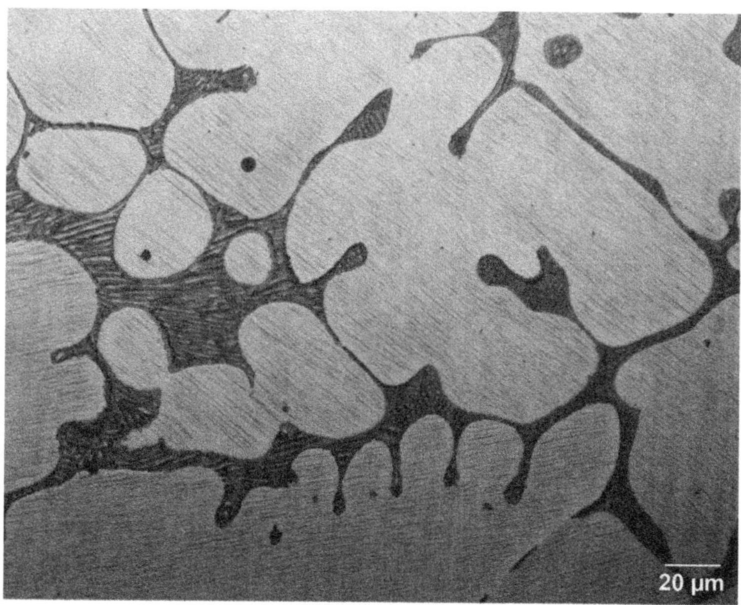

FIGURE 13.184 Optical microstructure of as-cast Mg-10Gd-6Zn alloy at high magnification. Lamellar eutectics of $(Mg,Zn)_3Gd$ at the grain boundaries and grain interiors can be seen. The etchant is Acetic Picral.

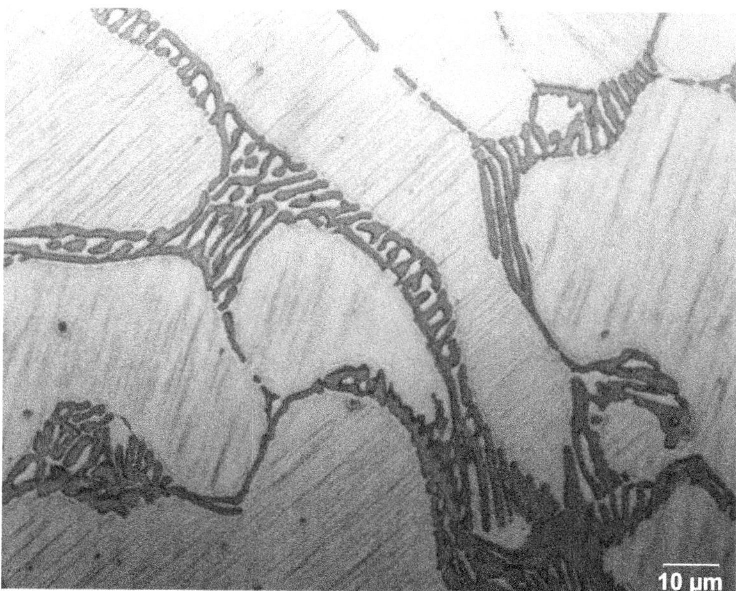

FIGURE 13.185 Optical microstructure of as-cast Mg-10Gd-6Zn alloy at high magnification. Lamellar eutectics of $(Mg,Zn)_3Gd$ at the grain boundaries can be seen. The etchant is Acetic Picral.

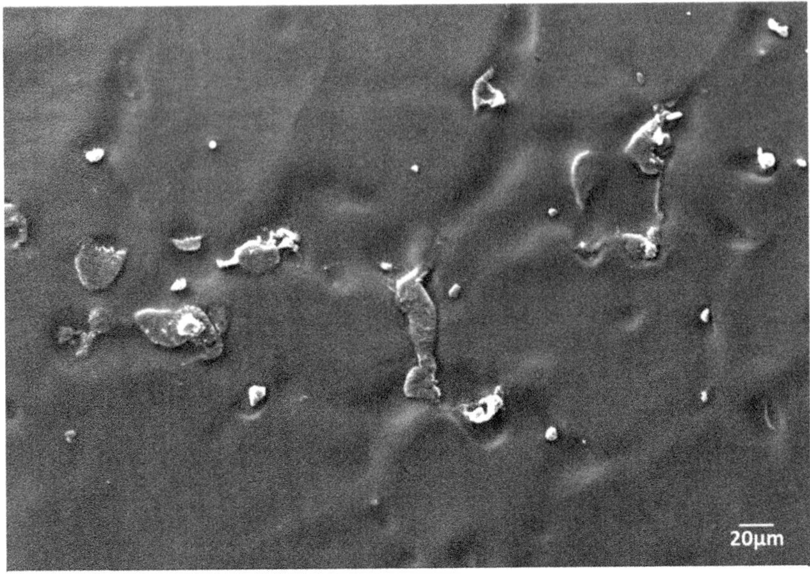

FIGURE 13.186 Microstructure of as-cast AZ31 alloy. The microstructure reveals eutectic of $Mg_{17}Al_{12}$ at the grain boundaries. The etchant is Acetic Picral.

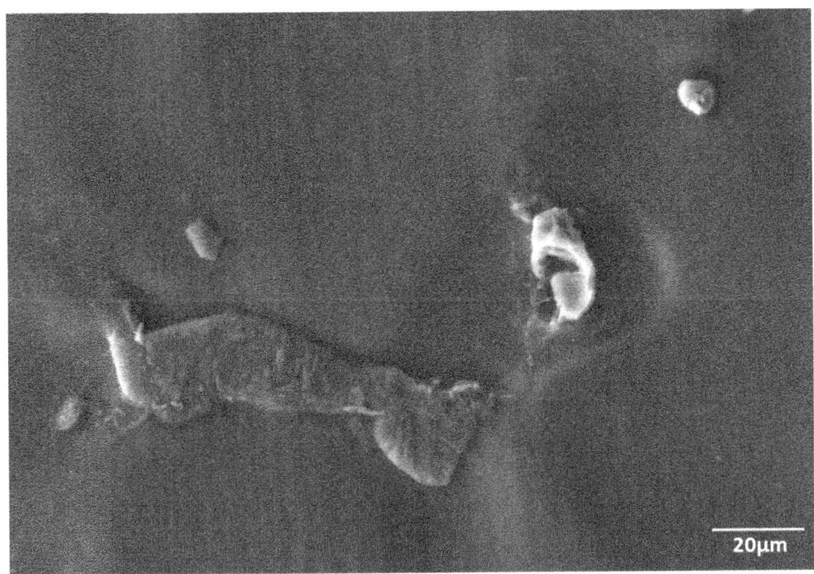

FIGURE 13.187 Microstructure of as-cast AZ31 alloy at high magnification. The microstructure reveals lamellar eutectic of $Mg_{17}Al_{12}$ at the grain boundaries. The etchant is Acetic Picral.

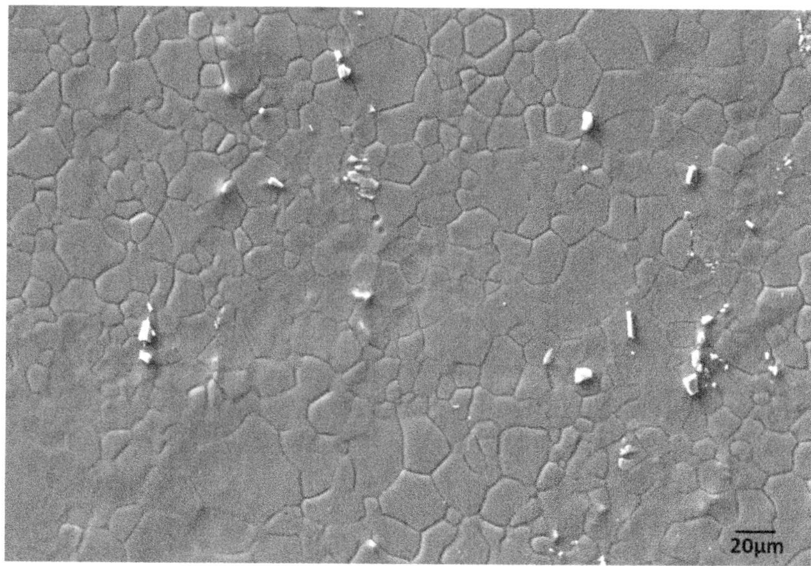

FIGURE 13.188 Microstructure of as-cast AZ31 hot rolled plate in H24 condition taken along the longitudinal direction. Fine equiaxed grain structure along with the fragmented particles of $Mg_{17}Al_{12}$ can be seen. The etchant is Acetic Picral.

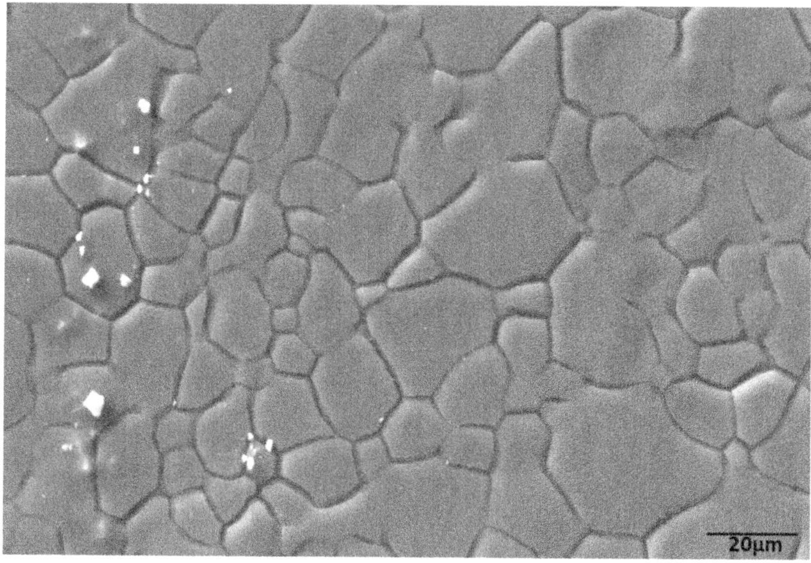

FIGURE 13.189 Microstructure of as-cast AZ31 hot rolled plate in H24 condition taken along the longitudinal direction at high magnification. Fine equiaxed grain structure along with the fragmented particles of $Mg_{17}Al_{12}$ can be seen. The etchant is Acetic Picral.

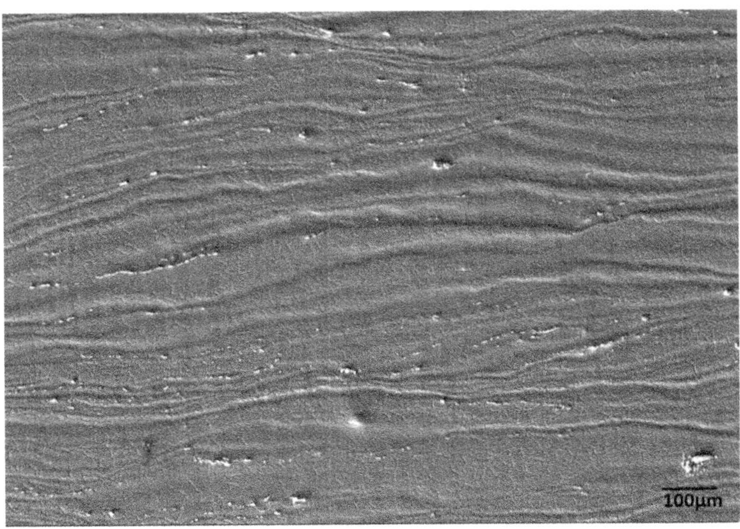

FIGURE 13.190 Microstructure of as-cast AZ31 hot rolled plate in H24 condition taken along the long transverse direction showing fine equiaxed grains with flow lines along rolling direction. Fragmented particles of $Mg_{17}Al_{12}$ can be seen along the flow lines. The etchant is Acetic Picral.

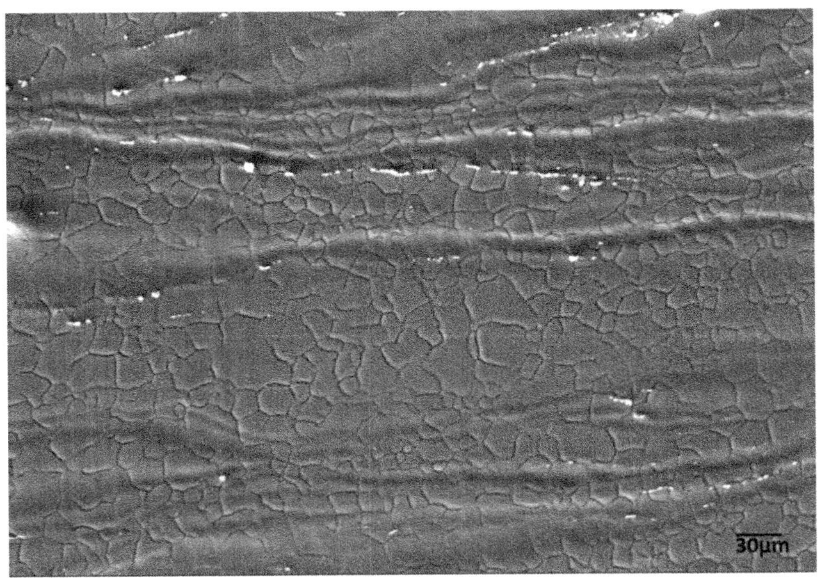

FIGURE 13.191 Microstructure of as-cast AZ31 hot rolled plate in H24 condition taken along the long transverse direction at high magnification showing fine equiaxed grains with flow lines along rolling direction. Fragmented particles of $Mg_{17}Al_{12}$ can be seen along the flow lines. The etchant is Acetic Picral.

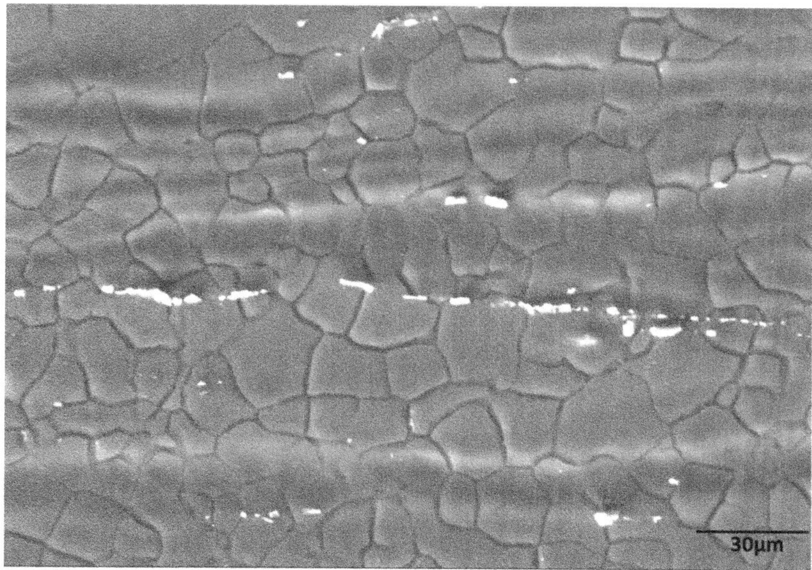

FIGURE 13.192 Microstructure of as-cast AZ31 hot rolled plate in H24 condition taken along the long transverse direction at high magnification showing fine equiaxed grains with fragmented particles of $Mg_{17}Al_{12}$ along the flow lines. The etchant is Acetic Picral.

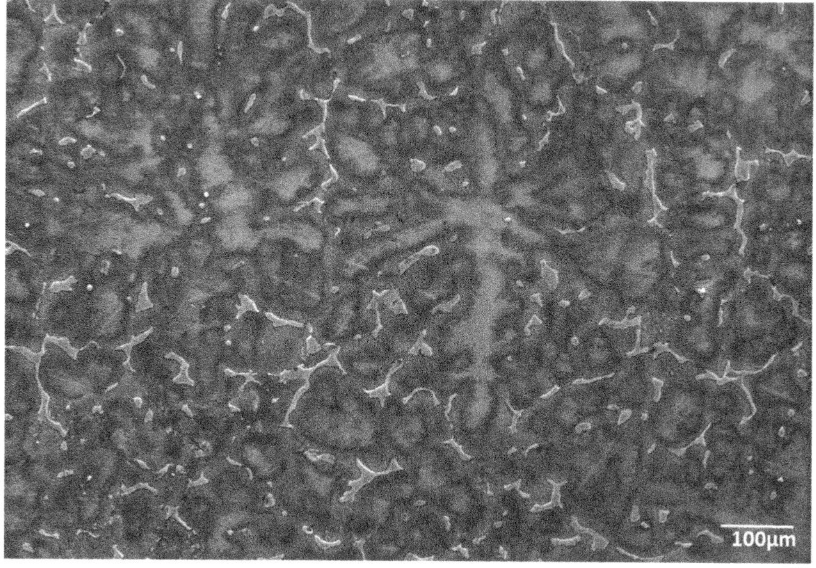

FIGURE 13.193 Microstructure of AZ91 magnesium alloy in as-cast condition. As-cast dendritic structure can be seen. The etchant is Acetic Picral.

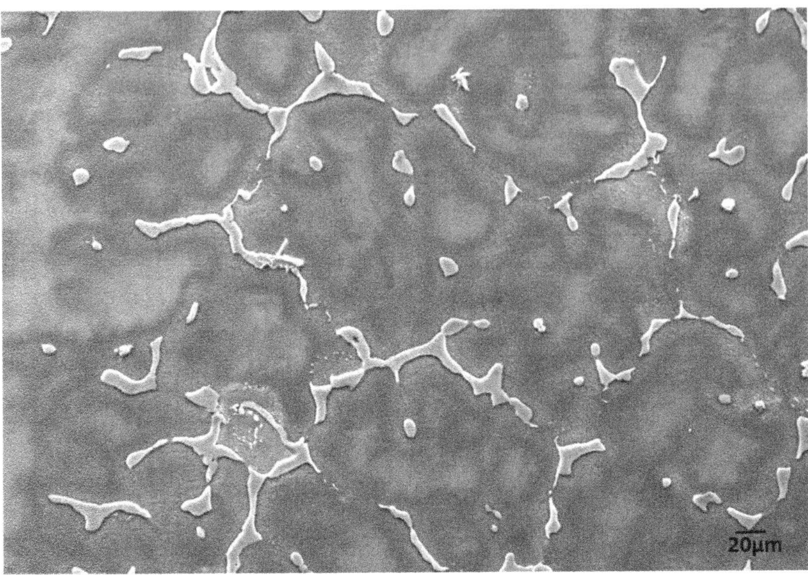

FIGURE 13.194 Microstructure of AZ91 magnesium alloy in as-cast condition at high magnification. Massive $Mg_{17}(Al,Zn)_{12}$ grain boundary phase along with fine lamellar eutectic of $Mg_{17}Al_{12}$ can be seen. The etchant is Acetic Picral.

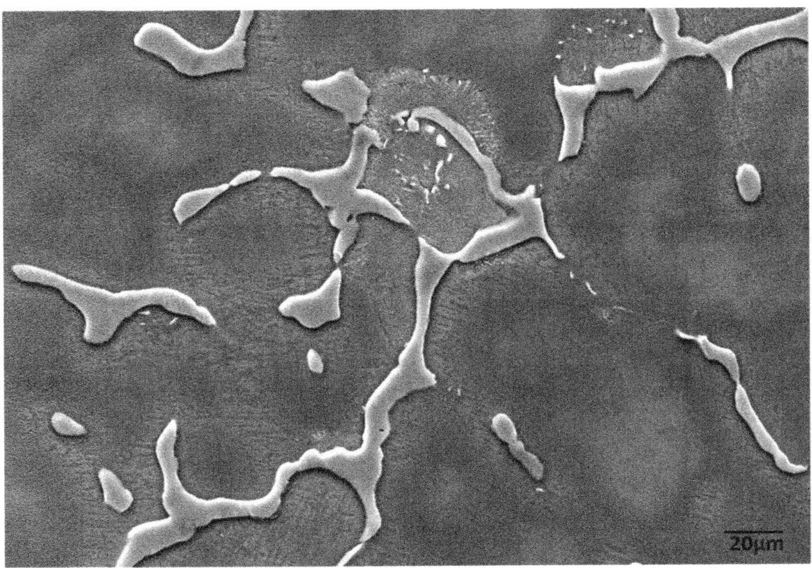

FIGURE 13.195 Microstructure of AZ91 magnesium alloy in as-cast condition at high magnification. Massive $Mg_{17}(Al,Zn)_{12}$ grain boundary phase along with fine lamellar eutectic of $Mg_{17}Al_{12}$ can be seen. The etchant is Acetic Picral.

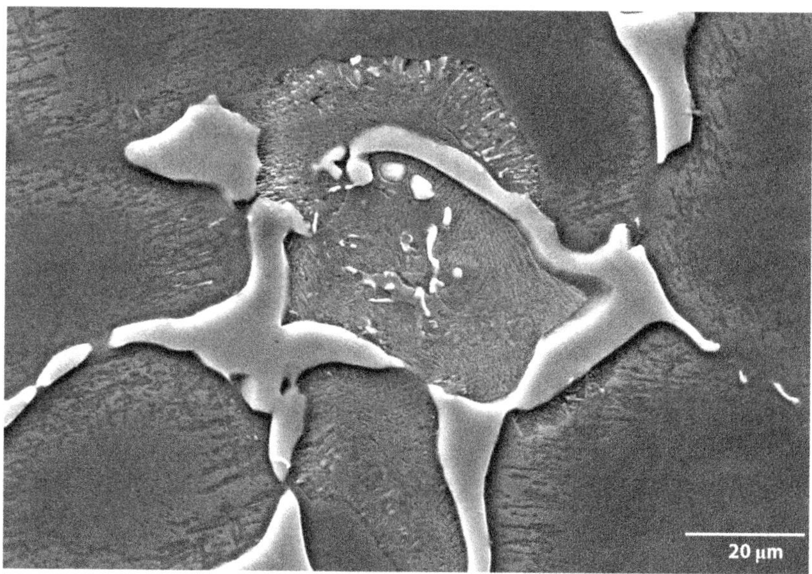

FIGURE 13.196 Microstructure of AZ91 magnesium alloy in as-cast condition at high magnification. Massive $Mg_{17}(Al,Zn)_{12}$ grain boundary phase along with fine lamellar eutectic of $Mg_{17}Al_{12}$ can be seen. The etchant is Acetic Picral.

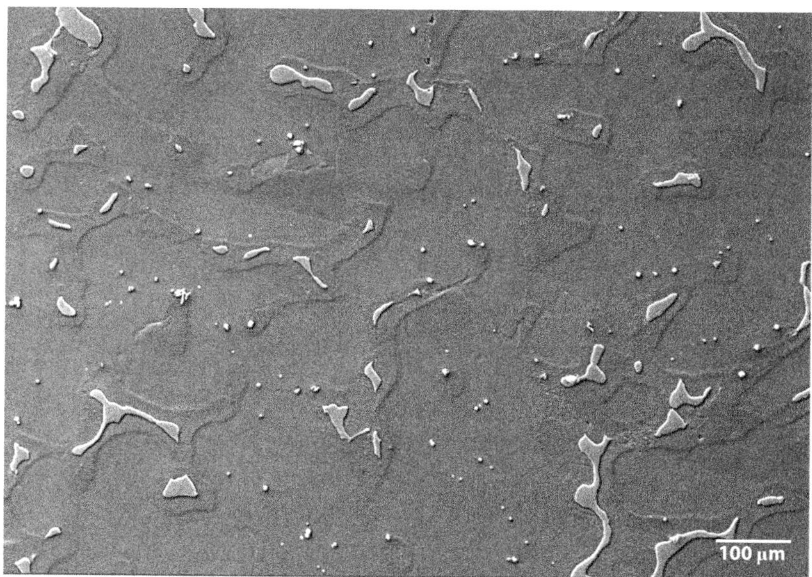

FIGURE 13.197 Microstructure of AZ91 magnesium alloy in F condition. $Mg_{17}(Al,Zn)_{12}$ grain boundary phase along with fine lamellar eutectic of $Mg_{17}Al_{12}$ can be seen. The etchant is Acetic Picral.

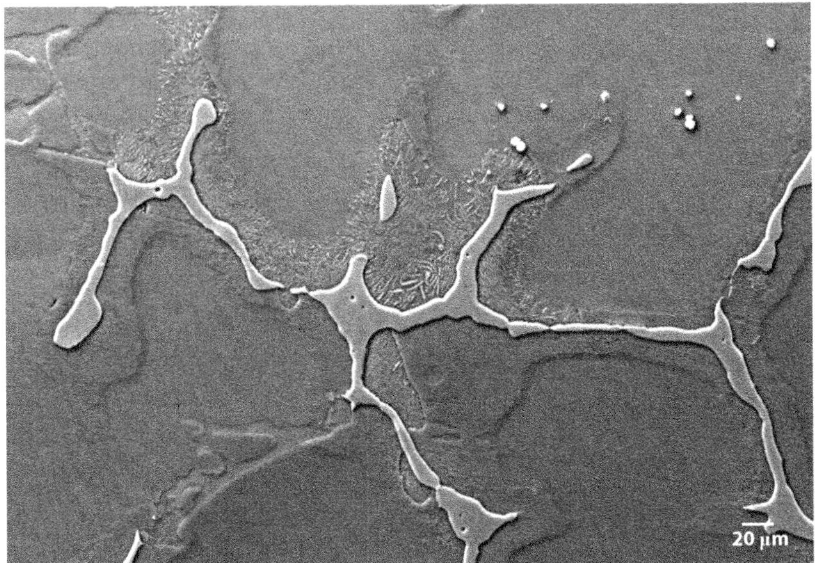

FIGURE 13.198 Microstructure of AZ91 magnesium alloy in F condition at high magnification. $Mg_{17}(Al,Zn)_{12}$ grain boundary phase along with fine lamellar eutectic of $Mg_{17}Al_{12}$ can be seen. The etchant is Acetic Picral.

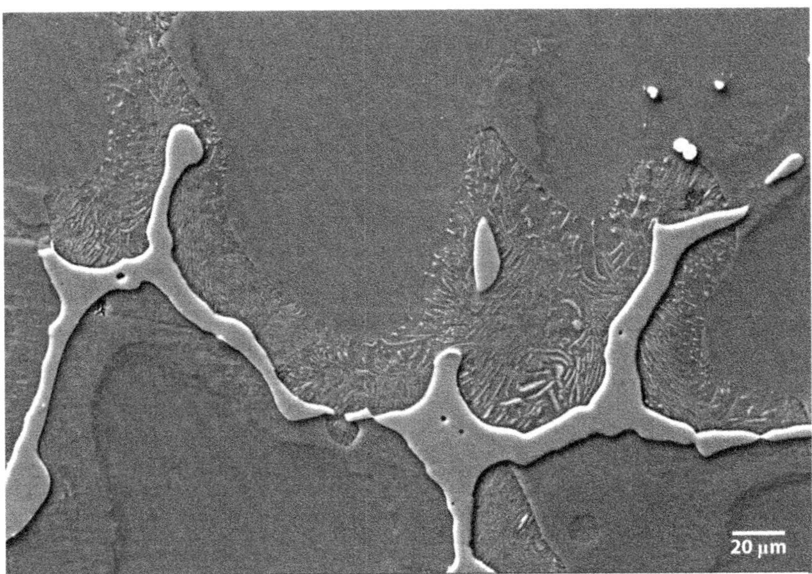

FIGURE 13.199 Microstructure of AZ91 magnesium alloy in F condition at high magnification. $Mg_{17}(Al,Zn)_{12}$ grain boundary phase along with fine lamellar eutectic of $Mg_{17}Al_{12}$. The etchant is Acetic Picral.

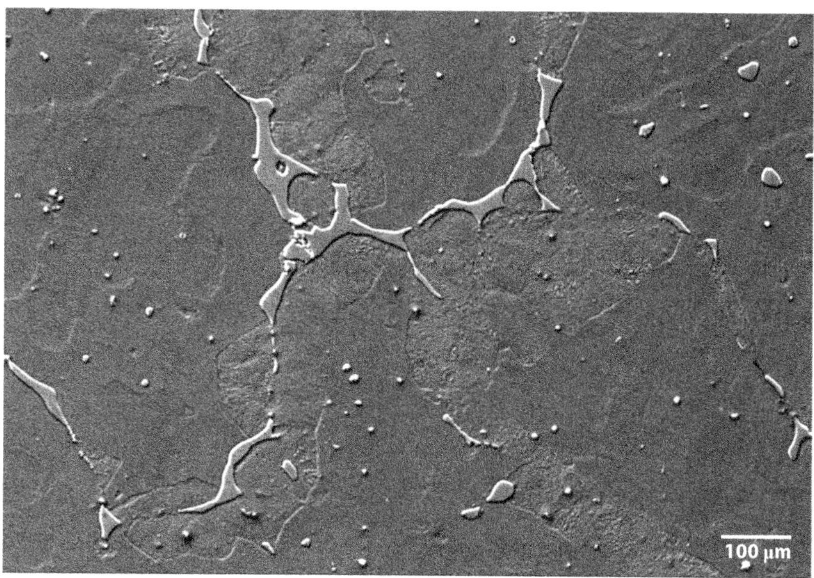

FIGURE 13.200 Microstructure of AZ91 magnesium alloy in T4 condition. $Mg_{17}(Al,Zn)_{12}$ grain boundary phase along with fine lamellar eutectic of $Mg_{17}Al_{12}$. The etchant is Acetic Picral.

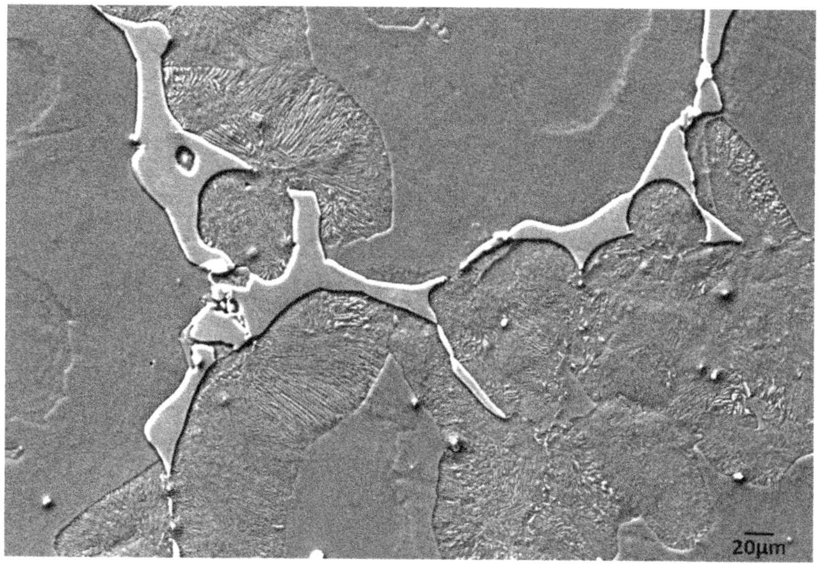

FIGURE 13.201 Microstructure of AZ91 magnesium alloy in T4 condition at high magnification. $Mg_{17}(Al,Zn)_{12}$ grain boundary phase along with fine lamellar eutectic of $Mg_{17}Al_{12}$. The etchant is Acetic Picral.

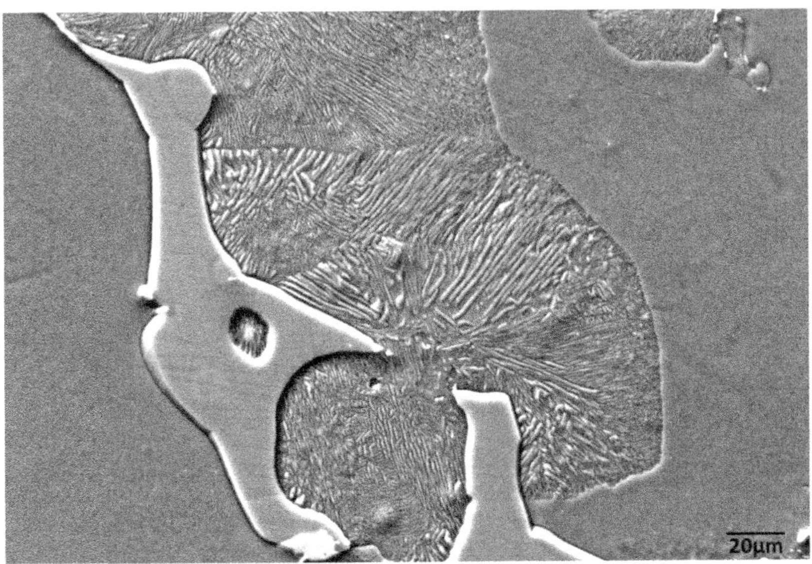

FIGURE 13.202 Microstructure of AZ91 magnesium alloy in T4 condition at high magnification. $Mg_{17}(Al,Zn)_{12}$ grain boundary phase along with fine lamellar eutectic of $Mg_{17}Al_{12}$ can be seen. The etchant is Acetic Picral.

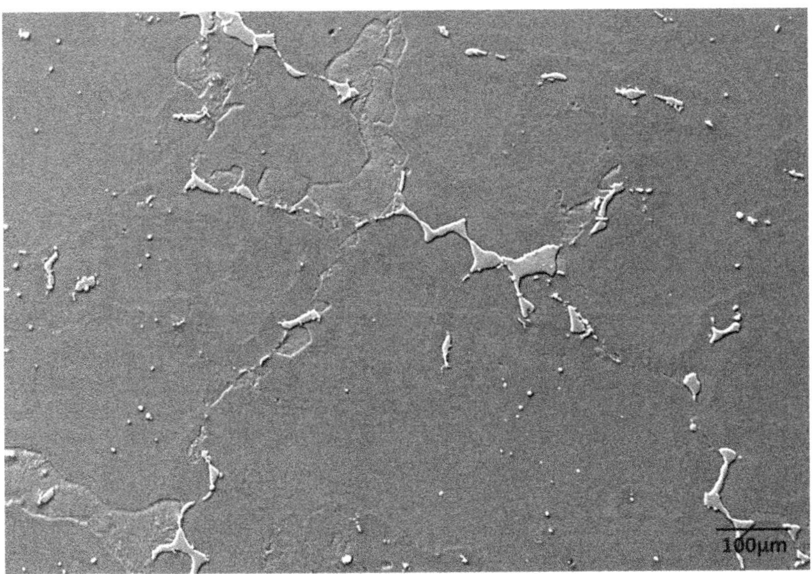

FIGURE 13.203 Microstructure of AZ91 magnesium alloy in T6 condition. $Mg_{17}(Al,Zn)_{12}$ grain boundary phase along with fine lamellar eutectic of $Mg_{17}Al_{12}$ can be seen. The etchant is Acetic Picral.

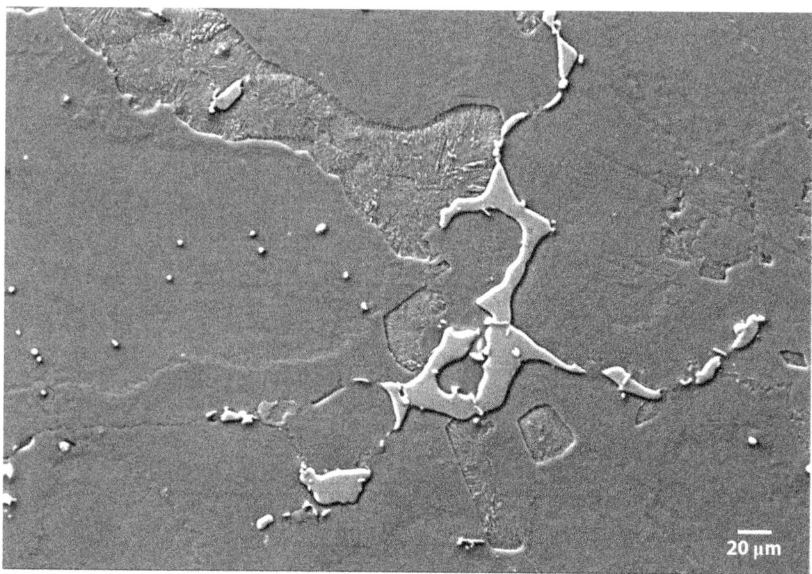

FIGURE 13.204 Microstructure of AZ91 magnesium alloy in T6 condition. at high magnification. $Mg_{17}(Al,Zn)_{12}$ grain boundary phase along with fine lamellar eutectic of $Mg_{17}Al_{12}$ can be seen. The etchant is Acetic Picral.

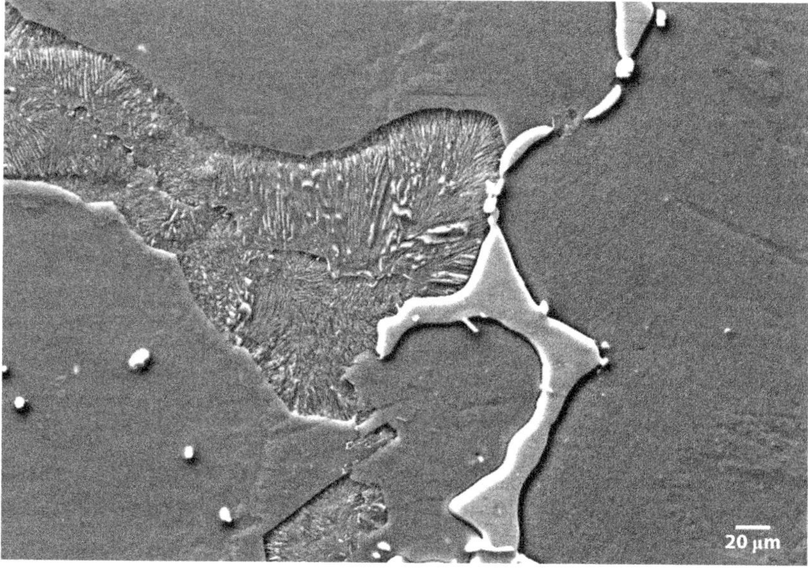

FIGURE 13.205 Microstructure of AZ91 magnesium alloy in T6 condition at high magnification. $Mg_{17}(Al,Zn)_{12}$ grain boundary phase along with fine lamellar eutectic of $Mg_{17}Al_{12}$ can be seen. The etchant is Acetic Picral.

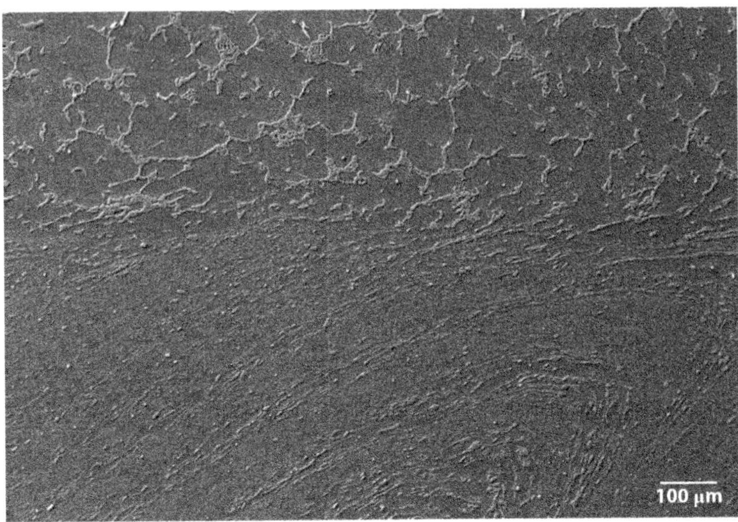

FIGURE 13.206 Microstructure of AZ91 magnesium cast plate that has been subjected to friction stir processing showing circular grain flow. The microstructure was obtained on the cross section at the interface between the parent material and stirred zone. The etchant is Acetic Picral.

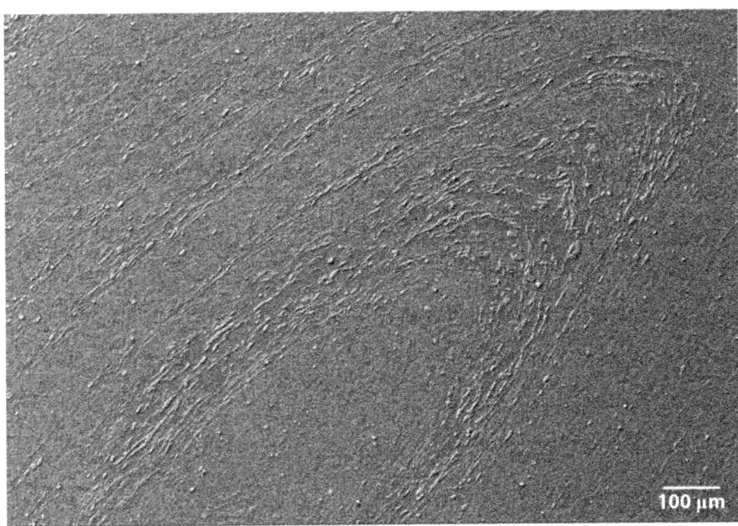

FIGURE 13.207 Microstructure of AZ91 magnesium alloy cast plate that has been subjected to friction stir processing showing circular flow lines. The microstructure was obtained on stirred zone. Fragmented particles along the circular flow lines can be seen. The etchant is Acetic Picral.

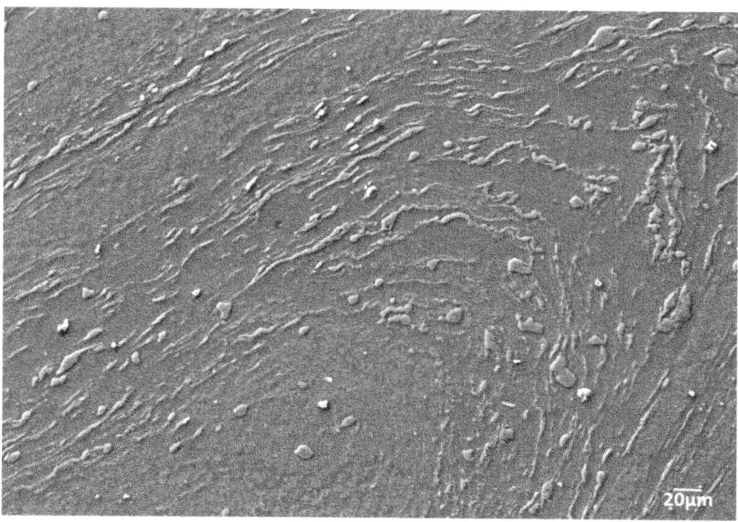

FIGURE 13.208 Microstructure of AZ91 magnesium alloy cast plate that has been subjected to friction stir processing showing circular flow lines at high magnification. The microstructure was obtained on stirred zone. Fragmented particles along the circular flow lines can be seen. The etchant is Acetic Picral.

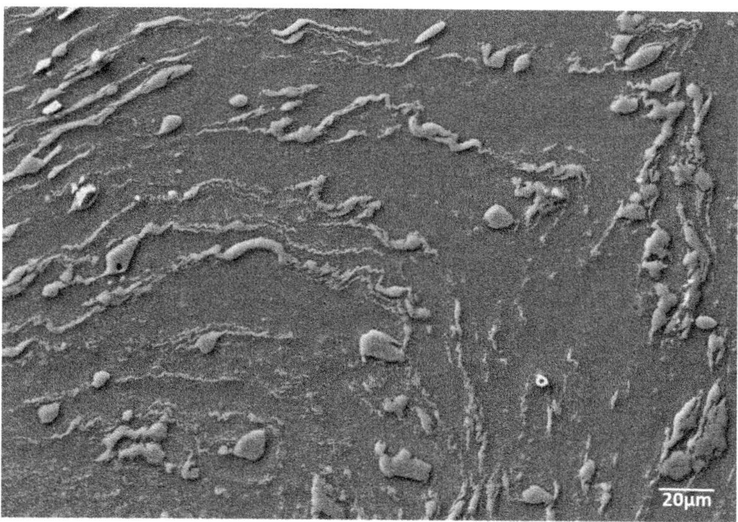

FIGURE 13.209 Microstructure of AZ91 magnesium alloy cast plate that has been subjected to friction stir processing at high magnification. The microstructure was obtained on stirred zone. Fragmented particles along the circular flow lines can be seen. The etchant is Acetic Picral.

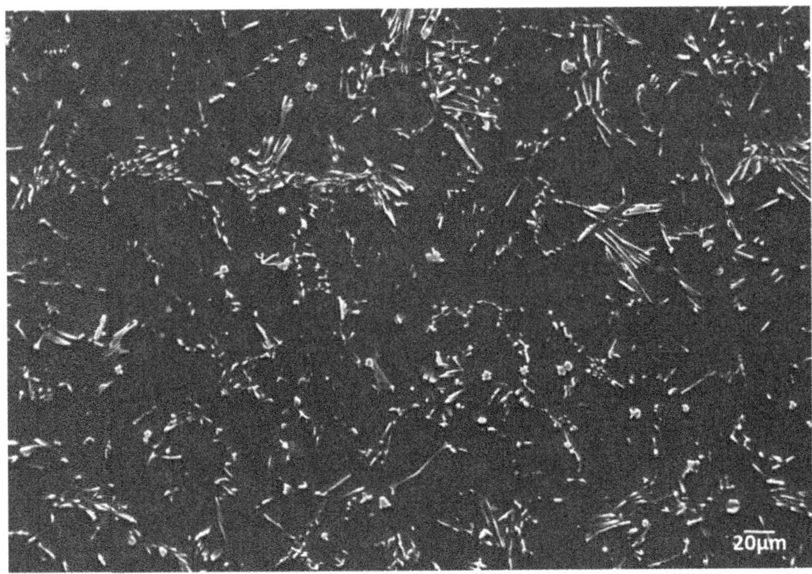

FIGURE 13.210 Microstructure of AE42 magnesium alloy in as-cast condition. Grain boundary $Al_{11}RE_3$ phase along with $Al_{10}Ce_2Mn_7$ particles can be seen. The etchant is Acetic Picral.

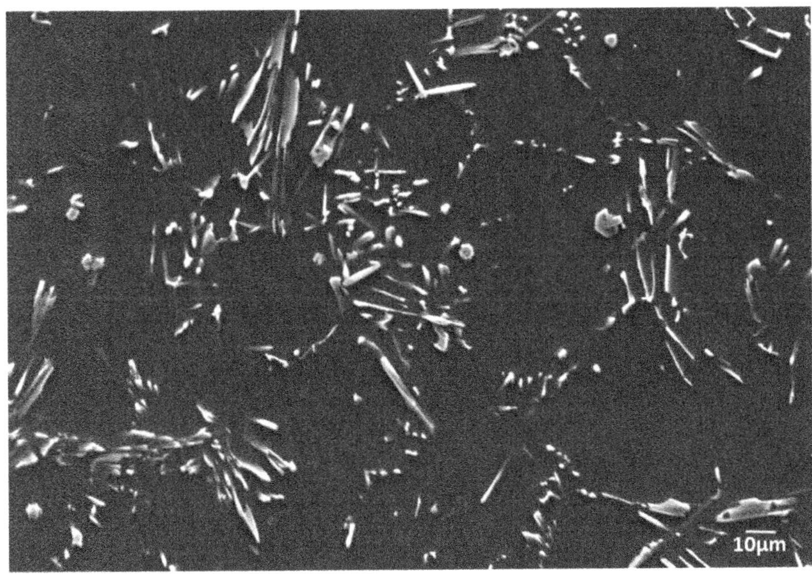

FIGURE 13.211 Microstructure of AE42 magnesium alloy in as-cast condition at high magnification. Grain boundary $Al_{11}RE_3$ phase along with $Al_{10}Ce_2Mn_7$ particles can be seen. The etchant is Acetic Picral.

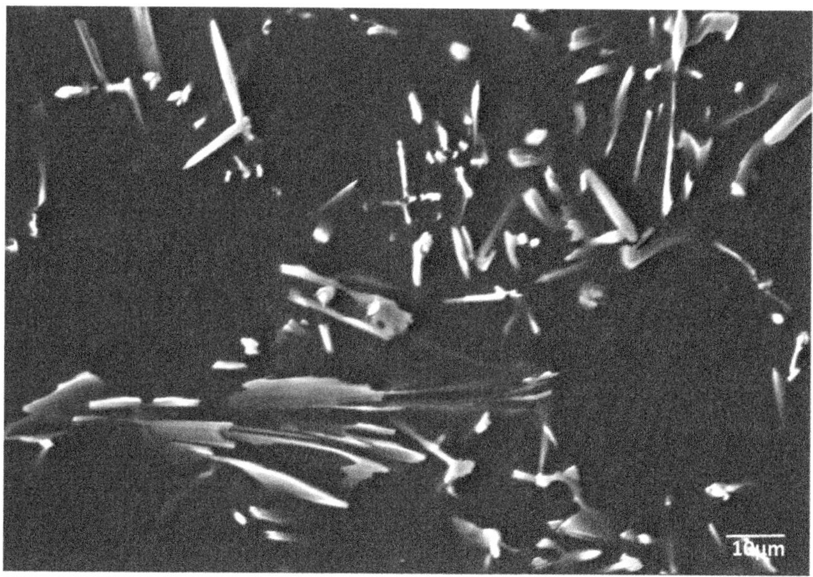

FIGURE 13.212 Microstructure of AE42 magnesium alloy in as-cast condition at high magnification. Grain boundary $Al_{11}RE_3$ phase along with $Al_{10}Ce_2Mn_7$ particles can be seen. The etchant is Acetic Picral.

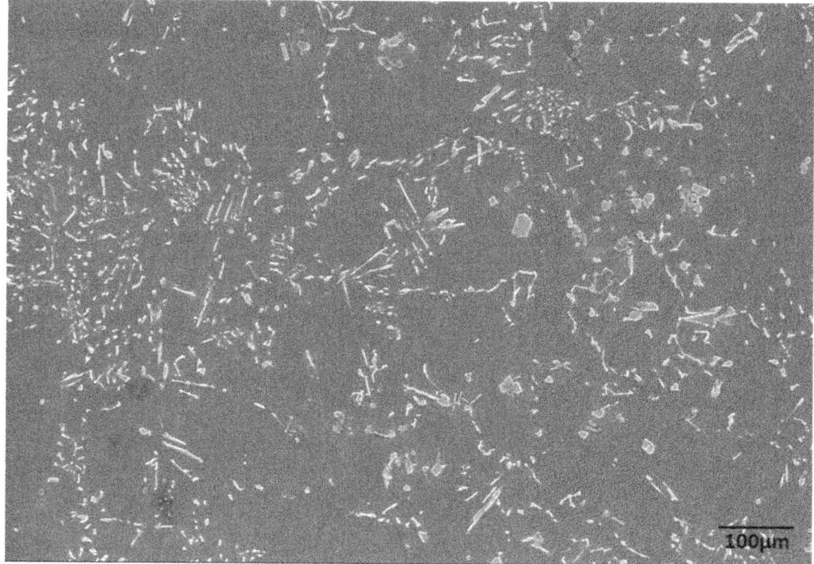

FIGURE 13.213 Microstructure of AE44 magnesium alloy in as-cast condition. The microstructure consists of grain boundary phase $Al_{11}(La,Ce,Nd)_3$, needles of $Al_{11}Ce_3$ and blocky particles of $Al_{10}(Ce,Nd)_2Mn_7$. The etchant is Acetic Picral.

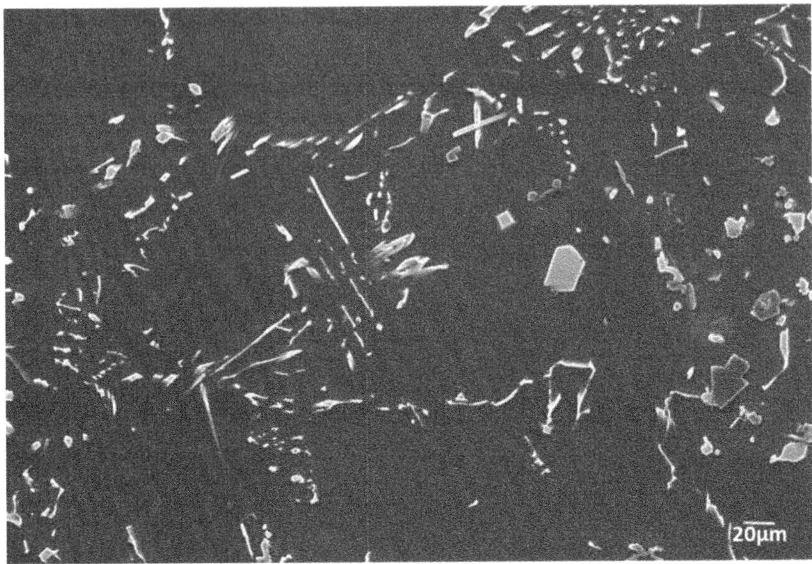

FIGURE 13.214 Microstructure of AE44 magnesium alloy in as-cast condition taken at high magnification. The microstructure consists of grain boundary phase $Al_{11}(La,Ce,Nd)_3$, needles of $Al_{11}Ce_3$ and blocky particles of $Al_{10}(Ce,Nd)_2Mn_7$. The etchant is Acetic Picral.

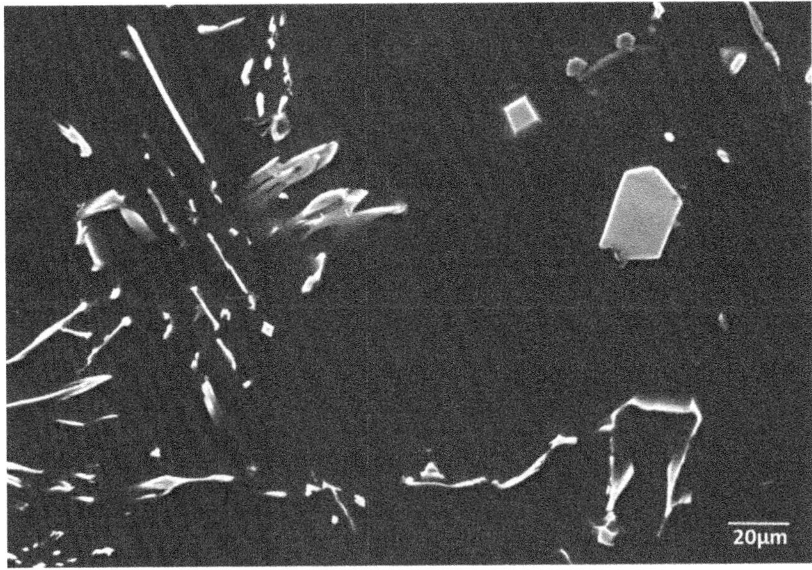

FIGURE 13.215 Microstructure of AE44 magnesium alloy in as-cast condition. The microstructure consists of grain boundary phase $Al_{11}(La,Ce,Nd)_3$, needles of $Al_{11}Ce_3$ and blocky particles of $Al_{10}(Ce,Nd)_2Mn_7$. The etchant is Acetic Picral.

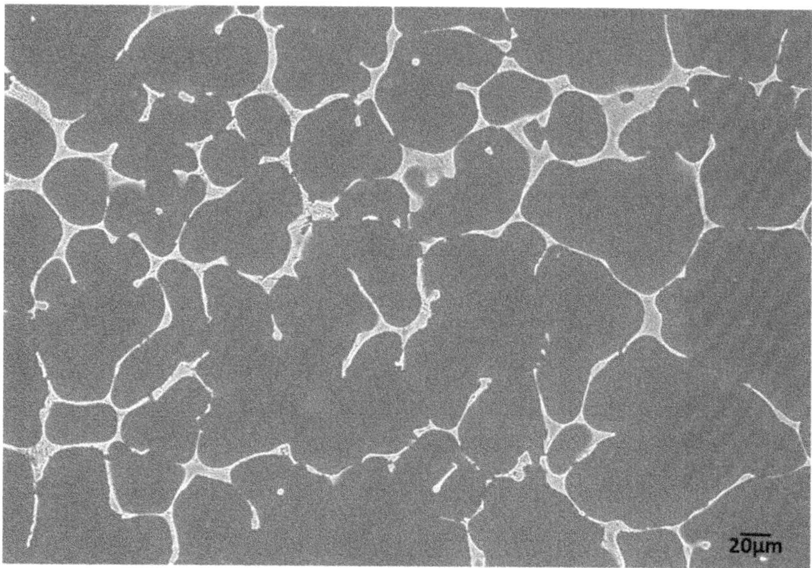

FIGURE 13.216 Microstructure of ZC63 magnesium alloy in as-cast condition. Fine lamellar eutectic of $Mg(Zn,Cu)_2$ can be seen at the grain boundaries along with the particles of Mg (Zn,Cu) inside the grains. The etchant is Acetic Picral.

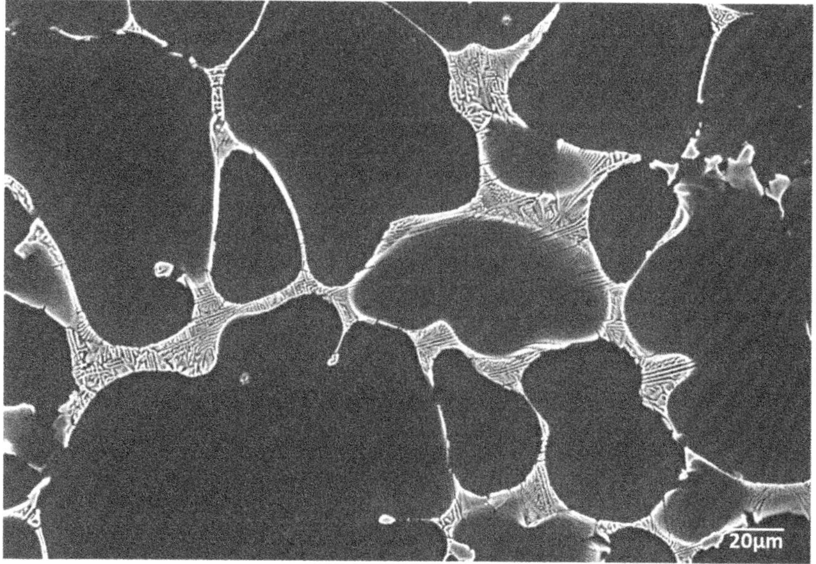

FIGURE 13.217 Microstructure of ZC63 magnesium alloy in as-cast condition. Fine lamellar eutectic of $Mg(Zn,Cu)_2$ can be seen at the grain boundaries along with the particles of Mg (Zn,Cu) inside the grains. The etchant is Acetic Picral.

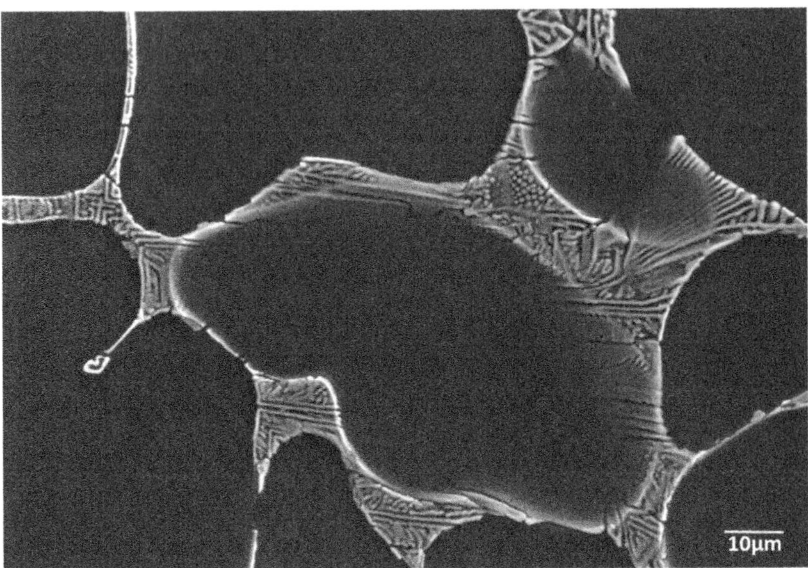

FIGURE 13.218 Microstructure of ZC63 magnesium alloy in as-cast condition taken at high magnification. Fine lamellar eutectic of $Mg(Zn,Cu)_2$ can be seen at the grain boundaries. The etchant is Acetic Picral.

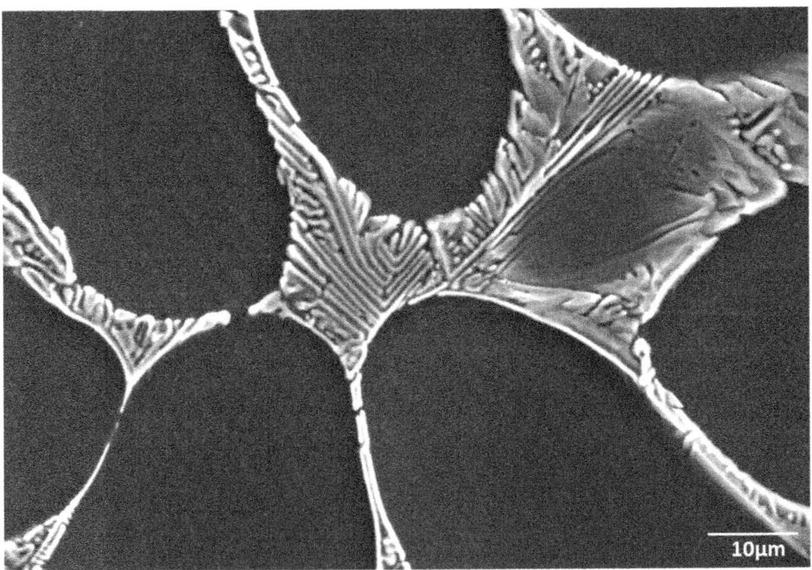

FIGURE 13.219 Microstructure of ZC63 magnesium alloy in as-cast condition at high magnification. Fine lamellar eutectic of $Mg(Zn,Cu)_2$ can be seen at the grain boundaries. The etchant is Acetic Picral.

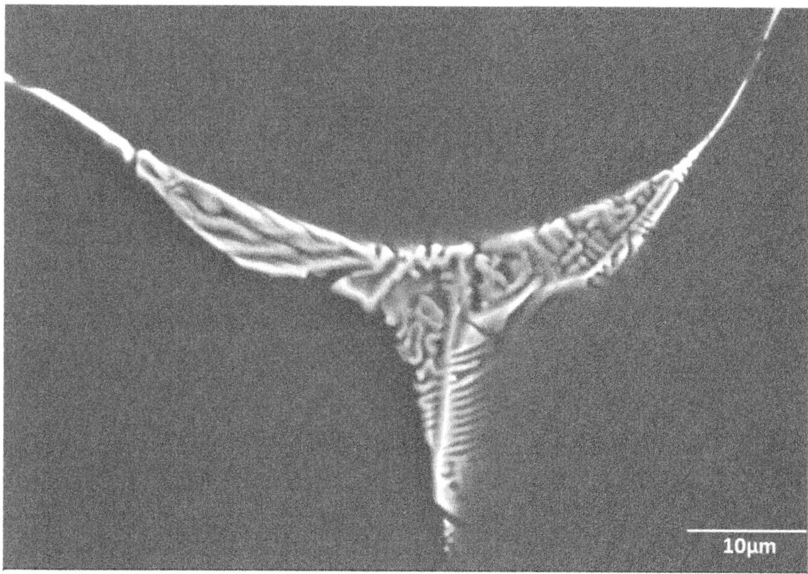

FIGURE 13.220 Microstructure of ZC63 magnesium alloy in as-cast condition at high magnification. Fine lamellar eutectic of $Mg(Zn,Cu)_2$ can be seen at the grain boundaries. The etchant is Acetic Picral.

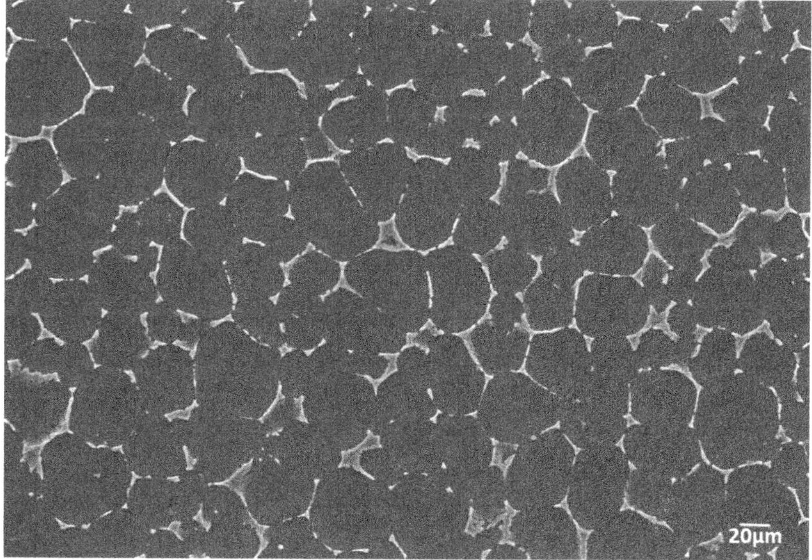

FIGURE 13.221 Microstructure of ZE41 magnesium alloy in as-cast condition. Fine equiaxed grain structure can be seen with Mg_7Zn_3RE eutectic at grain boundaries. The etchant is Acetic Picral.

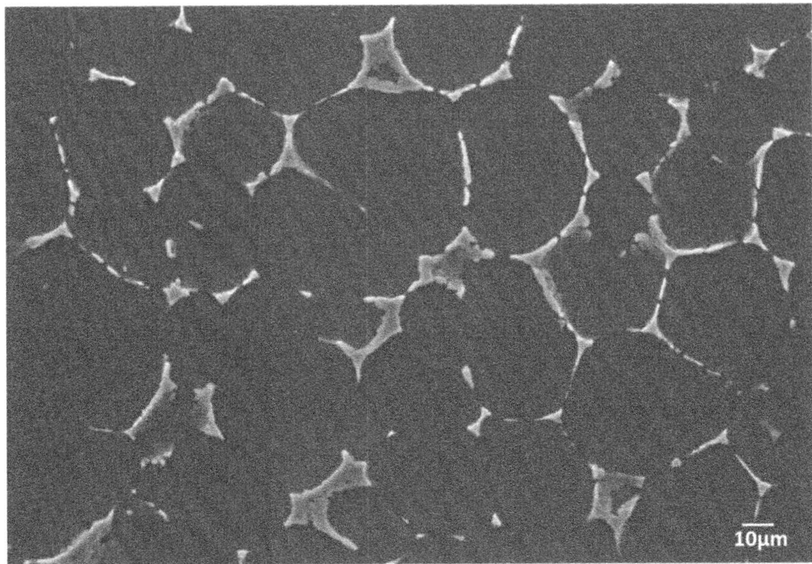

FIGURE 13.222 Microstructure of ZE41 magnesium alloy in as-cast condition taken at high magnification. Fine equiaxed grain structure can be seen with Mg_7Zn_3RE eutectic at grain boundaries. The etchant is Acetic Picral.

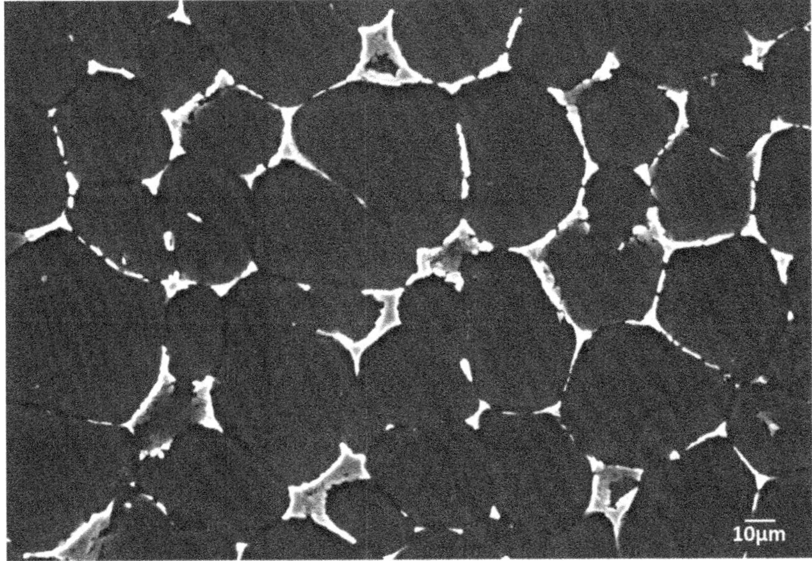

FIGURE 13.223 Microstructure of ZE41 magnesium alloy in as-cast condition taken at high magnification. Fine equiaxed grain structure can be seen with Mg_7Zn_3RE eutectic at grain boundaries. Note the presence of dendritic coring in the form of flowery pattern within the grains. The etchant is Acetic Picral.

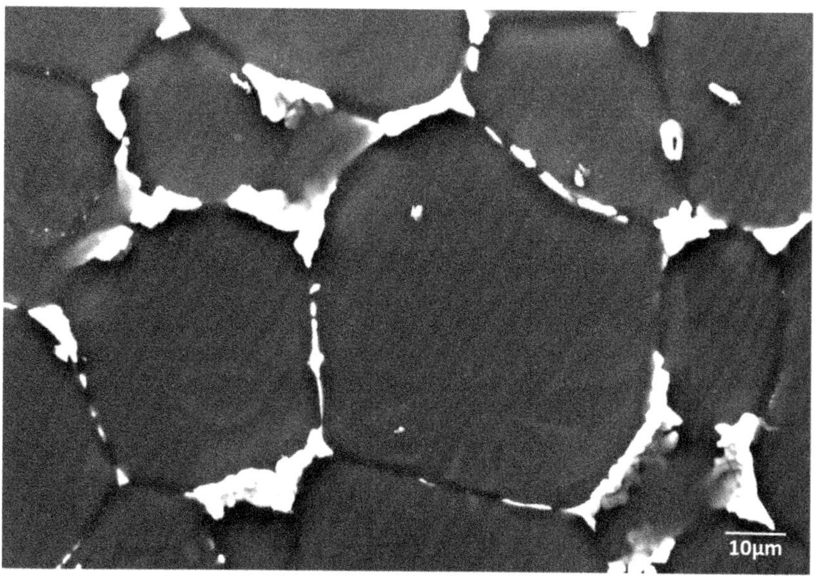

FIGURE 13.224 Microstructure of ZE41 magnesium alloy in as-cast condition taken at high magnification. Fine equiaxed grain structure can be seen with Mg_7Zn_3RE eutectic at grain boundaries. The etchant is Acetic Picral.

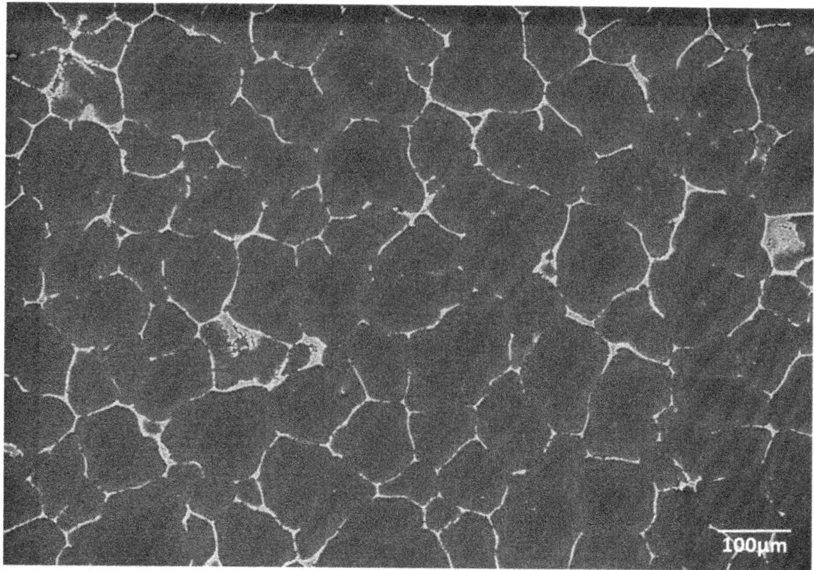

FIGURE 13.225 Microstructure of ZE41 R25 magnesium alloy in as-cast condition. Fine equiaxed grain structure can be seen with Mg_7Zn_3RE eutectic at grain boundaries. The etchant is Acetic Picral.

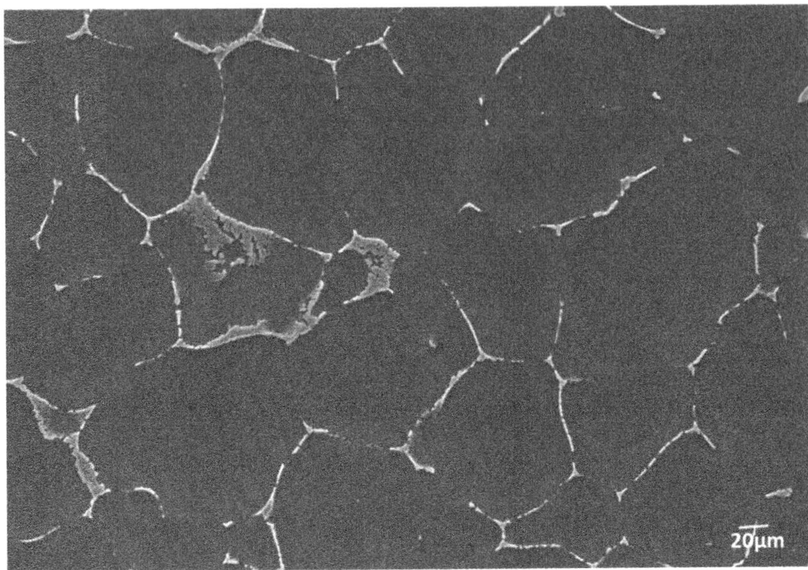

FIGURE 13.226 Microstructure of ZE41 R25 magnesium alloy in as-cast condition taken at high magnification. Fine equiaxed grain structure can be seen with Mg_7Zn_3RE eutectic at grain boundaries. The etchant is Acetic Picral.

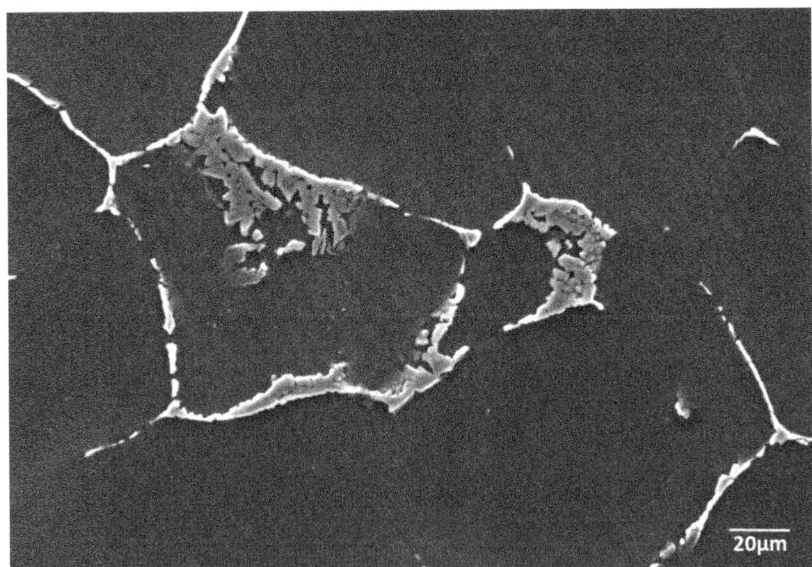

FIGURE 13.227 Microstructure of ZE41 R25 magnesium alloy in as-cast condition taken at high magnification. Fine equiaxed grain structure can be seen with Mg_7Zn_3RE eutectic at grain boundaries. The etchant is Acetic Picral.

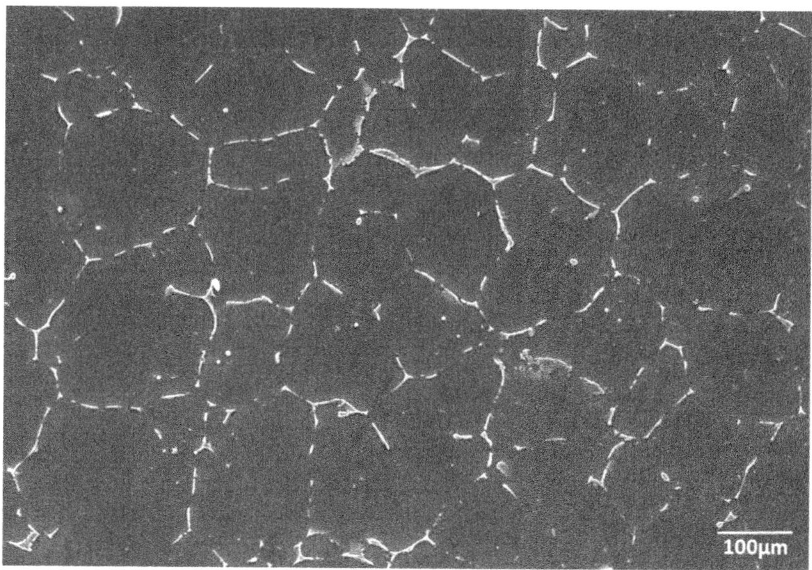

FIGURE 13.228 Microstructure of ZE10 magnesium alloy in as-cast condition. Grain boundary eutectic of Mg_7Zn_3RE can be seen. The etchant is Acetic Picral.

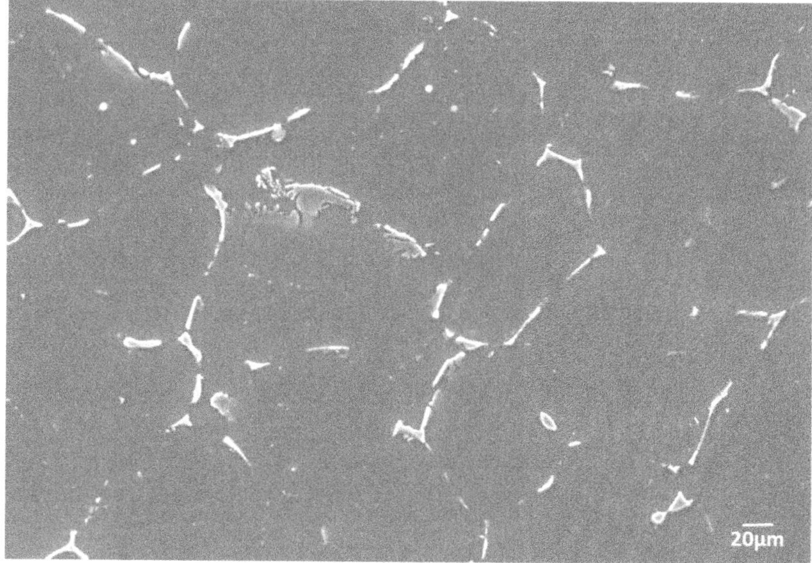

FIGURE 13.229 Microstructure of ZE10 magnesium alloy in as-cast condition taken at high magnification. Grain boundary eutectic of Mg_7Zn_3RE can be seen. The etchant is Acetic Picral.

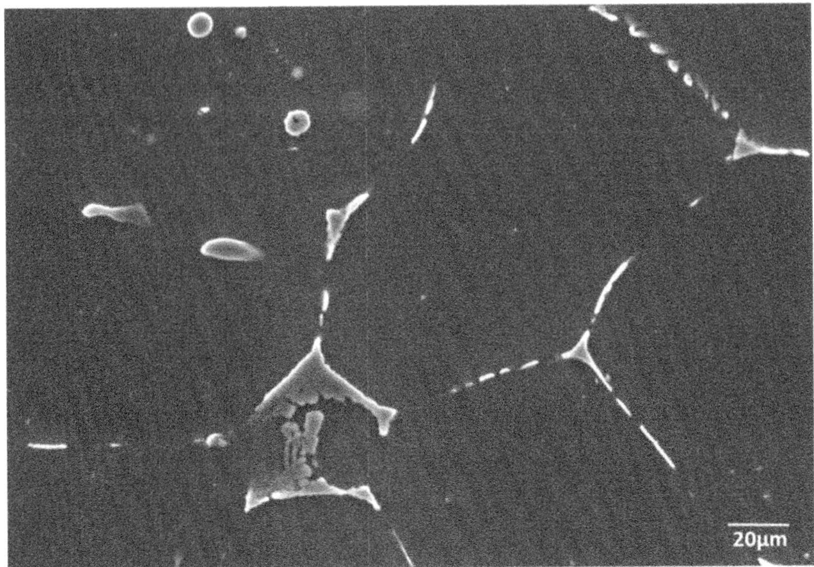

FIGURE 13.230 Microstructure of ZE10 magnesium alloy in as-cast condition taken at high magnification. Grain boundary eutectic of Mg_7Zn_3RE can be seen. The etchant is Acetic Picral.

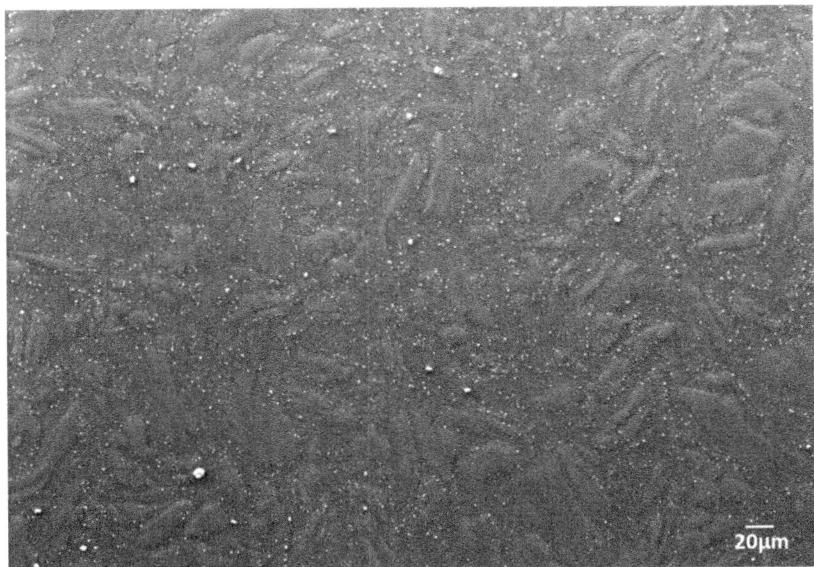

FIGURE 13.231 Microstructure of ZK60 magnesium alloy in as-cast condition. The specimen was etched with Acetic Picral.

FIGURE 13.232 Microstructure of ZK60 magnesium alloy in as-cast condition. The specimen was etched with Acetic Picral.

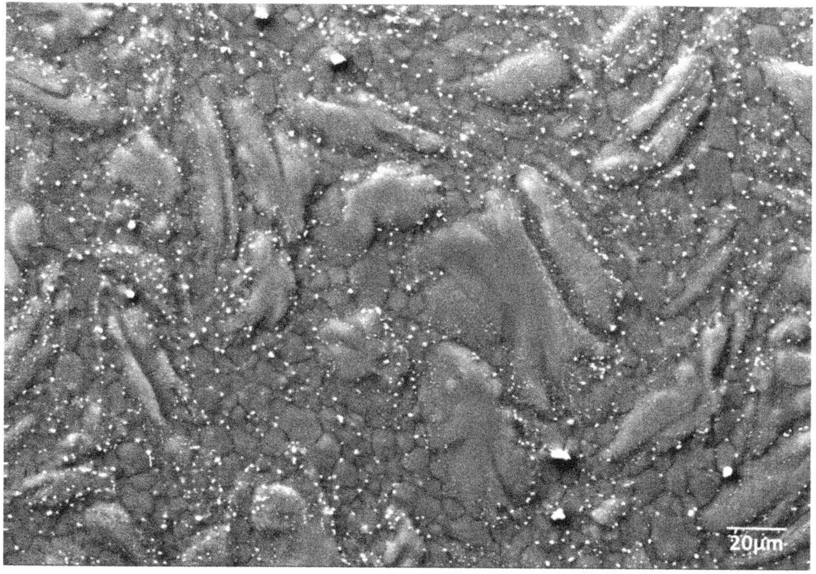

FIGURE 13.233 Microstructure of ZK60 magnesium alloy in as-cast condition. The specimen was etched with Acetic Picral.

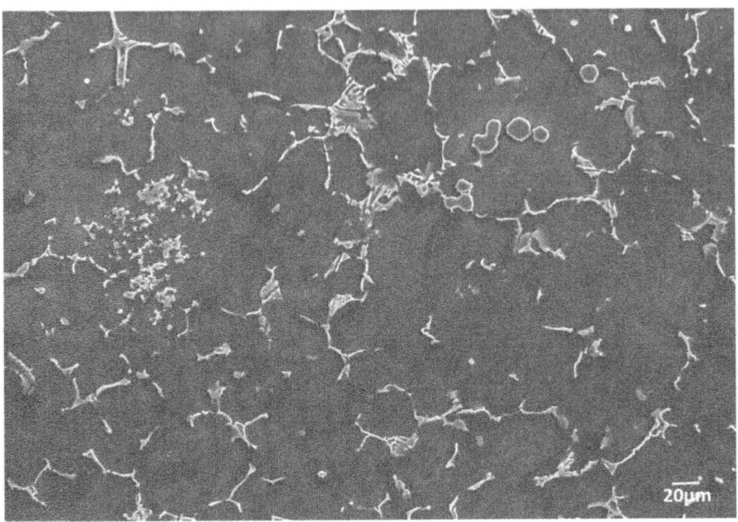

FIGURE 13.234 Microstructure of MRI153M magnesium alloy in as-cast condition. As-cast dendritic structure with grain boundary eutectic of $(Mg,Al)_2Ca$ phase can be seen. The etchant is Acetic Picral.

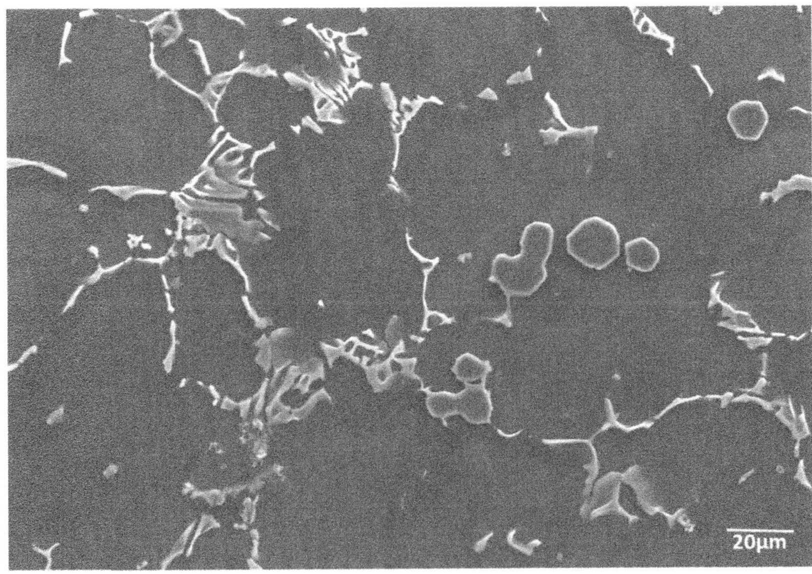

FIGURE 13.235 Microstructure of MRI153M magnesium alloy in as-cast condition. As-cast dendritic structure with grain boundary eutectic of $(Mg,Al)_2Ca$ phase can be seen. The etchant is Acetic Picral.

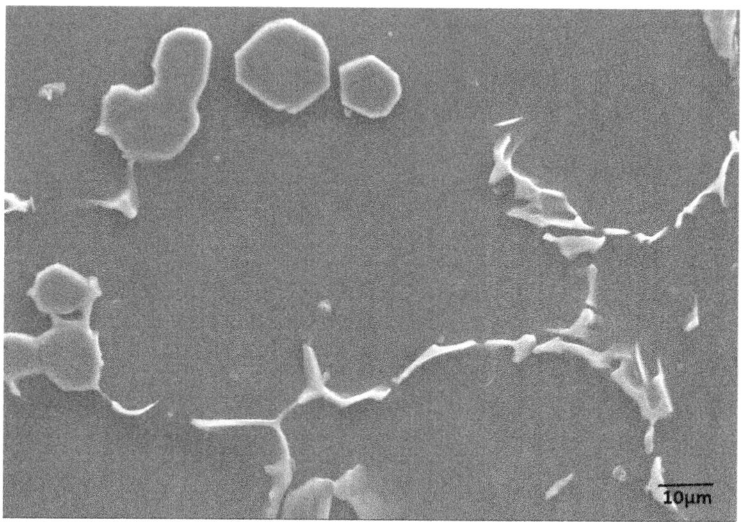

FIGURE 13.236 Microstructure of MRI153M magnesium alloy in as-cast condition taken at high magnification. As-cast dendritic structure with grain boundary eutectic of $(Mg,Al)_2Ca$ phase can be seen. The etchant is Acetic Picral.

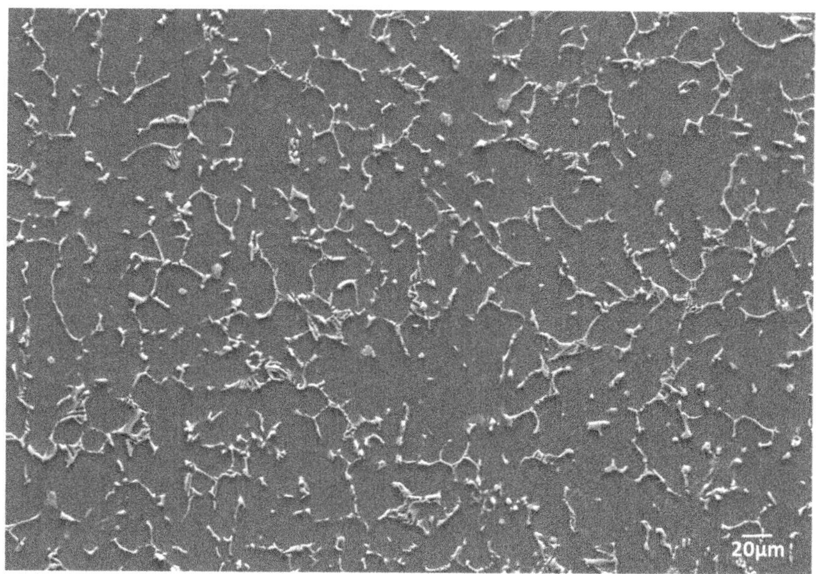

FIGURE 13.237 Microstructure of MRI230D magnesium alloy in as-cast condition. As-cast dendritic structure with grain boundary eutectic of $(Mg,Al)_2Ca$ phase can be seen. The etchant is Acetic Picral.

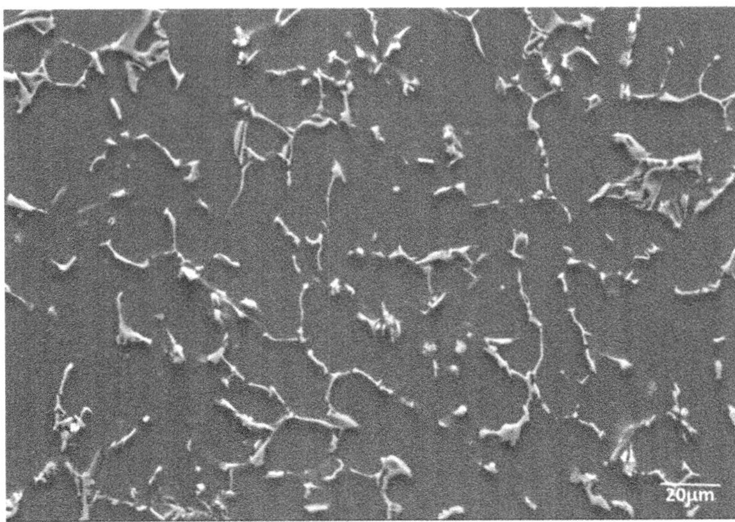

FIGURE 13.238 Microstructure of MRI230D magnesium alloy in as-cast condition taken at high magnification. As-cast dendritic structure with grain boundary eutectic of $(Mg,Al)_2Ca$ phase can be seen. The etchant is Acetic Picral.

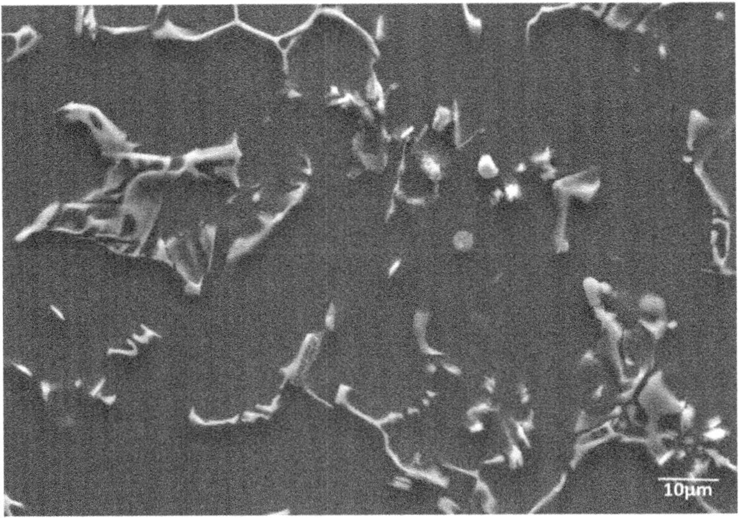

FIGURE 13.239 Microstructure of MRI230D magnesium alloy in as-cast condition taken at high magnification. As-cast dendritic structure with grain boundary eutectic of $(Mg,Al)_2Ca$ phase can be seen. The etchant is Acetic Picral.

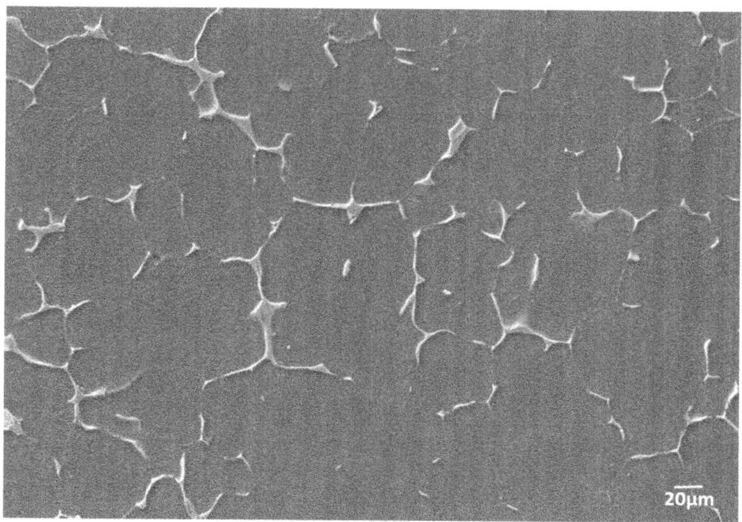

FIGURE 13.240 Microstructure of Elektron 21 magnesium alloy in as-cast condition. Dendritic solidification with Mg_{12}(Nd,Gd) grain boundary phase can be seen. The etchant is Acetic Picral.

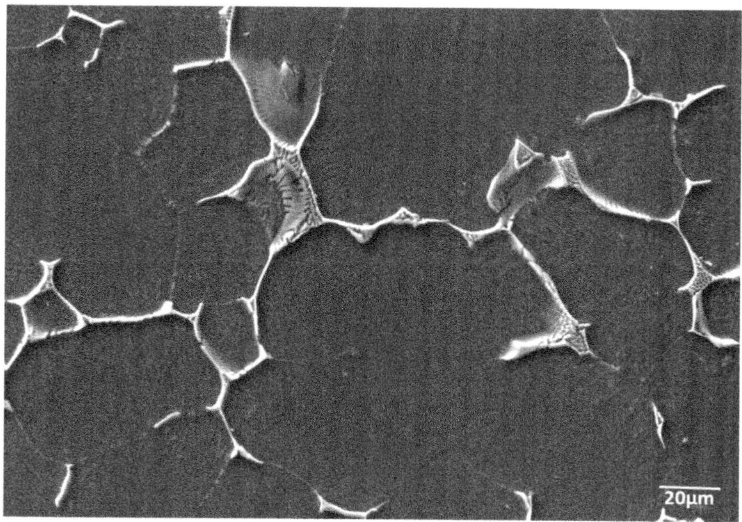

FIGURE 13.241 Microstructure of Elektron 21 magnesium alloy in as-cast condition taken at high magnification. Dendritic solidification with Mg_{12}(Nd,Gd) grain boundary phase can be seen. The etchant is Acetic Picral.

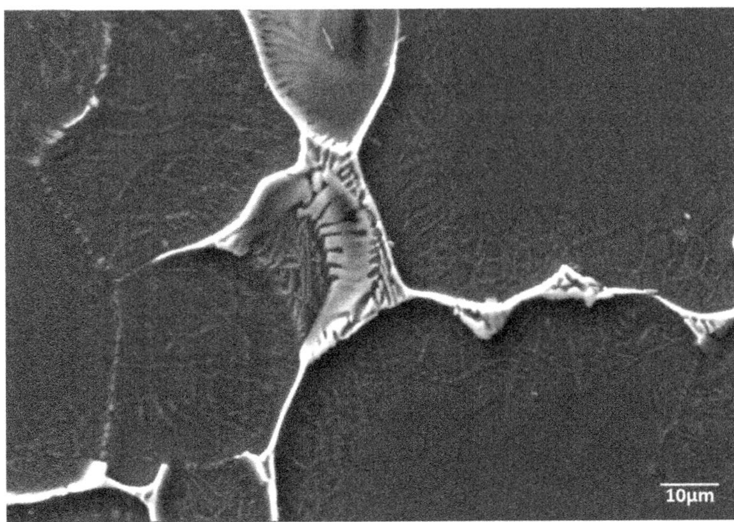

FIGURE 13.242 Microstructure of Elektron 21 magnesium alloy in as-cast condition taken at high magnification. $Mg_{12}(Nd,Gd)$ grain boundary phase can be seen. The etchant is Acetic Picral.

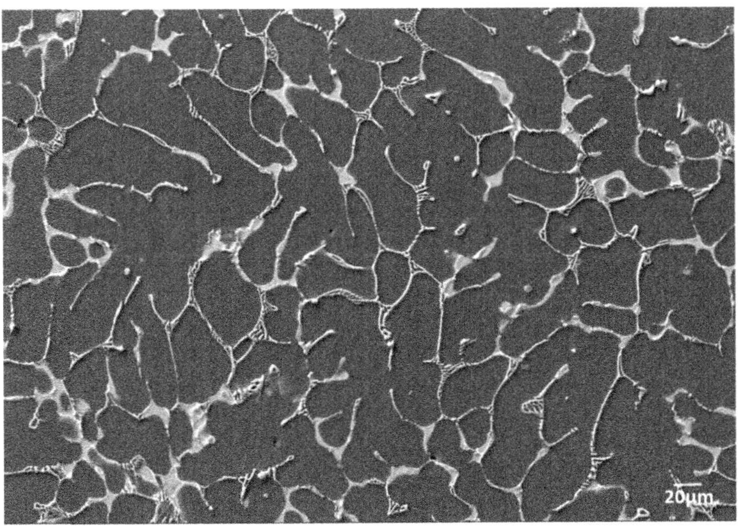

FIGURE 13.243 Microstructure of Diemag 422 magnesium alloy in as-cast condition. Compact Ba-rich phase $Mg_{21}Al_3Ba_2$ and lamellar Al_2Ca phase at grain boundaries can be seen. Dendritic solidification pattern can be seen. The etchant is Acetic Picral.

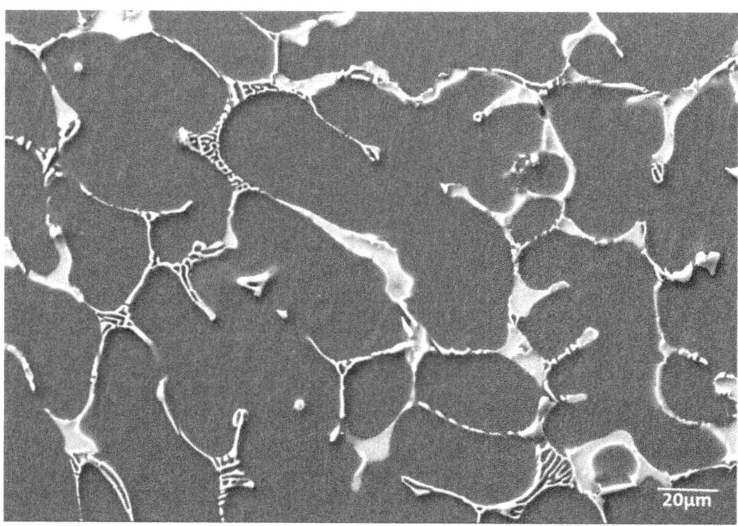

FIGURE 13.244 Microstructure of Diemag 422 magnesium alloy in as-cast condition taken at high magnification. Compact Ba-rich phase $Mg_{21}Al_3Ba_2$ and lamellar Al_2Ca phase at grain boundaries can be seen. Dendritic solidification pattern can be seen. The etchant is Acetic Picral.

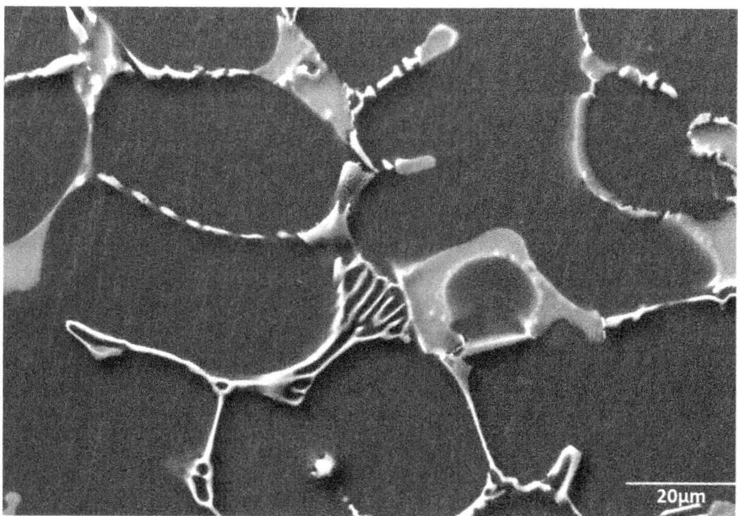

FIGURE 13.245 Microstructure of DieMag 422 magnesium alloy in as-cast condition taken at high magnification. Compact Ba-rich phase $Mg_{21}Al_3Ba_2$ and lamellar Al_2Ca phase at grain boundaries can be seen. Dendritic solidification pattern can be seen. The etchant is Acetic Picral.

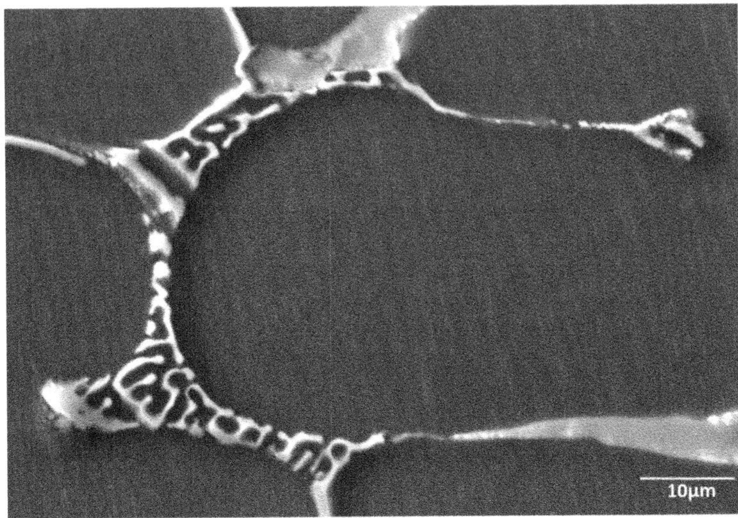

FIGURE 13.246 Microstructure of DieMag 422 magnesium alloy in as-cast condition taken at high magnification. Compact Ba-rich phase $Mg_{21}Al_3Ba_2$ and lamellar Al_2Ca phase at grain boundaries can be seen. Dendritic solidification pattern can be seen. The etchant is Acetic Picral.

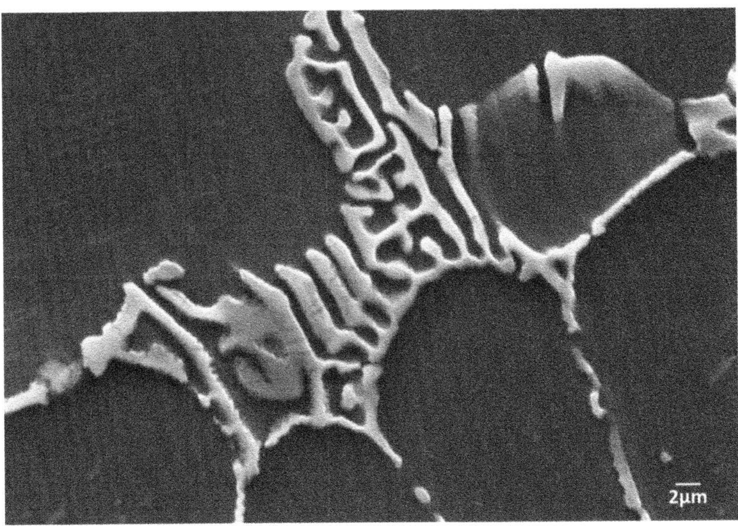

FIGURE 13.247 Microstructure of DieMag 422 magnesium alloy in as-cast condition taken at high magnification. Al_2Ca phase in the form of Chinese script at grain boundary can be seen. Dendritic solidification pattern can be seen. The etchant is Acetic Picral.

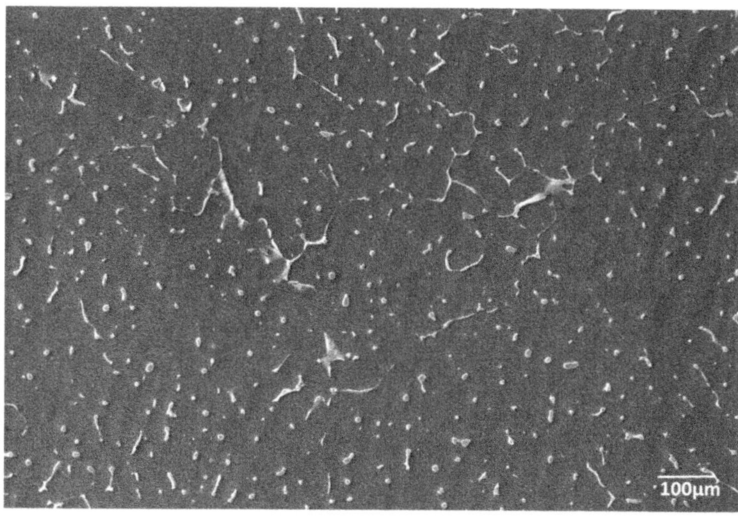

FIGURE 13.248 Microstructure of Mg-2Gd-6Zn magnesium alloy in as-cast condition. As-cast dendritic structure and $(Mg,Zn)_3Gd$ at the grain boundaries and grain interiors can be seen. The etchant is Acetic Picral.

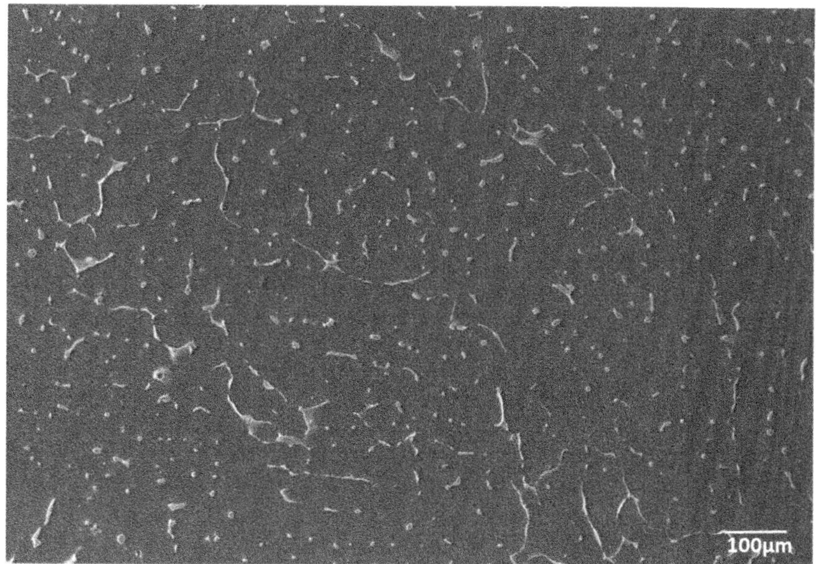

FIGURE 13.249 Microstructure of Mg-2Gd-6Zn magnesium alloy in as-cast condition taken at high magnification. $(Mg,Zn)_3Gd$ at the grain boundaries and grain interiors can be seen. The etchant is Acetic Picral.

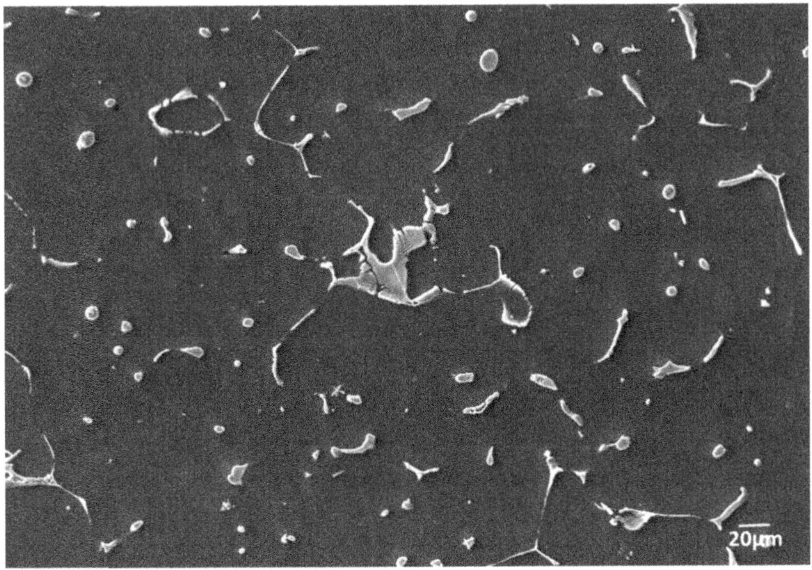

FIGURE 13.250 Microstructure of Mg-2Gd-6Zn magnesium alloy in as-cast condition taken at high magnification. $(Mg,Zn)_3Gd$ at the grain boundaries and grain interiors can be seen. The etchant is Acetic Picral.

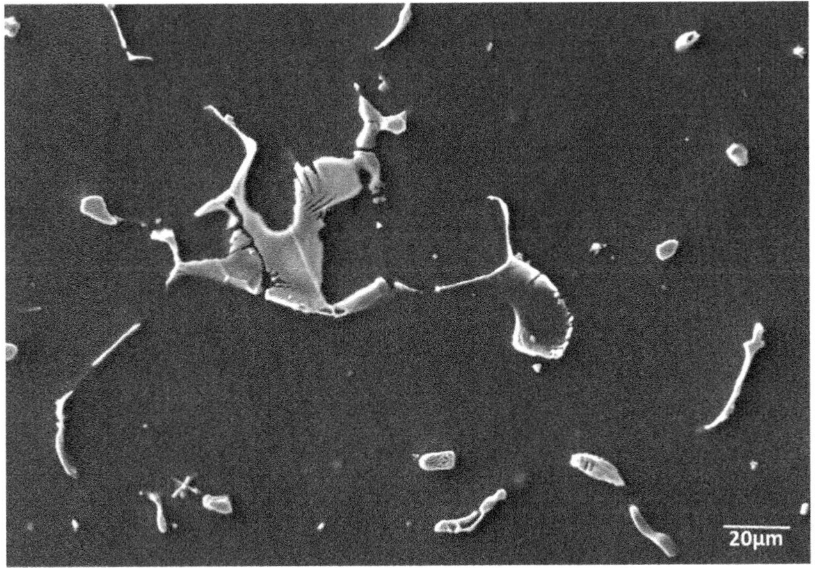

FIGURE 13.251 Microstructure of Mg-2Gd-6Zn magnesium alloy in as-cast condition taken at high magnification. $(Mg,Zn)_3Gd$ phase can be seen. The etchant is Acetic Picral.

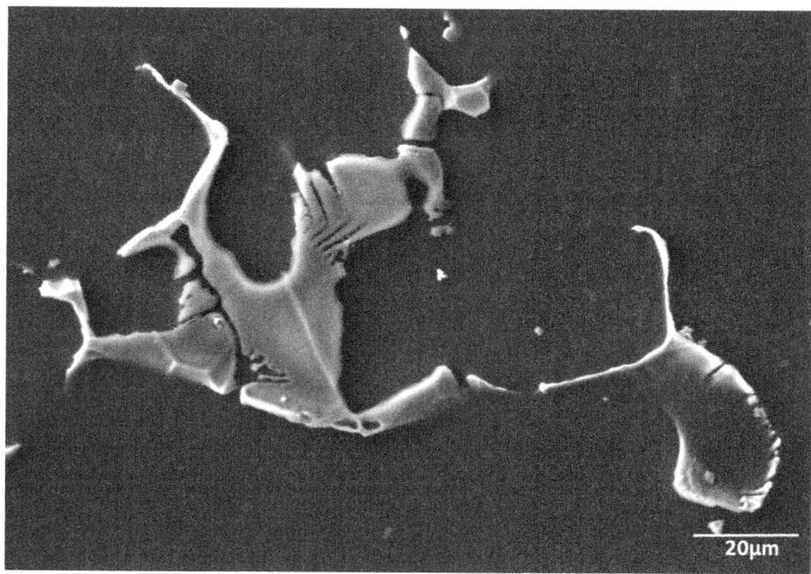

FIGURE 13.252 Microstructure of Mg-2Gd-6Zn magnesium alloy in as-cast condition taken at high magnification. $(Mg,Zn)_3Gd$ phase can be seen. The etchant is Acetic Picral.

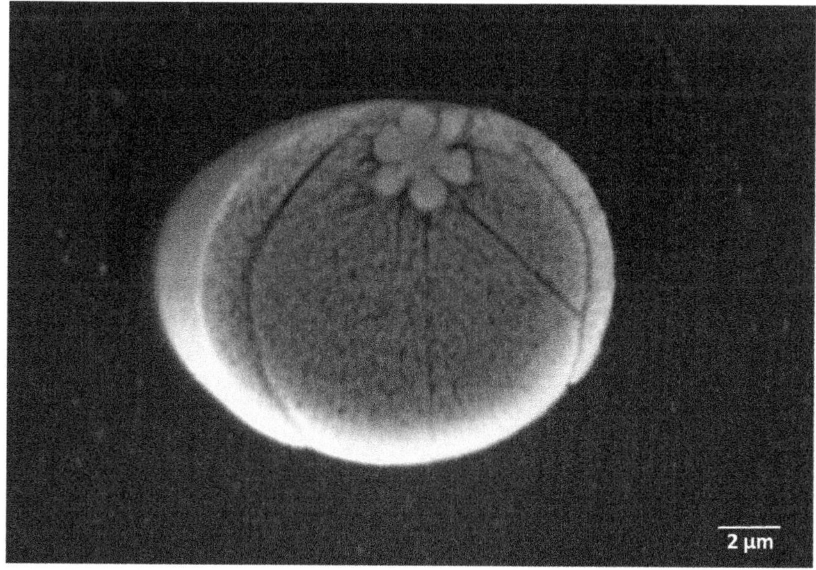

FIGURE 13.253 $(Mg,Zn)_3Gd$ particle seen as a flower in Mg-2Gd-6Zn magnesium alloy in as-cast condition. The etchant is Acetic Picral.

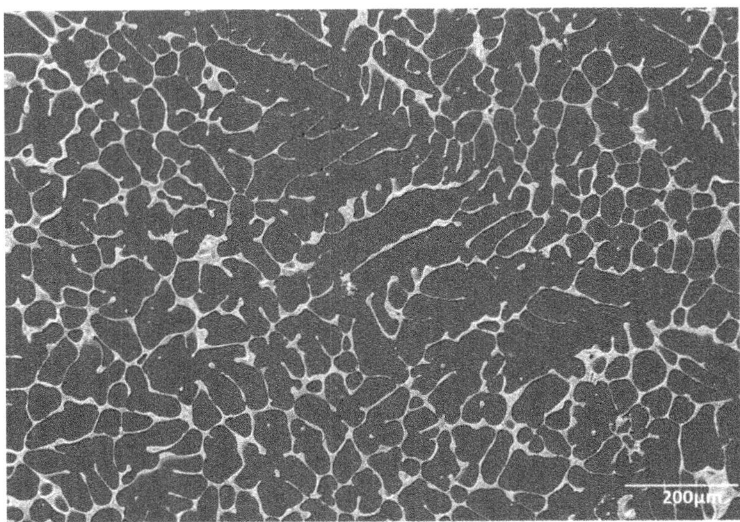

FIGURE 13.254 Microstructure of Mg-10Gd-6Zn magnesium alloy in as-cast condition. $(Mg,Zn)_3Gd$ at the grain boundaries and grain interiors can be seen. The etchant is Acetic Picral.

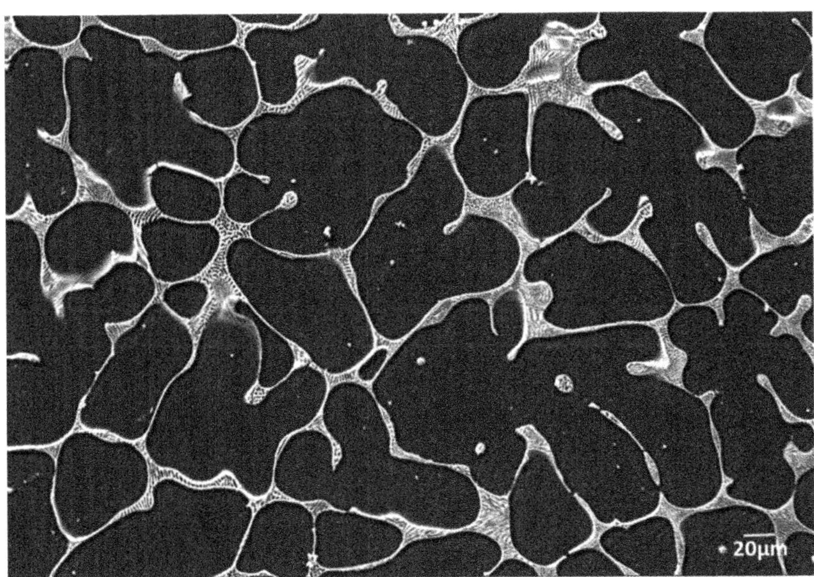

FIGURE 13.255 Microstructure of Mg-10Gd-6Zn magnesium alloy in as-cast condition taken at high magnification. $(Mg,Zn)_3Gd$ at the grain boundaries and grain interiors can be seen. The etchant is Acetic Picral.

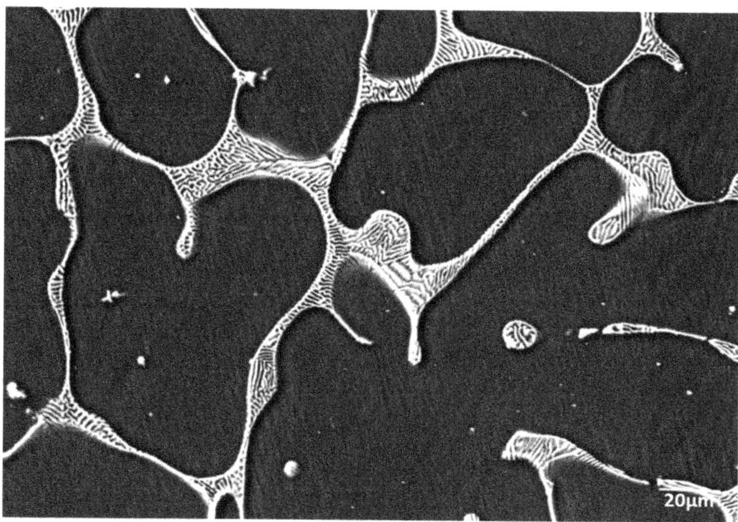

FIGURE 13.256 Microstructure of Mg-10Gd-6Zn magnesium alloy in as-cast condition taken at high magnification. $(Mg,Zn)_3Gd$ at the grain boundaries and grain interiors can be seen. The etchant is Acetic Picral.

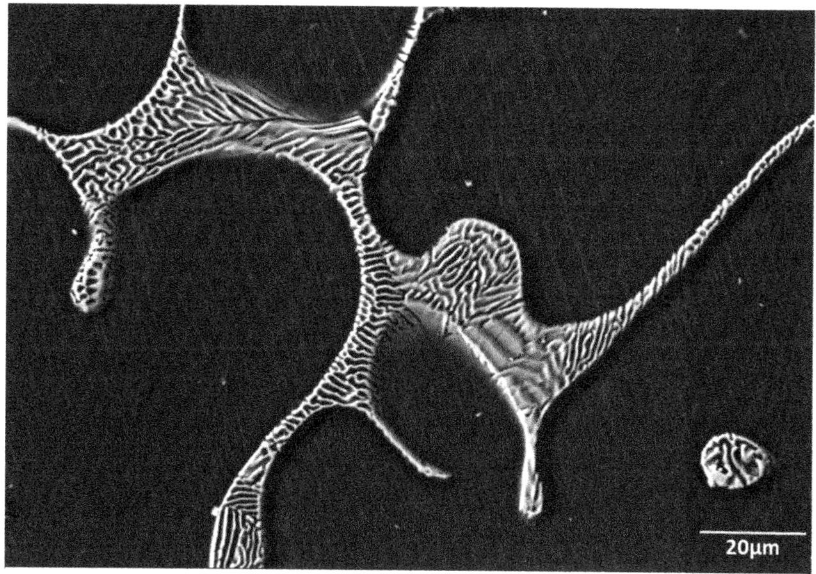

FIGURE 13.257 Microstructure of Mg-10Gd-6Zn magnesium alloy in as-cast condition taken at high magnification. $(Mg,Zn)_3Gd$ at the grain boundaries and grain interiors can be seen. The etchant is Acetic Picral.

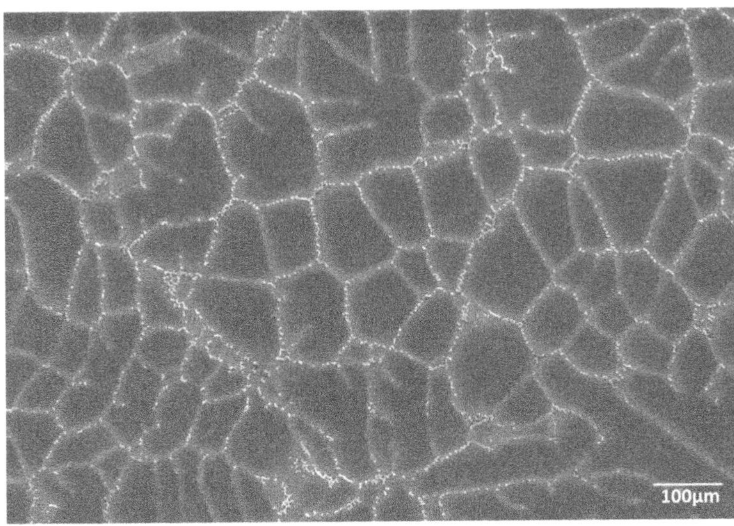

FIGURE 13.258 Microstructure of Mg-4Ag magnesium alloy in as cast condition. Precipitation of Mg_4Ag at the grain boundaries can be seen. The etchant is Acetic Picral.

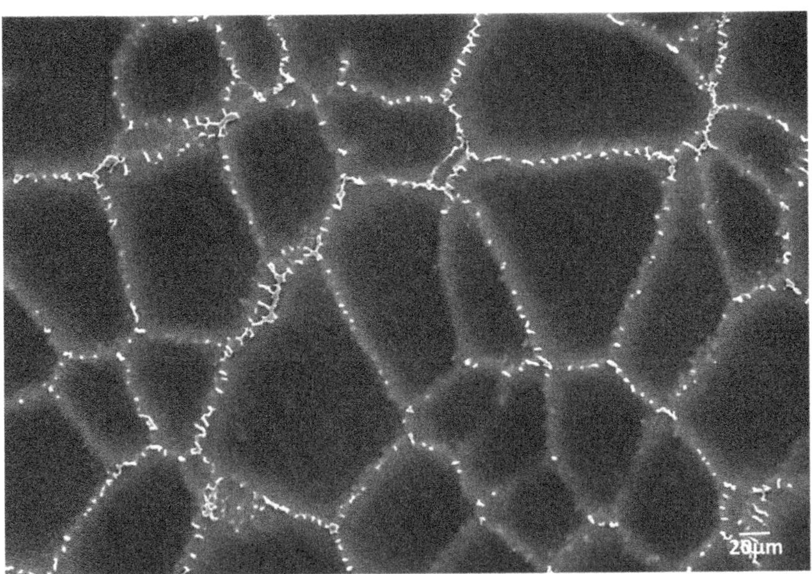

FIGURE 13.259 Microstructure of Mg-4Ag magnesium alloy in as-cast condition taken at high magnification. Precipitation of Mg_4Ag at the grain boundaries and silver rich light regions along the grain boundaries can be seen. The etchant is Acetic Picral.

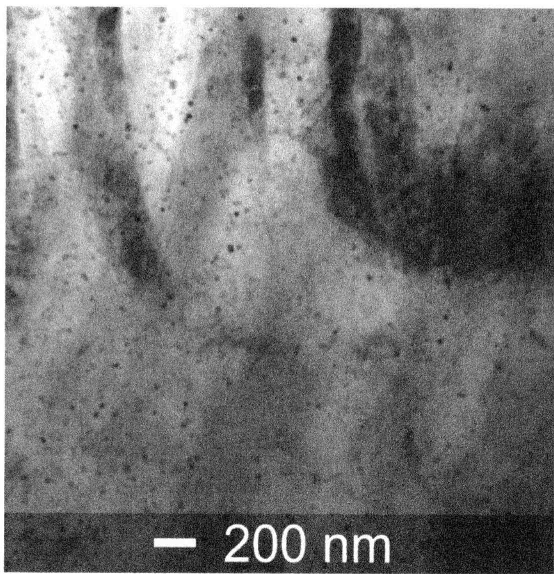

FIGURE 13.260 BFTEM image using objective aperture (10 μm) showing spherical/nodular Al_8Mn_5 precipitates in AZ31 alloy in the as-cast condition.

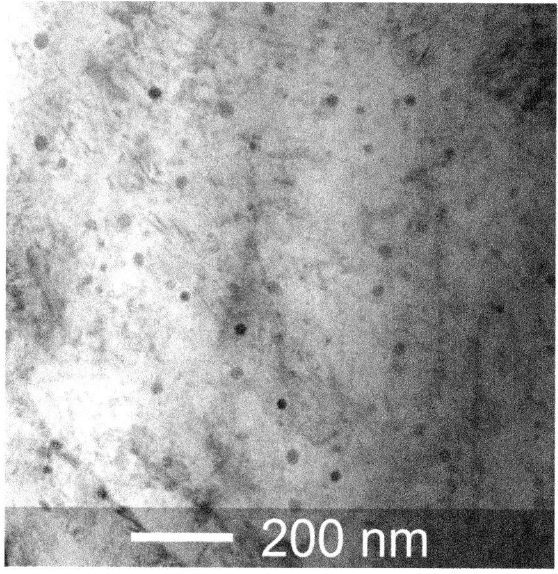

FIGURE 13.261 BFTEM image using objective aperture (10 μm) showing the precipitates of Al_8Mn_5 in AZ31 alloy in the as-cast condition.

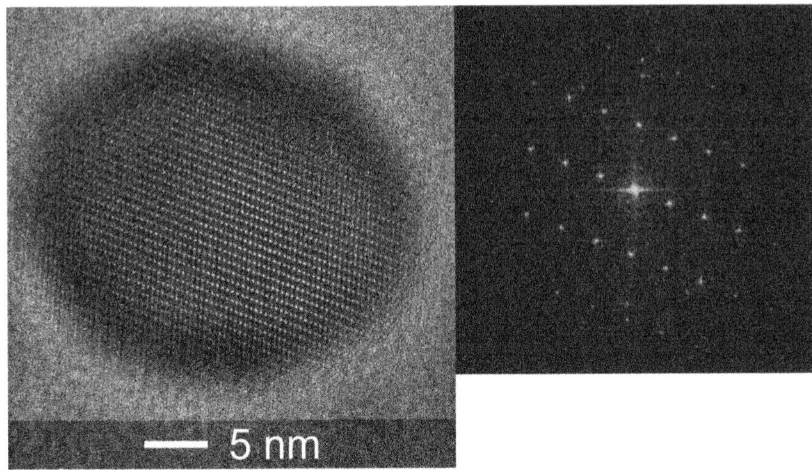

FIGURE 13.262 HRTEM micrograph with the corresponding fast Fourier transformation of an individual precipitate in AZ31 alloy in the as-cast condition.

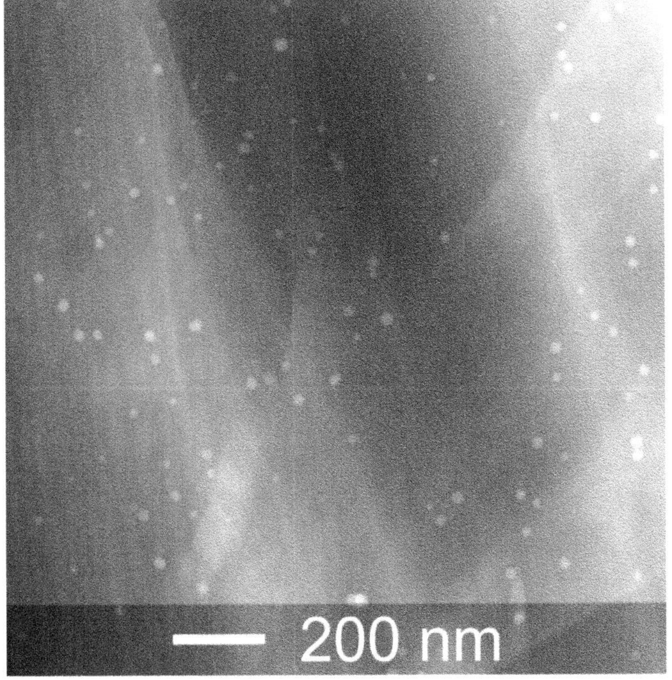

FIGURE 13.263 STEM-HAADF image showing the Z-contrast of Al_8Mn_5 precipitates in AZ31 alloy in the as-cast condition.

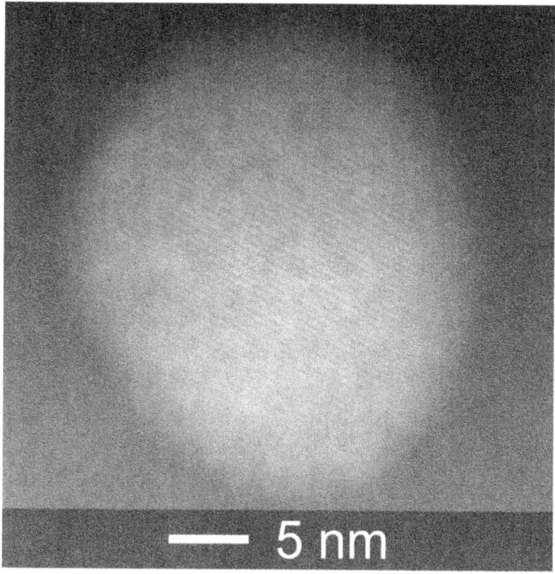

FIGURE 13.264 HRSTEM Image showing the lattice planes of an individual Al_8Mn_5 precipitate in AZ31 alloy in the as-cast condition.

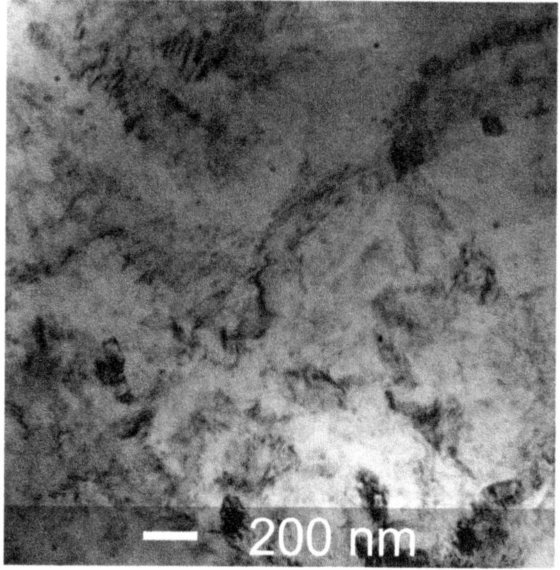

FIGURE 13.265 BFTEM image using objective aperture (10 μm) showing the Al_8Mn_5 precipitates and deformation bands in 4 mm thick-rolled AZ31 alloy sheet.

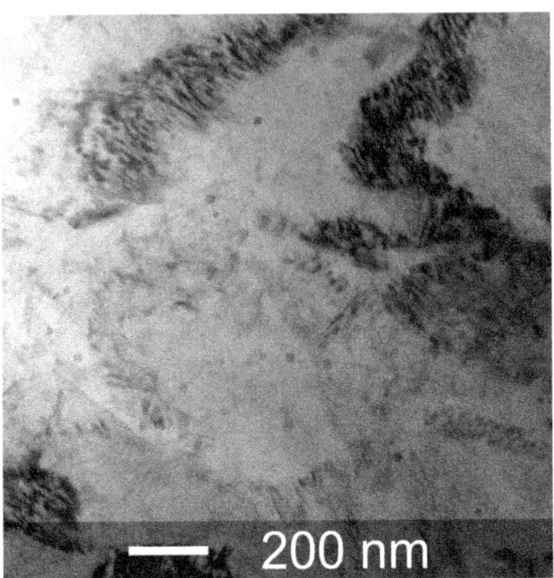

FIGURE 13.266 BFTEM image using objective aperture (10 μm) showing the Al_8Mn_5 precipitates in 4 mm thick-rolled AZ31 alloy sheet. Notice also the variation in contrast among the precipitates.

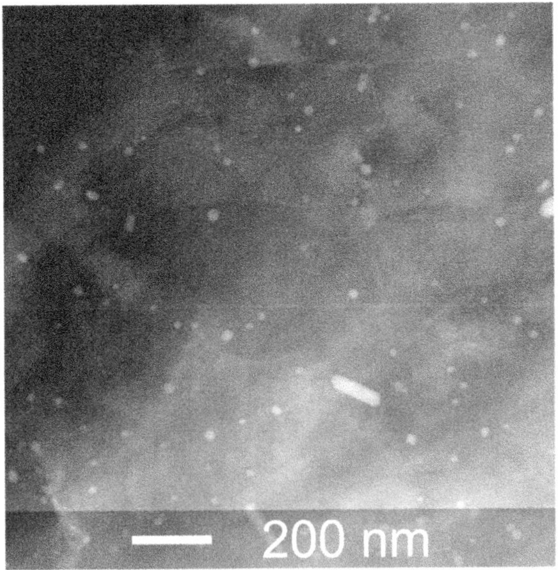

FIGURE 13.267 STEM-HAADF image showing the bright spherical/nodular Al_8Mn_5 precipitates in 4 mm thick-rolled AZ31 alloy sheet. Note that some of the spherical precipitates are on the grain boundary.

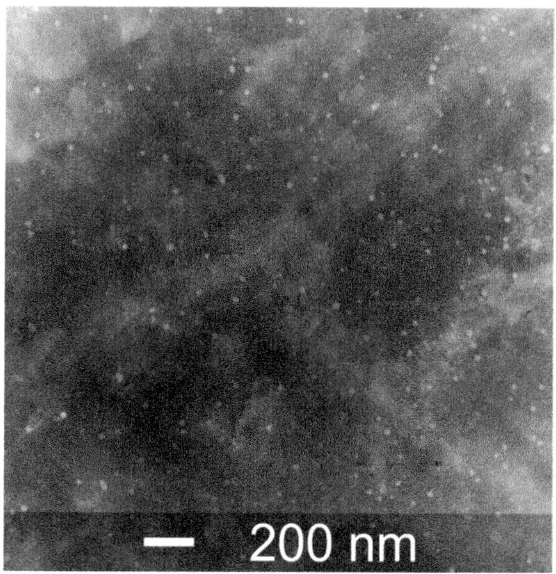

FIGURE 13.268 STEM-HAADF image showing the bright, spherical Al_8Mn_5 precipitates in 4 mm thick-rolled AZ31 alloy sheet.

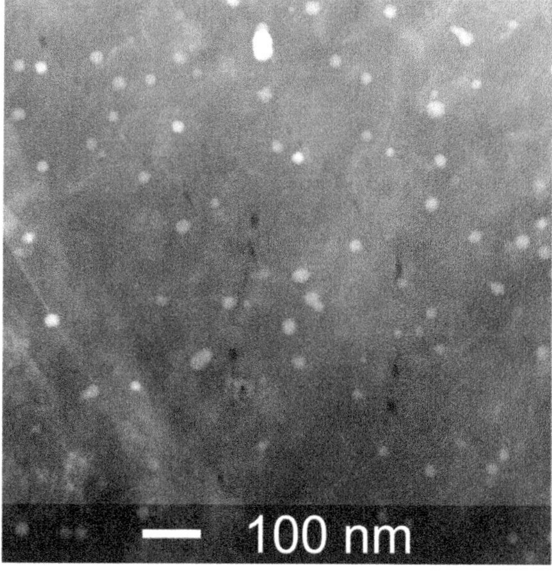

FIGURE 13.269 STEM-HAADF image showing the selective removal of precipitates from the dark pits (during preparation) in the 4 mm thick-rolled AZ31 alloy sheet.

FIGURE 13.270 STEM-HAADF image showing the overview of grains in AZ91 alloy in F condition. Average grain size 125 ± 10 nm.

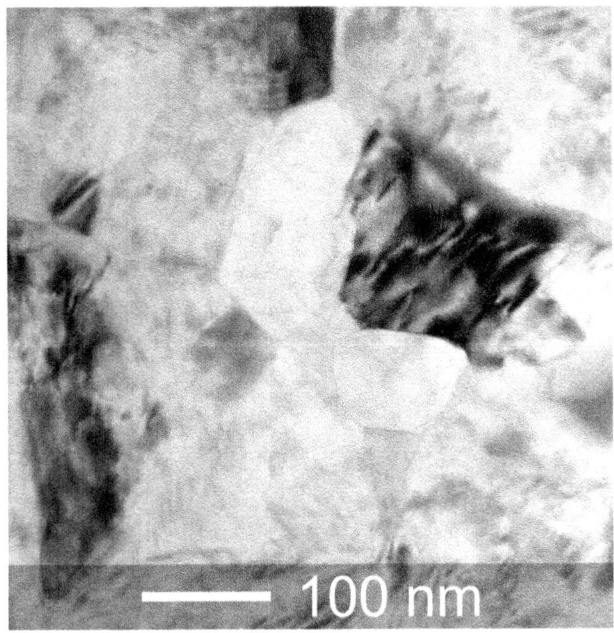

FIGURE 13.271 BFTEM image using objective aperture (10 μm) showing the grains of AZ91E alloy in F condition.

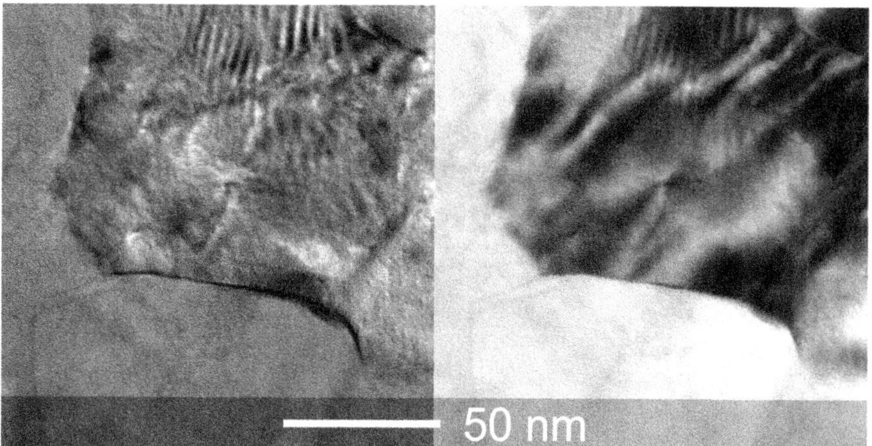

FIGURE 13.272 (Left) BFTEM image without objective aperture; (Right) BFTEM image using objective aperture (10 μm) showing the variation in contrast of the same grain of AZ91E alloy in F condition.

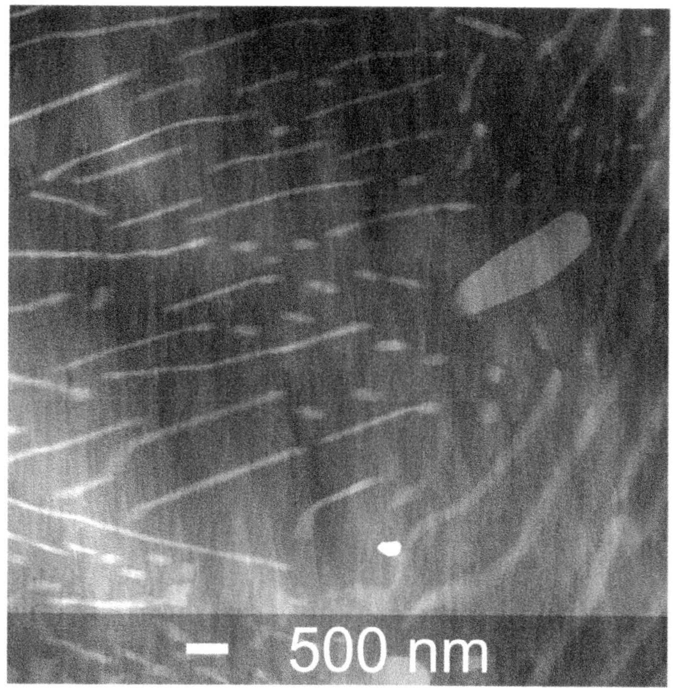

FIGURE 13.273 STEM-HAADF image showing continuous precipitates in AZ91E-T4 condition.

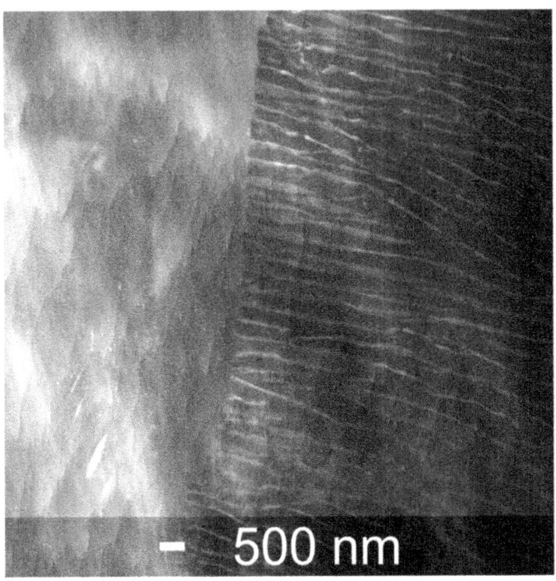

FIGURE 13.274 STEM-HAADF images showing lamellar eutectic of $Mg_{17}Al_{12}$, Al-Mn rich particles at an area where continuous precipitates are located along a grain boundary in AZ91E-T4 condition.

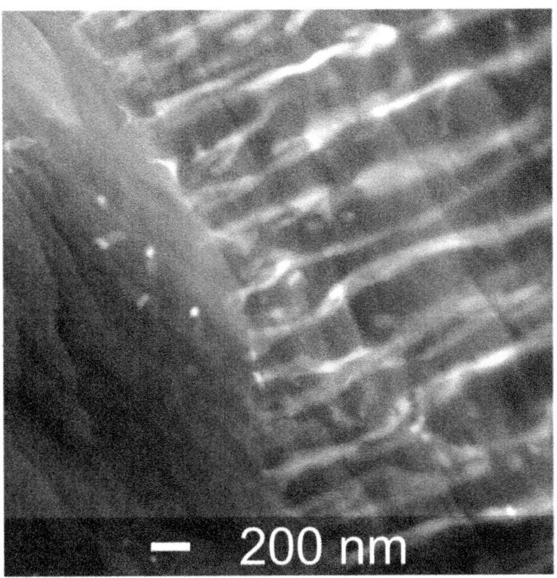

FIGURE 13.275 HAADF-STEM images showing lamellar eutectic of $Mg_{17}Al_{12}$ and Al-Mn rich particles at an area where continuous precipitates are located along a grain boundary in AZ91E-T4 condition.

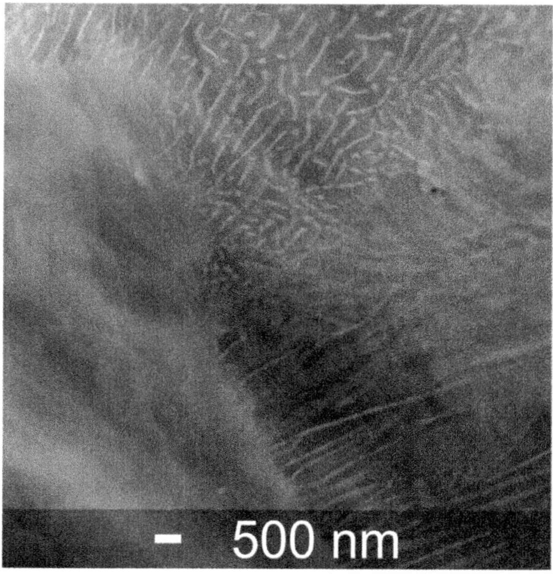

FIGURE 13.276 STEM-HAADF image showing the overview of the boundary, lamellar eutectic of $Mg_{17}Al_{12}$ according to the orientation of the precipitates.

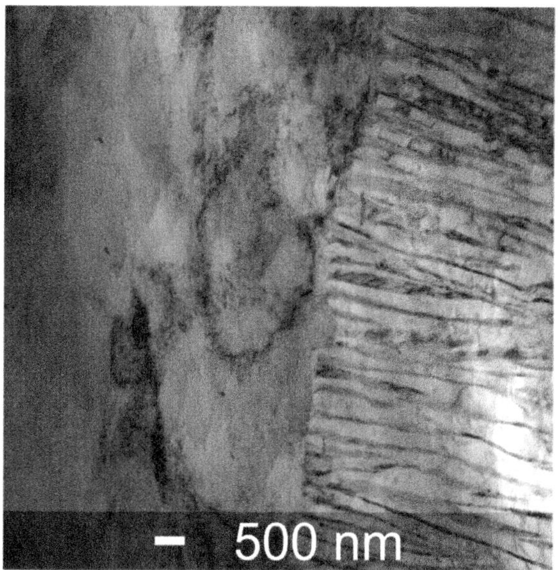

FIGURE 13.277 BFTEM image using objective aperture (10 μm) showing lamellar eutectic of $Mg_{17}Al_{12}$ in AZ91E-T4 condition. Also, note the presence of precipitates.

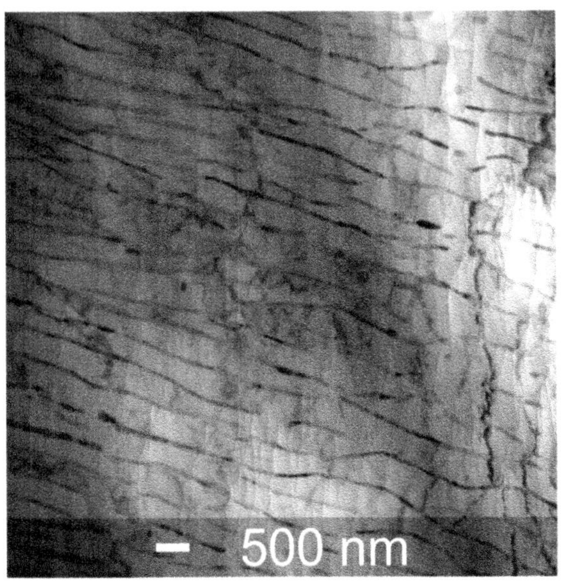

FIGURE 13.278 BFTEM image using objective aperture (10 μm) showing continuous precipitates in AZ91E-T4 condition.

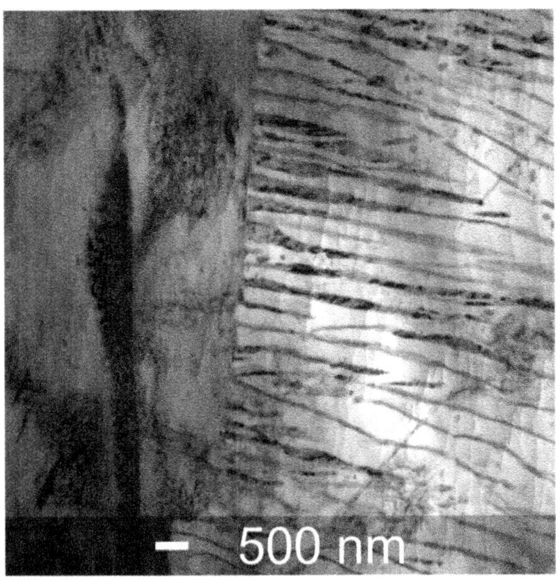

FIGURE 13.279 BFTEM image using objective aperture (10 μm) showing lamellar eutectic of $Mg_{17}Al_{12}$ in AZ91E-T4 condition.

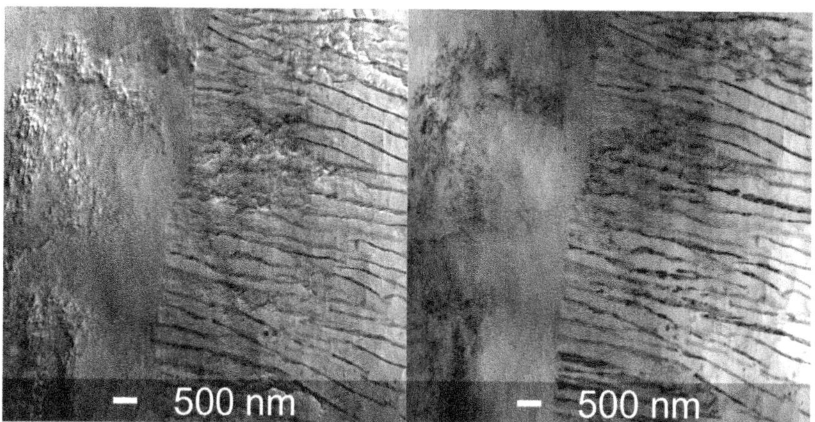

FIGURE 13.280 (Left) BFTEM image without objective aperture; (Right) BFTEM image using objective aperture (10 μm) showing the variation in contrast of the same lamellar eutectic area of $Mg_{17}Al_{12}$ in AZ91E – T4 condition.

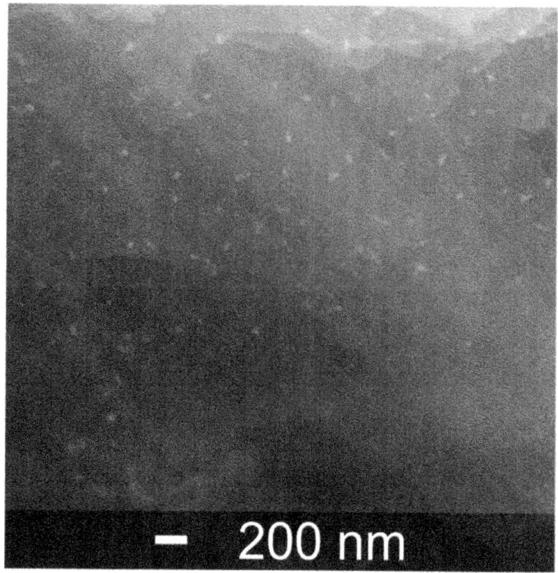

FIGURE 13.281 STEM-HAADF image showing the presence of irregular globular, short rod, long rod, precipitates in AZ91E – T6 condition.

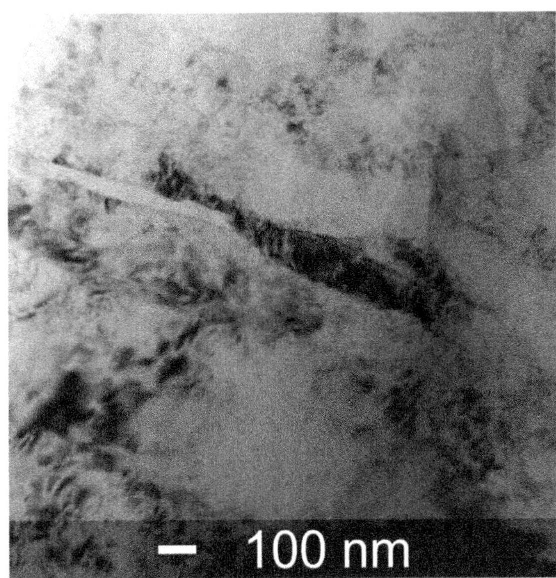

FIGURE 13.282 BFTEM image using objective aperture (10 μm) showing the twin ending at grain boundary in AZ91E – T6 condition.

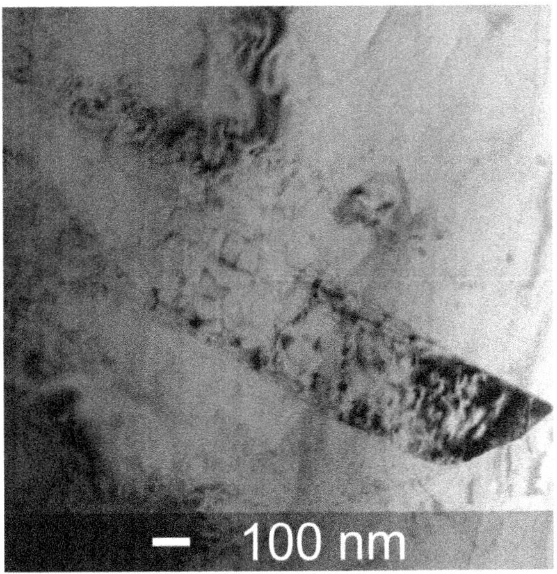

FIGURE 13.283 BFTEM image using objective aperture (10 μm) showing the twin ending at grain boundary in AZ91E – T6 condition.

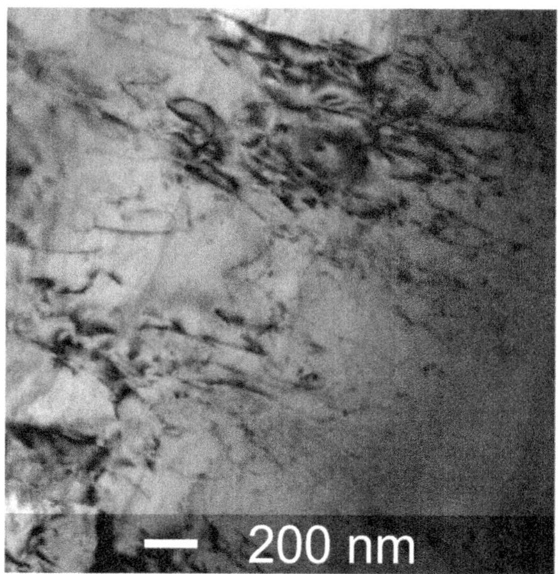

FIGURE 13.284 BFTEM image using objective aperture (10 μm) illustrates the dislocation network in the matrix of AZ91E – T6 condition.

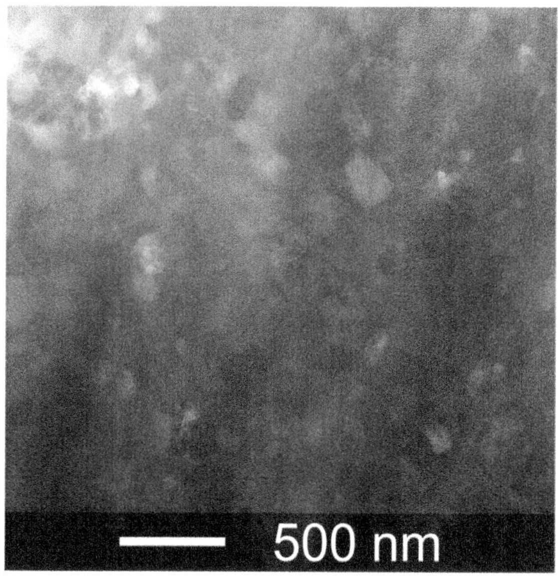

FIGURE 13.285 STEM-HAADF image showing the grain morphology along with the presence of fine precipitates in AE44. Average grain size 125 ∓ 10 nm.

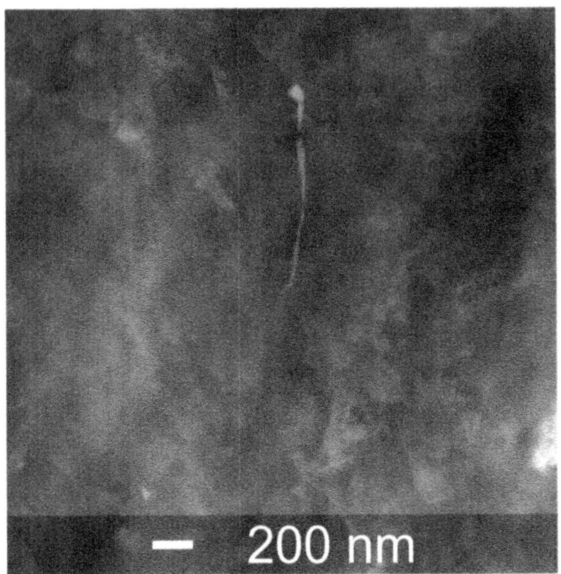

FIGURE 13.286 STEM-HAADF image showing Al-rich needle network along the grain boundary in AE44.

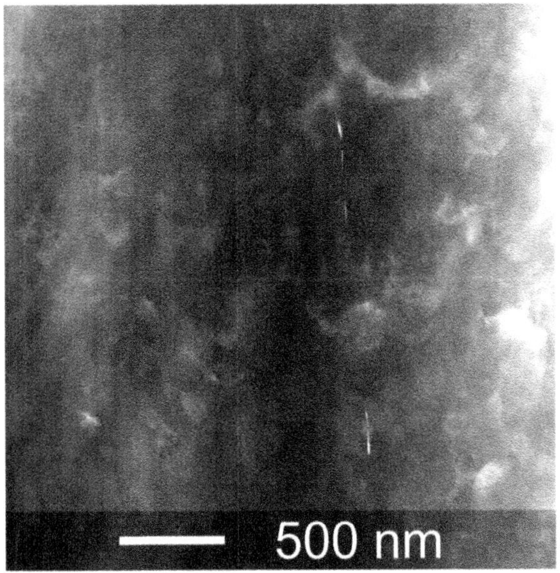

FIGURE 13.287 STEM-HAADF image showing Al-rich needle network along the grain boundary in AE44.

FIGURE 13.288 BFTEM image using objective aperture (10 μm) showing Al-rich needle network along the grain boundary in AE44.

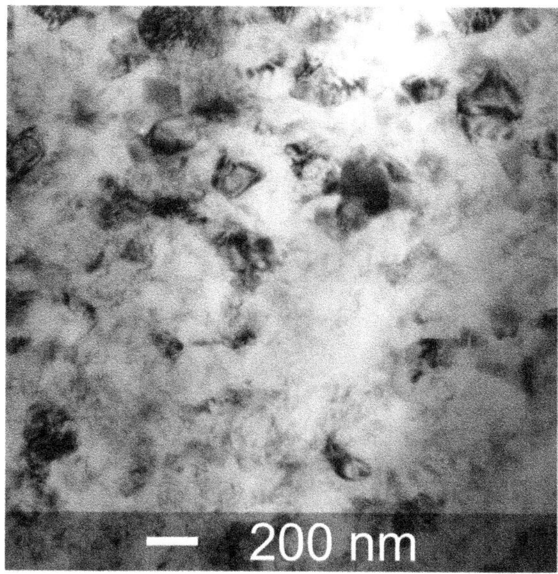

FIGURE 13.289 BFTEM image using objective aperture (10 μm) showing the typical grain morphology in AE44.

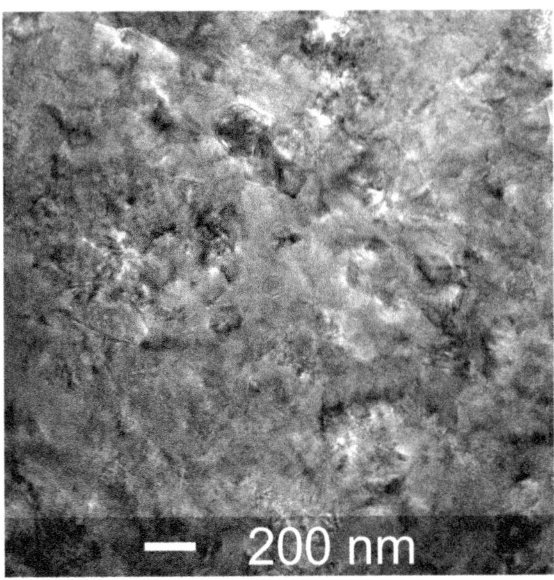

FIGURE 13.290 BFTEM image without using objective aperture showing the typical morphology in AE44.

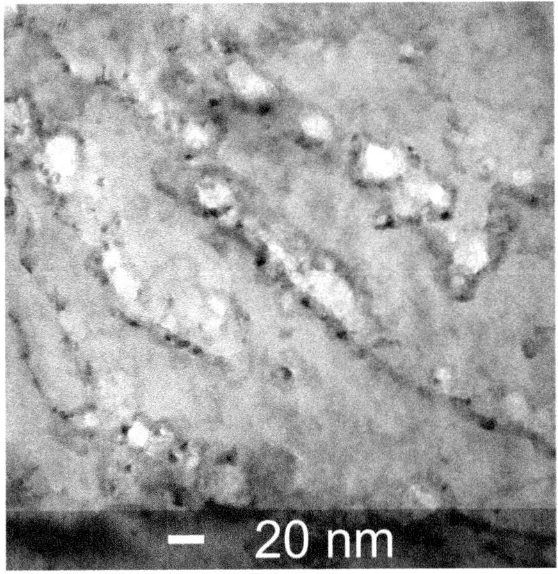

FIGURE 13.291 BFTEM image using objective aperture (10 μm) showing the typical morphology of fine, homogeneously distributed precipitates in DieMag422.

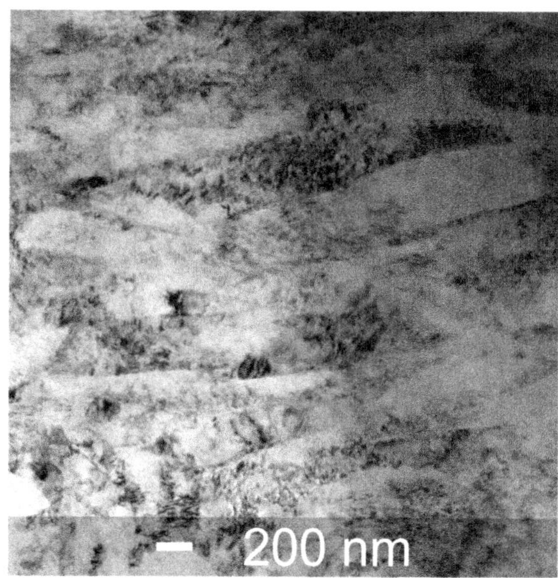

FIGURE 13.292 BFTEM image using objective aperture (10 μm) showing the typical grain morphology in DieMag422.

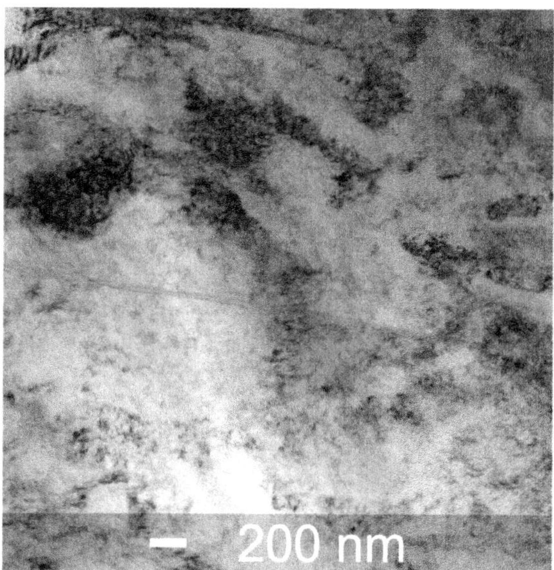

FIGURE 13.293 LM-TEM image using objective aperture (10 μm) showing long acicular precipitates in DieMag422. These long precipitates are prone to beam damage at high doses of electrons.

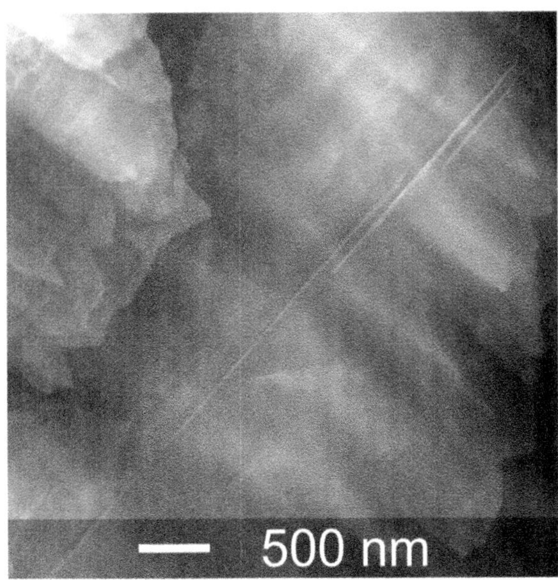

FIGURE 13.294 STEM-HAADF image showing elemental enrichment along the grain boundary in DieMag422.

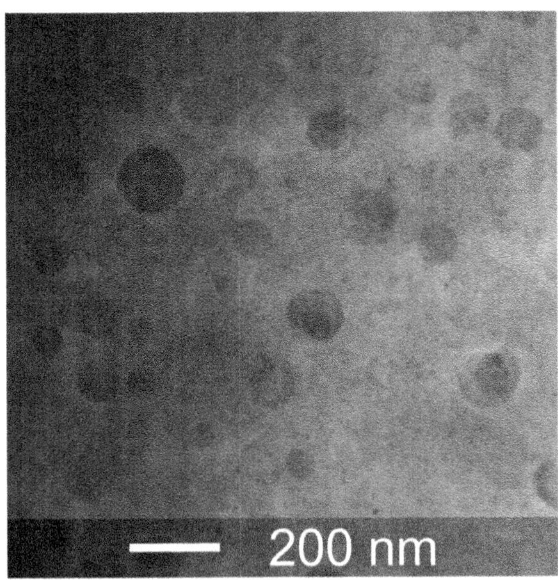

FIGURE 13.295 STEM-HAADF image showing the globular sample preparation artifacts (ion milling) in DieMag422.

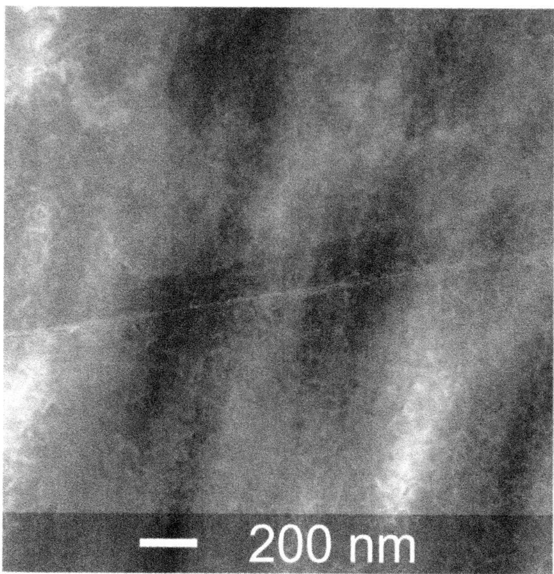

FIGURE 13.296 STEM-HAADF homogeneously distributed platelet-like precipitates in DieMag422.

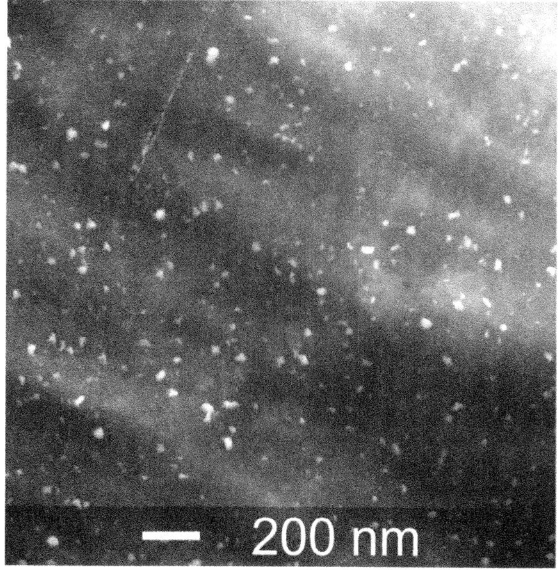

FIGURE 13.297 LM-STEM typically showing the morphology in DieMag422. Also, notice beam damage in the long precipitates.

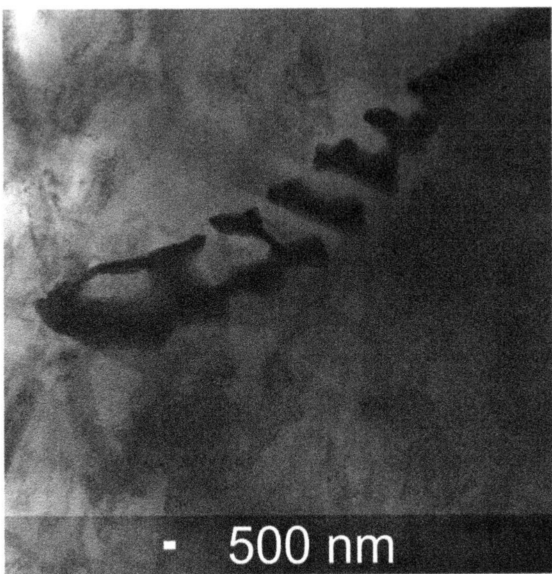

FIGURE 13.298 LM-STEM showing the Al_2Ca lamellar phase in DieMag422 alloy.

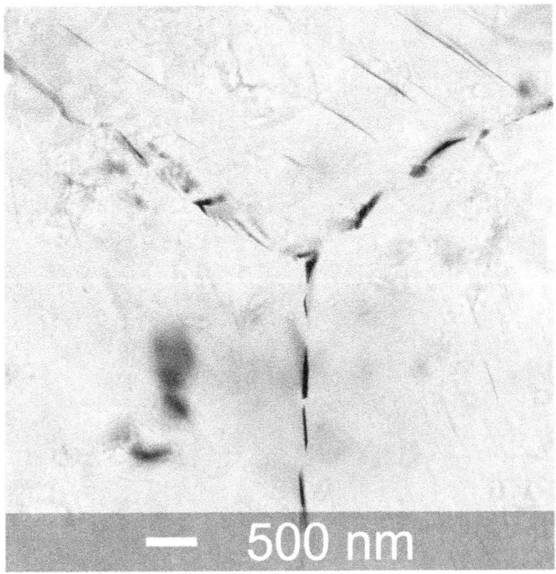

FIGURE 13.299 STEM-HAADF image showing enrichment along the grain boundaries of a triple junction as well as elongated platelet like precipitates in Elektron21.

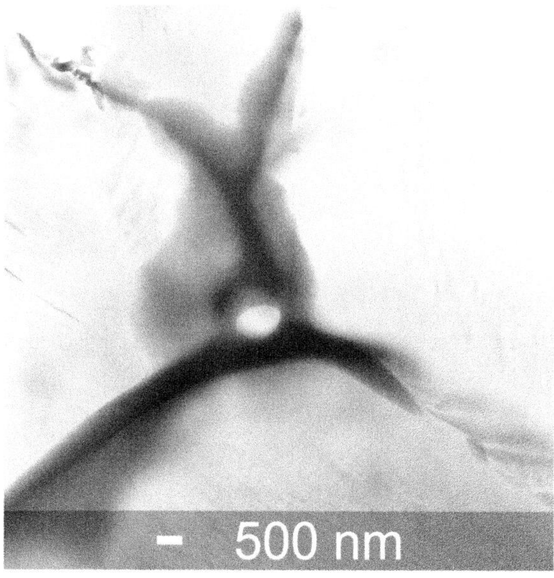

FIGURE 13.300 LM-STEM image showing enrichment along the grain boundaries of a triple junction as well as elongated precipitates in Elektron21.

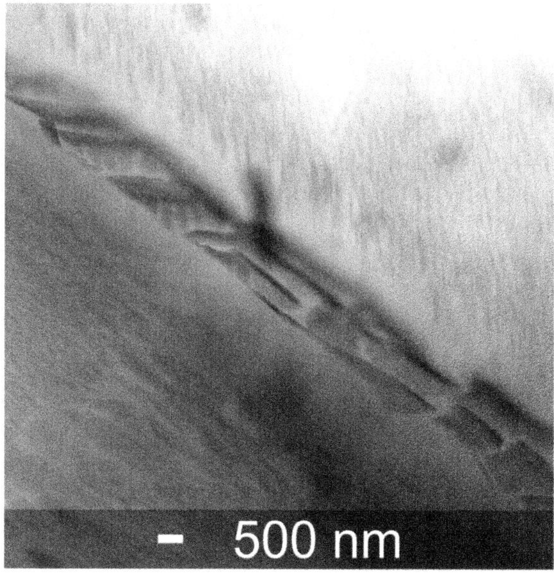

FIGURE 13.301 STEM-HAADF image of microstructure depicting the morphology at grain boundary in Elektron21 alloy.

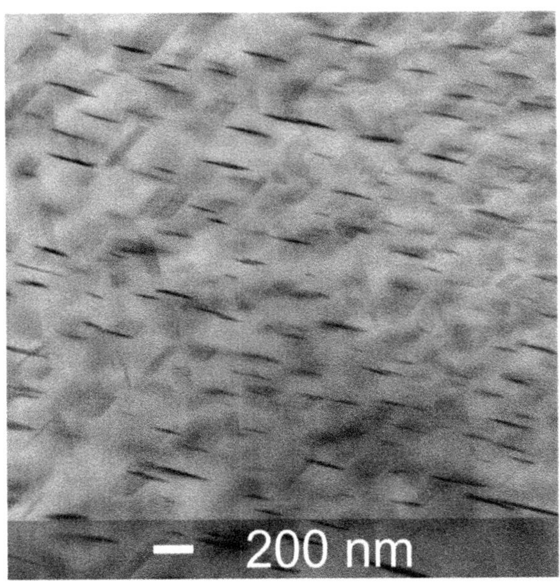

FIGURE 13.302 STEM-HAADF image of microstructure showing a family of high density of $Mg_{12}(Nd,Gd)$ precipitates in Elektron21 alloy.

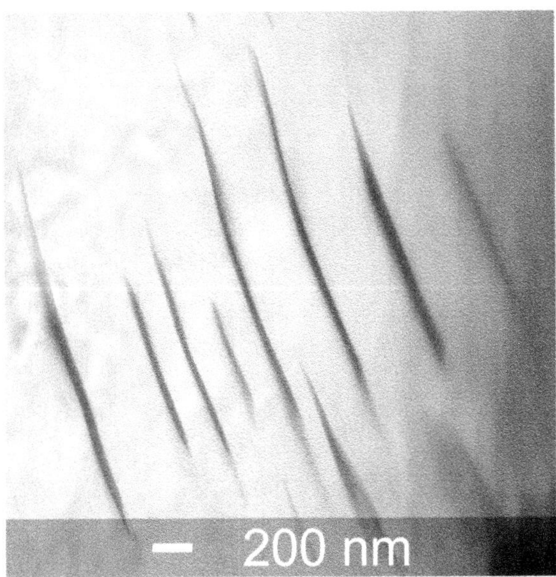

FIGURE 13.303 STEM-HAADF image of microstructure showing platelet like precipitates in Elektron21 alloy.

FIGURE 13.304 STEM-HAADF image of microstructure showing Mg_{12}(Nd,Gd) precipitates at grain boundary, two different orientations of Mg_{12}(Nd,Gd) precipitates is respective grains in Elektron21 alloy.

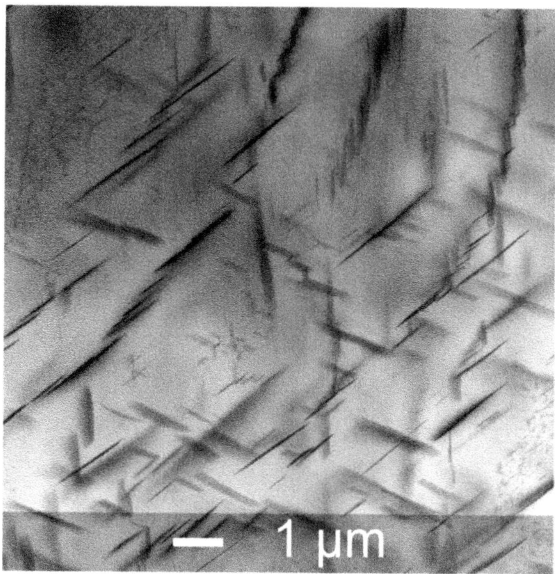

FIGURE 13.305 STEM-HAADF image depicting the presence of precipitates at dislocations in Elektron21 alloy.

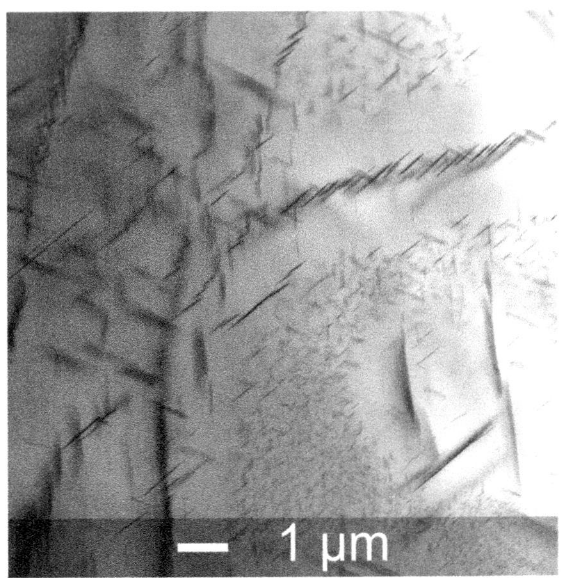

FIGURE 13.306 STEM-HAADF image depicting the presence of Mg_{12}(Nd,Gd) precipitates at dislocations in Elektron21. Also, notice the presence of precipitate free zone.

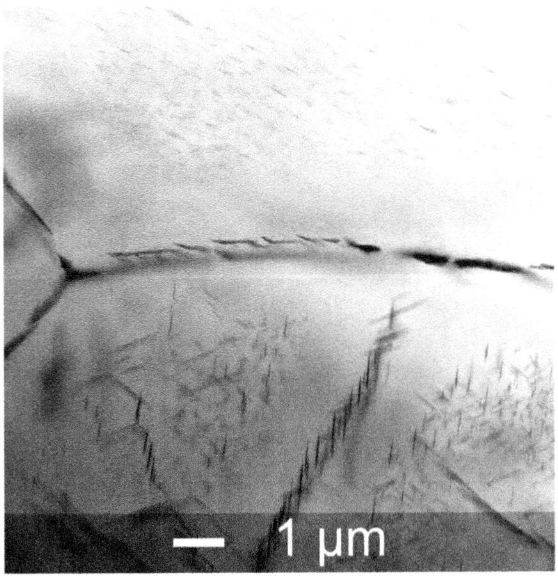

FIGURE 13.307 STEM-HAADF image depicting the presence of Mg_{12}(Nd,Gd) precipitates at dislocations in Elektron21. Also, notice the presence of precipitate free zone clearly.

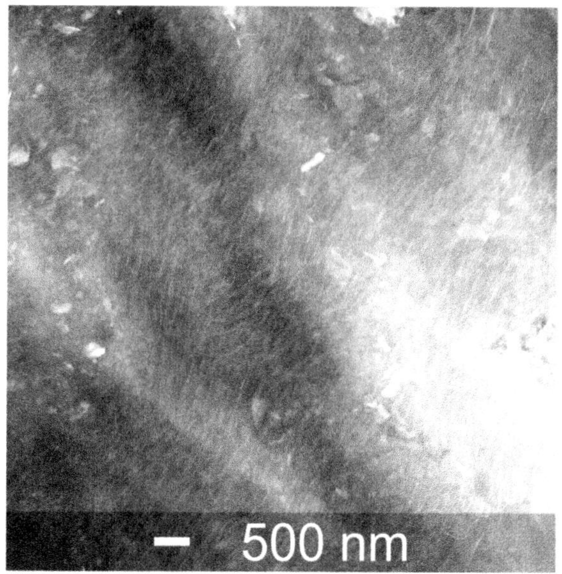

FIGURE 13.308 STEM-HAADF image depicting various precipitate morphologies in Elektron21 alloy.

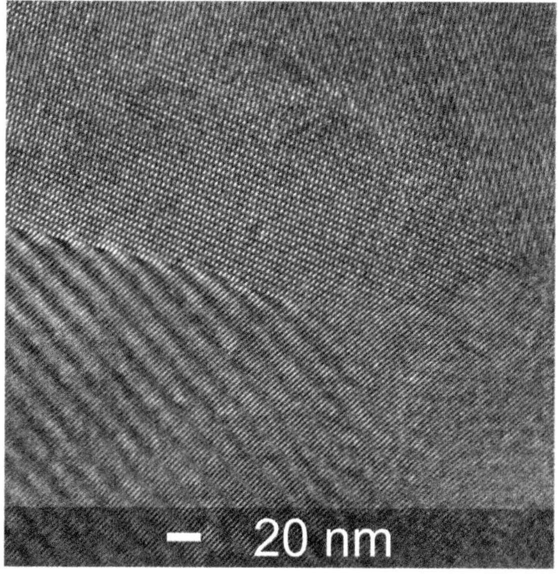

FIGURE 13.309 HRTEM micrograph of a grain boundary in Elektron21 alloy.

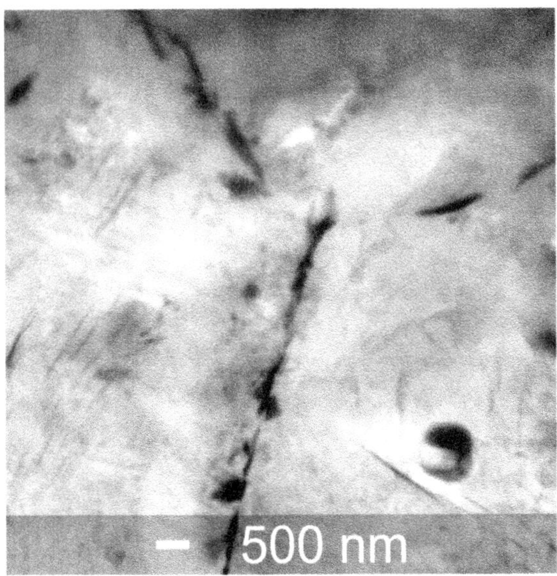

FIGURE 13.310 BFTEM image with objective aperture (10 μm) showing the presence of grain boundary precipitates at a triple junction in Elektron21 alloy.

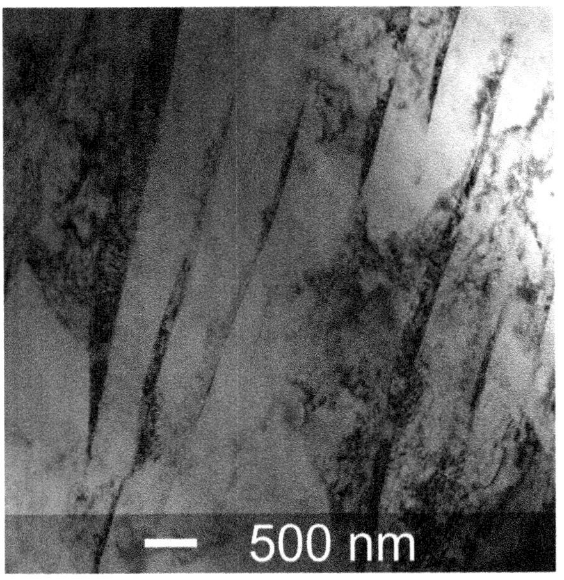

FIGURE 13.311 BFTEM image with objective aperture (10 μm) showing the typical morphology of MRI153M alloy.

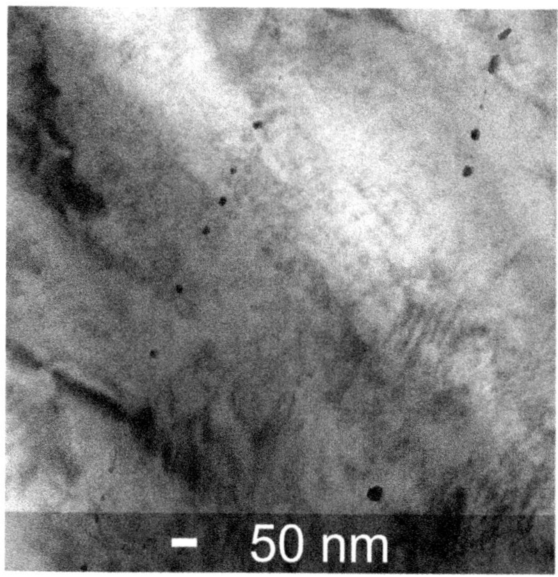

FIGURE 13.312 BFTEM image with objective aperture (10 μm) showing fine precipitates in MRI153M alloy.

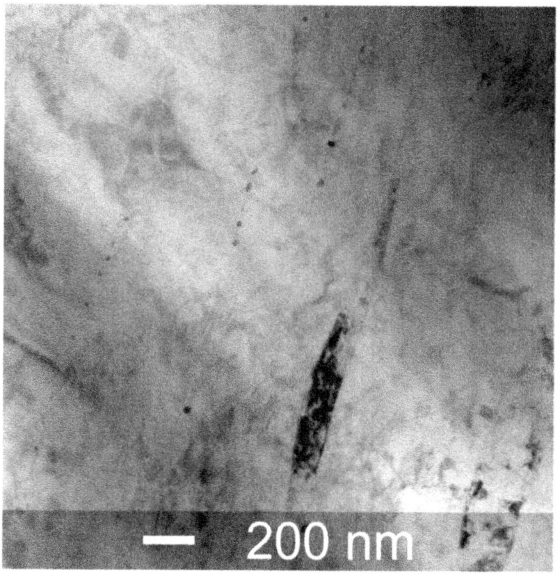

FIGURE 13.313 BFTEM image with objective aperture (10 μm) showing the presence of fine precipitates in MRI153M alloy.

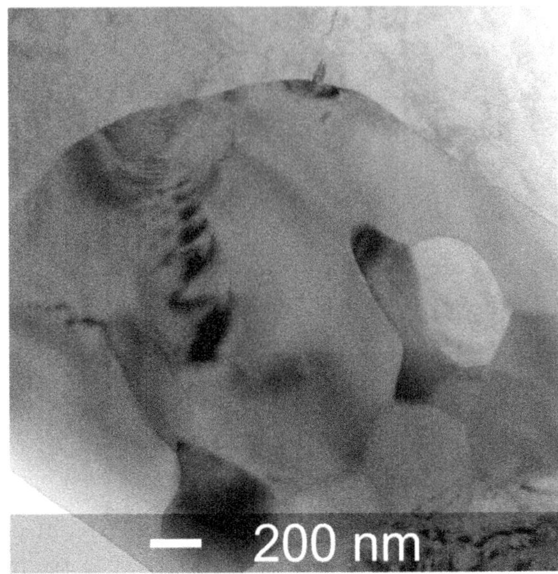

FIGURE 13.314 BFTEM image with objective aperture (10 μm) $(Mg,Al)_2Ca$ phase with equi-axed grains in MRI153M alloy.

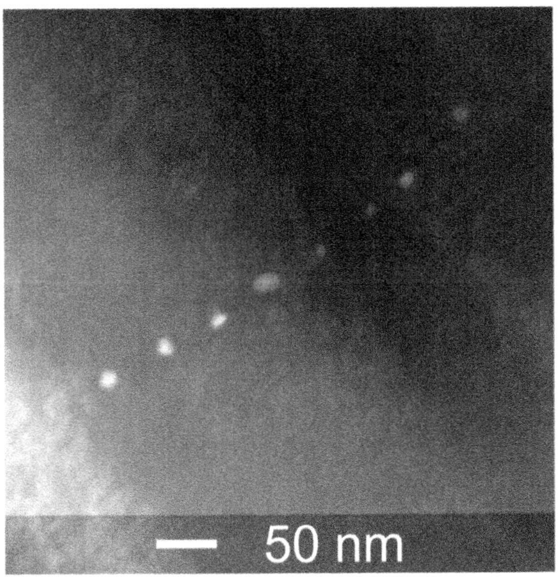

FIGURE 13.315 STEM-HAADF image showing fine spherical and elongated precipitates in MRI153M alloy.

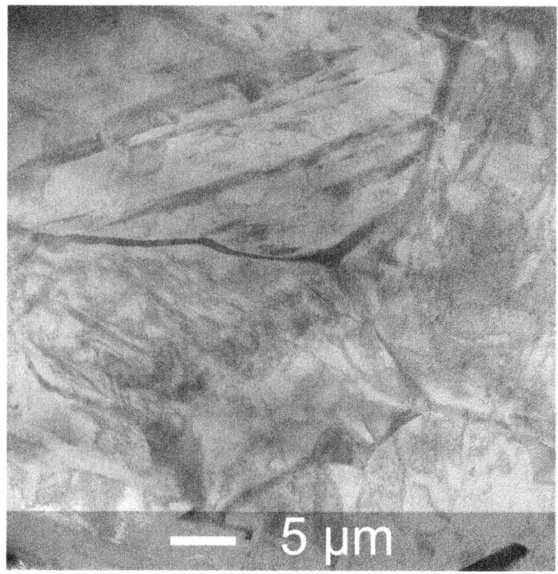

FIGURE 13.316 LM-STEM image showing the $(Mg,Al)_2Ca$ phase along the grain boundaries in MRI153M alloy.

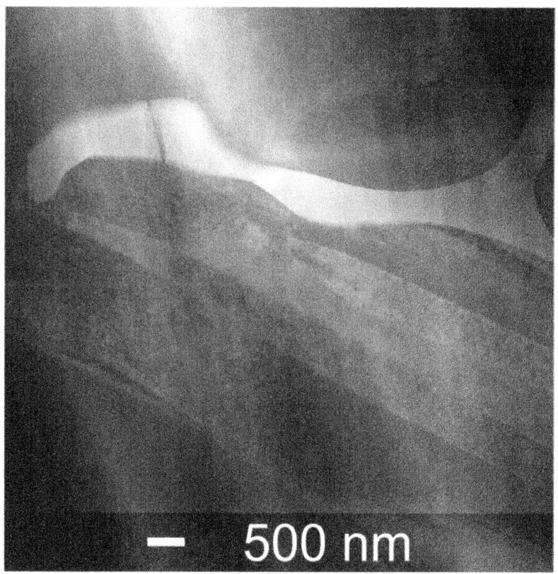

FIGURE 13.317 LM-STEM image showing deformation twins resulting from sample preparation along with $(Mg,Al)_2Ca$ phase at the grain boundary in MRI153M.

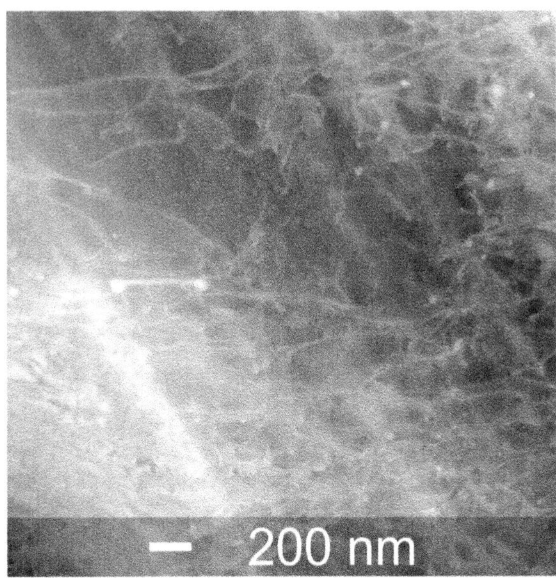

FIGURE 13.318 STEM-HAADF image showing forest of dislocations along with fine precipitates in MRI153M.

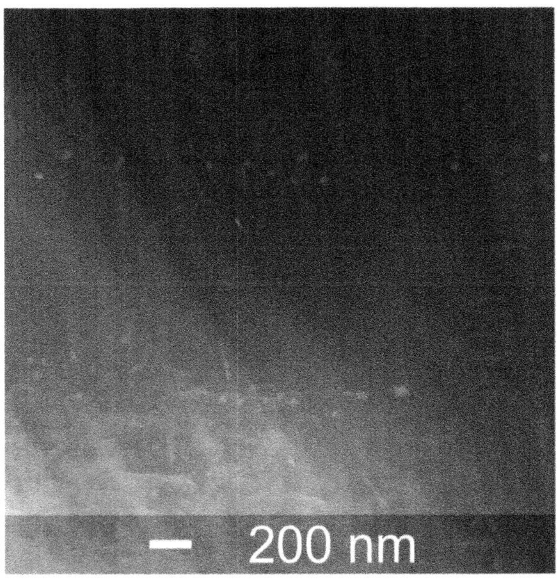

FIGURE 13.319 STEM-HAADF image of fine, near spherical and elongated precipitates in MRI153M alloy.

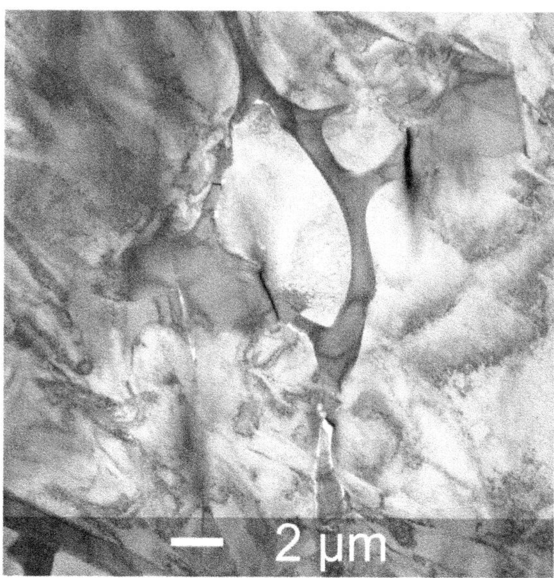

FIGURE 13.320 LM-STEM image showing network of $(Mg,Al)_2Ca$ phase at grain boundary in MRI230D alloy.

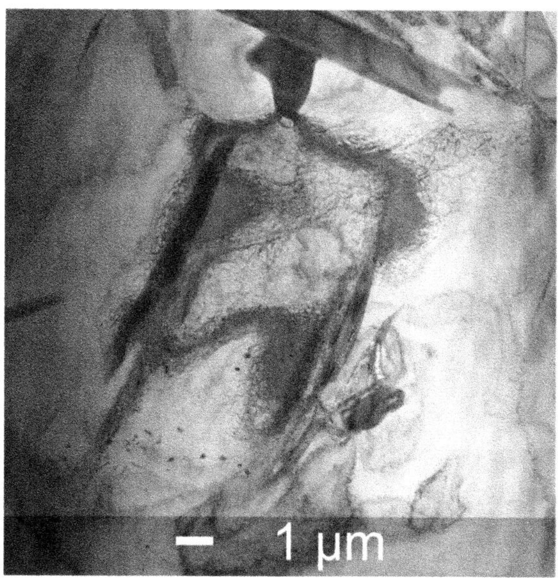

FIGURE 13.321 LM STEM image showing the distribution of fine precipitates in MRI230D alloy.

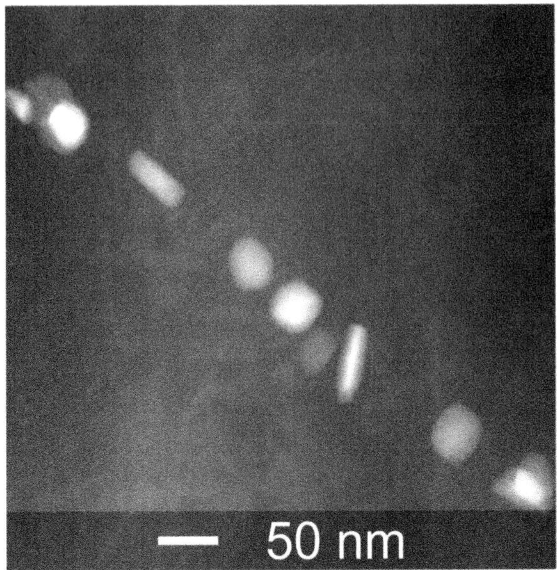

FIGURE 13.322 STEM-HAADF image showing fine spherical and elongated precipitates in MRI230D alloy.

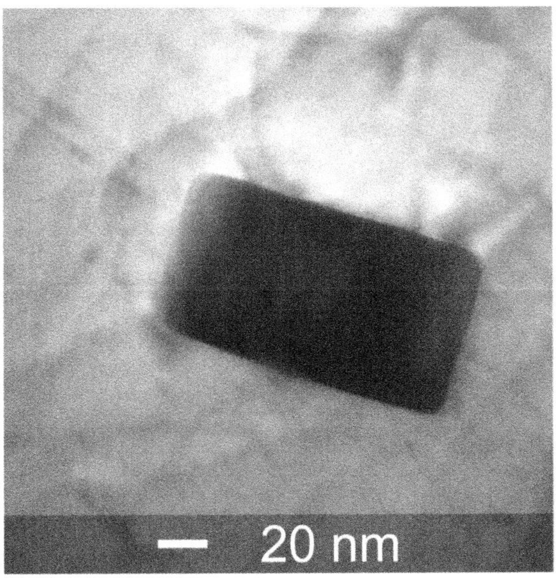

FIGURE 13.323 BFTEM image with objective aperture (10 μm) showing rectangular cross-section of a fine precipitate in MRI230D alloy.

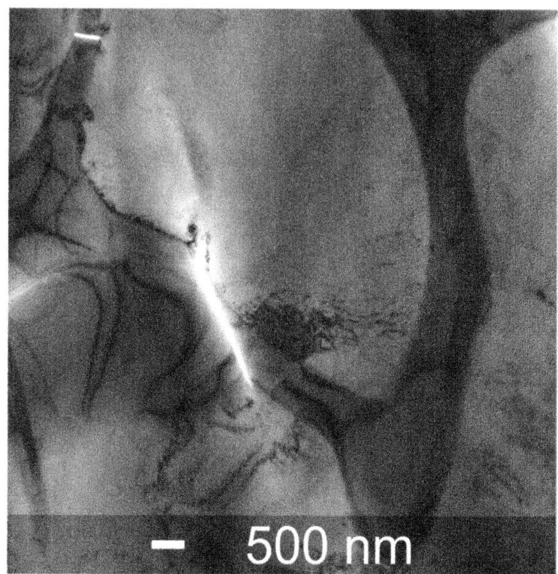

FIGURE 13.324 BFTEM image with objective aperture (10 μm) showing $(Mg,Al)_2Ca$ phase along the grain boundary in MRI230D alloy.

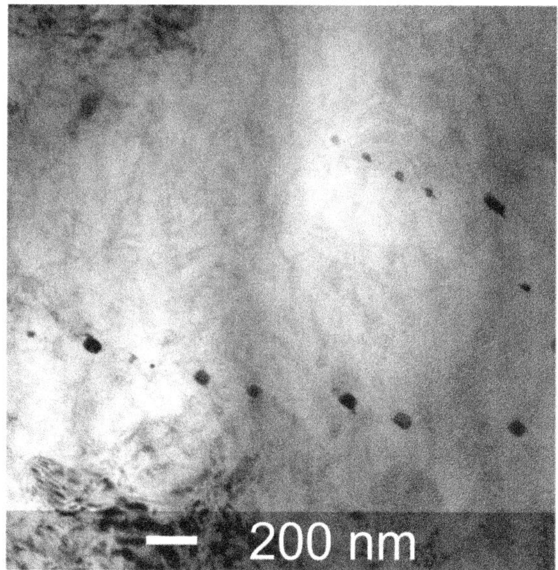

FIGURE 13.325 BFTEM image with objective aperture (10 μm) showing fine, near-spherical precipitates in MRI230D alloy.

FIGURE 13.326 HRTEM micrograph of an acicular precipitate in MRI230D alloy.

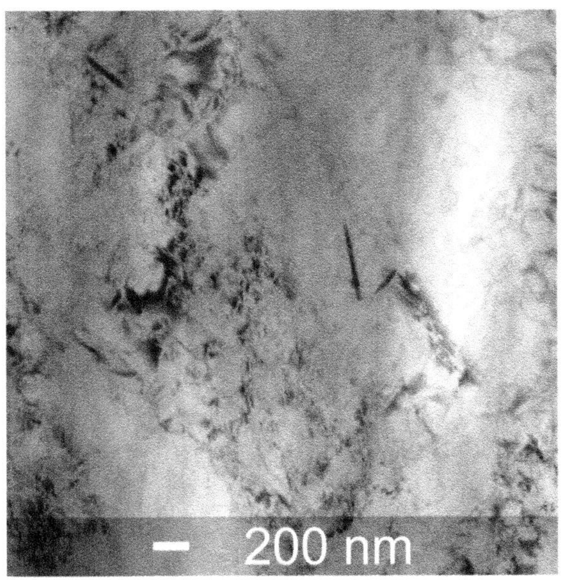

FIGURE 13.327 BFTEM image with objective aperture (10 μm) showing acicular precipitate in MRI230D alloy.

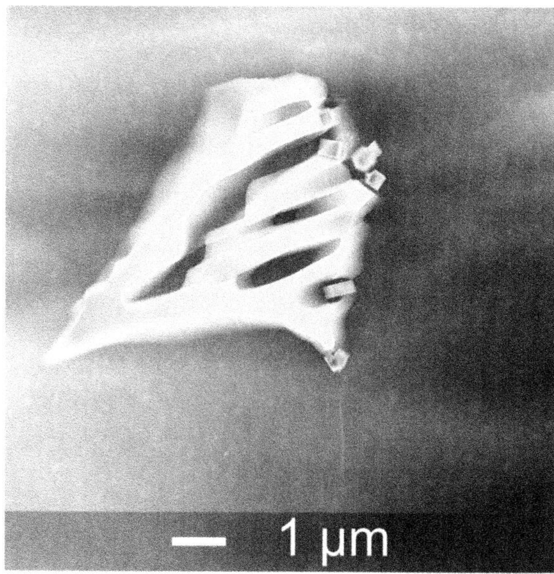

FIGURE 13.328 LM-STEM image showing secondary phase along with precipitates in WE54 alloy.

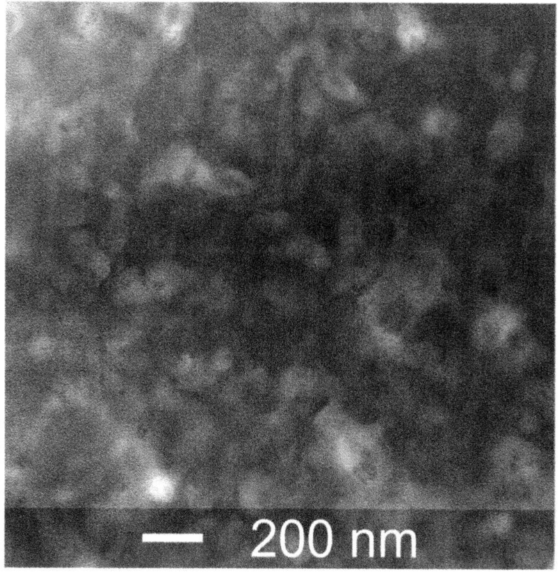

FIGURE 13.329 STEM-HAADF image showing beam damage in WE54 alloy.

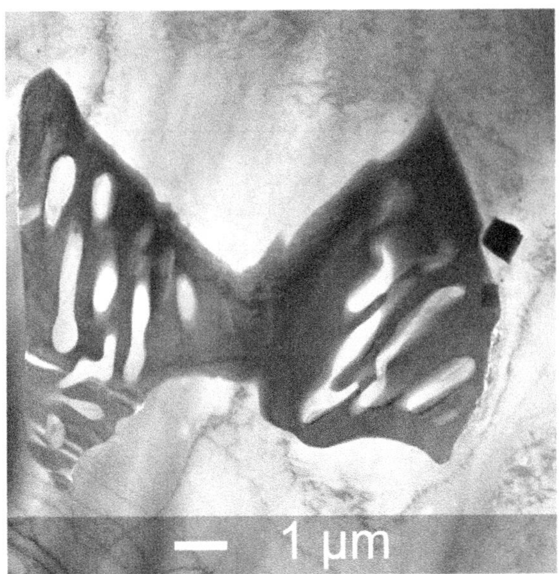

FIGURE 13.330 STEM-HAADF image showing the typical morphology of secondary phase in WE54 alloy.

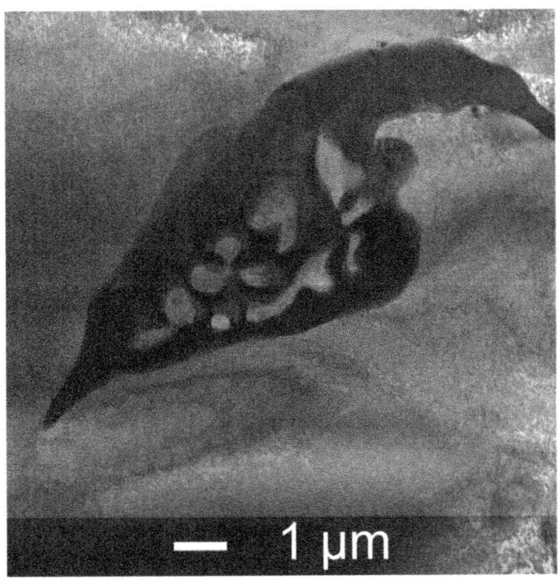

FIGURE 13.331 STEM-HAADF image showing the typical morphology of secondary phase in WE54 alloy.

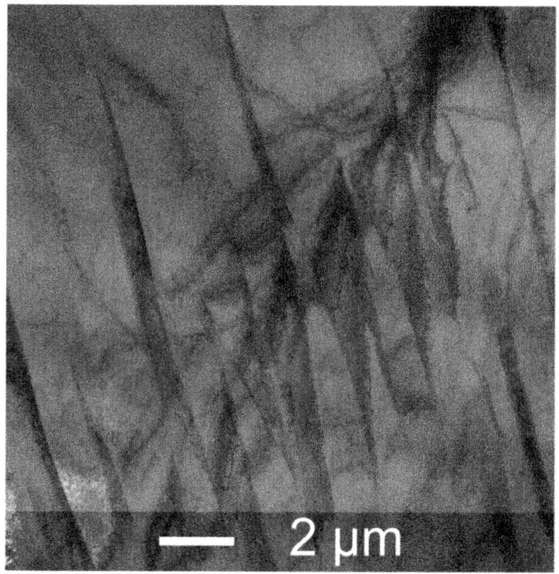

FIGURE 13.332 BFTEM image with objective aperture (10 μm) showing the typical morphology in WE54 alloy.

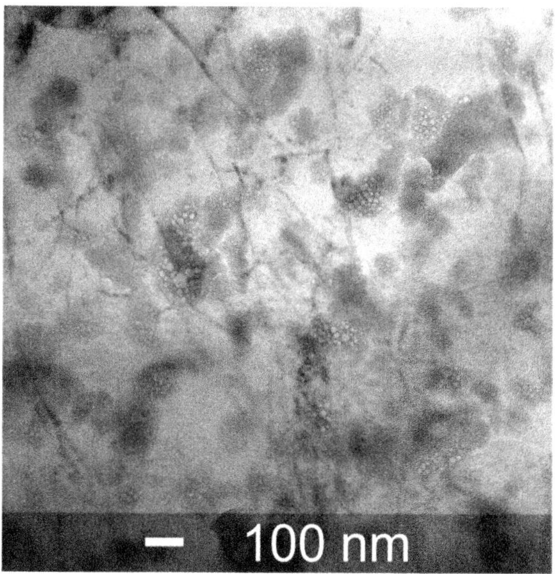

FIGURE 13.333 BFTEM image with objective aperture (10 μm) showing beam damage at an area with forest of dislocations.

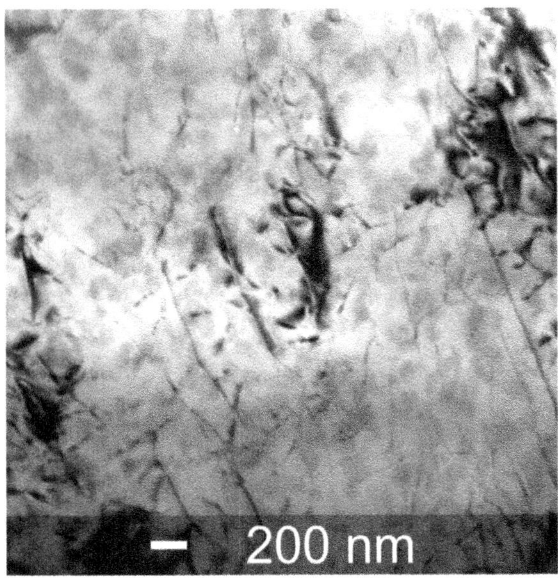

FIGURE 13.334 BFTEM image with objective aperture (10 μm) showing forest of dislocations along with precipitates on the dislocation loops.

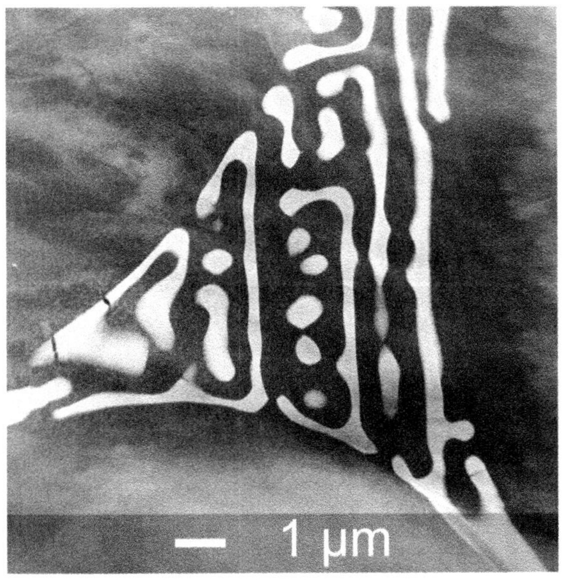

FIGURE 13.335 LM-STEM image showing Chinese script–type $Mg(Zn,Cu)_2$ phase in ZC63 alloy.

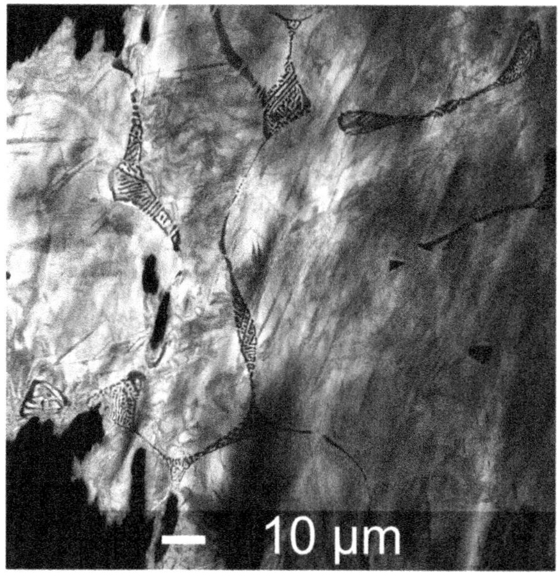

FIGURE 13.336 LM-STEM image showing an overview with various morphologies of Chinese script – $Mg(Zn,Cu)_2$ phases in ZC63 alloy.

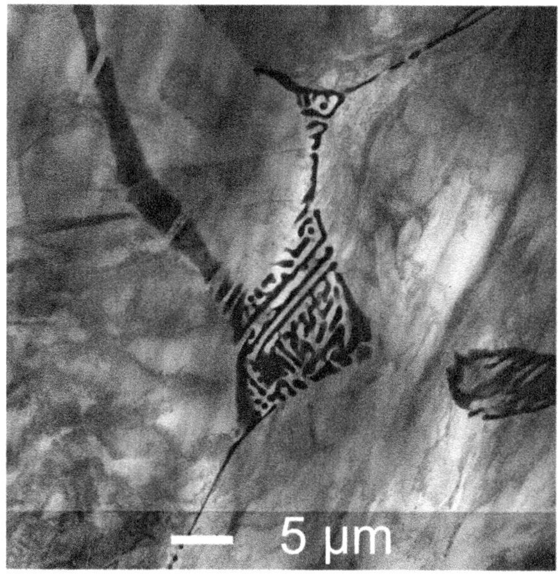

FIGURE 13.337 LM-STEM image showing Chinese script – $Mg(Zn,Cu)_2$ phase in ZC63 alloy.

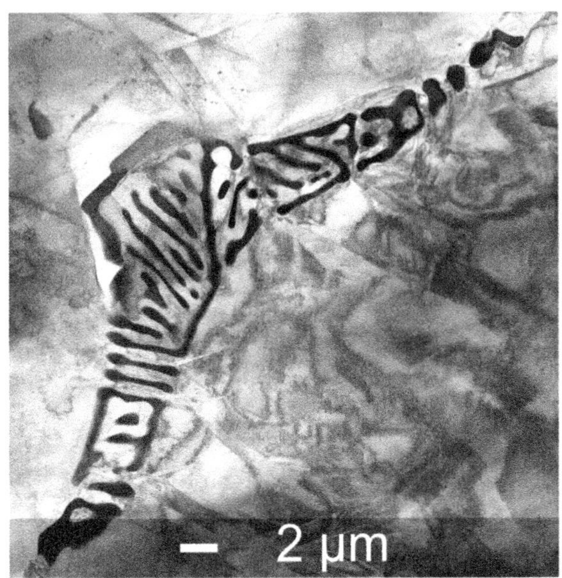

FIGURE 13.338 LM-STEM image showing Chinese script – $Mg(Zn,Cu)_2$ phase in ZC63 alloy.

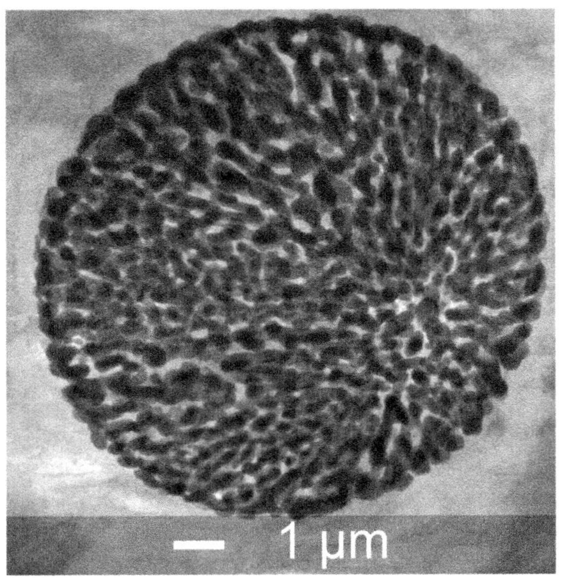

FIGURE 13.339 LM-STEM image showing circular morphology of the $Mg(Zn,Cu)_2$ phase in ZC63 alloy.

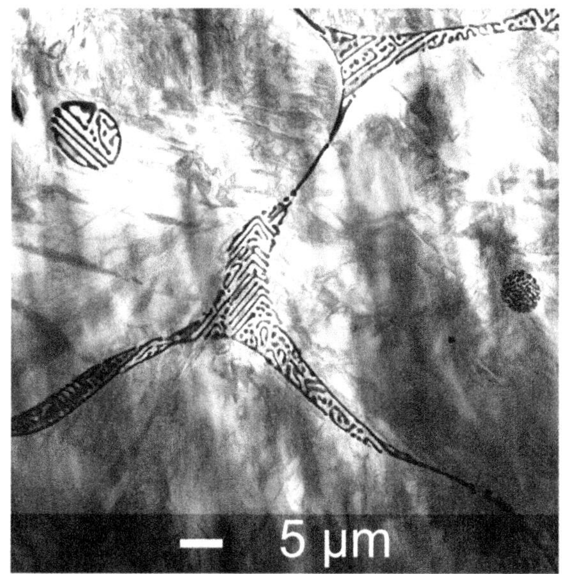

FIGURE 13.340 LM-STEM image showing Chinese script – $Mg(Zn,Cu)_2$ phase in ZC63 alloy.

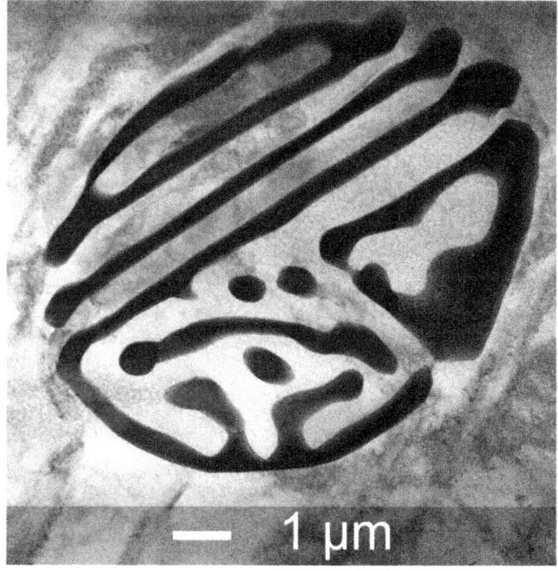

FIGURE 13.341 LM-STEM image showing Chinese script – $Mg(Zn,Cu)_2$ phase in ZC63 alloy.

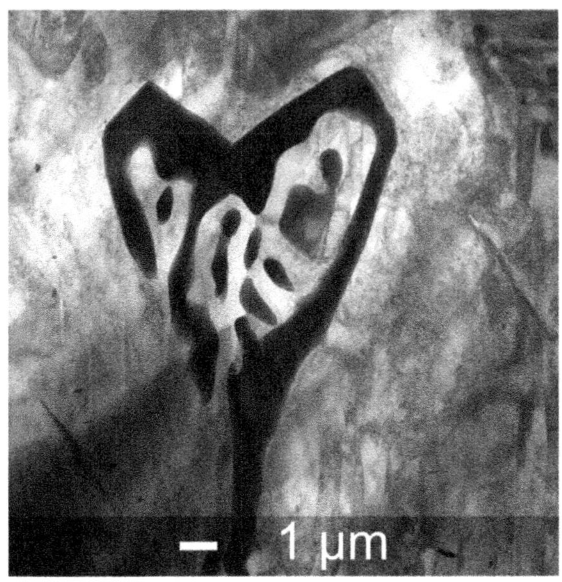

FIGURE 13.342 LM-STEM image showing Chinese script – $Mg(Zn,Cu)_2$ phase in ZC63 alloy.

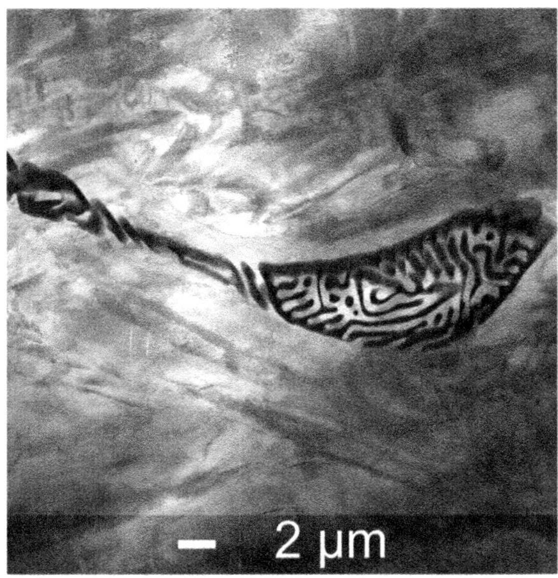

FIGURE 13.343 LM-STEM image showing Chinese script – $Mg(Zn,Cu)_2$ phase in ZC63 alloy.

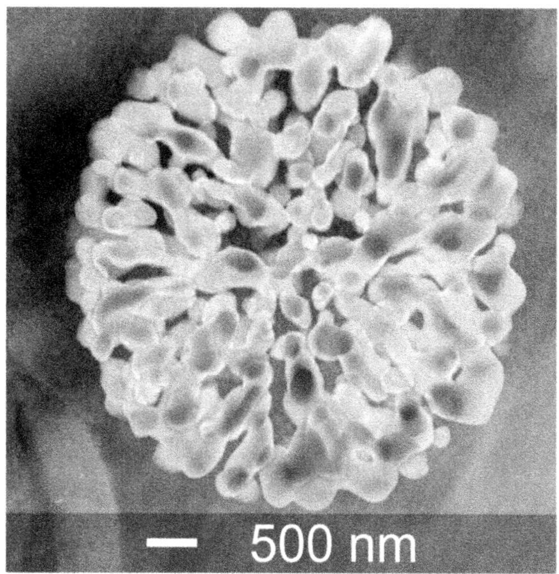

FIGURE 13.344 STEM-HAADF image showing $Mg(Zn,Cu)_2$ phase in the form of a flower in ZC63 alloy.

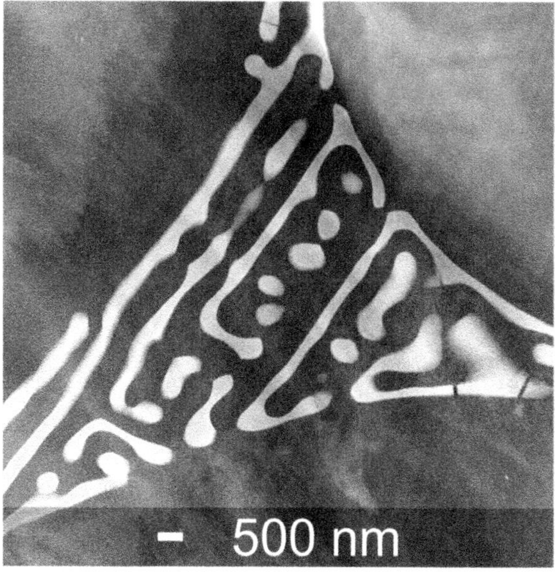

FIGURE 13.345 LM-STEM image showing Chinese script – $Mg(Zn,Cu)_2$ phase in ZC63 alloy.

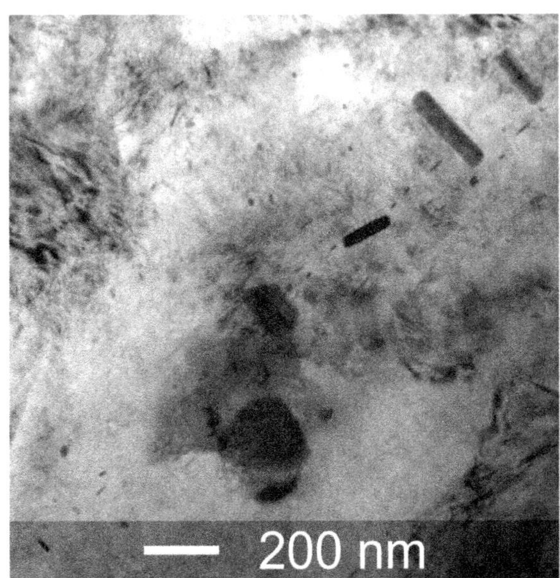

FIGURE 13.346 LM-STEM image showing various morphologies of precipitates in ZC63 alloy.

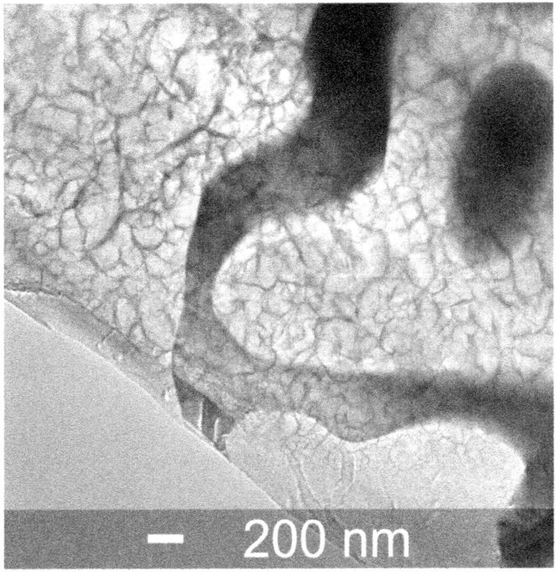

FIGURE 13.347 LM-STEM image showing the morphology between the Chinese script – $Mg(Zn,Cu)_2$ phase in ZC63 alloy.

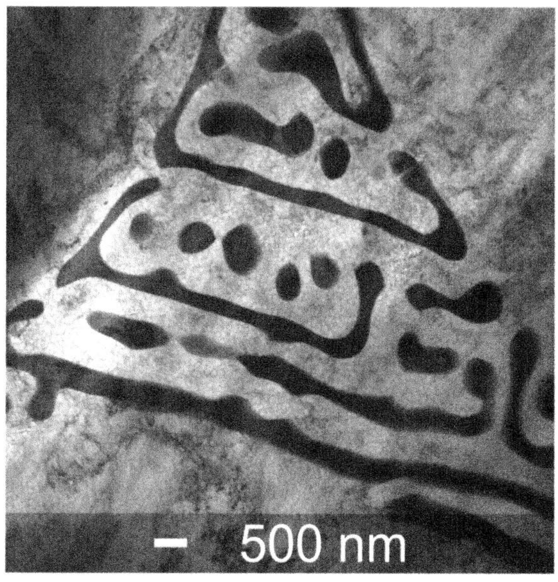

FIGURE 13.348 BFTEM image with objective aperture (10 μm) showing the Chinese script – $Mg(Zn,Cu)_2$ phase in ZC63 alloy.

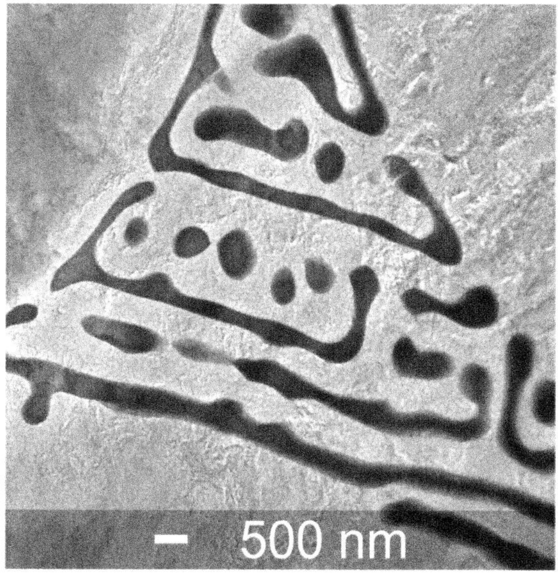

FIGURE 13.349 BFTEM image showing the Chinese script – $Mg(Zn,Cu)_2$ phase in ZC63 alloy.

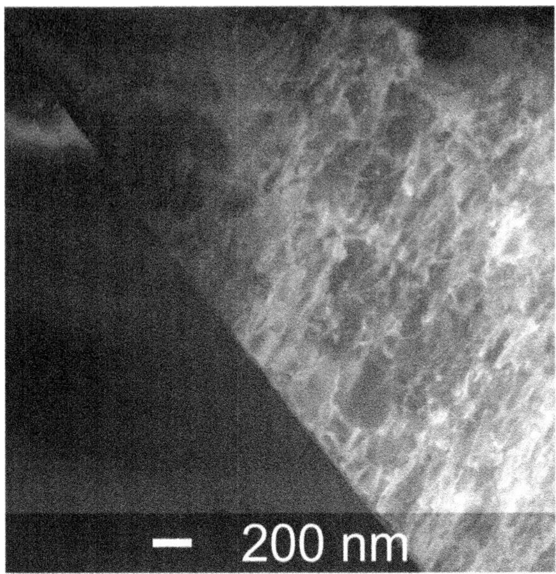

FIGURE 13.350 LM-STEM image showing a grain boundary along with dislocations and precipitates in ZE10 alloy.

FIGURE 13.351 LM-STEM image showing a grain boundary along with dislocations and precipitates in ZE10 alloy.

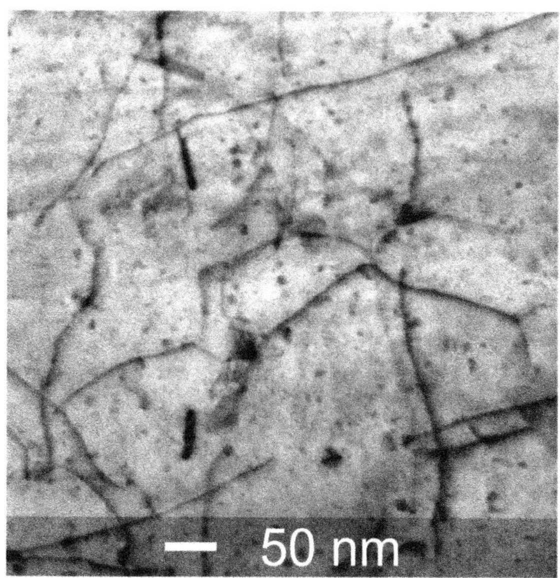

FIGURE 13.352 BFTEM image with objective aperture (10 μm) showing forest of dislocations with precipitates in ZE10 alloy.

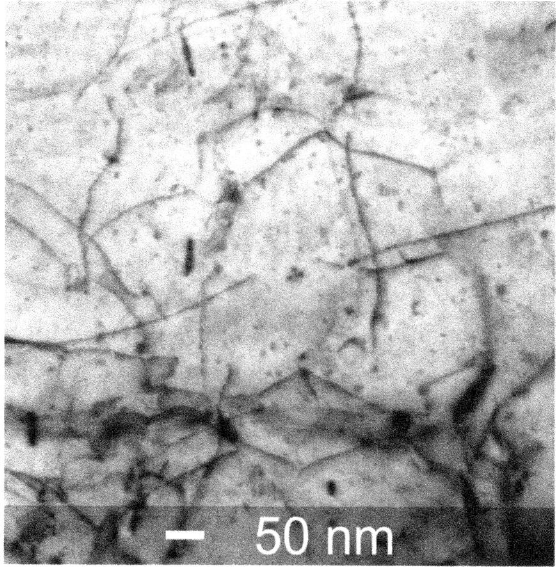

FIGURE 13.353 BFTEM image with objective aperture (10 μm) showing forest of dislocations with precipitates in ZE10.

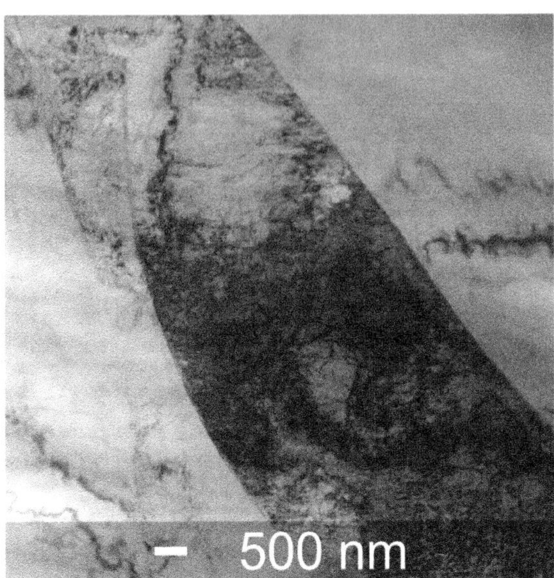

FIGURE 13.354 BFTEM image with objective aperture (10 μm) showing three large grains in ZE10 alloy.

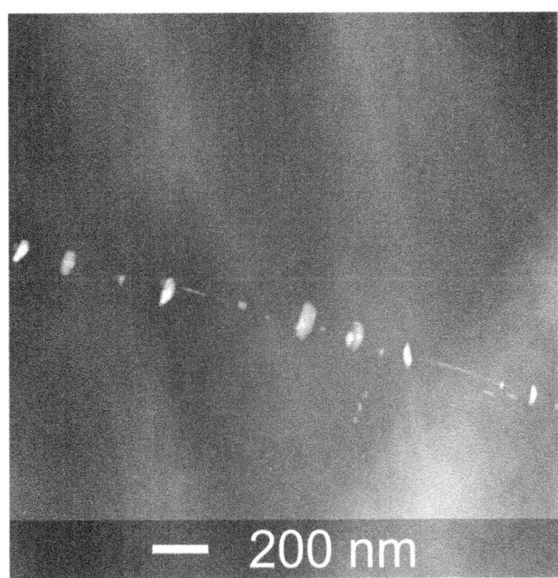

FIGURE 13.355 STEM-HAADF image of showing near spherical, acicular and elongated Zn_2Zr_3 precipitates in ZE41-R25 alloy.

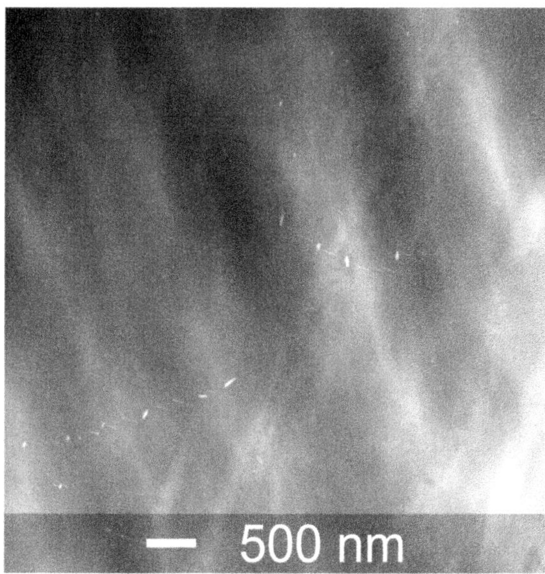

FIGURE 13.356 LM-STEM image of showing near spherical, acicular and elongated precipitates in ZE41-R25 alloy.

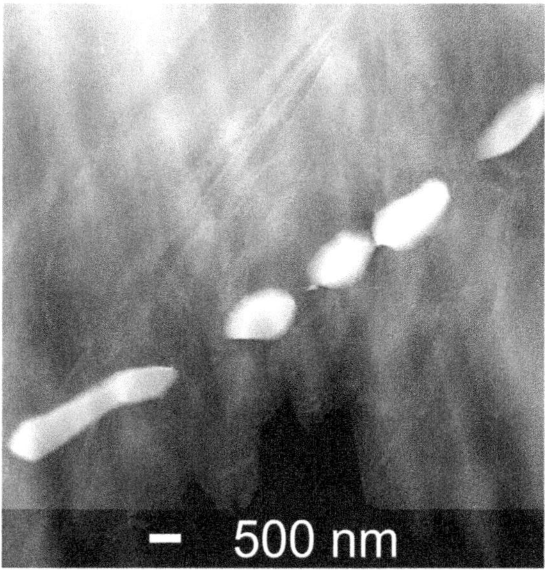

FIGURE 13.357 STEM-HAADF image of showing elongated Mg_7Zn_3RE phase precipitates along the grain boundary in ZE41-R25 alloy.

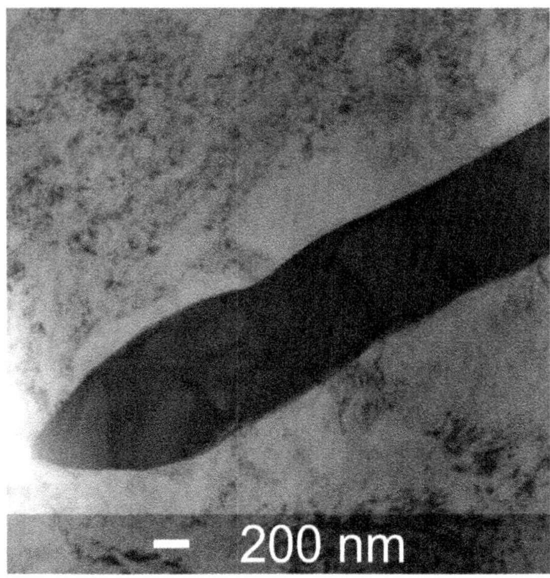

FIGURE 13.358 BFTEM image with objective aperture (10 μm) showing elongated Mg_7Zn_3RE phase precipitates in ZE41-R25 alloy.

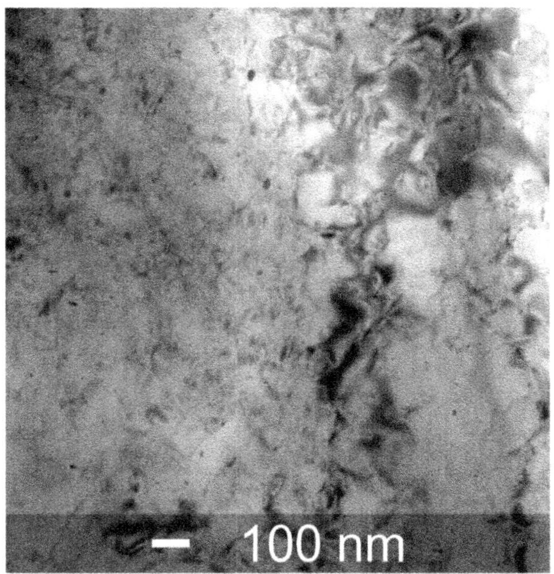

FIGURE 13.359 BFTEM image with objective aperture (10 μm) showing distribution of fine Zn_2Zr_3 precipitates in ZE41-R25 alloy.

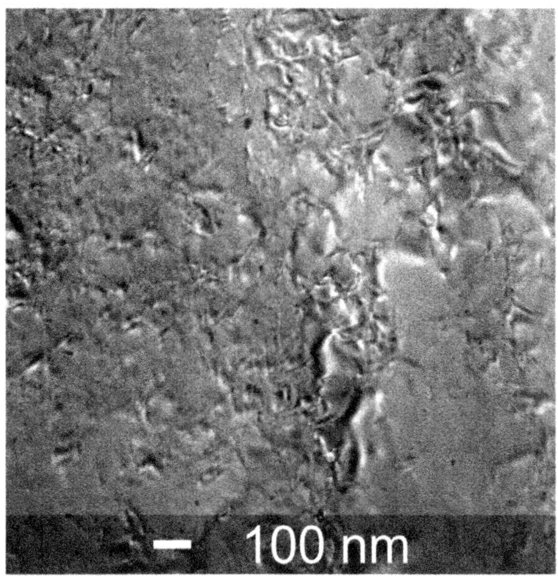

FIGURE 13.360 BFTEM image without objective aperture where the contrast of the precipitates is minimal in ZE41-R25 alloy.

Typical abbreviations used in the TEM section are:

STEM – Scanning Transmission Electron Microscopy
LM-STEM – Low Magnification STEM
HAADF – High Angle Annular Dark Field
TEM – Transmission Electron Microscopy
BFTEM – Bright Field Transmission Electron Microscopy

ACKNOWLEDGMENTS

Authors sincerely thank Prof. Karl Ulrich Kainer and Dr. Hajo Deringa of Helmholtz-Zentrum-Geesthacht for providing many of the magnesium alloys presented in this work. They also thank Dr. A. Srinivasan, CSIR-NIIST, Trivandrum; M/s Exclusive Magnesium, Hyderabad; and M/s Hindustan Magnesium, Hyderabad, for providing some of the commercial melt samples. Authors are thankful to Mrs. Aarcha UV and Mrs. Supaja Jasmin for their help towards sample preparation. Authors are grateful to Director, VSSC, for his kind permission to publish this work.

REFERENCES

1. Emley, E.F., *Principles of Magnesium Technology*, Pergamon Press, Oxford, 1996.
2. Evans, G.B. "Application of magnesium in aerospace", *Proceedings of the Conference On Magnesium Technology*, 1986, Institute of Metals, London, UK, 103.
3. Neite G. et al, "Magnesium base alloys." In *Structure and Properties of Nonferrous Alloys*, Vol. 8, Materials Science and Technology: A Comprehensive Treatment, Edited by Cahn R.W., Haasan P., and Krarner, E.J., 1996, VCH Publishers, New York, 113.
4. Vander Voort, G.F, Metallography, Principles and Practice, McGraw-Hill, New York, 1984.

Index

Note: Page numbers in italic and bold refer to figures and tables, respectively.

E

F

G

H

I

K

L

M

N

O

P

Q

R

S

Y

Z

Printed and bound by CPI Group (UK) Ltd, Croydon, CR0 4YY

24/10/2024

01778307-0019